Geophysical Monograph Series
(Including Maurice Ewing Volumes)

Geophysical Monograph Series
A. F. Spilhaus, Jr., Managing Editor

1 **Antarctica in the International Geophysical Year,** A. P. Crary, L. M. Gould, E. O. Hulburt, Hugh Odishaw, and Waldo E. Smith (Eds.)
2 **Geophysics and the IGY,** Hugh Odishaw and Stanley Ruttenberg (Eds.)
3 **Atmospheric Chemistry of Chlorine and Sulfur Compounds,** James P. Lodge, Jr. (Ed.)
4 **Contemporary Geodesy,** Charles A. Whitten and Kenneth H. Drummond (Eds.)
5 **Physics of Precipitation,** Helmut Weickmann (Ed.)
6 **The Crust of the Pacific Basin,** Gordon A. Macdonald and Hisashi Kuno (Eds.)
7 **Antarctic Research: The Matthew Fontaine Maury Memorial Symposium,** H. Wexler, M. J. Rubin, and J. E. Caskey, Jr. (Eds.)
8 **Terrestrial Heat Flow,** William H. K. Lee (Ed.)
9 **Gravity Anomalies: Unsurveyed Areas,** Hyman Orlin (Ed.)
10 **The Earth Beneath the Continents: A Volume of Geophysical Studies in Honor of Merle A. Tuve,** John S. Steinhart and T. Jefferson Smith (Eds.)
11 **Isotope Techniques in the Hydrologic Cycle,** Glenn E. Stout (Ed.)
12 **The Crust and Upper Mantle of the Pacific Area,** Leon Knopoff, Charles L. Drake, and Pembroke J. Hart (Eds.)
13 **The Earth's Crust and Upper Mantle,** Pembroke J. Hart (Ed.)
14 **The Structure and Physical Properties of the Earth's Crust,** John G. Heacock (Ed.)
15 **The Use of Artificial Satellites for Geodesy,** Soren W. Henriksen, Armando Mancini, and Bernard H. Chovitz (Eds.)
16 **Flow and Fracture of Rocks,** H. C. Heard, I. Y. Borg, N. L. Carter, and C. B. Raleigh (Eds.)
17 **Man-Made Lakes: Their Problems and Environmental Effects,** William C. Ackermann, Gilbert F. White, and E. B. Worthington (Eds.)
18 **The Upper Atmosphere in Motion: A Selection of Papers With Annotation,** C. O. Hines and Colleagues
19 **The Geophysics of the Pacific Ocean Basin and Its Margin: A Volume in Honor of George P. Woollard,** George H. Sutton, Murli H. Manghnani, and Ralph Moberly (Eds.)
20 **The Earth's Crust: Its Nature and Physical Properties,** John G. Heacock (Ed.)
21 **Quantitative Modeling of Magnetospheric Processes,** W. P. Olson (Ed.)
22 **Derivation, Meaning, and Use of Geomagnetic Indices,** P. N. Mayaud
23 **The Tectonic and Geologic Evolution of Southeast Asian Seas and Islands,** Dennis E. Hayes, (Ed.)
24 **Mechanical Behavior of Crustal Rocks: The Handin Volume,** N. L. Carter, M. Friedman, J. M. Logan, and D. W. Stearns (Eds.)
25 **Physics of Auroral Arc Formation,** S.-I. Akasofu and J. R. Kan (Eds.)
26 **Heterogeneous Atmospheric Chemistry,** David R. Schryer (Ed.)
27 **The Tectonic and Geologic Evolution of Southeast Asian Seas and Islands: Part 2,** Dennis E. Hayes, (Ed.)
28 **Magnetospheric Currents,** Thomas A. Potemra (Ed.)
29 **Climate Processes and Climate Sensitivity** (Maurice Ewing Volume 5), James E. Hansen and Taro Takahashi (Eds.)

Maurice Ewing Volumes

1 **Island Arcs, Deep Sea Trenches, and Back-Arc Basins,** Manik Talwani and Walter C. Pitman III (Eds.)
2 **Deep Drilling Results in the Atlantic Ocean: Ocean Crust,** Manik Talwani, Christopher G. Harrison, and Dennis E. Hayes (Eds.)
3 **Deep Drilling Results in the Atlantic Ocean: Continental Margins and Paleoenvironment,** Manik Talwani, William Hay, and William B. F. Ryan (Eds.)
4 **Earthquake Prediction—An International Review,** David W. Simpson and Paul G. Richards (Eds.)

Geophysical Monograph 30

Magnetic Reconnection
in Space and Laboratory Plasmas

Edward W. Hones, Jr.
Editor

American Geophysical Union
Washington, D.C.
1984

ASTRO
QC
809
.P5
M25
1984

Published under the aegis of the AGU
Geophysical Monograph Board:
Sean Solomon, Chairman; Francis Boyd,
Merle Henderschott, Janet Luhmann,
James Papike, and John Schaake, members.

Magnetic Reconnection
in Space and Laboratory Plasmas

Library of Congress Cataloging in Publication Data

Main entry under title:

Magnetic reconnection in space and laboratory plasmas.

(Geophysical monograph ; 30)
Papers presented at a Chapman Conference on Magnetic
Reconnection, held at Los Alamos National Laboratory,
Oct. 3-7, 1983.
Includes bibliographical references.
1. Magnetic reconnection--Congresses. 2. Space
plasmas--Congresses. 3. Plasma (Ionized gases)--
Congresses. 4. Magnetosphere--Congresses. I. Hones,
Edward W. II. American Geophysical Union. III. Chapman
Conference on Magnetic Reconnection (1983 : Los Alamos
National Laboratory) IV. Series.

QC809.P5M25 1984   538'.766   84-6376
ISBN 0-87590-058-5
ISSN 0065-8448

Copyright 1984 by the American Geophysical Union,
2000 Florida Avenue, N.W., Washington, D.C. 20009

Figures, tables, and short excerpts may be reprinted in
scientific books and journals if the source is properly
cited.

   Authorization to photocopy items for internal or personal
use, or the internal or personal use of specific clients, is
granted by the American Geophysical Union for libraries and
other users registered with the Copyright Clearance Center
(CCC) Transactional Reporting Service, provided that the base
fee of $1.00 per copy, plus $0.10 is paid directly to
CCC, 21 Congress St., Salem, MA 01970. 0065-8448/84/$01.+.10.
   This consent does not extend to other kinds of copying,
such as copying for creating new collective works or for
resale. The reproduction of multiple copies and the use of
full articles or the use of extracts, including figures and
tables, for commercial purposes requires specific permission
from AGU.

Printed in the United States of America

Ronald G. Giovanelli 1915–1984

"During the growth of a sunspot, there must be electric fields induced in its neighborhood... Apart from a general magnetic field, fields from other sunspots may still be of appreciable size in the neighborhood of the spot under consideration. It is thus to be expected that there will be places where actual neutral points exist and where conditions are thus suitable for the excitation of atoms by collision." R. G. Giovanelli, *Nature, 158,* 81, 1946.

HISTORICAL NOTE

A fundamental feature of plasmas is the interplay of energy forms that can occur within them, between the energy of fields on one hand and the kinetic or thermal energy of particles on the other. The origin of the concept that we now call magnetic reconnection (or magnetic field line merging or magnetic field annihilation) lies in an attempt by Ronald G. Giovanelli to explain solar flares in terms of such an interplay. Giovanelli was a member of the Division of Physics of the Commonwealth Council for Scientific and Industrial Research (later named the Commonwealth Scientific and Industrial Research Organization (CSIRO)) in Sydney, Australia. He had become an experienced observer of the sun and flares, working at the Mt. Stromlo Observatory in Canberra. Nearly four decades ago he advanced a theory [Giovanelli, 1946] that the flare optical emission is from atoms that are excited by electrons accelerated in induced electric fields near neutral points in the evolving magnetic fields of sunspots. In two subsequent papers [Giovanelli, 1947, 1948] he elaborated upon this theory, reporting detailed calculations of currents and fields in sunspots and of electron acceleration and atomic excitation in the sunspot environment.

Giovanelli remained with CSIRO throughout his career. He was primarily an experimental physicist but his continued interest in studying flares and other solar phenomena led him to a number of technological and instrumental developments for improved solar observing. He was very active in the international solar research community, publishing over 100 papers on solar physics and instrumentation. He was named a fellow of the Australian Academy of Science in 1962. He was largely responsible for establishing the Astronomical Society of Australia in 1966 and served as its president from 1968 to 1971. He was president of Commission 12 (Radiation and Structure of the Solar Atmosphere) of the International Astronomical Union from 1973 to 1976. He travelled abroad frequently and made extended visits to several observatories, including those at Kitt Peak, Sacramento Peak, Arcetri and Freiburg. In 1958 he was named head of the Division of Physics at CSIRO, a position he held until 1974. He retired from CSIRO in 1976 but continued to serve as a research fellow there. After a prolonged illness he died in Sydney on January 27, 1984 at the age of 69. Just before his death he completed a book, *Secrets of the Sun,* that will be published by Cambridge University Press.

The story of how Giovanelli's original idea began its evolution toward our present view of magnetic reconnection is not widely known and should be of interest particularly to magnetospheric scientists. In 1947 Giovanelli submitted his papers to the Australian National University in Canberra in partial fulfillment of the requirements for a Doctor of Science degree. Fred Hoyle, of Cambridge University, chosen as Giovanelli's external examiner, became interested in his neutral point theory and suggested an interesting variation of it. He proposed that the primary auroral particles that bombard the Earth's polar atmosphere are accelerated at neutral points in the combination of an interplanetary field and the geomagnetic field [Hoyle, 1949]. In 1947 Hoyle asked his student, James Dungey, for his thesis project, to develop Giovanelli's idea about the importance of neutral points and to apply it to the aurorae. Dungey [1983] has recorded interesting recollections of that period and ensuing events. "Giovanelli was in Sydney and, during my postdoctoral fellowship there, he was extremely kind to me both in my work and generally. My thesis in 1950 contained the essential of what I later called 'reconnection' and one major step in the application to what was later called the 'magnetosphere' . . . Figure 1 of my Citation Classic paper [Dungey, 1961] is one of the magnetospheric diagrams of my thesis, though it was never previously published . . . One reason against publishing was the rejection of my reconnection paper by the Royal Astronomical Society, though it was published in the *Philosophical Magazine* [Dungey, 1953] . . . Initially, the reconnection model seems to have been regarded as an entertaining fiction, but a prediction was successful and a controversy developed. I was enlightened by someone who was planning to make a movie about the magnetosphere and had visited most of the experts. He told me people get very emotional about reconnection. . . . Reconnection remained out of favor throughout the 1970s, but improved observations with the International Sun-Earth Explorers swung opinion the other way."

References

Dungey, J. W., Conditions for the occurrence of electrical discharges in astrophysical systems, *Phil. Mag., 44,* 725, 1953.

Dungey, J. W., Interplanetary magnetic field and the auroral zones, *Phys. Rev. Lett., 6,* 47, 1961.

Dungey, J. W., Citation classic, commentary on *Phys. Rev. Lett., 6,* 47, 1961, in *Current Contents, Physical Chemical and Earth Sciences, Vol. 23,* Number 49, page 20, December 5, 1983.

Giovanelli, R. G., A theory of chromospheric flares, *Nature, 158,* 81, 1946.

Giovanelli, R. G., Magnetic and electric phenomena in the sun's atmosphere associated with sunspots, *Mon. Not. Roy. Ast. Soc., 107,* 338, 1947.

Giovanelli, R. G., Chromospheric Flares, *Mon. Not. Roy. Ast. Soc., 108,* 163, 1948.

Hoyle, F., Magnetic storms and aurorae, in *Some Recent Researches in Solar Physics,* p. 102–104, Cambridge University Press, 1949.

CONTENTS

HISTORICAL NOTE	iii
PREFACE	xi

THEORY OF MAGNETIC RECONNECTION

MAGNETIC FIELD RECONNECTION W. I. Axford	1
SPONTANEOUS RECONNECTION K. Schindler	9
FAST SPONTANEOUS RECONNECTION BY THE RESISTIVELY COUPLED RADIATIVE INSTABILITY R. S. Steinolfson and G. Van Hoven	20
STEADY STATE ASPECTS OF MAGNETIC FIELD LINE MERGING Vytenis M. Vasyliunas	25
MAGNETIC RECONNECTION AND MAGNETIC ACTIVITY E. N. Parker	32
THE ELECTROMAGNETIC FIELD FOR AN OPEN MAGNETOSPHERE Walter J. Heikkila	39
QUASILINEAR EVOLUTION OF TEARING MODES DURING MAGNETIC RECONNECTION W. Horton, T. Tajima, and Ricardo Galvão	45
GLOBAL SINGLE ION EFFECTS WITHIN THE EARTH'S PLASMA SHEET Paul L. Rothwell and G. Kenneth Yates	51
MAGNETIC RECONNECTION AND ANOMALOUS TRANSPORT PROCESSES (A) J. F. Drake	61
COMPARISON OF AN ANALYTICAL APPROXIMATION FOR PARTICLE MOTION IN A CURRENT SHEET WITH PRECISE NUMERICAL CALCULATIONS (A) T. W. Speiser and L. R. Lyons	62

RECONNECTION IN ASTRONOMICAL OBJECTS

MAGNETIC RECONNECTION AT THE SUN E. R. Priest	63
MAGNETIC RECONNECTION IN COMETS Malcolm B. Niedner, Jr.	79
RECONNECTION IN THE JOVIAN MAGNETOSPHERE (A) A. Nishida	90
FINE-SCALE STRUCTURE OF THE JOVIAN MAGNETOTAIL CURRENT SHEET (A) K. W. Behannon	91

RECONNECTION AT EARTH'S MAGNETOPAUSE

MAGNETIC FIELD RECONNECTION AT THE MAGNETOPAUSE: AN OVERVIEW
B. U. Ö. Sonnerup — 92

EVIDENCE OF MAGNETIC MERGING FROM LOW-ALTITUDE SPACECRAFT AND GROUND-BASED EXPERIMENTS
Patricia H. Reiff — 104

PLASMA AND PARTICLE OBSERVATIONS AT THE MAGNETOPAUSE: IMPLICATIONS FOR RECONNECTION
Götz Paschmann — 114

RECONNECTION AT THE EARTH'S MAGNETOPAUSE: MAGNETIC FIELD OBSERVATIONS AND FLUX TRANSFER EVENTS
C. T. Russell — 124

RECENT INVESTIGATIONS OF FLUX TRANSFER EVENTS OBSERVED AT THE DAYSIDE MAGNETOPAUSE
R. P. Rijnbeek, S. W. H. Cowley, D. J. Southwood, and C. T. Russell — 139

A DUAL-SATELLITE STUDY OF THE SPATIAL PROPERTIES OF FTEs
M. A. Saunders, C. T. Russell, and N. Sckopke — 145

THE RELATION OF FLUX TRANSFER EVENTS TO MAGNETIC RECONNECTION (A)
J. D. Scudder, K. W. Ogilvie, and C. T. Russell — 153

FLUX TRANSFER EVENTS AND INTERPLANETARY MAGNETIC FIELD CONDITIONS (A)
J. Berchem and C. T. Russell — 154

SURVEY OF ION DISTRIBUTIONS IN FLUX TRANSFER EVENTS (A)
P. W. Daly, M. A. Saunders, R. P. Rijnbeek, N. Sckopke, and E. Keppler — 155

PATTERNS OF MAGNETIC FIELD MERGING SITES ON THE MAGNETOPAUSE (A)
J. G. Luhmann, R. J. Walker, C. T. Russell, N. U. Crooker, J. R. Spreiter, and S. S. Stahara — 156

ISEE-3 PLASMA MEASUREMENTS IN THE LOBES OF THE DISTANT GEOMAGNETIC TAIL: INFERENCES CONCERNING RECONNECTION AT THE DAYSIDE MAGNETOPAUSE (A)
J. T. Gosling — 157

RECONNECTION IN EARTH'S MAGNETOTAIL

RECONNECTION IN EARTH'S MAGNETOTAIL: AN OVERVIEW
A. Nishida — 159

MAGNETOTAIL ENERGY STORAGE AND THE VARIABILITY OF THE MAGNETOTAIL CURRENT SHEET
D. H. Fairfield — 168

PLASMA SHEET BEHAVIOR DURING SUBSTORMS
Edward W. Hones, Jr. — 178

STREAMING ENERGETIC ELECTRONS IN RECONNECTION EVENTS
John W. Bieber — 185

PARTICLE AND FIELD SIGNATURES OF SUBSTORMS IN THE NEAR MAGNETOTAIL
D. N. Baker — 193

IMPLICATIONS OF THE 1100 UT MARCH 22, 1979, CDAW 6 SUBSTORM EVENT
FOR THE ROLE OF MAGNETIC RECONNECTION IN THE GEOMAGNETIC TAIL
   T. A. Fritz, D. N. Baker, R. L. McPherron, and
   W. Lennartsson   203

SUBSTORM ELECTRIC FIELDS IN THE EARTH'S MAGNETOTAIL
   C. A. Cattell and F. S. Mozer   208

ENERGETIC IONS AND ELECTRONS AND THEIR ACCELERATION PROCESSES
IN THE MAGNETOTAIL
   Manfred Scholer   216

THE DISTANT GEOMAGNETIC TAIL IN THEORY AND OBSERVATION
   S. W. H. Cowley   228

ISEE 3 MAGNETIC FIELD OBSERVATIONS IN THE MAGNETOTAIL:
IMPLICATIONS FOR RECONNECTION
   G. L. Siscoe, D. G. Sibeck, J. A. Slavin, E. J. Smith,
   B. T. Tsurutani, and D. E. Jones   240

BOUNDARY LAYERS OF THE EARTH'S OUTER MAGNETOSPHERE
   T. E. Eastman and L. A. Frank   249

>35 keV ION OBSERVATIONS FROM ISEE-3 IN THE DEEP TAIL (A)
   P. W. Daly, T. R. Sanderson, and K.-P. Wenzel   263

COMPUTER MODELING

THREE-DIMENSIONAL COMPUTER MODELING OF DYNAMIC RECONNECTION IN THE
MAGNETOTAIL: PLASMOID SIGNATURES IN THE NEAR AND DISTANT TAIL
   Joachim Birn   264

EXTERNALLY DRIVEN MAGNETIC RECONNECTION
   Raymond J. Walker and Tetsuya Sato   272

COMPUTER MODELING OF FAST COLLISIONLESS RECONNECTION
   J. N. Leboeuf, F. Brunel, T. Tajima, J. Sakai,
   C. C. Wu, and J. M. Dawson   282

THE NONLINEAR TEARING MODE
   G. Van Hoven and R. S. Steinolfson   292

ON THE CAUSE OF X-LINE FORMATION IN THE NEAR-EARTH PLASMA SHEET:
RESULTS OF ADIABATIC CONVECTION OF PLASMA-SHEET PLASMA
   G. M. Erickson   296

NUMERICAL SIMULATION OF THE DAYSIDE RECONNECTION (A)
   M. Hoshino and A. Nishida   303

COMMENTS ON SIMULATION OF ANOMALOUS RESISTIVITY (A)
   J. W. Dungey   304

LABORATORY PLASMAS

RECONNECTION DURING THE FORMATION OF FIELD REVERSED CONFIGURATIONS
   Richard D. Milroy   305

DRIVEN MAGNETIC RECONNECTION DURING THE FORMATION OF A
TWO-CELL FIELD-REVERSED CONFIGURATION
   E. Sevillano and F. L. Ribe   313

RECONNECTION IN SPHEROMAK FORMATION AND SUSTAINMENT
    James H. Hammer   319

THE ROLE OF MAGNETIC RECONNECTION PHENOMENA IN THE
REVERSED-FIELD PINCH
    D. A. Baker   332

RECONNECTION IN TOKAMAKS
    V. K. Paré   341

A PLASMOID RELEASE MECHANISM THAT COULD EXPLAIN THE SUBSTORM'S
IMPULSIVE EARTHWARD DIVERSION OF CROSS-TAIL CURRENT
    A. Bratenahl and P. J. Baum   347

LABORATORY EXPERIMENTS ON CURRENT SHEET DISRUPTIONS, DOUBLE LAYERS,
TURBULENCE AND RECONNECTION (A)
    W. Gekelman and R. Stenzel   355

MAGNETIC RECONNECTION IN DOUBLETS (A)
    Torkil H. Jensen   356

APPRAISALS, UNANSWERED QUESTIONS, FUTURE DIRECTIONS

SOME COMMENTS ON SOLAR RECONNECTION PROBLEMS
    Ronald G. Giovanelli   357

DRIVEN AND NON-DRIVEN RECONNECTION; BOUNDARY CONDITIONS
    W. I. Axford   360

COMMENTS ON NUMERICAL SIMULATIONS
    T. Sato   362

DEFINITION OF SPONTANEOUS RECONNECTION
    K. Schindler   365

ASTROPHYSICAL IMPLICATIONS OF RECONNECTION
    Jonathan Arons   366

VALIDITY OF THE PETSCHEK MODEL
    D. Biskamp   369

SOLAR FLARES: AN EXTREMUM OF RECONNECTION
    Stirling A. Colgate   372

EVIDENCE FOR THE OCCURRENCE AND IMPORTANCE OF RECONNECTION BETWEEN
THE EARTH'S MAGNETIC FIELD AND THE INTERPLANETARY MAGNETIC FIELD
    S. W. H. Cowley   375

NOW CONSIDER DIFFUSION
    J. W. Dungey   379

DEFINITION OF A SUBSTORM, PHYSICAL PROCESSES IN A SUBSTORM
AND SOURCES OF DISCOMFORT
    G. Rostoker   380

THE LAST WORDS
    V. M. Vasyliunas   385

PREFACE

Magnetic reconnection is now widely believed to be a crucial process in cosmic plasmas, being the means by which their magnetic topologies are determined and by which energy stored in magnetic fields is released rapidly to produce such phenomena as solar flares and magnetospheric substorms. Furthermore, reconnection has been found to play important roles in several areas of fusion research.

This monograph reviews the basic concepts of the magnetic reconnection phenomenon. It describes the observational and theoretical evidence for the occurrence of reconnection in space plasmas, and it discusses the roles of reconnection in laboratory plasmas where the reconnection process offers practical benefits in some magnetic fusion experiments. The book is based on a Chapman Conference on Magnetic Reconnection and contains most of the papers presented at the conference as well as much of the discussion of them. The conference was held at the Los Alamos National Laboratory October 3–7, 1983 and was attended by 130 scientists, citizens of more than a dozen countries.

The Conference comprised six topical sessions of invited and contributed oral papers and a final half-day session for "Appraisals, Unanswered Questions, Future Directions." There was also one evening poster session of mixed topics. The sections of this book bear approximately the same titles as did the sessions of the conference. The poster session papers have been distributed among the sections according to their topics. The ISEE 3 satellite had made traversals of the distant magnetotail during the year preceding the conference and we were fortunate to be able to include initial reports of those very important observations in the sessions on reconnection in the magnetosphere. Those reports, given by D. N. Baker, P. W. Daly, J. T. Gosling, E. W. Hones, M. Scholer and G. L. Siscoe, provided dramatic new evidence for the importance of magnetic reconnection in the magnetosphere and are included in papers published here.

The objectives of the final session were summary, appraisal and evaluation of the rest of the conference and of our present understanding of magnetic reconnection. The following scientists, eminent in several research areas, were invited to give 15-minute accounts of their views, biases, impressions, or whatever, in that final session: J. Arons, W. I. Axford, D. Biskamp, S. A. Colgate, S. W. H. Cowley, J. W. Dungey, R. G. Giovanelli, G. Rostoker, and V. M. Vasyliunas. The remainder of the session was devoted to questions and comments from participants and two short commentaries by T. Sato and K. Schindler. The entire final session was tape-recorded so that it could be transcribed for this monograph.

Most of the papers that were presented in the topical sessions and the poster session were submitted for publication here. Some papers that had been, or were about to be, published in a journal were submitted only in abstract. The latter are indicated by (A) in the Table of Contents. All of the full-length papers underwent review by two referees in accordance with AGU policy for its monograph series; of the 44 such papers submitted 38 were accepted and are included here.

A transcript of the summary session taping was sent to each speaker and questioner of that session to correct errors made in transcribing and to convert from oral style of presentation to printed style; it was intended that no substantial changes of information content should be made. The "papers" in the last section of the book are those minimally altered transcripts. They contain essentially the authors' comments (and the ensuing discussions) about those topics and ideas brought up at the conference which most impressed and interested them. Ronald Giovanelli could not attend the conference because of illness so he very kindly prepared a video tape in which he presented some of his recent thoughts on magnetic reconnection in solar phenomena. The video tape was played for the conference participants at the final session and a transcript of it is included here.

The outstanding success of the conference resulted from the efforts of many people and organizations. I am pleased to acknowledge the efforts of the program committee members, D. N. Baker, J. U. Brackbill, F. V. Coroniti, D. H. Fairfield, A. Nishida, C. T. Russell, B. U. Ö. Sonnerup and V. M. Vasyliunas. I also thank J. Birn who, although not a member of the program committee, gave me important guidance during the formulation of the program and throughout the preparation of this book. I thank the session chairmen: W. I. Axford, E. N. Parker, B. U. Ö. Sonnerup, A. Nishida, T. Sato, R. K. Linford, and R. L. McPherron. I also thank the speakers and the ninety or so scientists who reviewed the papers submitted for publication. Finally, I am grateful for the sponsorship and support of the American Geophysical Union, the Los Alamos National Laboratory, the Institute of Geophysics and Planetary Physics of the University of California, the U.S. Department of Energy, the National Aeronautics and Space Administration and the National Science Foundation.

Edward W. Hones, Jr.

MAGNETIC FIELD RECONNECTION

W. I. Axford

Victoria University of Wellington
Private Bag, Wellington, New Zealand

Introduction

The suggestion that magnetic field line reconnection could be a significant cause of particle acceleration in cosmic plasmas, in particular in solar flares, was first made by Giovanelli [1947]. Hoyle [1949] and Dungey [1961] applied the idea to geomagnetic phenomena, notably concerning the interaction between the interplanetary and terrestrial fields (which we now call 'day-side' reconnection) and the acceleration of auroral particles ('night-side' reconnection). Various aspects of solar flare theory, including reconnection, were developed by Parker [1957], Sweet [1958], and Gold and Hoyle [1959], in particular the concepts of magnetic field configurational instability and energy release. For a brief period there was some concern that the possible rates of reconnection (defined in terms of a characteristic Alfvén Mach number) might not be adequate [Parker, 1963] but most doubts have been removed as a consequence of an important contribution by Petschek [1964], which has been essentially confirmed by subsequent analytical and numerical studies.

Magnetic field reconnection is a process involving a breakdown of the magnetohydrodynamic 'frozen field' theorem in which higher order effects (notably "resistivity"), normally negligible in the large, become locally dominant with dramatic consequences in the nature of the large-scale flow and magnetic field configuration which could not be achieved otherwise. Since such a process is mathematically singular and, in most situations highly non-linear, its implications are not easy to grasp conceptually, despite their essential importance. Unfortunately, this appears to have led to the emergence of a point of view in which it is asserted that reconnection, especially in the steady state, is misleading, often erroneous and unnecessary [e.g. Alfvén 1976, 1977, 1981]. For this reason, we have chosen in this paper to review some basic concepts and leave descriptions of the observational evidence for the occurrence of reconnection in natural plasmas to others. This evidence, which involves magnetic field configuration changes on the sun, the responsiveness of the magnetosphere to the sense of the interplanetary magnetic field, the penetration of energetic solar particles into the magnetosphere, escape of particles from the day side magnetosphere, the prevalence of energetic electrons/protons on the dusk/dawn side of the magnetotail, and the reconfiguration of the magnetosphere during substorms, is, in our view, overwhelming.

2. The Kelvin-Helmholtz Theorem

In order to understand the phenomenon of 'reconnection' it is useful to begin by understanding 'connection' and to this end we require the Kelvin-Helmholtz theorem describing the behaviour of the circulation and the vorticity vector in an inviscid, homentropic fluid [Batchelor, 1970]. In such a medium, the pressure (p) is isotropic with $p = p(\rho)$, where ρ is the mass density. Taking the curl of the equation of motion, we find the equation for vorticity ($\underline{\omega}$) transport:

$$\frac{\partial \underline{\omega}}{\partial t} = \nabla \times (\underline{u} \times \underline{\omega}) \ , \quad (2.1)$$

where $\underline{u}$ is the fluid velocity vector and $\underline{\omega} = \nabla \times \underline{u}$. Obviously, $\nabla \cdot \underline{\omega} = 0$.

Consider the flux of vorticity (Φ) at time t, through a material closed curve C spanning a simple surface S which becomes a new material curve C' and surface S' as the fluid moves during an interval δt (Figure 1a). The change in this flux ($\delta \Phi$) as the material curve moves from C to C' is given by

$$\begin{aligned}\delta \Phi &= \int_{S'}\int \underline{\omega}(\underline{r}', t + \delta t) \cdot \underline{dS} - \int_{S}\int \underline{\omega}(\underline{r}, t) \cdot \underline{dS} \\ &= \left\{\int_{S'}\int \underline{\omega}(\underline{r}', t + \delta t) \cdot \underline{dS} - \int_{S}\int \underline{\omega}(\underline{r}, t + \delta t) \cdot \underline{dS}\right\} \\ &\quad + \int_{S}\int \left\{\underline{\omega}(\underline{r}, t + \delta t) - \underline{\omega}(\underline{r}, t)\right\} \cdot \underline{dS} \end{aligned} \quad (2.2)$$

where $\underline{r}$ ($\underline{r}'$) is the position vector defining C(C') and $\underline{dS}$ is a surface element of S and S'. Using the divergence theorem on the volume element defined by C and C', which has surfaces comprising S, S' and the displacement surface defined by the surface elements $(\underline{u}\delta t) \times \underline{dl}$, where $\underline{dl}$ is a line elements on C, we obtain

$$\delta \Phi = \delta t \int_{S}\int \left\{\frac{\partial \underline{\omega}}{\partial t} - \nabla \times (\underline{u} \times \underline{\omega})\right\} \cdot \underline{dS} \ , \quad (2.3)$$

Since the integrand vanishes by virtue of equation (2.1), we have deduced that for an inviscid, homentropic fluid, $d\Phi/dt = 0$, i.e. the flux of vorticity through a closed loop moving with the fluid is conserved. The implication is that a vortex tube (see Figure 1b) moves with the fluid and its strength remains constant. In the limit of zero cross-section, we may consider a vortex tube to be a vortex "line" so that we may say, "a material line which initially coincides with a vortex line continues to do so." It is thus possible and convenient to regard a vortex line as having a continuing identity and as moving with (or being "frozen") to the fluid. We may also say that elements of fluid which are at one time *connected* by a common vortex line, remain so thereafter.

This result is of course valid only if equation (2.1) is a valid representation of the fluid motion, which is not the case if the stress tensor is not curl-free, or if non-conservative body forces are present. In particular, the presence of viscosity permits vortex lines to "diffuse" relative to the fluid since equation (2.1) takes on the form (for constant kinematic viscosity ν):

$$\frac{\partial \underline{\omega}}{\partial t} = \nabla \times (\underline{u} \times \underline{\omega}) + \nu \nabla^2 \underline{\omega} \ . \quad (2.4)$$

Viscosity can evidently promote the "reconnection" of vortex lines connecting different fluid elements. Further, if the fluid is electrically conducting, but otherwise inviscid and homentropic, equation (2.1) has the form

$$\frac{\partial \underline{\omega}}{\partial t} = \nabla \times (\underline{u} \times \underline{\omega}) + \nabla \times (\underline{j} \times \underline{B}/\rho) \ , \quad (2.5)$$

where $\underline{j}$ is the electric current and $\underline{B}$ the magnetic induction, indicating that the vorticity is not frozen to the fluid but is produced and transported by magnetic stresses. In effect, one might say that the angular momentum of a fluid element

(a)

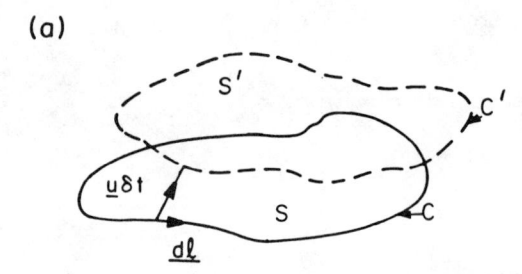

Fig. 1a. The closed material curve C moves to a new position C' in time δt. The surfaces S, S' span C C' respectively and the surface connecting C and C' is defined by the vector $\underline{u}\delta t$.

should be conserved during its motion if no torques are acting on it and that viscosity and Maxwell stresses can provide such torques.

3. The 'Frozen Field' Theorem

In a simple electrically conducting fluid, Ohm's law has the form

$$\underline{j} = \sigma(\underline{E} + \underline{u} \times \underline{B}), \quad (3.1)$$

where σ is the electrical conductivity and $\underline{E}$ the electric field. Using Faraday's and Amperes laws:

$$\frac{\partial \underline{B}}{\partial t} = -\nabla \times \underline{E}, \quad (3.2)$$

$$\mu\underline{j} = \nabla \times \underline{B}, \quad (3.3)$$

with μ the permeability of the medium, the curl of equation (3.1) can be written as

$$\frac{\partial \underline{B}}{\partial t} = \nabla \times (\underline{u} \times \underline{B}) + \eta\nabla^2\underline{B}, \quad (3.4)$$

where $\eta = 1/\mu\sigma$ has been assumed constant for convenience. This equation has the same form as equation (2.4) and since $\nabla \cdot \underline{B} = 0$, the same conclusions hold for the magnetic induction vector $\underline{B}$ as for the vorticity ω according to the Kelvin-Helmholz theorem in a non-conducting fluid. In particular, if the term $\eta\nabla^2\underline{B} = 0$ in equation (3.4) is negligible, the flux of the magnetic induction through a closed loop moving with the fluid is conserved and a material line which is initially coincident with a magnetic field line continues to be so. It is possible and convenient in these circumstances to regard a magnetic field line as having a continuing identity and as moving with the fluid (i.e. the magnetic field is "frozen" to the fluid).

Elements of fluid which are at one time connected by a single magnetic field line remain connected at subsequent times, unless the term $\eta\nabla^2\underline{B}$ or other terms neglected in equation (3.4) are at least locally sufficiently large as to cause the frozen field theorem to break down. Usually, if the characteristic speed (V) and length scale (L) of the configuration of fluid motion and magnetic field are such that the magnetic Reynolds number $VL/\eta \gg 1$, the magnetic field lines diffuse slowly relative to the fluid. However, since the characteristic length that must be considered pertains to the magnetic field rather than the entire configuration it is possible for situations to arise in which L is locally small (e.g. in a local region of magnetic field reversal, or current sheet). In this case, the time-scale for diffusion of magnetic field relative to the fluid is L^2/η and if this is comparable with the time-scale for removal of the field lines from the locality, say L/V_A (where $V_A = B/(\mu\rho)^{1/2}$), with B a characteristic magnitude of the magnetic induction, the frozen field theorem is locally invalid. The magnetic field lines may in these circumstances be described as "reconnecting." That is, the frozen field theorem and its implications hold generally, but fluid elements which are at one time connected by a common magnetic field line and which move in such a way that the field line between them passes through a diffusion-dominated region where the frozen field theorem breaks down completely, will become disconnected from

each other, and connected afterwards to different sets of fluid elements. This process of partner swapping or reconnection in a localized region is illustrated in Figure 2a. Note that the size of the region is of no immediate significance, since provided it is small compared with the scale of the whole system it can presumably adjust itself so that in combination with the magnetic diffusivity, the rate of reconnection demanded by the system (i.e. by distant and initial conditions) is achieved.

It is evident that the possibility that magnetic field lines can locally reconnect in this manner has important implications on the nature and stability of hydromagnetic configurations, both flowing and static. Since the tensile component of the Maxwell stress tensor (i.e. the tendency of curved magnetic field lines to contract like elastic strings) acts so as to remove fluid from a possible reconnection region (locally a two-dimensional X-type neutral point as illustrated in Figure 2b), it is possible for magnetic energy to be very easily transformed to fluid enthalpy and kinetic energy. The nature of the magnetic field configuration and flow is then likely to be quite different from that with η = 0. This cannot be prevented by a large electrical conductivity (small η) if this is compensated by a small value of L (i.e. the diffusion region is small), but it can be controlled, e.g., if the fluid is prevented from leaving the reconnection region by pressure gradients which counteract the effect of the Maxwell stresses. Such pressure gradients can only be produced as a consequence of the external conditions which determine the large scale configuration or flow; this is the basis of the assertion that reconnection rates are determined by boundary and/or initial conditions rather than the local properties of the medium such as electrical conductivity.

Finally, from this discussion, and the definition of 'reconnection' as a localized break-down of the requirement for 'connection' of elements of fluid at one time on a common magnetic field line, it should be evident that steady-state reconnection is perfectly possible. If one adopts the definition of reconnection as a change in topology of the magnetic field configuration (which may then presumably be controlled by sources of magnetic field external to the plasma under consideration) it might be difficult to conceive of a steady state. The alternative definition of reconnection as occurring when an electric field exists with a component parallel to a locally two-dimensional X-type magnetic neutral line is equivalent to break-down of connection and of course permits of a steady state. One must be careful not to be deceived by time-dependent X-type neutral line configurations described by several authors [Dungey, 1953, Chapman and Kendall 1963; Uberoi, 1963] in which an electric field is present generally but vanishes at the neutral line itself; it can be shown that connection is completely maintained in these cases [Yeh and Axford, 1970].

(b)

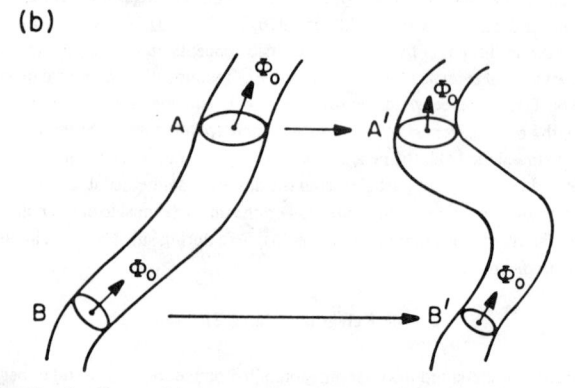

Fig. 1b. The closed material curves A and B initially defining a vortex tube with flux Φ_0 move to become the closed curves A', B' after a certain time. The Kelvin-Helmholtz theorem states that the fluxes of vorticity through A' and B' remain Φ_0 and since this is true for all closed curves A and B on the initial vortex tube the curves A' and B' also define a vortex tube. Alternatively, one may consider the surface connecting A and B through which there is initially no flux of vorticity: at any later time there can be no flux of vorticity through the surface connecting A' and B' so that the latter is again the surface of a vortex tube.

4. The Kelvin-Helmholtz Theorem in a Collisionless Plasma

If the velocity distribution functions for electrons and ions $[f_-(v), f_+(v)]$ in a collisionless plasma can be assumed to satisfy Boltzmann equations of the form

$$\frac{\partial f_\pm}{\partial t} + \underline{v} \cdot \frac{\partial f_\pm}{\partial \underline{x}} + \frac{e_\pm}{m_\pm}(\underline{E} + \underline{v} \times \underline{B}) \cdot \frac{\partial f_\pm}{\partial \underline{v}} = \left[\frac{\partial f_\pm}{\partial t}\right]_w \qquad (4.1)$$

where $\underline{E}$, $\underline{B}$ are 'smoothed' fields and $[\partial f_\pm/\partial t]_w$ represents the effects of 'microscopic' fluctuations, then one can derive moment equations in the usual way such that the first moment is

$$n_\pm m_\pm \left(\frac{\partial}{\partial t} + \underline{u}_\pm \cdot \nabla\right) \underline{u}_\pm = -\nabla \cdot \underline{\underline{S}}_\pm - n_\pm e_\pm (\underline{E} + \underline{u}_\pm \times \underline{B}) + \underline{K}_\pm . \qquad (4.2)$$

Here

$$n_\pm = \int f_\pm d\underline{v}, \qquad \underline{u}_\pm = \int \underline{v} f_\pm d\underline{v},$$

$$\underline{\underline{S}}_\pm = m_\pm \int (\underline{v} - \underline{u}_\pm)(\underline{v} - \underline{u}_\pm) f_\pm d\underline{v},$$

and

$$\underline{K}_\pm = m_\pm \int \underline{v} \left[\frac{\partial f_\pm}{\partial t}\right]_w d\underline{v}.$$

Equation (4.2) is exact if equation (4.1) is a useful representation of the situation, but the quantities $\underline{\underline{S}}_\pm$, $\underline{K}_\pm$ are not yet determined. However, if we can assume that $\nabla \cdot \underline{\underline{S}}_\pm = -\nabla p_\pm$ where $p_\pm = p_\pm(n_\pm)$, and that the terms involving momentum transfer via fluctuations are negligible, then taking the curl of equation (4.2) we obtain

$$\frac{\partial \underline{\Omega}_\pm}{\partial t} = \nabla \times (\underline{u}_\pm \times \underline{\Omega}_\pm), \qquad (4.3)$$

where

$$\underline{\Omega}_\pm = \nabla \times \underline{u}_\pm + e_\pm \underline{B}/m_\pm, \qquad (4.4)$$

(a)

Fig. 2a. At time t_1 the magnetic field line defined by fluid elements A and B approaches a more-or-less oppositely-directed field line defined by fluid elements C and D. At time t_2 the two field lines touch at some intermediate point between A and B and between C and D. At time t_3 the connection between A and B and between C and D is broken and the field line defined by fluid elements A and D moves away from the field line defined by fluid elements B and C.

(b)

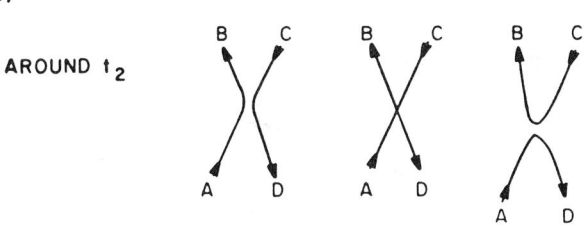

AROUND t_2

Fig. 2b. In the vicinity of the 'reconnection' event occurring around time t_2 in (a) the field lines A B, C D instantaneously form an X-type neutral point.

are generalized vorticities which take into account the gyrational motion of individual ions and electrons [Ferraro and Plumpton, 1966, Section 8.19]. From the Kelvin-Helmholtz analysis described in Section 2, we see immediately that $\underline{\Omega}_+$ and $\underline{\Omega}_-$ are 'frozen' to the ion and electron fluids respectively.

If we estimate the ordinary vorticities as

$$\nabla \times \underline{u}_+ \sim \nabla \times \underline{u}_- \sim \frac{V}{L} = \omega, \qquad (4.5)$$

with ω a characteristic vorticity or inverse time scale, then

$$\frac{m_\pm \underline{\Omega}_\pm}{e_\pm} \sim \underline{B}\left[1 + O\left(\frac{\omega}{\omega_{g_\pm}}\right)\right], \qquad (4.6)$$

where $\omega_{g_\pm} = e_\pm B/m_\pm$. Hence, if $\omega \ll \omega_{g_+} \ll \omega_{g_-}$ the right hand side of (4.6) is essentially $\underline{B}$ and the Kelvin-Helmholtz theorem implies that $\underline{B}$ is frozen to the plasma. The frozen field theorem is therefore to be regarded as the equivalent of the usual vorticity-freezing theorem in the presence of electric currents and magnetic fields. If $\omega_{g_+} \ll \omega \ll \omega_{g_-}$, $\underline{B}$ is frozen to the electron gas but not to the ions (the 'Hall' effect) and if $\omega \gg \omega_{g_-}$ then $\underline{B}$ is decoupled from the medium entirely. This suggests that for sufficiently small scales the effects of finite electron gyroradius should be sufficient to permit decoupling of the magnetic field, and therefore reconnection, regardless of the availability of other effects such as anomalous resistivity.

It is instructive to consider the same arguments using the generalized Ohm's Law as a starting point and following Section 3. On suitably subtracting the ion and electron first moment equations (4.2) and neglecting terms of order m_-/m_+, one obtains the generalized Ohm's Law (e.g. Rossi and Olbert, 1970):

$$(\underline{E} + \underline{u} \times \underline{B}) = \frac{1}{ne}\underline{j} \times \underline{B} - \frac{\nabla \cdot \underline{\underline{S}}_-}{ne} + \frac{m_-}{ne^2}$$
$$\times \left[\frac{\partial \underline{j}}{\partial t} + \nabla \cdot (\underline{u}\underline{j} + \underline{j}\underline{u})\right] + \frac{\underline{K}_-}{ne}, \qquad (4.7)$$

where n, $\underline{u}$ and $\underline{j}$ are bulk quantities as usually defined. Taking the terms on the left to be $O(1)$ and $\omega = V/L$ as before, one finds that the four terms on the right are $O(\omega/\omega_{g_+} M_A^2)$, $O(\omega/\omega_{g_-} M^2)$, $O(\omega^2/\omega_p^2)$ and $O(1/R_M)$ respectively, where $M_A = V/V_A$, $M^2 = 3\rho V^2/5p_-$, $\omega_p^2 = ne^2/\epsilon_0 m_-$ and $R_M = \mu\sigma_w VL$ with $\sigma_w = ne^2\tau_w/m_-$ and $e\underline{K}_-/m_- = \underline{j}/\tau_w$. Ignoring to begin with the effects of the term involving $\underline{K}_-$, it is evident that for $\omega \ll \omega_{g_+}$ equation (4.7) reduces to

$$\underline{E} + \underline{u} \times \underline{B} \approx 0, \qquad (4.8)$$

and hence the magnetic field is frozen to the bulk plasma. If $\omega_{g_+} < \omega \ll \omega_{g_-}$ then the first two terms on the right of equation (4.7) are important and Ohm's Law becomes

$$\underline{E} + \underline{u}_- \times \underline{B} \approx -\frac{\nabla \cdot \underline{\underline{S}}_-}{ne}, \qquad (4.9)$$

in which case, provided $\underline{\underline{S}}_- = \underline{\underline{S}}_-(n)$ the magnetic field is frozen to the electron gas. If $\omega_{g_-} \gtrsim \omega$ the third term on the right of equation (4.7) representing the effects of electron inertia is important and the magnetic field is decoupled from both the electron and the ion gases. Finally, the effects of anomalous resistivity represented by the fourth term on the right of equation (4.7) may become

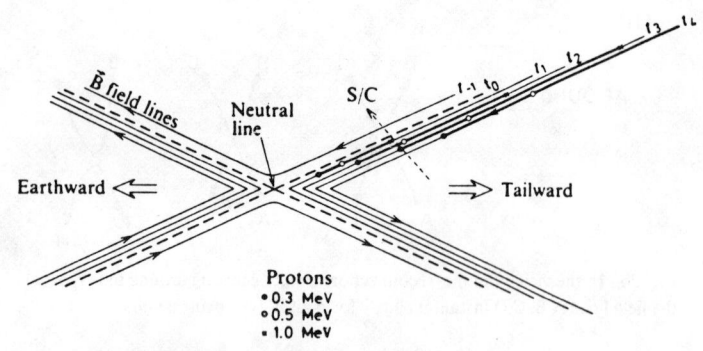

Fig. 3. A schematic diagram of the motions of magnetic field lines and energetic particles resulting from acceleration at an X-type neutral point. The positions of particles released from the neutral line at t_0 and at progressively later times t_1, t_2, t_3 and t_4 are shown more-or-less to scale, as is the motion of a spacecraft (s/c) which observes a progression of monoenergetic bursts of highly anisotropic protons with the lowest energy particles (0.3 MeV) being observed first and the highest energies (1.0 MeV) last [Sarris and Axford, 1979]. In this case the spacecraft moved from the post- to the pre-reconnection region.

important at any stage in this scheme provided there is sufficient wave turbulence present. However, for a reconnection region of finite size and therefore finite plasma transit time, unstable wave growth and the associated anomalous resistivity may be limited, whereas the effects of electron inertia must always become important at sufficiently small length scales.

A direct consequence of the decoupling of first ions and then electrons from the magnetic field in the vicinity of an X-type neutral point, is that the particles concerned are free to be accelerated directly by the electric field along the neutral line. The acceleration is limited ultimately by the total voltage drop available, EL_N, where L_N is the length of the neutral line. In addition, most particles reside in the region of de-coupling only for a time of order $1/\omega_{g\pm}$ and therefore the probable energy is limited. It seems reasonable to conclude that non-thermal, direct acceleration of particles at an X-type neutral point should lead to energy spectra which are exponential in form with a high energy cutoff at EL_N. The most distinctive characteristic of these energetic particles is that they must escape more-or-less along the separatrix of the magnetic field configuration since the dimensions of the acceleration region are so small. Consequently, when observed at large distances from the neutral line they should appear "velocity filtered" due to the fact that they must also take part in the field line motion across the separatrix. On crossing the separatrix from the pre-to the post-reconnection region an observer should see first the fastest particles and then progressively less fast particles until finally reaching the 'thermal' plasma flowing away from the neutral line or vice versa (Figure 3). This description was proposed by Sarris and Axford [1979] as an explanation of the dispersed energetic particle beams occasionally observed in the magnetotail in association with substorms [Sarris, et al., 1976]. The same effect was observed much earlier in the much lower energy particle beams produced as a consequence of dayside reconnection [Frank, 1971; Gurnett and Frank, 1972].

5. Solutions to Problems Involving Reconnection

It is useful before proceeding to detailed analyses of the reconnection process to consider the case of hydrostatic ($\underline{u} = 0$), time-dependent diffusion of magnetic field in the simple configuration shown in Figure 4a. The initial ($t = 0$) magnetic field distribution contains a discontinuity (current sheet) at $y = 0$ with $B_x = \pm B_0$ in $y \gtrless 0$. At later times "oppositely-directed" field lines diffuse towards each other and annihilate, with the energy of the field appearing as Joule heating:

$$\frac{B_n}{B_0} = \frac{2}{\sqrt{\pi}} \int_0^\xi e^{-u^2} du \,, \qquad \xi = z/(\eta t)^{1/2} \,. \quad (5.1)$$

Note that the thickness of the diffusion region increases as $t^{1/2}$ and so, in a sense, pure diffusion is self-defeating as a means of rapidly converting magnetic energy

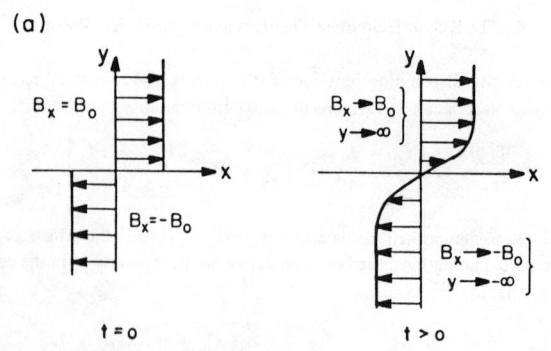

Fig. 4a. The distribution of magnetic field in a current sheet surrounding $y = 0$ for the case of pure diffusion ($\underline{u} = 0$). The current sheet has zero thickness at $t = 0$ and broadens for $t > 0$ according to equation (5.1).

to enthalpy since it simultaneously reduces strong field gradients. Here, and in the subsequent discussion, the addition of a uniform magnetic field in the z-direction does not affect the result which, although formally 2-dimensional, therefore refers to a 3-dimensional situation.

Parker [1957] and Sweet [1958] noted that the magnetic field diffusion rate could be kept at a high level provided the fluid moves towards the diffusion region (current sheet) to maintain the field gradient. This requires of course that the convergent flow be balanced by an equal outflow parallel to the current sheet as shown in Figure 4b. If the length of the current sheet is L and its thickness δ, mass conservation requires that the inflow and outflow speeds u_1, u_2 are related by by

$$u_1 L = u_2 \delta \,. \quad (5.2)$$

It can be shown from energy/momentum considerations that $u_2 = V_{A1}$ where V_{A1} is the Alfven speed in the inflow region immediately adjacent to the current sheet. Finally, since the thickness of the current sheet must be such that the diffusion and advection of the magnetic field balance (equation (3.4)),

$$u_1 \delta = \eta \,. \quad (5.3)$$

Eliminating δ between (5.2) and (5.3) we obtain

$$u_1 = \sqrt{V_{A1}/\mu\sigma L} = R_M^{-1/2} \, V_{A1} \,, \quad (5.4)$$

where R_M is the magnetic Reynolds number based on L and V_{A1}.

In applying this result to solar flares, Parker [1963] at first made the understandable assumption that L should be of the order of the scale of the system and found that the magnetic field annihilation rates to be expected on this basis were orders of magnitude too slow to account for the time scales typically observed. In fact, as pointed out earlier, L may have any value, but cannot be less than δ so that

$$0 \leq u_1 \leq V_{A1} \,, \quad (5.5)$$

Fig. 4b. The steady configuration considered by Parker [1957] and Sweet [1958]. Fluid moves with uniform speed u_1 towards the current sheet (hatched) carrying oppositely-directed magnetic fields B_1 over a length L. The fluid is ejected to either side with speed u_2 and the thickness of the current sheet is δ.

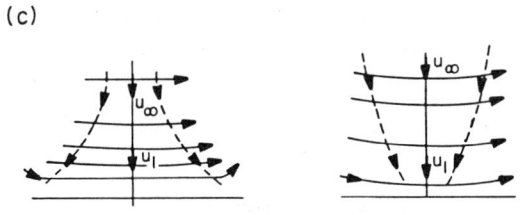

Fig. 4c. The relationship of u_1 and B_1 in (b) to u_∞ and B_∞ depends on whether the flow diverges ($u_1 < u_\infty$) or converges ($u_1 > u_\infty$) as it approaches the current sheet.

with the upper limit being independent of the electrical conductivity, however large.

A solar flare appears to be the result of a magnetic field configuration which has been brought to the point of instability and which reconfigures itself as fast as it can, so that one would expect that, initially anyway, $L \simeq \delta$, $u_1 \lesssim V_{A1}$. In this case, there is no difficulty with the observed time scales, since the size of the flare region is $\sim 10^6 - 10^7$ m and with $V_{A1} \sim 10^5$ m/s, time scales of 10-10^2 seconds are to be expected (comparable to the duration of the non-thermal flash phase of a flare). It is interesting to note that the corresponding maximum voltage drops available for the acceleration of particles are of the order of 10^9 volts if $B \sim 10^3$ gauss, consistent with the highest energy solar flare particles observed. The equivalent numbers for the case of magnetospheric substorms are $V_{A1} \sim 3 \times 10^6$ m/s, $B \sim 10 \gamma$ with length scales $\sim 10^7 - 10^8$ m leading to time scales of ~ 3-30 seconds and maximum voltages of $\sim 10^6$ volts, consistent with the highest energy particles observed in the magnetotail.

It is important in using the above result to remember that u_1 and V_{A1} refer to conditions just outside the current sheet in the region of inflow. There is no reason why these should reflect conditions at large distances (u_∞, $V_{A\infty}$) as indicated in Figure 4c. If the flow converges towards the current sheet $u_1 > u_\infty$, $B_1 < B_\infty$, $V_{A1} < V_{A\infty}$ whereas a divergent flow will make $u_1 < u_\infty$, $B_1 > B_\infty$, $V_{A1} > V_{A\infty}$. Both situations are found in more detailed analytic solutions of the reconnection problem, notably those of Petschek [1964] (see Figure 5) and Sonnerup and Priest [1975] (see Figure 6). Petschek's solution is essentially a singular perturbation treatment of the case of an unbounded medium in which the rate of reconnection is controlled by the rate at which fluid can be drawn towards the diffusing region as a consequence of the stress imbalance set up by an unimpeded outflow [Vasyliunas, 1975]. The corresponding "free" reconnection

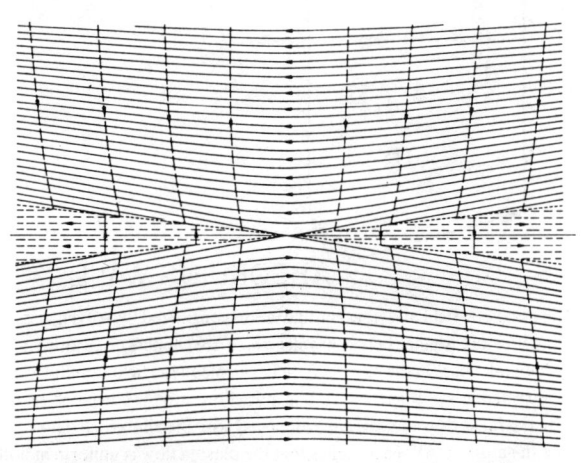

Fig. 5. The Petschek [1964] solution drawn to scale for $M_A = 0.1$ (from Vasyliunas [1975]). The magnetic field lines are solid and the plasma flow stream lines dashed. Slow shocks, which intersect at the reconnection region, are shown by dotted lines. Note that the flow converges towards the reconnection region so that $B_1 < B_\infty$ and $V_{A1} < V_{A\infty}$.

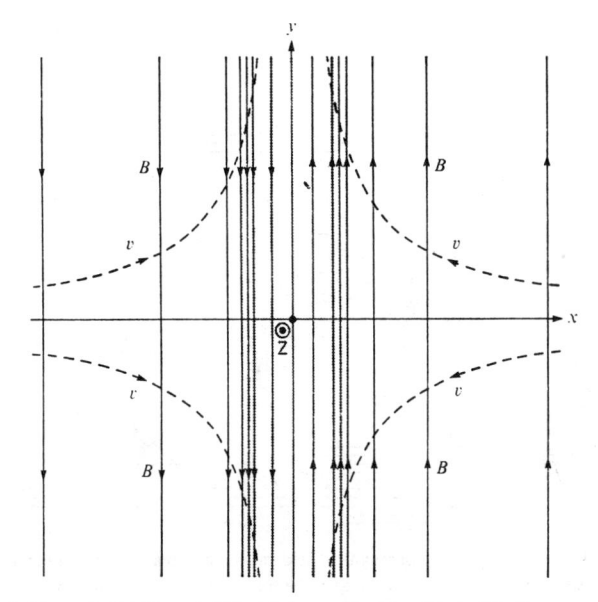

Fig. 6. Magnetic field lines (solid) and plasma flow lines (dashed) in the stagnation point flow solution found by Sonnerup and Priest [1975]. An intense current sheet is formed around $x = 0$ in which the field lines, no longer frozen to the plasma, merge and annihilate with magnetic energy being converted to heat. The flow diverges as it approaches the current sheet and the magnetic field strength correspondingly increases.

rate is

$$M_{A\infty} = \frac{u_\infty}{V_{A\infty}} \lesssim \left(\frac{\pi}{8}\right) \Big/ \log_e(2M_{A\infty}^2 R_M), \qquad (5.6)$$

where $R_M = V_{A\infty} L/\eta$ and L is the overall length of the system; the rate is only very weakly dependent on the electrical conductivity of the medium through the logarithmic term. The limitation on the rate, which differs from (5.5) is due to the fact that $B_1 < B_\infty$ and the flow converges ($u_1 > u_\infty$). In contrast, the exact stagnation point flow solution found by Sonnerup and Priest is such that the magnetic field is forced towards the current sheet from either side by fluid pressure gradients producing a diverging flow so that, in an incompressible medium the magnetic field gradients build up until the field annihilates through diffusion at exactly the rate at which it is being carried in from large distances; the annihilation rate is arbitrary and not affected in any way by the electrical conductivity of the medium.

A quite different case of driven reconnection has been discussed by Sonnerup [1970], the solution of which is the only non-singular member of the family of similarity solutions described by Yeh and Axford [1970]. The solution is obtained by matching regions of uniform flow to discontinuities and a diffusion region described approximately by the Pohlhausen technique of polynomial fitting. The configuration, in contrast to those already discussed, is bounded by plane walls with gaps on either side and the flow may be considered to be driven towards the reconnection region from either side by plane pistons which keep the flow uniform (see Figure 7). In this case, the direct influence of external boundary conditions on the reconnection region is made quite evident by the fact that it is situated at the point of intersection of waves emanating from the corners of the gaps in the walls. The rate of reconnection is again independent of the electrical conductivity of the medium and is determined entirely by the boundary conditions.

A number of authors have made numerical investigations of problems involving reconnection in finite regions with various initial and boundary conditions [e.g. Fukao and Tsuda, 1973; Ugai and Tsuda, 1977; Hayashi and Sato, 1978]. The essential results of these investigations are consistent with the deductions made from the analyses outlined above; that is, provided the conductivity is large enough for the diffusion region to be small compared with

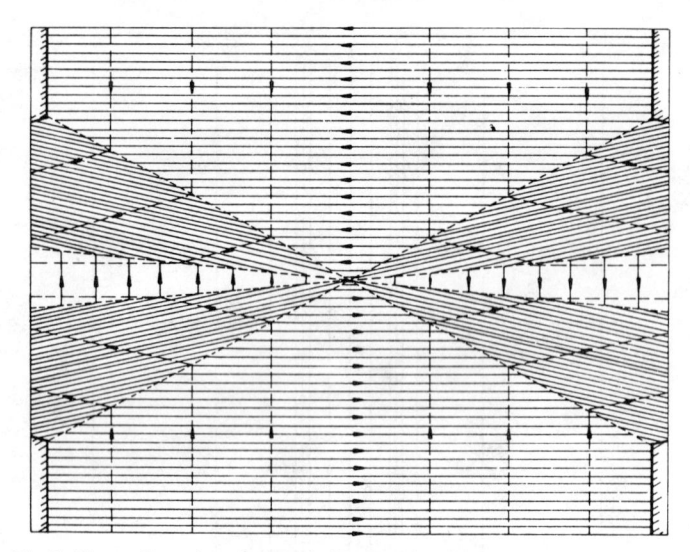

Fig. 7. The configuration of magnetic field lines (solid), plasma flow stream lines (dashed) and discontinuities (dotted) in the solution found by Sonnerup [1970] with the 'corners' required to set the proper boundary conditions corresponding to a merging rate $M_A = 0.5$ (from Vasyliunas [1975]).

the dimensions of the entire flow, the rate of reconnection is ultimately determined by boundary conditions and not by the electrical conductivity of the medium. Changes in the electrical conductivity simply produce compensating changes in the size of the diffusing region.

The essential problem in reconnection or magnetic field annihiliation is to maintain sharp gradients (i.e. high current densities) in the diffusion region by having the fluid carry the magnetic field towards it, which in turn requires the fluid must be removed from the region by some means. In all of the above situations the fluid is removed approximately along the magnetic field (x-direction) and perpendicular to the direction of the electric current (z-direction). The situation in which the current itself removes mass from the diffusion region has been considered by Alfven [1968], Dessler [1972] and Cowley [1973]. For a configuration in which uniform regions of oppositely-directed magnetic field ($B_y = \pm B_0$) are contained within parallel walls at $z = 0, L$ and with a current sheet in the plane $x = 0$ (see Figure 8), the current can potentially carry away mass at a sufficient rate to make the 'annihilation speed'

$$u_1 = \frac{2B_0}{en_0 L} = V_{A1}\left(\frac{\lambda_1}{L}\right), \quad (5.7)$$

where n_0 is the number density of the plasma and $\lambda_1 = (227 \text{ km})/(n_0 \text{ cm}^{-3})$ is the ion inertial length. Clearly this process can only compete with reconnection at an X-type neutral point if $L \sim \lambda_1$. More significantly, the process works only if the walls are effectively absorbers of the current-carrying electrons and ions, which may not be realistic in natural plasmas where the currents must close within the plasma and not on walls.

6. Configurational Instability

One may logically deduce from the previous discussion that, since the process of magnetic field line reconnection depends on the magnetic diffusivity η being non-zero, but not critically on its actual magnitude, the stability of a given configuration of plasma flow and magnetic field depends on its large-scale characteristics rather than the possible existence of localized current-driven instabilities which might produce a corresponding localized anomalously high electrical resistivity. Such localized instabilities might indeed occur and produce interesting effects, but they should not alone be able to produce a configurational instability and the latter should not be dependent on the presence of anomalous resistivity provided some form of resistivity is available, such as electron inertia as described in Section 4. For this reason, phenomena such as solar flares and magnetospheric substorms have been described as being examples of *gross resistive instabilities* [Axford, 1965], in which the whole configuration is involved, to emphasize the contrast with localized resistive instabilities which depend on the current intensity and local plasma parameters.

It has been proposed [e.g. Gold, 1964; Sweet, 1964] that a configurational instability is one in which the configuration can be imagined as evolving through a sequence of static states until a situation arises in which, as a result of magnetic field reconnection, a reconfiguration to a lower energy state occurs spontaneously. A useful mechanical analogue is the case of a ball in a valley in a flexible membrane which is continually and slowly changing shape (see Figure 9). Such configurations should also be *metastable* in the sense that disturbances of finite amplitude may induce rapid changes and energy release even when the configuration is stable against infinitesimal disturbances. Since ("sympathetic") solar flares and ("triggered") magnetospheric substorms can be initiated by finite perturbations (usually the passage of an externally produced shock) it appears that such metastable states occur, and since the magnetic field reconfigures as a result of such phenomena, both are evidently examples of configurational instabilities. It is important to note that impulsive events are observed in both cases, namely the X-ray flash phase of solar flares and the impulsive micropulsations in the initial phase of substorms, which suggest that a genuine instability is involved rather than a purely driven process as suggested by Akasofu [1981]. This argument is not completely watertight however as the impulsive phase of a substorm appears to coincide not with the commencement of reconnection, but with the time when reconnection first involves open ("lobe") magnetic field lines rather than previously closed plasma sheet field lines. At this instant, as pointed out by Sarris and Axford [1979], the Alfvén speed abruptly increases by a factor of about 10, with a corresponding abrupt increase in the reconnection rate and in the voltage available for the acceleration of particles in the vicinity of the neutral line.

The 'tearing' instability of plane current sheets separating regions of oppositely-directed magnetic fields is a case of a configurational instability. As indicated in Figure 10a, where the configuration is shown to be bounded by plane, non-conducting walls, a distance L apart, there is no unique mode of

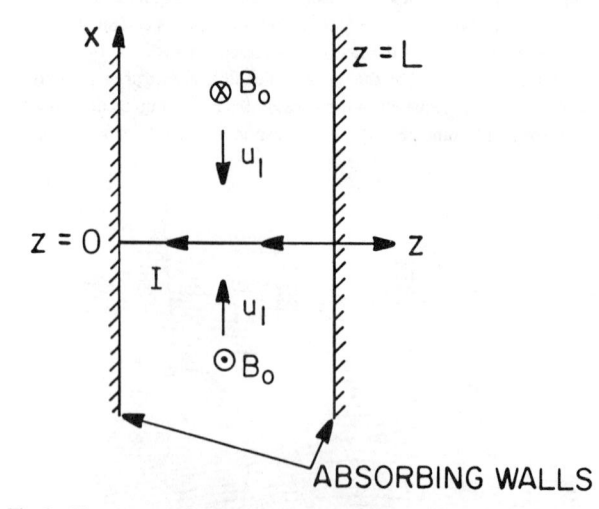

Fig. 8. The magnetic field and flow configuration considered by Alfvén and Cowley. The flow takes place between walls which are capable of absorbing particles which take part in the current flow I. The magnetic field is uniform but oppositely-directed in $x > 0$ and $x < 0$. The mass flow into the current sheet is balanced by the mass flow to the walls associated with the charge carriers constituting the current. Alfvén assumes that the plasma flow is uniform in both regions whereas Cowley treats the problem self-consistently and shows that, due to the different masses of electrons and ions, the stream lines converge to the end of the current sheet where ions are absorbed on the walls. The solution is changed if electrons and ions are emitted by the 'walls', which would be the case if the latter were another plasma in pressure equilibrium with the first.

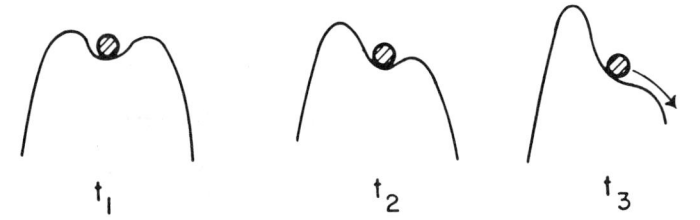

Fig. 9. Configurational instability of a heavy ball carried on a flexible membrane which changes shape with time ($t_1 < t_2 < t_3$). The configuration is stable at times t_1 and t_2 but unstable for $t = t_3$. At times such as t_2 when the valley in which the ball rests is shallow, the configuration may be metastable in the sense that the ball can escape as a result of a finite but small disturbance.

instability since any number of X-type neutral points can form; however the most unstable mode, releasing most energy, should be that with longest wavelength, i.e. containing a single X-type neutral point and half of an O-type neutral point at each wall. In all cases the new equilibrium configuration, if it can be achieved, involves conversion of magnetic energy to enthalpy in high pressure regions coinciding with the O-type neutral points. The instability will occur if the change in total energy $\iiint (B^2/2\mu + I)d^3x$ is negative (I being the enthalpy of the plasma) and there is somewhere for the excess energy to go. One can expect the above criterion to depend on the plasma β and to be easily satisified if $\beta \rightarrow 0$ and also the plasma is isothermal (when the excess energy can be radiated away). Plane current sheets are not necessarily unstable as is evidenced by the apparent stability (in general) of the current sheets defined by interplanetary sector boundaries; the plasma β and the diverging flow configuration appears to counteract instability in the inner ($\lesssim 5$ a.u.) solar system, although not necessarily at greater distances [Axford, 1972].

The nature of the tearing instability changes if the boundary conditions outlined above are changed, especially if the fluid is permited to escape through gaps in the walls as shown in Figure 10b. In this case, the configuration strives to reach a new equilibrium which is not hydrostatic, but involves a flow with a single X-type neutral point. If the walls are instead moved to infinity a similar flowing equilibrium with a single neutral point could be established, which should in fact be the Petschek solution. In all cases in which the initial configuration is static, the growth rate of the instability should be dependent on the electrical conductivity as well as on the wavelength, the structure of the current sheet and the plasma β. However, if the initial configuration involves a moving plasma with appropriate pressure gradients which force the configuration in the direction of instability the conductivity may not be important, since the flow may cause the current sheet to thin so much that it eventually tears at a high rate; it seems

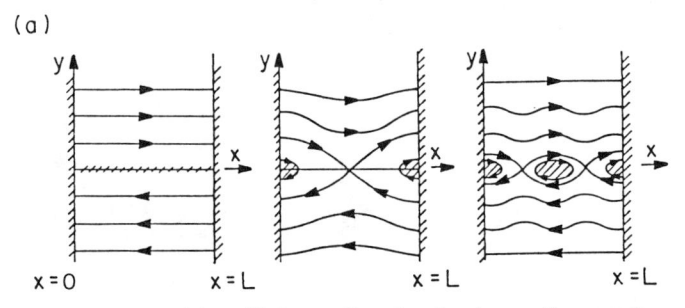

Fig. 10a. Some possible equilibrium configurations for plasma with a current layer near $y = 0$, contained between parallel, non-conducting walls at $x = 0, L$ and with the magnetic field becoming uniform at large y. The current (hatched) and its associated region of high pressure plasma may be contained in a sheet or concentrated in a series of O-type neutral points. If the plasma β is generally very small, the high pressure regions will be strongly localized and the magnetic field elsewhere almost stress-free. The configuration with a single X-type neutral point should in general correspond to the state with lowest total magnetic energy.

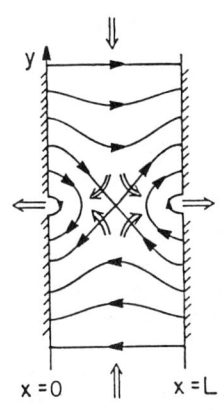

Fig. 10b. Development of the configuration shown in (a) when plasma is permitted to escape through holes in the walls at $y = 0$. In this case the new equilibrium is a steady flow with the reconnection rate at the single X-type neutral point being controlled by the rate at which plasma can escape through the holes.

possible that the initial tearing of the plasma sheet in the magnetotail leading to a substorm is of this nature.

It has often been asserted that substorms and flares are the consequence of current "disruption" with the implication that this must be caused by a sudden local enhancement of the resistivity. In fact, any reconfiguration of magnetic fluid and that due to reconnection in particular, represents current diversion or disruption, which as a description of what happened does nothing more than state the obvious without elucidating the cause. The essential question is to determine why a configuration should change rapidly, which is in turn connected with the question of why it was apparently stable beforehand. As we have argued throughout this paper, this cannot be a consequence of a decrease in electrical conductivity since the possibility of reconnection occurring is not affected provided the effective electrical conductivity is initially finite. Anomalous resistivity only changes the size of the 'diffusion' region surrounding an X-type neutral point which may appear, but in fact the conditions for its existence in such a region are to some extent disfavoured by a reduction in the current intensity and the short transit time of the plasma. The real cause of instability can only be ascribed to a change in the overall stress balance in the configuration (i.e. gradients of plasma pressure and Maxwell stresses) which as it evolves, becomes unstable so that reconnection (with or without anomalous resistivity) leads to rapid conversion of magnetic energy to plasma kinetic energy and enthalpy.

Finally, it is worth noting that, as a consequence of reconnection, magnetic field line configurations can arise which would otherwise be unexpected. In particular, reconnection in a sheared (three-dimensional) current sheet can lead to a twisted field which subsequently unwinds or straightens, generating Alfvén waves. An example occurs in the magnetotail, where an east-west component of the interplanetary magnetic field will cause the tail field lines to connect to their "wrong" partners; this is evidenced by "non-conjugacy" of high latitude aurorae [Mendis and Axford, 1970] and presumably by the generation of micropulsation activity in the magnetotail [Piddington, 1965, Axford, 1969]. If the configuration has two reconnection points, producing an O-type neutral point as shown in Figure 10a, the flux tube surrounding the latter will in general be twisted, forming a flux "rope", which has implications for flux tubes that become disconnected from the magnetotail for example. Wave-like structures in comet tails may have a similar explanation, rather than Kelvin-Helmholtz instabilities as is often proposed.

References

Akasofu, S. -I., Energy coupling between the solar wind and the magnetosphere, *Space Sci. Rev., 28,* 121, 1981.

Alfvén, H., Some properties of magnetospheric neutral surfaces, *J. Geophys. Res., 73,* 4379, 1968.

Alfvén, H., On frozen-in field lines and field-line reconnection, *J. Geophys. Res.*, *81* 4019, 1976.

Alfvén, H., Electric currents in cosmic plasmas, *Rev. Geophys. Space Phys. 15*, 271, 1977.

Alfvén, H., *Cosmic Plasma*, D. Reidel Publ. Co., Dordrecht, Holland, 1981.

Axford, W. I., Magnetic storm effects in the tail of the magnetosphere, *Proc. ESRO Symposium, Stockholm*, 1965 (see also *Space Sci. Rev., 7*, 149, 1967).

Axford, W. I., Magnetospheric convection, *Rev. Geophys., 7*, 421, 1969.

Axford, W. I., The interaction of the solar wind with the interstellar medium, in *Solar Wind, NASA Spec. Publ. SP-308*, 609, 1972.

Batchelor, G. K., *'An Introduction to Fluid Dynamics'*, Cambridge University Press, 1970.

Chapman, S., and P. C. Kendall, Liquid instability and energy transformation near a magnetic neutral line: A soluble non-linear hydromagnetic problem, *Proc. Roy. Soc. London, Ser. A, 271*, 435, 1963.

Coppi, B., G. Laval, and R. Pellat, Dynamics of the geomagnetic tail, *Phys. Rev. Lett, 16*, 1207, 1966.

Cowley, S. W. H., A self-consistent model of a simple magnetic neutral sheet system surrounded by a cold, collisionless plasma, *Cosmic Electrodynamics, 3*, 448, 1973.

Dessler, A. J., Vacuum merging: A possible source of the magnetospheric cross-tail electric field, *J. Geophys. Res., 76*, 3174, 1971.

Dungey, J. W., Conditions for the occurrence of electrical discharges in astrophysical systems, *Phil. Mag., Ser. 7, 44*, 725, 1953.

Dungey, J. W., *Cosmic Electrodynamics*, pp. 48-54, Cambridge University Press, London, 1958.

Dungey, J. W., Interplanetary magnetic field and the auroral zones, *Phys. Rev. Lett., 6*, 47-48, 1961.

Ferraro, V. C. A., and Plumpton, C., *An Introduction to Magneto-Fluid Mechanics*, Clarendon Press, Oxford, 1966.

Frank, L. A., Plasmas in the earth's polar magnetosphere, *J. Geophys. Res., 76*, 5202, 1971.

Fukao, S., and T. Tsuda, Re-connection of magnetic lines of force: Evolution in incompressible MHD fluids, *Planet. Space Sci., 21*, 1151, 1973.

Furth, H. I., J. Killeen and M. N. Rosenbluth, Finite-resistivity instabilities of a sheet pinch, *Phys. Fluids, 6* 459, 1963.

Giovanelli, R. G., Magnetic and electric phenomena in the sun's atmosphere associated with sunspots, *Mon. Notices Roy. Astron. Soc., 107*, 338, 1947.

Gold, T., Magnetic energy shedding in the solar atmosphere, in *AAS-NASA Symposium on the physics of Solar Flares, NASA Spec. Publ. SP-50.*, 389, 1964.

Gold, T., and F. Hoyle, On the origin of solar flares, *Mon. Notices Roy. Astron. Soc., 120*, 89, 1960.

Gurnett, D. A., and L. A. Frank, VLF hiss and related plasma observations in the polar magnetosphere, *J. Geophys. Res., 77*, 172, 1972.

Hayashi, T., and T. Sato, Magnetic reconnection-acceleration, heating and shock formation, *J. Geophys. Res., 83*, 217, 1978.

Hoyle, F., *Some Recent Researches in Solar Physics*, Cambridge University Press, London, 1949.

Mendis, D. A., and W. I. Axford, Geomagnetic Conjugacy Variations, *Comments on Astrophys. Space Phys., 2*, 99, 1970.

Parker, E. N., Sweet's mechanism for merging magnetic fields in conducting fluids, *J. Geophys. Res., 62*, 509, 1957.

Parker, E. N., The solar-flare phenomenon and the theory of reconnection and annihilation of magnetic fields, *Astrophys. J., suppl. Ser., 8*, 177, 1963.

Petschek, H. E., Magnetic field annihilation, in *AAS-NASA Symposium on the Physics of Solar Flares, NASA Spec. Publ. SP-50, 425*, 1964.

Piddington, J. H., The geomagnetic tail and magnetic storm theory, *Planet. Space Sci., 13*, 281, 1965.

Rossi, B., and S. Olbert, *Introduction to the Physics of Space*, McGraw-Hill, New York, 1970.

Sarris, E. T., S. M. Krimigis and T. A. Armstrong, Observations of magnetospheric bursts of high energy protons and electrons at $35R_e$ with IMP-7, *J. Geophys. Res., 81*, 2341, 1976.

Sarris, E. T., and W. I. Axford, Energetic protons near the plasma sheet boundary, *Nature, 277*, 460, 1979.

Sonnerup, B. U. Ö., Magnetic-field re-connection in a highly conducting incompressible fluid, *J. Plasma Phys., 4*, 161, 1970.

Sonnerup, B. U. Ö., and E. R. Priest, Resistive MHD stagnation-point flows at a current sheet, *J. Plasma Phys., 14*, 283, 1975.

Sweet, P. A., The neutral point theory of solar flares, in *Electromagnetic Phenomena in Cosmical Physics*, edited by B. Lehnert, Cambridge University Press, London, 1958.

Sweet, P. A., Instability problems in the origin of solar flares, in *AAS-NASA Symposium on the Physics of Solar Flares, NASA Spec. Publ. SP-50*, 409,1964.

Uberoi, M. S., Some exact solutions of magnetohydrodynamics, *Phys. Fluids, 6*, 1379, 1963.

Ugai, M., and T. Tsuda, Magnetic field-line reconnection by localized enhancement of resistivity, *J. Plasma Phys., 17*, 337, 1977.

Vasyliunas, V. M., Magnetic field line merging, *Rev. Geophys. Space Phys. 13*, 303, 1975.

Yeh, T., and W. I. Axford, On the re-connection of magnetic field lines in conducting fluids, *J. Plasma Phys., 4*, 207, 1970.

SPONTANEOUS RECONNECTION

K. Schindler

Ruhr-Universität Bochum, D-463 Bochum,
Federal Republic of Germany

Abstract. Spontaneous reconnection occurs when an originally stable system, which is gradually changed by external forces, passes over an onset point of an instability involving magnetic reconnection. The prototype of the instability is the tearing mode of a plane plasma sheet. Since plane plasma sheets are always unstable, more refined theories are being developed to explain the transition from stability to instability. In the case of the earth's magnetotail, during times when microturbulence is negligible, the main stabilizing factor is a finite amount of magnetic flux passing perpendicularly through the plasma sheet. If the motion of the electrons is adiabatic, the WKB-regime (small wavelengths along the direction of the main magnetic field) is stable. A recent numerical study has established an instability for wavelengths comparable with the equilibrium scale length along the main magnetic field. A sudden appearance of fluctuations or ionospheric feedback can lead to rapid growth. In the fluid approach the attention presently concentrates on the efficiency of the lower-hybrid-drift instability for generating the required resistivity. Several other factors that control spontaneous reconnection are discussed, among them shear in the magnetic field and in the plasma flow. Expressions for free energy available for reconnection-type instabilities are discussed, corresponding to situations with and without fluctuations. Large scale MHD-computations have largely confirmed and extended the analytical work, particularly for magnetotail configurations; there is also remarkable agreement with an empirical model. It appears that spontaneous reconnection plays a decisive role in controlling magnetic topology and in releasing previously stored free energy.

1. Introduction

Reconnection involves an electric field at a separator, i.e. at the intersection line of two surfaces separating magnetic fluxes of different topology. If this electric field occurs as an inherent part of an instability (an unstable linear eigenmode in the small perturbation limit) we speak of "spontaneous reconnection."

The prototype of spontaneous reconnection is the tearing instability of a one-dimensional plasma sheet with isotropic pressure separating antiparallel magnetic fields [Furth, 1962; Furth et al. 1963]. The instability is caused by the Lorentz force between parallel electrical currents leading to the formation of magnetic islands (Figure 1a). Tearing will not necessarily occur under all circumstances. One of the essential requirements for (spontaneous or driven) reconnection is that the medium must allow for changes in the magnetic field topology. Typically, this property holds for systems with dissipation caused either by fluctuations (Coulomb collisions, microturbulence generated by a small scale instability or fluctuations excited by an external source) or by inertia effects such as Landau-type resonance. Both cases involve violation of magnetic line and flux conservation, i.e.

$$\underline{E} + \underline{v} \times \underline{B} \neq 0. \tag{1.1}$$

For the prototype instability this condition follows directly from the fact that it generates separators with B=0 and E≠0 (Figure 1a). The r.h.s of (1.1) is directly related to the growth rate γ of the instability. If we perturb the magnetic flux function A(x,z,t) by a linear mode of the form $A_1 = \hat{A}_1(z)\exp(\gamma t)\sin kx$ one finds for z=0

$$(\underline{E}_1 + \underline{v}_1 \times \underline{B}_o)_y = -\gamma \hat{A}_1(o)\exp(\gamma t)\sin kx \tag{1.2}$$

Clearly, for a growing mode of the topological structure shown in Figure 1a the r.h.s. of (1.2) does not vanish.

If the equilibrium is such that a non-vanishing amount of magnetic flux passes through the neutral sheet the instability that causes spontaneous reconnection has the structure shown in Figure 1b. This case is of particular relevance for space plasmas and it will be addressed more in detail in sections 2 and 3.

Another instability that meets the definition

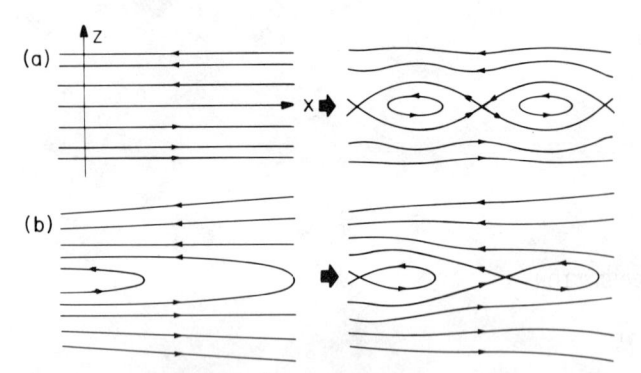

Fig. 1. Basic forms of spontaneous reconnection: (a) The tearing instability of a plane plasma sheet. (b) The generalized tearing instability in the presence of a normal magnetic field component.

of spontaneous reconnection is the coalescence mode where two magnetic islands spontaneously coalesce to form a larger island [Biskamp and Schindler 1971]. Also in that case a current sheet may be important, which forms between the contracting islands before significant magnetic flux merges [Finn and Kaw, 1977; Pritchett and Wu, 1979; Biskamp and Welter, 1980]. Figure 2 illustrates this process schematically. Coalescence typically occurs during the non-linear growth of the tearing mode [Tajima, 1981].

Spontaneous reconnection also occurs in configurations with magnetic shear. This effect is most pronounced in the resistive fluid case. We assume a two-dimensional equilibrium with translational invariance with respect to the y-direction, with shear generated by the B_y component. Figure 3a gives a one-dimensional example. It is convenient to split the magnetic field and the plasma velocity into a "toroidal," i.e. y-component and "poloidal" components, i.e. projection into the x,z-plane, deriving the poloidal components from flux functions A and D, respectively:

$$\underline{B} = \underline{B}_p + \underline{B}_T, \quad \underline{B}_p = \nabla A \times \underline{e}_y, \quad \underline{B}_T = B_y \underline{e}_y$$

$$\underline{v} = \underline{v}_p + \underline{v}_T, \quad \underline{v}_p = \nabla D \times \underline{e}_y, \quad \underline{v}_T = v_y \underline{e}_y \quad (1.3)$$

Assuming a resistive and incompressible (constant density ρ) fluid, one finds that the MHD equation may be cast into the following form

$$\frac{\partial A}{\partial t} = [D, A] + \frac{\eta}{\mu_o} \Delta A \quad (1.4)$$

$$\frac{\partial \Delta D}{\partial t} = [D, \Delta D] + \frac{1}{\mu_o \rho} [\Delta A, A] + \nu \nabla^4 D \quad (1.5)$$

$$\frac{\partial B_y}{\partial t} = [D, B_y] + [v_y, A] + \frac{\eta}{\mu_o} \Delta B_y \quad (1.6)$$

$$\frac{\partial v_y}{\partial t} = [D, v_y] + \frac{1}{\mu_o \rho} [B_y, A] + \nu \Delta v_y \quad (1.7)$$

where $[\alpha, \beta]$ denotes the Jacobian $\partial(\alpha, \beta)/\partial(x, z)$ and Δ denotes the Laplacian, ∇^2. For convenience we have included a kinematic viscosity ν.

We note that equations (1.4) and (1.5) decouple from the others in the sense that they contain only A and D. If A and D have been obtained from these equations, B_y and v_y are then found from (1.6) and (1.7). Thus it follows that the stability properties are determined by the poloidal fields only. Furthermore, one finds that the plasma pressure before it has been eliminated to form equations (1.4)-(1.7), enters only in the combination $p + B_y^2/2\mu_o + \rho(\nabla D)^2/2$. This means that in a static equilibrium, pressure, p, may be partly or wholly replaced by $B_y^2/2\mu_o$ without influencing the stability properties. Thus, the tearing instability shown in Figure 1a with $B_y = 0$ occurs in the same way in the force-free equilibrium of Figure 3a, leading to the field structure sketched in Figure 3b. The magnetic shear layer breaks up into flux tubes with helical field lines. Note that under more general conditions the asymptotic magnetic fields on both sides of the plasma/shear-layer may form any nonvanishing angle to guarantee instability. The asymptotic field angle may, however, influence the turbulence, giving rise to the resistivity. Huba et al. [1982] have shown that deviations from the antiparallel field orientation have an inhibiting effect on the penetration of the lower-hybrid-drift-mode (see section 2) toward the center plane of the plasma sheet. Figure 3c illustrates how a pair of sheared flux ropes can reconnect. Parker [1979] argues that flux rope reconnection is an important aspect of three-dimensional cosmical magnetic fields in their time-dependent evolution.

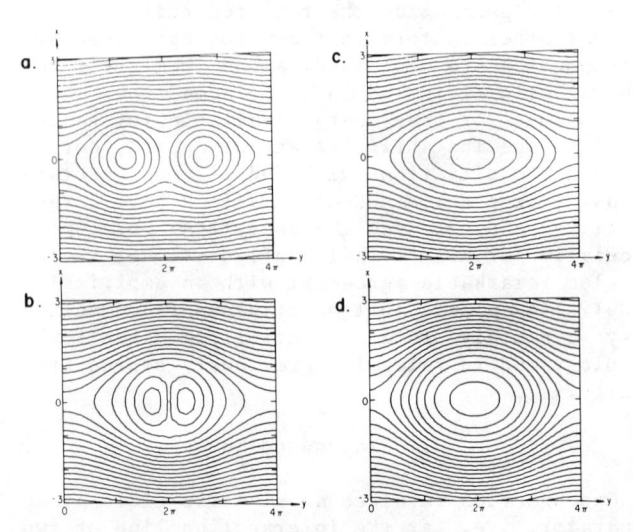

Fig. 2. Basic forms of spontaneous reconnection: Coalescence of magnetic islands [Pritchett and Wu, 1979].

The processes discussed so far may be considered as basic elements of spontaneous reconnection. Such configurations (e.g., current sheets) may be contained in various global field structures, that vary from case to case. Note that the instability exists in the basic structure without necessarily involving the larger scales. If, in addition, there is a global instability causing spontaneous reconnection, this process is of course much more difficult to describe theoretically. An important example is the possibility of ionospheric feed-back (section 3) into spontaneous reconnection in the magnetotail. An analysis which would fully include ionospheric effects would require analyzing eigenmodes of the entire magnetosphere.

Although this paper predominantly addresses theoretical results, it seems appropriate to list some of the main applications of spontaneous reconnection, without however aiming for details nor for completeness. Spontaneous reconnection has been inferred from in situ observations in the earth's magnetotail [Hones 1977 and 1979]. In this picture spontaneous reconnection initiates a process by which energy previously stored in the magnetotail is released partly toward the earth and partly away from the earth. The tailward energy flux includes the motion of a plasmoid. This picture is coincident with the observed delay of about 1 h between interplanetary and ionospheric parameters at intermediate activity [Bargatze et al. 1983]. Other investigations based on observations have also strengthened the conclusion that a neutral line forms in the near-earth-magnetotail [Nishida and Nagayama, 1973; Pytte et al., 1976]. Time-dependent changes of the magnetic field topology have also been observed at the magnetopause [Russell and Elphic 1978, Paschmann et al. 1978]. Time-dependent (as well as steady) reconnection at the magnetopause seem to play a key role in controlling energy influx from the solar wind into the magnetosphere. Solar activity is also widely associated with spontaneous reconnection [Kopp and Pneuman, 1976; see also papers in Priest 1981]. Further potential candidates for reconnection are stellar accretion disks and galactic jets. Spontaneous reconnection also occurs in appropriate laboratory plasma configurations, e.g. in the experiments by Stenzel and Geckelman (1981), in Θ-pinches and similar arrangements [Irby et al. 1976] and in Tokamaks [Kadomtsev, 1975; Wadell et al. 1979].

We emphasize that the term 'spontaneous' merely points at the presence of an instability causing the reconnection electric field. This term should not be taken as 'unpredictable'. In fact, if a stable plasma configuration is slowly changed such that it passes an onset point of instability, the onset point and the main characteristics of the growing mode are predictable. If the instability evolves into a turbulent state statistical averages will remain reproducible. In input-output correlation studies the storage

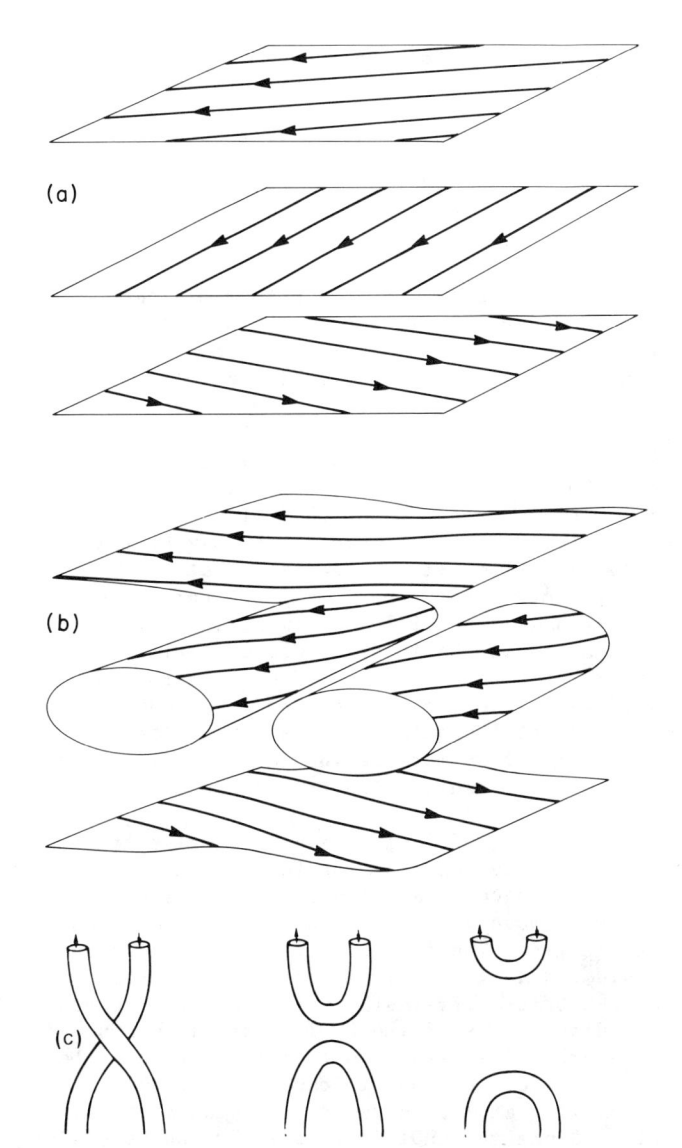

Fig. 3. Basic forms of spontaneous reconnection: The tearing instability (b) of a force free current sheet (a); spontaneous reconnection of magnetic flux ropes (c) [after Parker, 1979].

and unstable unloading will typically cause a characteristic time lag [Bargatze et al. 1983].

On the basis of its established and potential applications the process of spontaneous reconnection may be regarded as a fundamental plasma physics phenomenon. It seems to control the magnetic topology, to initiate the release and dissipation of previously stored energy in eruptive processes, to initiate steady state reconnection and accelerate particles. The relevance of spontaneous reconnection as a basic process of macroscopic plasmas is also apparent from the formal similarity of equations (1.4) and (1.5) with the Boussinesq equations describing Benard convection. Maschke and Saramito [1982] recently showed that both sets of equations possess similar bifurcation properties.

In the following sections we concentrate on particular aspects of spontaneous reconnection. Although we specifically address the tail of the magnetosphere, most of the results presented have a wider range of applicability.

2. Spontaneous Reconnection in the Earth's Magnetotail

Considerable theoretical effort has been devoted to the question whether the unloading process in the magnetotail can be explained by spontaneous reconnection [Schindler, 1974; Galeev and Zelenyi, 1976; Lembége, 1976; Galeev et al., 1978; Pellat, 1979; Coroniti 1980; Goldstein, 1981; Goldstein and Schindler, 1982]. Earlier work [Coppi et al. 1966], using the one-dimensional plasma sheet model, had already demonstrated that the magnetotail may be tearing unstable. Later it was realized that a satisfactory theory should explain both quiet and disturbed tail conditions. In fact the later refinements of the theory yielded stabilizing effects both for collision-free and turbulence based resistive modes. In both approaches the existence of the quiet tail is now at least plausible. Even more refinements, however, seem necessary to identify the precise condition of instability.

In the case of (collective) resistive tearing modes the occurrence of the instability depends on the presence of a suitable fluctuation field on the microscopic level generating a macroscopic resistivity. The instability that has been considered as the most promising one is the lower-hybrid-drift-instability (LHDI). From recent discussions of the LHDI in the literature it has emerged that two main conditions have to be satisfied for the LHDI to cause sufficient resistivity at the center of the plasma sheet:

(i) Since the LHDI is favored by low values of β (ratio of kinetic to magnetic pressure) there must exist an effective transport of the waves from the boundary of the plasma sheet to the center [Huba et al. 1977, Huba et al. 1980].

(ii) The turbulence must be such that it scatters electrons in an irreversible way [Drake and Lee, 1981; Huba et al. 1981]

At present, it seems that particularly the second condition imposes a rather severe limitation to the relevance of the LHDI for resistive tearing. As shown by Drake and Lee (1981) significant irreversibility (which is due to electron trapping) requires amplitudes that are available only if the plasma sheet width becomes smaller than an ion gyro-radius. This is a severe limitation because the quiet plasma sheet might only rarely reach that regime. In view of this difficulty, Huba et al. (1981) have suggested that the LHDI may be more relevant for externally driven than for spontaneous reconnection. Future more refined studies have to show, whether the requirements (i) and (ii) can be met under realistic conditions or whether or not perhaps there exists a more favorable instability.

In the case of collision-free modes the stabilization comes mainly from the normal magnetic field component $B_n = |B_z|$ (Figure 1b). It seems that B_n is always large enough to suppress electron Landau resonance. Therefore the attention concentrated on the role of non-adiabatic ions [Schindler 1974]. The work of Galeev and Zelenyi [1976], by Lembége [1976] and by Coroniti [1980] showed, however, that the ion-tearing mode is strongly stabilized by the adiabatic electrons, particularly for wavelengths in the WKB-regime $k_x L_x \gg 1$, where k_x and L_x denote wavenumber and equilibrium scale length along the direction of the tail axis (Figure 1b). The phenomenon of electron stabilization and recent results for arbitrary wavelengths are discussed separately in the following section.

The storage mode before onset of the instability seems to be a natural consequence of magnetotail convection due to the appearance of a crosstail electric field during periods with a southward interplanetary magnetic field component. This behavior was found in 2D and 3D self-consistent modelings of magnetotail convection, assuming that the outer closed flux tubes of the plasma sheet do not experience significant losses along the magnetic field lines [Schindler and Birn, 1982; Birn and Schindler, 1983]. Realistic steady convection states are inhibited by the pressure inconsistency due to the large compression ratios involved, as first pointed out by Erikson and Wolf [1980]. Figure 4 illustrates a typical time-development of the magnetic field during quiet periods between times of pronounced tail dynamics. An analysis of the evolution of the stability properties during time-dependent tail convection clearly shows considerable destabilization.

3. 2-D-Stability Theory of Plasma Sheets With $B_n \neq 0$ Against Perturbations of Arbitrary Wave-Length

3.1 Laminar Vlasov Theory

We first discuss the laminar collision-free case and then include fluctuations in section 3.2.

As pointed out above, an important stabilizing factor is due to adiabatic electrons. Following Lembége [1976], Coroniti [1980] obtained the following estimate for the variational functional δW [Laval et al. 1966; Schindler, 1966, Schindler et al. 1973]:

$$\delta W = C \int dA_0 |A_1|^2 \left[k_x^2 L_z^2 - 1 + \frac{\pi k_x^2 L_z^2 B_0}{8 B_n} \right] \quad (3.1)$$

Here A_0 denotes the equilibrium flux function, A_1 its perturbation, L_z the width of the plasma sheet, B_0 the asymptotic value of the magnetic field strength away from the neutral sheet and C a positive factor. (3.1 involves a special choice of a tearing type trial function A_1.) The WKB-approximation and additional simplification restrict the wavenumber k_x to the regime

$$k_x L_z \lesssim 1 \qquad (3.2)$$

for $k_x L_z$ of order 1 the last term in (3.1) clearly dominates, showing the stabilization for small wave-lengths.

Noting that in (3.1) as in similar expressions derived by Lembége [1976] the stabilizing effect decreases as k_x decreases, the question arose, whether the ion-tearing mode does exist outside the WKB-regime, e.g. at $k_x L_z \sim B_n/B_o$. This question was addressed by Goldstein [1981] and by Goldstein and Schindler [1982], taking into account self-consistent variations along the plasma sheet and ignoring only terms of order $[B_n/B_o]^2$. In that case, eigenmodes are no longer necessarily of the form $\exp(ik_x x)$ and a (partly) numerical approach is required. Goldstein [1981] used δW in the following rigorous form:

$$\delta W = \delta F + \delta W_i + \delta W_e + \delta W_\phi \qquad (3.3)$$

where

$$\delta F = \int dx dz \left(\frac{(\nabla A_1)^2}{2\mu_o} - \frac{\partial j_o}{\partial A_o} A_1^2 \right) \qquad (3.4)$$

$$\delta W_s = \frac{1}{2} e_s^2 \int d\Omega \left| \frac{\partial F_{os}}{\partial H_{os}} \right| < \frac{1}{m_s} (P - e_s A_o) A_1 >^2 \qquad (3.5)$$

$$\delta W_\phi = \sum_s \frac{1}{2} e_s^2 \int d\Omega \left| \frac{\partial F_{os}}{\partial H_{os}} \right| (\phi_1 - <\phi_1>)^2 \qquad (3.6)$$

where $s = i, e$ denotes particles species, j_o equilibrium electric current density, $F_{os}(H_{os}, P)$ equilibrium distribution functions as a function of Hamiltonian H_o and canonical momentum y-component P (used as a coordinate). The symbol $<\cdots>$ indicates a phase space average extended over the domain in phase space that is accessible to a particle of a given set of constants of the motion. The perturbation of the electric potential ϕ_1 is related to the trial flux function A_1 by the condition of quasi-neutrality:

$$\frac{\partial \rho_o}{\partial \phi_o} \phi_1 + \frac{\partial \rho_o}{\partial A_o} A_1 +$$

$$\sum_s e_s^2 \int dv_x d\frac{P}{m} dv_z \left| \frac{\partial F_{os}}{\partial H_{os}} \right| \times$$

$$< - \frac{1}{m_s}(P - e_s A_o)A_1 + \phi_1 > = 0 \qquad (3.7)$$

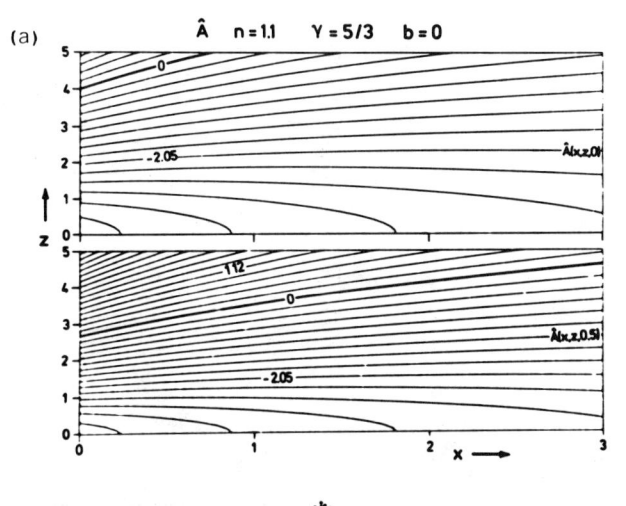

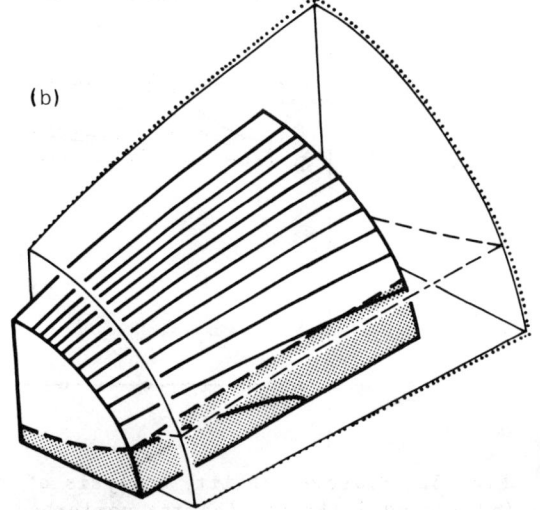

Fig. 4. Magnetic field evolution during time-dependent convection in the earth's magnetotail; (a) from two-dimensional theory [Schindler and Birn, 1982], (b) from three-dimensional theory [Birn and Schindler, 1983].

Here the electron-effects are contained both in δW_e and δW_ϕ; δF corresponds to the first two terms in (3.1). Figure 5 shows the results for a case with $b = B_n/B_o = 1/45$, $\varepsilon_i = a_{io}/L_z = 0.35$ (where a_{io} is the ion gyroradius evaluated at $B = B_o$, assuming protons). Figure 5a shows the equilibrium magnetic field lines, 5c the minimum of δW (curve d) as well as the contributions from individual terms in (3.3). The discontinuity of curves c and d is due to the change from adiabatic to nonadiabatic electron motion at $\varepsilon_e = b^2$ where the minimum radius of curvature of the magnetic field lines becomes comparable with the electron gyro-radius. Clearly, there exist significant intervals with $\delta W < 0$, corresponding to instability. Figure 5b shows the magnetic field lines of the minimizing mode for the unstable case $\varepsilon_e = 6.6 \times 10^{-4}$. In all cases that were considered, nonadiabatic ion motion was necessary for obtaining instability. The variational anal-

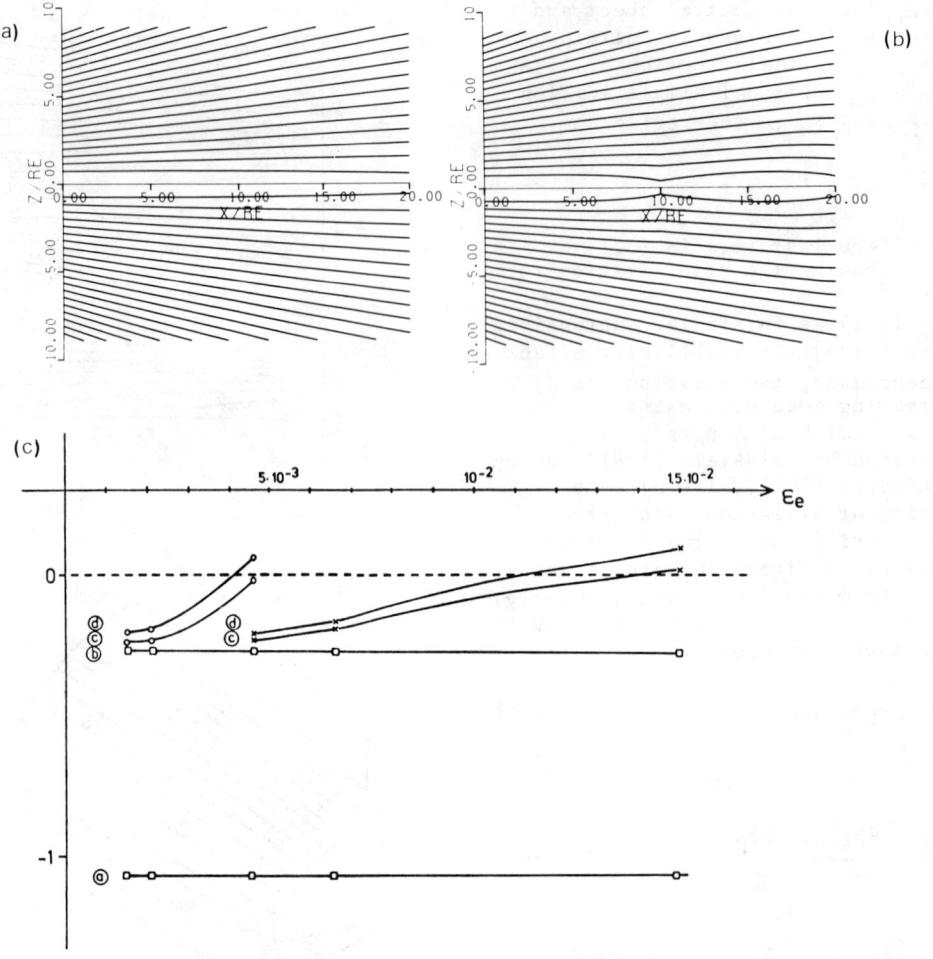

Fig. 5. Vlasov-stability analysis of the earth's magnetotail [Goldstein, 1981]; (a) the equilibrium, (b) the perturbed state, (c) the minimum of δW as a function of ε_e. (Curves a, b, c relate to individual contributions of δW.)

ysis used, of course, cannot identify details of the dynamics such as the possible role of bounce resonances. It should be noted, that it takes a finite amplitude before the topology of the field changes, because a neutral point requires $B_z = B_{zo} + B_{z1} = 0$, where B_{z1} is the perturbation $B_{z1} = \partial A_1/\partial x$. For nonadiabatic ions, the occurrence of neutral lines would not make an important difference. Electrons, however, will become non-adiabatically accelerated near the X-type neutral line(s). It is of interest to investigate this effect theoretically; however, at present, rigorous results are not available.

The point of view discussed in this section is in various ways an extreme case in which the constraints imposed on the particles are particularly strong. In the following two sections we discuss phenomena that would reduce the electron stabilization.

3.3 Destabilization by Fluctuations

Dissipative fluctuations caused by independent sources generally destabilize tearing modes. If the fluctuations are sufficiently strong and if the corresponding momentum exchange between electrons and ions can be described by an effective resistivity, resistive fluid theory applies (see section 2). In this case tearing is much less sensitive to B_n than in the framework of laminar Vlasov theory. In fact, the growth rate is not changed at all in the regime $b \lesssim S^{-1/2}$ where S denotes the magnetic Reynolds number [Janicke, 1980; 1982]. Here the main problem is the origin of the rather strong turbulence required.

If the fluctuations do not scatter particles sufficiently for fluid theory to apply, a kinetic description is required. Coroniti [1980] demonstrated that electron pitch angle scattering due to fluctuations may give rise to tearing for $B_n \neq 0$ and for wavelengths belonging to the WKB-regime. The scattering reduces the constraints that a purely laminar theory with adiabatic electrons imposes on the system. As a result the instability is shifted to smaller wavelengths. Applying these results to the geomagnetic tail, Coroniti finds a wavelength of the order of 15 R_E to be consistent with a magnetic field fluctuation spectrum with intensity $\delta B = 10^{-2}$ NT s$^{-1/2}$. The destabilization and the shift of the instability toward smaller wavelengths owing to fluctuations was recently confirmed by an approach based on statistical mechanics, computing the relaxed (maximum entropy) distribution function for a current carrying plasma [Kiessling, 1984]. On the basis of these results, we can conclude that the marginal wavelength, that is of the order of L_x without fluctuations, will be reduced to the order of L_z if fluctuations are effective.

3.3 Destabilization by Ionospheric Feedback

In almost all existing theoretical studies on tail dynamics one assumes that the mode decays away from the current sheet; in fact the tearing mode may be considered as a surface mode of a plasma sheet (magnetic vortex sheet) similar to the Kelvin-Helmholtz mode of a bulk flow vortex sheet. This approach is of course legitimate if one is interested in necessary conditions for instability. In the language of variational theory one looks at a restricted class of perturbations.

A more complete analysis must take the presence of neighboring regions into account. In fact, it seems that under suitable conditions, they can have a destabilizing effect.

Two suggestions have been made regarding the destabilization of the magnetotail by the ionosphere:

(i) Goldstein and Schindler [1978] suggested that electron relaxation in ion tearing modes might occur via field-aligned currents through the ionosphere. This process is similar to electron relaxation by fluctuations, except that the relaxation takes place in the ionosphere. The field-aligned currents lead to a redistribution of the electrons in the magnetotail.

(ii) Baker et al. [1982] have pointed at the fact that heavy ions such as oxygen ions provided by the ionosphere may reduce the onset threshold in the tail because heavier ions become nonadiabatic for a smaller degree of tail stretching than protons. From observed oxygen densities Baker et al. [1982] conclude that the maximum growth rate of the ion-tearing mode would appear in the range $-15 R_E < X_{GSM} < -10 R_E$ and $Y_{GSM} \approx 5 R_E$. (This study assumed absence of significant electron stabilization.)

These are first steps toward a global approach which seems to be required for a more complete understanding.

4. Further Results

Electron tearing is well-known to be sensitive to pressure anisotropy [e.g. Coppi and

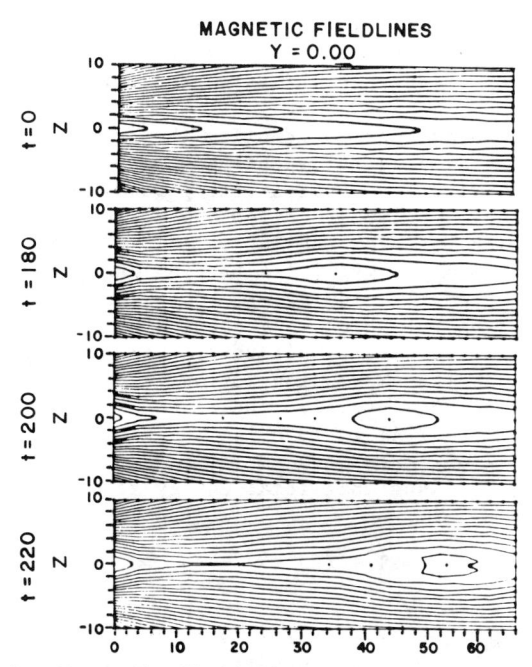

Fig. 6. Magnetic field lines of resistive MHD dynamics of the earth's magnetotail from a three-dimensional computer code [Birn and Hones, 1981].

Rosenbluth, 1968]. The effect of magnetic shear and of finite temperature gradients on this mode was investigated by Coppi et al. [1978] and considerable stabilization was found. The ion mode is less affected by anisotropy because the effect is determined by ε_i instead of ε_e. In the case of the Earth's magnetotail the pressure anisotropy during quiet times was measured to be of the order or less than 10% [Stiles et al., 1978]. The problem here seems not so much the stability of anisotropic plasmas, but rather to explain the large degree of isotropy of the steady state. A recent attempt in that direction has resulted in the concept of "non-local isotropization". This process is based on the small isotropy-threshold for anisotropy driven instabilities in the high-β domain near the center of the plasma sheet. A further reduction of anisotropy stems from adiabatic particle motion for suitable distribution functions such as double Maxwelleans [Nötzel et al. 1984].

Relativistic effects lead to an increase of the electron tearing growth rate because increasing particle inertia gives rise to more pronounced nonadiabaticity of the particles [Otto, 1983].

Shear flow may under suitable circumstances also increase the growth rate of the electron mode [Lakhina and Schindler, 1983].

Three-dimensional effects and non-linearity have been studied for the resistive tearing case by large scale MHD-computations [Birn, 1980; Birn and Hones, 1981]. These studies (Figure 6) clearly show the occurrence and the non-linear growth of spontaneous reconnection modes, largely in agreement with the empirical model by Hones [1977, 1979]. For further discussions of the MHD-computations see the survey by Birn in this volume.

In an attempt to identify the free energy that is available for spontaneous processes in a two-dimensional collision-free plasma, Schindler and Goldstein [1983] derived a non-linear expression W[A] for the "potential" or free energy. $W[A_o] - W[A_o^*]$ is the free energy available in an equilibrium A_o, A_o^* being the stable equilibrium for which W[A] assumes an absolute minimum. For small amplitudes $\delta W[A] - W[A_o]$ reduces to the expression for δW as given by (3.3) with (3.4)-(3.7). The method for constructing W[A] is to split the kinetic energy E_K into a non-adiabatic part T and an adiabatic part $E_K^{(ad)}$ such that total energy ready

$$E = E_K + E_M = T + (E_K^{(ad)} + E_M) = T + W \quad (4.1)$$

The definition of $E_K^{(ad)}$ is such that T vanishes in equilibrium and is positive otherwise. The explicit expression for W[A] is

$$W[A] = \sum_s \int H_s F_s(H_s, P) d\Omega + \frac{1}{2\mu_o} \int (\nabla A)^2 dxdz \quad (4.2)$$

where $F_s(H_s, P)$ is constructed from the equilibrium distribution function F_{os} by a phase space conserving mapping. This mapping is unique because one assumes monotonical decrease of F_{os} with H_s, where

$$H_s = \frac{m_s}{2}(v_x^2 + v_z^2) + \frac{1}{2m}(P - e_s A)^2 + e_s \phi \quad (4.3)$$

the potential ϕ being determined by the quasi-neutralizing condition

$$\sum_s e_s \int F_s(H, P) dv_x d\frac{P}{m} dv_z = 0. \quad (4.4)$$

That W[A] in fact describes the proper free energy is illustrated by the following property: That W[A] assumes a local minimum in equilibrium is necessary and sufficient for linear stability.

A free energy function F that allows for collisional relaxation in a current carrying plasma was derived by Brinkmann et al. [1984]:

$$F = \sum_s \left\{ -N_s kT \left[\ln \frac{V_s[A,\phi]}{N_s \lambda^3} + N_s \frac{m_s U_s^2}{2} \right] \right\}$$

$$+ \int \frac{(\nabla A)^2}{2\mu_o} dxdx \quad (4.5)$$

$$V[A,\phi] = \int \exp\left[\frac{-e_s}{kT}(\phi - U_s A)\right] dxdz \quad (4.6)$$

The potentials A and ϕ are again related by the quasi-neutrality condition. N_s denotes the total number of particles of species s, $\lambda_s = h/\sqrt{2\pi m_s T}$ for convenient normalization and U_s the y-component of bulk velocity of species s. The expression (4.6) represents a generalization of equilibrium thermodynamics. Note that for $U_s \equiv 0$ and $\phi \equiv 0$ the first term of F reduces to free energy of an ideal gas in thermodynamic equilibrium. The second variation δF of the functional F can be shown to be always smaller than δW. This is a formal basis for the destabilization by fluctuation discussed in the previous section.

5. Reconnection of Sheared Magnetic Fields and Parallel Electric Fields

Here we add a brief remark about the possible role of reconnection in causing parallel electric fields and double layers. The tearing instability involves a "toroidal" electric field, which in the case of sheared magnetic fields (Figure 3) has a significant component along the magnetic field. (The electric field at the separator in Figure 3b is parallel to $\underline{B}$). Thus, reconnection of sheared magnetic fields generates parallel electric fields. This parallel electric field might be sustained by either resistivity or by the formation of double layers. The shear layer would assume a filamentary structure (Figure 3b). By this process reconnection leads to enhanced dissipation of energy stored in magnetic shear. Note, however, that some dissipation may also occur in a stable sheared magnetic field due to field-aligned currents associated with that shear. Here we are pointing at the possibility that reconnection may enhance the dissipation rate. Dissipation in parallel electric fields seems to contribute significantly to the final ionospheric deposition of energy that had been involved in magnetospheric dynamics. It seems reasonable to assume that the parallel electric fields stem from magnetic shear. Whether or not reconnection enhancement plays a significant role is an open question.

6. Summary and Discussion

In this paper we discussed several aspects of spontaneous reconnection. It was argued that this process is a basic plasma physics phenomenon leading to dynamic states with important macroscopic consequences including control of magnetic topology, rapid dissipation of previously stored magnetic energy, formation of plasmoids, initiation of steady state reconnection, and particle acceleration. Evidence for spontaneous reconnection stems from in situ space observations, laboratory experiments, analytical theory, and large scale MHD-computations.

Although the prototype of spontaneous reconnection is the tearing instability of one-dimensional plasma sheets, more realistic systems had to be studied to explain the observations.

Regarding the magnetotail of the earth, it has become clear that the normal component B_n can account for the stabilization necessary for explaining the observed quiet states. In the collision-free case B_n stabilizes the electron tearing mode completely. Depending on the presence of nonadiabatic ions, the plasma sheet can become unstable at wavelengths of the order of the characteristic scale L_x (measured along the direction of the main tail magnetic field). Fluctuations extend the unstable regime to smaller wavelengths. A rapid growth of the ion tearing mode is expected if the stabilization by the adiabatic electrons is suddenly removed. Mechanisms that could cause such removal could be the sudden onset of fluctuations or field-aligned currents closing through the ionosphere. Another interesting ionospheric feed-back mechanism which may enhance unstable growth and extend the instability domain is the possibility of the emission of heavy ions from the ionosphere into the tail as a part of the overall dynamics. Ionospheric feed-back would be consistent with spontaneous reconnection occurring predominantly in the near-earth tail region. Note that strong ionospheric feed-back would imply that in a fully realistic description tail and ionosphere dynamics cannot be discussed separately. Possibly there exists a large scale unstable mode of the entire ionosphere-magnetosphere system.

The magnetic and flow fields in the tail obtained from large scale MHD-computations, simulating dissipation by an assumed resistivity y, are in remarkable agreement with in situ measurements.

The free energy that is available for release into non-adiabatic kinetic energy forms can be computed from an expression for ´potential´ energy W. A statistical mechanics approach that allows for fluctuations admits a considerably wider range of instability.

It is argued that there exists a close relationship between spontaneous reconnection and the formation of parallel electric fields. The required dissipation may be based on (collisional or collective) resistivity or on double layer.

Although considerable progress has been made in the field of spontaneous reconnection in recent years, it is also clear that we are still far from a complete theoretical understanding of all the different facets of this problem. The enormous difficulties arise from the fact that at least three major complicating factors of plasma theory are combined in this problem: strong spatial inhomogeneity, non-linearity and time-dependence. In spite of these difficulties, its high physical relevance asks for intensive future research in the area of spontaneous reconnection.

Acknowledgements. It is a pleasure for me to acknowledge the kind hospitality that I experienced during a visit to the Los Alamos National Laboratory, where part of this paper was prepared. Especially, I have profited from discussions with Drs. J. Birn and E. W. Hones, Jr. Furthermore, this work was supported by the Deutsche Forschungsgemeinschaft (SFB 162).

References

Baker, D. N., E. W. Hones, Jr., D. T. Young, and J. Birn, The possible role of ionospheric oxygen in the initiation and development of plasma sheet instabilities, Geophys. Res. Lett., 9, 1332, 1982.

Bargatze, L. F., D. N. Baker, R. L. McPherron and E. W. Hones Jr., Magnetospheric response for many levels of geomagnetic activity, Los Alamos National Laboratory, IGPP Publication No. 2457, Submitted to J. Geophys. Res., 1983.

Birn, J., Computer studies of the dynamic evolution of the geomagnetic tail, J. Geophys. Res., 85, 1214, 1980.

Birn, J. and E. W. Hones, Jr., Three-dimensional computer modeling of dynamic re-connection in the geomagnetic tail, J. Geophys. Res., 86, 6802, 1981.

Birn, J. and K. Schindler, Self-consistent theory of three-dimensional convection in the Earth´s magnetotail, J. Geophys. Res., in press, 1983.

Biskamp, D. and K. Schindler, Instability of two-dimensional collisionless plasmas with neutral points, Plasma Phys., 13, 1013, 1971.

Biskamp, D. and H. Welter, Coalescence of magnetic islands, Phys. Rev. Lett., 44, 1069, 1980.

Brinkmann, R.-P., M. Kiessling, and K. Schindler, A statistical approach to current carrying plasmas, in preparation at the Ruhr-Universität Bochum, 1984.

Coppi, B., G. Laval and R. Pellat, Dynamics of the geomagnetic tail, Phys. Rev. Lett., 16, 1207, 1966.

Coppi, B. and M. N. Rosenbluth in "Report ESRO-SP3" p. 1, Publ. European Space Research Organization, Paris, 1968.

Coppi, B., J. Mark, L. Sugiyama and G. Bertin, Magnetic reconnection in collisionless plasmas, Annals of Physics, 1978.

Coroniti, F. V., On the Tearing mode in quasi-neutral sheets, J. Geophys. Res., 85, 6719, 1980.

Drake, J. F. and T. T. Lee, Irreversibility and transport in the lower hybrid drift instability, Phys. Fluids, 24, 1115, 1981.

Erickson, G. M. and R. A. Wolf, Is steady convection possible in the Earth´s magnetotail? Geophys. Res. Lett., 7, 897, 1980.

Finn, J. M. and P. V. Kaw, Coalescence instability of magnetic islands, Phys. Fluids, 20, 72, 1977,

Furth, H. P., The "mirror instability" for finite particle gyro-radius, Nucl. Fusion Suppl. Pt. 1, 169, 1962.

Furth, H. P., J. Killeen and M. N. Rosenbluth, Finite-resistivity instabilities of a sheet pinch, Phys. Fluids, 6, 459, 1963.

Galeev, A. A., and L. M. Zelenyi, Tearing instability in plasma configurations, Sov. Phys. JETP, 43, 1113, 1976.

Galeev, A. A., F. V. Coroniti, and M. Ashour-Abdalla, Explosive tearing mode reconnection in the magnetospheric tail, Geophys. Res. Lett., 5, 707, 1978.

Goldstein, H., Die Ionen-tearing-Instabilität zweidimensionaler Schichtgleichgewichte und ihre Bedeutung für die Dynamik der Magnetosphäre, Dissertation, Ruhr-Universität Bochum, 1981.

Goldstein, H. and K. Schindler, On the role of the ionosphere in substorms: generation of field-aligned currents, J. Geophys. Res., 83, 2574, 1978.

Goldstein, H. and K. Schindler, Large scale collision-free instability of two-dimensional plasma sheets, Phys. Rev. Lett., 48, 1468, 1982.

Hones, E. W., Jr., Substorm processes in the magnetotail: Comments on 'On hot tenuous plasmas, fireballs, and boundary layers in the Earth's magnetotail,' by L. A. Frank, K. L. Ackerson, and R. P. Lepping, J. Geophys. Res., 82, 5633, 1977.

Hones, E. W. Jr., Plasma flow in the magnetotail and its implications for substorm theories, in Dynamics of the Magnetosphere (S. I. Akasofu ed.), D. Reidel Publ. Co., p. 545, 1979.

Huba, J. D., N. T. Gladd, and K. Papadopoulos, The lower-hybrid-drift instability as a source of anomalous field line reconnection, Geophys. Res. Lett., 4, 125, 1977.

Huba, J. D., J. F. Drake, and N. T. Gladd, lower hybrid-drift-instability in field reversed plasmas, Phys. Fluids, 23, 552, 1980.

Huba, J. D., N. T. Gladd, and J. F. Drake, On the role of the lower hybrid drift instability in substorm dynamics, J. Geophys. Res., 86, 5881, 1981.

Huba, J. D., N. T. Gladd and J. F. Drake, The lower hybrid drift instability in nonantiparallel reversed field plasmas, J. Geophys. Res., 87, 1692, 1982.

Irby, J. H., J. F. Drake and H. R. Grilin, University of Maryland, reprint No. 808P003, 1976.

Janicke, L., The resistive tearing mode in weakly two-dimensional neutral sheets, Phys. Fluids, 23, 1843, 1980.

Janicke, L., Resistive tearing mode in coronal neutral sheets, Solar Phys., 76, 19, 1982.

Kadomtsev, B. B., Disruptive instability in tokamaks, Fiz Plazmy 1, 710, 1975 [Sov. J. Plasma Phys., 1, 389, 1975].

Kiessling, M., 2D thermodynamic stability of the earth's magnetotail, to be published, 1984.

Kopp, R. A. and G. W. Pneuman, Magnetic reconnection in the corona and the loop prominence phenomenon, Solar Phys., 50, 85, 1976.

Lakhina, G. S., and K. Schindler, Tearing modes in the magnetopause current sheet, Astrophys. Space Sci., in press, 1983.

Laval, G., R. Pellat, and M. Vuillemin, Instabilite-electromagnetiques des plasmas sans collisions, in Plasma Physics and Controlled Fusion Research, vol. II, p. 259, International Atomic Energy Agency, Vienna, 1966.

Lembége, B., Stabilité d'un modéle bidimensionel de la couche quasineutre de la queue magnétosphèrique terrestre, vis-á-vis du mode de "cisaillement" (tearing mode) linéaire, thése du doctorat 3e cycle, Univ. of Paris XI, 1976.

Maschke, E. K. and B. Saramito, On the transition to turbulence in magneto-hydrodynamic models of confined plasmas, Physica Scripta, T2/2, 410, 1982.

Nishida, A., and N. Nagayama, Synoptic survey for the neutral line in the magnetotail during the substorm expansion phase, J. Geophys. Res., 78, 3782, 1973.

Nötzel, A., J. Birn, and K. Schindler, On the cause of approximate pressure isotropy in the quiet near-earth plasma sheet, submitted to J. Geophys. Res., 1984.

Otto, A., Ein Energieprinzip für stossfreie relativistische Plasmen, Diplomarbeit, Ruhr-Universität, Bochum, 1983.

Parker, E. N., Cosmical Magnetic Fields, Clarendon Press, Oxford, 1979.

Paschmann, G., N. Sckopke, G. Haerendel, J. Papamastorakis, S. J.Bame, J. R. Asbridge, J. T. Gosling, E. W. Hones, Jr., and E. R. Tech, ISEE plasma observations near the subsolar magnetopause, Space Sci. Rev., 22, 717, 1978.

Pellat, R., About 'reconnection' in a collisionless plasma, Space Sci. Rev., 23, 359, 1979.

Priest, E. R., Solar Flare Magnetohydrodynamics, Gordon and Breach, London, 1981.

Pritchett, P.L., and C. C. Wu, Coalescence of magnetic islands, Phys. Fluids, 22, 2140, 1979.

Pytte, T., R. L. McPherron, M. G. Kivelson, H. I. West, Jr., and E. W. Hones, Jr., Multiple-satellite studies of magnetospheric substorms: Radial dynamics of the plasma sheet, J. Geophys. Res., 81, 5921, 1976.

Russell, C. T. and R. C. Elphic, Initial ISEE magnetometer results: magnetopause observations, Space Sci. Rev., 22, 681, 1978.

Schindler, K., A variational principle for one-dimensional plasmas, in Proceedings of the Seventh International Conference on Phenomena in Ionized Gases, vol. II, p. 736, Gradevinska Knjiga, Belgrad, Yugoslavia, 1966.

Schindler, K., A theory of the substorm mechanism, J. Geophys. Res., 79, 2803, 1974.

Schindler, K., D. Pfirsch, and H. Wobig, Stability of two-dimensional collision free plasmas, Plasma Phys., 15, 1165, 1973.

Schindler, K., and J. Birn, Self-consistent theory of time-dependent convection in the Earth's magnetotail, J. Geophys. Res., 87, 2263, 1982.

Schindler, K., and H. Goldstein, A non-linear kinetic energy principle for two-dimensional collision-free plasmas, Physics of Fluids, 26, 2222, 1983.

Stenzel, R. L., and W. Geckelman, Magnetic field line reconnection experiments 1. Field topologies, J. Geophys. Res., 86, 649, 1981.

Stiles, G. S., E. W. Hones, Jr., S. J. Bame and J. R. Asbridge, Plasma sheet pressure anisotropies, J. Geophys. Res., 83, 3166, 1978.

Tajima, T. Nonlinear collisionless tearing and reconnection, The University of Texas, Report DOE/ET/53088-38, 1981.

Wadell, B. V., B. Carreras, H. R. Hicks, and J. A. Holmes, Non-linear interaction of tearing mode in highly resistive Tokamaks, Phys. Fluids, 22, 896, 1979.

Questions and Answers

Moore: What do you make of the observed correlation of the onset of substorms with northward turning of the IMF (from south-pointing during buildup) (e.g., as found in the paper by Caan, McPherron, and Russell, 1975)? Do you see that this could trigger the onset of reconnection in the tail?

Schindler: Any external perturbation is a potential trigger for a metastable state, assuming that it contains appropriate Fourier components. I am not aware of any quantitative analysis addressing especially the northward turnings of the IMF.

Vasyliunas: There have been several different computer simulations that give different results, presumably because they use different boundary conditions. Is it really established mathematically what is a well-posed problem, what boundary conditions can properly be imposed?

Schindler: I am not aware of a mathematical existence theorem for the full nonlinear resistive MHD-systems. In practice, one assumes a more pragmatic point of view and chooses the boundary conditions on the basis of the requirements imposed by the difference scheme one uses. Another obvious method to keep the influence of unphysical boundary conditions small is to choose sufficiently large systems. Present computer experiments clearly demonstrate the difference between spontaneous (homogeneous boundary conditions) and driven (inhomogenous boundary conditions) processes.

FAST SPONTANEOUS RECONNECTION BY THE RESISTIVELY COUPLED RADIATIVE INSTABILITY

R. S. Steinolfson and G. Van Hoven

Department of Physics, University of California, Irvine, California 92717

Abstract. The thermal and tearing instabilities give rise to the development of filaments and flares in sheared magnetic fields. We are investigating the coexistence of these physical mechanisms in the case when Coulomb resistivity $\eta_c(T)$ couples the energy evolution to the plasma dynamics. We find that the analogue of the thermal mode, which still develops on the radiative time scale, involves significant magnetic field reconnection. When compared on the basis of equal magnitudes of nonlinear terms, the fast radiative mode provides some 30% of the reconnected magnetic flux associated with the much slower ($<10^{-2}$ in growth rate for solar coronal conditions) tearing mode. This finding opens the possibility for a more rapid initiation of magnetic reconnection, with resulting energy release, than had previously been thought feasible.

Introduction

Theories of the thermal instability (Field, 1965; Chiuderi and Van Hoven, 1979), which is driven by a radiation output that decreases with temperature, have usually ignored the resistive magnetohydrodynamic effects of the resulting, very collisional, relatively low-temperature plasma. Treatments of the tearing instability (Furth, Killeen, and Rosenbluth, 1963; Steinolfson and Van Hoven, 1983), which is catalyzed by finite conductivity, have usually ignored the energy-flux consequences of the resulting magnetic energy release. Since, in astrophysical situations, radiation is a strong effect and the relevant resistivity is the temperature-dependent Coulomb value, these two dynamic processes should be treated in a unified way (Van Hoven, Steinolfson, and Tachi, 1983; Steinolfson, 1983). This paper describes the outcome of such a coupled-instability study, in which the relevant linearized equations are solved numerically over a range of parameters, and interprets the most important results, including the existence of a reconnection component of the fast radiative/thermal mode.

Formulation

We model a current-carrying volume of plasma by the sheared-field form

$$\vec{B}_o = B_o[\hat{e}_z \tanh(y/a) + \hat{e}_z \mathrm{sech}(y/a)], \quad (1)$$

which is force-free and thus consistent with an initially uniform temperature and density. We describe the temporal development of this system by using the resistive MHD equations, relating flow velocity to magnetic induction,

$$\rho d\vec{v}/dt = -\nabla p + \vec{J} \times \vec{B}, \quad (\mu_o \vec{J} = \nabla \times \vec{B}), \quad (2)$$

$$d\vec{B}/dt = (\vec{B} \cdot \nabla)\vec{v} + (\eta(T)/\mu_o)\nabla^2 \vec{B} + \vec{J} \times \nabla \eta(T), \quad (3)$$

along with the incompressible energy-transport equation

$$K'\rho dT/dt = \nabla \cdot [\vec{B}\kappa_\parallel B^{-2}(\vec{B} \cdot \nabla)T] - R\rho^2 T^r + \eta(T)J^2 \quad (4)$$

and the ideal gas law. The notation here is the same as that in Van Hoven, Steinolfson and Tachi (1983), except that K' is Boltzmann's constant per unit mass, divided by $\gamma-1$. The transport coefficients in the above 2 equations are the resistivity η and the parallel thermal conductivity $\kappa_\parallel$. These equations are solved by assuming small-amplitude perturbations of the form $T_1(\vec{r},t) = T_1(y,t)\exp(ikx)$, for example, where $T_1 \ll T_0$, the equilibrium value.

If we temporarily ignore the underlined resistive terms in (2) - (4), we obtain two uncoupled dynamic systems. The first, from (2) and (3), describes the frozen-in-field condition, and variations on the hydromagnetic time scale $\tau_{hm} = a(\mu_o\rho)^{\frac{1}{2}}/B_o$. The second from (4) provides unstable, incompressible, radiative cooling at $\tau_{ra} \cong \Omega_\rho^{-1} = [-rR\rho_o T_o^{r-1}/K']^{-1}$, when $r < 0$ so that the radiation falls with temperature (Hildner, 1974). This unstable cooling only occurs, however, near the center of the magnetic-shear layer $y \sim 0$, where $\vec{B} \cdot \vec{\nabla} = ikB_z = ikB\tanh(y/a) \sim 0$, so that the much stronger parallel (to B) thermal conduction is ineffective (Chiuderi and Van Hoven, 1979).

Let us now restore the underlined resistive terms to (2) - (4), using the Coulomb value $\eta(T) = \bar{\eta}T^{-3/2}$ (Spitzer, 1962) as the relevant form for astrophysical applications. This added level of fidelity has two important coupling effects, beyond the superposition of (slow) resistive diffusion and heating on the time

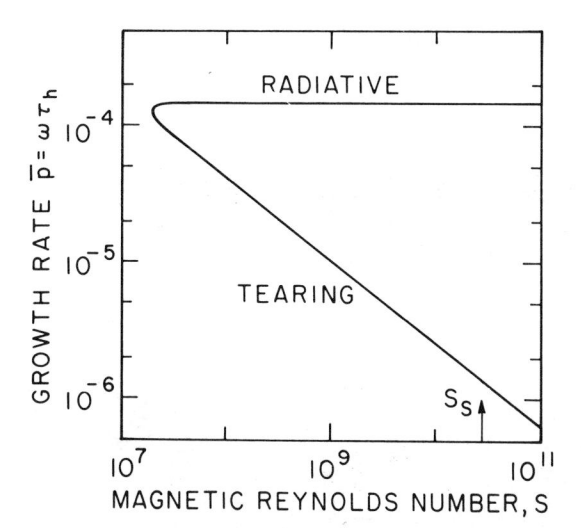

Fig. 1. Growth rates at α = 0.1.

scale $\tau_{re} \sim \mu_o a^2/\eta$. First, as is well known (Furth, Killeen, and Rosenbluth, 1963), Eqs. (2) and (3) now exhibit a new reconnection, or magnetic-tearing, excitation which grows on the hybrid time scale $\tau_{te} \sim \tau_{hm}^\nu \tau_{re}^{1-\nu}$, where $2/5 < \nu < 2/3$ (Steinolfson and Van Hoven, 1983). The tearing mode, as with the radiative-cooling perturbation of (4), exists primarily in the layer $y \sim 0$, where the usually dominant first term on the right of (3) is ineffective. This local frustration of the normal frozen-in-field constraint allows the magnetic field to reconnect into a topology from which it can release its magnetic energy (Van Hoven, 1979). [For completeness, the resistive term has no significant influence (for astrophysical conditions) on the radiative phenomena described by (4) alone.]

The second primary effect of a temperature-dependent resistivity is the coupling together of the magnetodynamic system (2,3) and the energy transport of (4) through the underlined Ohmic diffusion/heating terms. The consequences of this new interaction form the principal topic of this paper.

Growth Rates

The coupled dynamic system (2) - (4) exhibits two, growing, small-amplitude excitations, a (mainly) radiative mode and a tearing-like mode. We solve the linearized equations by using a finite-difference scheme (Killeen, 1970), with the addition of variable grid spacing (Steinolfson and Van Hoven, 1983) to resolve the sharp central gradients which arise in this problem. The growth rate ω is measured from the various perturbation quantities, such as $\partial T_1/T_1 \partial t$, when they become uniform throughout the x-y grid. At this point, various eigenfunction profiles and topographic plots are produced.

The input parameters and output quantities are expressed in normalized terms. These include the wavenumber $\alpha = ka$, the magnetic Reynolds number τ_{re}/τ_{hm}, and the ratio of the equilibrium radiation to Ohmic heating $\epsilon = R\rho_o^2 T_o r/\eta_o J_o^2 = -\beta\Omega_\rho \tau_{re}/3r$ (with $\gamma = 5/3$ and β the ratio of plasma to magnetic pressures).

Growth rates ω are given either in terms of the hydromagnetic time, $\bar{p} = \omega\tau_{hm}$, or of the resistive time, $p = \omega\tau_{re}$, convertible by a factor of S. The magnetic Reynolds number S is artificially scaled (Van Hoven, Steinolfson, and Tachi, 1983) by the coefficient $\bar{\eta}$ of the Coulomb resistivity, with the equilibrium temperature T_o held constant. Thus, $S = S_c \bar{\eta}_c/\bar{\eta}$ where the subscript c denotes the classical value. Since the parameter $\epsilon \propto \eta^{-1}$, it also varies as $\epsilon_c S/S_c$.

An illustrative run of growth rates $\bar{p}$ vs. S, for $\alpha = 10^{-1}$ is displayed in Fig. 1. Under conditions of large $S(S_s = S_c$ is the classical solar coronal value), the radiative (filamentation) mode grows $\sim 100 \times$ more quickly than the tearing (flare) mode, and the character of the eigenfunctions $(\tilde{B}, T, \tilde{v})$ is quite different, as will be described in what follows. As the value of S drops, the ratio of radiation loss to Ohmic heating falls to unity with ϵ, and the mode profiles become quite similar. Below $S \sim 10^{7.3}$, where the curves meet, the growth rates are complex.

A more general plot of the growth rate, normalized to the resistive-diffusion time, is shown in Fig. 2, where the α (wavenumber) dependence is emphasized. [The physical conditions here, which can affect the normalized magnitudes, are $T = 10^6$K, $n = 10^{10}$ cm^{-3}, $B = 83$ G and $a = 10^7$cm.] The solid curves specify the growth rates of the tearing mode, and the dashed line segments those of the radiative mode (Van Hoven, Tachi and Steinolfson, 1984). The growth of the latter mode is dispersionless over this range of (relatively long) wavelengths and so is not shown in

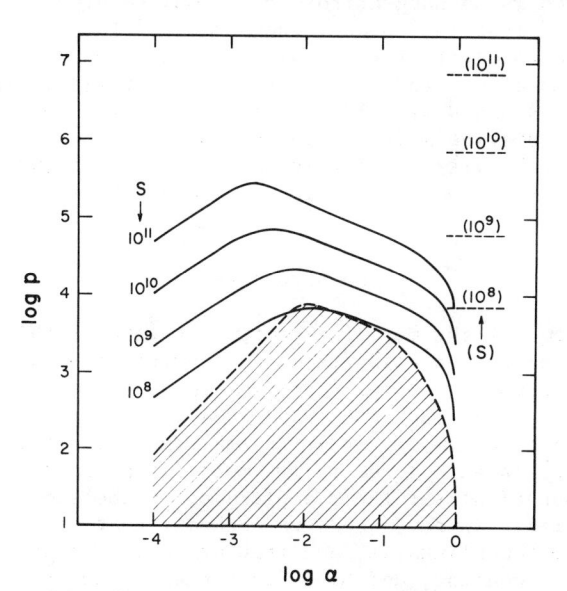

Fig. 2. Growth rates vs. wavenumber.

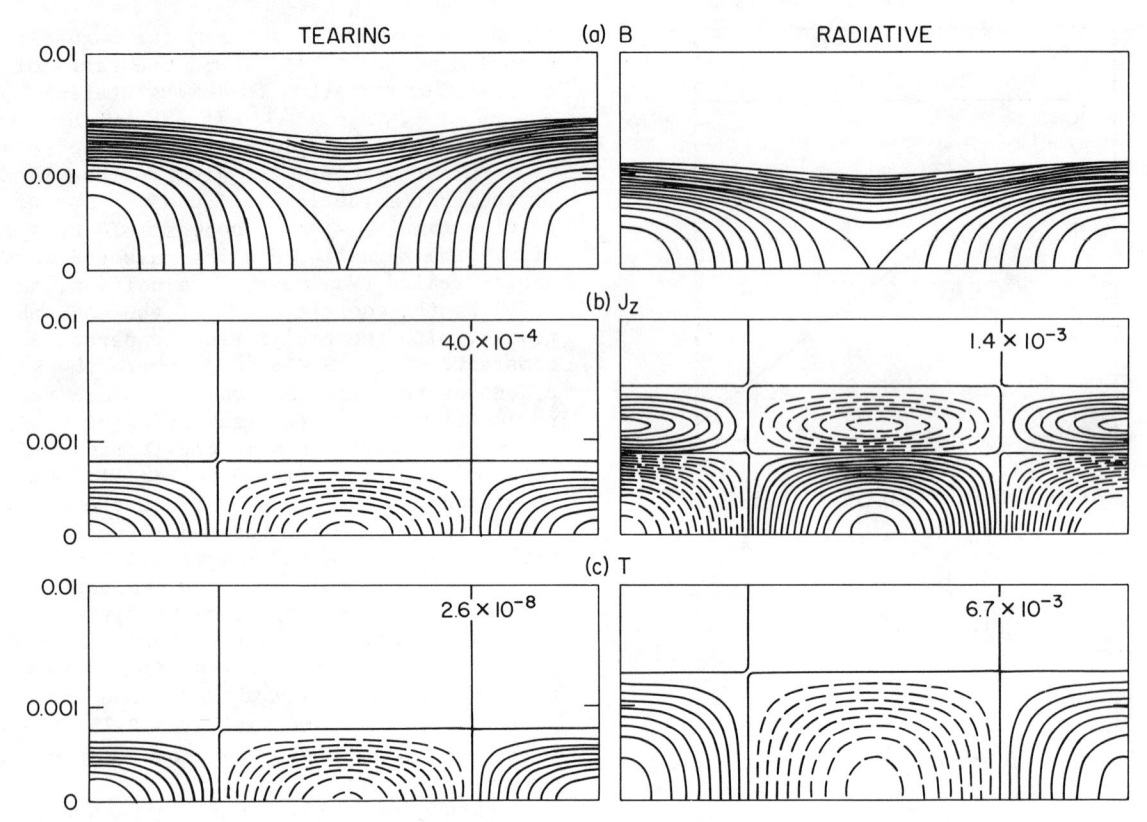

Fig. 3. Mode structures at equivalent levels.

detail. [For solar coronal parameters, the growth rate is independent of parallel (to $\vec{B}$) heat flow and does not drop until $\alpha \sim 10^2$ where perpendicular thermal conduction becomes important.] The cross-hatched area is the regime of strong radiative-to-tearing mode coupling (as in the $S < 10^{7.3}$ range of Fig. 1), in which the growth rate becomes complex.

For S values above 10^9, the tearing mode growth is unchanged from the normal (nonradiative) case. That is, $p \sim S^{-\nu}$, where $\nu = 2/3$ for (fixed) α to the left of the peak, $\nu = \frac{1}{2}$ at the peak values of α, and $\nu = 2/5$ to the right (Steinolfson and Van Hoven, 1983). The growth rate of the radiative mode can be expressed as $p \sim -2(\gamma-1) r \epsilon/\beta$, which is equivalent to the well-known result $\omega \sim \Omega_\rho$ (Chiuderi and Van Hoven, 1979).

Mode Structures

Some of the more interesting aspects of this coupled reconnection/radiation system arise when one looks at the forms of the eigenfunctions (Steinolfson and Van Hoven, 1984). In particular we wish to compare their thermal and reconnection performances. To do so, we must (in a linear computation) set their amplitudes on the same footing. We have used two methods, that of incipient nonlinearity and that of equal linear energy content, and have found these methods to be approximately equivalent.

Figure 3 shows a situation in which each of the two modes (at S_s) has individually (orthogonally) grown to a level in which nonlinearity would just start to become important. [The sum of the nonlinear terms, computed but not used in the dynamic equations, is a fixed fraction of the largest linear term.] The plots of Fig. 3 refer to the coordinate system of Eq. (1), and show the x-y halfplane (on its side) as seen from the z direction. The horizontal (x) scale is linear and shows one wavelength, and the vertical scale ($y/a = 10^{-2}$ at the top) is expanded near the origin to better show the details in this narrow layer.

The top plot (a) of Fig. 3 shows the magnetic flux function, or $B_x - B_y$ field lines, artificially amplified by a factor of 10^2 so that the fine structure can be appreciated. On the left is the well-known X-point field pattern of the tearing mode (Furth, Killeen, and Rosenbluth, 1963). A new result is shown at the right. At an equivalent incipient nonlinear level, the fast radiative mode provides 30% as much magnetic reconnection as the tearing mode, with a similar potential for magnetic energy release. Part (b) displays the z-directed current density which, when multiplied by η_0, provides the parallel (to $\vec{B}$) electric field intrinsic to the reconnection process. The peak levels shown (with respect to J_0) are small, at this amplitude, but the radiative-mode $E_\parallel$ field is relatively 35× larger.

Part (c) of Fig. shows the isotherms of the perturbations, with the dotted curves indicating negative values. The strong temperature reduction in the center of the radiative mode was expected, but the moderate cooling of the X point of the tearing mode is significant. The behavior is opposite to that in the absence of radiation (Tachi, Steinolfson, and Van Hoven, 1983), when the locally concentrated Ohmic dissipation heats the plasma and can reduce the level of Coulomb resistivity which catalyzes the reconnection.

Discussion

This paper has described a unified treatment of the resistive filamentation/radiation and flare/reconnection instabilities. The growth rates and excitation structures of the coupled linear modes in a sheared magnetic field (Van Hoven, Steinolfson, and Tachi, 1983), as determined from numerical-computations, have been detailed. An alternative, approximate, analytic calculation of the growth rates has been given by Steinolfson (1983). In particular, we have demonstrated a significant reconnection component to the quasi-hydromagnetic ($\tau_{ra} \sim 10^4 \tau_{hm}$) radiative/thermal mode.

The growth-rate results show a strong coupling of the modes below a minimum value of the magnetic Reynolds number. However, for solar coronal S conditions, the growth rates are similar to their uncoupled values, which are separated by a factor of order 10^2, with the radiative mode being the faster. For $T = 10^6$K, $B = 10^2$G, $a = 10^7$ cm and $n = 10^9 - 10^{10}$, the growth time of the radiative mode is 3 - 15 minutes (Van Hoven, Tachi, and Steinolfson, 1984).

Although the growth behavior is not much modified under high-S, sheared-field, astrophysical conditions, the individual mode structures are significantly changed. The tearing-like mode, as shown in Fig. 3, exhibits a negative temperature perturbation at the X point. This is the result of the strong radiation dominance at high S, and indicates that the ηJ^2 heating accompanying tearing cannot lower the local Coulomb resistivity. Thus, the magnetic reconnection can proceed to higher levels, and further energy release, without the self-quenching or slowing-down that a thermally reduced resistivity would provide.

What is more important is the fact that the radiative mode has a previously unknown magnetic-reconnection component (Steinolfson and Van Hoven 1984). This cooling mode is 30X times faster than, and provides 30% of the flux reconnection of, the normal tearing mode. This radiative excitation, therefore, has the potential of providing magnetic energy release on a much shorter time scale than previously believed possible.

In addition, for the radiative instability, the linear energy in the perturbed magnetic fields exceeds the perturbed thermal energy by about a factor of five. Consequently, the Coulomb-resistivity-coupled radiative instability clearly does not have the physical characteristics of the condensation mode of the thermal instability (Field, 1965; Chiuderi and Van Hoven, 1979), despite the fact that it occurs on the same time scale. Instead it is more of a hybrid mode with the growth rate and temperature structure of a thermal instability, yet with the field-line reconnection of a resistive tearing instability.

It is significant that, for both instabilities, a temperature reduction occurs at the X point, with an equal temperature increase inside the magnetic islands. This circumstance has an important consequence. The Coulomb resistivity increase at the reconnection site (opposite to what occurs in normal nonradiative tearing; Steinolfson and Van Hoven, 1983) indicates a potential for increased reconnection, and this condition is accentuated in the case of the radiative mode. Normal resistive reconnection is also aided by radiative cooling, so that it may go beyond the limitations due to self-heating. The final characterization of the connection between these two instabilities must come from nonlinear computations, which are under development. These simulations are also necessary in order to establish a time scale for the energy release. Since it is well-known that the linear growth rate of the tearing mode cannot be extrapolated to obtain nonlinear energy release times, there is no reason to expect that this can be done for the radiative mode either.

Acknowledgments. This research was supported in part by the Solar-Terrestrial Theory Program of NASA and the Atmospheric Science Section of NSF. This work was completed while R. S. Steinolfson was a visiting scientist at the High Altitude Observatory of the National Center for Atmospheric Research, Which is sponsored by the National Science Foundation.

References

Chiuderi, C., and G. Van Hoven, The Dynamics of Filament Formation: The Thermal Instability in a Sheared Field, Astrophys. J. 232, L69, 1979.

Field, G. B., Thermal Instability, Astrophys. J. 142, 531, 1965.

Furth, H. P., J. Killeen, and M. N. Rosenbluth, Finite-Resistivity Instabilities of a Sheet Pinch, Phys. Fluids 6, 459, 1963.

Hildner, E., The Formation of Solar Quiescent Prominences by Condensation, Solar Physics 35, 123, 1974.

Killeen, J., Computational Problems in Plasma Physics and Controlled Thermonuclear Research, in Physics of Hot Plasmas, eds. B. J. Rye and J. C. Taylor (New York: Plenum), 212-219, 1970.

Spitzer, L., Encounters between Charged Particles,

in *Physics of Fully Ionized Gases* (New York: Wiley Interscience), 136-145, 1962.

Steinolfson, R. S., Energetics and the Resistive Tearing Mode: Effects of Joule Heating and Radiation, *Phys. Fluids* 26, 2590, 1983.

Steinolfson, R. S., and G. Van Hoven, The Growth of the Tearing Mode: Boundary and Scaling Effects, *Phys. Fluids* 26, 117, 1983.

Steinolfson, R. S., and G. Van Hoven, Radiative Tearing: Magnetic Reconnection on a Fast Thermal Instability Time Scale, *Astrophys. J.* 276, 391, 1984.

Tachi, T., R. S. Steinolfson, and G. Van Hoven, The Effects of Ohmic Heating and Stable Radiation on Magnetic Tearing, *Phys. Fluids* 26, 2976, 1983.

Van Hoven, G., The Energetics of Resistive Magnetic Tearing, *Astrophys. J.* 232, 572, 1979.

Van Hoven, G., R. S. Steinolfson, and T. Tachi, Energy Dynamics in Stressed Magnetic Fields: The Filamentation and Flare Instabilities, *Astrophys. J.* 268, 860, 1983.

Van Hoven, G., T. Tachi, and R. S. Steinolfson, Radiative and Reconnection Instabilities: Filaments and Flares, *Astrophys. J.* (in press) 1984.

STEADY STATE ASPECTS OF MAGNETIC FIELD LINE MERGING

Vytenis M. Vasyliunas

Max-Planck-Institut für Aeronomie, D-3411 Katlenburg-Lindau, Federal Republic of Germany

Abstract. Although a true steady state does not seem ever to occur in the magnetosphere, there are topics profitably discussed as steady-state aspects, including (1) description of phenomena whose essential features are the same with or without time variations, and (2) description of the time-averaged configuration, important for understanding and classifying the basic structures which may then serve as a framework for the more complex time-dependent effects. Magnetic field line merging (I consider this term preferable to its synonym "magnetic reconnection," for linguistic reasons) exhibits such steady-state aspects on both a local and a global scale. The local aspects are primarily those concerned with the region centered about the magnetic X line (also called the separator) formed by the intersection of two branches of the separatrix surface (between volumes occupied by topologically different magnetic field lines). Bulk flow of the plasma across the separatrix surface implies an electric field along the X line, a conclusion that can be derived independently of the presence or absence of time variations by applying Faraday's law to a loop lying within the separatrix surface, whether steady or not. The pivotal point of the argument is that the global length scales of the system (e.g. the radius of curvature of the X line) are very large compared to the microscopic length scales (e.g. gyroradii, resistive lengths) at which the MHD approximation breaks down, so that intermediate length scales, macroscopic but local, exist. On a global scale, the complex three-dimensional geometrical configuration may be visualized with the help of several diagrams, including in particular representations of (1) magnetic field lines and flow streamlines on electric equipotential surfaces and (2) potential contours on a surface passing through the magnetic X lines. The large-scale magnetic topology associated with the open magnetosphere is characterized by a single X line that closes on itself to form a complete ring (nonplanar in general); it can be divided, on the basis of the relative direction of the electric field, into two distinct segments generally referred to as the "dayside" and "nightside" merging lines, plus possibly some additional "inactive" segments where the electric field along the X line vanishes. Despite the names, the actual locations of these segments even in a time-averaged configuration are very uncertain, and a number of distinct models can be proposed without conflicting with the (as yet rather limited) observations.

Introduction

In this paper I use the term "magnetic field line merging" rather than "magnetic field line reconnection" found in most of the other papers in this volume, for purely linguistic reasons and with no implication of any physical distinction whatsoever: "merging" is shorter than "reconnection," it obviates the question "why reconnection?" (the answer to which is that "reconnection" is really an abbreviation of "cutting and reconnection"), and I have consistently used it in previous papers including a comprehensive review of the theoretical models of the process (Vasyliunas, 1975). The models discussed in that review describe systems that have been approximated, for the most part, as two-dimensional, steady-state configurations. The real magnetosphere is three-dimensional and time-varying, and that fact has been invoked to question the applicability not only of the various specific models but sometimes also of the concept of magnetic field line merging itself. The purpose of this paper is to discuss some major aspects of magnetic merging, showing that they remain valid in a broader context, as well as to present some graphical tools for describing the global configuration associated with magnetic merging in more complex geometries.

Even though no true steady state seems ever to occur in the magnetosphere, a discussion of steady-state aspects is useful for at least two reasons. First, there are phenomena whose essential features remain the same regardless of whether time variations occur or not. Second, an adequate description of the time-averaged configuration (which is steady-state by definition) is often useful or even necessary in order

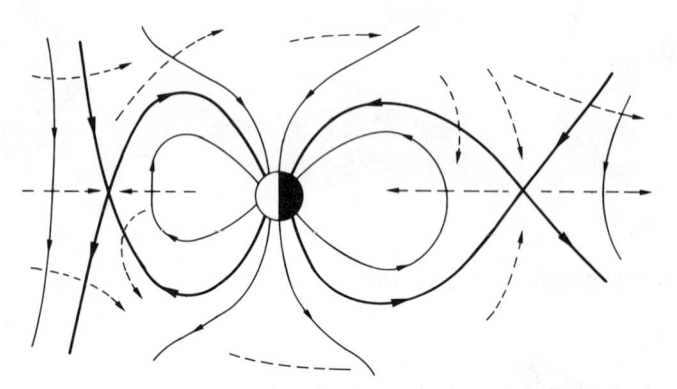

Fig. 1. Sketch of the magnetic field lines (solid lines) and plasma flow (dashed arrows) in a simple model of the open magnetosphere. The separatrix is shown by thick lines.

to understand and classify the basic structures which may then serve as a framework for the more complex time-dependent effects. It also enables one to recognize phenomena that may be time-dependent only accidentally, so to speak - those that could well be imagined in principle to be steady even if in actuality they are not - and to distinguish them from others that are intrinsically time-varying and cannot be fitted into any steady-state framework.

The Electric Field Along the Separator

A prerequisite for magnetic merging is a magnetic field configuration that is topologically complex, containing more than one topological class of field lines. The regions of space threaded by field lines of the various distinct topologies are then separated by a surface, or possibly several surfaces; such a bounding surface is known as the separatrix. As an example, the simple open magnetosphere sketched in Figure 1 contains three topological classes of magnetic field lines: (1) closed field lines, contained inside a volume topologically identical to a torus, (2) interplanetary magnetic field lines not connected to the earth, lying outside a volume topologically identical to a cylinder, and (3) open field lines, connected to the earth in one direction and extending into interplanetary space in the other, occupying the volume outside the torus but inside the cylinder. It is evident that the surface of the torus touches the surface of the cylinder, and thus the separatrix in this case has two branches which intersect along a closed line known as the X line or the separator. At the X line, all magnetic field components perpendicular to it vanish, but there may be (and, in non-idealized geometries, usually is) a non-zero component along the X line itself (hence the name "magnetic neutral line," although fairly widely used, is not really appropriate).

Magnetic field line merging, or reconnection, has been defined as the process of plasma flow across the separatrix (e.g. Vasyliunas, 1975). An alternative definition is that it is the process associated with a non-zero electric field component along all or part of the magnetic X line or separator (e.g. Baum and Bratenahl, 1980); this definition has the advantage of an obvious relation to magnetic flux transfer between open and closed field lines (the intuitive root of the magnetic merging concept), but it also forces one to admit as instances of magnetic merging what are mere time variations of a moderately complex but straightforward magnetostatic system (e.g. waving a bar magnet near a current-carrying coil). The first definition, on the other hand, makes magnetic merging a specifically plasma-physical process. It does not really matter which definition one adopts as long as one is dealing with large-scale plasma systems, because in that case the two definitions are equivalent: plasma flow across the separatrix surface implies a non-zero electric field along the separator line, and conversely. In simplified models this equivalence is often presented as the result of assuming a two-dimensional time-independent configuration, but in fact it can be established in complete generality.

Consider a closed loop which lies at all times in the separatrix surface and includes a segment of the X line (Figure 2). The time rate of change of the magnetic flux Φ_M through the loop can be expressed as a line integral around the loop, by a well-known theorem:

$$d\Phi_M/dt = \oint d\vec{\ell} \cdot (\vec{E} + \vec{U} \times \vec{B}/c) \qquad (1)$$

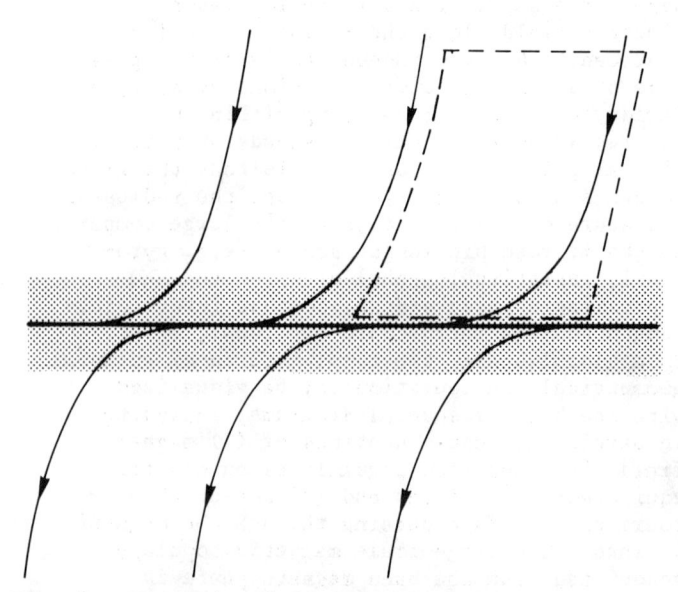

Fig. 2. Magnetic field lines on a branch of the separatrix surface seen face-on, near the X line (thick line) surrounded by the non-MHD region (shaded). The dashed line is a path of integration.

where $\vec{U}$ at any point is the local velocity of the separatrix surface, defined purely geometrically by the position of the separatrix as a function of time, with no reference to the bulk flow velocity $\vec{V}$ of the plasma; in a steady state, $\vec{U} = 0$ everywhere. By definition, no field lines cross the separatrix; hence $\Phi_M = 0$ at all times and equation (1) becomes

$$\oint d\vec{l} \cdot (\vec{E} + \vec{U} \times \vec{B}/c) = 0 \qquad (2)$$

The line integral around the loop can be written as the sum of two terms, the contribution from the X line segment plus the contribution from the rest of the loop. In the first term, $d\vec{l} \cdot \vec{U} \times \vec{B} = 0$ because $\vec{B}$ is either zero or aligned with $d\vec{l}$. In the second term, the MHD approximation

$$\vec{E} + \vec{V} \times \vec{B}/c \simeq 0 \qquad (3)$$

may be applied at all parts of the loop where the spatial scales of the field and plasma structures are large compared to a characteristic microscopic length scale λ_{ch} of the plasma, with λ_{ch} a gyroradius, or an inertial length, or a resistive diffusion length, etc., depending on the relative importance of the various terms in the generalized Ohm's law (see, e.g., Axford, 1984 - this volume); for a large-scale system, (3) thus holds over most of the separatrix and breaks down only at distances from the X line comparable to or less than λ_{ch} (see, e.g., Vasyliunas, 1975). The contribution from this non-MHD region can be reduced to a negligible fraction of the second term by choosing the loop to be sufficiently large, which is possible as long as λ_{ch} is very small compared to typical macroscopic length scales (such as the linear dimension of the magnetosphere or the radius of curvature of the X line); equation (2) can then be rewritten as

$$c \int_{\text{x line}} d\vec{l} \cdot \vec{E} \approx \int_{\substack{\text{rest of} \\ \text{loop}}} d\vec{l} \times \vec{B} \cdot (\vec{U} - \vec{V}) \qquad (4)$$

Noting that $\vec{V} - \vec{U}$ is the plasma flow velocity relative to the separatrix and that $d\vec{l}$ and $\vec{B}$ both lie in the separatrix surface and hence $d\vec{l} \times \vec{B}$ is normal to it, we recognize the RH side of (4) as proportional to the plasma flow through the separatrix.

Equation (4) thus displays the equivalence between the two defining phenomena of magnetic field line merging, electric field along the separator and plasma flow across the separatrix. Its derivation presupposes only a macroscopic scale size sufficiently large so that significant departures from the MHD approximation are confined to small localized regions. No other restrictions need be assumed on the geometry or on time dependence.

Representations of the Merging Geometry

The global geometry of the open magnetosphere can be readily visualized and illustrated for the case of a southward interplanetary magnetic field, as in Figure 1. A similar illustration, but only in the noon-midnight meridian plane, can be drawn for the case of an interplanetary field with no component out of that plane. In the general case of an arbitrarily oriented interplanetary field, the basic topology remains the same, with a two-branched separatrix surface that consists of a topological torus touching a topological cylinder at a closed separator line (see, e.g., Cowley, 1973), but visualizing the field and flow patterns or representing them in a flat two-dimensional figure is no longer a simple matter. However, as long as the magnetosphere can be described or approximated as a steady-state system, a method of graphical representation applicable for an arbitrarily oriented interplanetary field exists.

This method is based on the fact that, in the MHD approximation (3), the electric field $\vec{E}$ is perpendicular to both $\vec{B}$ and $\vec{V}$. In a steady state, $\vec{E}$ is derivable from a scalar potential,

$$\vec{E} = -\nabla \Phi \qquad (5)$$

and hence both the magnetic field lines and the streamlines of the plasma bulk flow lie on equipotential surfaces $\Phi = $ constant. We may thus treat the global configuration in two steps, visualizing first the electric equipotential surfaces in space and secondly the two-dimensional pattern of magnetic field lines and plasma streamlines on any given equipotential surface.

Figure 3 shows a sketch of equipotentials on the surface of an earth-centered sphere somewhat outside of the earth; normally the surface is taken at the ionosphere, but for our purposes it may be placed somewhat higher, above the non-MHD regions where auroral particle acceleration occurs (the difference, however, is not qualitatively important since the associated localized potential drops along $\vec{B}$ are generally small compared to global electric potential differences in the magnetosphere). The corresponding equipotential surfaces in space can be constructed by mapping the equipotential contours from the sphere outward along magnetic field lines; within the region of open field lines, identified with the polar cap, the mapping leads ultimately to the equipotentials in the solar wind (see, e.g., review by Stern, 1977). A quantitative construction requires a unified model for the magnetic fields in the magnetosphere, its

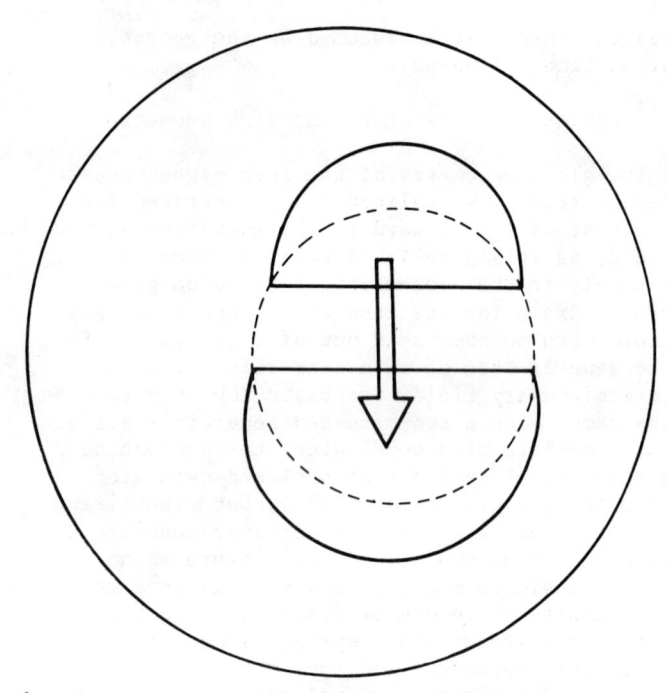

Fig. 3 Sketch of two equipotential contours on an earth-centered sphere, looked at from above the north pole with the sun to the left. Dotted line: boundary of the polar cap. Open arrow: direction of the electric field.

boundary regions (taking into account interconnection of interior and exterior fields), and the magnetosheath. The development of such a model remains a task for the future. A qualitative sketch of the equipotential surfaces can, however, be drawn for any orientation of the interplanetary magnetic field, by a method that is illustrated with a particular example in Figure 4.

This figure shows the equipotentials in the dawn-dusk cross-section of the magnetosphere, drawn for an interplanetary field orientation dawn-to-dusk ($B_y > 0$) and slightly northward ($B_z > 0$), but the generalization for other planes normal to the sun-earth line and for other field orientations will be obvious. (The case $B_y < 0$, $B_z > 0$ is simply the mirror image). In the solar wind outside the bow shock, the equipotential surfaces are well known: they are equidistant parallel planes containing the vectors $\vec{V}$ and $\vec{B}$. In the low-altitude regions of the magnetosphere, the equipotentials may be assumed known from observations complemented by the theory of magnetosphere-ionosphere coupling (e.g. Stern, 1977, Vasyliunas, 1979, Wolf, 1983, and references therein). The magnetic field component normal to the magnetopause is relatively small and its effects on the local geometry (as distinct from the topology) of the magnetic field are largely confined to a relatively thin boundary region on either side of the magnetopause. Thus, to map the equipotentials in the space from low altitudes to the bow shock, one may rely on current models of the magnetic field in the magnetosphere as long as one is well inside of the boundary region, and one may use the well-known "draping" of the magnetic field in the magnetosheath (Fairfield, 1967; Spreiter, Alksne, and Summers, 1968) as long as one is well outside of the boundary region. Only in extending the equipotentials across the boundary region itself does one have little or no guidance at present and has to draw them essentially arbitrarily, subject only to the condition that the required interconnection of a magnetic field line between outside and inside be possible (this determines how, in Figure 4, the north and south sections of a magnetospheric equipotential are to be paired with the dawn and dusk sections of the corresponding solar-wind equipotential). As a result of the relative thinness of the boundary region, however, this arbitrariness has little effect on the qualitative over-all shape of the equipotentials in Figure 4.

Evidently, an equipotential surface that extends into the open field line region has a complicated curved and folded shape which moreover depends strongly on the orientation of the interplanetary magnetic field. Topologically, however, the surface (except possibly in the case of a few isolated potential values) is identical to a plane, and of course beyond the bow shock it actually is a plane. Thus, to represent the configuration of magnetic field lines and plasma streamlines which lie on a particular equipotential surface it is convenient to deform the surface into a plane

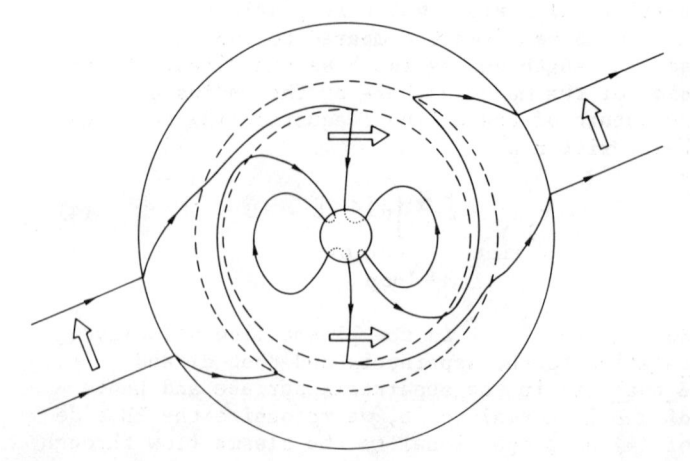

Fig. 4. Equipotential surfaces, corresponding to the two equipotentials of Figure 3, in the dawn-dusk meridian looked at from the sun. Arrows on the contours indicate the direction of the projected magnetic field. Open arrows: direction of the electric field. Outermost circle: the bow shock. The magnetosphere-magnetosheath boundary region lies between the two dashed circles (not to scale).

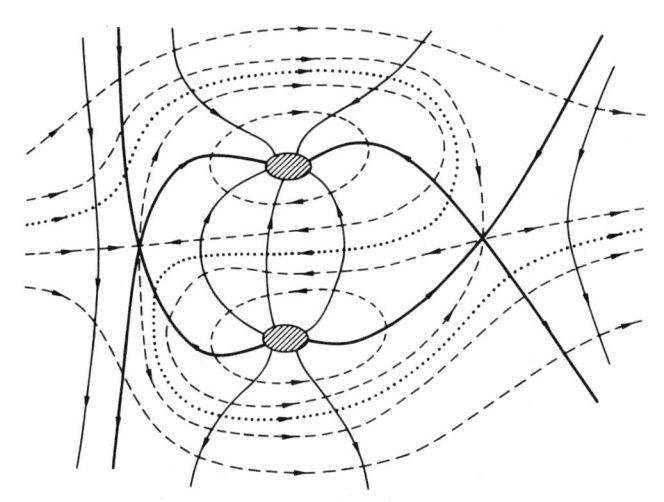

Fig. 5. Sketch of the magnetic field lines (solid lines, separatrix shown as thick line) and plasma bulk flow streamlines (dashed lines) on an electric equipotential surface. The dotted line is a streamline of the "plasma-entry" topological class (see text).

and simply draw the lines on it, as illustrated in Figure 5. In this representation, the upper and lower quasi-circular "holes" represent the parts of the equipotential surface that dip below the MHD region of the magnetosphere (into the auroral acceleration region and/or the neutral atmosphere below the ionosphere) in the northern and southern hemispheres, respectively; the equipotential contour in Figure 3 that corresponds to the equipotential surface under discussion may be taken as the boundary of the northern (upper) hole in Figure 5. Closed field lines run from one hole to the other; open field lines extend from one of the holes into the interplanetary space at the edges of the figure. The magnetic topology of Figure 5 is very similar to that of Figure 1 (in fact, Figure 1 can be viewed as a degenerate limit of Figure 5 where the two holes have merged into one).

Of particular interest is the topology of the plasma streamlines in Figure 5. Three topological classes of streamlines are apparent: (1) "solar-wind" flow, from interplanetary to open magnetic field lines and back to interplanetary, (2) "magnetospheric-convection" flow, circulating about the one or the other of the two holes, flowing between open and closed field lines, and (3) "plasma-entry" flow, crossing all branches of the magnetic separatrix, from interplanetary to open to closed and out again. The first two are obvious and well known, but the third requires a north-south asymmetry and can exist in two topologically distinct variants: the streamlines from the dayside and from the nightside intersections of the X line with the equipotential surface go around (a) the northern and the southern holes, respectively (shown in Figure 5), or (b) vice versa. (If a streamline joins the two intersections, as in a case of ideal symmetry, this type of flow disappears.) Under what conditions (a) or (b) occurs is not yet well understood; presumably both the orientation of the interplanetary magnetic field and the location of the equipotential surface (dawn or dusk) play a role. The B_y-dependent asymmetric appearance of magnetotail lobe plasma (Hardy, Hills, and Freeman, 1975; Gosling, 1984 - this volume) may possibly be associated with the two variants of this flow.

Except for the possible difference between (a) and (b), the topological pattern of the flow and the field shown in Figure 5 should not depend particularly on the orientation of the interplanetary magnetic field. Thus the merging geometry assumes a relatively simple and nearly universal form when viewed in the equipotential surface representation, all the complexity having been transferred to the description of the equipotential surface itself in three-dimensional space.

Global Magnetic Merging Configurations

A further aspect of the global geometry of magnetic field line merging is the large-scale pattern of plasma flow and the associated distribution of the electric field along the separator. This can be represented by equipotential contours on a surface that bisects both branches of the separatrix (the cylindrical and the toroidal) and contains the X line or separator which, as previously noted, is a closed loop. If the surface is constructed with its normal antiparallel to $\vec{B}$ in the region exterior to the separator and parallel to $\vec{B}$ everywhere in the interior (except inside the earth, through which the surface must pass in order to satisfy the latter requirement), then the direction of plasma flow along each equipotential is unambigously determined, although this is in general only a projection of the flow (the velocity vector need not lie in the surface itself, unlike the case of Figure 5). In the ideal case of a purely southward interplanetary magnetic field, this surface is the magnetospheric equatorial plane; for more general orientations, it can be considerably curved and folded, but it remains topologically identical to a plane and thus easily representable by a flat drawing.

On such a surface (which one may call the "topological equator"), the separator line and the projected plasma flow are shown sketched in Figure 6 for three different models of the global magnetic merging configuration. In all three cases one recognizes two types of "active" segments of the separator, one (the dayside merging line) with plasma flow toward it from both sides and the other (the nightside merging line) with plasma flow away from it on both sides, plus some "inactive" segments (which may

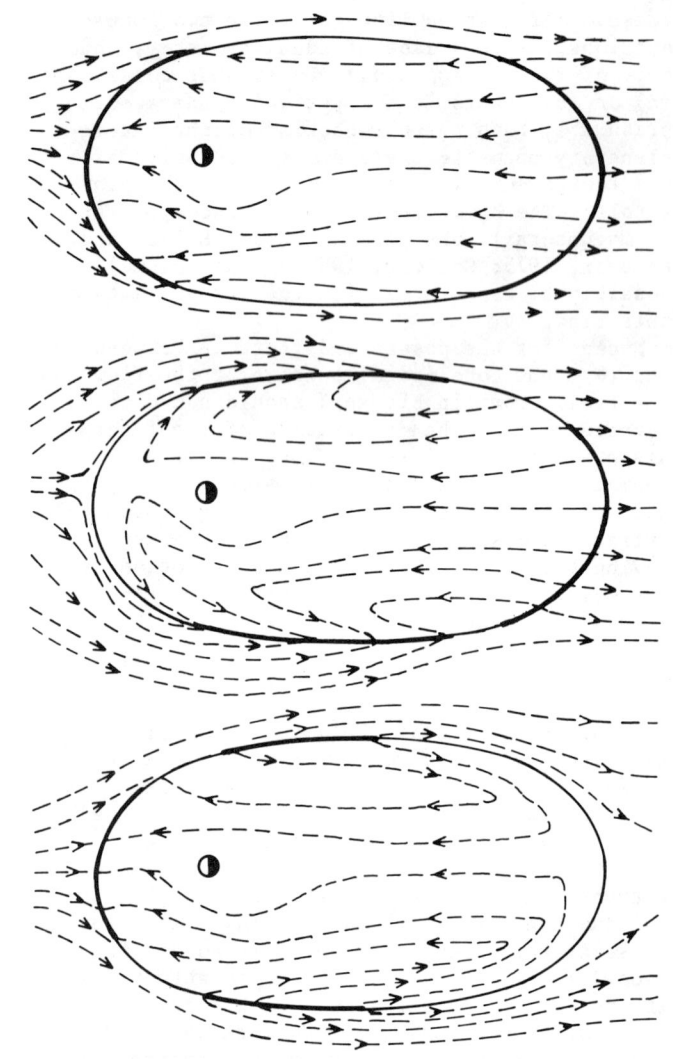

Fig. 6. (a) top, (b) middle, (c) bottom. Equipotential contours on the topological equator surface (dashed lines, arrows indicate direction of the projected plasma flow) and the separator or magnetic X line (solid line, active segments shown by thick line).

or may not be of nonnegligible length) where the electric field component tangent to the separator line vanishes. The three models differ in their assumptions about the location of these various segments and the implied configuration of plasma flow in the magnetospheric boundary regions; the equipotential contours in the interior of the magnetosphere, where the main features of the large-scale electric field plasma flow are reasonably well established (e.g. Wolf, 1983), have been drawn identical in all three.

Figure 6a represents what may be considered the conventional view: the dayside merging line is on the day side of the magnetosphere, the nightside merging line is somewhere well back in the magnetotail, and the plasma flow is generally sunward throughout the closed field line region in the outer magnetosphere. Figure 6b is a model based on the hypothesis (discussed as an alternative by Wolf, 1970, and strenuously argued by Heikkila, 1975) that the electric field along the dayside magnetopause is negligible: the "dayside" merging line has been split into two and placed at the flanks of the magnetotail, leaving the dayside part of the separator as an inactive segment, with the result that the magnetospheric plasma flow turns antisunward while still on closed field lines in the dayside boundary region. Figure 6c is a model that is implicit in the discussion of Frank, Ackerson, and Lepping (1976): here the "nightside" merging line has been placed at the flanks, and there is antisunward flow on closed field lines in the boundary regions of the magnetotail. More complex variants of these models as well as combinations of them can be readily imagined.

Which of these models, if any, can be considered as an approximation to the magnetosphere in a steady state (viewed as the result either of time averaging or of quiet solar wind conditions)? That question has at present no clear answer, whether on observational or on theoretical grounds. There are major aspects of the global geometry of magnetic merging whose specific configuration (particularly in the boundary regions) and possible dependence on the orientation of the interplanetary magnetic field remain to be determined.

References

Axford, W.I., Magnetic field reconnection, this volume, 1984.
Baum, P.J., and A. Bratenahl, Magnetic reconnection experiments, Adv. Electronics and Electron Phys., 54, 1-67, 1980.
Cowley, S.W.H., A qualitative study of the reconnection between the Earth's magnetic field and an interplanetary field of arbitrary orientation, Radio Sci., 8, 903-913, 1973.
Fairfield, D.H., The ordered magnetic field of the magnetosheath, J. Geophys. Res., 72, 5865-5877, 1967.
Frank, L.A., K.L. Ackerson, and R.P. Lepping, On hot tenuous plasmas, fireballs, and boundary layers in the earth's magnetotail, J. Geophys. Res., 81, 5859-5881, 1976.
Gosling, J.T., ISEE 3 plasma measurements in the lobes of the distant geomagnetic tail: inferences concerning reconnection at the dayside magnetopause, this volume, 1984.
Hardy, D.A., H.K. Hills, and J.W. Freeman, A new plasma regime in the distant geomagnetic tail, Geophys. Res. Lett., 2, 169-172, 1975.
Heikkila, W.J., Is there an electrostatic field tangential to the dayside magnetopause and neutral line?, Geophys. Res. Lett., 2, 154-157, 1975.
Spreiter, J.R., A.Y. Alksne, and A.L. Summers, External aerodynamics of the magnetosphere, in

Physics of the Magnetosphere, edited by R.L. Carovillano, J.F. McClay, and H.R. Radoski, pp. 301-375, D. Reidel, Dordrecht-Holland, 1968.

Stern, D.P., Large-scale electric fields in the earth's magnetosphere, Rev. Geophys. Space Phys., 15, 156-194, 1977.

Vasyliunas, V.M., Theoretical models of magnetic field line merging, 1, Rev. Geophys. Space Phys., 13, 303-336, 1975.

Vasyliunas, V.M., Interaction between the magnetospheric boundary layers and the ionosphere, in Proceedings of Magnetospheric Boundary Layers Conference, edited by B. Battrick, pp. 387-393, ESA SP-148, Noordwijk, The Netherlands, 1979.

Wolf, R.A., Effects of ionospheric conductivity on convective flow of plasma in the magnetosphere, J. Geophys. Res., 75, 4677-4698, 1970.

Wolf, R.A., The quasi-static (slow-flow) region of the magnetosphere, in Solar-Terrestrial Physics, edited by R.L. Carovillano and J.M. Forbes, pp. 303-368, D. Reidel, Dordrecht-Holland, 1983.

MAGNETIC RECONNECTION AND MAGNETIC ACTIVITY

E. N. Parker

Dept. of Physics, Dept. of Astronomy and Astrophysics
University of Chicago, Chicago, Illinois 60637

Abstract. A large-scale magnetic field extending through a highly conducting tenuous fluid may become distorted on a small-scale as a consequence of slow small-scale shuffling of the magnetic lines of force at the boundaries of the tenuous fluid. Any slow wrapping and winding introduced at the boundaries is distributed along the field (at the Alfven speed). It is a curious and little known fact that such wrapping and winding possesses no static equilibrium (except for a set of solutions of extreme symmetry). The result is neutral point reconnection of the strains in the field, rapidly dissipating the wrapping and winding. We suggest that this is the principal cause of the extreme heating that produces the active corona of the sun and other stars. The shuffling of the footpoints of the magnetic field in the photospheric turbulence introduces small-scale wrapping and twisting into the coronal loops. The work done by the turbulence in twisting the fields is dissipated within a matter 10-20 hours by neutral point reconnection, introducing heat into the corona at a rate $\sim 10^7$ ergs/cm^{2}sec for photospheric turbulence of 0.5km/sec. We suggest that this is the basic cause of the X-ray corona.

Introduction

The salient feature of magnetized plasma is its activity. The laboratory plasma, the planetary magnetosphere, and the ordinary star exhibit continual activity, punctuated by particularly intense outbursts or flares. The activity involves plasma turbulence and waves, shocks, superheated gases, the production of fast particles, and the associated radio, X-ray, and γ-ray emissions. The phenomenon of activity is universal, occurring whenever and wherever a magnetic field in a tenuous plasma is subject to externally imposed strains.

It should be noted that the existing state of universal activity is forced upon us by observations of the magnetically confined plasma in the laboratory, by observations of the plasmas and fields in space, by observations of the sun and by observations of the distant stars and galaxies. Theory anticipated none of it and has been successful in understanding activity only where detailed observations are available to define the problem. For the fact is that a magnetized plasma has too many degrees of freedom to allow direct deduction of its behavior from the basic dynamical equations for an aggregate of charged particles. Even in the fluid approximation (the magnetohydrodynamic equations) there is so much freedom of motion as to prohibit general deductive solutions. The magnetic Reynolds number for a system of scale L is $N=Lv/\eta$, and, roughly speaking, the number of independent states is proportional to N^3. We note that N is of the order of 10^{15} for the solar corona and $1-10^3$ for the laboratory plasma.

Consequently, a subtler approach has been necessary, sparked by observations of discovery, followed by mapping and probing along with the development of theoretical understanding of each special situation. The special cases illustrate the many effects that collectively make up the real activity in the world around us. Only some "supercomputer" could combine the interacting effects to mock up the total situation. As a matter of fact, nature has already done that for us, the results of which we read out through our observations, which points out the problem in the first place.

This Chapman Conference focusses attention on neutral point reconnection of magnetic fields, long recognized as a central dynamical effect in magnetic activity. Indeed, it seems fair to say now that neutral point reconnection may be the central effect.

The basic observational fact is that when a magnetic field embedded in a highly conducting gas or fluid is subjected to external strains which force the field out of its natural potential form, then the field and its fluid become active, subject to small-scale fluctuations, rapid transport of fluid across the field, acceleration of particles, etc. In those cases where a genuine equilibrium state exists the activity is in response to a complex repertoire of dynamical instabilities. But in many circumstances the dynamical activity is in response to the absence of any equilibrium for the field topology at hand. The reverse-field pinch seems to be an example. It looks, too, that the active corona of the sun is heated as a consequence of an absence of equilibrium.

Reconnection turns up in too many forms to be covered comprehensively in a single review, or even in a whole conference. So we limit the present exposition to the circumstance in which a magnetic field is anchored in a dense plasma ($\beta \gg 1$) and extends through a region of hot tenuous gas, where the gas pressure p is, at most, comparable to the field pressure $B^2/8\pi$. This situation contains the essential physics of the active region on the sun, composed of magnetic lines of force that arch up from the surface into the corona. The field is manipulated by the convective motions beneath the visible surface, introducing wrapping and twisting into the field in the corona. The example provides a well defined situation in which we can examine the mathematical properties and restrictions imposed by the equation

$$\nabla(p+B^2/8\pi)=(\vec{B}\cdot\nabla)\vec{B}/4\pi \qquad (1)$$

for static equilibrium. The point is that when the field is subjected to essentially any strain-wrapping or twisting—that takes it away from a potential form ($\vec{B}=-\nabla\phi$), the field develops regions of internal dynamical nonequilibrium, in the form of neutral point reconnection. It is this ubiquitous property of neutral point reconnection that makes it the driving force, directly converting the internal strains of the magnetic field into fluid motion, heat, and fast particles.

In nature the distortion of the magnetic field from a potential form is created in a variety of ways. Planetary magnetospheres are distorted by the pressure and drag of the solar wind, the rotation of the planet, the pressure of the internal gases, etc. In stars it is the convective motions beneath the photosphere that are primarily responsible.

Before launching into the theory for the general occurrence of the neutral point reconnection, it is appropriate to say a few words about the development of the concept of reconnection itself. The idea that X-type neutral points in magnetic fields in a highly conducting fluid are of special interest may be traced back at least as far as Giovanelli (1947), who suggested that an electrical discharge would develop along the neutral line, producing a solar flare. The idea was developed and elaborated by Dungey (1953,1958) who pointed out the reconnection of the lines of force associated with the thinning of the current sheet. Sweet (1958) suggested that the neutral point reconnection was the cause of the solar flare and further illustrated the thinning of the current sheet between opposite fields.

The interest in the X-type neutral point was motivated by the fact that a magnetic field can be consumed (dissipated) at a speed u comparable to the resistive diffusion coefficient $\eta=c^2/4\pi\sigma$ divided by the characteristic scale ℓ of the field gradient (thickness of the current sheet). If ℓ is equal to the general dimension L of the field, the rate is $u \sim \eta/L$, which can be written as V/N, where V is the characteristic Alfven speed in the field and N is the magnetic Reynolds number, defined in terms of $V(N=VL/\eta)$. Hence in the solar corona, where $V \sim 10^3$km/sec, we have u equal to 10^{-7}cm/sec, i.e. 3cm/year. Even a continental ice sheet progresses more rapidly.

The work of Dungey and Sweet indicated that the thickness ℓ of the current sheet may be greatly reduced below the dimension L of the field as a whole. Unfortunately, however, the dynamics of the problem, involving the expulsion of the fluid from between the two opposing fields, limited ℓ to $L/N^{1/2}$ and u to $V/N^{1/2}$ (Parker, 1957, 1963). This is larger by the large factor $N^{1/2}$ than the result L/N for passive diffusion, but it is still too small to play a role in the observed magnetic activity. In the solar corona, one estimates $V/N^{1/2} \sim$1cm/sec, with generally smaller values elsewhere.

The breakthrough came with Petschek's (1964) point that opposite fields of scale L need not press together over more than some narrow region at their opposing apexes, so that the effective scale is small compared to L. He went on to demonstrate a self-consistent condition in which the reconnection proceeds at a rate u of the order of V/lnN. Since logarithms are generally of the order of only 20-100, the result is a rapid cutting of the lines of force at a rate of the order of 0.1-0.01V. Neutral point reconnection had come into its own, cutting across fields at speeds of 1-100km/sec in the sun and planetary magnetosphere. Indeed, the speed u may lie anywhere in the range $(V/N^{1/2}, V/lnN)$ depending upon the boundary conditions. Hence opposite fields may lie quietly together at times, and fiercely consume each other at other times. Sonnerup (1970) provided exact solutions to the magnetohydrodynamic equations showing that u may be as large as V, for arbitrarily large N, if one exerts suitable external pressures on the system (see also the similarity solutions of Yeh and Axford, 1970).

In recent years these conditions have been explored and elaborated by many authors, reviewed by other speakers in these proceedings (see Vasyliunas, 1975; Parker, 1979, pp. 392-439; Priest, 1981, 1982). The resistive tearing mode instability (Furth, Killeen, and Rosenbluth, 1963) is a particularly important parallel development of neutral point reconnection into an instability of the equilibrium of opposite fields, as distinct from the absence of equilibrium when progressive reconnection is the result.

The more difficult question is under what

circumstances neutral point reconnection occurs in nature. Clearly when two opposite fields are pressed together, the fluid squeezes out from between (unless especially prevented from doing so by suitably applied external pressures) so that there is generally no equilibrium for the fields. The field gradients increase without bound and rapid reconnection arises. But if this is a sufficient condition for rapid reconnection, what is a necessary condition?

Syrovatsky (1978,1981,1982; Bobrova and Syrovatsky, 1979) pointed out that when an X-type neutral point is distorted by externally applied forces, it forms a current sheet which is subject to neutral point reconnection. We realize how generally this suggests reconnection when we recall that the projection of any field $\vec{B}(\vec{r})$ onto a plane perpendicular to $\vec{B}$ at any point $\vec{r}_o$ has a neutral point at $\vec{r}=\vec{r}_o$. Since $\vec{B}(\vec{r})$ is a well behaved function of position, it can be expanded about $\vec{r}_o$, showing that the neutral point is either an O or X-type (see discussion in Parker, 1979, pp. 383-391). There are finite regions throughout which the neutral point is of the X-type at every point $\vec{r}_o$. It is obvious that there cannot be a continuum of current sheets throughout the volume—that is a contradiction in terms—so we need to understand more about the process.

Fortunately, there is another way to approach the problem, recognizing that neutral-point rapid reconnection is a nonequilibrium state. That is to examine the equilibrium equation (1) to see for what field topologies it has solutions. When the equations have no solution, the result is nonequilibrium reconnection. To pursue this inquiry we employ the special state already described, illustrated in Fig. 1, where a unidirectional magnetic field extending through a simply connected volume V of highly conducting fluid is anchored in the infinitely conducting surface S enclosing V. The field within V can be manipulated by moving the footpoints of the field lines on S. We shall suppose, too, that the pressure of the fluid can be controlled at the surface S, so as to make the best possible case for equilibrium throughout V.

Denote by h the very large distance across the volume V in the direction of the field, and suppose that the breadth of V is very much larger in the directions perpendicular to the field. To fix ideas denote the upper surface of S by z=h in the region sketched in Fig. 1 and denote the lower surface by z=0. Maintain the anchor points fixed in the upper surface, while we manipulate the anchor points on z=0 in some arbitrary manner, subject to the limitation that the velocity field $\vec{v}(x,y,t)$ of the footpoints is incompressible and contains no sources or sinks ($\nabla\cdot\vec{v}=0$). We also assume that the path of any particular footpoint never strays farther from its starting point than some distance comparable to the characteristic scale L of variation of $\vec{v}$. That is to say, the individual

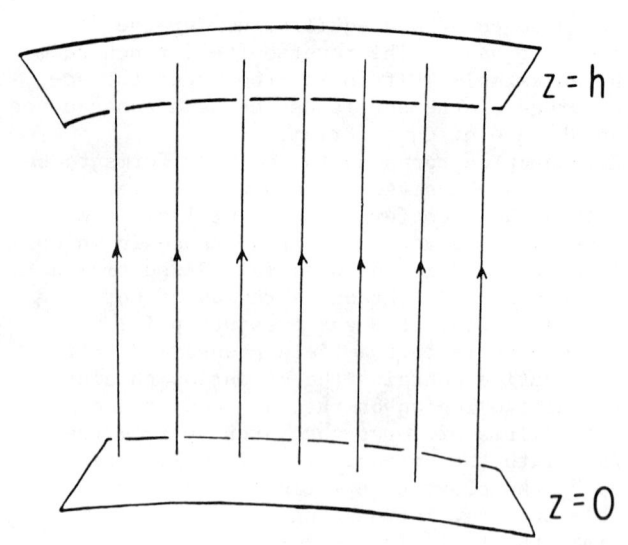

Fig. 1. A sketch of the unidirectional field extending between fixed surfaces z=0, h through a highly conducting fluid.

footpoints may go round and round in complicated ways, traversing a total pathlength s>>L, but never going beyond a distance of the order of L from their starting place. We suppose that the length h of the lines of force is so large that h>>s>>L, so that the lines of force are inclined no more than some small angle $\varepsilon=s/h$ to the general direction of the field across V. Fig. 2 is a sketch of the resulting wrapping and twisting of the field.

The purpose of this ordering (h>>s>>L) is to ensure that the interior of V is free of the effects of the boundaries (which effects decline inward at least as fast as exp(-z/L). We are concerned with the static equilibrium of field and fluid without the interference of rigid boundaries (which can always be introduced in such a way as to support an equilibrium). The boundaries are necessary to control and define the topology of the field, and otherwise play no role.

Now Fig. 2 sketches some special cases of the wrapping and twisting of the field that might arise as a consequence of the manipulation of the footpoints at z=0. The simplest is the twisted flux tube sketched on the left side of the figure. The topology of the field in the twisted tube is invariant along the tube and the field itself is invariant ($\partial\vec{B}/\partial z=0$) once one is well away from the boundaries (z=0,h). Next to it is a more complicated twisted tube made up of two twisted tubes wrapped in a larger twisted envelope. The topology is invariant along the tube. The third from the left is a braid, in which the direction of wrapping of the field alternates first left and then right along the lines of force. The field is clearly not invariant along the tube, nor is its topology. The right hand configuration is intended to

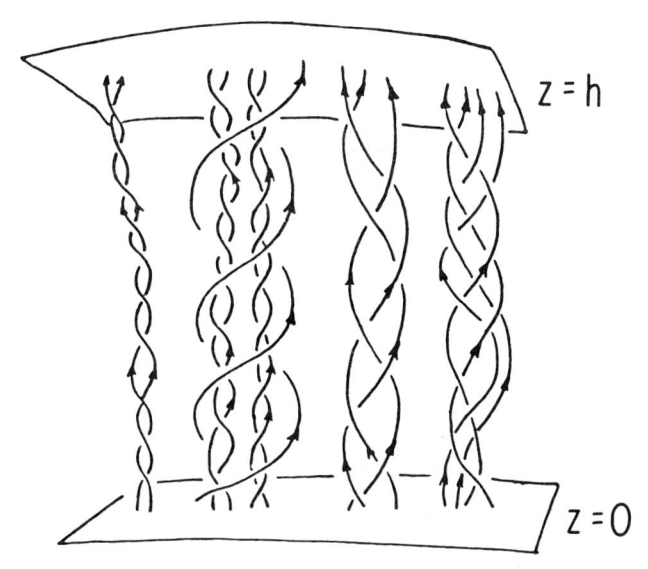

Fig. 2. A sketch of some of the forms of twisted and wrapped field lines that may occur in the unidirectional field as a consequence of shuffling the footpoints of the lines of force (at z=0).

illustrate a more general wrapping of the lines of force in which there is no repeatable pattern at all.

Now, as a consequence of the small inclination ε of the field to the z-direction, the field sketched in Fig. 2 can be written as

$$\vec{B}(\vec{r}) = \vec{e}_z B_o + \varepsilon \vec{b}(\vec{r}), \qquad (2)$$

where $\vec{b}$ represents the perturbation arising from the shuffling of the footpoints and ε is a number small compared to one. The fluid pressure is similarly perturbed, to the form $p_o + \varepsilon p(\vec{r})$. Then, to first order in ε, (1) becomes

$$\nabla(p + B_o b_z / 4\pi) = (B_o/4\pi) \partial \vec{b}/\partial z. \qquad (3)$$

The divergence of this equation yields

$$\nabla^2 (p + B_o b_z / 4\pi) = 0$$

as a consequence of $\nabla \cdot \vec{b} = 0$. We require solutions with the perturbations p and b_z bounded over the entire region no matter how large is h/L. The only bounded solution to Laplace's equation over an arbitrarily large space is a constant. But if $p + B_o b_z / 4\pi$ is a constant, it follows from (3) that

$$\partial \vec{b}/\partial z = 0. \qquad (4)$$

The calculation can be carried to all orders in ε, and can be repeated for perturbations about any field that is independent of z, from which we conclude that equilibrium requires the winding and wrapping of the field about itself to be invariant along the field (Parker, 1972, 1979, pp. 359-378; Yu, 1973).

One may conjecture on exceptions to this theorem, in the form of families of solutions for which $\partial \vec{b}/\partial z \neq 0$, that contain no invariant field ($\partial \vec{b}/\partial z = 0$) as a member. Such a family is not analytic in ε at $\varepsilon = 0$ and would lie outside the above considerations, based on expansion in ascending powers of ε. No such families of solution have been discovered up to the present (see discussion in Rosner and Knobloch, 1982). It would be exceedingly interesting to study their properties if they exist. Tsinganos, Distler, and Rosner (1984) have extended the above proof of the necessity for invariance by showing the correspondence between the magnetic lines of force and the trajectories of a Hamiltonian system in phase space, to which they apply the Kolmogoroff-Arnold-Moser theorem. They show that there are no equilibrium solutions within a finite region ε about invariance ($\partial \vec{b}/\partial z = 0$).

The physical basis for invariance in equilibrium is readily understood. It follows from (1) that the fluid pressure is uniform along the lines of force ($\vec{B} \cdot \nabla p = 0$). Suppose that the winding pattern varies along the field. There is an immediate difficulty, for if the pressure is adjusted to provide equilibrium in one winding pattern, say at the lower end of the field, that pressure is projected along the lines of force into different winding patterns where it does not suffice for equilibrium. In particular, the fluid pressure is inadequate to keep the opposite fields (facing each other across the X-type neutral point at each point $\vec{r}_o$) from pressing together and reconnecting at various points in other patterns.

Suppose, then, that we carefully tailor the shuffling of the footpoints, by choosing a fixed pattern for the streamlines of v, so that (4) is satisfied throughout the interior of the volume. In that case, it is well known that (1) reduces to

$$p = p(A), \quad B_z = B_z(A)$$

where $A = A(x,y)$ with

$$B_x = +\partial A/\partial y, \quad B_y = -\partial A/\partial x.$$

The vector potential A is determined by the exact field equation

$$\nabla^2 A + 4\pi F'(A) = 0 \qquad (5)$$

where

$$F(A) \equiv p(A) + B_z(A)^2/8\pi \qquad (6)$$

It is obvious that with arbitrary F(A) there are infinitely many forms of (5), many of which have infinitely many well behaved solutions. Clearly there is a variety of equilibrium forms for the field.

However, the infinity of solutions to (5) have special properties, so that we do not expect to find them in nature. That is to say, the physical world generally violates the special equilibrium requirement (5). The simplest way to see the difficulty is to note that, if $\partial \vec{B}/\partial z=0$ so that the equilibrium (5) exists, then the shuffling of the footpoints $\vec{v}$, (a) has no sources or sinks, (b) the shuffling pattern is time independent, and (c) the shuffling is limited to excursions of L or less from the starting point. Together these mean that the streamlines of $\vec{v}$ form closed-packed cells of circulation. Hence the projection of the resulting $\vec{B}(x,y)$ onto any plane z=constant gives lines of force (A=constant) forming closed, close-packed cells, which is just another way of saying that the field is composed of distinct twisted flux tubes, all packed tightly together. Each elastic tube is squashed out of round, into some sort of polygonal cross section, being jammed between its nearest neighbors. Fig. 3 is a sketch of what a cross section might look like. Each twisted tube would like to assume a circular cross section, but is prevented from doing so by the larger pressure exerted on it by its neighbor near the center of their common face. Note, then, that if at any point two tubes with the same twist are pressed together, they present opposite transverse fields across their common face. The fluid is squeezed from between the opposite fields by the higher pressure near the middle of the common face, so that neutral point reconnection (i.e. dynamical nonequilibrium) is the result (Parker, 1982,1983a,c).

To put it in different terms, the common vertices of three or more twisted flux tubes must be four-fold (or 2n-fold) rather than three-fold, for if only three twisted tubes meet, as sketched in Fig. 4, at least two of the three tubes must have the same sense of twist, and therefore present opposite transverse fields to each other. But a four-fold

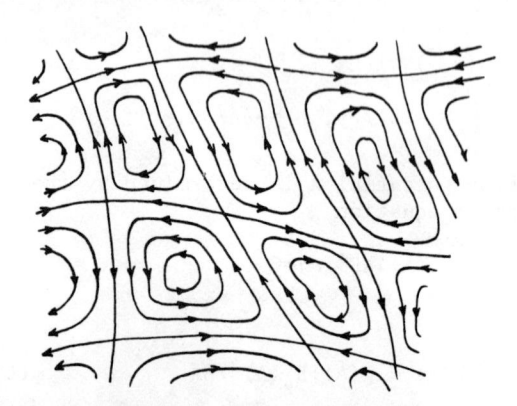

Fig. 3. A sketch of the field configuration (projected onto a surface z=constant) of close-packed twisted flux tubes.

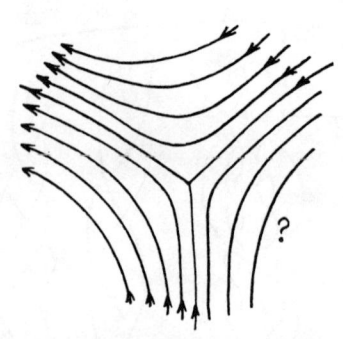

Fig. 4. A sketch of the expected three-fold vertex between contiguous tubes, illustrating the fact that at least one of the tubes must have the same sense of twist as one of the others, so that opposite fields meet across their common interface.

vertex is a special construction, requiring perfect balance between the strength (stiffness and diameter) of all four tubes in order that each tube meet the other three at the common point of the other three. Four-fold vertices are only mathematical constructions (i.e. solutions to (5)) and do not occur in nature, where something akin to hexagonal close packing is expected.

And so we are generally defeated in our quest for equilibrium in strained magnetic fields in nature. Any strains introduced into the fields by shuffling the footpoints through more than a couple of revolutions produces nonequilibrium in the form of reconnection.

It is on this basis, then, that we can understand the universal appearance of reconnection of the strained fields in nature, subject to the bulging plasmas of the laboratory, the rushing solar wind around the magnetosphere, and the continual shuffling of the footpoints of the stellar magnetic field in the convective photosphere. Thus, for instance, most of the work done by the convection of the magnetic loops extending above the active regions of the sun is soon dissipated into fluid motion and heat in the corona above (Parker, 1983b). Solar flares are the most spectacular result (Priest, 1981). The active X-ray corona of the sun and other stars is another.

We should not fail to note that the dynamical nonequilibrium of the close-packed twisted flux tubes is precisely the phenomenon of two-dimensional magnetohydrodynamic turbulence (Parker, 1983d) that has been studied in detail by several authors (Fyfe and Montgomery, 1976; Fyfe, Montgomery, and Joyce, 1977; Montgomery and Vahala, 1979; Matthaeus and Montgomery, 1981; Matthaeus, 1982) in recent years. Their numerical simulations provide a direct illustration of dynamical nonequilibrium (and the related neutral point reconnection) in action. We can understand from the nonequilibrium of such configurations why the turbulence cannot cease until the entire field is reduced to two oppositely twisted flux tubes.

As a final remark, the discussion in this paper has been in terms of a fluid, ignoring the small-scale plasma effects. The fact is that the dynamical behavior of a plasma differs significantly from a fluid only on small scales, in the thin current sheets between opposite fields, where the finite cyclotron radius, the steep magnetic gradients, the large drift velocity of the individual particles, and the large conduction velocities may provide plasma turbulence and other effects. The rate of reconnection is controlled primarily by the conditions outside the current sheet where the dynamical behavior is accurately represented by a fluid with perhaps an anisotropic pressure in extreme cases (see the review by Axford and by Vasyliunas in these proceedings). The complex plasma and particle effects within the current sheet have relatively little effect on the rate, although they may have profound effects upon the disposition of the free energy into particle acceleration to high energy, etc. Basically it is the large-scale dynamical nonequilibrium of the field and fluid that sets up the situation for the intense plasma effects around the neutral point, and not vice versa.

Acknowledgements. This work was supported by the National Aeronautics and Space Administration through NASA grant NGL-14-001-001.

References

Bobrova, N.A. and S.I. Syrovatsky, Singular lines of one-dimensional force-free magnetic field, Solar Phys. 61, 379-388, 1979.

Dungey, J.W., Conditions for the occurrence of electrical discharges in astrophysical systems, Phil. Mag. 44, 725-738, 1953.

Dungey, J.W., Cosmic Electrodynamics, Cambridge, Cambridge University Press, pp. 98-102, 1958.

Furth, H.P., J. Killeen and M.N. Rosenbluth, Finite-resistivity instabilities of a sheet pinch, Phys. Fluid 6, 459-484, 1963.

Fyfe, D. and D. Montgomery, High-beta turbulence in two-dimensional magnetohydrodynamics, J. Plasma Phys. 16, 181-191, 1976.

Fyfe, D., D. Montgomery and G. Joyce, Dissipative, forced turbulence in two-dimensional magnetohydrodynamics, J. Plasma Phys. 17, 369-398, 1977.

Giovanelli, R.G., Electric phenomena associated with sunspots, Mon. Not. Roy. Astron. Soc. 107, 338-355, 1947.

Matthaeus, W.H., Magnetic reconnection in two dimensions: Localization of vorticity and current near magnetic X-points, Geophys. Res. Letters 9, 660-667, 1982.

Matthaeus, W.H. and D. Montgomery, Nonlinear evolution of the sheet pinch, Plasma Phys. 25, 11-22, 1981.

Montgomery, D. and G. Vahala, Two-dimensional magnetohydrodynamic turbulence, J. Plasma Phys. 21, 71-83, 1979.

Parker, E.N., Sweet's mechanism for merging magnetic fields in conducting fluids, J. Geophys. Res. 62, 509-520, 1957.

Parker, E.N., The solar flare phenomenon and the theory of reconnection and annihilation of magnetic fields, Astrophys. J. Suppl. 8, 177-212, 1963.

Parker, E.N., Topological dissipation and the small-scale fields in turbulent gases, Astrophys. J. 174, 499-510, 1972.

Parker, E.N., Cosmical Magnetic Fields, Oxford, Clarendon Press, 1979.

Parker, E.N., The rapid dissipation of magnetic fields in highly conducting fluids, Geophys. Astrophys. Fluid Dyn. 22, 195-218, 1982.

Parker, E.N., Magnetic neutral sheets in evolving fields. I. General theory, Astrophys. J. 264, 635-641, 1983a.

Parker, E.N., Magnetic neutral sheets in evolving fields. II. Formation of the solar corona, Astrophys. J. 264, 642-647, 1983b.

Parker, E.N., Absence of nonequilibrium among close-packed twisted flux tubes, Geophys. Astrophys. Fluid Dyn. 23, 85-102, 1983c.

Parker, E.N., The hydrodynamics of magnetic nonequilibrium, Geophys. Astrophys. Fluid Dyn. 24, 79-108, 1983d.

Petschek, H.E., Magnetic field annihilation, AAS-NASA Symposium on the Physics of Solar Flares, US Government Printing Office, pp. 425-437, ed. by W.N. Hess, 1964.

Priest, E.R., Solar Flare Magnetohydrodynamics, New York, Gordon and Breach, pp. 139-212, 1981.

Priest, E.R., Solar Magnetohydrodynamics, Dordrecht, D. Reidel, pp. 345-365, 1982.

Rosner, R. and E. Knobloch, On perturbations of magnetic field configurations, Astrophys. J. 262, 349-357, 1982.

Sonnerup, B.U.O., Magnetic-field reconnection in a highly conducting incompressible fluid, J. Plasma Phys. 4, 161-174, 1970.

Sweet, P.A., The neutral point theory of solar flares, in Electromagnetic Phenomena in Cosmical Physics, I.A.U. Symp. No. 6, pp. 123-134, ed. by B. Lehnert, Cambridge, Cambridge University Press, 1958.

Syrovatsky, S.I., On the time evolution of force-free fields, Solar Phys. 58, 89-94, 1978.

Syrovatsky, S.I., Pinch sheets and reconnection in astrophysics, Ann. Rev. Astron. Astrophys. 19, 163-229, 1981.

Syrovatsky, S.I., Model of flare loops, fast motions, and opening of magnetic field in the corona, Solar Phys. 76, 3-20, 1982.

Tsinganos, K.C., J. Distler, and R. Rosner, On the topological stability of magnetostatic equilibrium, Astrophys. J. (in press) 1984.

Vasyliunas, V.M., Theoretical models of magnetic field line merging, Rev. Geophys. Space Sci. 13, 303-336, 1975.

Yeh, T. and W.I. Axford, On the reconnection of

magnetic field lines in conducting fluids, *J. Plasma Phys.* 4, 207-229, 1970.

Yu, G., Hydrostatic equilibrium of hydromagnetic fields, *Astrophys. J.* 181, 1003-1008, 1973.

Questions and Answers

Moore: It seems that you are seconding the remarks of Schindler this morning, i.e., you seem to be showing that the real problem is the stability allowing energy to build up before a flare or substorm. Why doesn't the energy just continually burn away and never build up for a large explosion?

Parker: *The answer is that neutral point reconnection may proceed at any speed in the interval ($V/N^{1/2}$, $V/\ln N$) as noted above. If, for instance, one presses together two oppositely directed magnetic fields, they flatten against each other, forming a broad thin magnetic neutral sheet between them. The excess pressure gradually expels the fluid from the neutral sheet, with magnetic reconnection between the two opposite fields progressing at the slow rate $V/N^{1/2}$. The energy burns away so slowly ($N \sim 10^{10}$ or more) that the magnetic strains may build up in spite of the dissipation. One expects that eventually various instabilities, e.g. the slow resistive tearing mode, develop localized X-type neutral points, which reconnect more rapidly, along the lines of the Petschek configuration, at a rate nearer $V/\ln N$. Thay may provide a local rapid burn, which quickly relieves local strains in the field. However, rapid reconnection on a small local scale may be self-quenching because the fluid expelled from the neighborhood of each local neutral point is still in residence between the large-scale opposite fields, and must be expelled over the large-scale before further reconnection can take place. On the other hand, if the Petschek configuration is set up on a large scale, so that the fluid is expelled from the whole neutral sheet, then the burn may consume most, if not all, of the magnetic free energy at a rapid rate.*

It is not possible to determine by any quantitative method the rate of reconnection for a given initial configuration of opposite fields in a fluid of large magnetic Reynolds number N. That is why, in connection with heating the solar corona, I could do no more than state the average rate of reconnection on the assumption that the active corona is heated largely by neutral point dissipation of strained (i.e. nonequilibrium) magnetic fields. The rate, you may recall, turned out to be of the same general order of magnitude as the harmonic mean of the two extreme values $V/N^{1/2}$ and $V/\ln N$. That is to say, the mean rate of reconnection is very large compared to the slow rate $V/N^{1/2}$ and very small, by the same large factor, compared to the fast rate $V/\ln N$. The actual reconnection rate in the active corona presumably sputters along at various rates above and below the mean. One may imagine that when the strains in the magnetic field get above some critical level, the reconnection is prone to something akin to the Petschek configuration, which quickly reduces the strains to more modest levels.

THE ELECTROMAGNETIC FIELD FOR AN OPEN MAGNETOSPHERE

Walter J. Heikkila

Center for Space Sciences
The University of Texas at Dallas
Richardson, Texas 75080

Abstract. Two-dimensional steady-state theories of reconnection are based on an electric field that is constant across the separator line; consequently, curl $\underline{E}$ is assumed to vanish. However, a finite curl is required so that stored magnetic energy can be tapped, since $\partial(B^2/2\mu_0)/\partial t = \underline{H}\cdot\partial\underline{B}/\partial t$, and $\partial\underline{B}/\partial t = -$ curl $\underline{E}$. With reversal of the electric field at the magnetopause (which implies a finite curl), magnetosheath plasma can feed the boundary layers, just inside the magnetopause. Since the boundary layer plasma cannot all flow into the plasma sheet, it must continue flowing tailward, still on closed field lines. The topology of the assumed electric field must be revised from being everywhere in the dawn-dusk direction (in the reconnection model) to the reverse direction within the boundary layers. Even though the mechanism for creating this electric field may be transient, the polarization charge in the boundary layer is not lost immediately, and the electric field will be quasi-steady state, especially toward the flanks where there is always antisunward flow. This revision to the electric field profile implies that steady state reconnection may not be important in powering magnetospheric phenomena such as large scale circulation; localized and transient processes are more important, including impulsive transport of magnetosheath plasma to the boundary layers; and that the boundary layers provide the plasma, momentum, and energy to the plasma sheet, in a new kind of viscous interaction.

Introduction

In 1975 I predicted that a drastic revision might be required in our concepts of the open magnetosphere, and of the associated electric field structure in particular (Heikkila, 1975). Many persons have apparently misunderstood my point, thinking that I favored a closed magnetosphere (Cowley, 1982); this is not true, as pointed out in a comment on Cowley's article (Heikkila, 1983a). My view is that the reality is more complex than can be incorporated in steady-state models, at least analytically; nevertheless, it is possible that conceptually the basic physics can be understood by appealing directly to first principles.

Dungey's (1961) model for an open magnetosphere has a dawn-to-dusk electric field everywhere for a southward interplanetary magnetic field (IMF). He proposed that solar wind plasma flow, which would be associated with an electric field as in Fig. 1(a), would convect flux tubes into the dayside X-type separator line, tailward on open polar cap field lines, and then into the nightside X-line. He assumed that the electric field remained finite across the magnetopause, and that a reconnection process would occur at both the dayside and nightside X-lines, as reviewed for example by Vasyliunas (1975).

My criticism of reconnection models and theories, for example as voiced in the article in Planet. Space Sci. (1978), is that the steady state reconnection models are too simplistic; these models cannot possibly explain the essential physics, primarily because the electric field has no curl. This was spectacularly illustrated in the study of airfoils. Zhukovskii (see Landau and Lifshitz, 1959) tried very hard to see how an airfoil could provide lift; he was unsuccessful at first, until he introduced a flow pattern which had a curl. Similarly, without a curl in the electric field stored magnetic energy is not accessible. This is obvious upon careful inspection of Poynting's energy conservation theorem:

$$\int \underline{J}\cdot\underline{E}\, d\tau = -\int_s \underline{E}\times\underline{B}\cdot d\underline{S}/\mu_0 - \int \underline{B}\cdot\partial\underline{B}/\mu_0\partial t\cdot d\tau$$
$$- \int \varepsilon_0\underline{E}\cdot\partial\underline{E}/\partial t\cdot d\tau \qquad (1)$$

With steady state models based on ideas similar to those of Dungey, the only source of energy is energy that is entering through the boundary surface (by means of the Poynting flux), or alternatively by means of a current delivering power from an outside source, which is exactly

ELECTROMAGNETIC FIELD

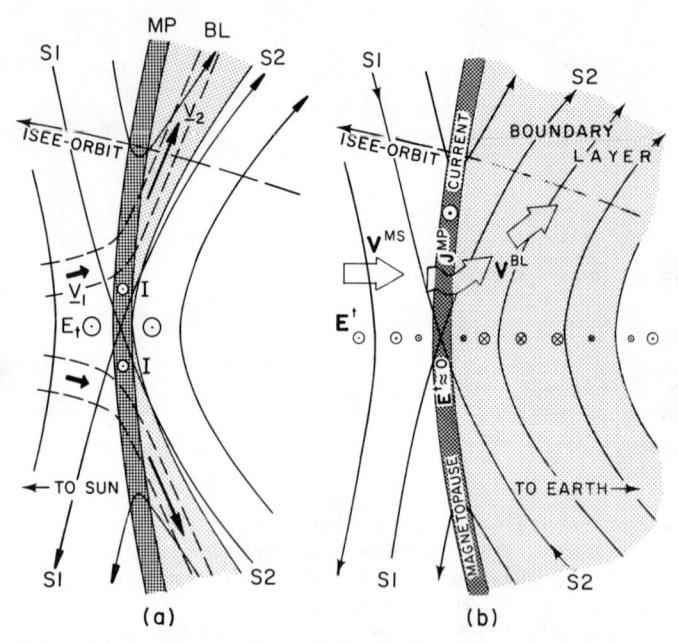

Fig. 1(a) Dungey's model of the dayside magnetopause. The $\underline{J} \times \underline{B}$ force produces a jet of plasma on open field lines, and a positive $\underline{J} \cdot \underline{E}$ implies power dissipation of about 5×10^{11} watts over the entire frontside magnetopause. Fig. 1(b) Heikkila's model with the same magnetic field topology, but with the tangential electric field reversing direction. The $\underline{J} \times \underline{B}$ force also produces a jet of plasma, but this time toward closed field lines. Here $\underline{J} \cdot \underline{E}$ is positive on one side, but negative on the other; the current perturbation can be regarded as a dynamo followed by a load.

equivalent (Vasyliunas, 1968; Heikkila et al., 1979, p. 1385).

When the solar wind changes suddenly there is no time for steady state conditions to be set up; what happens in the first instance must be entirely local, as pointed out by Heikkila (1982). When we include time-dependent fields we have a new source of the electric field through the time variation of the magnetic field curl $\underline{E} = -\partial \underline{B}/\partial t$, by Faraday's law. Notice that $\underline{H} \cdot \partial \underline{B}/\partial t = \partial/\partial t(B^2/2\mu_0)$, showing that we can add to, or tap, the energy that is stored in the magnetic field only through the curl of the electric field.

The more recent discovery of the low-latitude boundary layer on closed field lines (Hones et al., 1972; Akasofu et al., 1973; Eastman et al., 1976; Eastman and Hones, 1979) makes even more urgent a revision to theories of the electric field structure. The electric field is essentially normal to the magnetopause at the dawn and dusk flanks for tangential flow, being in the dusk-dawn direction within the boundary layers. Further toward the dayside, the field must still be so directed in order that the magnetosheath plasma, which is observed inside

the sub-solar magnetopause, can convect tailward, as indicated in Fig. 2. In the magnetosheath the electric field will be in the reverse direction for a southward IMF. Thus, at the dayside magnetopause in the subsolar region the tangential component must be as indicated by Figure 1(b), from Heikkila (1982). This simultaneous reversal of both the electric and magnetic fields causes the plasma particles to be convected across the magnetopause onto closed field lines by a transient process to form the boundary layer.

It should be noted that this can be only a transient process, as the magnetopause can convect only a relatively short distance toward the earth; it must eventually return, in a breathing motion, as has been reported by Williams et al.

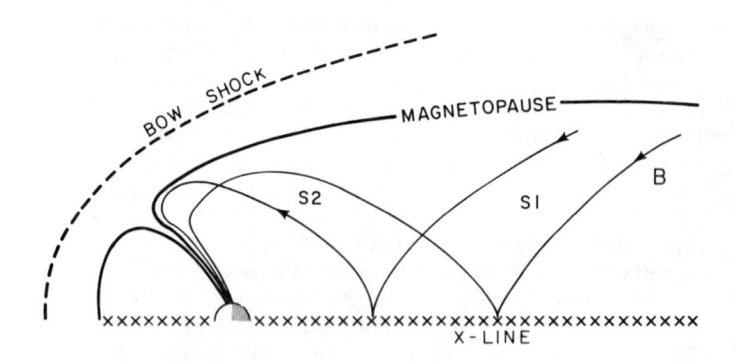

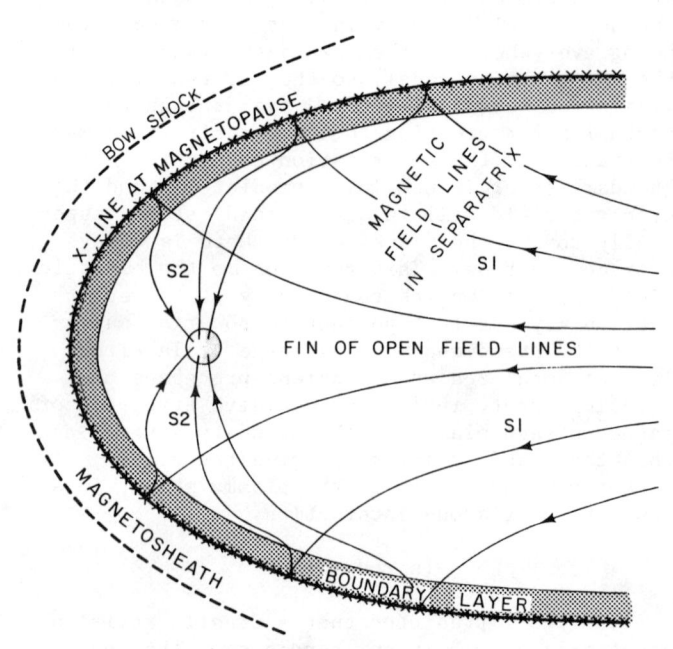

Fig. 2 The separatrix consists of two sheets S_1 and S_2, going to the solar wind and the earth respectively from the X-type separator line. The X-line is at the magnetopause in the equatorial plane for a strictly southward interplanetary magnetic field. It will continue to be at magnetopause in the distant tail, as shown in Figure 3.

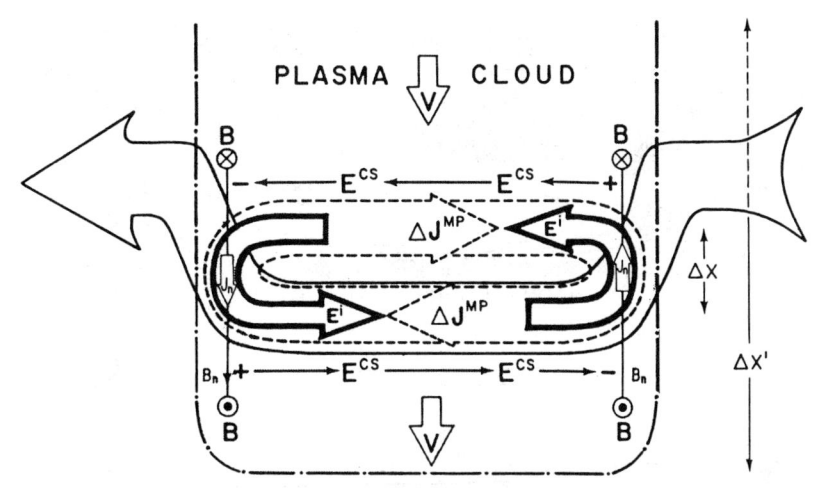

Fig. 3 Impulsive penetration is produced by an electrostatic field due to a charge distribution created by an induction electric field. Charged particles from the old plasma cloud go through the current sheet along B_n and form a new plasma cloud which can extend to closed magnetic field lines.

(1979). The conditions depicted in Figure 1(b) are meant to apply only briefly (as explained by Heikkila, 1982), during the pseudo-steady state conditions of inward motions of the magnetopause. However, the polarization charge that is formed in the boundary layer takes some time to be lost (especially on closed field lines), and a continued barrage of plasma injection events will create an electric field that has a steady state component.

The magnetopause is too thin for MHD conditions to be satisfied; instead, we should take the view that the plasma on the outside supplies particles to create a new plasma on the inside, as indicated in Figure 3, again only in the subsolar region. We can view the moving magnetopause current as a travelling perturbation $\Delta \underline{J}^{MP}$, to create an induction electric field everywhere opposing the current perturbation, with a finite curl since the vector has no beginning or end. In the forward part of the current $\underline{E} \cdot \underline{J}$ is negative, showing that the plasma particles lose energy. In the trailing part $\underline{E} \cdot \underline{J} > 0$, indicating that particles gain energy. The complete perturbation can be regarded as a localized circuit, with the forward part a dynamo supplying energy to the trailing part. Thus cause (with $\underline{E} \cdot \underline{J} < 0$) and effect (a load with $\underline{E} \cdot \underline{J} > 0$) are in close proximity.

Akasofu et al. (1973) showed that the magnetopause boundary layer is about 1 R_e thick and about 20 R_e high at the Vela orbit. Assuming an average particle density of 10 cm^{-3} and a flow speed of 300 km/s yields a flux of 4×10^{27}/s for both the dawn and dusk boundary layers. Further toward the dayside Eastman (1979) estimated a flux of 6×10^{26}/s, possibly indicating that additional entry occurs well back toward the flanks. This massive flow cannot all go into the plasma sheet; Hill (1974) has estimated that $10^{25} - 10^{26}$ is a sufficient source. I have proposed (Heikkila, 1983b) that the dawn and dusk boundary layers become joined together in the distant magnetotail, with the plasma still flowing tailward on closed magnetic field lines, inside the magnetopause. This would require that the electric field have the profile shown in Figure 4, being in the dusk-to-dawn sense throughout most of the distant magnetotail. This combination of fields acts as a giant pump forcing particles tailward. The plasma particles can cross the distant nightside magnetopause by a process similar to that occurring on the dayside.

Again, this is a transient process, implying that the distant magnetotail changes shape, possibly by forming new x-lines as the old ones are swept back, beyond the end of the plasma sheet (this is not related to substorms, which occur much closer to the earth).

Conclusions

The electric field profile shown in Figure 4 is the suggested drastic revision to the electric field, as compared to what has been assumed in reconnection models (as indicated in my Figure 5, from Cowley, 1980) which is everywhere from dawn to dusk.

In fact, the reconnection model shown in Figure 5 has three serious difficulties. First, in Figure 5 there is no indication of how the nightside X-line begins or ends. The dayside X-line is at the magnetopause, and it will continue at the magnetopause toward the dawn and dusk flanks (see Figure 2). If it is assumed that this is continued into the nightside X-line, in a steady-state feature, then Figure 5 is inconsistent in that the magnetopause is shown extending tailward over the lobes past the

42   ELECTROMAGNETIC FIELD

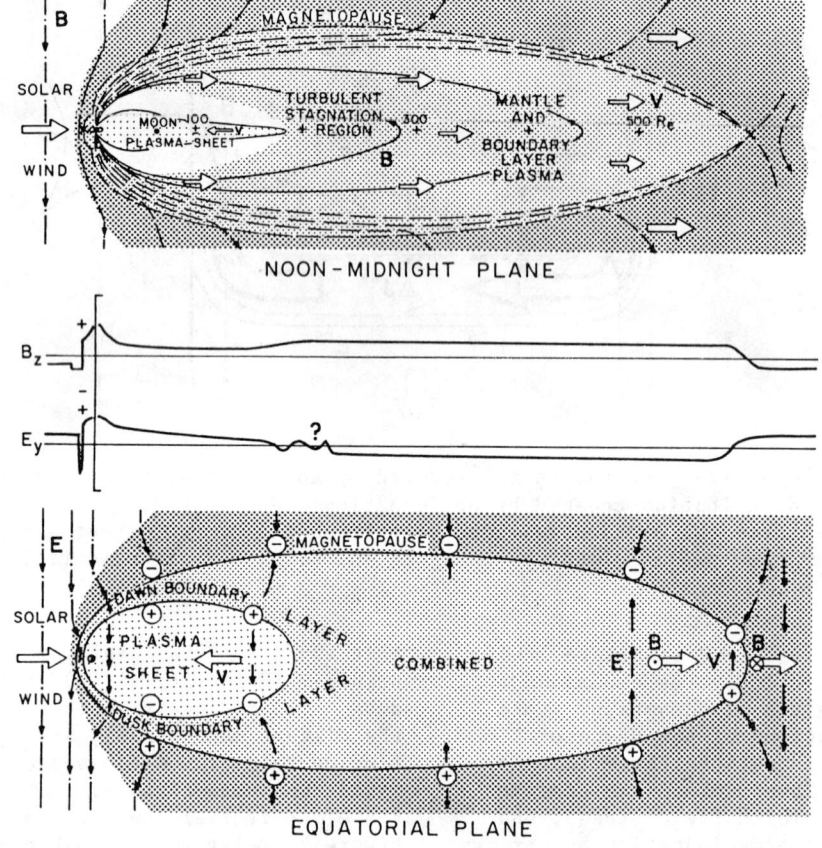

Fig. 4 Two cuts of the entire magnetosphere showing that the plasma sheet is a small cavity of low density plasma within a sea of boundary layer and mantle plasma. The flow pattern requires that the electric field reverses direction in the boundary layer.

X-line. Where does the X-line leave the magnetopause? If, on the other hand, the X-line in the tail is local, not extending to the dawn and dusk flanks, then it must be joined to an O-line; such a feature is a magnetic island or plasmoid entirely within the magnetotail, and is related to transient substorm phenomena; that happens only occasionally.

Second, it is commonly thought (e.g. Russell and McPherron, 1973) that reconnection causes such things as magnetospheric convection, through flux erosion from the dayside. This is commonly thought to be described by a two-dimensional steady state reconnection process (Russell and McPherron, 1973), which is necessarily a load (Vasyliunas, 1975). A process which is a load cannot be the cause of anything; a dynamo is needed to be the cause of something, i.e., plasma must lose energy. Furthermore, there is the question of the source of energy to power the reconnection load. The suggestion is that it is a dynamo over the high latitude lobes of the magnetopause, where the current opposes the electric field. The difficulty with that view is that it would take considerable time (an Alfvén wave travel time,

of tens to hundreds of minutes, assuming mantle densities of 10 cm^{-3}) for the power to be delivered to the reconnection load in response to a transient process. The model shown in Figure 1(b) and 3 avoids this by having a local dynamo instead.

Third, the very existence of the boundary layer on closed field lines shows that the electric field must reverse for anti-sunward flow on both sides of the magnetopause.

Figure 4 avoids all these difficulties by the assumption of a new electric field profile. We must be equally concerned with the topology of the electric field as with the topology of the magnetic field if we are to understand the essential physics.

In the new scheme it is the boundary layer that causes most internal magnetospheric phenomena as a by-product, such as the formation of the plasma sheet by diffusion of boundary layer plasma, sunward convection within the plasma sheet, and the production of auroras (see Heikkila, 1983c; Piddington, 1979). Losses of particles, momentum, and energy of the boundary layer plasma are involved; broadly speaking, the boundary layer acts basically as a viscous pro-

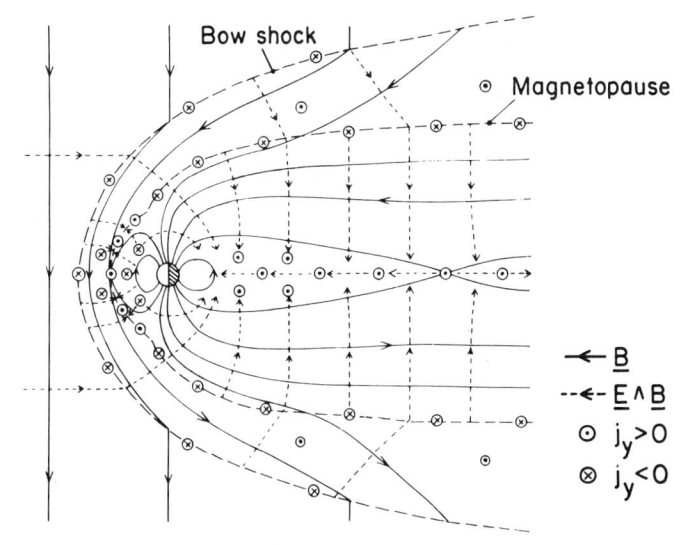

Fig. 5 Sketch of the conventional magnetosphere model (from Cowley, 1980) in the noon-midnight meridian plane for a southward directed interplanetary magnetic field, showing the magnetic field lines (solid lines), the bow shock and magnetopause boundaries (long dash lines), and the direction of the ExB drift or Poynting vector (short dash lines). The electric field is directed everywhere out of the plane of the diagram. The circled dots indicate regions in which the current is directed out of the plane of the diagram such that $\mathbf{j}\cdot\mathbf{E}>0$ and energy is transferred from the field to the plasma, while circled crosses indicate current flow into the plane of the diagram such that $\mathbf{j}\cdot\mathbf{E}<0$ and energy is lost by the plasma to the field. This should be compared with the topology proposed here in Fig. 4

cess, by any conceivable definition of "viscous" phenomena. Thus, boundary layer processes provide a mechanism for the "viscous" interaction" between the solar wind and the magnetosphere, a concept that was first postulated by Axford and Hines (1961).

Acknowledgements. I acknowledge support from NASA grant NAG5-226 and the National Science Foundation, grant ATM-8025194.

References

Axford, W.I. and C.O. Hines, A unifying theory of high-latitude geophysical phenomena and geomagnetic storms, Can. J. Phys. 39, 1433, 1961.

Akasofu, S.-I., E.W. Hones, Jr., S.J. Bame, J.R. Asbridge, and A.T.Y. Lui, Magnetotail and boundary layer plasmas at a geocentric distance of $-18R_E$ Vela 5 and 6 Observations, J. Geophys. Res., 78, 7257, 1973.

Cowley, S.W.H., Plasma populations in a simple open magnetosphere, Space Sci. Rev. 26, 217, 1980.

Cowley, S.W.H., The causes of convection in the Earth's magnetosphere: A review of developments during the IMS, Rev. Geophys. Space Phys. 20, 531, 1982.

Dungey, J.W., Interplanetary field and the auroral zones, Phys. Rev. Lett. 6, 47, 1961.

Eastman, T.E., The plasma boundary layer and magnetopause layer of the earth's magnetosphere, written in partial fulfillment of requirements for degree of Doctor of Philosophy, Press of Los Alamos Scientific Laboratory, 1979.

Eastman, T.E., E.W. Hones, Jr., S.J. Bame, and J.R. Asbridge, The magnetospheric boundary layer: site of plasma, momentum, and energy Transfer from the Magnetosheath into the magnetosphere, Geophys. Res. Lett., 3, 685, 1976.

Eastman, T.E., and E.W. Hones, Jr., Characteristics of the Magnetospheric boundary layer and magnetopause layer as observed by Imp 6, J. Geophys. Res., 84, 2019, 1979.

Heikkila, W.J., Is there an electrostatic field tangential to the dayside magnetopause and neutral line? Geophys. Res. Lett., 2, 954, 1975.

Heikkila, W.J., Criticism of reconnection models of the magnetosphere, Planet. Space Sci., 26, 121, 1978.

Heikkila, W.J., Impulsive plasma transport through the magnetopause, Geophys. Res. Lett., 9, 159, 1982.

Heikkila, W.J., Comment on 'The Causes of Convection in the Earth's Magnetosphere: A Review of Developments during the IMS' by S.W.H. Cowley', Rev. Geophys. Space Phys., 21, 1787, 1983a.

Heikkila, W.J., Exit of boundary layer plasma from the distant magnetotail, Geophys. Res. Lett., 10, 218, 1983b.

Heikkila, W.J., Magnetospheric topology of fields and currents, published in the Geophysical Monograph 28, Magnetospheric Currents, p. 365, by the American Geophysical Union, 1983c.
Proceedings of the AGU Chapman Conference on Magnetospheric Boundary Layers, p. 365, 1983c.

Heikkila, W.J., R.J. Pellinen, C.-G. Fälthammar, and L.P. Block, Potential and inductive electric fields in the magnetosphere during auroras, Planet. Space Science, 27, 1383, 1979.

Hill, T.W., Origin of the plasma sheet, Rev. Geophys. Sp. Phys. 12, 379, 1974.

Hones, E.W., Jr., J. R. Asbridge, S.J. Bame, M.D. Montgomery, S. Singer, and S.-I. Akasofu, Measurements of magnetotail plasma flow made with Vela 4B, J. Geophys. Res. 77, 5503, 1972.

Landau, L.D. and E.M. Lifshitz, Fluid Mechanics, Vol. 6, p. 139, 1959.

Piddington, J.H., The closed model of the earth's magnetosphere, J. Geophys. Res., 84, 95, 1979.

Russell, C.T. and R. McPherron, The magnetotail

and substorms, Space Sci. Rev. 15, 205-266, 1973.

Vasyliunas, V., Theoretical models of magnetic field line merging, 1, Rev. Geophys. Space Phys. 13, 303, 1975.

Vasyliunas, V., Discussion of paper by Harold E. Taylor and Edward W. Hones, Jr., "Adiabatic Motion of Auroral Particles in a Model of the Electric and Magnetic Fields Surrounding the Earth", J. Geophys. Res, 73, 5805, 1968.

Williams, D.J., T.A. Fritz, B. Wilken, and E. Keppler, An energetic particle perspective of the magnetopause, J. Geophys. Res., 84, 6385-6396, 1979.

QUASILINEAR EVOLUTION OF TEARING MODES
DURING MAGNETIC RECONNECTION

W. Horton and T. Tajima

Institute for Fusion Studies
University of Texas, Austin, Texas 78712

and

Ricardo Galvão

Instituto de Estudos Avancados, Centro Tecnico Aerospacial
São Jose dos Campos, SP-12200 BRAZIL

Abstract. Particle simulations of magnetic reconnection show that the presence of a magnetic field parallel to the current sheet changes the character of reconnection from a laminar process to a weakly turbulent one. An analysis of the quasilinear spectrum of tearing modes is presented which may account for some features of the 2-1/2D simulations. The turbulent, incompressible plasma flow carrying reversed magnetic flux to the tearing layer is shown to produce a positive definite anomalous resistivity. The relationship to the negative anomalous resistivity of Biskamp is discussed.

I. Neutral Current Sheet and Tearing Instability

Extensive 2-1/2D particle simulations [Tajima, 1982] of collisionless magnetic reconnection show that there are two regimes depending on the ratio of the reconnecting poloidal magnetic field to the toroidal magnetic field. In the absence of the toroidal magnetic field parallel to the current sheet, the reconnection involves a compressible, singular laminar flow leading to the Sweet-Parker process [Parker, 1963] followed by a faster second phase [Brunel et al., 1982]. In the presence of a strong enough toroidal or parallel guide field, however, the flow becomes essentially incompressible and the simulations [Tajima, 1982] show that the magnetic reconnection is a turbulent process. Such a case may become relevant to the reconnection process associated with the coalescence instability [Pritchet and Wu, 1979] or the Kadomtsev [1975] disruption model. In general, any reconnection process with a strong toroidal field driven by other instabilities, other island activities, or flow activities may involve the present process. We, however, exclude consideration of any three dimensional effects such as multiple helicity.

Tajima, [1982] points out that the transition of the laminar to turbulent reconnection is observed only in the collisionless simulation but not in the magnetohydrodynamic (MHD) particle simulation. This suggests that the turbulence involves electron dynamics in an essential way, since the MHD smears out the electron dynamics. This onset of turbulence may be related to the onset of magnetized electron dynamics. In the present work, we concentrate on analysis of the resultant turbulence; we do not discuss the onset of or transition to the turbulence.

The flow and magnetic activity are characterized by a spectrum of k_x, k_y fluctuations with time scales comparable to the maximum growth rate for collisionless tearing modes $\Delta\omega \sim \gamma_{max} \lesssim k_y |\Delta B_y| v_e/B$. Figure 1 shows typical examples of the contours of constant poloidal flux in the turbulent phase [Tajima, 1982]. In the following, we analyze some aspects of the quasilinear evolution of a spectrum of tearing mode turbulence.

The geometry we consider is a neutral current sheet $j_z(x)$ of thickness $\Delta x = x_0(t)$. The reversed poloidal field $B_y(x) = \partial\psi/\partial x$ vanishes at x=0 and is held fixed at $B_y = \pm B_y^0$ at $x = \pm L$. The electric and magnetic fields in the plasma are

$$\underset{\sim}{E} = -\nabla\phi(x,y,t) + \frac{1}{c}\frac{\partial\psi}{\partial t}\hat{z} \qquad (1)$$

$$\underset{\sim}{B} = \hat{z} \times \nabla\psi + B\hat{z}, \qquad (2)$$

and thus the component of the electric field

46 QUASILINEAR EVOLUTION

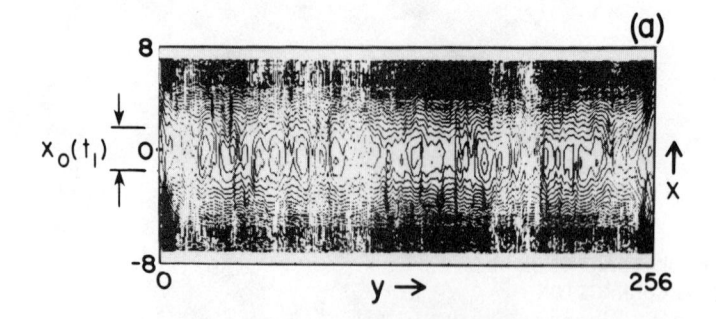

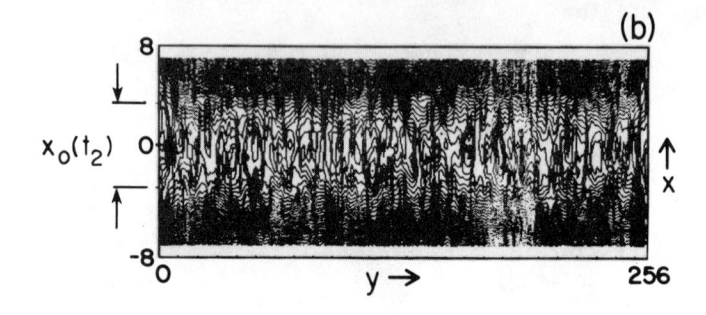

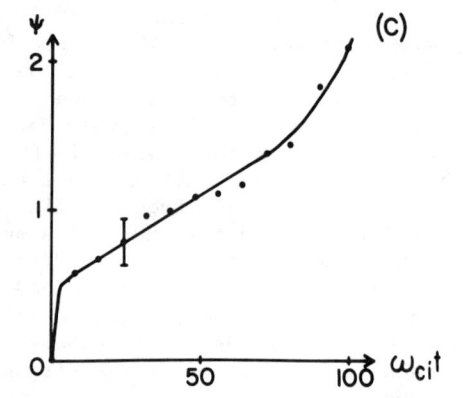

Fig. 1. (a) and (b) Contour of constant poloidal magnetic flux ψ at times t_1 and $t_2 > t_1$ showing the growth of the turbulent layer $x_0(t)$. (c) The approximately linear increase of the flux ψ trapped in the turbulent layer.

parallel to $\underset{\sim}{B}$ is

$$E_\| = \frac{\underset{\sim}{B} \cdot \underset{\sim}{E}}{B} = \frac{1}{c}\left(\frac{\partial \psi}{\partial t} + \underset{\sim}{v} \cdot \nabla \psi\right) \quad (3)$$

with

$$\underset{\sim}{v} = \frac{c\hat{z} \times \nabla \phi}{B} \quad (4)$$

and the current density is

$$\underset{\sim}{j} = \frac{c}{4\pi} \nabla \times \underset{\sim}{B} = \frac{c}{4\pi} \nabla^2 \psi \hat{z} . \quad (5)$$

In the turbulent motion the fields are split into their mean and fluctuating components by

$$\langle F \rangle = \int \frac{dy}{L_y} \int \frac{dt}{T} F \text{ and } \delta F = F - \langle F \rangle$$

where L_y is the period of the system in the y direction and $T\Delta\omega \gg 1$ and yet T is short compared with the time scale of the nonlinear evolution of the background.

The space-time scales of the fluctuations are determined by the width of the current sheet x_0 and the poloidal Alfvén speed $v_{Ay} = B_y^o/(4\pi n m_i)^{1/2}$ with

$$\Delta k_\perp \lesssim \frac{1}{x_0(t)}, \quad \Delta k_\| = \frac{\Delta k_y B_y(x)}{B} \quad \Delta\omega \lesssim \frac{v_{Ay}}{x_0} \quad (6)$$

in the laboratory frame. The thermal electrons see a correlation time $\tau_c^e = 1/\Delta k_\| v_e$ and the ions a correlation time $\tau_c^i = \min(\gamma_k^{-1}, 1/\Delta k_\perp \tilde{v})$ where $\tilde{v} = \langle v^2 \rangle^{1/2}$.

For the mean profiles $\langle \psi \rangle$ and $\langle \phi \rangle$ we take $\langle \phi \rangle = 0$ and

$$\langle \psi \rangle = x_0 B_y^o \ln\left[\cosh\left(\frac{x}{x_0}\right)\right]$$

$$\langle B_y \rangle = B_y^o \tanh\left(\frac{x}{x_0}\right) \quad (7)$$

$$\langle j_z \rangle = \frac{B_y^o}{x_0} \text{sech}^2\left(\frac{x}{x_0}\right) .$$

The analysis of the linear stability of the system at large magnetic Reynolds numbers $S = x_0 v_{Ay}/\eta$ is given by Drake and Lee [1977]. The linear outside solution of $\underset{\sim}{B} \cdot \nabla j_\| = \hat{z} \cdot \nabla \psi \times \nabla \nabla^2 \psi = 0$ gives

$$\psi_{k_y}(x,y) = \psi_{k_y}(0) e^{\mp k_y x}$$

$$\times \left[1 \pm \frac{\tanh\left(\frac{x}{x_0}\right)}{k_y x_0}\right] \cos(k_y y) \quad (8)$$

with

$$\Delta'_{k_y} = \frac{2\partial_x \psi_{k_y}}{\psi_{k_y}}\bigg|_{x=0} = \frac{2}{x_0}\left[\frac{1}{k_y x_0} - k_y x_0\right] . \quad (9)$$

The collisionless growth rate is

$$\gamma_{k_y} = \frac{B_y^o k_y v_e c^2 x_0 \Delta'(k_y)}{2\pi^{1/2} B x_0^2 \omega_{pe}^2} \quad (10)$$

and the collisional growth rate is

$$\gamma_{k_y} = \left(\frac{v_{Ay}}{x_0}\right)(Cx_0 \Delta'_{k_y})^{4/5}(k_y x_0)^{2/5} S^{-3/5} \quad (11)$$

with $C = \Gamma(1/4)/\pi\Gamma(3/4)$.

In the tearing layer the magnetic fluctuations $\psi_k(x)$ drive an electrostatic potential $\phi_k(x)$ through Poisson's equation or in the quasi-neutral limit valid for $k_\perp \lambda_{De} \ll 1$ through $\nabla \cdot \underset{\sim}{J} = -\partial_t \rho_\phi = 0$.

II. Nonlinear Electron Dynamics and Ohm's Law

Even at a low amplitude of the fields ψ_k, ϕ_k the electron motion becomes nonlinear. The distribution $f(\underset{\sim}{x},v,t)$ of magnetized or adiabatic electrons evolves by

$$\frac{\partial f}{\partial t} + v_\parallel \frac{\underset{\sim}{B}}{B} \cdot \frac{\partial f}{\partial \underset{\sim}{x}} + \frac{c\underset{\sim}{E} \times \underset{\sim}{B}}{B^2} \cdot \frac{\partial f}{\partial \underset{\sim}{x}} - \frac{e}{m_e} E_\parallel \frac{\partial f}{\partial v_\parallel} = \hat{C}f \quad (12)$$

where we include a weak electron-ion collision operator $\hat{C}$ to define the classical resistivity η_o. With the 2D spatial dependence of the simulations and the electromagnetic fields in Eqs. (1)-(3) the kinetic equation (12) becomes

$$\frac{\partial f}{\partial t} + \frac{\hat{z}}{B} \cdot \nabla(v_\parallel \psi + c\phi) \times \nabla f - \frac{e}{m_e c} \frac{d\psi}{dt} \frac{\partial f}{\partial v_\parallel} = \hat{C}f \quad (13)$$

with

$$\frac{d\psi}{dt} = \frac{\partial \psi}{\partial t} + \frac{c}{B} \hat{z} \cdot \nabla\phi \times \nabla\psi \equiv \frac{\partial \psi}{\partial t} + \frac{c}{B}[\phi,\psi] . \quad (14)$$

In Eqs. (13) and (14) it is useful for the nonlinear analysis to introduce the Poisson brackets $[f,g]$ defined by

$$[f,g] = \hat{z} \cdot \nabla f \times \nabla g = \frac{\partial f}{\partial x}\frac{\partial g}{\partial y} - \frac{\partial f}{\partial y}\frac{\partial g}{\partial x}$$

because of their obvious conservation and symmetry properties. We note that $\int dx\, f[f,g] = \int dx\, g[f,g] = 0$ and $\langle[\tilde{f},g]\rangle_y = -\partial_x \langle \tilde{g}\, \partial f/\partial y\rangle = \partial_x \langle \tilde{f}\, \partial g/\partial y\rangle$.

One of the most important aspects of the electron motion during field line reconnection is the form of Ohm's law parallel to the fluctuating magnetic field. Thus, we take the parallel current moment $j_\parallel = -e\int v_\parallel f\, dv$ of Eq. (13) and then average over the fluctuations to obtain the generalized Ohm's law

$$-\frac{1}{e}\frac{\partial \langle j_\parallel\rangle}{\partial t} + \frac{1}{m_e B}\frac{\partial}{\partial x}\langle -\frac{\partial \psi}{\partial y}\delta p_\parallel\rangle + \frac{c}{eB}\frac{\partial}{\partial x}\langle \frac{\partial \phi}{\partial y}\delta j_\parallel\rangle$$
$$+ \frac{e\langle n\rangle}{m_e}\langle E_\parallel\rangle + \frac{e}{m_e}\langle \delta n\, \delta E_\parallel\rangle = \frac{\nu_e}{e}\langle j_\parallel\rangle . \quad (15)$$

There are three important quasilinear transport processes described by Eq. (15).

1. Radial Transport of $\langle j_\parallel\rangle$

The fluctuations in the electron pressure $\delta p_\parallel$ combine with the cross-field tilting of the magnetic field $\delta B_x = -\partial\psi/\partial y$ to produce a net flow of current across the neutral sheet. In addition, the current is convected by the $E\times B$ flow across the current sheet. The net current flux F_j is given by (second and third terms of Eq. (15))

$$F_{j_\parallel} = \frac{c}{B}\langle -\frac{e}{m_e c}\frac{\partial \psi}{\partial y}\delta p_\parallel + \frac{\partial \phi}{\partial y}\delta j_\parallel\rangle$$
$$= -\frac{c}{B}\sum_{\underset{\sim}{k}} ik_y[\langle \phi_k^*\, \delta j_{\parallel k}\rangle - \frac{e}{m_e c}\langle \psi_k^*\, \delta p_{\parallel k}\rangle] . \quad (16)$$

A lengthy calculation of $\delta p_\parallel$ and $\delta j_\parallel$ using renormalized turbulence theory [Horton and Choi, 1979] shows that the principal contribution to $F_{j_\parallel}$ reduces to an anomalous electron viscosity

$$F_{j_\parallel} = -\mu_A \frac{d\langle j_\parallel\rangle}{dx} \quad (17)$$

with

$$\mu_A = -\frac{c^2}{B^2}\sum_{\underset{\sim}{k}}\int dv$$
$$\langle f_e\rangle \frac{k_y^2 v_\parallel^2}{v_e^2}\Big|\phi_k + \frac{v_\parallel}{c}\psi_k\Big|^2 \text{Im}\, G_k(v_\parallel) \quad (18)$$

where $G_k(v_\parallel) = (\omega - k_\parallel(x)v_\parallel + i\nu_k)^{-1}$ and

$$\nu_k = i\nu_o - \sum_{\underset{\sim}{k}_1}(\underset{\sim}{k}\times\underset{\sim}{k}_1\cdot\hat{z}/B)^2|v_\parallel \psi_{k_1} + c\phi_{k_1}|^2 G_{k-k_1}(v_\parallel) .$$

For the regime $\nu_o \leq \nu_k \leq \Delta k_\perp v_e \langle \delta B^2\rangle^{1/2}/B \leq (v_e/x_o)(B_y^o/B)$ we estimate that the maximum anomalous viscosity is given by

$$\mu_A \equiv \sum_k \frac{v_e^2|\delta B_k|^2}{\nu_k B^2} \leq \frac{v_e|B_y^o|}{x_o B} \quad (19)$$

where the last formula applies for large $\langle \delta B_\perp^2\rangle^{1/2}$ and $\nu_k \sim k v_e \langle \delta B_\perp^2\rangle^{1/2}/B$.

2. Mean Parallel Electric Field $\langle E_\parallel\rangle$

The convection of the magnetic flux $\underset{\sim}{v}\cdot\nabla\psi$ produces an average inductive change in the electric field parallel to the neutral sheet current. The total mean field is given by

$$\langle E_\parallel\rangle = \frac{\partial\langle\psi\rangle}{\partial t} + \frac{\partial F_\psi}{\partial x} \quad (20)$$

where F_ψ is the flux of B_y-magnetic flux into the turbulent layer given by

$$F_\psi = \langle v_x \psi\rangle = \sum_{\underset{\sim}{k}}\frac{ick_y}{B}[\langle\tilde{\phi}_k^*\tilde{\psi}_k\rangle - \langle\tilde{\phi}_k\tilde{\psi}_k^*\rangle] . \quad (21)$$

This flux F_ψ depends on the cross-correlation function of the magnetic and kinetic fluctuations. In Sec. III we calculate this flux in detail for a simple dynamical model. In general $\Delta E_z = \partial F_\psi/\partial x$ can have either sign -- ($\Delta E_z > 0$) reinforcing the ambient resistive E_z or ($\Delta E_z < 0$) reacting against it.

3. Parallel Friction from Density Fluctuations

The correlation function

$$\langle \delta n \delta E_\parallel \rangle = \sum_{\underset{\sim}{k}} \langle \delta n_{\underset{\sim}{k}}^* E_{\parallel \underset{\sim}{k}} \rangle \qquad (22)$$

produces a parallel friction to the passage of the electron current. The most common mechanisms giving rise to this type of anomalous resistivity are the lower hybrid and ion-acoustic waves. These mechanisms for producing anomalous resistivity are reviewed by Papadopoulos, [1977].

III. Evolution of the Mean Flux and the Effective Resistivity

In this section we neglect processes (1) anomalous electron viscosity and (3) parallel friction from density fluctuations in Sec. II and calculate process (2) the turbulent transport of magnetic flux.

For $\partial_t \langle j_\parallel \rangle \ll (e^2 \langle n \rangle / m_e) \langle E_\parallel \rangle$ the generalized Ohm's law (15) and Eq. (20) yield

$$\frac{\partial \langle \psi \rangle}{\partial t} + \frac{\partial F_\psi}{\partial x} = \frac{c^2 \eta_o}{4\pi} \frac{\partial^2}{\partial x^2} \langle \psi \rangle \qquad (23)$$

where η_o is the collisional resistivity and F_ψ is given by Eq. (21). The value of F_ψ requires knowledge of the ψ-ϕ cross-correlation function.

The electrostatic potential is determined by Poisson's equation in the quasi-neutral limit

$$\nabla \cdot \underset{\sim}{J}_\perp^i = -\underset{\sim}{B} \cdot \nabla (J_\parallel^e / B) = -\frac{1}{B} [\psi, \nabla^2 \psi]$$

where in the cross-field ion current we include the effect of a weak ion viscous force $\mu_o \nabla^2 v$. The magnetic flux ψ is determined by Ohm's law parallel to magnetic field $E_\parallel = \eta_o j_\parallel$ using Eqs. (3) and (5). The model nonlinear equations for ψ and ϕ are

$$\frac{\partial \psi}{\partial t} + \frac{c}{B}[\phi, \psi] = \frac{\eta_o c^2}{4\pi} \nabla^2 \psi \qquad (24)$$

$$\frac{1}{v_A^2} \nabla^2 \frac{\partial \phi}{\partial t} = \frac{c}{B}[\psi, \nabla^2 \psi] + \frac{\mu_o}{v_A^2} \nabla^4 \phi \ . \qquad (25)$$

Although the system (24) and (25) is an overly simplified description of the reconnection dynamics, it possess the correct conservation laws and describes the tearing mode instability, and thus is a useful model for calculating the ψ-ϕ correlation function.

Small amplitude fluctuations $\delta \psi_{\underset{\sim}{k}}(x)$, $\delta \phi_{\underset{\sim}{k}}(x) \exp(i k_y y + \gamma t)$ in Eqs. (24)-(25) satisfy the linear eigenvalue problem

$$AX = \lambda BX \qquad (26)$$

with the eigenvalue $\lambda = \gamma$ and eigenvector $X^T = (\delta \psi_k(x), \delta \phi_k(x))$ and the matrix components

$A_{11} = \eta_o(\partial_x^2 - k_y^2)$, $A_{12} = i k_y \frac{\partial \psi}{\partial x}$

$A_{21} = i k_y \frac{\partial \psi}{\partial x}(k_y^2 - \partial_x^2) + i k_y \frac{\partial^3 \psi}{\partial x^3}$

$A_{22} = -\mu_o(\partial_x^2 - k_y^2)^2$

$B_{11} = 1$, $B_{12} = B_{21} = 0$, $B_{22} = k_y^2 - \partial_x^2$

The sixth order system (26) with suitable boundary conditions is an eigenvalue problem for $\gamma = \gamma(S, M, k_y x_o)$ with $S = x_o v_{Ay}/\eta_o$ and $M = x_o v_{Ay}/\mu_o$. The same equation is solved by Dalhburg et al., 1983, for the problem of 2D incompressible resistive-viscous magnetohydrodynamics. For $SM > 2$ the system is unstable for all modes with $k_y x_o < 1$ and the growth rate is given by Eq. (11) for large $S = x_o v_{Ay}/\eta_o$.

The linear solutions of Eq. (26) determine the phase relation between $\psi_k(x)$ and $\phi_k(x)$ or $\psi(k_x, k_y)$ and $\phi(k_x, k_y)$ by a Fourier transform in x. From Eq. (26) we write the quasilinear phase relations as

$$\delta \psi_{\underset{\sim}{k}} = \frac{i k_y \psi'(x)}{\gamma_{\underset{\sim}{k}} + k_\perp^2 \eta_o} \delta \phi_{\underset{\sim}{k}} \qquad (27)$$

$$\delta \phi_{\underset{\sim}{k}} = \frac{i k_y (k_\perp^2 \psi' + \psi''')}{k_\perp^2 (\gamma_{\underset{\sim}{k}} + k_\perp^2 \mu_o)} \delta \psi_{\underset{\sim}{k}} \ . \qquad (28)$$

Substituting relationships (27) and (28) into Eq. (21) we obtain the flux formula

$$F_\psi = -\sum_{\underset{\sim}{k}} \left[\frac{k_y^2 \langle |\delta \phi_{\underset{\sim}{k}}|^2 \rangle}{\gamma_{\underset{\sim}{k}} + k_\perp^2 \eta_o} - \frac{k_y^2(k_\perp^2 + \psi'''/\psi')}{k_\perp^2(\gamma_{\underset{\sim}{k}} + k_\perp^2 \mu_o)} \right.$$

$$\left. \times \langle |\delta \psi_{\underset{\sim}{k}}|^2 \rangle \right] \frac{\partial \psi}{\partial x} \ . \qquad (29)$$

The first term in the square bracket in Eq. (24) describes turbulent convection by the $\underset{\sim}{E} \times \underset{\sim}{B}$ drift; the second term originates from the electron meandering in the turbulent magnetic fields. Formula (29) is instructive for several purposes.

First we consider the short wavelength, dissipationless limit, $k_\perp^2 \gg \psi'''/\psi' \simeq x_o^{-2}$ and $\eta_o = \mu_o = 0$. Formula (29) then reduces to the Biskamp-Welter [1983] formula

$$F_\psi = -\eta_A \frac{\partial \psi}{\partial x} \qquad (30)$$

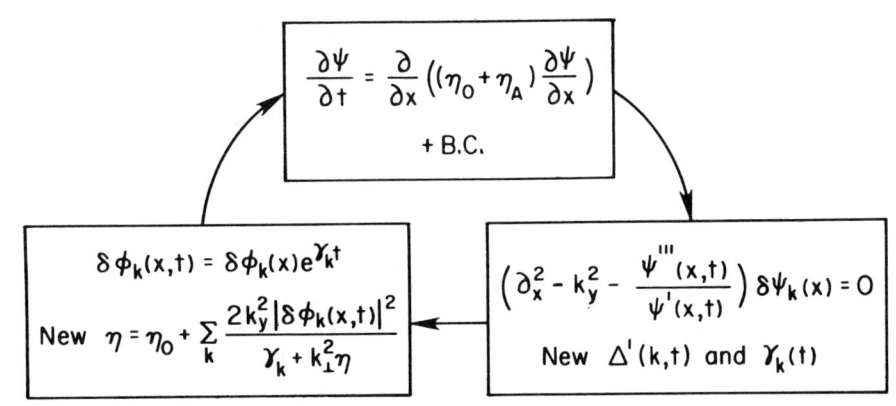

Fig. 2. Schematic of the evolution described by the quasilinear system of equations.

with

$$\eta_A = \sum_k \gamma_k^{-1} k_y^2 [\langle |\delta\phi_k|^2\rangle - \langle |\delta\psi_k|^2\rangle]$$

$$= \tau_c [\langle v_x^2\rangle - \langle \delta B_x^2\rangle] \qquad (31)$$

with τ_c the average correlation time in the laboratory frame. Biskamp uses formula (31) along with the observation that the 3D simulations of tokamaks show the buildup of a short scale magnetic turbulence with $\langle \delta B^2\rangle \gg \langle \delta v^2\rangle$ to propose an explanation of the major disruption based on the negative value of η_A.

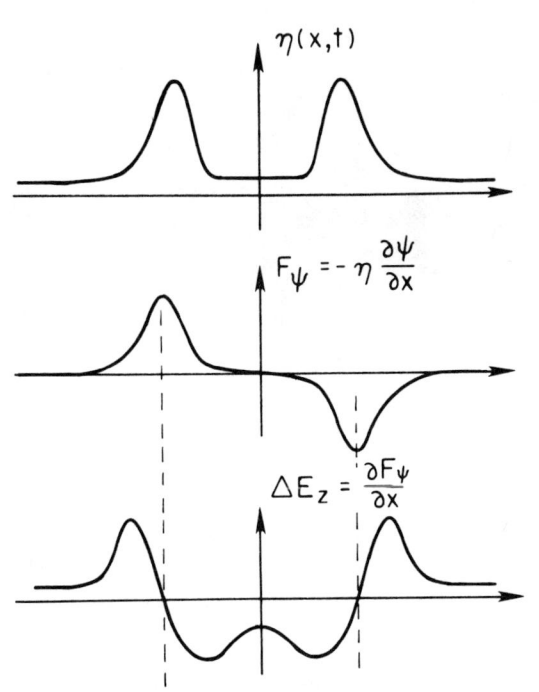

Fig. 3. Evolution of the quasilinear effective anomalous resistivity, the inward flux F_ψ of the poloidal magnetic flux and the change ΔE_z in the parallel electric field produced by the inward turbulent convection of magnetic flux.

Formula (31) shows that a spectrum of Alfvén wave turbulence does not produce a transport of magnetic flux since $|\delta\phi_k|^2 = |\delta\psi_k|^2$ for Alfvén waves and thus $F_\psi = 0$.

Finally, we observe that for tearing mode turbulence, where formula (29) must be used, the magnetic and kinetic contributions in Eq. (29) are related through the local dispersion relation

$$k_\perp^2(\gamma_k + k_\perp^2\mu_0)(\gamma_k + k_\perp^2\eta_0) + k_y^2(k_\perp^2\psi'^2 + \psi'\psi''') = 0$$

from Eq. (26). Along this dispersion relation (k,γ) the magnetic and kinetic contributions are equal and combine to yield

$$F_\psi = -\sum_k \frac{2k_y^2\langle|\delta\phi_k|^2\rangle}{\gamma_k + k_\perp^2\eta_0} \frac{\partial\psi}{\partial x} \qquad (32)$$

giving a positive definite anomalous η_A from tearing mode turbulence.

A complete proof of the positivity of η_A requires use of the boundary layer solutions of Eq. (26) and is given in a recent report [Horton and Galvao, 1984].

IV. Summary and Discussion

To complete the quasilinear calculation shown in Fig. 2 we carry out the integration of

$$\frac{\partial\psi}{\partial t}(x,t) = \frac{\partial}{\partial x}\left(\eta(x,t)\frac{\partial\psi}{\partial x}\right) \qquad (33)$$

with

$$\eta(x,t) = \eta_0 + \sum_{k_y} \frac{2k_y^2\langle|\delta\phi_{k_y}(x,t)|^2\rangle}{\gamma_{k_y} + k_\perp^2\eta} \qquad (34)$$

with $\delta\phi_k(x,t)$ given by the inner solution of the tearing mode equations and $\gamma_k = \gamma_k(\Delta'_k)$ calculated from the solution of the outer equation for $\delta\psi_k(x)$ using $\psi'''(x,t)/\psi'(x,t)$ for the local potential.

The results are shown schematically in Fig. 3 and will be reported in detail in a later work.

The principal features are that there is a strong inward convection of the magnetic flux to the turbulent reconnection layer. This flux induces an electric field $E_o + \Delta E_z$ with $\Delta E_z = \partial F_\psi / \partial x$ which has $E_o \Delta E_z < 0$, inhibiting the current flow in the center of the current sheet, and $E_o \Delta E_z > 0$, strengthening the current flow, in the outer regions of reconnection. As the current layer broadens the fluctuation spectrum shifts to longer wavelengths. If the turbulence remains weak then we expect the turbulent broadening to continue until the last mode $k_y = 2\pi/L_y$ is stabilized.

Acknowledgments. Useful conversations with D. Biskamp and P. Diamond are gratefully appreciated. This work was supported by Department of Energy Contract #DE-FG05-80ET-53088 and National Science Foundation grant ATM82-14730.

References

Biskamp, D., and H. Welter, Negative anomalous resistivity - a mechanism of the major disruption in tokamaks, Phys. Letters, 96A, 25, 1983, and preprint Max-Planck-Institut fur Plasmaphysik IPP6122, 1983.

Brunel, F., T. Tajima, and J. M. Dawson, Fast magnetic reconnection processes, Phys. Rev. Lett., 49, 323, 1982.

Dahlburg, R. B., Z. A. Zang, D. Montgomery and M. Y. Hussaini, Viscous, resistive MHD stability computed by spectral techniques, NASA Report 17129, ICASE, Hampton, Virginia 23665, 1983.

Drake, J. F. and Y. C. Lee, Kinetic theory of tearing instabilities, Phys. Fluids, 20, 1341, 1977.

Horton, W., and R. Galvao, Quasilinear evolution of collisional tearing modes, Institute for Fusion Studies Report #121, 1984.

Horton, W., and D. I. Choi, Renormalized turbulence theory for the ion acoustic problem, Physics Reports, 44, 273, 1979.

Kadomtsev, B. B., Disruptive instability in tokamaks, Fiz. Plazmy, 1, 710, 1975 [Sov. Phys. Plasma Phys., 1, 389, 1975].

Papadopoulos, K., Dynamics of magnetosphere, (eds. S.-I. Akasofu, D. Reichel), p. 289, 1979.

Parker, E. N., The solar-flare phenomenon and the theroy of reconnection and annihilation of magnetic fields, App. J. Suppl. Sec., 77, 177, 1963.

Pritchett, P. L., and C. C. Wu, Coalescence of magnetic islands, Phys. Fluids, 22, 2140, 1979.

Tajima, T., Tearing and reconnection, in Fusion - 1981 (International Centre for Theoretical Physics, Treiste, 1982), p. 403.

GLOBAL SINGLE ION EFFECTS WITHIN THE EARTH'S PLASMA SHEET

Paul L. Rothwell and G. Kenneth Yates

Air Force Geophysics Laboratory
Hanscom Air Force Base, Bedford, Massachusetts 01731

Abstract. Two global properties of single ion motion in the magnetotail are examined. The first effect is caused by the magnetic field in the plasma sheet directing boundary ions to the neutral sheet. Exact solutions to the Lorentz equation indicate that these ions can have sufficient energy to trigger the ion tearing mode if $B_0/aB_z > 6.0$, where B_0 is the tail lobe magnetic field, B_z is the magnetic field in the north-south direction and 'a' is a parameter related to the growth of the ion tearing instability. It is found that this effect occurs at a lower energy for oxygen than for protons. The second global property is related to the thinning or expansion of the plasma sheet. The results indicate that in the absence of reconnection the plasma sheet adiabatically maintains equilibruim by allowing plasma and magnetic flux to cross the boundaries. The presence of reconnection modifies the flow across the boundaries as well as the spatial distribution of the induced electric field.

Introduction

Single ion motion has been shown to play an important role in the dynamics of the magnetotail [Lyons and Speiser, 1982; Rothwell and Yates, 1979]. Here we first solve the Lorentz equation for single ions in a shear magnetic field and determine how the topology of the ion motion scales with the plasma sheet thickness and ion mass. The tail magnetic field, for example, focuses specific ions from the plasma sheet boundaries onto the neutral sheet. Comparison with the ion tearing mode [Galeev, 1979; Galeev and Zelenyi, 1976] implies that these focused ions are sufficiently energetic to trigger the instability. It is found that boundary ions may trigger substorm onsets at larger plasma sheet thicknesses ≈ 2,000 km (half-thickness) but that the ion tearing mode is always excited when the plasma sheet half-thickness is less than ≈ 500 km. The latter effect is simply due to the self consistent current density at the neutral sheet exceeding the ion tearing mode stability criteria.

The second effect addresses particles near the plasma sheet boundaries as the plasma sheet thins and expands. First, we take the results of Sonnerup [1971] and calculate the guiding center location based on the conservation of the adiabatic invariants. These results agree very well with the much simpler approach of using A_y = constant to define a field line. For a linearly varying magnetic field A_y is quadratic.

$$A_y = B_0(L/2 - z) \qquad z \geq L$$
$$A_y = -B_0 z^2/2L \qquad L \geq z \geq -L \qquad (1)$$
$$A_y = B_0(L/2 + z) \qquad z \leq -L$$

$$z_f/z_i = (L_f/L_i)^{1/2} \qquad (2)$$

The location, z, of a given field line moves proportional to the square root of the plasma sheet thickness, L. This means that with no reconnection at the neutral sheet a field line (with attached particles) will move outside the boundary for a thinning plasma sheet and into the boundary for an expanding plasma sheet. In this way the plasma sheet adiabatically responds to external perturbations. The presence of reconnection modifies the results since magnetic flux will then be lost at the neutral sheet. A particle tracing computer code is used to examine four different cases corresponding to different boundary conditions imposed on A_y. These boundary conditions define the value of A_y at specific locations and determine the transport of particles across the boundary, the induced electric field spatial distribution, the presence of reconnection and the possible displacement of the tail field lines. The transport of particles perpendicular to the magnetic field is similar to the observations reported by Parks et al [1979].

Single Ion Motion

Single ion motion in a shear field geometry has been investigated by Speiser [1965, 1967],

Sonnerup [1971], Eastwood [1972, 1974], Stern and Palmadesso [1975], Cowley [1978], Rothwell and Yates [1979] and Lyons and Speiser [1982]. All of these efforts assumed a shear magnetic field of the form:

$$B_x = B_0 \quad z \geq L$$
$$B_x = B_0 z/L \quad L \geq z \geq -L \quad (3)$$
$$B_x = -B_0 \quad z \leq -L$$

where positive x is towards the sun, y towards dusk and z north and the half-thickness of the plasma sheet is L. The addition of a small (≈ 1 nT) B_z component causes the ions to be ejected from the plasma sheet [Lyons and Speiser, 1982; Speiser, 1965, 1967; Stern and Palmadesso, 1975; and Cowley, 1978]. Near the neutral sheet where $B_x = 0$ the guiding center approximation for the particle motion breaks down [Sonnerup, 1971]. For particles closer to the plasma sheet boundaries the approximation hold at larger plasma sheet thicknesses. In this paper, we show that as the plasma sheet thins the guiding center approximation breaks down even for ions with turning points on the plasma sheet boundaries. These boundary ions travel across the entire plasma sheet in the z direction and dynamically connect the north-south tail lobes. It is claimed that this coupling plays a role in regard to the periodicity of substorms [Rothwell and Yates, 1979]. In addition, the tail magnetic field causes selected boundary ions to approach the neutral sheet asymptotically forming a localized current antiparallel to the bulk current. It will be shown below that the energy at which these focused ions leave the plasma sheet boundary is comparable to the threshold energy required for the ion tearing mode [Galeev, 1979].

Equations of Motion

The ratio B_x/B_z is usually greater than ten so that the motion caused by B_x is characteristically ten times faster than the motion from B_z. This suggests that to a first approximation one can treat the effect of B_x independently to that of B_z [Speiser, 1965; Eastwood, 1972]. We, therefore, first treat the one-dimensional magnetic field given by equation (3) and then show that the results are consistent with a weak B_z. The Lorentz equation in component form is:

$$\ddot{x} = 0$$
$$\ddot{y} = (eB_0/Lmc) z \dot{z} \quad (4)$$
$$\ddot{z} = -(eB_0/Lmc) z \dot{y}$$

Integrating $\ddot{y}$,

$$\dot{y}(t) = (eB_0/2Lmc)(z^2 - z_0^2) + \dot{y}_0 \quad (5)$$

which when inserted into the $\ddot{z}$ equation gives

$$\ddot{z} = (eB_0/Lmc)\left[(eB_0/2Lmc)z_0^2 - \dot{y}_0\right] z - 2(eB_0/2Lmc)^2 z^3 \quad (6)$$

which can be multiplied by $2\dot{z}$ and integrated exactly.

$$\dot{z}^2(t) = \dot{z}_0^2 + (eB_0/Lmc)\left[(eB_0/2Lmc)z_0^2 - \dot{y}_0\right](z^2-z_0^2) - (eB_0/2Lmc)^2(z^4-z_0^4) \quad (7)$$

where y_0, $\dot{y}_0$, z_0 and $\dot{z}_0$ are the initial conditions. The equation is simplified if we take the initial conditions to be the turning point. Here $z_0 = z_m$, $\dot{z}_0 = 0$, $\dot{y}_0 = v$, and y_0 is arbitrarily taken as zero. The results can be written in dimensionless form by setting

$$z \rightarrow z/z_m$$
$$t \rightarrow t(v\omega_c/L)^{1/2} \quad (8)$$
$$k^2 = \omega_c z_m^2/4Lv$$

$$\dot{z}^2 = (1 - k^2 + k^2 z^2)(1 - z^2) \quad (9)$$

which is satisfied by the Jacobi elliptic function $cn[t(v\omega_c/L)^{1/2}, k]$ [Byrd and Friedman, 1954]. This is more conveniently expressed as $cn(2kvt/z_m, k)$, where k is the modulus of the Jacobi function. The angular gyrofrequency, ($\omega_c = eB_0/mc$), is defined with the tail lobe magnetic field, B_0, and v, the ion velocity in the y-z plane.

The topology of the individual ion orbit is defined by the numerical value of k^2 [Sonnerup, 1971; Rothwell and Yates, 1979]. For $k^2 > 1$, the orbits are confined to one side of the neutral sheet. For $k^2 = 1$ the orbits correspond to trajectories that originate in the plasma sheet and asymptotically approach the neutral sheet, while the $k^2 < 1$ orbits cross the neutral sheet. Figure 1 shows proton orbits in the dawn-dusk plane with a full plasma sheet thickness of 1,500 km. Protons with turning points on the boundary ($z_m = 750$ km) will cross the neutral sheet provided their energy is greater than 0.2 keV. As argued by Rothwell and Yates [1979] this dynamically couples both tail lobes with the plasma sheet and could be a source for multiple onsets.

Our attention, however, turns to Figure 1b. The tail magnetic field acts as a magnetic lens that focuses the $k^2 = 1$ trajectories onto the neutral sheet forming a localized beam antiparallel to the bulk current. The solutions to (9) for this case are:

$$y = z_m \tanh(2vt/z_m) - vt$$
$$z = \pm z_m \cosh(2vt/z_m) \quad (10)$$

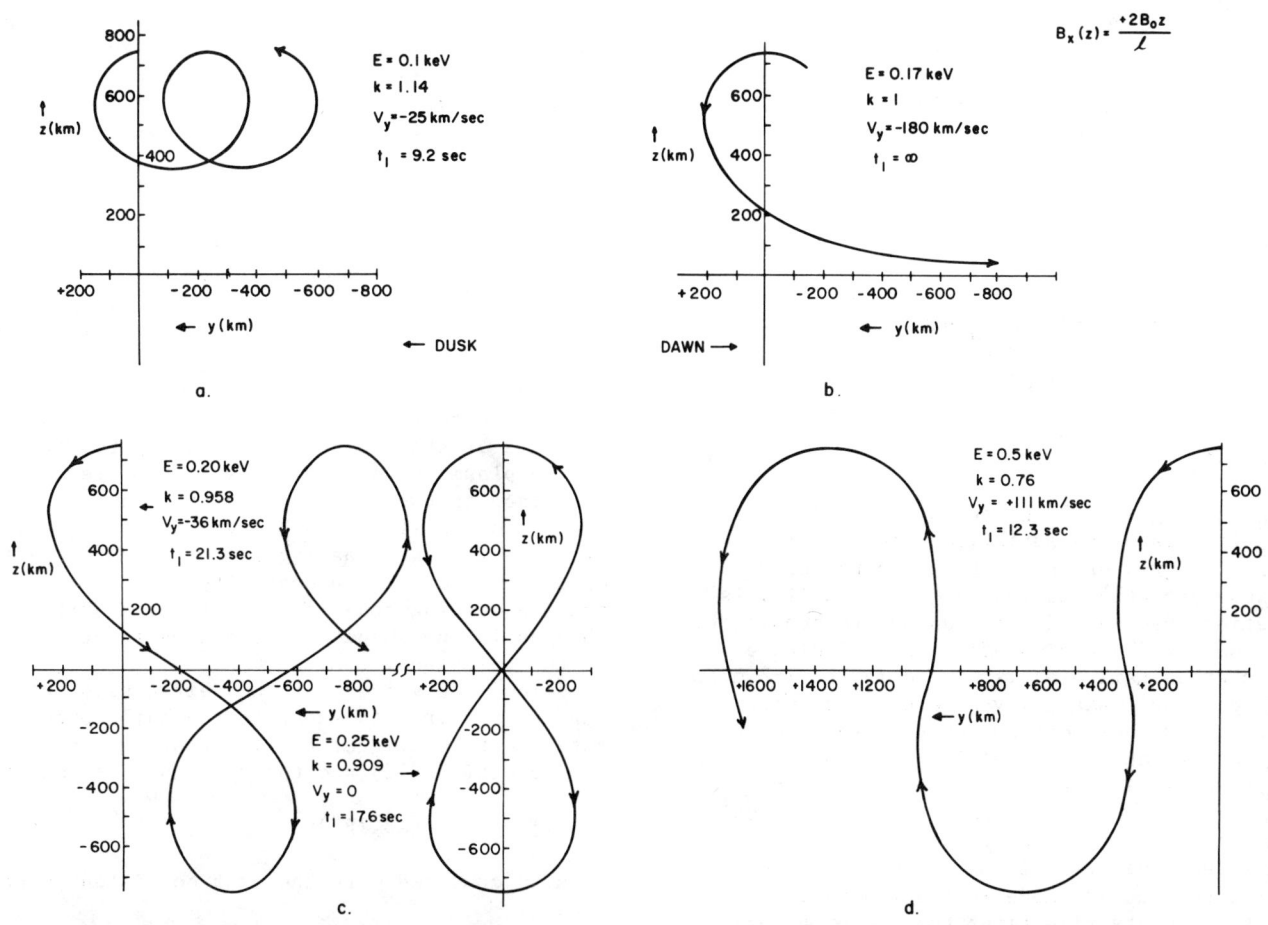

Fig. 1. A dusk-dawn cross section of the magnetotail showing various types of ion orbits. Fig. 1b shows the focusing effect of the magnetic field for ions with $k^2 = 1$ (see text). Here script ℓ refers to the full plasma sheet thickness.

The energy, E_c, and corresponding velocity, v_c, at which ions satisfy $k^2 = 1$ with turning points on the boundary ($z_m=L$) are from the definition of k^2:

$$E_c = m\omega_c^2 L^2/32 \qquad (11)$$

$$v_c = \omega_c L/4$$

which for protons and $B_0 = 10$ nT equals 17.5 keV (L=7335 km) and 0.2 keV (L= 750 km).

If the plasma sheet has a temperature of 5 keV, then the larger L value corresponds to few protons. However, for the thinned case, this energy is near the peak of the Maxwellian distribution. We note that E_c scales as m^{-1}. At the same thickness, lower energy, higher mass particles connect the boundary and the neutral sheet.

Ion Tearing Mode

The linear growth rate for the ion tearing mode is given by Galeev [1979], Galeev and Zelenyi [1976], Schindler et al [1973] and Coppi et al [1966].

$$\gamma = (v_i/L\pi^{1/4})(\rho_i/L)^{3/2}(1 + T_e/T_i)(1-\bar{k}^2L^2) \qquad (12)$$

Here $\bar{k}$ is the perturbation wave number in the z direction. Based on equation 13 from Galeev [1979] we take the $\bar{k}^2L^2$ term as being much less than unity (i.e. long wavelength limit). The ion gyroradius, ($\rho_i = v_i/\omega_c$), is defined in the uniform field region. We take $T_e/T_i \approx 1/2$. The parameter L is again the plasma sheet half-thickness. Inverting (12) in terms of a required threshold velocity for a predetermined growth rate we have:

$$E_c' = 1/2 mv_i^2 = K_s mL^2 \omega_c^{6/5} \gamma^{4/5}$$
$$\propto B_0^{6/5} B_z^{4/5} L^2/m \qquad (13)$$

where $K_s = 1/2 (2\pi^{1/2}/3)^{4/5} = 0.57$. The

growth rate can be defined as being proportional to the ion gyrofrequency, ω_z, about B_z, the normal magnetic component. Hence, $\gamma = a\omega_z/2\pi$, where a is the constant of proportionality. Note that both E_c (Equation 11) and E_c^* (Equation 13) scale as L^2 and m^{-1} suggesting an underlying relationship. The scaling of E_c^* with m is implied by the work of Baker et al [1982].

The ratio between the two energies is

$$E_c/E_c^* = 0.24 \, (B_0/aB_z)^{4/5} \qquad (14)$$

which must be greater than one if the focused boundary ions ($k^2=1$) are to have sufficient kinetic energy to exceed the threshold energy, E_c^*, for the ion tearing mode. This will occur when

$$B_0/aB_z > 6.0 \qquad (15)$$

Given that $B_0 \approx 10\text{-}60$ nT and $B_z \approx 1\text{-}4$ nT this inequality is often satisfied even if a is substantially greater than one. Computer simulations show that as long as equation (15) is satisfied then beam dispersion due to bending by B_z is small. These simulations also indicate that a cross tail electric field in the 1-2 mv/m range does not materially affect the results. Galeev [1979] has pointed out that the linear mode must reach a critical B_z value in order for the explosive, nonlinear mode to be initiated. Terasawa [1981] found from a numerical simulation that the linear mode can saturate before reaching the critical value if the scale size of the tearing mode in the x direction is less than twenty five times the proton gyroradius. These subtle questions are not addressed in this analysis.

We conclude that single ions from the plasma sheet boundaries form a beam along the neutral sheet with a kinetic energy that can exceed the threshold energy for the linear ion tearing mode. This energy is inversely proportional to the ion mass and, therefore, oxygen ions are more efficient than protons for triggering the ion tearing mode by this method [Baker et al, 1982].

The triggering of the ion tearing mode must also be dependent on the current density of the ions reaching the neutral sheet. This density depends on the rate at which the ions are diffusing into the $k^2 = 1$ neighborhood of the distribution function on the boundary and the location of this neighborhood in the distribution function itself. The $k^2 = 1$ orbits are not adiabatic and, therefore, they are lost to the boundary. This produces a "hole" in the distribution function that is filled by diffusion driven by boundary turbulence. This diffusion is probably rapid enough to saturate the loss rate. However, if it is not then the single ion triggering mechanism proposed here will be modulated by boundary turbulence. The situation is analogous to the loss cone formed by trapped protons in the Van Allen radiation belts. There significant ion loss cone depletions are maintained in a quasi-steady state. One expects the $k^2 = 1$ boundary ions to trigger the ion tearing mode when the current density is a maximum. This will occur when $v_c = v_t$, where v_t is the ion thermal velocity. From equation (11) this corresponds to a plasma sheet half-thickness of

$$L = 4v_t/\omega_c \qquad (16)$$

A proton temperature of 5 keV and $B_0 = 20$ nT yields $L \approx 2,000$ km. A colder plasma sheet and a stronger lobe field requires substantially more thinning to trigger a substorm. Ions originating on the boundary and for which $k^2 = 1$ contribute to the neutral sheet current density dawnward of that location. This implies that if the plasma sheet has a bowtie profile in the y-z plane that these duskward originating $k^2 = 1$ ions will have a higher energy (equation 11) than ions originating near the narrow part of the bowtie. The probability of triggering the ion tearing mode is, therefore, sensitive to changes in the shape of the plasma sheet profile in the y-z plane.

Below some value of L the self consistent current required to separate the tail lobes is sufficient to trigger the ion tearing mode. From equation (12) and the relation $\gamma = a\omega_z/2\pi$,

$$v_i > 0.51 \, (aB_z/B_0)^{2/5} \, \omega_c L \qquad (17)$$

This is compared with the self consistent current density

$$j = nev_i = cB_0/4\pi L \qquad (18)$$

which is consistent with equation (1) and also with the Harris [1962] equilibrium distribution at $z = 0$. Combining equations (17) and (18) gives the plasma sheet half-thickness below which the ion tearing mode is triggered by the self consistent current density

$$L \leq 1.40(c/\omega_{pi})(B_0/aB_z)^{1/5} \qquad (19)$$

where $\omega_{pi} = (4\pi n e^2/m)^{1/2}$ is the ion plasma frequency. For protons with a density of $n \approx 1$ cm^{-3} and $\omega_{pi} = 1.32 \times 10^3$ sec^{-1} have

$$L \leq 319 \, (B_0/aB_z)^{1/5} \text{ km} \qquad (20)$$

The question arises whether the boundary ions can trigger the ion tearing mode before the plasma sheet thins to the levels indicated by equation (20). From equation (15), we find that the focused boundary ions equaled or exceeded the necessary threshold velocity if $B_0/aB_z \geq 6$. Inserting this value into equation (20) implies $L \leq 456$ km. From equation (16) we saw that the maximum flux of boundary ions focused on the neutral sheet probably occurs when their velocity is equal to the thermal velocity which occurs when L ≈ 2,000 km. The physical picture is that the ion tearing mode may be triggered by focused boundary ions at larger plasma half-thicknesses (≈ 2,000 km), but that below ≈ 500 km the ion tearing mode is definitely triggered by the self consistent current.

Ion Motion in a Thinning or Expanding Plasma Sheet

In this section we examine the effect of moving plasma sheet boundaries and look for appropriate magnetic configurations consistent with magnetic flux conservation. Within the guiding center approximation, we first show that predictions based on Sonnerup's [1971] results agree with those obtained by simpler methods. The displacement of the guiding centers or field lines as a function of the boundary velocity in the z direction is found to depend on the boundary conditions imposed on the vector potential, $A_y(z)$. The magnetic field, $\bar{B}$, itself is independent of these boundary conditions.

The results of a time dependent computer particle tracing code are presented for four selected boundary conditions on $A_y(z)$. Individual field lines as well as selected proton orbits are tracked as a function of time. We include the effects of the induced electric field on particle trajectories. The field geometry given by equation (3) leads to the adiabatic invariant, J, as given by Sonnerup [1971], for $|z| \leq |L|$.

$$J = (32/3)(Lv_\perp E_\perp/\omega_c)^{1/2} f_1(k) \quad k \leq 1$$
$$J = (16/3)(Lv_\perp E_\perp/\omega_c)^{1/2} f_2(k) \quad k \geq 1$$
$$f_1(k) = [(1-k^2)K(k)-(1-2k^2)E(k)]$$
$$f_2(k) = k[2(1-k^2)K(1/k)-(1-2k^2)E(1/k)]$$
(21)

where $E_\perp$ and $v_\perp$ are the kinetic energy and the velocity perpendicular to the magnetic field respectively. K and E are the complete elliptic integrals of the first and second kind respectively with modulus k. Another invariant of the motion is is the canonical momentum in the y direction:

$$P_y = mv_n = mv_y + eA_y(z)/c = \text{constant} \quad (22)$$

which can be expressed in terms of k^2.

$$v_n = v_\perp(1-2k^2) = \text{constant} \quad (23)$$

Averaging equation (23) over one oscillation period leads to

$$\tilde{v}_n = \tilde{v}_\perp(1-2k^2) = \text{constant} \quad (24)$$

where the tilde implies the time averaged quan-

CHANGE IN TURNING POINT DUE TO ADIABATIC THINNING OF THE PLASMA SHEET

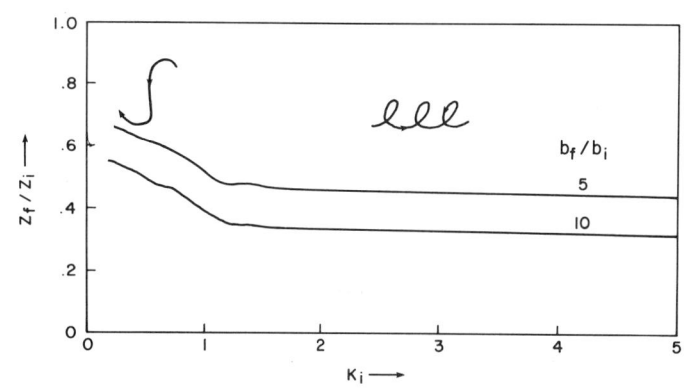

Fig. 2. The ratio of the proton's turning points on the boundary as the plasma sheet is adiabatically thinned by factors of five and ten respectively. The figure is based on the results of Sonnerup [1971].

tities. Combining (24) and (21) leads to the total invariant.

$$Lf_1^2(k)/[B_0(1-2k^2)^3] = \text{constant} \quad k \leq 1$$
$$Lf_2^2(k)/[B_0(1-2k^2)^3] = \text{constant} \quad k \geq 1$$
(25)

Instead of defining k^2 in terms of v_n and $v_\perp$, we express it in terms of the particles turning point, z_m.

$$k^2 = \omega_c z_m^2/(4Lv) \quad (26)$$

(see equation 8). The perpendicular subscript on $v_\perp$ has been dropped and $v = v_\perp$ is understood. For a given tail configuration defined by (B_0, L_i) and a given ion trajectory defined by its mass, z_i and v_i, we can calculate k_i^2 by equation (26). If the plasma sheet adiabatically thins or expands to a new thickness, L_f, then a new k^2 (i.e. k_f^2) can be found from equation (25). Equation (23) is then used to find the new velocity, v_f, and equation (26) the new turning point distance, z_f.

The results are shown in Figures 2 and 3. In these examples, the plasma sheet was thinned by factors of five and ten, increasing the magnetic field gradient, B_0/L, by the same amounts. Note that the crossing orbits (k<1) show a significant dispersion in velocity and location as the plasma sheet boundaries change. The noncrossing orbits, (k≥1), on the other hand, show little dispersion. This is to be expected since the noncrossing orbits satisfy the guiding center approximation. In the following analysis we treat only noncrossing orbits. The results, therefore, are strictly applicable only at larger plasma sheet thicknesses. (See equation 14).

These orbits should adiabatically follow the motion of the field lines in the x-z plane. This is seen more clearly by noting that in a

VELOCITY INCREASE DUE TO ADIABATIC THINNING OF THE PLASMA SHEET

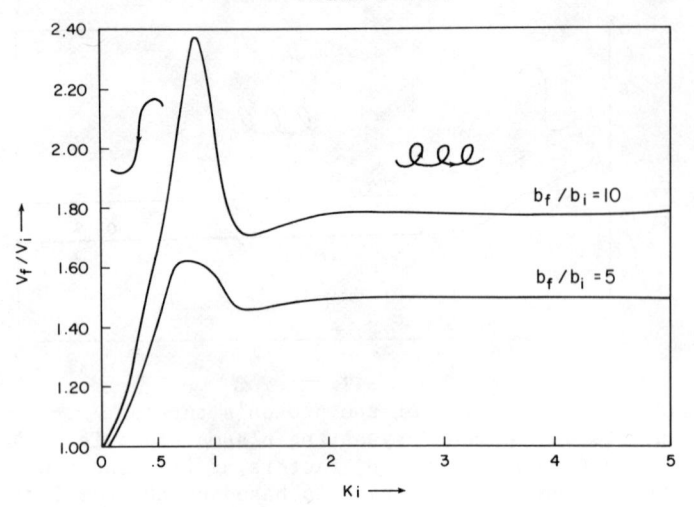

Fig. 3. The ratio of the proton's initial and final velocities as the plasma sheet is thinned as in Figure 2. This figure is also based on the results of Sonnerup [1971].

two-dimensional geometry, A_y = constant defines a magnetic field line. The displacement of a specific field line can be located by simply finding where A_y has the same numerical value. The constancy of equation (1), for example, implies

$$z_f = z_i(L_f/L_i)^{1/2} \qquad (27)$$

where i and f denote the initial and final values of the plasma sheet thickness and the turning point distance from the neutral sheet. Thinning the plasma sheet by a factor of five implies $z_f = 0.44 z_i$. This agrees very well with Figure 2. Similarly, a factor of ten implies $z_f = 0.32 z_i$. This is also in good agreement with Figure 2. Conservation of the first adiabatic invariant, ($\propto v^2/B$), implies

$$v_f = v_i(L_i/L_f)^{1/4} \qquad (28)$$

The ratios obtained from equation (28) are in good agreement with those shown in Figure 3 for k > 1. The simpler approach of requiring A_y = constant is equivalent to following the adiabatic motion of the k > 1 particles.

The main point we want to emphasize, however, is that in both approaches the field lines and the attached particles do not move linearly with the boundary displacements. Magnetic field lines move relative to the plasma sheet boundaries. Field lines initially inside a thinning plasma sheet move outside the boundary and field lines initially outside an expanding plasma sheet move inside. The boundary marks a transition in the spacing between field lines and is not necessarily identified with any one field line. We will now show that these phenomena are related to boundary conditions imposed on A_y.

Equation (1), for example, is incompatible with the presence of magnetic reconnection at the neutral sheet. The $A_y = 0$ field line remains fixed at z = 0 independent of boundary motion. The other field lines change location in order to conserve magnetic flux. The lack of reconnection at the neutral sheet requires the passage of magnetic flux through the boundaries.

With this insight we now modify equation (1) so that the $A_y = 0$ field line always coincides with the boundaries (z=±L). Because $A_y(-z) = A_y(z)$ throughout the following and for simplicity, we ignore the region z < 0.

$$A_y = B_o(L^2-z^2)/L \qquad z \leqslant L$$
$$A_y = B_o(L-z) \qquad z \geqslant L \qquad (29)$$

where L is a function of time. Note that B_x has not changed although the numerical value of $A_y(z=0)$ depends on L. That is, which field line is at the neutral sheet now depends on L(t). Field lines will be annihilated or created to preserve magnetic flux.

Another possible boundary condition is $\partial A_y/\partial t = 0$, for z > L. For that case:

$$A_y = -B_o(z^2+L^2)/2L \qquad z \leqslant L$$
$$A_y = -B_o z \qquad z \geqslant L \qquad (30)$$

which implies magnetic reconnection at the neutral sheet.

Equation (1) we label as case A, equation (29) as case B and equation (30) as case C. Case D is the Harris [1962] equilibrium vector potential.

$$A_y = -LB_o \ln(\cosh z/L) \qquad (31)$$

which holds for all values of z.

The induced electric fields for the four cases are ($v_L \equiv dL/dt$)

Case A
$$E_y = -B_o z^2 v_L/2cL^2 \qquad z \leqslant L$$
$$E_y = -B_o v_L/2c \qquad z \geqslant L \qquad (32)$$

Case B
$$E_y = -B_o v_L (1+z^2/L^2)/2c \qquad z \leqslant L$$
$$E_y = -B_o v_L/c \qquad z \geqslant L \qquad (33)$$

Case C
$$E_y = B_o v_L (1-z^2/L^2)/2c \qquad z \leqslant L$$
$$E_y = 0 \qquad z \geqslant L \qquad (34)$$

Case D $$E_y = B_o v_L \left[\ln(\cosh z/L) - z/L \tanh(z/L)\right]/c \qquad (35)$$

Case A predicts no induced electric field at the neutral sheet. The induced field, however, increases quadratically to a constant value throughout the tail lobes. Case B predicts a finite electric field at the neutral sheet which points in the dawn-dusk direction for a thinning plasma sheet. The electric field is a minimum at the neutral sheet and also quadratically increases to a maximum value at the boundary and is constant in the tail lobes. Case C predicts a dusk-dawn electric field (opposite to case B) which quadratically decreases to zero at the boundaries and remains zero throughout the tail lobes. Case D is similar to case A except that $E_y \to -v_L B_0 \ln(2)/c$, as z goes to infinity.

For $B_0 = 20$ nT and $v_L = 10$ km/sec [DeCoster and Frank, 1979 and Frank et al, 1981], we find $E_y(z=0) = 0.1$ mv/m for cases B and C. A reasonable upper limit to the boundary velocity, v_L, is the local Alfvén speed. The Alfvén speed is $v_A = 436$ km/sec for $B_0 = 20$ nT and for a local mass density equivalent to one proton per cubic centimeter. Boundary motion at this speed would induce an electric field of 4.4 mv/m at the neutral sheet, a magnitude comparable to that of the cross tail field during active periods. The importance of locally induced electric fields has been stressed by Heikkila and Pellinen [1977].

Figures 4-7 show the results of a computer code for the four cases. This code traces magnetic field lines and particles including the effects of the induced electric field. In each case the boundary, which is represented by two closely spaced lines, moves from 15,000 km to 10,000 km at 4.2 km/sec. The time history of field lines selected at equal flux intervals are shown as are selected proton orbits. The protons near the neutral sheet initially start with turning points at 3,000 km and velocities of 140 km/sec (0.10 keV) in the y direction. The protons near the boundary start at 14,000 km and have a velocity of 1,000 km/sec (5.2 keV) in the y direction. Protons in the tail lobe start at 30,000 km and also have a velocity of 1,000 km/sec in the y direction. The paired symbols correspond to the upper and lower turning points.

Figure 4, which portrays case A, shows no reconnection present at the neutral sheet. Protons also move across the boundary, and the tail lobe field lines are affected. The later effect arises because of the terms in equation (1) required to make A_y continuous at the boundary.

Figure 5 corresponds to case B. Field lines are now being annihilated at the neutral sheet. Protons do not cross the boundary, and particles in the tail lobe are affected. Particles that are noncrossing before the thinning takes place become crossing during the thinning. By forcing the boundary to remain identified with the same field line ($A_y=0$) reconnection must take place at the neutral sheet to maintain overall flux conservation.

Figure 6 represents case C. Here field lines are created at the neutral sheet. Protons move across the boundary, and the tail lobe field

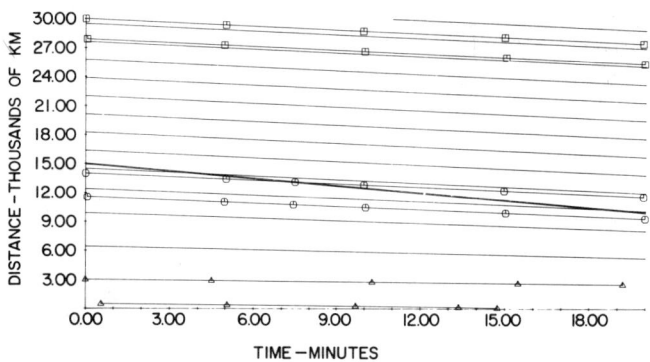

Fig. 4. Case A. A time history of selected field lines and proton orbits in the magnetotail. This figure was produced by a computer program using the boundary condition $A_y(z=0)=0$. The double line is the plasma sheet boundary which is moving towards the neutral sheet at 4.2 km/sec. The triangles near the neutral sheet correspond to the upper and lower turning points of 0.1 keV protons starting at 3,000 km. They become crossing particles under thinning. The circles show the upper and lower turning points of 5.2 keV protons near the boundary which is initially at 15,000 km. Note that the protons cross the boundary. The squares represent 5.2 keV protons initially at 30,000 km. The position of the boundary after twenty minutes is 10,000 km.

lines are not affected by the thinning plasma sheet. Field lines are created because their velocities are zero at the boundary. That is, the boundary is moving across the stationary field lines. Since the magnetic field outside the plasma sheet is greater than inside, field lines must be created in a thinning plasma sheet to provide the additional flux. Note that the

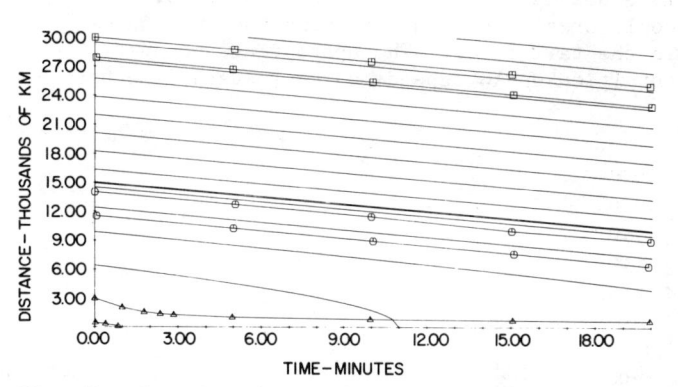

Fig. 5. Case B. A time history similar to Figure 4 except now a field line is is fixed relative to the plasma sheet boundary and reconnection at the neutral sheet is required to maintain overall flux conservation. Note that particles do not cross the boundary. The single lines in Figures 4-7 correspond to individual magnetic field lines spaced at equal flux intervals.

58 GLOBAL SINGLE ION EFFECTS

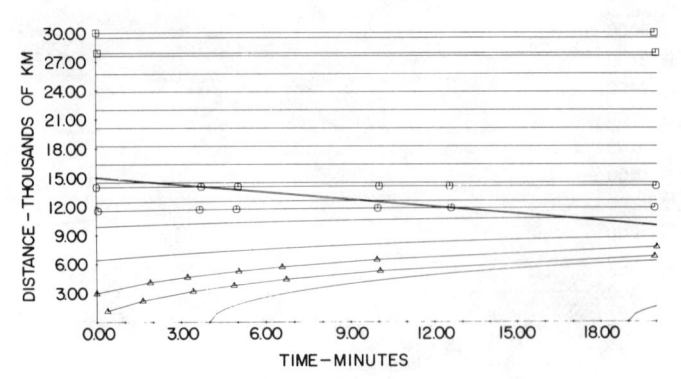

Fig. 6. Case C. Similar to Figures 4 and 5 except now the boundary condition is that the induced electric field be zero in the tail lobes. Note that flux is being created at the neutral sheet in order to maintain flux conservation and protons are crossing the boundary.

protons near the neutral sheet move to larger z, opposite to that shown for the other cases and are significantly accelerated.

Finally, Figure 7 shows the results for case D. As seen in Table 1, the results are similar to those in Figure 4.

The computer code results have highlighted three distinct responses to plasma sheet thinning. Each response corresponds to different boundary conditions on A_y. These conditions reflect whether the plasma sheet and tail lobes are both moving or whether the plasma sheet is thinning independent of the tail lobes. Cases A, B and D correspond to the former and case C to the later. Cases A and B imply that reconnection may or may not be present when both move.

Plasma sheet thinning accompanied by tail lobe movement implies that the thinning is driven by enhanced solar wind pressure on the external lobe boundaries. Thinning with static lobes could arise from standing compressional waves in the tail lobes. The appropriate boundary condition on A_y depends on the external mechanisms that drive the thinning and, therefore, may even be time dependent. Note that for each case the corresponding graph can be traced backwards in time to obtain the effect of an expanding plasma sheet. Indeed, the computer code was run in this manner and replicated the original conditions.

Table 1 summarizes the proton motion for the four cases. Note that protons near the neutral sheet can be significantly accelerated by the induced electric field. Case B is somewhat deceptive because the acceleration region exceeds the nominal tail width of 40 R_E.

If the plasma sheet is thinning tailward of an observation point then Figures 4, 6 and 7 suggest that ions and electrons would be transported from the plasma sheet into the uniform field region. They would be then guided by the boundary field lines to the observation point and appear as earthward boundary flows. Ions would be predominant at the observation point since electrons are more susceptable to pitch angle scattering by boundary fluctuations. Ion boundary fluxes, which are proportional to v_L times the plasma sheet particle density, can be significant. The flux crossing a boundary moving at 5 km/sec is 5 x 10^5 particles/(cm^2-sec) for a plasma sheet density of one proton per cubic centimeter. This phenomena provides a possible source for ion boundary flows but does not explain their high energies (>100 keV) [Williams, 1981].

Parks et al [1979] have noted the existence of a layer of energetic ions and electrons ($\approx$ 1.5 keV) just outside the plasma sheet boundary. These particle structures had velocities from a few kilometers per second to > 60 km/sec. They were associated with the thinning and expansion of the plasma sheet. Parks et al argue that the velocity of these structures was perpendicular to the magnetic field. This is compatable with the results presented here.

Table 2 lists the velocity of the boundary field line and that of the turning points in the north-south direction. All velocities are expressed in units of the boundary velocity. This table was constructed by setting the total time derivative of A_y in each of the four cases to zero and solving for the field line velocity. The differences between this velocity and the boundary velocity gives the velocity of the turning points across the boundary.

Summary

In summary, two features of single ion motion in the plasma sheet have been shown. The first feature is the directing of ions onto the neutral sheet by the tail magnetic field. These ions have sufficient energy to trigger the ion tearing mode when the tail magnetic field becomes sufficiently elongated or tail-like (i.e. $B_0/aB_z > 6$). This triggering mechanism, therefore, becomes enhanced during extended periods of

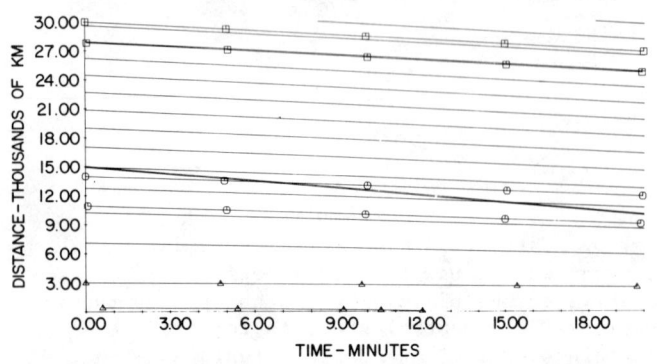

Fig. 7. Case D. A time history of selected field lines and proton orbits using the Harris [1962] magnetic field distribution. This case is very similar to Case A.

TABLE 1. The Effect of Induced Electric Fields on Magnetotail Protons, Initial Boundary at 15,000 Kilometers, Final Boundary at 10,000 Kilometers, Boundary Velocity = -4.2 Kilometers per Second

	Proton in tail lobe		Proton just inside boundary		Proton near neutral sheet	
	Position[1] 10^3 km	Energy keV	Position[1] 10^3 km	Energy keV	Position[1] 10^3 km	Energy keV
Initial conditions	30.00	5.21	14.00	5.21	3.00	0.102
Final conditions:						
Case A $A_y(0)=0$	27.50	5.21	11.60	6.08	2.51	0.116
Ratio[2]	0.92	1.00	0.83	1.17	0.84	1.14
Case B $A_y(L)=0$	25.03	5.21	8.93	4.70	0.895	39.8[3]
Ratio[2]	0.83	1.00	0.64	0.90	0.30	390.[3]
Case C $\partial A_y/\partial t(z>L)=0$	30.00	5.21	14.12	6.08	7.75	0.576
Ratio[2]	1.00	1.00	1.01	1.17	2.58	5.65
Case D $A_y \propto \ln(\cosh)$	26.78	5.37	11.88	5.96	2.50	0.117
Ratio[2]	0.89	1.03	0.85	1.14	0.83	1.15

[1] location of upper turning point.
[2] final condition divided by the initial condition.
[3] a proton in this case would have exited the plasma sheet before obtaining this energy.

southward directed IMF. The second feature uses ions satisfying the guiding center approximation to follow individual field lines when the plasma sheet boundaries are moving. It was found found that the boundary conditions imposed on A_y determines how particles move relative to the boundaries, whether the tail lobes are affected by plasma sheet motion and if reconnection takes place. The boundary conditions also determine how the induced electric field is distributed in the plasma sheet when it is thinning or expanding. It is suggested that the various boundary conditions on A_y are related to the external mechanisms that drive the plasma sheet thinning and to the presence of reconnection at the neutral sheet. For large L the results indicate that thinning is a necessary but not sufficient condition for reconnection. That is, if reconnection is not present then the plasma sheet can adiabatically adjust to external perturbations through the flow of plasma and field lines across the boundaries. For small L ($L<<4v_t/\omega_c$) there are few noncrossing ions and the crossing ions dominate the plasma sheet dynamics. [Rothwell and Yates, 1979]. Our results regarding the flow of electrons and ions across the boundary are consistent with the observations of Parks et al [1979].

Acknowledgements. We wish to acknowledge stimulating discussions with A. A. Galeev and L. R. Lyons.

TABLE 2. Normalized[1] Key Velocities in a Thinning/Expanding Plasma Sheet

Case	Velocity of field lines at boundary	Particle velocity across boundary
A	0.5	-0.5
B	1.	0.
C	0.	-1.
D	0.43	0.57

[1] so that the boundary velocity, $dV_L/dt = 1$.

References

Baker, D. N., E. W. Hones, D. T. Young and J. Birn, The possible role of ionospheric oxygen in the initiation and development of plasma sheet instabilities, Geophys. Res. Letts., 9, 1337-1340, 1982.

Coppi, B., G. Laval and R. Pellat, Dynamics of the geomagnetic tail, Phys. Rev. Letts., 16, 1207-1210, 1966.

DeCoster, R. J., and L. A. Frank, Observations pertaining to the dynamics of the plasma sheet, J. Geophys. Res., 84, 5099-5142, 1979.

Byrd, Paul F., and Morris D. Friedman, Handbook of elliptic integrals for engineers and physicists, Springer-Verlag Publ, Berlin, 1954.

Cowley, S. W. H., A note on the motion of charged particles in one-dimensional current sheets, Planet Space Sci., 26, 539-545, 1978.

Eastwood, J. W., Consistency of fields and particle motion in the "Speiser" model of the current sheet, Planet and Space Sci., 20, 1555-1568, 1972.

Eastwood, J. W., The warm current sheet model and its implications on the temporal behavior of the geomagnetic tail, Planet and Space Sci., 22, 1641-1668, 1974.

Frank, L. A., R. L. McPherron, R. J. DeCoster, B. G. Burek, K. L. Ackerson, and C. T. Russell, Field-aligned currents in the earth's magnetotail, J. Geophys. Res., 86, 687-700, 1981.

Galeev, A. A., and L. M. Zelenyi, Tearing instability in plasma configurations, Sov. Phys. JETP, 43, 1113-1123, 1976.

Galeev, A. A., Reconnection in the magnetotail, Space Science Rev., 23, 411-425, 1979.

Harris, E. G., On a plasma sheath separating regions of oppositely directed magnetic field, Il Nuovo Cimento, 23, 115-121, 1962.

Heikkila, W. J., and R. J. Pellinen, Localized induced electric field within the magnetotail, J. Geophys. Res., 82, 1610-1614, 1977.

Lyons, L. R., and T. W. Speiser, Evidence for current sheet acceleration in the geomagnetic field, J. Geophys. Res., 87, 2276-2286, 1982.

Parks, G. K., C. S. Lin, K. A. Anderson, R. P. Lin and H. Reme, ISEE 1 and 2 particle observations of outer plasma sheet boundary, J. Geophys. Res., 84, 6471-6476, 1979.

Rothwell, P. L., and G. K. Yates, A dynamical model for the onset of magnetospheric substorms, Dynamics of the Magnetosphere, D. Reidel Publ. Co., ed. SI. Akasofu, 497-518, 1979.

Schindler, K., D. Pfirsch and H. Wobig, Stability of two-dimensional collison-free plasmas, Plasma Physics, 15, 1165-1184, 1973.

Sonnerup, B. U. O., Adiabatic particle orbits in a magnetic null sheet, J. Geophys. Res., 76, 8211-8222, 1971.

Speiser, T. W., Particle trajectories in model current sheets 1. Analytic solutions, J. Geophys. Res., 70, 4219-4226, 1965.

Speiser, T. W., Particle trajectories in model current sheets 2. Applications to auroras using a geomagnetic tail model, J. Geophys. Res., 72, 3919-3932, 1967.

Stern, David P., and Peter Palmadesso, Drift-free magnetic geometries in adiabatic motion, J. Geophys. Res., 80, 4244-4248, 1975.

Terasawa, Toshio, Numerical study of explosive tearing mode instability in a one-component plasmas, J. Geophys. Res., 86, 9007-9019, 1981.

Williams, D. J., Energetic ion beams at the edge of the plasma sheet: ISEE 1 observations plus a simple explanatory model, J. Geophys. Res., 86, 5507-5518, 1981.

MAGNETIC RECONNECTION AND ANOMALOUS TRANSPORT PROCESSES

J. F. Drake

Department of Physics, University of Maryland
College Park, MD 20742

Mechanisms for the generation of anomalous resistivity and its role in reconnection in collisionless plasmas are discussed here. Theoretical models of reconnection are often based on model Ohm's laws,

$$\underline{E} + \underline{V} \times \underline{B} = \eta \underline{J} , \qquad (1)$$

where η is the resistivity. In the vicinity of the field null or x-line in reversed field configurations, $\underline{V} \times \underline{B} \approx 0$ and the electric field driving the reconnection produces a current $\underline{J} \simeq \underline{E}/\eta$. When the classical resistivity is small, as in many applications in space plasmas, $\underline{J}$ is quite large and reconnection is inhibited unless an instability is excited which limits $\underline{J}$. Candidate instabilities must (1) be causally related to large drift velocity, $V_d = J/n_e$, (larger current → stronger instability), (2) lead to momentum exchange between electrons and ions, (3) reach significant amplitude before saturating, and (4) go unstable above reasonable thresholds in V_d. Most instabilities fail to satisfy one or more of these criteria and must therefore be discarded [Papadopoulos, 1979]. The ion-acoustic and Buneman instabilities are only unstable for $V_d \gtrsim V_e$ for $T_e \sim T_i$, corresponding to sheet thicknesses $\lambda \sim \rho_e = V_e/\Omega_e$, where ρ_e is the electron Larmor radius. The beam cyclotron mode saturates at extremely small amplitudes. The magnetized ion-ion instability is driven by the relative drift of two ion species, is therefore not obviously causally related to $\underline{J}$, and is probably not important for reconnection. The lower-hybrid-drift instability (LHD) is the only known candidate which satisfies all of these criteria [Huba and Drake, 1981].

The lower hybrid drift instability is a flute mode ($\underline{k} \cdot \underline{B} = 0$) which is driven unstable by relative drift of electrons and ions when $V_d/V_i > (m_e/m_i)^{1/4}$ [Krall and Liewer, 1971]. The frequency is given by $\omega \gtrsim k_y V_d$ with a growth rate which peaks at $k_y \rho_i \sim (m_i/m_e)^{1/2}$. The frequency satisfies $\Omega_i << \omega << \Omega_e$, so that the ions behave as if they are unmagnetized. In finite β plasma both species of particles resonate with the wave, the electrons through their ∇B drift and the ions directly through the Landau resonance. This resonant interaction allows electrons and ions to irreversibly exchange energy and momentum [Huba and Drake, 1981].

The nonlinear saturation of the LHD instability is now well understood [Drake et al., 1983a, 1984]. A nonlinear wave equation has been derived which describes 2-D LHD turbulence ($\underline{k} \cdot \underline{B} = 0$) with realistic sources and sinks of energy. In detailed numerical computations the instability saturates by transferring energy from growing, long-wavelength to damped, short-wavelength modes. Both particle simulations [Brackbill et al., 1984] and laboratory observations [Fahrbach et al., 1981] are consistent with this saturation mechanism. The anomalous resistivity has been calculated self-consistently with the wave spectrum and is represented by the effective collision frequency.

$$\nu_{an} = 2.4 \, (V_d/V_i)^2 \omega_{lh} , \qquad (2)$$

where $\omega_{lh} = (\Omega_e \Omega_i)^{1/2}$ is the lower hybrid frequency.

The remaining unresolved issue in understanding the role of LHD turbulence in magnetic reconnection is the problem of localization. Electrons strongly absorb the wave energy through their ∇B resonance when $\beta_e = 8\pi n T_e/B^2 > 1$. This strong absorption prevents the wave from propagating into the high β region of the neutral sheet and results in a resistivity profile across the sheet which peaks away from $B = 0$ [Huba et al., 1980]. It has been shown that such a resistivity profile causes magnetic flux to be transported toward the neutral line, dramatically increasing the current density J and V_d near the center of the neutral sheet [Drake et al., 1981]. Once V_d locally exceeds V_e the wave turbulence is expected to spread across the entire sheet. This process has been observed in computer simulations [Winske, 1981; Tanaka and Sato, 1981] of the LHD instability and is still under active study.

References

Brackbill, J. U., D. W. Forslund, K. Quest, and D. Winske, The nonlinear evolution of the lower hybrid drift instability, *Phys. Fluids*, submitted, 1984.

Davidson, R. C., N. T. Gladd, C. S. Wu and J. D. Huba, Effects of finite plasma beta on the lower-hybrid-drift instability, *Phys. Fluids 20*, 301, 1977.

Drake, J. F., N. T. Gladd and J. D. Huba, Magnetic field diffusion and dissipation in reversed field plasmas, *Phys. Fluids 24*, 78, 1981.

Drake, J. F., P. N. Guzdav and J. D. Huba, Saturation of the lower-hybrid-drift instability by mode coupling, *Phys. Fluids 26*, 601, 1983a.

Drake, J. F., J. D. Huba and N. T. Gladd, Stabilization of the lower-hybrid-drift instability in finite-β plasmas, *Phys. Fluids 26*, 2247, 1983b.

Drake, J. F., P. N. Guzdav, A. B. Hassam and J. D. Huba, Nonlinear mode coupling theory of the lower-hybrid-drift instability, *Phys. Fluids*, to be published, 1984.

Fahrbach, H. U., W. Köppendörfer, M. Münich, J. Neuhauser, H. Röhr, G. Schramm, J. Sommer and E. Holzhauer, Measurement of lower hybrid drift fluctuations in the boundary layer of a high-beta plasma by collective CO_2 laser light scattering, *Nucl. Fusion 21*, 257, 1981.

Huba, J. D., J. F. Drake and N. T. Gladd, Lower-hybrid-drift instability in field reversed plasmas, *Phys. Fluids 23*, 552, 1980.

Huba, J. D. and J. F. Drake, Physical mechanism of wave-particle resonances in an inhomogeneous magnetic field. 1. linear theory, *Phys. Fluids 24*, 1650, 1981.

Krall, N. A. and P. C. Liewer, Low-frequency instabilities in magnetic pulses, *Phys. Rev. A, 4*, 2094, 1971.

Papadopoulos, K., the role of microturbulence in collisionless reconnection, in *Dynamics of the Magnetosphere* (S.-I. Akasofu, ed.) p. 289, D. Reidel Publ. Co., Dordrecht, Holland, 1979.

Tanaka, M. and T. Sato, Simulations on lower hybrid drift instability and anomalous resistivity in the magnetic neutral sheet, *J. Geophys. Res. 86*, 5541, 1981.

Winske, D., Current-driven microinstabilities in a neutral sheet, *Phys. Fluids 24*, 1069, 1981.

Questions and Answers

Dungey: What do you mean by reconnection time scale?

Drake: To carry out a full 3-D kinetic simulation of the complete reconnection process, calculating self-consistently the growth and saturation of the LHD turbulence, is very difficult because the time scales required to follow the LHD turbulence $[t \sim (r_e r_i)^{1/2}]$ are so disparate from those required to follow the reconnection process. The time required to simulate the reconnection process (reconnection time scale) is at least several Alfvén times $(t \gg \lambda/c_A)$ so that the macroscopic flows can build up to their steady (or non-steady) levels.

Priest: How would extra magnetic field components, both in the plane of the sheet and normal to it, affect the anomalous lower hybrid resistivity qualitatively?

Drake: A component of the magnetic field B_y in the plane of the sheet (East-West in the magnetotail) produces magnetic shear which is generally stabilizing. For $B_y \gtrsim .15 B_x$ the instability becomes more strongly localized away from the center of the sheet than when $B_y = 0$ [see Huba, et al., J. Geophys. Res. 87, 1697 (1982)]. I do not think that a small normal component B_z would strongly affect the instability although no calculations have been carried out for this case.

COMPARISON OF AN ANALYTICAL APPROXIMATION FOR PARTICLE MOTION IN A CURRENT SHEET WITH PRECISE NUMERICAL CALCULATIONS

T. W. Speiser[1] and L. R. Lyons[2]

Space Environment Laboratory, NOAA
Boulder, CO 80303

Approximate analytic solutions exist for charged particle motion in a one-dimensional current sheet with constant magnetic field component (Speiser, 1965, 1967; Cowley, 1973, 1978; Eastwood, 1972, 1974). The equations of motion in a frame representing the geomagnetic tail are:

$$\ddot{x} = c_2 \eta V_y \tag{1}$$

$$\ddot{y} = c_3 + c_1 z V_z - c_2 \eta V_x \tag{2}$$

$$\ddot{z} = - c_1 V_y z \tag{3}$$

where $\eta = B_z/B_{xT}$; $c_1 = (q/m)B_{xT}/d$; $c_2 = c_1 d$; $c_3 = (q/m)E$; B_{xT} is the uniform x-component of the lobe field; B_z is the uniform normal component; E the uniform cross tail electric field; d is the current sheet half-width; and q and m are particle charge and mass. A transformation to a frame moving with speed E/B_z in the +x direction eliminates the electric field in equation (2) (the De-Hoffman-Teller frame). For positive c_1V_y, equation (3) indicates oscillatory motion about the sheet. A simple approximation therefore ignores the second term in (2) during the oscillation interval. Therefore, final velocity components can be given in terms of initial values as:

$$V_{xF} = -V_{xI} + 2E/B_z; \; V_{yF} \approx V_{yI} \approx 0; \; |V_{zF}| \approx |V_{zI}| \tag{4}$$

where final and initial values are chosen at "ejection" and "injection" points, defined as those turning points in the particle orbit where V_y (see (3)) goes through zero. The neglect of the $c_1 z V_z$ term is justified not because it is small at any time, but because its average value is small, averaged over an oscillation. Direct comparison of these results with numerical integration, in the earth's frame, shows that these simple results are valid over a fairly wide range of current sheet parameters:

$$3nT \left[\frac{E d_o}{E_o d}\right]^{1/2} \gtrsim B_z \gtrsim 1/4nT \left[\frac{E}{E_o}\right] \left[\frac{W_o}{W}\right]^{1/2} \tag{5}$$

Here E is the electric field strength, d is the current sheet half-thickness, B_z is the normal field component, and W is the scale width (in the y-direction) of the tail or the scale width over which the model applies (constant E and B_z). E_o, d_o, W_o have nominal values of 1/4mV/m, 500 km, and 10 R_e, respectively. Using (4), initial distributions can be mapped into accelerated (ejection) particle distribution functions. Such mappings compare favorably with observations of streaming ions on the edge of the plasma sheet in the geomagnetic tail. The analytic solutions can also be used to construct consistent equilibrium models for the fields and particles for the tail current sheet, as has been done, for example, by Eastwood (1972), Hill (1975), and Cowley (1980).

References

Cowley, S. W. H., A self-consistent model of a simple magnetic neutral sheet system surrounded by a cold, collisionless plasma, *Cosmic Electrodyn., 3*, 448, 1973.

Cowley, S. W. H., A note on the motion of charged particles in one-dimensional magnetic current sheets, *Planet. Space Sci., 26*, 539, 1978.

Cowley, S. W. H., Plasma populations in a simple open model magnetosphere, *Space Sci. Rev., 26*, 217, 1980.

Eastwood, J. W., Consistency of fields and particle motion in the 'Speiser' model of the current sheet, *Planet. Space Sci., 20*, 1555, 1972.

Eastwood, J. W., The warm current sheet model and its implication on the temporal behavior of the geomagnetic tail, *Planet. Space Sci., 22*, 1641, 1974.

Hill, T. W., Magnetic merging in a collisionless plasma, *J. Geophys. Res., 80*, 4689, 1975.

Speiser, T. W., Particle trajectories in model current sheets, 1, analytical solutions, *J. Geophys. Res., 70*, 4219, 1965.

Speiser, T. W., Particle trajectories in model current sheets, 2, Aplications to auroras using a geomagnetic tail model, *J. Geophys. Res., 72*, 3919, 1967.

Speiser, T. W. and L. R. Lyons, Comparison of an analytical approximation for particle motion in a current sheet with precise numerical calculations, *J. Geophys. Res.*, in press, 1984.

Questions and Answers

Sonnerup: How do the particle pitch angles transform during the current sheet interaction?

Speiser: In the frame where the electric field is zero, the x-component of velocity is simply reversed, with z-component remaining about the same (in absolute value) and the y-component is very small at the injection and ejection points. Therefore pitch angle is preserved in this frame. Transforming back to the earth's frame, we find small pitch angles for initially low energy particles and pitch angle preservation for initially high energy particles.

---

[1] Also at the Department of Astrophysical, Planetary and Atmospheric Sciences, Campus Box 391, University of Colorado, Boulder, CO 80309.

[2] Now at Space Sciences Laboratory, Marshall Space Flight Center, Huntsville, AL 35812.

MAGNETIC RECONNECTION AT THE SUN*

E. R. Priest

Applied Mathematics Department, The University, St Andrews, Scotland

Abstract. Recent advances in reconnection theory include the realisations that reconnection may be initiated in many different ways and the tearing mode instability may develop nonlinearly along several different pathways. This resolves the dichotomy between the "forced" and "spontaneous" reconnection schools of thought, since both types of reconnection can occur, depending on the application. Also, a new regime has been discovered of *impulsive bursty reconnection* when the Petschek-Sonnerup regime goes unstable to tearing and rapid coalescence as pairs of X- and O-points are repeatedly created and annihilated.

In the Sun reconnection may be responsible for a wide range of phenomena. Prominences are huge sheets of plasma supported in a magnetic configuration that is thought to evolve slowly through a series of equilibria containing neutral points. The solar corona may be heated by small-scale tearing in Alfvén waves that have been phase-mixed by magnetic field inhomogeneities. Alternatively, the heating may be caused by a continuous generation of tearing turbulence as the coronal magnetic field evolves through a series of equilibria subject to a constraint on the evolution of magnetic helicity.

In solar flares reconnection may play several roles including the initial triggering of an eruptive magnetic instability by new flux emergence. Also the main phase of energy release occurs as open magnetic field lines (that are line-tied to the solar surface) close down. This process has been modelled numerically and reveals a linear tearing phase followed by a quasi-steady phase of nonlinear Petschek reconnection with slow shocks and below the reconnection site a fast shock. Later, it evolves into the impulsive bursty reconnection regime.

1. Introduction

The Sun represents a cosmic laboratory where one can observe the basic properties of a plasma at high magnetic Reynolds number. In particular, it is a testing ground for theories of the reconnection process, which is inevitable in such a plasma and which is important in a rich variety of solar phenomena.

Our understanding of the Sun's atmosphere has changed dramatically over the past 10 years. The old view of the Sun was of a spherically symmetric atmosphere with the magnetic field being negligible except in sunspots. The corona was thought to be heated by sound waves and to be expanding uniformly as the solar wind. Now we appreciate that the photospheric magnetic field is concentrated to intense values of 1-2 kG at the edges of supergranule cells. The corona too is highly structured and is instead believed to be heated magnetically. Soft X-ray images from Skylab (Figure 1) reveal much of this structure. Open fields show up as coronal holes from which much of the solar wind is escaping, while closed field lines are seen as loops of higher pressure plasma. The small X-ray bright points scattered over the whole disc represent regions where new magnetic flux is emerging from below the solar surface and creating microflares.

Our new view of the Sun is dominated by the magnetic field and its relation with the plasma atmosphere, in which, as we shall see, magnetic reconnection plays a prime role (e.g. Priest 1981, 1984). There are many solar phenomena where reconnection is important. For instance, in the convective zone magnetic fields are wound up by thermal convection until reconnection relieves the magnetic stresses. In the dynamo mechanism reconnection allows the field reversals to take place. In the corona large-scale reconfigurations of the magnetic field are allowed by courtesy of reconnection. Also prominences, the corona itself, and flares would not exist without reconnection. Before describing some of these varied solar phenomena, however, I wish to make some points about reconnection in general.

2. Some Properties of Magnetic Reconnection

2.1 Reconnection May Be Initiated in Different Ways

In a current sheet or a sheared magnetic field (such as a twisted flux tube) reconnection may be

---
*Invited review.

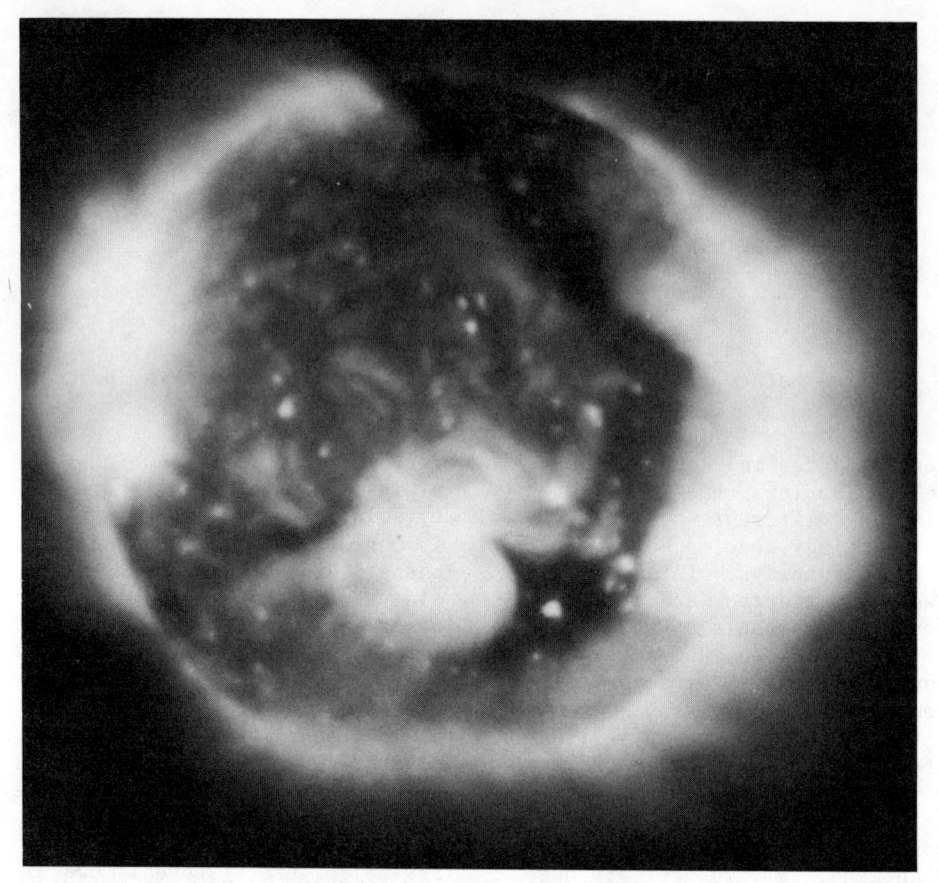

Fig. 1. An X-ray picture of the highly structured corona, showing coronal holes, coronal loops and bright points. (Courtesy D. Webb, American Science and Engineering.)

driven locally by a sudden enhancement of the magnetic diffusivity (η) due to, for example, the onset of microturbulence (Ugai and Tsuda, 1977). Alternatively, it may be driven from outside when topologically separate flux systems are pushed together (Sato and Hayashi, 1979). Furthermore, reconnection may occur spontaneously by a resistive instability such as the tearing mode. In particular, for a twisted toroidal flux tube having field components $\underline{B}_0 = (0, B_\phi(r), B_p(r))$ that depend on the distance r from the magnetic axis, a radial perturbation of the form

$$\xi = \xi(r) e^{i(m\theta - n\phi)}$$

produces a shape like sausage if m = 0 or a single helix if m = 1 or a double helix if m = 2 (Figure 2). The perturbed induction equation for the departure ($\underline{B}_1$) from the equilibrium ($\underline{B}_0$) then takes the form (for incompressible flow)

$$\frac{\partial \underline{B}_1}{\partial t} = \underline{v}_1 (\underline{k} \cdot \underline{B}_0) + \eta \nabla^2 \underline{B}_1,$$

which (for $R_m = vL/\eta \gg 1$) implies that the diffusion of magnetic field lines (represented by $\eta \nabla^2 \underline{B}_1$) is negligible except in a thin layer (called a resonant surface) where

$$\underline{k} \cdot \underline{B}_0(r) = 0. \quad (1)$$

At this layer the orientation of the perturbation matches that of the field, so that the crests and troughs of the helix follow the field lines. The magnetic field can reconnect easily here, just as it can at a neutral current sheet, where $\underline{B}_0 = 0$ (a particular case of (1)). For a given field profile ($\underline{B}_0(r)$), (1) is satisfied at different radii for different values of m.

2.2 Nonlinear Development of Tearing

Several nonlinear pathways are possible for the growth of the tearing mode instability out of its linear phase. Modes with m > 1 may *saturate* at a very low amplitude when the islands have only grown as wide as the resistive layer (Rutherford, 1973). (For the m = 1 mode saturation occurs somewhat later when the current profile flattens within the resonant surface (Kadomtsev, 1975).) This pathway appears to be especially relevant for tokamaks with their strong axial fields and conducting walls that generally inhibit more dramatic behaviour. An alternative is that modes on neighbouring surfaces may *couple* to one another

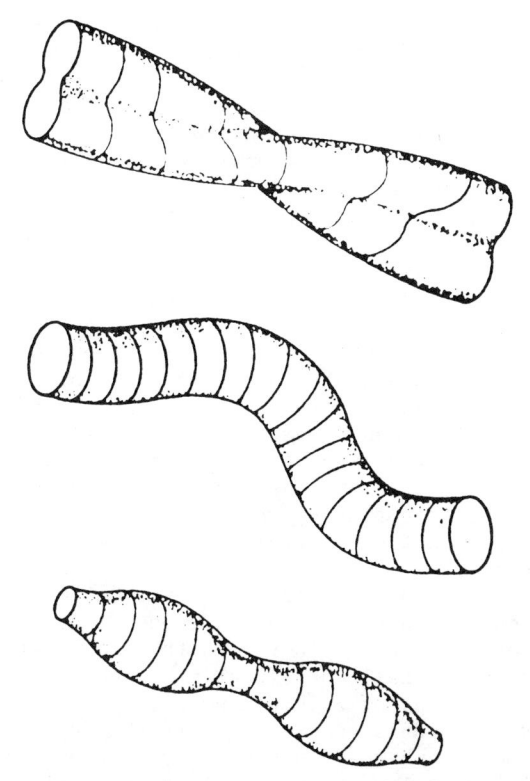

Fig. 2. The distortions of a magnetic flux tube produced by perturbations of type m = 0 (sausage), m = 1 (kink) and m = 2 (double helix). (After Bateman, 1978.)

if they are close enough (Waddell et al, 1978).

When several islands have formed along a neutral sheet or resonant surface they may also attract one another due to an ideal mode known as the *coalescence instability* (Finn and Kaw, 1977; Pritchett and Wu, 1979). Two neighbouring islands approach one another on an Alfvénic time-scale, creating a current sheet between them (Figure 3c) and driving reconnection so that the two islands combine to give a single island.

A fourth possibility is that tearing grows to give fast reconnection with inflow speeds that are a significant fraction of the Alfvén speed. In this *Petschek-Sonnerup regime* (Figure 4a) slow MHD shock waves propagate away from the reconnection point (Vasyliunas, 1975). Depending on the external boundary conditions there is a range of possible inflow behaviour from the pure fast-mode expansion of Petschek (1964) to the pure slow-mode expansion of Sonnerup (1970). This regime tends to develop when the outflow boundary conditions are free enough that the Alfvénic jets are not inhibited from developing. However, numerical codes will only exhibit the regime if the numerical dissipation is low enough that they can cope adequately with the shocks. Especially when they are weak, the shocks tend to show up more clearly in plots of current density or plasma velocity than in the magnetic field lines. Another point to note is that the Petschek-

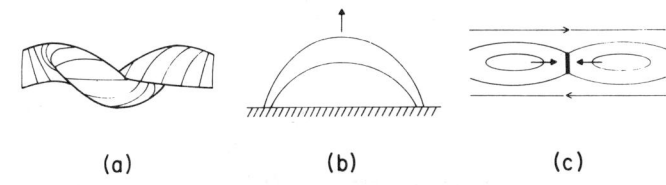

Fig. 3. Ideal magnetic instabilities that create sheets: (a) kink, (b) eruptive, (c) coalescence.

Sonnerup regime can develop from a driven reconnection scenario (§2.1) as well as a linear tearing mode.

2.3 Ideal Versus Resistive Instability

There is a subtle interplay between ideal and resistive instabilities, with each being able to lead into the other. For example, an ideal instability such as the kink mode (Figure 3a) or the eruptive mode (§4.3) may create current sheets, within which the tearing mode will subsequently develop. Also, as we have seen above a primary tearing mode may produce many islands which are then attracted towards one another by the coalescence instability. The current sheet at the interface between two approaching islands may then go unstable to *secondary tearing* on a much smaller scale (Biskamp, 1982).

2.4 Regimes of Fast Reconnection

When separate flux systems are approaching one another at a steady speed v (at large distances from the neutral point) the resulting type of reconnection depends on the value of v. If

$$\frac{v_A}{R_m^{1/2}} < v < v_{max}, \qquad (2)$$

where $v_A = B/(\mu\rho)^{1/2}$ is the Alfven speed at large distances and $R_m = L_e v_A/\eta$ is the large-scale Lundquist number based on the overall length-scale (L_e) of the system, one finds the *Petschek-*

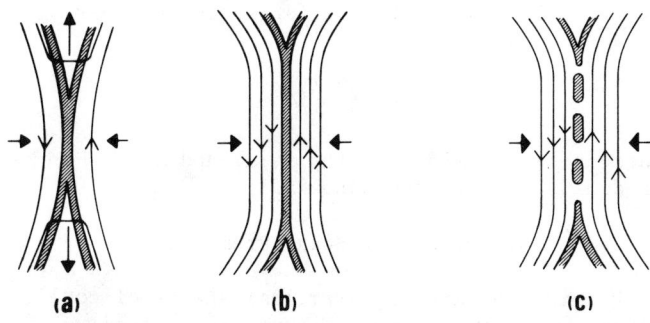

Fig. 4. Regimes of fast reconnection: (a) Petschek-Sonnerup, (b) supercritical with flux pile-up, (c) impulsive bursty.

Sonnerup regime. For the pure Petschek mode

$$v_{max} = \frac{\pi v_A}{8 \log R_m}, \quad (3)$$

which typically lies between 0.01 v_A and 0.1 v_A. The central current sheet bifurcates into two pairs of slow shocks because the inflow is supersonic with respect to the slow mode speed. If there are magnetic obstacles in the downstream flow one will also find fast shocks in the outflowing jets (Yang & Sonnerup, 1976; Forbes and Priest, 1983).

The *flux pile-up regime* (Figure 4b) occurs when

$$v_{max} < v \quad \text{or} \quad R_m > R_{m_{max}} \quad (4)$$

(Biskamp, 1982; Forbes and Priest, 1982) so that a steady state is impossible with the imposed values of v and B at large distances. Instead the *flux piles up* and the central sheet or diffusion region grows in length since all the incoming flux cannot be reconnected. The inflow region is highly time-dependent with fast mode waves propagating back to the distant sources and interacting with them in some way that depends on the particular application. Either the velocity and magnetic field strength at the sources are modified or shocks appear in the inflow.

A new regime of *impulsive bursty reconnection* (Birn and Hones, 1981; Forbes and Priest, 1982; Biskamp, 1982; Tajima et al, 1982; Bhattacharjee et al, 1983; Brecht et al, 1982; Park et al, 1983) can develop from either of the above two regimes when the diffusion region length (L) becomes so great relative to its width (or $R_m > R_{m_{crit}}$) that the Petschek mode becomes unstable to secondary tearing (Figure 4c). The islands form relatively slowly (on the tearing mode time-scale) and then coalesce very rapidly (in a few Alfvén times). The process repeats sporadically and gives an impulsive energy release. If the islands are allowed to be swept along the sheet and carried out, the net effect is to increase the reconnection rate but otherwise it is slowed down by the secondary tearing. According to Bulanov et al (1978) the criterion for secondary tearing in a sheet with flow is

$$L > \ell S^{3/7},$$

where ℓ is the width of the sheet and S the Lundquist number (see also §5.3).

3. Quiescent Solar Prominences

Prominences are huge vertical sheets of cool plasma up in the corona (Figure 5), typically 200,000 km long, 50,000 km high and 6,000 km wide, with a density that is a hundred times that of the ambient coronal medium and a temperature that is only one-hundredth. They can remain in a stable configuration for months, presumably supported against gravity by a magnetic field (of typical strength 10 G). The Sun is able to perform such a feat with ease, while man finds confinement of similar plasma in the laboratory for only a second almost impossible!

A prominence contains much internal structure in the form of thin vertical threads only 300 km wide, which may be created by a resistive instability modified by gravity and thermal effects. It is also found to lie above a reversal in the line-of-sight component of the photospheric field. Two models have been proposed for prominence magnetic fields by Kippenhahn and Schluter (1957) and Kuperus and Raadu (1974), depending on the sign of the prominence magnetic field. In the former case the field consists of a simple magnetic arcade, whereas in the latter case there is an X-type neutral point below the prominence.

Recently, some dramatic observations of prominence magnetic fields have been reported by Leroy et al (1983). They find that in two-thirds of the cases (tall, quiescent prominences with field strengths of 5-10 G) the sign of the observed field is consistent with the Kuperus-Raadu model, whereas in one-third of the cases (low-lying (below 30,000 km altitude) active-region prominences with fields of about 20 G) the observations suggest the classical Kippenhahn-Schlüter configuration.

Malherbe and Priest (1983) have shown how to use complex variable techniques (with $z = x + iy$) to set up simple models of both types (Figure 6). For example,

$$B_y + i B_x = -B_0 \frac{[(p^2+z^2)(q^2+z^2)]^{1/2}}{z(z+ih)^2} - \frac{B_1}{z} \quad (5)$$

gives a simple Kippenhahn-Schlüter field, with a current sheet (the prominence) represented by a cut in the complex plane from $z = ip$ to $z = iq$ (Figure 6a). Adding the field

$$B_y + i B_x = \frac{B_1}{z} + \frac{B_1}{z} \frac{(z+iH)}{(z-iq)}$$

to this introduces an X-point above the prominence (Figure 6b). Similarly, Kuperus-Raadu configurations may be modelled by

$$B_y + i B_x = \frac{B_0[(p^2+z^2)(q^2+z^2)]^{1/2}}{z} + B_1(z-ip) \quad (6)$$

or

$$B_y + i B_x = -\frac{B_0[(p^2+z^2)(q^2+z^2)]^{1/2}}{z(z+ih)^2} + B_1 \frac{z-ip}{z(z-iq)},$$

as sketched in Figures 6c and 6d, respectively. Another important observation, however, is of

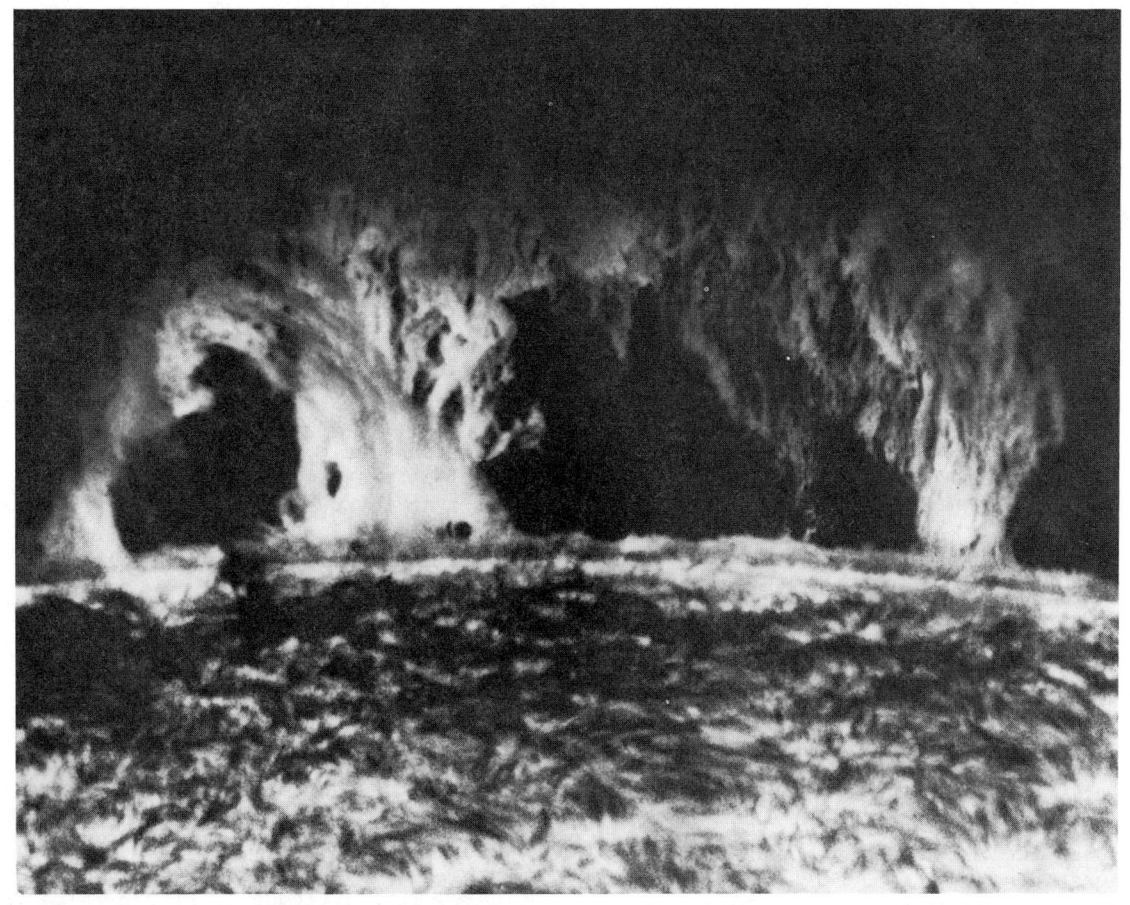

Fig. 5. A quiescent prominence seen from the side at the limb of the Sun. (Courtesy H. Zirin, Big Bear Solar Observatory.)

slow steady upflows at 0.5 km s^{-1} in prominences seen from above on the solar disc (Malherbe et al, 1983). This may be explained by a dynamic prominence model in which the coronal field evolves through a series of equilibria, reconnecting if necessary, in response to slow motions of the photospheric footpoints, as indicated in Figure 6. This suggests that Kuperus-Raadu prominences may be situated along large-scale fault lines (possibly giant cell boundaries) in the solar surface where the plasma is converging and flowing down. By contrast, Kippenhahn-Schlüter prominences would tend to form where the surface plasma is upwelling and diverging.

4. Coronal Heating

The solar corona is heated to a few million degrees, compared with the Sun's surface temperature of only 6000 K. In active regions 5000 W m^{-2} is needed and this was at one time thought to be provided by acoustic waves which are generated by granulation at the photosphere and

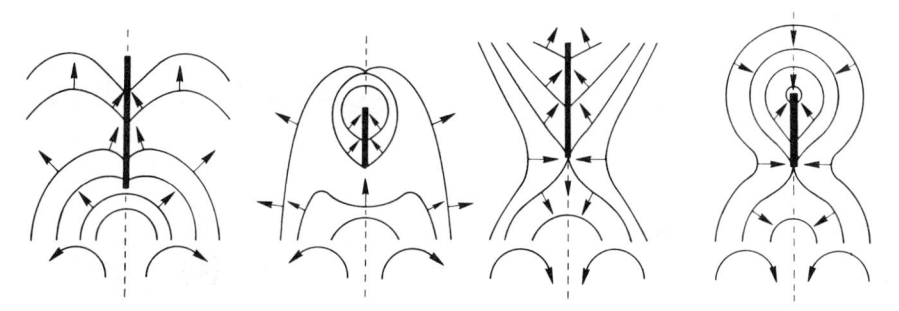

Fig. 6. Dynamic prominence models of Kippenhahn-Schluter type (a and b) and Kuperus-Raadu type (c and d). The arrows indicate motions perpendicular to field lines.

68   MAGNETIC RECONNECTION AT THE SUN

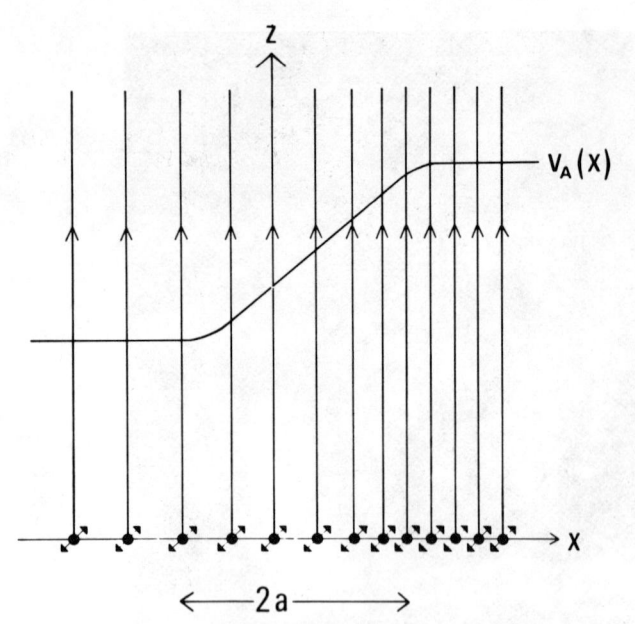

Fig. 7a. The equilibrium field and Alfvén profile.

then steepen into shocks. However, the observed acoustic power is only 10 W m^{-2} (Athay and White, 1978; Mein and Schmieder, 1981) and the heating mechanism is now thought to be magnetic in nature (Chiuderi, 1981; Hollweg, 1981; Kuperus et al, 1981; Priest, 1982b). Nevertheless, the suspect has not yet been identified and is at present a topic of intense debate, although my own feeling is that magnetic reconnection is probably an important ingredient. The response to motions of photospheric footpoints depends on the time-scale (τ_f) of those motions relative to the time (L/v_A, say) it takes for waves to propagate along a closed field line of length L, where the Alfvén speed is v_A. This has led to two distinct types of mechanism, as exemplified below.

4.1 Wave Heating ($\tau_f < L/v_A$)

The energy generated by rapid footpoint motions propagates upwards as waves (e.g. Habbal et al, 1979; Zweibel, 1980; Hollweg, 1981; Leroy, 1980). In particular, the difficulty with Alfvén waves has been to suggest how they give up their energy, since they tend to damp extremely slowly under coronal conditions. The answer appears to lie in the effect of strong inhomogeneities in the coronal magnetic field. For waves that are polarised in the direction of the inhomogeneity, resonant absorption and surface wave effects are crucial (Hasegawa and Chen, 1974; Rae and Roberts, 1982; Ionson, 1978; Wentzel, 1979; Roberts, 1981, 1982). For the other sense of polarization, the *phase mixing of shear Alfvén waves* has been recently proposed, as outlined below (Heyvaerts and Priest, 1983; Browning and Priest, 1983a; Nocera et al, 1983).

Consider a unidirectional magnetic field ($B_0(x)\hat{z}$) at rest with an Alfvén speed profile ($v_A(x)$) varying over a width 2a (Figure 7a). Linear perturbations with velocity $v(x,z,t)\hat{y}$ and magnetic field $b(x,z,t)\hat{y}$ satisfy the equation

$$\frac{\partial^2 v}{\partial t^2} = v_A^2(x) \frac{\partial^2 v}{\partial z^2} + \nu \left(\frac{\partial^2}{\partial x^2} + \frac{\partial^2}{\partial z^2} \right) \frac{\partial v}{\partial t} \quad (7)$$

where ν is the diffusion coefficient (magnetic or viscous). For travelling waves with $\nu = 0$ each surface x = constant oscillates independently with a wavenumber $k_z(x) = \omega/v_A(x)$, and so the waves on neighbouring surfaces become increasingly out of phase as they propagate, since $\partial/\partial x \sim (dk_z/dx)z$ increases with altitude z (Figure 7b). The effect of dissipation ($\nu \neq 0$) in the limits of long waves, weak damping and strong phase mixing is to give the solution of (7) as

$$v(x,z,t) = v_0 e^{-z^3/\Lambda^3} e^{i(\omega t - k_z(x)z)}, \quad (8)$$

where the damping length is

$$\Lambda = \frac{(6R_e)^{1/3}}{k_z(x)} = 5000 \; B_G^{5/3} \left(\frac{\tau}{10s} \right)^{2/3} \text{km}$$

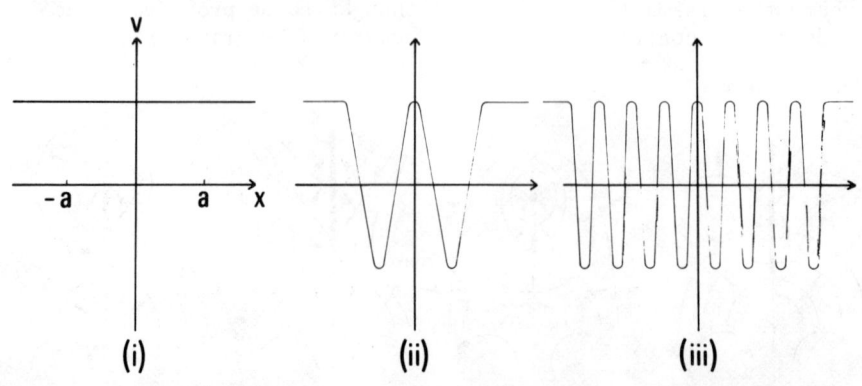

Fig. 7b. The x-variation of v_y at different altitudes (in a propagating wave) or times (in a standing wave).

in terms of the local Reynolds number $R_e = \omega a^2/\nu$. Thus short-period waves (~10s) in a weak field (~1 G) damp efficiently over 5000 km.

For standing waves on loops there are reflections at the transition regions and each surface oscillates with a frequency $\omega(x) = k_z v_A(x)$. This produces phase mixing in time since $\partial/\partial x \sim (d\omega/dx)t \equiv q$ for disturbances proportional to $\exp i(\omega(x)t - k_z z)$. The damping time is $\tau_d = (R_e)^{1/3}/\omega$, which is typically 20 wave periods and so phase mixing enhances the dissipation by building up strong local gradients. However, the dissipation is enhanced still further by the creation of even smaller scale structure by local instabilities in the standing phase-mixed Alfvén waves. In particular, Kelvin-Helmholtz instabilities are active near magnetic field nodes (Browning and Priest, 1983a). Also, near the velocity nodes where $\underline{v} = 0$ and $\underline{B} = B_0(\hat{z} - A \sin qx \, \hat{y})$ one finds tearing modes with a growth time $\tau = (\tau_A^4 \tau_d^3)^{1/7}$ for periodic boundary conditions (Bobrova and Syrovatsky, 1980), or even faster for the newly discovered radiative tearing mode (Steinolfson and Van Hoven, 1983; Steinolfson, 1983; Van Hoven et al, 1983). This is faster than a wave period for phase mixing times longer than a few periods, and our conclusion is that phase mixing appears to be a very efficient means of dissipating Alfvén energy, especially in standing waves.

4.2 Tearing Turbulence in Slowly Evolving Fields ($\tau_f > L/v_A$)

When the footpoint motions are slow enough the coronal magnetic field tries to evolve through a series of equilibria, which are force-free for the low-beta plasma of an active region above sunspots. Parker (1972, 1983) has suggested that, if the magnetic field consists in its fine-scale structure of a set of closely packed flux tubes, the tubes will tend to form current sheets at their interfaces when they are simply twisted and will tend to have no equilibrium at all when they are braided, both cases leading to enhanced dissipation.

If the large-scale force-free equilibria do exist they are in general unstable to tearing and so the coronal magnetic field is likely to be in a state of tearing turbulence as it evolves due to slow photospheric motions and continuously relaxes to the lowest available (linear) force-free state. Heyvaerts et al (1984) have recently set up the technical apparatus for estimating this evolution and the resulting heat input by tearing turbulence.

Previously, the evolution of the coronal field in active regions had been thought to be through nonlinear force-free fields subject to the constraint of imposing the connections of each field line from one footpoint to another (e.g. Birn and Schindler, 1981; Low, 1982). When such fields went unstable they were thought to reduce

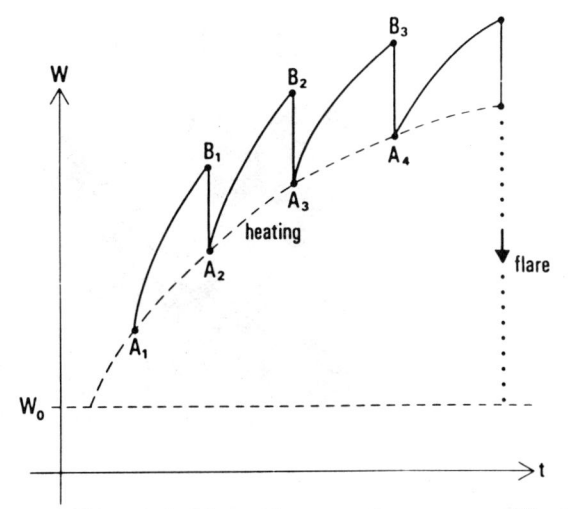

Fig. 8. The evolution of magnetic energy (W) in time due to a series of discrete footpoint motions. A_i represent linear force-free states and B_i nonlinear ones.

to their lowest energy state subject to no constraint (except imposing the normal field component at the photosphere), namely a potential field. We are instead suggesting that coronal fields evolve in a more subtle manner through force-free equilibria that are subject to a *new constraint*. The field is continually unstable to tearing, but in general this has the effect of reducing the field to an equilibrium with an energy well above potential (except in the unusual and catastrophic case of a major flare). The magnetic helicity is defined as

$$K = \int_V \underline{A} \cdot \underline{B} \, dV \qquad (9)$$

in terms of the magnetic vector potential ($\underline{A}$) such that $B = \nabla \times A$. The new constraint is then that the magnetic helicity evolves in time according to the relation

$$\frac{dK}{dt} = \int_C (\underline{A} \cdot \underline{v}) B_n \, dS \qquad (10)$$

due to motions ($\underline{v}$) on the boundary C where the normal magnetic field is B_n. This is a generalisation of Taylor's (1974) hypothesis that K = constant for the evolution of a reversed field pinch in which $B_n \equiv 0$.

The evolution of the field may then be understood in terms of the following thought experiment for the response of the corona to a series of small discrete footpoint motions. Suppose the footpoints make a sudden small motion and cause the coronal magnetic field to change from an initial linear force-free state to a new nonlinear force-free state while preserving the field line connections and not allowing any reconnection (i.e. from point A_1 to B_1 in Figure 8). Then the

3B FLARE OF JULY 29, 1973

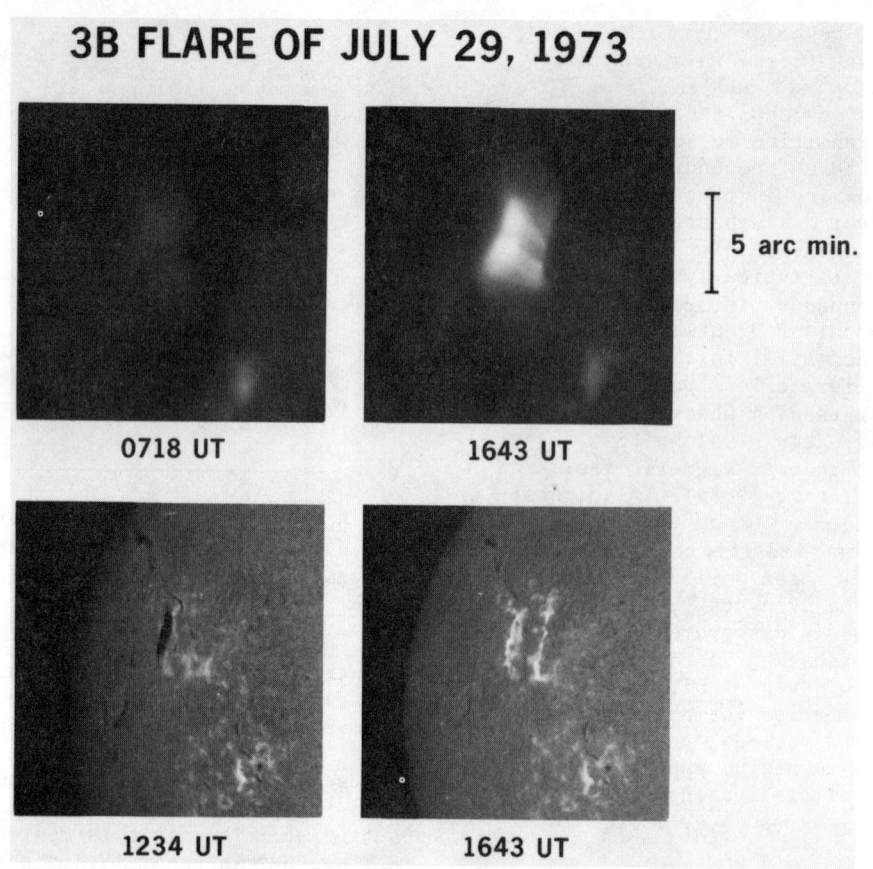

Fig. 9a. Hα (bottom) and soft X-ray (top) pictures before (left) and after (right) the start of a large flare, showing X-ray loops joining Hα ribbons. (Courtesy D. Webb, American Science and Engineering.)

coronal field relaxes (from B_1 to A_2) due to small-scale tearing to a new linear force-free state whose parameter $\alpha = \nabla \times \underset{\sim}{B}/B$ is determined by the helicity evolution constraint (10). Next, a new footpoint motion causes the field to evolve from A_2 to a nonlinear state at B_2 and then to relax to A_3, and so the process repeats. The net effect is an evolution through the sequence of linear force-free states $A_1, A_2, A_3, \ldots$. In Figure 8 the energy is shown to be continuously increasing up to a point of catastrophic energy release (a flare) as the linear force-free state itself goes unstable or ceases to exist and the energy does fall back to potential. However, once W is above a potential value it may of course increase or decrease in its evolution through linear force-free states, depending on the details of the footpoint motion. Furthermore, the difference between the energy build-up due to the Poynting flux through the photosphere and the energy of the sequence of linear states gives the amount of energy dissipated as heat during the tearing processes (B_1A_2, B_2A_3, B_3A_4, ...). This is found to be sufficient to heat the corona.

5. Solar Flares

5.1 Observations

A large flare involves the release of 3×10^{25} J of energy and consists of three phases (e.g. Svestka, 1976; Sturrock, 1980). During the *preflare phase* (~ ½ hr) a large flux tube (an active-region filament) moves up slowly and there is a weak brightening in soft X-rays. At the *rise phase* (between 5 mins and 1 hr in duration) the flux tube suddenly erupts much more rapidly, while there is a steep rise in the Hα and X-ray emission, and two ribbons of emission form in the chromosphere. During the *main phase* (1 hr - 1 day) the intensity declines slowly, while the ribbons move apart and are joined by a rising arcade of hot loops up to 100,000 km high and with a density and temperature that decline from $10^{17} m^{-3}$ and $2 \times 10^7 K$ to $10^{16} m^{-3}$ and $5 \times 10^6 K$ later in the event (Figure 9a). The ribbon and loop velocities are rapid at first (20 - 50 km s^{-1}) and much slower later on (0.5 km s^{-1}) and have been measured in several events (e.g. Figure 9b). The motion is thought to indicate a reconnection of the field lines with the X-point rising and new loops being energised sequentially rather than representing the rise of a single loop. Indeed, the observed motion has been used to deduce the reconnection rate, as measured by the electric field at the X-point (Forbes and Priest, 1982b). It is found to be nearly constant through the main phase and implies a magnetic Reynolds number of 10^8 for reconnection at the maximum Petschek rate. Major theoretical problems were to explain how to heat plasma up to $2 \times 10^7 K$ and how to account for the enormous amount of mass that is seen to be falling down (as cool loops (10^4K) below the hot ones) and which would be sufficient to fill the whole corona normally.

5.2 Small Flares - Emerging Flux

Many theories have been put forward for solar flares (see e.g. Priest, 1981, 1982c for reviews), but one possibility for small events is that new flux is emerging from below the photosphere and creating a current sheet at height h at the interface between the new and overlying flux (Heyvaerts et al, 1977). In this *emerging flux model* the onset of the flare takes place when the sheet reaches a critical height (h_{crit}) for the current density to exceed the threshold for strong microturbulence. h_{crit} has been calculated by solving the energy balance in the sheet (Milne and Priest, 1981).

Some preliminary results of a numerical simulation of emerging flux by Forbes and Priest (1984) are shown in Figure 10. The normal 2D resistive MHD equations are solved in a box with free boundary conditions on the sides and top and with initially a uniform horizontal field at rest. The effect of forcing new flux in through the base at a speed of $v_A/8$ and with a magnetic Reynolds number of 400 can be seen. The new flux at first emerges and reconnects easily with the overlying field. After t = 4 (measured in units of the Alfvén travel time across a unit distance) no more new flux is forced through the base, but the flux continues to rise through its own inertia and pinches off near the base to form a plasmoid. This disappears and the field relaxes to a potential field based on the new photospheric boundary condition.

Some results at a higher magnetic Reynolds number (2000) that we have not yet analysed in detail are shown in Figure 11. The magnetic field reconnects much less readily, and the current density contours on the left of the Figure reveal that the current structure is quite complex. The curved current sheet at the interface between new and old flux is much longer than one would expect from steady reconnection theory due to *flux pile-up*. At its ends the curved sheet bifurcates to give what appear to be a slow expansion fan and a slow shock (c.f. Levy et al, 1964; Hoshino and Nishida, 1983; Sonnerup, 1979). The reconnection is reminiscent of the Sonnerup mechanism, with a slow mode expansion in the inflow above the curved sheet since the magnetic field strength increases slightly and the flow diverges as it approaches the sheet. After t = 4, when no more flux emerges, these sheet pairs die away. At the X-point which forms later below the plasmoid one finds a bifurcation of a new vertical current sheet into pairs of slow shocks, but the plasmoid is more persistent than before and decays away much more slowly. It should be noted that the wiggles in the current density imply that there is real fine structure (Matthaeus and Montgomery, 1981). We are running the code at the highest

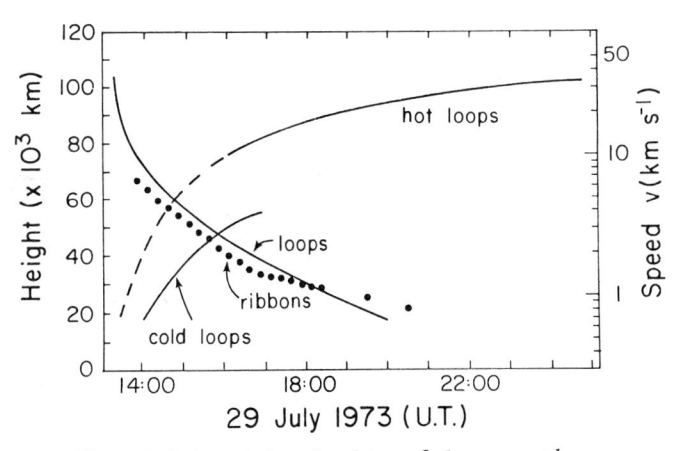

Fig. 9b. Height and velocity of loops and ribbons as functions of time. (After Moore et al, 1980.)

magnetic Reynolds number that gives acceptable accuracy on our present machine. By halving the grid-spacing we have estimated that the errors are only a few percent.

5.3 Large (Two-Ribbon) Flares

The overall behaviour of a large flare is believed to be as follows (Figure 12). During the preflare phase the large flux tube with its overlying magnetic arcade rises slowly because of: either an ideal eruptive *MHD instability* (Hood and Priest, 1980) when the flux tube twist or height are too great; or *magnetic nonequilibrium* (Parker, 1979; Browning and Priest, 1984b) when the buoyancy by magnetic tension; or nearby *emerging flux*, which may trigger the eruption by locally enhancing the resistivity and so driving reconnection (§2.1).

The onset of the rapid eruption may take place when the field lines have been stretched out so much by the slow rise that they begin to reconnect by the tearing mode instability.

At the main phase, the reconnection continues nonlinearly and creates hot loops and ribbons as the field closes down, the erupting flux tube having long since disappeared from view. Kopp and Pneuman (1976) and Pneuman (1981) attempted to explain the tremendous mass source by showing how plasma can flow up along open field lines and then become trapped and heated as the field lines reconnect and create closed field. Subsequently, the hot plasma cools radiatively and conductively and falls back down. Furthermore, Cargill and Priest (1982) demonstrated that the slow Petschek shocks can heat the plasma to the observed temperatures.

A numerical simulation of such a reconnection process in stretched-out field lines has been carried out by Forbes and Priest (1982a, 1983), including the line tying of the footpoints at the base of the numerical box (Figure 13). Along the upper and right-hand edges the boundary conditions are free-floating, whereas along the left-hand

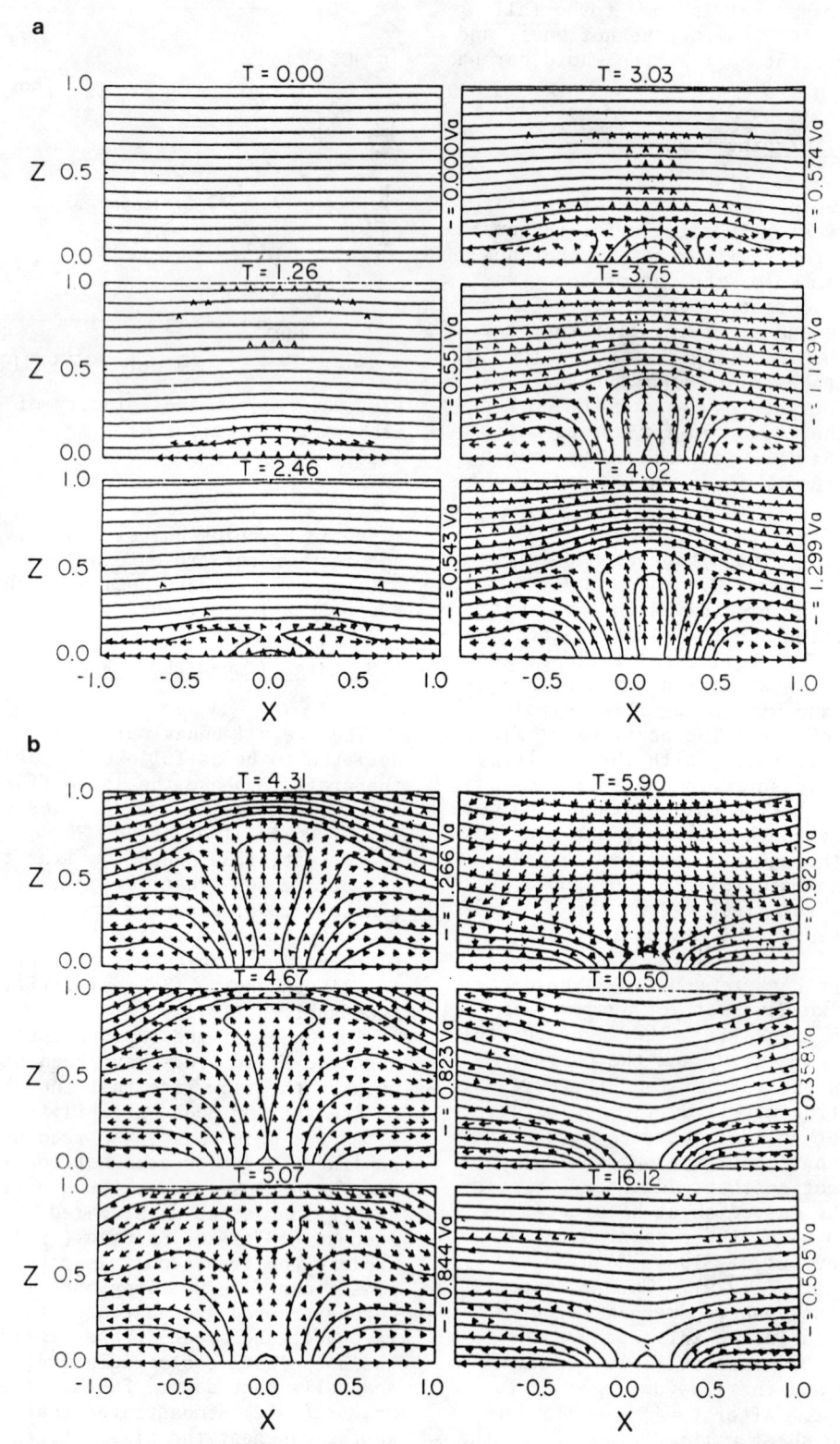

Fig. 10. Numerical simulation of emerging flux at a magnetic Reynolds number of 400 (Forbes and Priest, 1984).

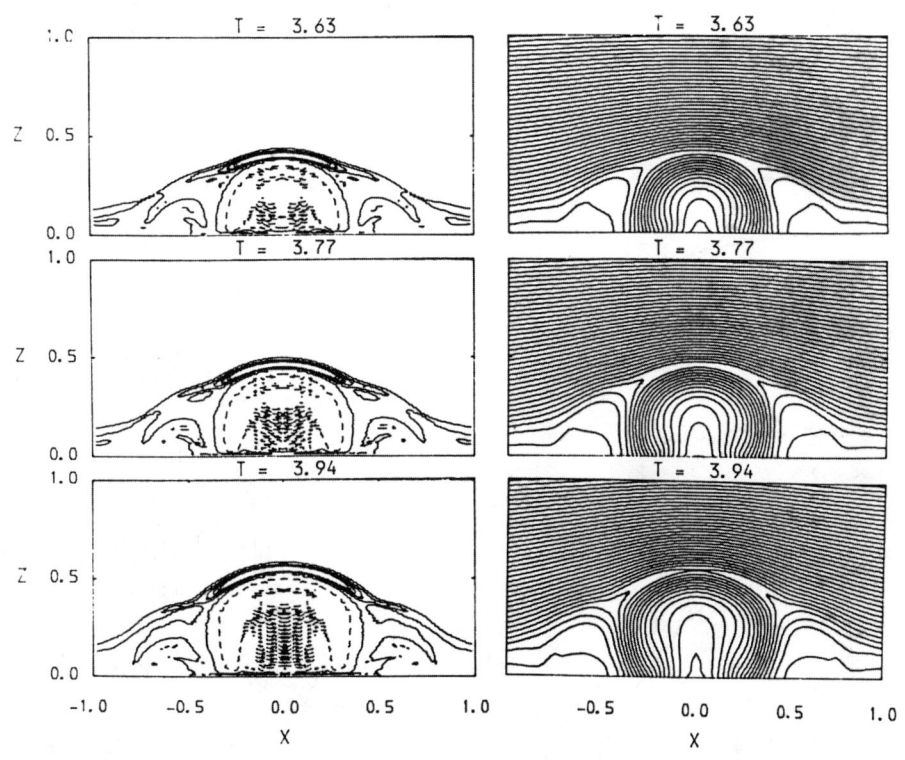

Fig. 11. Numerical simulation of emerging flux at a magnetic Reynolds number of 2000 showing current density on the right and magnetic field lines on the left (Forbes and Priest, 1984).

edge they are symmetry conditions. The initial state consists of a vertical field in equilibrium, and the magnetic Reynolds number (R_m) based on the height of the box is 2000, while the ambient plasma beta (β) is 0.1 and the width (a) of the initial vertical current sheet is 0.075 of the box width. A flux-corrected algorithm with a variable grid is used to minimise numerical diffusion and discover any shocks that should be present. Time is measured in units of the Alfvén travel time (a/v_A).

The behaviour of the numerical experiment was initially as expected. The sheet tears near the base and in the nonlinear regime a quasi-steady state of Petschek reconnection is set up, with the magnetic field continuing to close down while the X-type neutral point rises and a plasmoid is ejected upwards. In Figure 14 the plot of current density contours indicates the presence of two *slow Petschek shocks* extending up and down from the neutral point. Also, the inflow region to the right of the current sheet is characteristic of a fast-mode expansion, with the field strength decreasing and the flow converging as it approaches the X-point. Another feature is the presence of a *fast-mode shock* (Yang and Sonnerup, 1976) which slows down the jet of plasma that is being accelerated down towards the obstacle of closed magnetic flux near the base.

The later development (t > 90) was a complete surprise to us. The Petschek mode goes unstable and enters the new regime of *impulsive bursty reconnection*. The current sheet thins and secondary tearing creates a new pair of O- and X-points. Reconnection at the upper X dominates, so that the O is shot down and coalesces with the lower X very rapidly by the coalescence instability (§2.2). Meanwhile, a new pair is created and the process repeats in a sporadic manner that we found rather fun. The result is that energy is released in bursts faster than the steady Petschek mode. The time of energy release is roughly the Alfvén time (τ_A), while the time for the islands to grow is the tearing time $(\tau_A \tau_d)^{\frac{1}{2}}$, where τ_d is the diffusion time. However, the interval between bursts may be less than the tearing time because several islands may be present at once.

The impulsive bursty regime is found to occur more readily as R_m, β^{-1} or a^{-1} are increased. In order of magnitude, the condition for its occurrence is that the sheet length (L) be so great that the time (L/v_A) for newly created islands to be swept out of the sheet exceed the time (roughly $(\ell^3/(v_A\eta))^{\frac{1}{2}}$) it takes to create them by tearing, where ℓ is the sheet width. In other words, the onset of impulsive bursty regime takes place approximately when

$$L \gg S^{\frac{1}{2}} \ell, \qquad (11)$$

where $S = \ell v_A/\eta$. For our numerical experiment, $S = 10$ locally (since $\ell v_i/\eta \approx 1$ and $v_i/v_A \approx 0.1$ in terms of the inflow speed v_i) and the sheet width

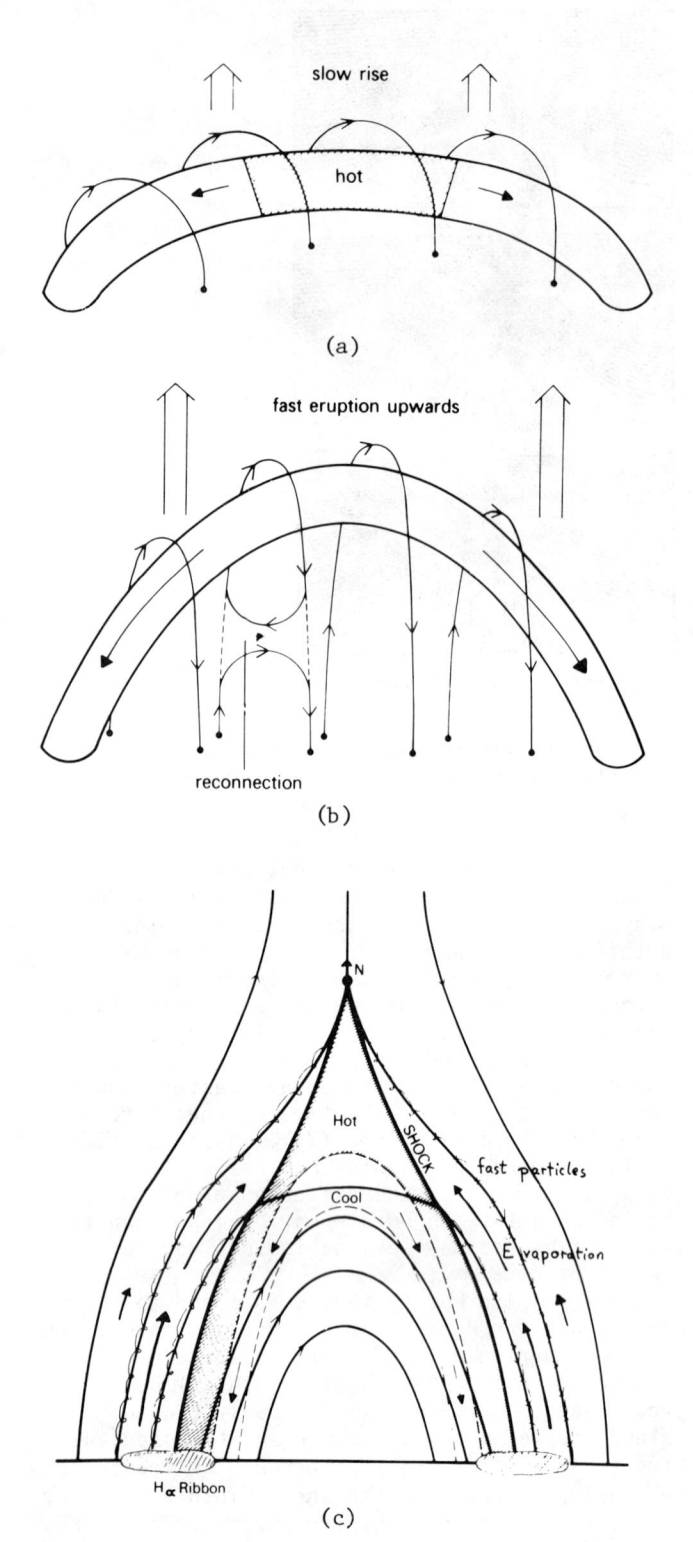

Fig. 12. The overall behaviour of a large flare: (a) preflare phase, (b) rise phase, (c) a section across the arcade during the main phase.

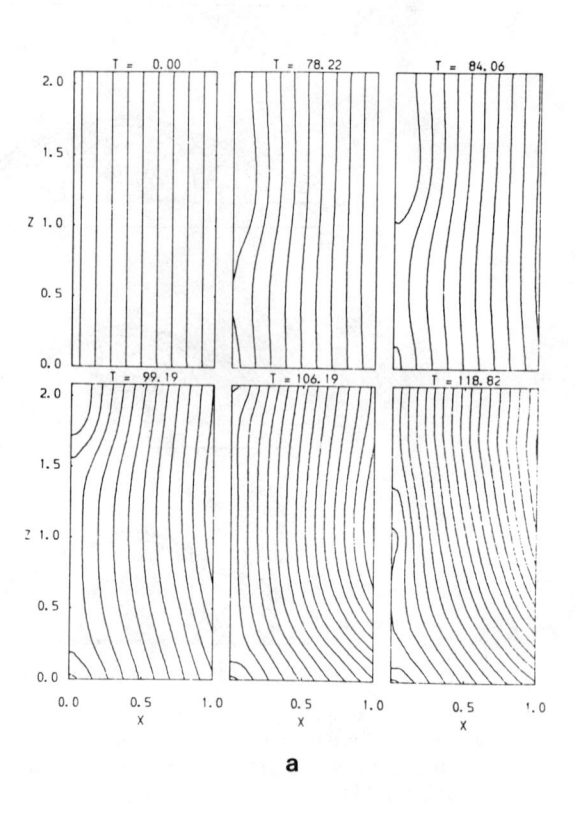

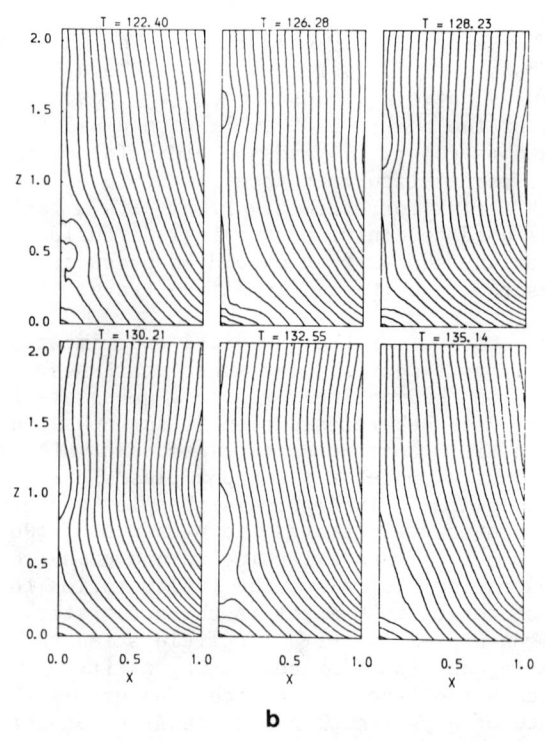

Fig. 13. Numerical simulation of line-tied reconnection (Forbes and Priest, 1983).

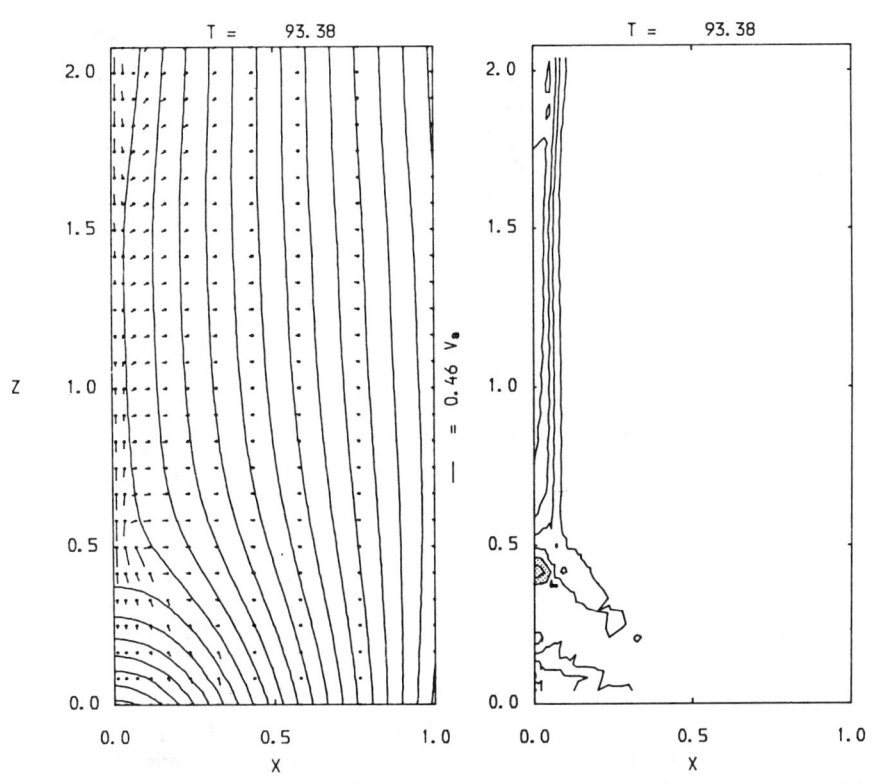

Fig. 14. Line-tied reconnection, showing (a) magnetic field lines and flow velocity vectors, (b) current density contours with slow and fast mode shocks present.

$\ell = 0.01$, which is resolved numerically and is the Petschek width ($\ell = \eta/v_i$) for $R_m = v_A d/\eta = 10^3$ in terms of the box width (d) of unity and $v_i/v_A \approx 0.1$. Thus (11) gives $L \gg 0.03$, which is consistent with the observed length for onset of roughly 1. However, the onset criterion needs to be investigated much more carefully, including the stabilizing effects of $v_y(y)$ (Bulanov et al, 1978) and B_x (Bulanov et al, 1979; Nishikawa and Sakai, 1982) and the destablizing effect of $v_y(x)$ (Sato and Walker, 1982; Dobrowolny et al, 1983) on the tearing mode in a reconnecting sheet.

6. Conclusion

A few years ago there was a clear division in reconnection theory between the linear tearing mode instability and the nonlinear steady state of fast Petschek-Sonnerup reconnection. Now, however, by studying nonlinear time-dependent reconnection, we are beginning to understand the subtle link between the two, whereby each can give rise to the other. Also, numerical experiments are presenting us with new surprises such as the presence of the fast shock and flux pile-up and the new regime of impulsive bursty reconnection, which occurs faster than steady reconnection. At still higher magnetic Reynolds numbers this may well lead to a regime of *fully turbulent reconnection*!

Some of the story of solar reconnection that I have been telling may have rung bells for magneto- sphericists. Emerging flux reconnection seems similar to dayside reconnection (Figure 15a) (although the Alfvén Mach number is much less than unity for the Sun and about 8 for the dayside magnetosphere). Also, two-ribbon flares appear similar to geomagnetic substorms. In the flash phase of such a flare an X-line forms below a rising filament, just as in the expansion phase (lasting 10 mins) of a substorm a new X-line forms earthward of the escaping plasmoid (Figure 15b). In the flare main phase the Hα ribbons move apart and the X-line rises, just as in the recovery phase (for 30 min) the aurorae leap poleward and the new X-line moves tailward. However, there are important differences in geometry and parameter regime, and different physical effects are present. The build-up to flares and substorms appears to be quite different, with a coronal arcade being sheared until it goes eruptively unstable and the flux building up in the geomagnetic tail until it tears. The solar case is collisional with a density of $10^{16} m^{-3}$, a mean-free path of 5 km, an iongyroradius of 1 cm, a plasma beta of $10^4 - 10^{-1}$, an Alfvén speed of 100 km s^{-1} and a magnetic Reynolds number of $10^8 - 10^{10}$, whereas the mangeto- sphere is collisionless and the corresponding geomagnetic values are $10^5 m^{-3}$, $10^{10} m$, $10^4 m$, 10^{-3} in the tail lobes (or 1 in the plasma sheet), 10^4 km s^{-1} in the tail lobes (or 10^3 km s^{-1} in the plasma sheet), and 10^{16} or more. Furthermore, in flares radiative and thermal cooling, gravity and efficient photospheric line-tying are important

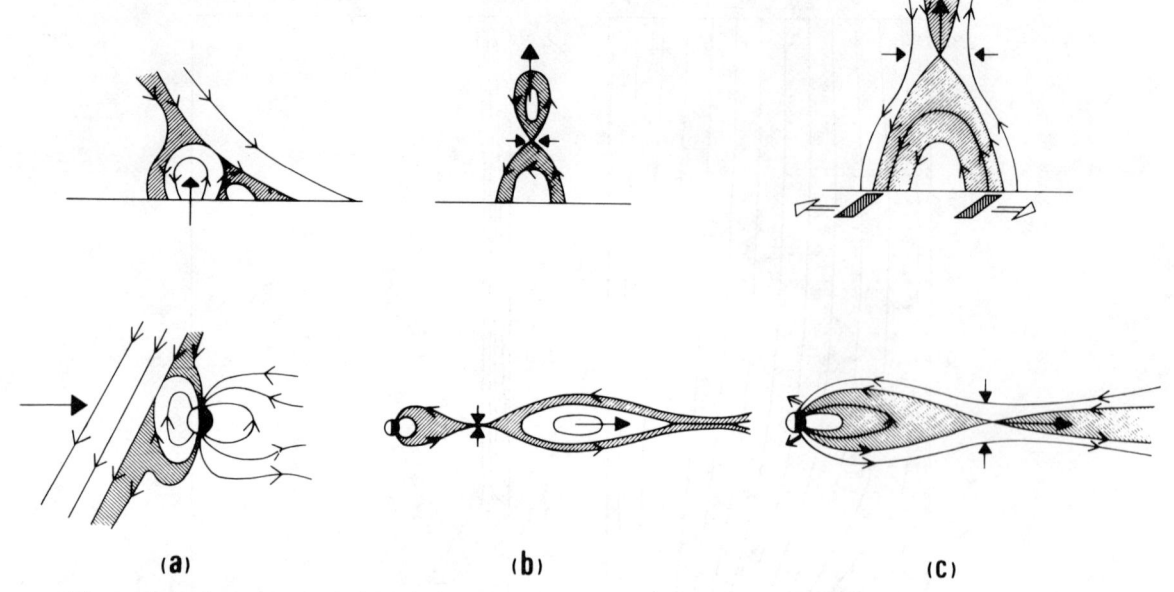

Fig. 15. Apparent similarities between solar flares (top) and the magnetosphere (bottom).

ingredients, whereas in the magnetosphere there are no such efficient cooling mechanisms, atmospheric ions are injected by parallel electric fields and the ionospheric line tying is only partial. Magnetic reconnection is certainly the key ingredient in both phenomena, but they are by no means identical and shed different light on its operation. Together they are allowing us to understand more of this beautiful process and of the intimate relation between the magnetic field and a cosmical plasma.

Acknowledgements. I am most grateful to the Los Alamos Laboratory for financial support, to Ed Hones for his hospitality and efficiency in organising the conference, and to T. Forbes, J. Heyvaerts and B. Roberts for invaluable discussions.

References

Athay, R.G. and White, O.R., Chromospheric and coronal heating by sound waves, Astrophys. J. 226, 1135, 1978.

Bateman, G. MHD instabilities, MIT Press, 1978.

Birn, J. and Schindler, K., Two-ribbon flares: magnetostatic equilibria, Ch.6 of Solar flare MHD (ed. E.R. Priest), 1981.

Bhattacharjee, A., Brunel, F. and Tajima, T., Magnetic reconnection driven by the coalescence instability, preprint, 1983.

Birn, J. and Hones, E.W., Three-dimensional computer modelling of dynamic reconnection in the geomagnetic tail, J. Geophys. Res. 86, 6802, 1981.

Biskamp, D., Dynamics of a resistive sheet pinch, Z. Naturforsch 37a, 840, 1982a.

Biskamp, D., Effect of secondary tearing instability on the coalescence of magnetic islands, Phys. Letters 87A, 357, 1982b.

Bobrova, N.A. and Syrovatsky, S.J., Dissipative instability of a one-dimensional force-free magnetic field, Sov. J. Plasma Phys. 6, 1, 1980.

Browning, P.K. and Priest, E.R., Kelvin-Helmholtz instability of a phase-mixed Alfvén wave, Astron. Astrophys., in press, 1984a.

Browning, P.K. and Priest, E.R., The magnetic non-equilibrium of buoyant flux tubes, Solar Phys., in press, 1984b.

Bulanov, S.V., Syrovatsky, S.I. and Sasarov, P.V., Stabilizing influence of plasma flow on dissipative tearing instability, JETP Lett. 28, 117, 1978.

Bulanov, S.V., Sakai, J. and Syrovatsky, S.I., Tearing-mode instability in approximately steady MHD configurations, Sov. J. Plasma Phys. 5, 157, 1979.

Cargill, P.J. and Priest, E.R., Slow-shock heating and the Kopp-Pneuman model for 'post'-flare loops, Solar Phys. 76, 357, 1982.

Chiuderi, C., Magnetic heating in the Sun, in R.M. Bonnet and A.K. Dupree (eds), Solar Phen. in Stars and Stellar Systems, D. Reidel, p.269, 1981.

Dobrowolny, M., Veltri, P. and Mangeney, A., Dissipative instabilities of magnetic neutral layers with velocity shear, preprint, 1983.

Finn, J.M. and Kaw, P.K., Coalescence instability of magnetic islands, Phys. Fluids 20, 72, 1977.

Forbes, T.G. and Priest, E.R., Numerical study of line-tied magnetic reconnection, Solar Phys. 81, 303, 1982a.

Forbes, T.G. and Priest, E.R., Neutral line motion due to reconnection in two-ribbon solar flares and magnetospheric substorms, Planet. Space Sci. 30, 1982b.

Forbes, T.G. and Priest, E.R., A numerical experiment relevant to line-tied reconnection in two-ribbon flares, Solar Phys. 84, 169, 1983.

Forbes, T.G. and Priest, E.R., A numerical simulation

of reconnection in an emerging magnetic flux region, Solar Phys., submitted, 1984.

Habbal, S.R., Leer, E. and Holzer, T.E., Heating of coronal loops by fast mode MHD waves, Solar Phys. 64, 287, 1979.

Hasegawa, A. and Chen, L., Plasma heating by Alfvén wave phase-mixing, Phys. Rev. Lett. 32, 454, 1974.

Heyvaerts, J. and Priest, E.R., Coronal heating by phase-mixed shear Alfvén waves, Astron. Astrophys. 117, 220, 1983.

Heyvaerts, J., Priest, E.R. and Chiuderi, C., Coronal heating by reconnection in D.C. current systems: a theory based on Taylor's hypothesis, Astron. Astrophys., in press, 1984.

Heyvaerts, J., Priest, E.R. and Rust, D.M., An emerging flux model for the solar flare phenomenon, Astrophys. J. 216, 123, 1977.

Hollweg, J.V., Mechanisms of energy supply, in F. Orrall (ed), Solar Active Regions, Colo. Ass. Univ. Press, 1981.

Hood, A.W. and Priest, E.R., Magnetic instability of coronal arcades as the origin of two-ribbon flares, Solar Phys. 66, 113, 1980.

Hoshino, M. and Nishida, A., Numerical simulation of the dayside reconnection, J. Geophys. Res. 88, 6926, 1983.

Ionson, J.A., Resonant absorption of Alfvénic surface waves and the heating of solar coronal loops, Astrophys. J. 226, 650, 1978.

Janiche, L., Resistive tearing mode in weakly two-dimensional neutral sheets, Phys. Fluids 23, 1843, 1980.

Kadomtsev, B.B., Fiz. Plasmy 1, 710, 1975.

Kippenhahn, R. and Schluter, A., Eine Theorie der solaren Filamente, Zs. Ap. 43, 36, 1957.

Kopp, R.A. and Pneuman, G.W., Magnetic reconnection in the corona and the loop prominence phenomenon, Solar Phys. 50, 85, 1976.

Kuperus, M. and Raadu, M.A., The support of prominences formed in neutral sheets, Astron. Astrophys. 31, 189, 1974.

Kuperus, M., Ionson, J.A. and Spicer, D.S., On the theory of coronal heating mechanisms, Ann. Rev. Astron. Astrophys., 19, 7, 1981.

Leroy, B., Propagation of waves in an atmosphere in the presence of a magnetic field II, Astron. Astrophys. 91, 136, 1980.

Leroy, J.L., Bommier, V. and Sahal-Brechot, S., The magnetic field in the prominences of the polar crown, Solar Phys. 83, 135, 1983.

Levy, R., Petschek, H. and Siscoe, G., Aerodynamic aspects of magnetospheric flow, AIAA J. 2, 2065, 1964.

Low, B.C., Nonlinear force-free magnetic fields, Rev. Geophys. Space Phys., 20, 145, 1982.

Malherbe, J.M. and Priest, E.R., Current sheet models for solar prominences I, Astron. Astrophys. 123, 80, 1983.

Malherbe, J.M., Priest, E.R., Forbes, T.G. and Heyvaerts, J., Current sheet models for solar prominences II, Astron. Astrophys. 127, 153, 1983a.

Malherbe, J.M., Schmieder, B., Ribes, E. and Mein, P., Dynamics in filaments II, Astron. Astrophys. 119, 197, 1983b.

Matthaeus, W.H. and Montgomery, D., Nonlinear evolution of the sheet pinch, J. Plasma Phys. 25, 11, 1981.

Mein, N. and Schmieder, B., Mechanical flux in the solar chromosphere III, Astron. Astrophys. 97, 310, 1981.

Milne, A. and Priest, E.R., Internal structure of reconnecting current sheets and the emerging flux model for solar flares, Solar Phys. 73, 152, 1981.

Moore, R. et al, in Solar flares (ed. P. Sturrock) p.341, 1980.

Nishikawa, K. and Sakai, J., Stabilizing effect of a normal magnetic field on the collisional tearing mode, Phys. Fluids, 25, 1384, 1982.

Nocera, L., Leroy, B. and Priest, E.R., Phase mixing of propagating Alfvén waves, IAU Symp. No.107, 1983.

Park, W., Monticello, D.A. and White, R.B., Two and three-dimensional reconnection in tokamaks, preprint, 1983.

Parker, E.N., Topological dissipation and the small-scale fields in turbulent gases, Astrophys. J. 174, 499, 1972.

Parker, E.N., Cosmical magnetic fields, Oxford University Press, 1979.

Parker, E.N., Dissipation of inhomogeneous magnetic fields, Astrophys. J., 265, 468, 1983.

Petschek, H.E., Magnetic field annihilation, AAS-NASA Symp. Solar Flares, NASA SP-50, p.425, 1964.

Pneuman, G.W., Two-ribbon flares: (post)-flare loops, in Solar flare MHD, (ed. E.R. Priest) p.379, 1981.

Priest, E.R., Solar flare MHD, Gordon and Breach, 1981.

Priest, E.R., Solar MHD, D. Reidel, 1982a.

Priest, E.R., Coronal loop structure and its heating mechanism, Commission 12 report of Rep. Astron., D. Reidel, 1982b.

Priest, E.R., Theories for simple-loop and two-ribbon solar flares, Fund. Cosmic Phys., 8, 1982c.

Priest, E.R., The magnetohydrodynamics of current sheets, Rep. Prog. Phys., in press, 1984.

Pritchett, P.L. and Wu, C.C., Coalescence of magnetic islands, Phys. Fluids, 22, 2140, 1979.

Rae, I.C. and Roberts, B., On MHD wave propagation in inhomogeneous plasmas and the mechanism of resonant absorption, Mon. Not. Roy. Astron. Soc., 201, 1171, 1982.

Roberts, B., Wave propagation in a magnetically structured atmosphere I & II, Solar Phys., 69, 39, 1981.

Roberts, B., Waves in magnetic structures, Physics of Sunspots, (ed. L. Cram & J.H. Thomas) p.369, 1982.

Roberts, B. and Mangeney, A., Solitons in solar magnetic flux tubes, Mon. Not. Roy. Astron. Soc., 178, 7P, 1982.

Rutherford, P.H., Nonlinear growth of the tearing mode, Phys. Fluids, 16, 1903, 1973.

Sato, T. and Hayashi, T., Externally driven magnetic reconnection and a powerful magnetic energy converter, Phys. Fluids, 22, 1189, 1979.

Sato, T. and Walker, R.J., Magnetotail dynamics

excited by the streaming tearing mode, J. Geophys. Res., 87, 7453, 1982.

Sonnerup, B.U.O., Magnetic field reconnexion in a highly conducting incompressible fluid, J. Plasma Phys., 4, 161, 1970.

Sonnerup, B.U.O., Magnetic field reconnection, Ch. III 1.2 of Solar System Plasma Physics (ed. L. Lanzerotti, C. Kennel and E.N. Parker), 1979.

Steinolfson, R.S., Energetics and the resistive tearing mode: effects of joule heating and radiation, Phys. Fluids, in press, 1983.

Steinolfson, R.S. and Van Hoven, G., The growth of the tearing mode: boundary and scaling effects, Phys. Fluids, 26, 117, 1983.

Sturrock, P., Solar flares, Colo. Ass. Univ. Press, Boulder, 1980.

Svestka, Z., Solar Flares, D. Reidel, 1976.

Tajima, T., Brunel, F. and Sakai, J., Loop coalescence in flares and coronal X-ray brightening, Ap. J., 258, L45, 1982.

Taylor, J.B., Relaxation of toroidal plasma and generation of reverse magnetic field, Phys. Rev. Lett., 33, 1139, 1974.

Ugai, M. and Tsuda, T., Magnetic field line reconnection by localized enhancement of resistivity, J. Plasma Phys., 17, 337, 1977.

Van Hoven, G., Simple-loop flares: magnetic instabilities, Ch.4 of Solar flare MHD, (ed. E.R. Priest), 1981.

Van Hoven, G., Steinolfson, R.S. and Tachi, T., Energy dynamics in stressed magnetic fields: the filamentation and flare instabilities, Ap. J. Letts., in press, 1983.

Vasyliunas, V.M., Theoretical models of magnetic field line merging I, Rev. Geophys. Space Phys., 13, 303, 1975.

Waddell, B.V., Carreras, B., Hicks, H.R., Holmes, J.A. and Lee, D.K., Mechanism for major disruptions in tokamaks, Phys. Rev. Letts., 41, 1386, 1978.

Wentzel, D.G., Hydromagnetic surface waves on cylindrical flux tubes, Astron. Astrophys., 76, 20, 1979.

Yang, C.K. and Sonnerup, B.U.O., Compressible magnetic field reconnection; a slow wave model, Astrophys. J., 206, 570, 1976.

Zweibel, E., Thermal stability of a corona heated by fast mode waves, Solar Phys., 66, 305, 1980.

Questions and Answers

Birn: Looking at your simulation pictures it seemed to me that the earlier upward motion of the plasmoid is related to diverging field lines whereas the later downward motion of neutral lines is related to converging field lines. Is that impression correct?

Priest: Yes, I agree. The motion of the neutral lines and plasmoids is very sensitive to both the boundary conditions and the particular initial state that is adopted. We would very much like to repeat the line-tied experiment in the future with different initial states, particularly two dimensional ones that represent a stretched out arcade, similar to the one you have adopted yourself.

Moore: Based on your nice results for reconnection in flares, do you have any clues or suggestions for how the large production of energetic ($\gtrsim 10$ keV) electrons is accomplished in flares?

Priest: The prime candidates for high energy particle acceleration in flares are the MHD shock waves which arise naturally in the MHD simulations. These include the slow mode Petschek shocks and the fast mode shock below the reconnection site that I have described in the numerical simulations of line-tied reconnection and which move very rapidly and are strong during the early stages of the process. Also, the fast shock ahead of the erupting filament may play a role. For an excellent review of particle acceleration I can recommend the chapter by Heyvaerts in the book "Solar Flare MHD." Also the recent work of M. Lee (Durham, New Hampshire) looks most promising.

Speiser: You talk about erupting prominences along with flares. Do you imply they are in fact the same? (One to one?)

Priest: There are two types of filaments (or prominences), namely large quiescent ones far away from active regions and much smaller active-region (or plage) filaments. When an active-region filament erupts, it gives a large solar flare with an x-ray brightening and two ribbons of emission down in the chromosphere. When a quiescent filament erupts it gives an x-ray brightening but usually no chromospheric ribbons, presumably because its magnetic field is much less than that of an active-region filament and so much less energy is released. However, the basic magnetic instability in both types may well be the same, and so we can probably learn a lot about two ribbon flares by studying these erupting quiescent prominences which are much larger and lie within much simpler magnetic field configurations.

MAGNETIC RECONNECTION IN COMETS

Malcolm B. Niedner, Jr.

Laboratory for Astronomy and Solar Physics
NASA Goddard Space Flight Center, Greenbelt, MD 20771

Abstract. We are at a point today where many of the traditionally-puzzling phenomena in the cometary plasma-tail environment can plausibly be linked to magnetic reconnection occurring in several regions of a comet [Niedner and Brandt, 1978, 1980]. The turn-on of these various reconnection sites appears to follow a cyclic pattern in which the plasma-tail disconnection event (DE) is the primary feature and the periodic sector structure of the solar wind is the external driver. The purpose of this review is to discuss these different classes of cometary activity, to state the justifications for linking them to reconnection, to discuss proposed alternate (non-reconnection) models, and to suggest future tests of the hypotheses presented.

Introduction

Although no spacecraft has yet to visit a comet, the study of magnetic fields in these objects enjoys one distinct advantage over the study of planetary magnetospheres and magnetotails: the global magnetic structure of comets is made visible by ultraviolet fluorescence radiation emitted by ions (principally CO^+) trapped on field lines. Whereas the study of Earth's magnetosphere and magnetotail often depends critically on a spacecraft's timing and position, and on the presence of other spacecraft to provide multi-point observations, a comet's entire magnetic tail may often be recorded in a single wide-field photograph. When they can be obtained, frequently-spaced images then provide the means to study the dynamics of a diverse array of constantly-evolving structures in the tail.

A rich heritage extending back to the 1890's exists in cometary photography. Many images have been published; most are stored in archival observatory plate vaults. Even the oldest images are extremely useful today for studies both of cometary structure and of the solar wind (using comets as "natural probes"). Although this material has been available for analysis for many decades, only in the last 5-6 years has a systematic attack been made on the body of data as a whole to recognize and understand the different classes of cometary plasma activity, their solar-wind associations, and their physical causes. The present review examines observational evidence concerning magnetic reconnection in comets.

The General Comet/Solar-Wind Interaction

Because of the large amounts of gas sublimated off its small, several-km wide icy nucleus, a bright comet with a gas production rate of $10^{29} < Q < 10^{30}$ molecules s^{-1} is a very extended obstacle to the solar wind. The $\geq 10^6$ s photoionization timescales of species such as H, OH, and CO, combined with mean outflow speeds ≥ 1 km s^{-1}, results in mass loading of the solar wind at distances $>10^6$ km upstream of the nucleus. A weak M $\sim$ 2 shock is predicted to form at distances $\geq 2 \times 10^5$ km upstream for bright comets [Brosowski and Wegmann, 1972; Wallis, 1973; Schmidt and Wegmann, 1980, 1982]. Figure 1, from Brandt and Mendis [1979], shows the shock and other cometary features discussed below.

Downstream of the shock, the heated flow continues to decelerate, with a resultant compression of the interplanetary magnetic field (IMF). If plentiful enough, cometary ions created close to the nucleus will have sufficient streaming pressure $\varrho_i V_i^2$ to balance the streaming pressure of the incoming solar wind. At the point of balance, a contact surface (or tangential discontinuity) forms which separates shocked and contaminated solar wind from purely cometary gas. If an inner shock exists [Houpis and Mendis, 1980], then the solar wind is balanced by the thermal pressure of ionosphere ions. Houpis and Mendis [1981] have discussed the role in pressure balance played by the neutrals, which are collisionally coupled to the ions in the inner coma. For medium-bright comets (e.g., Halley's Comet), the subsolar contact surface distance is presumed to be $10^3 - 10^4$ km from the nucleus [Brandt and Mendis, 1979; Houpis and Mendis, 1981]. Ion densities near the contact surface should be several x 10^3 cm^{-3}.

The IMF becomes draped over the contact surface with a field strength presumably dictated by the solar-wind dynamic pressure; B $\sim$ 50−100γ is expected at 1 AU. The extent to which the ionosphere is field-free will depend on the stability of the contact surface

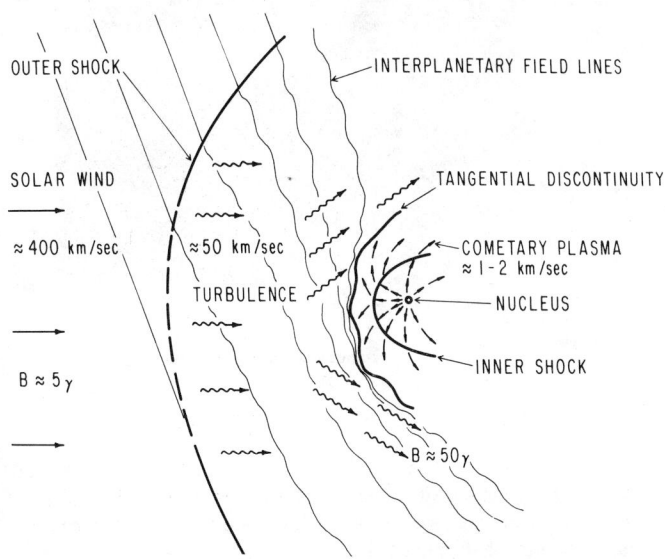

Fig. 1. Schematic of general structure in the sunward comet/solar-wind interaction region.

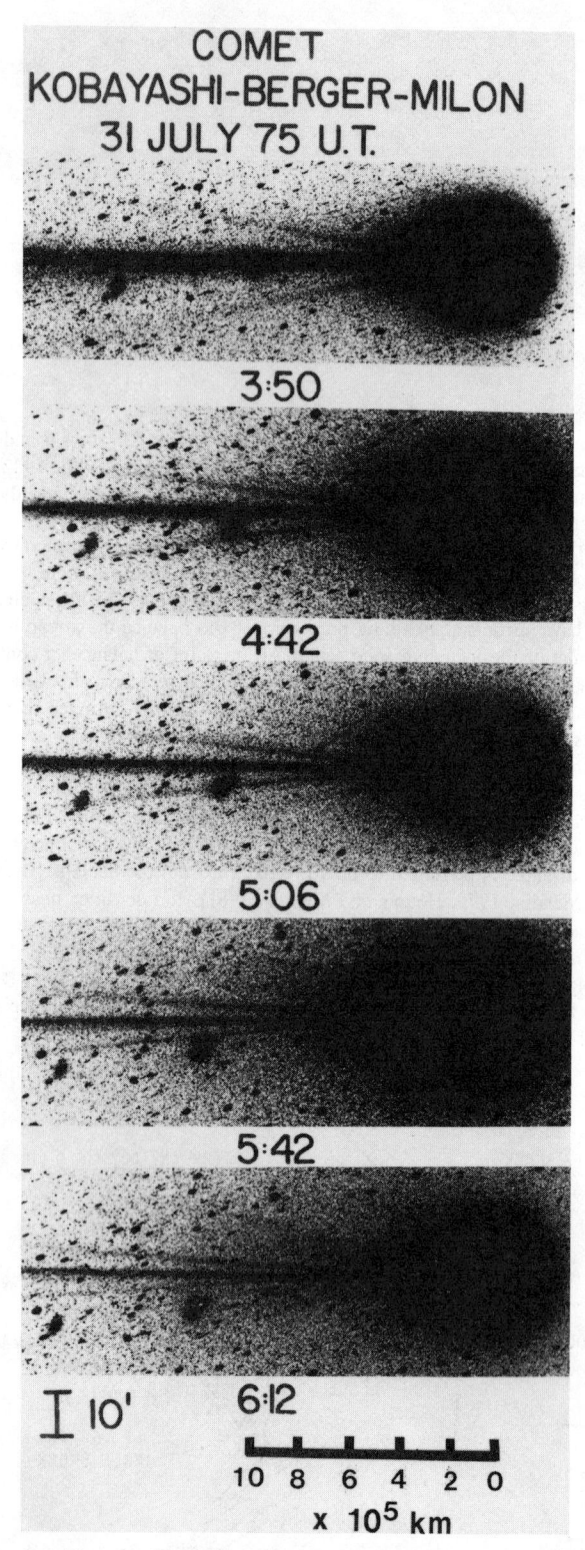

Fig. 2. Photographic time sequence showing the folding of a tail ray pair into the main tail of comet Kobayashi-Berger-Milon 1975 IX. Times below the photographs are in UT (Joint Observatory for Cometary Research photographs).

[Ershkovich and Mendis, 1983]. The properties and even the existence of the contact surface have been debated for years [Wallis and Ong, 1976; cf. Schmidt and Wegmann, 1982; Ip and Axford, 1982]. The long lifetime ($\geq$ 1 day) of tail rays and streamers argues for the IMF becoming effectively hung-up in the head, however. The contact surface in comets Bennett 1970 II and West 1976 VI may have been observed spectroscopically by Delsemme and Combi [1979] and Combi and Delsemme [1980].

A consequence of field capture is the creation of a magnetic tail due to the continued propagation of solar wind on either side of the comet. In this model, first advocated by Alfvén [1957], the tail consists of two oppositely-polarized magnetic lobes channeling CO^+ and other cometary ions created in the head; the tail lobes are expected to be separated by a neutral sheet. Alfvén's concept of a plasma tail was 2-dimensional, but is readily adaptable to a 3-D IMF: while the capture plane of a field line is essentially its original plane [Schmidt and Wegmann, 1982], the tail is made up of flux spanning a range of initial planes. Figure 2 is a photographic time sequence of comet 1975 IX which shows the folding of a tail-ray pair into the tail axis; tail rays have folding timescales of 0.5-1.0 days and are interpreted as IMF capture [cf. Brandt and Mendis, 1979; Brandt, 1982].

For additional details about the cometary plasma environment, consult the reviews by Mendis and Ip [1977] and Brandt and Mendis [1979].

Plasma-Tail Morphology and Activity Cycles

Disconnection Events (DE's)

Disconnection Events (or "DE's") are characterized by the entire plasma tail being uprooted from the cometary head, convected away in the solar wind, and replaced by a new tail constructed from folding tail rays. DE's are arguably the most spectacular phenomenon exhibited by comets, and they constitute the observational foundation of reconnection studies in these objects.

Figure 3 shows a highly visible DE which occurred in comet Borrelly 1903 IV on 1903 July 24 [Barnard, 1903]. The detached tail is separated from an attached tail and the near end of the former is well-defined at a downstream distance of 2.7 x 10^6 km. A 16-hour sequence of Halley's Comet in 1910 is shown in Figure 4. Note the recession of a very conspicuous disconnected tail and the lengthening and strengthening of the attached tail through the sequence; the mean speed of recession during the 16-hour period was 57 km s^{-1} [Niedner, 1981]. Although the sequence in Figure 4 does not definitively locate the spatial origin of the disconnection, examination of the photograph in Figure 5 shows that the seat of the disturbance was in the head. The photograph was

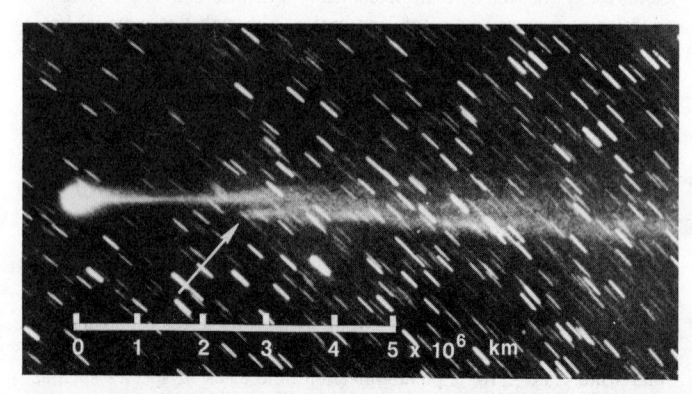

Fig. 3. Wide-field photograph showing a disconnection event (DE) in comet Borrelly 1903 IV on 1903 July 24 (Yerkes Observatory photograph).

shows that the ray pair, which is still considerably inclined, has become more prominent and it encloses a disconnected tail, the near end of which is indicated by the arrow at a downstream distance of 3×10^6 km. Since the tail in Figure 6a is still attached (or else just detached), a lower limit of ~ 125 km s^{-1} can be put on the recession speed during the 7 hr interval [Niedner and Brandt, 1980; Niedner, 1981]. It is important to note in Figure 6b that there is no ion structure rooted into the head besides the folding rail rays, i.e., no tail proper. Hence, the head of the comet is where the disconnection occurred, and in virtually every DE where photographic coverage is complete enough to allow a judgement, this is the case [Niedner, 1981].

The presumption in the reconnection model of Niedner and Brandt [1978] is that events major enough to be classified as "DE's" originate in the head. This is not to deny the existence of small, detached tail streamers in the middle of an otherwise attached and undisturbed tail. These relatively unspectacular structures (an example of which will be discussed later) may be caused by reconnection in the tail or by "leakage processes" in the head such as Venusian magnetic flux ropes [Russell and Elphic, 1979].

The longevity of DE's is shown in Figure 7, a 3-day sequence of comet Morehouse 1908 III in which consecutive images are spaced by 24 hours. In the first photograph, the tail is attached although flaring and disturbed [Barnard, 1908a]. In the second, the tail is disconnected and a new, weak attached tail has taken its place. The disconnected tail is still visible in the third photograph although very diffuse; note the growth of the new tail between the second and third images.

A summary of the basic DE sequence of events is as follows. In the pre-disconnection phase, which lasts some 12-24 hours (18 hours was adopted by Niedner et al. [1981] on the basis of pre-disconnection tail- and tail-ray activity, and was supported by Niedner [1982]), the comet becomes very active with the generation of prominent tail rays (Figures 5 and 6a), and the tail often brightens and narrows down (i.e., becomes wedge-shaped; Figure 5). In the immediate post-disconnection phase, the detached tail is often clearly visible in the space between the still widely-inclined tail rays (Figure 6b). The detached tail continues its recession from the head and the rays close to define a new tail (Figures 4 and 7). Not all features are observed in all DE's, due both to often poor observational coverage and to differences between events. The disconnection of a plasma tail is, however, always accompanied by the construction of a new one.

DE's are not rare. From a search of the literature and of archival observatory photographs, Niedner [1981] has catalogued 72 DE's in 29 comets during the years 1892-1976; comet Morehouse 1908 III had 9 DE's, followed by Halley's Comet with 5 events in 1910. DE's occur over wide ranges in heliocentric distance ($r = 0.47 - 3.1$ AU), heliographic latitude ($b = -61° - +53°$), and solar cycle phase; refer to Figure 1 of Niedner [1982]. The speeds of recession measured for those of the 72 events which had multiple observations are typically in the 50-100 km s^{-1} range [Niedner, 1981].

Post-DE "Arcades" in the New Plasma Tail

The best known example of "arcades" occurred in comet Kohoutek. The photograph in Figure 8a was taken 24 hours after that in Figure 6b, and shows the result of the two rays of Figure 6b coming together to form a new tail following the tail disconnection. The new tail is heavily structured with cross-tail "arcades" (indicated by the bracket), the convex sides of which point toward the head region (to the right). Also, there is a gap in the tail (arrow) which is virtually structureless. For additional details of these features, see Niedner and Brandt [1980].

Figure 8b shows a computer-enhanced enlargement of a photograph taken 17 hours after that in Figure 8a. The two left-most arrows point to cross-tail structures which resemble the earlier arcades, although they are much more diffuse and less conspicuous. The right-most

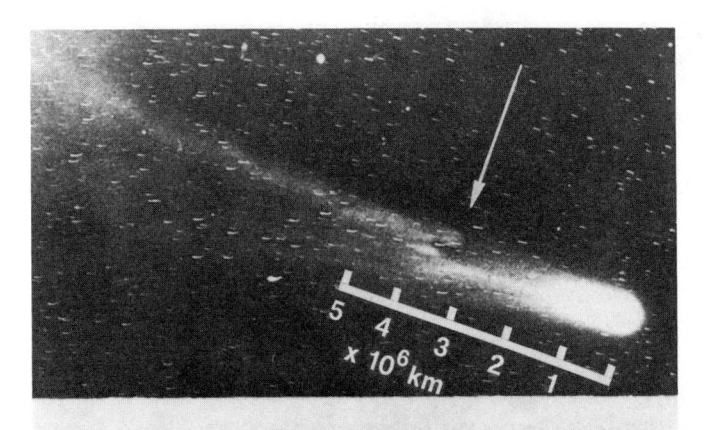

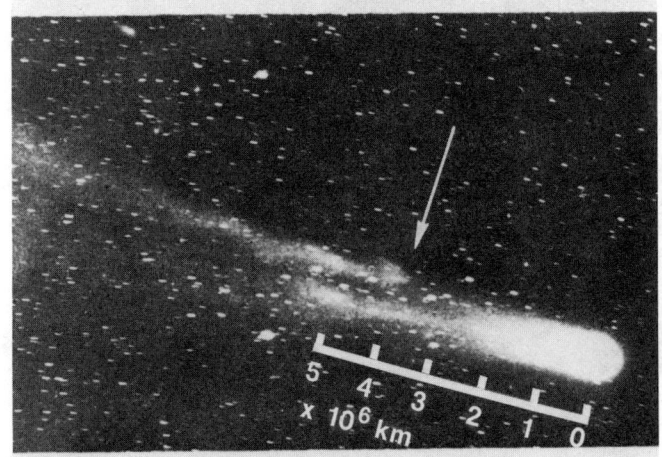

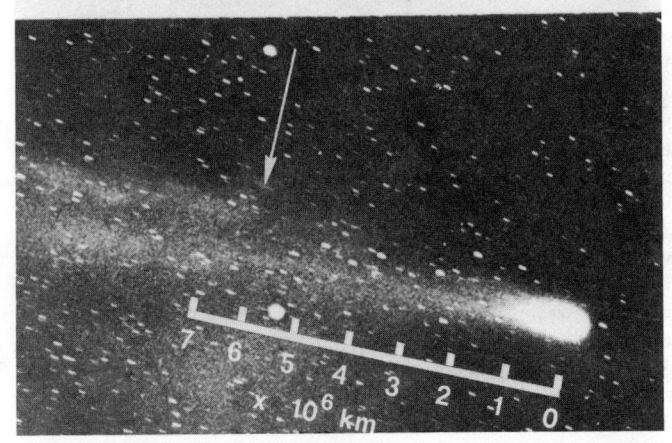

Fig. 4. 16-hour time sequence of Halley's Comet showing the recession of a disconnected plasma tail and the growth of a new attached tail. Sequence photographs were taken (top to bottom) 1910 June 6.66, 6.84, and 7.29 GMT (Yerkes Observatory photographs).

taken ~ 22 hours before the first image in Figure 4; note the strongly wedge-shaped tail which is attached and flanked by a pair of widely-inclined tail rays. The combination of a flaring (i.e., wedge-shaped) inner tail and prominent tail rays is a precursor to tail disconnection which is frequently seen in DE's [Niedner and Brandt, 1979].

Another example of plasma-tail behavior immediately before and after a disconnection is provided by Figures 6a and 6b. In Figure 6a, an attached tail flanked by a ray pair is seen in comet Kohoutek 1973 XII. Only ~ 7 hours later, the photograph in Figure 6b was taken; it

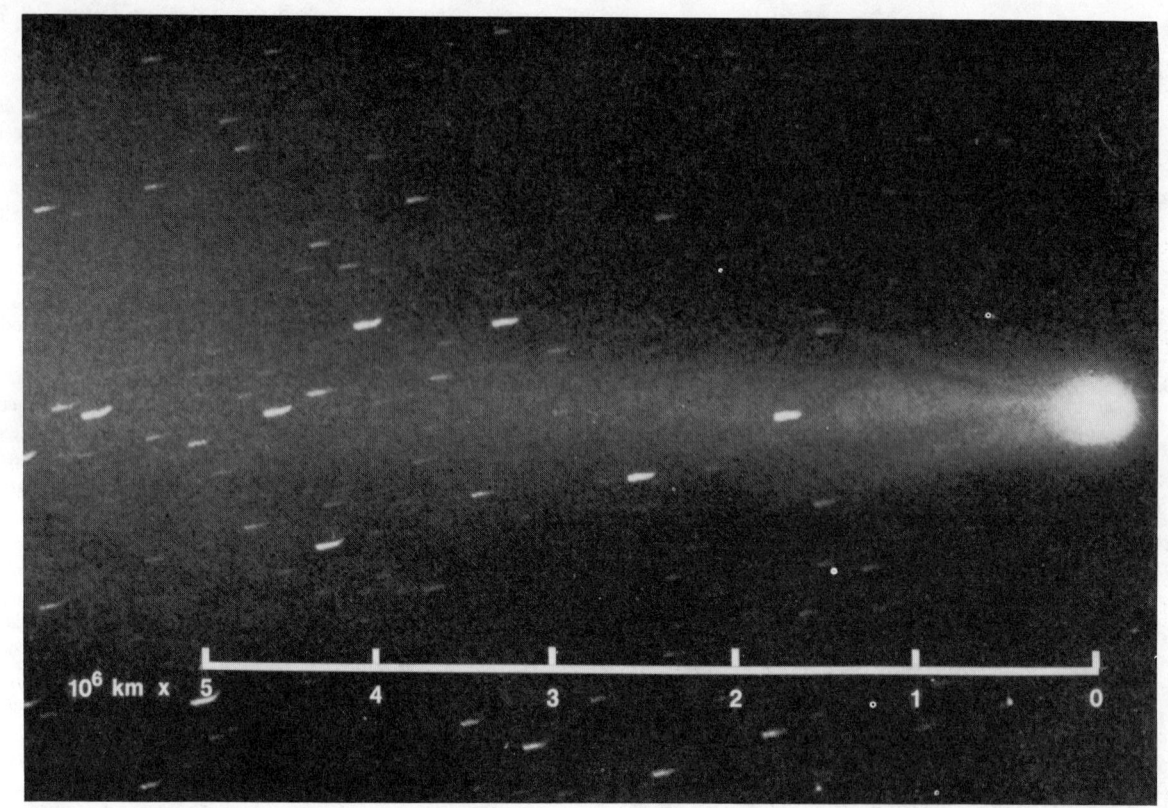

Fig. 5. Lick Observatory photograph showing Halley's Comet on 1910 June 5.746 GMT, ∼22 hours before the first image in Figure 4. The tail is attached, although strongly wedge-shaped (flaring), and is flanked by a pair of prominent tail rays.

arrow indicates a gap structure which could be a later manifestation of the tail gap shown in Figure 8a.

As discussed by Niedner and Brandt [1980], the arcade phenomenon is closely related to DE's when it is seen (e.g., the comet Kohoutek example discussed above), but it is observed much less frequently than DE's. Either arcades rarely occur or selection effects hinder their observation.

Loops Between Colliding Plasma Tails

As in the case of tail arcades discussed above, there is in this category a well-defined "best example": it occurred in comet Bennett 1970 II on 1970 April 4 and is shown in Figure 9. Note the large-scale loop joining the disconnected end of a rejected tail with one or more of the narrow streamers making up a fledging new tail. The structureless strong emission near the head is a strong dust tail which partially masks the visibility of this event. Nonetheless, from imagery obtained earlier, before the loop formed, the event was classified as a "disconnection" by Wurm and Mammano [1972]. As in the case of tail arcades, the frequency of spectacular loops and bridges between rejected and attached tails is much less than the frequency of DE's.

A schematic summary of the sequence of events during a DE cycle is shown in Figure 10. As indicated earlier, well-observed comets often are seen to undergo this cycle several times.

A Reconnection Model

Construction of a model of the structures discussed above requires knowledge of what solar-wind feature (if any) tends to be associated with DE sequences. This exercise has been carried out by Niedner and Brandt [1978, 1979], and by Niedner [1982], using *in situ* solar-wind data (1960's onward, with gaps), aa geomagnetic indices (1868 onward, continuous), and geomagnetically-inferred interplanetary polarities (1926 onward, continuous). The procedure uses a corotation technique which shifts solar-wind or geomagnetic data from the Earth to a comet. The net result (consult cited references for details) for post-1926 DE's is that a close correlation exists with high-speed streams on whose leading edges sector boundaries are found. For pre-1926 DE's, the correlation is restricted to streams (polarity data are lacking). This latter result is still consistent with a sector boundary correlation, however, because of the location of sector boundaries on the leading edges of corotating streams [but not all streams; Gosling et al., 1976]. Niedner and Brandt [1979] suggested on statistical grounds that it is the sector boundary, and not the high-speed stream, with which DE's are more strongly correlated. Sector boundary crossings are the foundation of the reconnection model of DE's developed by Niedner and Brandt [1978], which is discussed below (crossings of local, small-scale polarity reversals can explain DE's not obviously associated with any particular sector boundary).

Figure 11 illustrates the sector boundary model; the IMF is for simplicity assumed to be perpendicular to the direction of solar-wind flow. Panel A shows the capture of magnetic flux within a sector of uniform polarity; this is basically Alfvén's model. After sector boundary crossing (panel B), reconnection occurs on the dayside between previously captured fields and the first fields past the sector boundary, which are of opposite polarity. The reconnected fields are not bound to the comet and after leaving the reconnection site they are pushed back into the tail. Because reconnection starts at the "top" of the captured fields draped over the ionosphere and works itself toward the contact surface, reconnection could be expected to bring a narrowing of the

COMET KOHOUTEK ON 1974 JANUARY 9

A $2^h\ 18^m\ 30^s$ UT (Joint Observatory for Cometary Research)

B $8^h\ 59^m\ 30^s$ UT (Tokyo Astronomical Observatory)

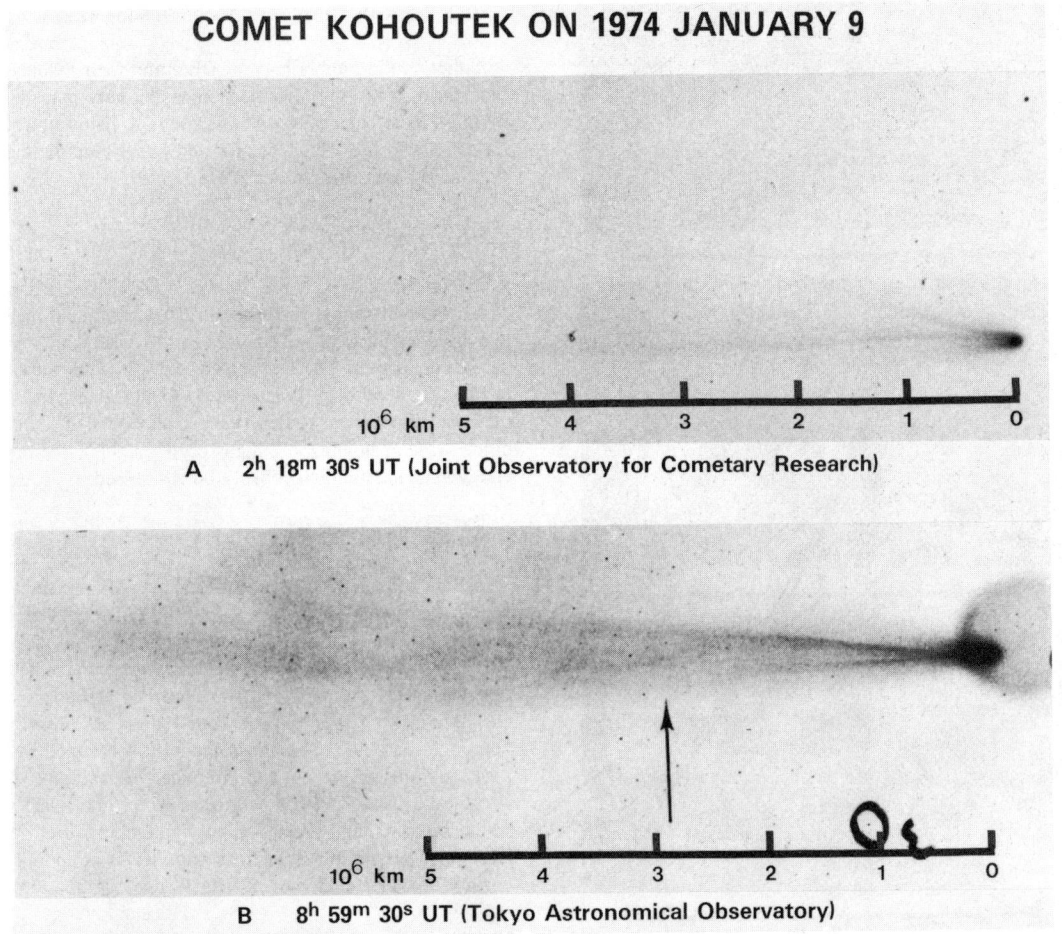

Fig. 6. a) Comet Kohoutek 1973 XII on 1974 January 9, $2^h18^m30^s$ UT (midexposure). The main tail is flanked by a pair of tail rays and is about to disconnect (Joint Observatory for Cometary Research photograph). b) Comet Kohoutek 1973 XII on 1974 January 9, $8^h59^m30^s$ UT (midexposure). The tail of ~7 hours earlier (Fig. 6a) has disconnected and is in the space between two strong tail rays (Tokyo Astronomical Observatory photograph).

inner plasma tail. A variety of circumstantial evidence (see earlier discussion) suggests that the duration of reconnection is about 18 hours, after which all of the draped flux has been reconnected. At that point the tail has been disconnected (panel C). Capture of flux from the new sector, which has been occurring since the last of the reconnection phase (panel B), results in the formation, first, of prominent tail rays, and second—when these tail rays have closed—a new plasma tail (panel D). The comet then awaits the next sector boundary crossing.

Even its qualitative form this model has some appealing features. Most importantly, it reproduces the observed morphology of an entire DE sequence: tail narrowing accompanied by strong tail rays (Figures 5, 6a), tail disconnection and folding of tail rays (Figure 6b), and generation of a new tail via ray closure (Figures 4, 7). Moreover, it suggests that the reason some comets have five or more DE's is because over the course of several months of observation, a comet could cross that many sector boundaries. The possible role of reconnection in pre-disconnection brightenings of the plasma tail has been considered by Niedner [1980]. The arcades observed between tail rays closing to define a new plasma tail (Figure 8) are interpretable as flux tubes reconnected across the tail between the oppositely-polarized rays. The comet Bennett loop (Figure 9) also has a reconnection interpretation as the result of oppositely-polarized magnetic elements in the two tails coming into contact.

Examination of the Dayside Reconnection Process

A more quantitative treatment of possible dayside reconnection has been performed by Niedner et al. [1981]. Observations were used as much as possible to estimate the size and shape of the diffusion region, the reconnection speed, and the role of anomalous transport processes. Probably the most uncertain parameter was the width of the magnetic barrier at the onset of reconnection. 10^4 km was assumed for two reasons: first, tail rays near the head have characteristic widths of $\sim 10^3$ km and several rays are often visible at one time; second, the width of the barrier seemed unlikely to be greater than the contact surface distance, which should be $\sim 10^4$ km for a comet in Halley's brightness range. Using a reconnection duration of $\tau_{rec} = 0.75$ days, the mean inflow speed came to 0.15 km s^{-1}. The outflow speed was assumed to be the Alfvén speed measured in the inflow region [Vasyliunas, 1975]; the assumed magnetic field of 100γ and ion density of 5000 then yielded $V_{out} = 4.5$ km s^{-1}. The merging speed was thus estimated as $M_A = V_{in}/V_A \sim 0.04$.

Application of standard equations of reconnection theory showed that the assumed value of τ_{rec} and the quasi-derived value of M_A did not unambiguously set either the diffusion region dimensions x^*-z^* or the resistivity η. It was suggested, however, that the diffusion region thickness set by classical resistivity was much lower than the "critical"

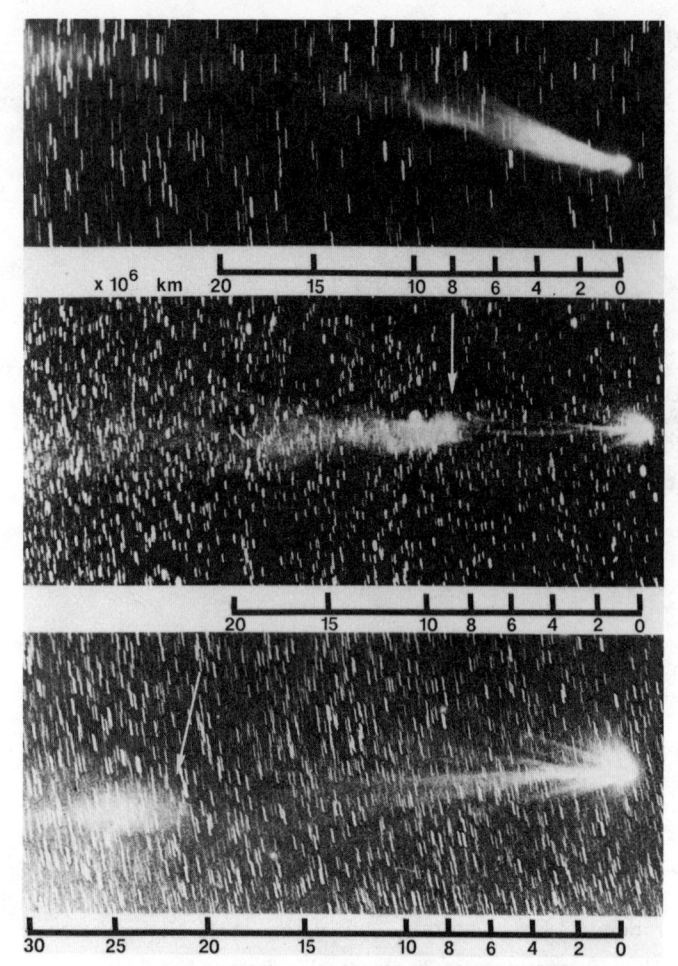

Fig. 7. Three-day sequence of comet Morehouse 1908 III, 1908 September 30-October 2. The photographs show the disconnection and recession of the comet's plasma tail, and the growth of a new attached tail (Yerkes Observatory photographs).

value set by marginal stability criteria for both the ion-acoustic and lower hybrid drift instabilities. Exactly what values x^*, z^*, and η achieve in cometary reconnection could not, however, be determined.

Niedner et al. [1981] suggested that tearing mode instabilities operate in at least some DE's. Because it is a macroscopic process, break-up of the reconnecting current sheet into filaments and islands could load the tail with observable non-uniformities prior to the actual tail disconnection. Some 12 hours before one of its tail disconnections, comet Morehouse's innermost tail became distinctly brighter and loaded with clumps and condensations [Barnard, 1908b]. The clumpiness Barnard referred to cannot be reproduced well in a photographic print, but the situation is shown schematically in Figure 12.

Fedder et al. [1981, 1983] used a 3-D MHD computer code to study the solar-wind interaction with comets and derived some interesting results of relevance to the proposed dayside reconnection process. They found that 180° rotations in the magnetic polarity (i.e., sector boundaries) produce tail rays and a depletion in the inner plasma tail density which they interpret as a possible disconnection event. Hence, Fedder et al. are in agreement with Niedner and Brandt [1978] that DE's are likely produced by cometary crossings of magnetic field polarity reversals.

Alternative (Non-Reconnection) Models of DE's

Although the morphology of DE's and their association with sector boundaries make the reconnection model very plausible, other models of DE's have been proposed and they are listed in Table 1. Each is discussed briefly below, but for additional remarks the reader should consult Niedner and Brandt [1978, 1979].

Internal Model

The model of Wurm and Mammano [1972] was based on the DE in comet Bennett shown in Figure 9. They attributed the tail disconnection to a cessation of ion sources in the head which fed the tail before the disconnection. The stoppage of ion production was presumed due to physical conditions (unspecified) totally internal to the comet. Given the correlations now known to exist between DE's and solar-wind conditions (sector boundaries, high-speed streams), the model of Wurm and Mammano should probably be reassessed.

Shock Interaction Model

Jockers and Lüst [1973] examined two comet Bennett events, one of which was the DE in Figure 9, but did not recognize them as disconnected tails. They attributed the events to the formation of secondary tails in response to the passage of flare-generated interplanetary shocks. For one event, a flare shock was observed at Earth which had a plausible association with comet Bennett; for the other one, no such association could be made. The shock model is weakened by Jockers and Lüst's non-recognition of the disconnection nature of the events and also by the fact that DE's as a class tend to correlate with sector boundaries and high-speed streams, not shocks. Both comet Bennett DE's have plausible associations with sector boundaries observed at Earth [Niedner and Brandt, 1978; Niedner, 1982].

High-Speed Stream/Flute Instability Model

Ip and Mendis [1978], whose model invokes an interruption of ion production as a result of the contact surface becoming unstable to the flute instability, felt that the DE reconnection model of Niedner and Brandt placed sector structure at unrealistically high solar latitudes (>30°). In a study by Niedner [1982], the view was advocated that sector structure can extend up to latitudes of 45° and higher for much of a solar cycle, and hence the latitude distribution of DE's does not contradict the sector boundary model. Central to this argument was a re-examination of the Rosenberg-Coleman predominant polarity effect with latitude (Rosenberg and Coleman, 1969), and specifically, the degree to which it controls the tilt and latitude extent of all sector boundaries (or current sheet warps) at any given time. In short, the predominant polarity effect need not effect all sectors equally, which

TABLE 1. Proposed DE Models

Internal Cessation of Ion Tail Sources [Wurm and Mammano, 1972]

Interplanetary Shock [Jockers and Lüst, 1973]

Sector Boundary/Magnetic Reconnection [Niedner and Brandt, 1978, 1979]

High-Speed Stream/Flute Instability [Ip and Mendis, 1978]

High-Speed Stream/Differential Acceleration Mechanism [Jockers, 1981]

COMET KOHOUTEK ON 1974 JANUARY 10–11

A 9^h 05^m UT, January 10 (Tokyo Astronomical Observatory)

B 2^h 42^m UT, January 11 (Joint Observatory for Cometary Research)

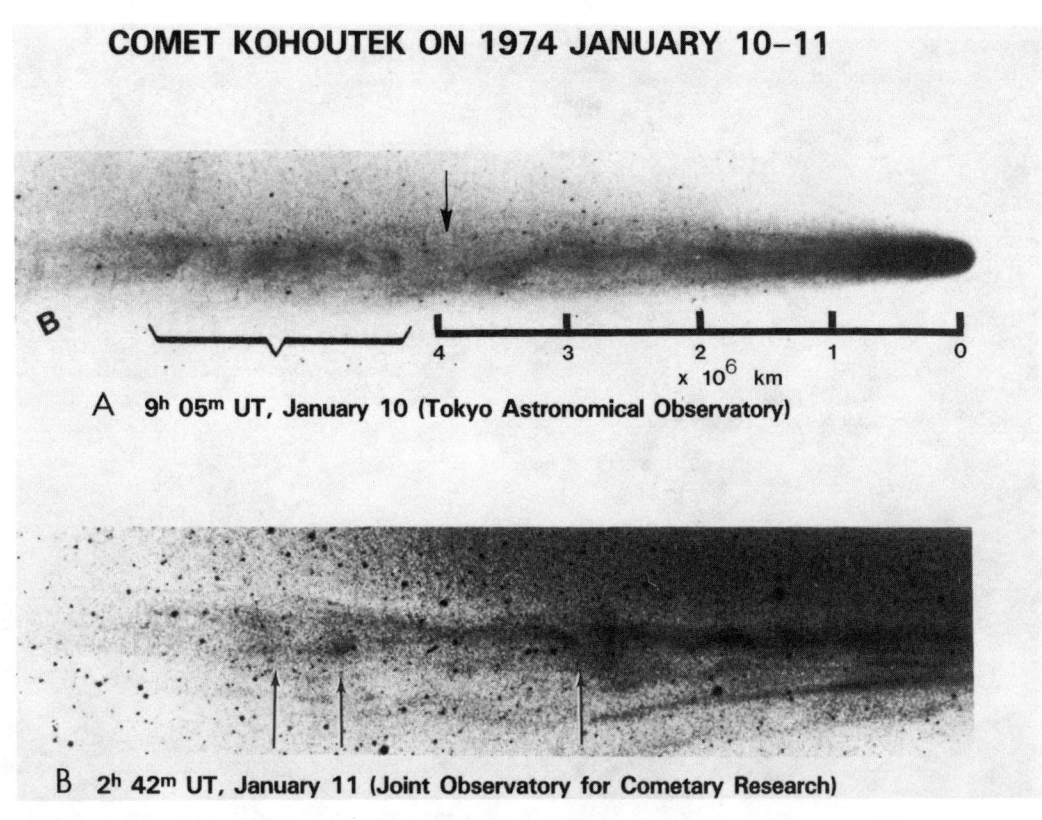

Fig. 8. a) Photograph taken 24 hours after that in Fig. 6b, showing arcades in comet Kohoutek on 1974 January 10 (Tokyo Astronomical Observatory photograph). b) Computer-enhanced image showing a possible later phase of the arcades seen 17 hours earlier on January 10 (Fig. 8a) (Joint Observatory for Cometary Research photograph).

vitiates, *ab initio*, attempts to determine the latitudinal extent of sector structure by extrapolation of the low-latitude Rosenberg-Coleman effect. Niedner [1982] presented a phenomenological model based on coronal hole evolution throughout the solar cycle which can produce a measureable Rosenberg-Coleman effect while at the same time allowing some steeply tilted sector boundary surfaces to extend to high latitudes to produce the DE's. The subsequent work of Hoeksema et al. [1982, 1983] using computed coronal magnetic fields appears to confirm the hypothesis of sector structure at high latitudes ($\geq 45°$) for much of the solar cycle.

Niedner and Brandt [1979] pointed out that the Ip-Mendis model does not account for the morphology of DE's as explicitly as the reconnection model does, nor does it explain why DE's correlate more closely with sector boundaries than with high-speed streams.

High-Speed Stream/Differential Acceleration Model

The recent model of Jockers [1981] invokes high-speed streams and a "differential acceleration effect". It is not obvious how the mechanism produces the disconnection morphology and in any event, it has the same problem with the higher correlation of DE's with sector boundaries (vs. high-speed streams) as that mentioned above in connection with the Ip-Mendis model.

Reconnection Processes in Attached Plasma Tails

All of the examples of cometary reconnection presented so far have dealt with disturbed conditions in which DE's are the first and major feature in time sequences lasting 1-3 days. A natural question concerns the cometary analog of the geomagnetic substorm, and specifically, tail plasmoids [Hones, 1977, 1979]. Ip and Mendis [1975, 1976] considered the stability of the cross-tail current sheet in comets. In their substorm or "auroral" model, the sheet occasionally becomes unstable with a resultant field-aligned current discharge into the head. Although Ip and Mendis' emphasis was not on the formation of observable tail features, but rather on ionizing properties of discharging currents, Morrison and Mendis [1978] returned to the question of current sheet instabilities and their effect on tail structure. They showed several examples of fine-scale tail structure in comet Kohoutek (kinks, waves) and attributed them to the tearing and rippling mode instabilities.

Are there observed cases where a quiescent, attached tail reconnects across its neutral sheet with the ejection of a plasmoid? The answer is a guarded "yes", but the phenomenon may not be very spectacular. A good example occurred in comet West 1976 VI and is shown in Figure 13. The plasma tail is quiet and made up of the usual streamer structure, but there is a broken section in the tail as indicated by the arrow. Possibly reconnection or tearing in the tail isolated this bundle of magnetic flux, but it is hard to be sure because imagery is not available in the hours preceding this photograph.

Present Limitations and Prospects for the Future

The weak points in our understanding of cometary magnetic reconnection specifically, and of cometary plasma phenomena generally, are fairly obvious. First, no space probe has ever visited a comet and our remote methods for inferring physical conditions are not perfect by any means. Second, temporal coverage of past comets has been poor. Third, crucial plasma structures in the head, such as the contact surface, have not been seen with certainty. Especially troubling is that for the proposed dayside reconnection which produces disconnected tails,

86   MAGNETIC RECONNECTION IN COMETS

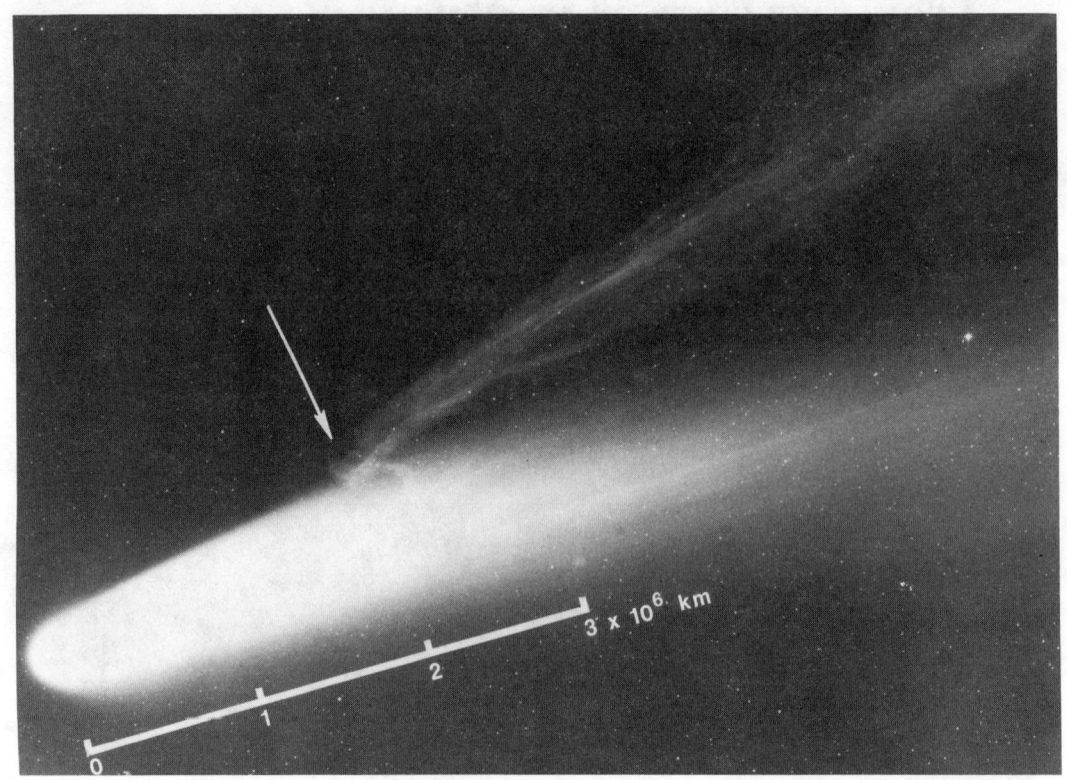

Fig. 9. Spectacular loop structure which formed in comet Bennett 1970 II on 1970 April 4, apparently as a result of the collision of a disconnected tail and the young new tail which formed soon after the disconnection (Hamburg Observatory photograph).

PLASMA TAILS: A MORPHOLOGICAL SEQUENCE

Phase I:
Narrowing tail ("Streaming")
"Condensations" in tail (Sometimes)
Strong ray system

Phase II:
Disconnection of Tail
Helical Structures in Disconnected Tail (Sometimes)
Turning of Rays

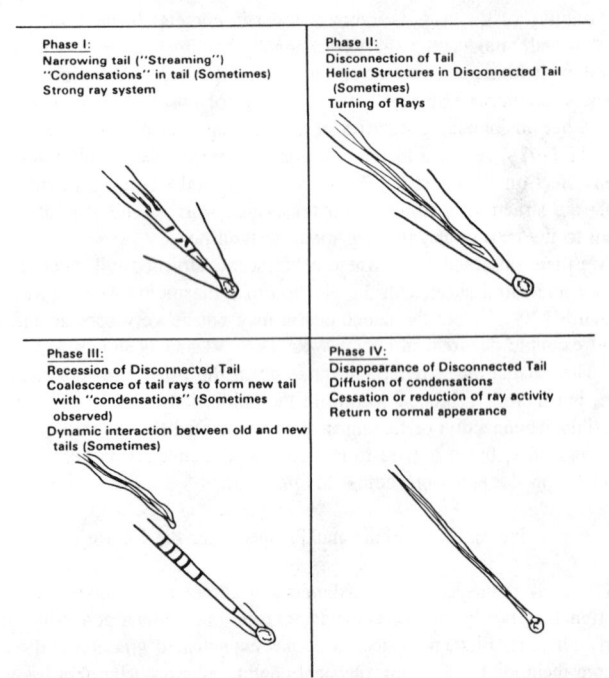

Phase III:
Recession of Disconnected Tail
Coalescence of tail rays to form new tail with "condensations" (Sometimes observed)
Dynamic interaction between old and new tails (Sometimes)

Phase IV:
Disappearance of Disconnected Tail
Diffusion of condensations
Cessation or reduction of ray activity
Return to normal appearance

Fig. 10. Schematic diagram showing morphological sequence of events during a DE cycle. Not all features are observed in all DE's; the DE which perhaps comes closest is the 1974 January 9-11 event in comet Kohoutek (Figs. 6 and 8).

almost no data—imaging, spectroscopy, or otherwise—exists on the actual reconnection region in the head when reconnection was occurring.

Responding to each of these limitations in order, the following can be said concerning future progress:

1) Five spacecraft from three international space agencies are being sent to Halley's Comet to carry out *in situ* measurements in 1986 March [Reinhard, 1982], and NASA's ISEE-3 satellite has been retargeted away from the Earth for a 1985 September encounter with comet Giacobini-Zinner [ISEE Working Team, 1982]. As a result of these pioneering space efforts, the bow shock and contact surface could for the first time become "observables" to use in comet models.

2) The International Halley Watch [IHW; Rahe and Newburn, 1982] is committed to avoiding the usual past mistakes of long data gaps and poor coordination among observers. It will be by far the most concentrated attack on a comet. Of relevance here is the possibility that unprecedented completeness in the wide-field imagery, combined with high-level monitoring of the solar wind, could definitively "nail down" the association of DE's and sector boundaries and thereby provide an important verification of the basic foundation of the reconnection model. Or, the results may favor another (non-reconnection) mechanism.

3) The reconnection jets operating in the head during DE onset may be observed on the ASTRO-1 Space Shuttle mission, which will take high spatial resolution images and spectrophotometry of Halley's Comet for a 7-10 day period in 1986 March. The chances of at least one DE occurring during the mission are very good.

In summation, the arrival of Halley's Comet in 1985-1986 presents an unparalleled opportunity to test the reconnection and alternate DE models with much more thoroughness than has been possible in the past. As a result of the extensive plans now being made to obtain uni-

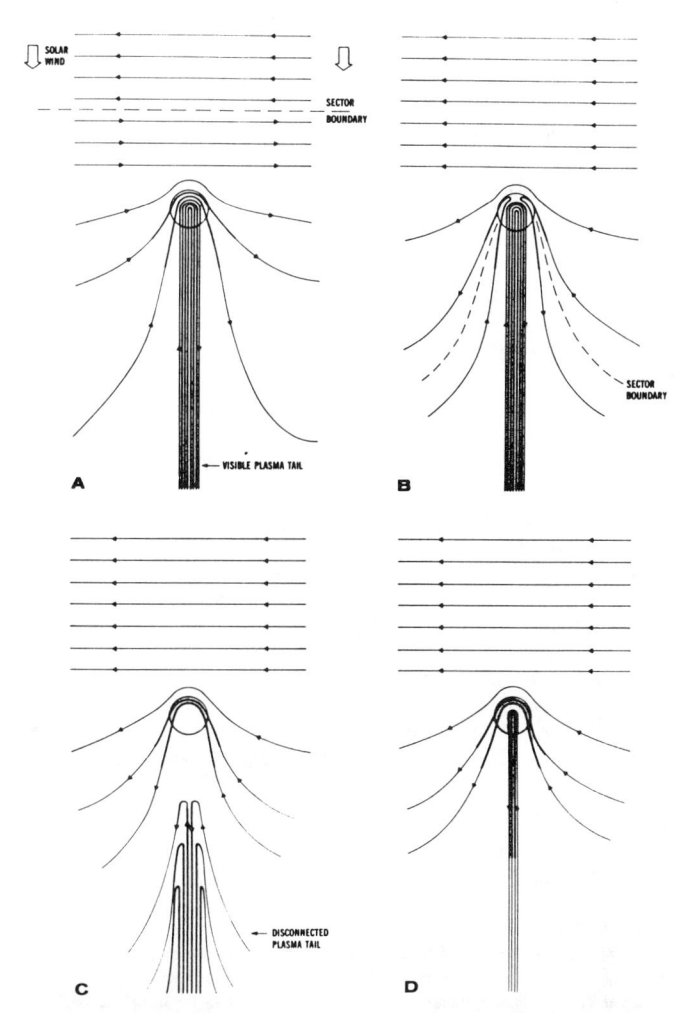

Fig. 11. Sector boundary/magnetic reconnection model of DE's [from Niedner and Brandt, 1978].

TEARING MODE INSTABILITY AND PRE-DISCONNECTION CLUMPING IN TAIL

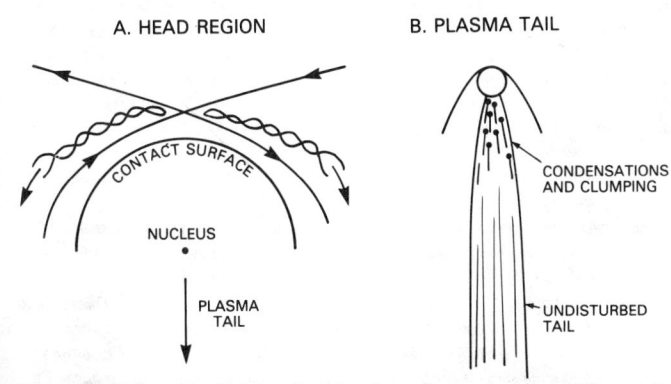

Fig. 12. Schematic diagram showing how the tearing instability might load the plasma tail with clumpiness or condensations in the hours immediately preceding a DE. Such condensations in a disconnecting tail were observed in comet Morehouse.

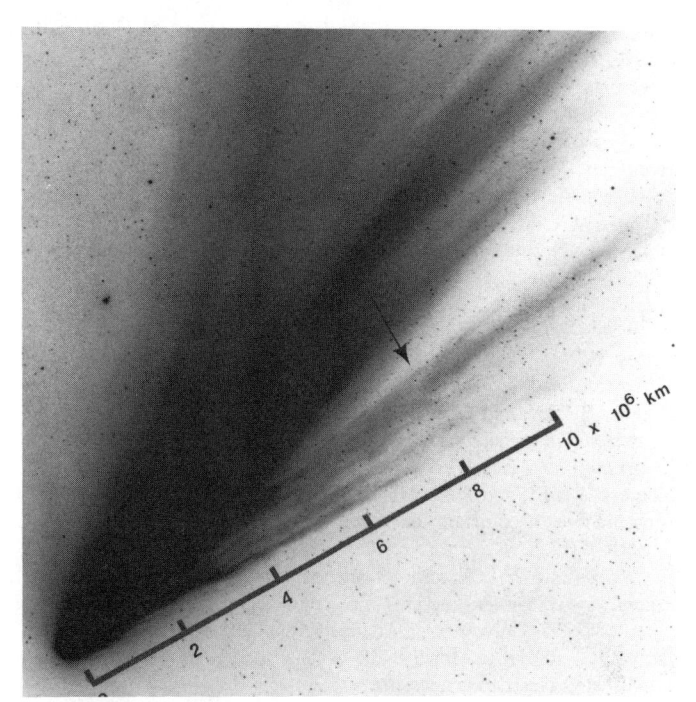

Fig. 13. Photograph of comet West 1976 VI showing (arrow) a possible cometary analog to plasmoids in the Earth's magnetotail (Joint Observatory for Cometary Research photograph).

que observations of the comet, we have every reason to expect a great deal of progress in our understanding.

References

Alfvén, H., On the theory of comet tails, *Tellus, 9,* 92-96, 1957.

Barnard, E.E., Photographic observations of Borrelly's comet and explanation of the phenomenon of the tail on July 24, 1903, *Astrophys. J., 18,* 210-217, 1903.

Barnard, E.E., Comet c 1908 (Morehouse), *Astrophys. J., 28,* 292-299, 1908a.

Barnard, E.E., Photographic observations of comet c 1908. Second Paper, *Astrophys. J., 28,* 384-388, 1908b.

Brandt, J.C., Observations and dynamics of plasma tails, in *Comets,* edited by L.L. Wilkening, pp. 519-537, University of Arizona Press, Tucson, Arizona, 1982.

Brandt, J.C., and D.A. Mendis, The interaction of the solar wind with comets, in *Solar System Plasma Physics,* Vol. II, edited by C.F. Kennel *et al.*, pp. 253-292, North Holland Publ. Co., Amsterdam, 1979.

Brosowski, B., and R. Wegmann, Numerische behandlung eine kometenmodells, *Max-Planck Institut Astrophysik Publ.* MPI/PAE-Astro. 46, 1972.

Combi, M.R., and A.H. Delsemme, Brightness profiles of CO^+ in the ionosphere of comet West, *Astrophys. J., 238,* 381-387, 1980.

Delsemme, A.H., and M.R. Combi, $O(^1D)$ and H_2O^+ in comet Bennett 1970 II, *Astrophys. J., 228,* 330-337, 1979.

Ershkovich, A.I., and D.A. Mendis, On the penetration of the solar wind into the cometary ionosphere, *Astrophys. J., 269,* 743-750, 1983.

Fedder, J.A., S.H. Brecht, and J.G. Lyon, Comet magnetospheres, *EOS, 62,* 367, 1981.

Fedder, J.A., S.H. Brecht, and J.G. Lyon, submitted to *Icarus,* 1983.

Gosling, J.T., J.R. Asbridge, S.J. Bame, and W.C. Feldman, Solar wind speed variations: 1962-1974, *J. Geophys. Res., 81,* 5061-5070, 1976.

Hoeksema, J.T., J.M. Wilcox, and P.H. Scherrer, Structure of the heliospheric current sheet in the early portion of sunspot cycle 21, *J.Geophys. Res., 87,* 10,331-10,338, 1982.

Hoeksema, J.T., J.M. Wilcox, and P.H. Scherrer, The structure of the heliospheric current sheet: 1978-1982, *J. Geophys. Res., 88,* 9910-9918, 1983.

Hones, E.W., Substorm processes in the magnetotail: comments on 'On hot tenuous plasmas, fireballs, and boundary layers in the earth's magnetotail' by L.A. Frank, K.L. Ackerson, and R.P. Lepping, *J. Geophys. Res., 82,* 5633-5640, 1977.

Hones, E.W., Transient phenomena in the magnetotail and their relation to substorms, *Space Sci. Rev., 23,* 393-410, 1979.

Houpis, H.L.F., and D.A. Mendis, Physicochemical and dynamical processes in cometary ionospheres. I. The basic flow profile, *Astrophys. J., 239,* 1107-1118, 1980.

Houpis, H.L.F., and D.A. Mendis, On the development and global oscillations of cometary ionospheres, *Astrophys. J., 243,* 1088-1102, 1981.

Ip, W.-H., and W.I. Axford, Theories of physical processes in the cometary comae and ion tails, in *Comets,* edited by L.L. Wilkening, pp. 588-634, University of Arizona Press, Tucson, Arizona, 1982.

Ip, W.-H., and D.A. Mendis, The cometary magnetic field and its associated electric currents, *Icarus, 26,* 457-461, 1975.

Ip, W.-H., and D.A. Mendis, The generation of magnetic fields and electric currents in cometary plasma tails, *Icarus, 29,* 147-151, 1976.

Ip, W.-H., and D.A. Mendis, The flute instability as the trigger mechanism for disruption of cometary plasma tails, *Astrophys. J., 223,* 671-675, 1978.

ISEE Science Working Team, Report of the Comet Subcommittee, 1982 June.

Jockers, K., Plasma dynamics in the tail of comet Kohoutek 1973 XII., *Icarus, 47,* 397-411, 1981.

Jockers, K., and Rh. Lüst, Tail peculiarities in comet Bennett caused by solar wind disturbances, *Astron, Astrophys., 26,* 113-121, 1973.

Mendis, D.A., and W.-H. Ip, The ionospheres and plasma tails of comets, *Space Sci. Rev., 20,* 145-190, 1977.

Morrison, P.J., and D.A. Mendis, On the fine structure of cometary plasma tails, *Astrophys. J., 226,* 350-354, 1978.

Niedner, M.B., Interplanetary Gas. XXV. A solar wind and interplanetary magnetic field interpretation of cometary light outbursts, *Astrophys. J., 241,* 820-829, 1980.

Niedner, M.B., Interplanetary Gas. XXVII. A catalog of disconnection events in cometary plasma tails, *Astrophys. J. (Supplement), 46,* 141-157, 1981.

Niedner, M.B., Interplanetary Gas. XXVIII. A study of the three-dimensional properties of interplanetary sector boundaries using disconnection events in cometary plasma tails, *Astrophys. J. (Supplement), 48,* 1-50, 1982.

Niedner, M.B., and J.C. Brandt, Interplanetary Gas. XXIII. Plasma tail disconnection events in comets: evidence for magnetic field line reconnection at interplanetary sector boundaries?, *Astrophys. J., 223,* 655-670, 1978.

Niedner, M.B., and J.C. Brandt, Interplanetary Gas. XXIV. Are cometary plasma tail disconnections caused by sector boundary crossings or by encounters with high-speed streams?, *Astrophys. J., 234,* 723-732, 1979.

Niedner, M.B., and J.C. Brandt, Structures far from the head of comet Kohoutek. II. A discussion of the swan cloud of January 11 and of the general morphology of cometary plasma tails, *Icarus, 42,* 257-270, 1980.

Niedner, M.B., J.A. Ionson, and J.C. Brandt, Interplanetary Gas. XXVI. On the reconnection of magnetic fields in cometary ionospheres at interplanetary sector boundary crossings, *Astrophys. J., 245,* 1159-1169, 1981.

Rahe, J., and R.L. Newburn, The International Halley Watch, in *The Need for Coordinated Ground-Based Observations of Halley's Comet,* Proceedings of ESO Workshop, edited by P. Veron et al., p. 301, 1982.

Reinhard, R., Space missions to Halley's Comet and related activities, *ESA Bull., No. 29,* 68-83, 1982.

Rosenberg, R.L., and P.J. Coleman, Heliographic latitude dependence of the dominant polarity of the interplanetary magnetic field, *J. Geophys. Res., 74,* 5611-5622, 1969.

Russell, C.T., and R.C. Elphic, Observation of magnetic flux ropes in the Venus ionosphere, *Nature, 279,* 616-618, 1979.

Schmidt, H.U., and R. Wegmann, MHD-calculations for cometary plasmas, *Computer Phys. Commun., 19,* 309-326, 1980.

Schmidt, H.U., and R. Wegmann, Plasma flow and magnetic fields in comets, in *Comets,* edited by L.L. Wilkening, pp. 538-560, University of Arizona Press, Tucson, Arizona, 1982.

Vasyliunas, V.M., Theoretical models of magnetic field line merging, 1., *Rev. Geophys. Space Phys., 13,* 303-336, 1975.

Wallis, M.K., Weakly-shocked flows of the solar wind plasma through atmospheres of comets and planets, *Planet. Space Sci., 21,* 1647-1660, 1973.

Wallis, M.K., and R.S.B. Ong, Cooling and recombination processes in cometary plasmas, in *The Study of Comets,* edited by B. Donn et al., pp. 856-876, NASA SP-393, Washington, 1976.

Wurm, K., and A. Mammano, Contributions to the kinematics of type 1 tails of comets, *Astrophys. Space Sci., 18,* 273-286, 1972.

Questions and Answers

Reiff: It appears to me that there are two other ways to get disconnected plasma tails besides the dayside reconnection model that you showed.

The first mechanism is a combination of diffusion and convection that allows a draped interplanetary field line to partially penetrate while slipping over the ionosphere, accelerating away some ionospheric plasma that had been gravitationally bound. This apparently occurs at Venus [Cloutier, Venus book]. Another way which might occur involves nightside reconnection. If the field lines get hung up on the dayside and are prevented from convecting over the pole by collisions, they could fold up, becoming antiparallel on the nightside. Reconnection there could result in a ring of magnetic field around the comet and a V-shaped plasma tail flowing downstream.

It seems that the observational data are not good enough yet to rule out the occurrence of these kinds of disconnection events.

Niedner: There is almost no question that field line slippage over the ionosphere and nightside reconnection do occur in comets. The most important consideration then becomes: how major are the observed effects, and can DE's be produced by either of these two mechanisms? Presumably, field line slippage is occurring all the time for IMF incident at moderate-to-high "magnetic latitudes", i.e., well away from the subsolar point. These fields which reside only briefly in the coma could drag some ions downstream as you suggest, but unless the strongly hung-up fields at the subsolar point (which constitute the roots of the plasma tail) are also uprooted, there probably will be no large-scale tail disconnection: recall that in the early phase of a DE (Figures 6b, 10, and 11c), there is no tail structure rooted into the head, only the inclined tail rays. I do think, however, that the slippage mechanism is a good candidate for explaining some smaller-scale structures—clouds, fine rays, and "spray"—which are occasionally seen along the outer borders of attached tails.

The occurrence of nightside reconnection is of course favored by the 2-lobe structure of the tail and it almost certainly does occur, but unless the reconnection continues for the entire width of the tail (in effect cutting it in two), any observed feature may only be substructure contained within the envelope of a "normal tail", and not a DE. This is not to say that nightside reconnection doesn't operate in comets. There may be cases where nightside reconnection does break the tail, giving the appearance of a DE. If so, however, we've never had an observational sequence showing some location in the tail to be the site of a DE, and it is important to note this point. However, a myriad of moderate- and fine-scale tail structure has been observed in many past comets, and tail reconnection is an exciting and plausible candidate for some of these kinds of features. The severed structure in the tail of comet West (Figure 13) may be an example of tail reconnection.

D. N. Baker: Why do you not have cometary Disconnection Events (D.E.'s) due to reconnection on the nightside of the comet rather than invoking dayside reconnection? This sort of "tail reconnection" would be in closer analogy with what happens during terrestrial magnetospheric substorms and could give rise to the loss of the cometary tail in the same way as plasmoids are formed during substorms.

Niedner: This question is identical to the second part of Dr. Reiff's question—the result of the questions being written—but since this issue of the nightside is on the minds of the reconnection community, it might be worth making a few additional points. From an observational viewpoint, the reason for making the statement that DE's are a dayside process is that in every case where photographic observations are made frequently enough, events classified "DE's" are seen to originate in, and emerge from, the coma, and not the tail proper. The **physical reason** *why spectacular tail reconnection and plasmoids are not observed is not certain, but the differences between the cometary and geomagnetic tails must be kept in mind: 1.)because it is an induced (captured) field, the comet-tail field is approximately cylindrical, not a distorted dipolar field as in the case of the Earth's near-tail, and 2.) the comet tail has dense concentrations (n_i = 100-1000 cm^{-3}) of heavy CO^+ ions (28 amu) which may restrict motion of the fields.*

Hopefully, detailed and continuous observations of Halley's Comet in 1985-1986 will show us what kinds of reconnection structures are produced in the tail region. The goal is that extended photographic sequences will show the site of origin and the temporal development of structures like that seen in Figure 13. Until then, the importance of tail reconnection must remain a somewhat open question.

RECONNECTION IN THE JOVIAN MAGNETOSPHERE

A. Nishida

Institute of Space and Astronautical Science
4-6-1 Komaba, Meguro, Tokyo 153, Japan

While the plasma in the Jovian magnetosphere usually flows in the direction of corotation with the planet, an intense anisotropy indicating outflows in the direction away from Jupiter were sometimes encountered near the magnetopause. Krimigis et al. [1980] have designated them as magnetospheric wind.

Figure 1 is the magnetic field data (9.6 s averages) that correspond to the magnetospheric wind event reported by Krimigis et al. In the 80 ~137 keV ion data the event lasted from about 0800 to 1200 SCET. During this event Voyager 2 was located around 0300 LT at a radial distance of about 150 R_J. It is seen in Figure 1 that the inclination δ and the z component of the magnetic field (in the spacecraft centered heliographic coordinates) were northward while the anti-Jupiterward flow was encountered. The simultaneous occurrence of the outflow and the northward magnetic inclination suggests that field-line reconnection was in progress upstream of the observing site.

From a statistical survey of similar events in Voyager 1 data obtained beyond 100 R_J, we have found indications that reconnection operates Jupiterward of the 0430 LT meridian in more than 15% of the time.

While reconnection in the earth's magnetotail is basically a temporal feature, reconnection in the Jovian magnetodisc can be envisaged as a basically spatial feature. When field lines rotate with the planet from the nightside toward the dayside, their length has to be shortened to meet the shorter radius of the dayside magnetosphere. However, the centrifugal force acting on the plasma opposes the shortening, so that the tips of the field lines have to be separated by reconnection. Our observation is essentially consistent with this idea, but it adds a point that the Jovian reconnection is not entirely a steady-state process since tailward flow and northward field do not always occur together.

This work was published in full in Nishida [1983].

Acknowledgments. High-resolution magnetic field data of Vogager 2 have been provided by the WDC-A, Rockets and Satellites, in NASA Goddard Space Flight Center.

References

Krimigis, S. M., T. P. Armstrong, W. I. Axford, C. O. Bostrom, C. Y. Fan, G. Gloeckler, L. J. Lanzerotti, D. C. Hamilton, and R. D. Zwickl, Energetic (~100 keV) tailward-directed ion beams outside the Jovian plasma boundary, *Geophys. Res. Lett., 7*, 13, 1980.

Nishida, A., Reconnection in the Jovian magnetosphere, *Geophys. Res. Lett., 10*, 451, 1983.

Questions and Answers

Behannon: Your statistical study was carried out with hourly averages. It was seen in some of the early work that attempted to locate the neutral line in the earth's magnetotail that there was not a very close correspondence between hourly averaged data around the neutral sheet and high resolution measurements within the sheet as far as the polarity of the field component perpendicular to the sheet was concerned. That appears to be the case also at Jupiter, since your results infer a high percentage of southward-pointed field at large distances and our detailed sheet-crossing data show predominantly northward fields beyond 70-80 R_J. Can you comment on this?

Nishida: In fact, the data I used have a much better resolution. Anisotropy data are given at every 15 min, and the magnetic field data are 48 sec averages. From these data predominant features of flow and field polarity have been read out every hour.

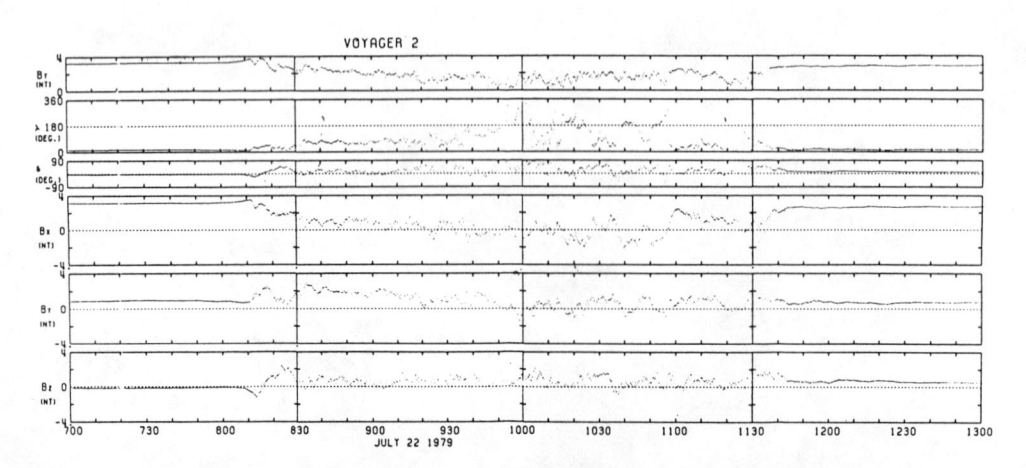

Figure 1. Voyager 2 magnetic field data obtained in the Jovian magnetodisc.

FINE-SCALE STRUCTURE OF THE JOVIAN MAGNETOTAIL CURRENT SHEET

K. W. Behannon

NASA/Goddard Space Flight Center
Laboratory for Extraterrestrial Physics
Greenbelt, MD 20771

Detailed study of the Voyager 1 and 2 magnetometer data taken outbound in the Jovian magnetosphere show that the magnetic structure near and within the current sheet was variable with time and distance from Jupiter, but generally corresponded to one of the following four types: 1. Simple rotation of field across the sheet, with an approximately southward direction in the sheet (generally northward beyond a distance from Jupiter of ~ 84 R_J); 2. Field having a southward component in a broad region near the sheet, but northward in a restricted region at the sheet itself; 3. A clear bipolar variation of the sheet-normal field component as the sheet was crossed (i.e., the field became northward and then southward, or vice versa, in crossing the sheet); 4. Large amplitude fluctuations in all field components near and in the sheet, with alternating northward and southward polarities.

The different types are illustrated schematically in Figure 1 in terms of the observed characteristic variation of the north-south (B_z) component of the magnetic field across the current sheet. These magnetic structures are all morphologically similar to those observed at the current sheet in the earth's magnetotail at different times. The question mark next to the type 2 sketch indicates that the interpretation of that variation or structural form is not yet well understood. It could be the result of a thinning of the plasma sheet with distance from the planet, with a corresponding gradient in the diamagnetic effect, and a field component in the current sheet that is directed oppositely to that in the broader, surrounding plasma sheet because of the existence of a cross-tail neutral line planetward of the observation point [Dessler and Hill, 1970]. This could be a single, large-scale neutral line or one of multiple neutral lines or points within the sheet fine structure. Type 3 geometry can also be interpreted in terms of sheet fine structure. One possible explanation is that it is the signature of crossing a magnetic loop or bubble, perhaps a tearing mode magnetic island, with the motion of the current sheet past the spacecraft unsteady as a result either of flapping motion of the sheet or wavelike oscillations. Related to these are the type 4 variations, which are signatures of multiple traversals of the sheet in which a combination of the other three types was sometimes seen; most notably it consisted of successive occurrences of type 3. Such dynamic, complex geometries as types 3 and 4 may be indicative of the occurrence of magnetic merging associated with either the spontaneously-excited resistive tearing mode instability or Petschek-type externally-driven reconnection. Signatures of both types of processes have been found in the geomagnetic tail [Nishida et al., 1981; Nishida and Hones, 1981]. The Voyager results presented here strongly suggest that these processes occur at least sporadically in Jupiter's magnetotail.

The occurrence distributions of the various types of sheet traversals are shown in Figure 2 along the Voyager 1 and 2 trajectories projected on the plane of Jupiter's orbit about the sun. Satellite position as a function of time is given by day-of-year numbers along the trajectory curves (circled points). The model dawnside magnetopause (MP) is also shown. Sheet crossing events are identified where they occurred by type (see key in the figure). Considering Voyager 1 and 2 observations together, twice as many of type 1 signatures were seen as types 2 and 4, which occurred approximately in equal numbers, whereas type 3's were only half as frequent as the latter types. The strong tendency can be seen for the more complex forms (types 3 and 4) to dominate at greater distances from Jupiter (for $X < -70$ R_J or $r > 80$ R_J).

Calculations using the limited data available from the Voyager encounters show that it is plausible for the tearing mode and hence reconnection to occur in the Jovian magnetotail environment, since instability growth times and plasmoid motion times are of the same order and also are similar to those derived for Earth's tail sheet by Schindler [1974], Galeev et al. [1978] and Speiser and Schindler [1981]. Additional data are needed to establish this conclusively, as well as to demonstrate that such processes are related to the occurrence of substorms at Jupiter. See Behannon [1984] for a more complete description and discussion of these results.

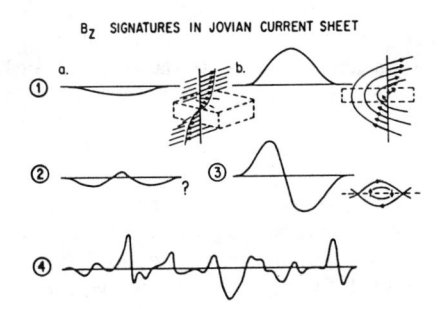

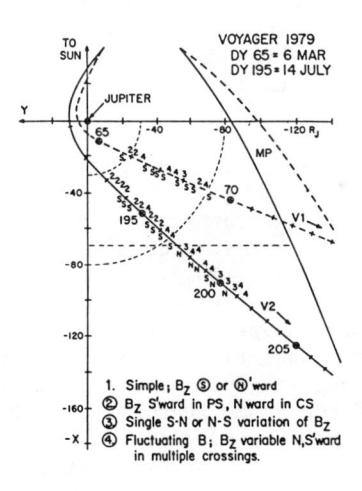

References

Behannon, K. W., Fine-scale structure of the Jovian magnetotail current sheet, *J. Geophys. Res.*, to be published, 1984.

Dessler, A. J., and T. W. Hill, Location of neutral line in magnetotail, *J. Geophys. Res.*, 75, 7323, 1970.

Galeev, A. A., F. V. Coroniti, and M. Ashour-Abdalla, Explosive tearing mode reconnection in the magnetospheric tail, *Geophys. Res. Lett.*, 5, 707, 1978.

Nishida, A., H. Hayakawa, and E. W. Hones, Jr., Observed signatures of reconnection in the magnetotail, *J. Geophys. Res.*, 86, 1422, 1981.

Nishida, A. and E. W. Hones, Jr., Undulation of field line loops in the neutral sheet, *Program and Abstracts, 4th IAGA Scientific Assembly*, p. 435, 1981.

Schindler, K., A theory of the substorm mechanism, *J. Geophys. Res.*, 79, 2803, 1974.

Speiser, T. W., and K. Schindler, Magnetospheric substorm models: comparison with neutral sheet magnetic field observations, *Astrophys. Space Sci.*, 77, 443, 1981.

MAGNETIC FIELD RECONNECTION AT THE MAGNETOPAUSE: AN OVERVIEW

B.U.Ö. Sonnerup

Thayer School of Engineering, Dartmouth College, Hanover, NH 03755

Abstract. A brief summary is presented of the basic qualitative and quantitative aspects of reconnection in its magnetopause setting. First, the basic morphological and dynamic features of asymmetric reconnection are examined with emphasis on the important role played by the rotational discontinuity in these geometries. Second, the structure and other properties of rotational discontinuities are discussed. Third, the manner in which individual particles are energized or de-energized during their interaction with current layers in general, and rotational discontinuities in particular, is examined. Finally, the question of nonsteady, localized reconnection and its relation to flux transfer events is discussed and a qualitative model is proposed to describe these phenomena.

1. Introduction

Magnetic field reconnection in a plasma may in principle occur wherever the magnetic field exhibits strong shear. In planetary magnetospheres, the two principal active sites that have been considered are the magnetopause and the magnetotail. In this paper, attention is focussed on reconnection in its magnetopause setting. An overview is presented of the basic features and local signatures of the process predicted by existing theory. Important new information concerning magnetopause reconnection has been obtained during the last few years, principally as a result of the ISEE mission. These results, along with the magnetospheric consequences of magnetopause reconnection, will not be dealt with in detail, since they form the topic of several papers to follow in this volume.

2. Basic Morphology and Dynamics

In the early closed model of the magnetosphere (Johnson, 1960), the magnetopause was that surface, usually marked by an electric current sheet, which separated the earth's magnetic field from the solar-wind plasma and the interplanetary magnetic field (IMF) embedded in it. No interconnection between the two fields was included so that the magnetopause was a tangential discontinuity. When the concept of magnetic field reconnection was introduced into magnetospheric physics by Dungey [1961], the meaning of the term magnetopause became blurred. Indeed, in Dungey's [1961] drawing of the open magnetosphere for purely southward IMF (Figure 1), the magnetopause seems to be entirely absent, except perhaps near the subsolar point. A geometry of this type suggests that most or all of the interplanetary magnetic flux impinging upon the magnetosphere is reconnected, but it must be remembered that the drawing was intended only as a qualitative illustration of the "open" magnetic-field topology.

The first quantitative analysis of the magnetopause reconnection process was carried out by Levy et al. [1964] who concluded that only some 10-20% of the incident magnetic flux reconnects while the remainder is carried past the magnetosphere without becoming interconnected with the earth's field. In such a situation, the effect of reconnection on the dayside magnetopause can be thought of as a small perturbation. In other words, the magnetopause remains a well defined current sheet as shown in Figure 2. However, the physical character of this sheet changes drastically: instead of being a tangential discontinuity (TD), i.e., a layer with a vanishing normal magnetic field component, B_n, it is now a rotational discontinuity (RD) and has a small but significant B_n, amounting to 10-20% of the total field.

A rotational discontinuity is a large-amplitude Alfvén wave across which the interplanetary plasma flows with the Alfvén speed based on B_n. During its passage through the dayside portion of this current layer, this plasma is accelerated away from the subsolar point by the $\underline{I} \times \underline{B}_n$ force, where $\underline{I}$ is the magnetopause current, to form a boundary layer of jetting plasma immediately inside the magnetopause. Tailward of the two cusp regions, the direction of $\underline{I}$ is reversed so that the plasma is decelerated instead (a feature not included in the Levy et al. model). It is in these latter regions that mechanical energy is extracted from the solar wind and stored as magnetic energy in the geomagnetic tail [e.g., Swift, 1980]. However, the presence of such deceleration regions is merely a consequence of ongoing reconnection somewhere on the subsolar magnetopause. For this reason, we shall focus attention on this latter region.

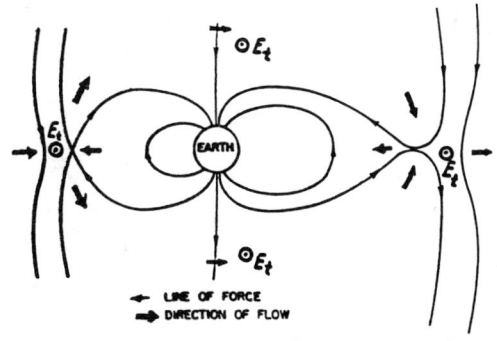

Figure 1. The first reconnection model of the magnetosphere [Dungey, 1961]. The reconnection electric field, E_t, leads to the flow pattern shown.

In the MHD model of Levy et al. (Figure 2), it was assumed that no plasma is present in the magnetosphere. As a result, the inner edge of the plasma boundary layer inside the magnetopause consists of a narrow slow-mode expansion fan in which the plasma expands to zero pressure and is accelerated further [see Yang and Sonnerup, 1977]. In reality the dilute but hot magnetospheric plasma has a substantial pressure so that the expansion fan may be absent or even replaced by a slow shock. It is not clear how collision-free slow shocks and expansion fans manifest themselves in the narrow magnetopause-boundary layer region.

The rotational discontinuity is expected to be present at the magnetopause in all but the most unusual circumstances: strictly antiparallel fields and identical or nearly identical plasma and field states on the two sides of the magnetopause in which case the usual symmetric Petschek [1964] model containing pairs of slow shocks applies. Even a small increase in density outside the magnetopause over that inside is sufficient to bring the outer slow shocks in that model to their maximum strength (switch-off of the tangential field). After that, an RD in which the tangential field rotates by 180° will appear. If the reconnecting magnetic fields are not antiparallel, an RD is always needed. Thus, the principal magnetic and plasma signatures of the magnetopause region away from the separator (reconnection line; X line) should be those associated with an RD.

The separator is commonly assumed to pass through the subsolar point on the magnetopause and to be oriented approximately along the net magnetopause current, $\underline{I}$, as illustrated in Figure 3a. Figure 3b shows that an orientation strictly along $\underline{I}$ would not permit reconnection between fields $\underline{B}_1$ and $\underline{B}_2$, where $|\underline{B}_1| < |\underline{B}_2|$, for an angle $\Delta\theta$ such that $\cos\Delta\theta \geq B_1/B_2$, because the magnetic field component perpendicular to the separator must reverse sign for reconnection to be possible [Sonnerup, 1974; Gonzalez and Mozer, 1974; Hill, 1975]. Cowley [1976] has argued persuasively that the separator orientation along $\underline{I}$, while consistent with incompressible MHD theory [see Petschek, 1964], is not required by it. Thus, the separator orientation may depend not only on local conditions but also on the global reconnection configuration. Nevertheless, all so-called "component-merging" hypotheses have a separator tilt in qualitative agreement with Figure 3a. In other words, a positive (negative) B_Y component in the magnetosheath (in GSM coordinates) corresponds to a separator location north (south) of the equatorial plane on the afternoon side and south (north) of that plane on the forenoon side of the magnetopause. An alternate hypothesis has been advanced [Crooker, 1979] to the effect that reconnection occurs only at those locations on the magnetopause where $\underline{B}_1$ and $\underline{B}_2$ are antiparallel or nearly antiparallel ("antiparallel merging"; see J. Luhmann, this volume). The merits of this suggestion have yet to be fully evaluated.

Direct observational proof of the occurrence of reconnection would consist of a measured electric field along a separator since that is a standard definition of reconnection. However, such information is difficult to obtain, not only because electric-field measurements in the magnetopause plasma environment are difficult, but also because the diffusion region, i.e., the narrow channel around the separator in which the frozen magnetic field condition is violated, has small

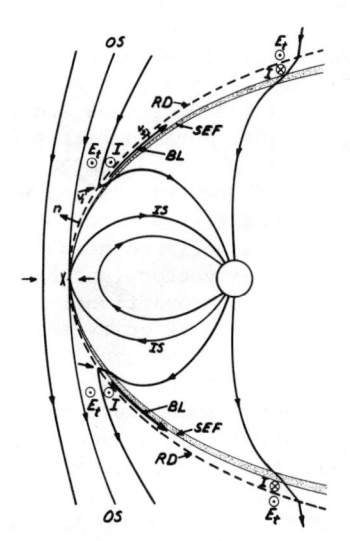

Figure 2. Magnetopause reconnection due to Levy et al. [1964] in which only a small portion of the incident magnetic flux is reconnected. The principal geometrical features of this model are: the separator (X); the magnetopause current layer in the form of a rotational discontinuity (RD); a thin high-velocity plasma boundary layer (BL) immediately inside the magnetopause and, at its innermost edge, a slow-mode expansion fan (SEF, shown shaded). The inner and outer separatrix surfaces are marked by (IS) and (OS), the reconnection electric field by E_t, and the magnetopause current by I.

physical dimensions and therefore is difficult to identify. The difficulty is compounded by the fact that our theoretical understanding of the important plasma processes in the diffusion region is poor (J.F. Drake, this volume). In effect, we do not know what plasma signatures to look for. A suggestion concerning a magnetic signature produced by Hall currents may be found in Sonnerup [1979] [see also Terasawa, 1983; 1984].

In principle, it is possible to obtain persuasive evidence for reconnection even from satellite traversals of the magnetopause away from the reconnection site itself, i.e., away from the separator. An electric field E_t tangential to the magnetopause or a magnetic field component B_n normal to it, if present over a region of linear dimensions much greater than the magnetopause thickness, indicates that reconnection is occurring or has occurred in the recent past. According to simple steady-state 2D reconnection theory, E_t is equal to the reconnection rate and B_n is proportional to E_t. But both E_t and B_n are difficult to determine: over most parts of the magnetopause, the largest electric field component is normal to the magnetopause (see T. Aggson, this volume) and the largest magnetic field component is tangential to it. Thus a reliable determination of E_t or B_n depends critically upon the accurate knowledge of the normal vector, $\underline{n}$. Because of wave motion and other irregularities in the magnetopause surface, a model normal is not useful for this purpose. The remaining possibility is to obtain a normal vector (and B_n) from minimum-variance analysis of the magnetic field data [Sonnerup and Cahill, 1967]. Occasionally, reliable normal vectors and B_n values are obtained from this process and there is little doubt [e.g., Sonnerup and Ledley, 1979a,b] that a few magnetopause crossings have been identified where B_n was significantly different from zero. But more often than not, the minimum variance method fails to give a reliable normal vector (and B_n). A third quantity which is also proportional to the reconnection rate is the flow velocity component, v_n, across the magnetopause. However, it too is difficult to measure, not only because an accurate $\underline{n}$ vector is needed but also because the magnetopause is usually not stationary. In addition, a normal flow velocity component is in itself not convincing evidence of reconnection since it could be the result of rapid diffusion of plasma across a TD.

The difficulties described above usually preclude a reliable estimate of the reconnection rate. However, a simple check on the presence or absence of reconnection can be obtained by examination of the behavior of the tangential plasma velocity across the magnetopause. According to the reconnection model, the interplanetary plasma experiences a change in tangential momentum caused by the $\underline{I} \times \underline{B}_n$ force as it crosses the current layer. On the dayside, and with the interplanetary magnetic field due essentially south, the net effect is plasma acceleration to veloci-

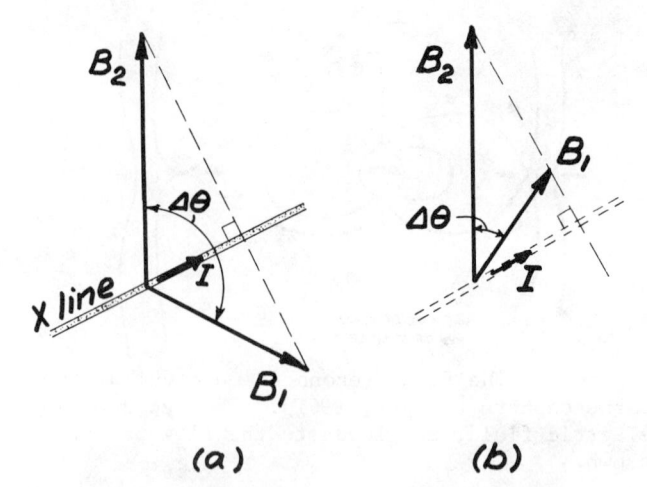

Figure 3. (a) Illustration of the hypothesis that the separator (X line) is aligned with the net magnetopause current I. (b) According to this hypothesis no reconnection is possible for $\cos\Delta\theta > B_1/B_2$.

ties of about twice the Alfvén speed, $2v_A$, leading to the plasma jets shown in Figure 2, just inside the magnetopause. These jets have indeed been observed [Paschmann et al., 1979; Sonnerup et al., 1981; Gosling et al., 1982], and the observed detailed agreement with the theoretical tangential momentum change makes it unlikely that these jets were the product of processes other than reconnection. Recently (G. Paschmann, this volume) the energy balance across the magnetopause has also been examined.

In spite of the fact that the $\underline{I} \times \underline{B}_n$ force is proportional to the reconnection rate via B_n, the plasma momentum change is independent of that rate. The reason is that the mass flow rate, ρv_n, across the magnetopause is also proportional to the reconnection rate. Thus, observations of this type cannot be used to establish the reconnection rate.

There are a number of other direct consequences of the reconnection field topology which are observable in the region between the inner and outer separatrix surfaces, shown in Figure 2. These surfaces intersect in the diffusion region and they might therefore bear the signature, in the form of heat flow, of any electron heating in that region (see J.D. Scudder, this volume). In addition, the outer separatrix might be traced by escaping magnetospheric electrons.

Between the outer separatrix and the magnetopause one might expect to find magnetospheric and ionospheric ions that have leaked across the magnetopause and have been given a tangential velocity change of up to $2v_A$ by the $\underline{I} \times \underline{B}_n$ force [Scholer et al., 1981; Sonnerup et al., 1981], as well as magnetosheath ions that have been reflected in the magnetopause and have been similarly influenced by the $\underline{I} \times \underline{B}_n$ force [Sonnerup et al., 1981]. Between the magnetopause and the inner separatrix, one expects to find, not only the jetting plasma boundary layer, but also any magnetospheric ions that have not leaked across the magnetopause but have been reflected against it [Scholer and Ipavich, 1983]. Detailed consideration of these various signatures may be found in G. Paschmann's article (this volume). Finally, it is noted that the inner separatrix comprises the last set of closed field lines in the magnetosphere. This surface and the region just outside it project into the high-latitude ionosphere so that many of the consequences of magnetopause reconnection should be observable there (see P. Reiff, this volume).

3. Rotational Discontinuity

The basic properties of an RD are: (i) it has a nonvanishing magnetic field component, B_n, normal to the layer; (ii) the plasma flows toward and away from the discontinuity with normal speed, v_n, equal to the Alfvén speed based on B_n and corrected for nonisotropic pressure; (iii) the tangential magnetic field can change direction by an arbitrary angle $\Delta\theta$ across the sheet; indeed, it is the only discontinuity having this capability. The fact that the flow speeds into and out of the discontinuity are equal to the corresponding Alfvén-wave speeds guarantees that no wave steepening or broadening of the usual type is present in a uniform medium.

Jump Conditions

In a collision-free plasma with nonisotropic pressure the jump conditions across an RD are [Hudson, 1970; 1971; 1973]:

$$\Delta\{\rho(1-\kappa)\} = 0 \qquad \text{(Mass Cons.)} \qquad (1)$$

$$\Delta\{p_\perp + B^2/2\mu_o\} = 0 \qquad \text{(Normal Mom. Cons.)} \qquad (2)$$

$$\Delta\{\underline{v}_t - \underline{v}_{At}\} = 0 \qquad \text{(Tang. Mom. Cons.)} \qquad (3)$$

$$\Delta\{(\kappa + \tfrac{1}{2})B^2/\mu_o\rho + 5p_\perp/2\rho\} = -q \qquad \text{(Energy Cons.)} \qquad (4)$$

In these equations, the symbols, ρ, p, $\underline{B}$, and $\underline{v}$ have their usual meaning and the delta symbol is defined by $\Delta F \equiv F_2 - F_1$, where the subscripts 1 and 2 denote conditions upstream and downstream of the discontinuity. Also, q is the amount of energy per unit mass which leaves the system via heat conduction or radiation. The pressure anisotropy, κ, is defined by

$$\kappa \equiv (p_\| - p_\perp)\mu_o/B^2 \qquad (5)$$

Finally, $\underline{v}_{At} \equiv \underline{B}_t[(1-\kappa)/\mu_o\rho]^{1/2}$ is the tangential Alfvén speed. In the frame of the RD, the normal plasma flow speeds towards and away from the layer are $v_{n1} = v_{An1} = |B_n|[(1-\kappa_1)/\mu_o\rho_1]^{1/2}$ and $v_{n2} = v_{An2}$, respectively. In the firehose limit, $\kappa_1 = \kappa_2 = 1$, the plasma ceases to flow across the discontinuity.

From the above formulas with $q=0$ and $\kappa_1=\kappa_2=0$, the well known MHD results [e.g., Landau and Lifshitz, 1960] are recovered: ρ, p, and B remain unchanged across the RD while the direction of $\underline{B}_t$ can change by an arbitrary angle, this change being accompanied by a corresponding change in $\underline{v}_t$ as indicated by (3). If the tangential field $\underline{B}_t$ rotates by an angle $\Delta\theta$, (3) yields $|\Delta\underline{v}_t| = 2|v_{At}|\sin(\Delta\theta/2)$ so that the maximum value of $|\Delta\underline{v}_t|$ is $2v_{At}$, a result already mentioned in section 2.

If $q=0$ and $\kappa_1=\kappa_2$, then ρ, p, and B again remain constant but the Alfvén speed now contains the correction factor $(1-\kappa)^{1/2}$. This situation arises if the double adiabatic relations $p_\perp/\rho B$ = const. and $p_\| B^2/\rho^3$ = const. hold. If only the first of these conditions, representing the conservation of the magnetic moment of a particle, is valid, a curious situation arises [Hudson, 1973]. One possible root is $\kappa_2=\kappa_1$ in which case ρ, p, and B all remain unchanged. But there also exist parameter ranges in which a second root $\kappa_2 \neq \kappa_1$ occurs so that ρ, p, and B all change across the RD. In discontinuities of this type the entropy usually also changes: for $q=0$ only solutions that bring about an entropy increase, $\Delta s>0$, are physically acceptable. If neither of the adiabatic relations hold and/or if $q \neq 0$, even more general behavior appears possible.

In summary, contrary to the case of shocks, the jump conditions across a rotational discontinuity do not uniquely specify the change of state of the plasma and field. As long as $q=0$, one possibility always is that the plasma state and field magnitude remain unchanged but other types of behavior exist as well. These have been explored to some extent by Hudson [1971, 1973]. However, recent experimental evidence (G. Paschmann, this volume) indicates that substantial ion heat flow away from the magnetopause RD occurs occasionally so that the assumption $q=0$ is not always a good one.

Structure

The structure and thickness of the rotational discontinuity must be such that the changes implied by the jump conditions are achieved. For example, one must presume that the two conditions $q=0$ and $\Delta s=0$ together imply a sufficient thickness of the layer so as to permit laminar, non-dissipative behavior of the plasma. Conversely, limitations derived from the RD structure may eliminate certain downstream states that are allowed by the jump conditions.

In dissipationless MHD, the RD structure is simple: the plasma state and the magnetic field magnitude remain constant throughout the layer while the angle θ, defined by $\tan\theta = B_z/B_y$, of the tangential field changes by the desired amount, $\Delta\theta$. The function $\theta(x)$, x being the normal coordinate ($\hat{x} = -\underline{n}$), can be specified arbitrarily. If a small amount of dissipation is present, the width of the layer increases gradually [Landau and Lifshitz, 1960]. Our theoretical knowledge of the structure of rotational discontinuities in a collision-free plasma is limited to laminar structures with $|B|$ = const., as discussed below. However, a vast body of observations [e.g., Sonnerup and Ledley, 1979b;

Berchem and Russell, 1982b; Paschmann et al., 1979] indicates that in reality magnetopause RDs are usually turbulent and have $|\underline{B}| \neq$ const.

In analyzing the structure of current layers, it is convenient to use the so-called de Hoffmann-Teller (dHT) frame in which the external electric field vanishes. In this frame, the inflow and outflow velocities at an RD are field aligned and equal to the Alfvén speed on the two sides. The following basic statements can be made:

(1) It appears that the structure of a rotational discontinuity involves an electric field $E_n(x)$ normal to the layer and an associated potential barrier $\phi(x)$. The first-order orbit theory model examined by Su and Sonnerup [1968] as well as the double-adiabatic model by Lee and Kan [1982] contain such an electric field. On the other hand, Kirkland and Sonnerup [1982] failed to find RD solutions in a two-beam plasma model with $\phi(x) \equiv 0$. Su and Sonnerup recognized the possibility of trapping particles electrostatically in the layer but Lee and Kan were the first to elucidate the role played by trapped electrons in satisfying charge neutrality. Their calculation predicts E_n values of the order of one millivolt per meter for typical magnetopause conditions. However, recently Wang and Sonnerup [1984] have developed a model in which large deviations from charge neutrality occur in two narrow layers (electrostatic "shocks") of width equal to a few Debye lengths, one at each edge of the RD. Between the shocks, charge neutrality is at hand and the magnetic field rotates at a constant rate. The electric fields in the shocks are of the order of volts per meter.

Rotational discontinuities which have the same plasma state (and therefore the same field magnitude) on its two sides will also have the same electric potential, in the dHT frame, on the two sides. If the plasma state changes across the discontinuity, one would also expect the potential to change. It should also be noted that both $E_n(x)$ and $\phi(x)$ change when one transforms from the dHT frame to some other frame of reference (the "laboratory" frame) moving parallel to the discontinuity. In the laboratory frame, the potential will in general not have the same value on the two sides of the RD. This effect must be kept in mind when comparing electric field measurements to theory.

(2) For an RD in a plasma consisting of electrons with isotropic pressure tensor and one specie of ions of charge q, the tangential magnetic field, $B_t = B_y + i B_z$, where $i^2 = -1$, can be shown to obey the equation

$$\frac{dB_t}{dx} = iB_t \frac{\sqrt{1-\kappa_1}}{\lambda_{i1}} \left[1 - \frac{N_i}{N_{i1}} \left(1 - \frac{\mu_o P_{ixt}}{B_x B_t} \right) \frac{1}{1-\kappa_1} \right] \frac{B_x}{|B_x|} \quad (6)$$

where $\lambda_{i1} \equiv (m_i / \mu_o N_{i1} q^2)^{1/2}$ is the ion inertial length, N_{i1} is the ion number density ($\rho_1 \equiv N_{i1} m_i$), and κ_1 is the pressure anisotropy factor, upstream of the RD. Also $P_{ixt} \equiv P_{ixy} + i P_{ixz}$ is the tangential ion stress. Note that because $N_{i1}(1-\kappa_1) = N_{i2}(1-\kappa_2)$ all upstream conditions (subscript 1) can be replaced by downstream ones (subscript 2). It is easy to show that for the upstream or downstream gyrotropic state $P_{ixt} = B_x B_t \kappa / \mu_o$ so that $dB_t/dx = 0$ there.

In the absence of viscous stresses, the ion pressure tensor P_i is diagonal in a coordinate system with one axis along $\underline{B}$. In that case P_{ixt} is proportional to B_t so that the ratio P_{ixt}/B_t is purely real. Equation (6) is then of the form $dB_t/dx = ig(x) B_t$, where $g(x)$ is a real function of x, indicating that the field magnitude $|B_t|$ is constant. Examples of this situation are the models by Lee and Kan [1982] and Wang and Sonnerup [1984].

In observed magnetopause rotational discontinuities, the magnetic field magnitude often has a minimum in the center of the layer. The above results indicate that such an effect is likely to be caused by viscous stresses in the layer (although nonisotropy of the electron stress tensor and/or the presence of more than one ion specie may be contributing factors as well). Viscous stresses are expected to be important only in thin layers so that strong deviations from $|B_t|$ = const. in an RD should be an indication that the layer width is small. When viscous stresses are important, the use of the jump conditions (1) and (3) across only a portion of the magnetopause layer [Paschmann et al., 1979; Sonnerup et al., 1981] is inaccurate.

(3) Equation (6) also forms a suitable basis for discussing the sense of polarization of the RD. The right-hand (electron) and left-hand (ion) polarizations are obtained when the rectangular bracket is positive and negative, respectively. For example, if the ions are cold so that $P_{ixt} = 0$, $\kappa_1 = 0$, then the ion polarization is obtained for a positive potential barrier, $\phi(x) > 0$, since such a barrier will lead to $N_i > N_{i1}$. Similarly, $\phi(x) < 0$ yields the electron polarization [see Wang and Sonnerup, 1984]. For hot ions, the situation is less well understood. For example, in the double-adiabatic ion fluid description employed by Lee and Kan [1982] the square bracket reduces to $[1 - N_i/N_{i1}]$ but for $\phi(x) > 0$ (the only case dealt with by Lee and Kan) the density ratio N_i/N_{i1} is found to be less than unity except for cold or almost cold ions, the result being the electron polarization. The model by Su and Sonnerup [1968], in which electron inertia drift is important, permits the ion polarization only in special circumstances and then only for thick layers.

Sonnerup and Ledley [1979a,b] have argued that in a warm plasma only the electron polarization should occur for layers that are sufficiently thin so that only electrons but not ions are capable of moving across the layer by sliding along the magnetic field lines so as to provide the required field-aligned current distribution in the layer. This situation should arise when the layer thickness is comparable to, or less than, the ion gyroradius. On the basis of Explorer 12 and OGO 5 data, these authors also state that both polarizations have been observed but with a preference for the electron sense. Magnetopause thicknesses are, for the most part, unavailable for these data sets. On the other hand, Berchem and Russell [1982a,b] (see also C.T. Russell, this volume) have found from ISEE data that the magnetopause thickness is usually in the range 400-1000 km, i.e., several to many gyroradii, and that the basic polarization rule is for the field rotation angle $\Delta\theta$ in the layer to obey the inequality $|\Delta\theta| \leq \pi$. In other words, in rotating from its magnetospheric to its magnetosheath orientation, the tangential magnetic field vector chooses the shortest route, for RDs as well as TDs. Reexamination of published Explorer 12 and OGO 5 RDs basically supports the Berchem and Russell rule: only one RD structure has been published in which $|\Delta\theta| > \pi$ [Sonnerup and Ledley, 1974]. This crossing had the electron polarization. Furthermore, all of the reconnection events in Sonnerup et al., [1981] have $|\Delta\theta| \leq \pi$, as noted by Cowley et al. [1983].

Additional support for the rule $|\Delta\theta| \leq \pi$ comes from a recent computer simulation of RDs by Swift and Lee [1983] in which structures with $|\Delta\theta| > \pi$ were observed to be unstable. The reason for this effect is not clear although it is evident that the looped field line configurations for $|\Delta\theta| > \pi$ entail higher free energy than the nonlooped ones for $|\Delta\theta| \leq \pi$.

Further observational studies of the polarization question should focus on magnetopause crossings that can be unambiguously identified as rational discontinuities either by having a nonvanishing normal magnetic field, B_n, obtained from minimum variance analysis, or by having the appropriate plasma signatures. Also, the magnetopause thickness should be available as is the case for the ISEE data set [Berchem and Russell, 1982a]. On the theoretical side, it must be concluded that all existing analytical models of the rotational discontinuity rest on assumptions that are too restrictive to permit general conclusions concerning the sense of polarization. We need to examine in what general circumstances, if any, only one polarization sense is allowed and to establish whether the observed paucity of structures with $\pi < |\Delta\theta| < 2\pi$ is the result of nonexistence or of instability of steady-state solutions.

4. Particle Orbits

Motion in de Hoffmann-Teller Frame

In this section we review certain general results concerning particle orbits in one-dimensional current sheets with $B_x = -\underline{B} \cdot \underline{n} \neq 0$. These results are relevant not only to the magnetopause but to the geomagnetic tail and the earth's bow shock as well. It is advantageous to study the motion in the dHT frame because in that frame particles have simple helical orbits as they travel towards and away from the current sheet. In the sheet itself their motion may be extremely complicated but it remains constrained by the conservation of total energy and the two generalized tangential momenta:

$$mv^2/2 + q\phi(x) = \text{const.} \quad (7)$$

$$mv_y - qB_x z + qA_y(x) = \text{const.} \quad (8)$$

$$mv_z + qB_x y + qA_z(x) = \text{const.} \quad (9)$$

where $B_y \equiv -dA_z/dx$ and $B_z \equiv dA_y/dx$ are the field components tangential to the layer.

Cowley [1978] has shown that (7), (8), and (9) constrain the particle orbit to lie inside a surface in space, the intersection of which with any plane parallel to the yz plane is circular with radius $R_{Lx} = mv/qB_x$. As one moves from one plane parallel to the yz plane to another, R_{Lx} changes as the particle speed v changes in response to $\phi(x)$. The center of the circle moves in such a way that it remains located on one and the same field line.

It is instructive to arrive at these results by use of the diagrams in Figure 4. As shown in part (a) of that figure, at a given fixed x value the position of a particle in velocity space is constrained to lie on a sphere of radius v where $v = v_1\sqrt{1 - 2q\phi(x)/m}$, v_1 being the particle speed upstream of the current layer where $\phi = 0$. On this sphere the circles which represent constant pitch angles, α, with respect to the local magnetic field are shown for convenience. If the geometry shown in Figure 4a is projected onto the $v_y v_z$ plane, the result is the circular disk $v_y^2 + v_z^2 \leq v^2$ shown in Figure 4b. The circles of constant pitch angle now appear as ellipses inside the disk. The cases $\alpha = 0$, corresponding to particle motion along $\underline{B}$ towards increasing x, and $\alpha = \pi$, corresponding to motion along $\underline{B}$ towards decreasing x, appear as two points labeled, 0 and π, respectively. For a fixed x value, (8) and (9) provide a linear transformation from the tangential velocity map in Figure 4b to a plane in configuration space parallel to the yz plane and located at the assumed x value. This transformation consists of a clockwise rotation of the disk and a scaling by the factor m/qB_x, as shown in Figure 4c, so that the disk radius becomes the gyroradius, $R_{Lx} = mv/qB_x$, based on the normal magnetic field component B_x.

The terms $A_y(x)$ and $A_z(x)$ in the transformation lead to a shift of the disk location with changing x value such that the center remains on one and the same field line. To see this, we differentiate (8) and (9) to obtain

$$B_x dz = (dA_y/dx)dx = B_z dx \quad (10)$$

$$B_x dy = (-dA_z/dx)dx = B_y dx \quad (11)$$

which are the differential equations for a field line.

In the uniform field region on either side of the current layer, the surface traced by the disk (Figure 4d) is a cylinder of elliptical cross section as shown in Figure 4e for the upstream side, say, of the layer. In the figure, the symbol φ denotes the angle between the incident magnetic field, assumed to lie in the xy plane, and the

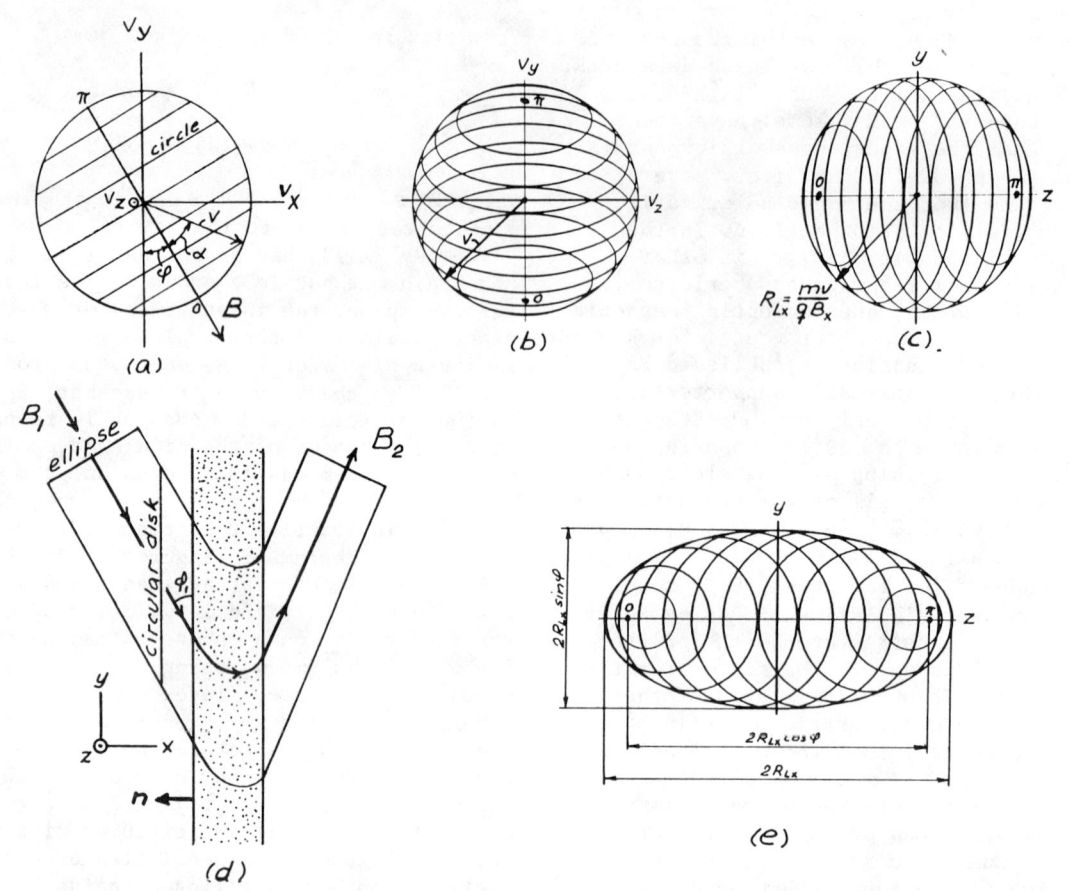

Figure 4. (a) In the de Hoffmann-Teller frame, and at a fixed x value, a particle is located on a sphere of radius v in velocity space. Circles of constant pitch angle on this sphere are shown. (b) Projection of the sphere and the constant-pitch-angle circles on the $v_y v_z$ plane. (c) The generalized tangential momenta provide a linear transformation from the $v_y v_z$ plane to the yz plane. In assigning the location of the points $\alpha=0$ and $\alpha=\pi$ in the yz plane, a positively charged particle, q>0, has been selected. (d) Surface in space traced by the disk (c) as x is changed. (e) Elliptic cross section of the surface (d) on left-hand side of current layer.

surface of the discontinuity. Inside the ellipse are constant-pitch-angle circles, each labeled by a corresponding pitch angle, α. It is geometrically evident from Figure 4 that these circles have their centers at y=0 and $z_\alpha = -(R_{Lx}\cos\alpha)\cos\varphi$. For $0 \leq \alpha < \pi/2$ the particle moves in a helical orbit toward the layer while for $\pi/2 < \alpha < \pi$ it moves away from it. The two foci of the ellipse correspond to $\alpha=0$ and $\alpha=\pi$. For pitch angles near 0 and π, the circles do not touch the ellipse: they correspond to helices in which v_x does not change sign. For circles that touch the ellipse, v_x changes sign at the points of osculation.

As a simple illustration of Figure 4e, consider a particle incident upon the current sheet with pitch angle α_1 ($<\pi/2$) which is reflected and leaves with pitch angle α_2 ($>\pi/2$). Upon use of the expression for z_α given in the previous paragraph, Figure 4e shows that during the interaction with the current sheet, the guiding center of the particle is displaced by the distance

$$\Delta z = z_{\alpha 2} - z_{\alpha 1} = (mv/qB_x)\cos\varphi \, (\cos\alpha_1 - \cos\alpha_2) \quad (12)$$

purely in the z direction. The maximum displacement occurs when $\alpha_1=0$, $\alpha_2=\pi$, in which case the orbit moves from one focus of the ellipse to the other (a result that may be compared to the approximation used by T.W. Speiser, this volume).

A second illustration is provided in Figure 5 which shows a view along the negative normal of a magnetopause rotational discontinuity with $\Delta\theta=135°$. A field line bends as shown in the figure, as it passes through the current layer. Disks are shown at the entrance to (#1), in the middle of (#1½), and at the exit from (#2) the layer. Note that the constant pitch angle pattern in the disk rotates as one progresses through the current layer. Disks #1 and #2 are shown with the same diameter, corresponding to an assumed equal value, $\phi=0$, of the electric potential on both sides of the layer in the dHT frame. Disk #1½ has a smaller diameter corresponding to the presence of a

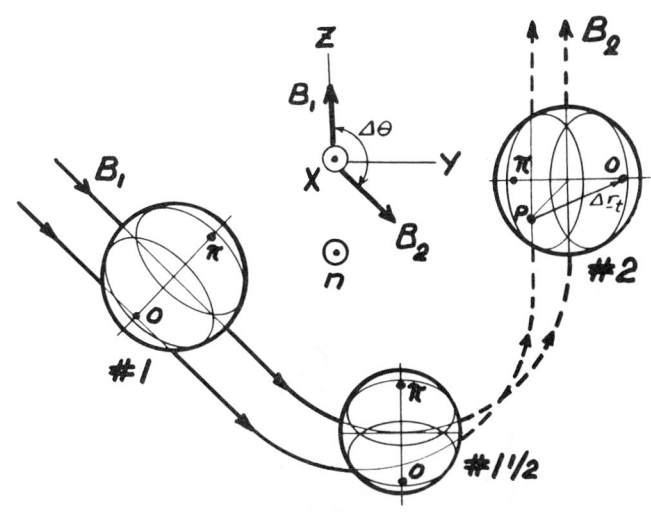

Figure 5. View of the magnetopause from the sun (the coordinates X, Y, and Z are GSM so that X = -x). The interplanetary field B_1 connects with the magnetospheric field B_2 across the magnetopause. Cowley's [1978] disks are shown outside (#1), in the middle of (#1½), and inside (#2) the magnetopause.

potential barrier $\phi(x)$. As an example, a positively charged particle with zero pitch angle enters the layer at the point $\alpha=0$ in disk #1. If the magnetic moment μ is preserved, it leaves at the point $\alpha=0$ in disk #2. Note that because of inertia drift it does not follow the field line through the entry point which leaves the layer at the point labeled P. If the particle were reflected rather than transmitted, it would leave at the point $\alpha=\pi$ in disk #1 instead (assuming constant μ). In reality, the magnetic moment is not necessarily preserved for either reflected or transmitted particles and numerical orbit calculations are needed to obtain the exit pitch angle.

Particle Energization

Equation (3) describes the tangential momentum change of the plasma as it crosses an RD. Depending on circumstances, this change will lead either to energization or to deenergization of the plasma. These effects can be conveniently examined by transforming velocities from the dHT frame to the "laboratory" frame, a procedure that is now in common use for magnetopause RDs as well as for the tail current sheet [e.g., Cowley, 1980; 1982] and for the earth's bow shock [e.g., Paschmann et al., 1980]. As illustrated in Figure 6, this transformation consists of adding a constant velocity $\underline{v}_o$ parallel to the sheet. Energization of transmitted and reflected particles is illustrated in Figures 6b and 6c, while an example of deenergization is shown in Figure 6d. For simplicity, only the case where $\underline{B}_1$, $\underline{B}_2$, and $\underline{v}_o$ all lie in the xy plane is shown.

In the laboratory frame a constant tangential electric field, $\underline{E}_t = -\underline{v}_o \times \underline{B}_n$, will be present, the direction of which is shown in Figure 6. This is the reconnection electric field, and it now becomes clear that $\underline{E}_t$ and B_n are indeed proportional. At the magnetopause $\underline{E}_t$ is a remnant of the interplanetary electric field (IEF) and is directed from dawn to dusk for southward interplanetary magnetic field (IMF). Note that $\underline{E}_t \cdot \underline{I} > 0$ for energization and $\underline{E}_t \cdot \underline{I} < 0$ for deenergization. The former case occurs on the frontside magnetopause where the magnetopause current, $\underline{I}$, is directed from dawn to dusk for southward IMF. The latter case occurs on portions of the magnetopause tailward of the cusp where $\underline{I}$ is directed from dusk to dawn. The suggestion has been made [Crooker, 1979] that it may also occur on the frontside magnetopause as a consequence of cusp reconnection. But since $\underline{I}$ is directed from dawn to dusk on the frontside, $\underline{E}_t$ would then have to be directed from dusk to dawn, i.e., it would have to oppose the IEF. It is difficult to see how such a situation would arise.

The energization or deenergization of individual particles occurs as a result of displacement of the particle guiding center along $\underline{E}_t$. As an example, consider a particle reflected at the current layer. It is easy to show from the velocity triangles in Figure 6c that the kinetic energy increase is

$$\Delta \mathcal{E} = m v v_o \cos\varphi (\cos\alpha_1 - \cos\alpha_2) \equiv q E_t \Delta z \qquad (13)$$

(note that $\alpha_2 > \pi/2$) where Δz is the guiding center displacement. Since $v_o = E_t/B_x$ we can solve this expression for Δz, the result being exactly the displacement given by (12). In other words, orbit considerations based on the conservation of energy and generalized tangential momenta lead to exactly the same result as the frame transformation method described in Figure 6. This is not surprising but it serves to illustrate that the reconnection model which predicts plasma acceleration by the $\underline{I} \times \underline{B}_n$ force is internally consistent: the reconnection electric field is precisely the field needed to energize the plasma particles by the requisite amount. Thus experimental checks of the tangential momentum balance automatically provide a check of the electromechanical part of the energy balance. However, the remaining, thermal part is in general not negligible (see G. Paschmann, this volume).

Equation (12) indicates that the particle displacement Δz along E_t becomes very large as B_x approaches zero. However, at the same time E_t itself becomes small, the result being that the energy gain $\Delta \mathcal{E}$ remains the same. Therefore, as stated earlier, the plasma energization is independent of the reconnection rate. It is also evident from (12) that the energization of electrons is negligible compared to that of ions.

Assuming for simplicity that no net potential change $\Delta\phi$ occurs across the magnetopause in the dHT frame, the energy change of a particle transmitted through an RD can be shown by the frame transformation method to be

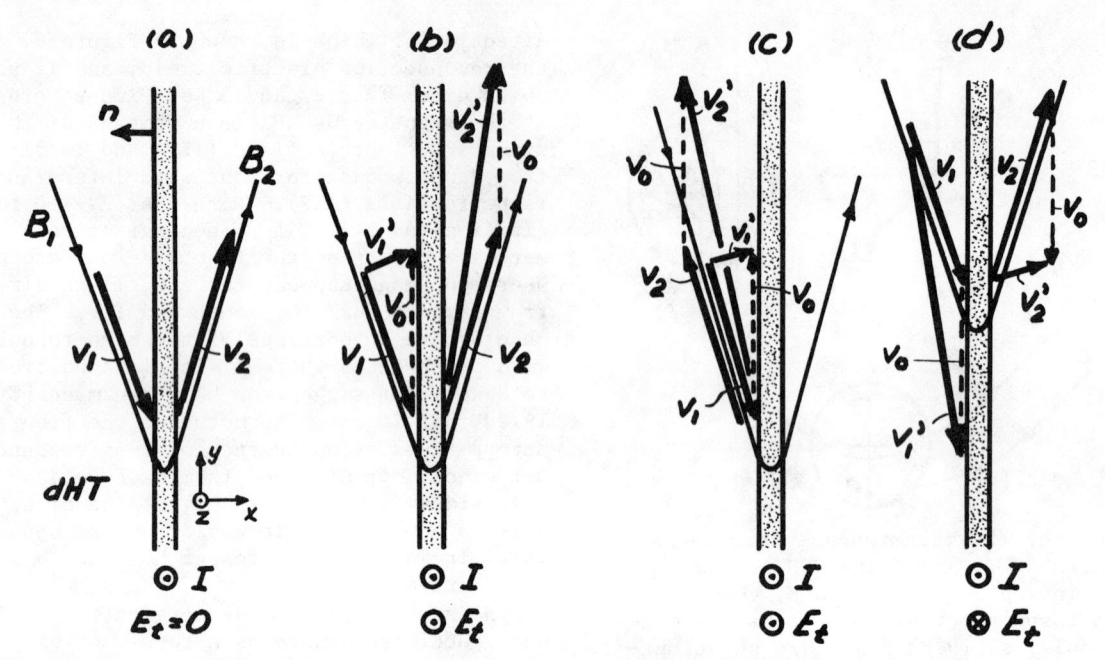

Figure 6. Velocity change of a particle as it interacts with the magnetopause current layer (shaded). (a) In the de Hoffmann-Teller frame, a transmitted particle has velocities $v_1 = v\cos\alpha_1$ and $v_2 = v\cos\alpha_2$ before and after the interaction. (b) In the laboratory frame a transformation velocity $\underline{v}_o$ has been added and the particle now has velocities v_1' and v_2' instead where $v_2' \gg v_1'$. (c) A reflected particle is similarly energized. (d) Deenergization is achieved by reversing the direction of $\underline{v}_o$. The magnetopause current $\underline{I}$ and the electric field $\underline{E}_t = -\underline{v}_o \times B_n$ are parallel for energization, antiparallel for deenergization.

$$\Delta\mathcal{E} = m v \underline{v}_o \cdot [(\underline{B}_{t2}/B_2)\cos\alpha_2 - (\underline{B}_{t1}/B_1)\cos\alpha_1] \quad (14)$$

After simple manipulations, this result can also be interpreted in terms of a displacement of the particle guiding center along $\underline{E}_t$, as illustrated in Figure 5. For example, a particle which crosses the layer while retaining $\alpha=0$ will undergo an effective tangential displacement $\Delta\underline{r}_t$ from point P to point $\alpha=0$ in disk #2 so that its energy increase will be $\Delta\mathcal{E} = q\underline{E}_t \cdot \Delta\underline{r}_t$. The location of point P is such that the vector from the center of disk #2 to P is equal to the vector from the center of disk #1 to the point $\alpha=0$ in that disk. The reason that the distance from the point $\alpha=0$ in disk #1 to the point P is not included in $\Delta\underline{r}_t$ is that these two points have the same electric potential: the total transformation electric field $\underline{E}_o = -\underline{v}_o \times \underline{B}$ has no component along $\underline{B}$ and the two points are located on one and the same field line. As mentioned already in section 3, $\underline{E}_o$ does have a substantial component along the normal direction so that in the laboratory frame the electric potential changes across the layer. However, the energy change of a particle as it moves across this potential difference is exactly equal and opposite to the energy change associated with the displacement of the particle in the tangential field $\underline{E}_t$ from the point $\alpha=0$ in disk #1 to point P in disk #2. Note also that the latter displacement is different for $\Delta\theta = 135°$ and $\Delta\theta = 215°$ whereas $\Delta\underline{r}_t$ is the same.

5. Nonsteady Localized Reconnection

In the preceding sections, we have dealt with reconnection as a steady-state process. However, there is mounting evidence [Russell and Elphic, 1978; Cowley, 1982] that a nonsteady patchy version of reconnection, referred to as flux transfer events (FTEs), may be the dominant reconnection mode at the magnetopause. The FTE is envisaged as a pair of flux tubes, one in each hemisphere, each passing through a "hole" in the magnetopause and connecting to the earth and each being accelerated along it away from the subsolar region by the magnetic tension associated with the sharp kink in the tube at the hole (see Figure 7a). Such tubes seem to have typical cross-sectional dimensions of the order of 1 R_E, implying a hole dimension of a few R_E since $B_n < B$, but in reality there may be a continuum of sizes ranging from dimensions comparable to the ion gyroradius up to the scale of quasisteady reconnection. Detailed discussion of the morphology and statistics of FTEs may be found elsewhere in this volume. Here a few general remarks are offered concerning these structures and their relation to reconnection.

(1) As illustrated in Figure 7b, the pair of

holes in the magnetopause associated with FTEs is the result of reconnection, limited to a narrow longitude interval, which started near the equatorial region and then ceased before a steady state could be established, the method of cessation being conversion of the reconnection line into two passive or almost passive X lines and an O line.

(2) If quasisteady reconnection takes place in a limited longitude segment then, for $\Delta\theta \neq \pi$, there are regions on the magnetopause surface, located to the left or to the right of the reconnection site, where one would observe open field lines but not the plasma jetting discussed previously. Signatures of this type should be looked for in the data and their relation to FTEs should be examined in detail.

(3) Whenever the IMF is due south, the magnetopause situation would seem to be an ideal one for quasisteady reconnection to occur at its maximum permitted rate. Yet, this is apparently not what happens. Even with a southward IMF, quasisteady reconnection is observed only occasionally, indicating that the process occurs sporadically and/or that it is limited to narrow longitude segments. If the currently popular interpretation of FTE signatures is correct, then what is observed most of the time is therefore a patchy time-dependent version of reconnection (for a detailed discussion of occurrence rates, see Cowley, 1982). This set of circumstances indicates the existence of a threshold, other than a southward IMF, for the onset of the process. Either by design or by accident the magnetopause plasma state seems to hover around this threshold.

(4) The detailed nature of the threshold is not understood but the following scenario illustrates how FTEs might be generated. Assume that a flux tube containing interplanetary magnetic field lines drapes over the subsolar magnetopause and gets hung up there, perhaps in a preexisting indentation. The plasma in this tube will then escape by flowing tailward along the lines of force, the result being a lowering of plasma density, N, β value (in particular $\beta_\| \equiv p_\| \mu_0/B^2$), and Alfvén Mach number, M_A, of the flow. It may be argued that low values of these factors should be conducive to the onset of reconnection between the field lines in the flux tube and the geomagnetic field. For example, the tearing mode growth rate is inversely proportional to N [e.g., Quest and Coroniti, 1981]. As soon as reconnection has been initiated, two developments occur: (i) a deepening of the magnetopause indentation associated with the flux tube must take place as reconnection

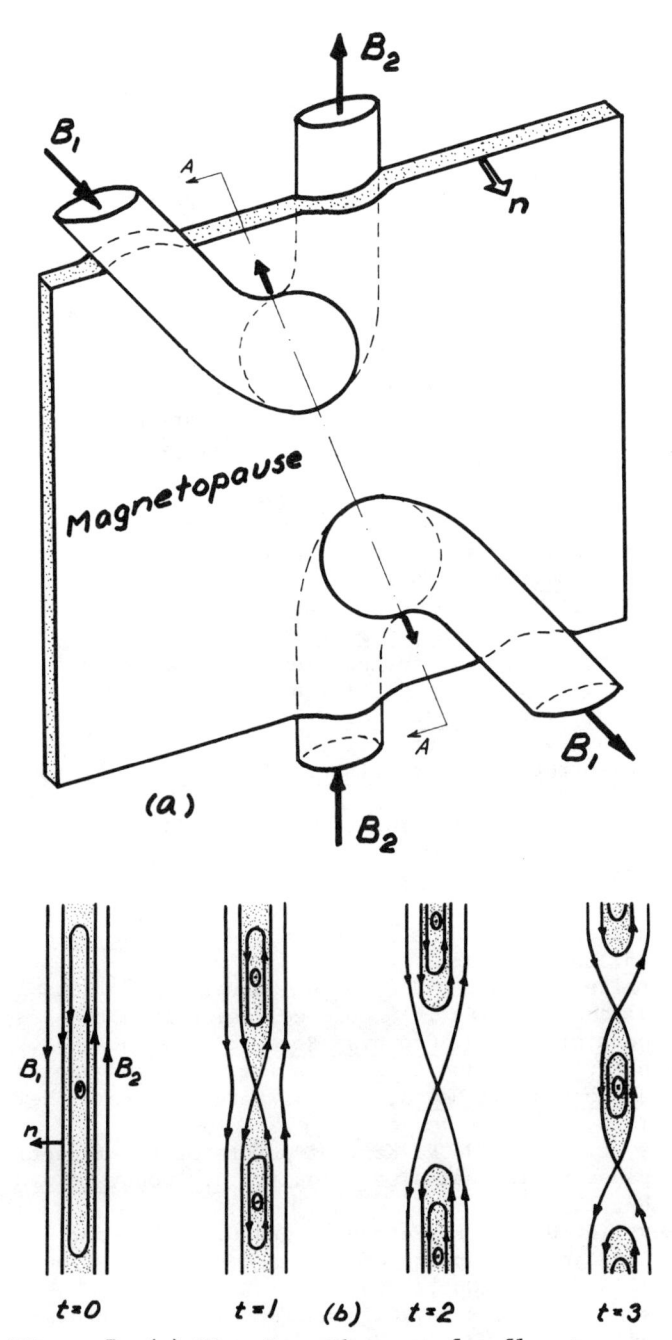

Figure 7. (a) View from the sun of a flux transfer event (FTE) on the magnetopause. (b) Side view of the equatorial magnetopause (shaded) during the development of an FTE. At t=0 the frontside magnetopause is closed, i.e., it is very nearly a one-dimensional TD. However, the northern and southern magnetic cusps (not shown) lead to weak two-dimensional effects which produce an O-type magnetic null line at the subsolar point. Reconnection starts, splitting the O-line into two O-lines and an X line, the latter remaining at the subsolar point. At t=1 all the magnetic flux comprising the original magnetopause has been reconnected. At t=2 some interplanetary magnetic field, B_1, has become connected to the earth's field, B_2. At this time, the X line is converted to an O-line and two X lines, the former remaining at the subsolar point. A small amount of reconnection occurs at the two X lines, the result being the formation of a closed magnetopause near the subsolar point at t=3, along with two separate holes in the magnetopause. The situation at t=3 corresponds to the section A-A in part (a) of the figure.

erodes flux locally from the magnetosphere; (ii) the reconnection region gets replenished with fresh solar-wind plasma in which N, β, and M_A return to their original values. Via the threshold, the latter effect may lead to the cessation of reconnection. The former effect creates a suitable trap for new interplanetary magnetic flux tubes. These indentations and reconnection regions may also get swept away over the flanks of the magnetopause so that FTEs should be observable over the entire frontside magnetopause and not just near local noon. And on occasion, the O-type null line created when reconnection ceases (see Figure 7b) may be swept over one or the other of the poles so that FTE signatures characteristic of the southern hemisphere may occasionally be observed in the northern hemisphere and vice versa.

It also seems reasonable to assume that for suitable plasma conditions a quasisteady reconnection configuration may be established which either remains limited to a narrow longitude segment or spreads over a substantial part of the frontside magnetopause.

The above scenario is by no means unique, but it may serve as a useful guide for future theoretical and observational studies of FTEs. In particular, it illustrates the importance of developing a better understanding of the threshold conditions for onset of magnetopause reconnection. It seems clear that these thresholds must be associated with local conditions at the reconnection site rather than with global boundary conditions. A complete theory is needed for the onset and evolution of the tearing mode in a streaming collision-free plasma when the magnetic field exhibits strong shear as it usually does in the magnetopause current layer.

Acknowledgements. The research was supported by the Division of Atmospheric Sciences, National Science Foundation, under Grant ATM-8201974, and by the National Aeronautics and Space Administration by Grant NSG5348 to Dartmouth College. The geometrical derivation of Cowley's disks in Figure 4, including the constant-pitch-angle curves on each disk, was developed in collaboration with C. Thron.

References

Berchem, J., and C. T. Russell, The thickness of the magnetopause current layer: ISEE 1 and 2 observations, J. Geophys. Res., 87, 2108, 1982a.

Berchem, J., and C. T. Russell, Magnetic field rotation through the magnetopause, J. Geophys. Res., 87, 8139, 1982b.

Cowley, S. W. H., Comments on the merging of non-antiparallel fields, J. Geophys. Res., 81, 3455, 1976.

Cowley, S. W. H., A note on the motion of charged particles in one-dimensional magnetic current sheets, Planet. Space Sci., 26, 539, 1978.

Cowley, S. W. H., Plasma populations in a simple open model magnetosphere, Space Sci. Rev., 26, 217, 1980.

Cowley, S. W. H., The causes of convection in the earth's magnetosphere: a review of developments during the IMS, Revs. Geophys. Space Phys., 20, 531, 1982.

Cowley, S. W. H., D.J. Southwood, and M. A. Saunders, Interpretation of magnetic field perturbations in the earth's magnetopause boundary layers, Planet. Space Sci., 31, 1237, 1983.

Crooker, N. U., Dayside merging and cusp geometry, J. Geophys. Res., 84, 951, 1979.

Dungey, J. W., Interplanetary magnetic field and the auroral zones, Phys. Rev. Lett., 6, 47, 1961.

Gonzalez, W. D., and F. S. Mozer, A quantitative model for the potential resulting from reconnection with an arbitrary interplanetary magnetic field, J. Geophys. Res., 79, 4186, 1974.

Gosling, J. T., J. R. Asbridge, S. J. Bame, W. C. Feldman, G. Paschmann, N. Sckopke, and C. T. Russell, Evidence for quasi-stationary reconnection at the dayside magnetopause, J. Geophys. Res., 87, 2147, 1982.

Hill, T. W., Magnetic merging in a collisionless plasma, J. Geophys. Res., 80, 4689, 1975.

Hudson, P. D., Discontinuities in an anisotropic plasma and their identification in the solar wind, Planet. Space Sci., 18, 1611, 1970.

Hudson, P. D., Rotational discontinuities in an anisotropic plasma, Planet. Space Sci., 19, 1693, 1971.

Hudson, P. D., Rotational discontinuities in an anisotropic plasma - II, Planet. Space Sci., 21, 475, 1973.

Johnson, F. S., The gross character of the geomagnetic field in the solar wind, J. Geophys. Res., 65, 3049, 1960.

Kirkland, K. B., and B. U. Ö. Sonnerup, Contact discontinuities in a cold collision-free two-beam plasma, J. Geophys. Res., 87, 10355, 1982.

Landau, L. D., and E. M. Lifshitz, Electrodynamics of Continuous Media, p. 224, Pergamon Press, New York, 1960.

Lee, L. C., and J. R. Kan, Structure of the magnetopause rotational discontinuity, J. Geophys. Res., 87, 139, 1982.

Levy, R. H., H. E. Petschek, and G. L. Siscoe, Aerodynamic aspects of the magnetospheric flow, AIAA J., 2, 2065, 1964.

Paschmann, G., B. U. Ö. Sonnerup, I. Papamastorakis, N. Sckopke, G. Haerendel, S. J. Bame, J. R. Asbridge, J. T. Gosling, C. T. Russell, and R. C. Elphic, Plasma acceleration at the earth's magnetopause: evidence for reconnection, Nature, 282, 243, 1979.

Paschmann, G., N. Sckopke, J. R. Asbridge, S. J. Bame, and J. T. Gosling, Energization of solar wind ions by reflection from the earth's bow shock, J. Geophys. Res., 85, 4689, 1980.

Petschek, H. E., Magnetic field annihilation, in The Physics of Solar Flares, W. N. Hess, ed., NASA SP-50, 425, 1964.

Quest, K. B., and F. V. Coroniti, Tearing at the dayside magnetopause, J. Geophys. Res., 86, 3289, 1981.

Russell, C. T., and R. C. Elphic, Initial ISEE

magnetometer results: magnetopause observations, Space Sci. Rev., 22, 681, 1978.

Scholer, M., F. M. Ipavich, G. Gloeckler, D. Hovestadt, and B. Klecker, Leakage of magnetospheric ions into the magnetosheath along reconnected field lines at the dayside magnetopause, J. Geophys. Res., 86, 1299, 1981.

Scholer, M., and F. M. Ipavich, Interaction of ring current ions with the magnetopause, J. Geophys. Res., 88, 6937, 1983.

Sonnerup, B. U. Ö., Magnetopause reconnection rate, J. Geophys. Res., 79, 1546, 1974.

Sonnerup, B. U. Ö., Magnetic field reconnection, in Solar System Plasma Physics, L. J. Lanzerotti, C. F. Kennel, and E. N. Parker, eds., p. 45, North Holland, Amsterdam, 1979.

Sonnerup, B. U. Ö., and L. J. Cahill, Jr., Magnetopause structure and attitude from Explorer 12 observations, J. Geophys. Res., 72, 171, 1967.

Sonnerup, B. U. Ö., and B. G. Ledley, Magnetopause rotational forms, J.Geophys.Res., 79, 4309, 1974.

Sonnerup, B. U. Ö., and B. G. Ledley, OGO 5 magnetopause structure and classical reconnection, J. Geophys. Res., 84, 399, 1979a.

Sonnerup, B. U. Ö., and B. G. Ledley, Electromagnetic structure of the magnetopause and boundary layer, in Magnetospheric Boundary Layers, B. Battrick, ed., p. 401, ESA Scientific and Technical Publ. Branch, Noordwijk, 1979b.

Sonnerup, B. U. Ö., G. Paschmann, I. Papamastorakis, N. Sckopke, G. Haerendel, S. J. Bame, J. R. Asbridge, J. T. Gosling, and C. T. Russell, Evidence for magnetic field reconnection at the earth's magnetopause, J. Geophys. Res., 86, 10049, 1981.

Su, S.-Y., and B. U. Ö. Sonnerup, First-order orbit theory of the rotational discontinuity, Phys. Fluids, 11, 851, (Correction, 13, 1423), 1968.

Swift, D. W., Substorms and magnetospheric energy transfer processes, in Dynamics of the Magnetosphere, S.-I. Akasofu, ed., p. 327, D. Reidel Publ. Co., Dordrecht, 1980.

Swift, D. W., and L. C. Lee, Rotational discontinuities and the structure of the magnetopause, J. Geophys. Res., 88, 111, 1983.

Terasawa, T., Hall current effect on tearing mode instability, Geophys. Res. Lett., 10, 475, 1983.

Terasawa, T., Hall tearing instability in dissipationless plasmas, submitted to Phys. Rev. Lett., 1984.

Wang, D.-J., and B. U. Ö. Sonnerup, Electrostatic structure of the rotational discontinuity: trapped particles, to be submitted to Phys. Fluids, 1984.

Yang, C.-K, and B. U. Ö. Sonnerup, Compressible magnetopause reconnection, J. Geophys. Res., 82, 699, 1977.

EVIDENCE OF MAGNETIC MERGING FROM LOW-ALTITUDE SPACECRAFT AND GROUND-BASED EXPERIMENTS

Patricia H. Reiff

Department of Space Physics and Astronomy, Rice University, Houston, Texas 77251

Abstract. Because dayside merging occurs at the magnetopause (either near the nose or at the cusps), evidence of merging obtained by low altitude spacecraft or ground-based instruments must necessarily be indirect. Nevertheless, the evidence is compelling. Much of the evidence consists of correlations between magnetospheric parameters and the interplanetary magnetic field (IMF). For example, geomagnetic perturbations near the noon cusp are affected by the sign of the east-west component of the IMF, both on the ground (the "Svalgaard-Mansurov effect") and at spacecraft altitudes; this is most easily explained in terms of the east-west stress transmitted along interconnected field lines. Other lines of evidence include the expansion of the auroral zone and the intensification of the auroral electrojets with increases in the southward component of the IMF, IMF control of the access of energetic solar particles to the polar cap, and the change in the sign of magnetic perturbations as the IMF changes from southward to northward, indicating a region of sunward flow in the polar cap. Using low-altitude spacecraft data, one can also infer the existence of magnetic merging from the significant correlations of the cross-polar-cap potential difference, and of the Birkeland current intensity, with the southward component of the IMF. The observed energy-latitude dispersion of cusp ions, wherein the characteristic energy of precipitating ions decreases with increasing latitude, is most readily explained by magnetic merging. At mid-altitude (3-4 R_E), one can even discern the corresponding merging dispersion signature within a single pitch-angle scan. By analyzing these dispersion signatures, one can infer that acceleration of magnetosheath ions by about 1 keV occurs at a geocentric field-aligned distance ~10 R_E, as one expects from magnetic merging. During periods of northward IMF, one can also observe dispersions of the reverse type, which may be caused by merging between northward magnetosheath field lines and geomagnetic field lines tailward of the cusp. Such tail merging may also explain features of the "theta" arc phenomenon.

Introduction

The concept of magnetic merging was first applied to the magnetosphere by Dungey [1961], who proposed magnetic interconnection as a specific mechanism whereby the solar wind drives magnetospheric and ionospheric convection. A southward Interplanetary Magnetic Field (IMF) connects to northward magnetospheric magnetic fields through the formation of a magnetic X-line at the dayside magnetopause (line 1 in Figure 1b, from Cowley [1982]). Except very near the X-line, the concept of "frozen-in-flux" applies [Alfvén and Fälthammar, 1963], and the solar wind drags the open magnetospheric field lines antisunward (lines 2-4). At another magnetic X-line in the magnetotail (line 5), magnetospheric field lines from the northern and southern polar caps reconnect to form closed field lines that return at lower latitudes (along the flanks, outside the plane of this figure) to the dayside magnetopause (lines 6-8). Thus there are three topological classes of magnetic field lines: interplanetary field lines that do not intersect the earth's surface (line 1); "open" magnetospheric field lines that intersect the earth at one point and extend into the solar wind (lines 2-4); and "closed" magnetospheric field lines that intersect the earth at two places (lines 6-8). Flow of plasma across a "separatrix" separating magnetic fields of different topology is, by definition, magnetic merging [Vasyliunas, 1975].

The ionospheric footprint of the magnetospheric convection pattern is shown in Figure 1a. Shown is the classic "two-cell" convection pattern, with antisunward flow over the polar caps and return flow at lower latitudes. It should be emphasized that the existence of the convection pattern is by itself not unambiguous evidence of magnetic merging, because quasiviscous processes in a near-equatorial boundary layer can also, in principle, produce a flow pattern of this type [Axford and Hines, 1961]. However, it can be shown [Hill, 1983] that cross-field diffusion of magnetosheath particles into the magnetosphere (probably the dominant quasiviscous process) can be expected to yield polar-cap potential drops

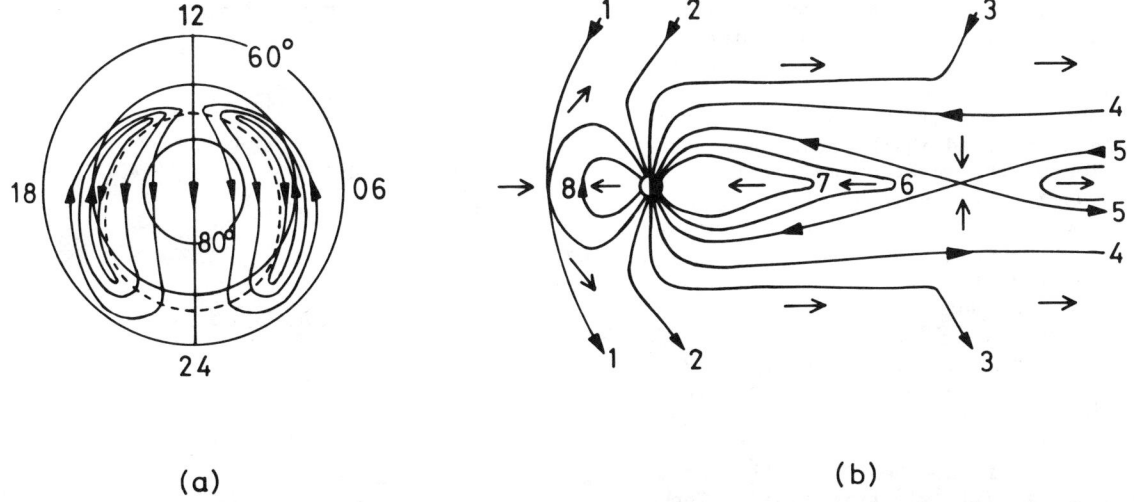

Fig. 1. (a) Sketch of the basic two-cell convection pattern observed under usual conditions, when the IMF B_z is not strongly positive. The dashed line is the boundary of open field lines, and the antisunward convection just equatorward of it presumably maps to the magnetopause boundary layers on closed field lines. The flow in the vicinity of noon is represented as a narrow "throat" structure in conformity with the observations of Heelis et al. [1976], but no attempt has been made to represent the "Harang discontinuity" on the nightside. The coordinates are magnetic latitude and local time. (b) Sketch of the convection cycle proposed by Dungey [1961] initiated by dayside merging between the earth's magnetic field and a southward-directed interplanetary magnetic field. The numbers indicate the successive positions of a flux tube in the cyclic flow [from Cowley, 1982].

of, at most, only about 15 kV, whereas the observed polar-cap potential drop is typically ~50 kV and can increase to nearly ~200 kV at times [Reiff et al., 1981]. Thus the boundary between open and closed magnetic field lines (shown dashed in Figure 1a) is shown slightly poleward of the convection reversal (the boundary between sunward and antisunward-flowing plasma). Although the open-closed boundary is notoriously difficult to identify, several spacecraft studies have suggested that this separation (1°-2°) is real and persistent [Heelis et al., 1980; Reiff, 1979; McDiarmid et al., 1978; Klumpar et al., 1976]. This corresponds to a hybrid model in which magnetospheric convection is driven by both magnetic merging and quasi-viscous drag with the former mechanism dominating in general [Burch et al., 1984; Reiff, 1979; Oberc, 1979]. Although the existence of ionospheric convection per se is not incontrovertible evidence for the existence of magnetic merging, one can build a strong case for magnetic merging by examining the IMF control of the strength and direction of the convective flow. Evidence of B_z control of the strength, and B_y control of the direction, of the flow is available from both ground-based instrumentation and low altitude spacecraft data.

Another class of evidence for magnetic merging is associated with the net transfer of magnetic flux from the dayside to the nightside that is observed whenever the IMF becomes more southward. Evidence for IMF control of the polar cap size derives from studies of auroral, convection, and cusp boundaries, as well as from observed motions of auroral electrojets.

A magnetic merging model also predicts effects on the magnetosphere of the x-component of the IMF (principally interhemispheric differences); these have been reviewed recently by Cowley [1981b] and will not be discussed here.

Finally, the effects of magnetic merging can be observed in the resulting particle distributions, producing a convective dispersion that can be observed from both low- and mid-altitude spacecraft.

Effects of the y-Component of the IMF

The case shown in Figure 1 is the simplest: the IMF is due southward, and the resulting flow is symmetric about the noon-midnight meridian. The X-line in this case is in the equatorial plane, and in fact extends completely around the earth, joining with the tail neutral line [Vasyliunas, 1976]. In the more realistic case where the IMF has significant x- and y-components, the location and orientation of the X-line is more difficult to predict. In some models, the X-line still passes through the subsolar point, but has a tilt depending on the sign of B_y [Gonzales and Mozer, 1974; Stern, 1973]. In other models, merging occurs preferentially at the dayside cusps [Crooker, 1979, 1980; Haerendel et al., 1978] and may favor one side (dawn or dusk) or the other, depending on the sign of B_y (and the hemisphere in question) [Luhmann et al., 1984;

also this volume]. Regardless of the location of the X-line, all models agree on two fundamental predictions: (1) the interconnection geometry is such that an arbitrary electric field in the solar wind maps along field lines to a generally dawn-dusk electric field in the polar cap; and (2) the tension on the interconnected field lines causes the plasma in the northern hemisphere to flow westward (eastward) if the y-component of the IMF is positive (negative) and vice versa in the southern hemisphere [Russell and Atkinson, 1973; see reviews by Burch and Heelis, 1980; Cowley, 1981a; Nishida, 1983; Burch, 1983]. Both of these effects are illustrated in Figure 2 [from Cowley, 1973]. In this figure, a -y component of the IMF corresponds to a north-to-south interplanetary electric field. If the field lines are equipotentials, then this will map to a dawn-to-dusk ionospheric electric field. The tension on the field lines will correspond to a pull towards the dusk in the northern hemisphere, and towards the dawn in the southern. Although the magnetic field configuration is more complicated than this simple vacuum superposition of a dipole and a horizontal IMF [Longenecker and Roederer, 1981], the qualitative nature of this east-west flow asymmetry was the one of earliest established unambiguous consequences of magnetic merging.

The east-west flows associated with merging have been observed in many ways: in horizontal magnetic perturbations observed at dayside auroral zone stations and vertical magnetic perturbations at the magnetic pole (the "Svalgaard-Mansurov effect"; see Matsushita and Trotter [1980] and references therein); in direct measurements of plasma flows and associated electric fields from low- and mid-altitude spacecraft [Heelis, 1984; Heppner, 1972, 1977], and indirectly from east-west magnetic perturbations (which are opposite to the flow direction in the northern hemisphere, and in the same direction in the southern) [see reviews by Burch and Heelis, 1979; Meng, 1979; Cowley, 1981a; Burch, 1983]. One can interpret the magnetic perturbations in two equivalent ways: one is the concept of "line-tying," wherein the solar wind tugs on the earth's field, causing the field to tilt [McDiarmid et al., 1978]; the other is to consider the current system that is required to cause the ionosphere to flow. For example, to cause eastward flow in the northern hemisphere, one requires two matching Birkeland current sheets, one upward on the equatorward side, one downward on the poleward side, closing by an equatorward Pedersen current in the ionosphere. The resulting magnetic perturbation for that current system, as measured above the ionosphere between the Birkeland current sheets, is westward (opposite the plasma flow), as is observed [Saflekos and Potemra, 1980; McDiarmid et al., 1978].

The dawn-dusk asymmetry of the access of low- and high-energy solar electrons to the polar cap as a function of B_y is also best explained by the open model [Akasofu et al., 1981; McDiarmid et al., 1980; Meng, 1979; Meng et al., 1977].

Other magnetospheric asymmetries consistent with merging can be found in the magnetotail [see Cowley, 1981a, and others in this volume].

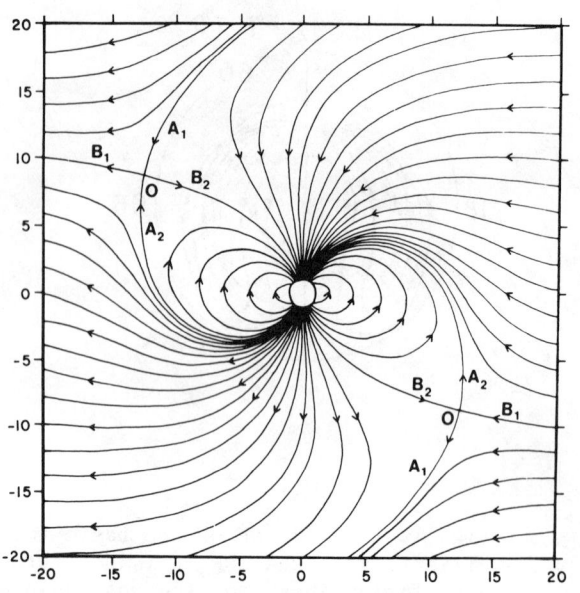

Fig. 2. Magnetic field line configuration in the dawn-dusk meridian looking from the sun for a vacuum superposition of a dipole and a $-B_y$ interplanetary magnetic field. Note that, if field lines are equipotentials, a north-south solar-wind electric field maps to a dawn-dusk polar-cap field. Note also that the tension on the field lines causes duskward flow in the northern hemisphere [from Cowley, 1973].

Effects of the z-Component of the IMF

Another major line of evidence for magnetic merging involves the effects of the southward component of the IMF. First, erosion of flux from the dayside is expected to increase the size of the polar cap when the IMF turns southward, and this has been widely confirmed from cusp measurements [e.g., Reiff et al., 1980; Burch, 1972], from the size of the auroral zone [e.g., Burch, 1979] and of open regions for solar particle access [Akasofu et al., 1981; see review by Meng, 1979] and from other particle boundaries [e.g., Gussenhoven et al., 1981]. One example is shown in Figure 3, from Meng [1979], which shows a ten-degree decrease in polar cap radius as the IMF B_z goes from -4 to + 4 nT.

Perhaps of greater importance to the magnetosphere and ionosphere is the increase in the magnitude of the convection (as quantified by the cross-polar-cap potential drop). It is clear that increases in the southward component of the IMF lead to larger polar cap potentials, and that specific merging models [e.g., Hill 1975; Sonnerup 1974] predict the polar cap potential better than simply a fixed fraction of the east-

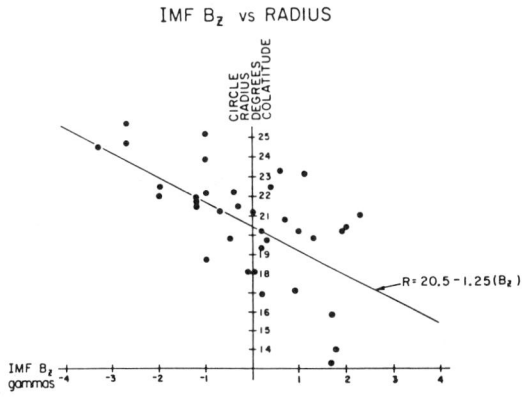

Fig. 3. Response of the auroral oval size to changes in the hourly IMF B_z component [from Meng, 1979].

west solar-wind electric field $V_{sw}B_{sw}$; however, it is not clear yet which merging model is the best predictor [Wygant et al., 1983; Doyle and Burke, 1983; Reiff et al., 1981; Sergeev and Kuznetsov, 1981; Lei et al., 1981]. It is also clear that the potential drop appears to saturate at a value of ≲ 200 kV, but the reason for that saturation is not clear. It may be caused by an amplification process in the magnetosheath; alternatively, it may be a consequence of finite ionospheric conductivity limiting the potential drop [Hill et al., 1976].

As the polar cap potential increases, other measurable quantities also increase, contributing to the confirmation. For example, as the ionospheric Pedersen currents increase, so must the Birkeland currents feeding them, and this has been observed [Figure 4, from Iijima and Potemra, 1982]. Also, as the Hall current increases, the electrojet intensity increases [Sergeev and Kuznetsov, 1981]. Another tool involves using the global pattern of ionospheric currents to remotely infer the dayside merging rate. Through an inversion technique combining a conductivity model with the global horizontal current pattern inferred from magnetometer chain data, one can deduce not only the electric field pattern, but the Birkeland current pattern as well [Kamide et al., 1981].

Another consequence of the increase in potential drop and transfer of magnetic flux to the magnetotail is the general correlation of geomagnetic activity (as quantified, e.g., by the AL index) with B_z [see, e.g., Clauer et al., 1983; Akasofu, 1980]. Again, there is still controversy about which function of the IMF parameters is the best predictor of geomagnetic activity, and whether the tail response is "driven" or "unloading" in nature (or a combination); however, it is clear that the evidence for IMF control of auroral-zone activity is quite strong.

Magnetospheric Convection for Northward IMF

There is some evidence that the magnetosphere is qualitatively, not just quantitatively, different during times of northward IMF, and many of the differences are attributable to a different magnetic configuration caused by magnetic merging. In some ways the magnetosphere appears to be a "half-wave rectifier" — that is, certain magnetospheric parameters are functions only of B_s, where $B_s = -B_z$ for $B_z < 0$, and $B_s = 0$ for $B_z \geq 0$ [Crooker, 1980; Russell and McPherron, 1973; Arnoldy, 1971]. For example, the amount of magnetic flux transferred from closed to open field lines goes to zero as B_s decreases to zero, and remains zero for any value of $B_z > 0$. Other functions do not have a cutoff at $B_z = 0$, however: for example, the polar cap size continues to shrink as the IMF becomes increasingly northward, but, as far as we know, remains of finite size; the polar cap potential continues to decrease as B_z increases, and so on.

Perhaps the most interesting qualitative difference is in the direction of the polar cap convection. When B_z is southward, the flow expected from merging is in the same direction as that from viscous processes. When the IMF turns northward, merging with closed or open tail field lines can occur (Figure 5, from Cowley [1982]). Depending on the configuration (and that, in turn, is dependent on both the y- and x-components of the IMF), sunward flow can occur in one or both polar caps, opposite to the viscous flow. This reversed flow was first inferred from magnetic perturbations [Maezawa, 1976], long after its prediction from magnetic merging theory [Russell, 1972; Dungey, 1963]. Thus a three-cell [Crooker, 1980] or four-cell [Burch et al., 1980; Burke et al., 1979; Reiff, 1979] pattern can occur, with a viscous cell along each edge and one or two merging cells driving sunward convection in the polar cap.

Evidence for reversed flow in the polar cap comes both from ground magnetometer measurements [Maezawa, 1976], in situ flow and magnetic field measurements [Reiff, 1982; Banks et al., 1981;

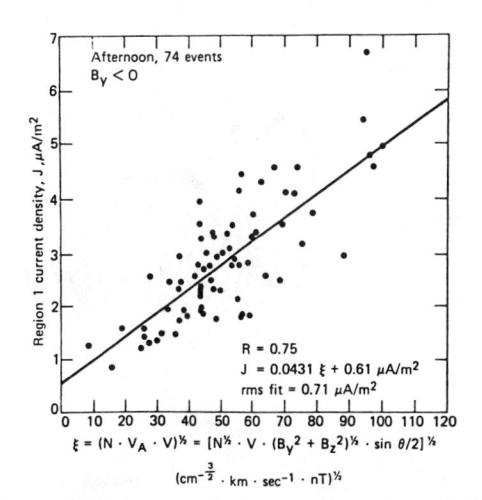

Fig. 4. Dependence of the region-1 Birkeland current intensity on the epsilon parameter [from Iijima and Potemra, 1982].

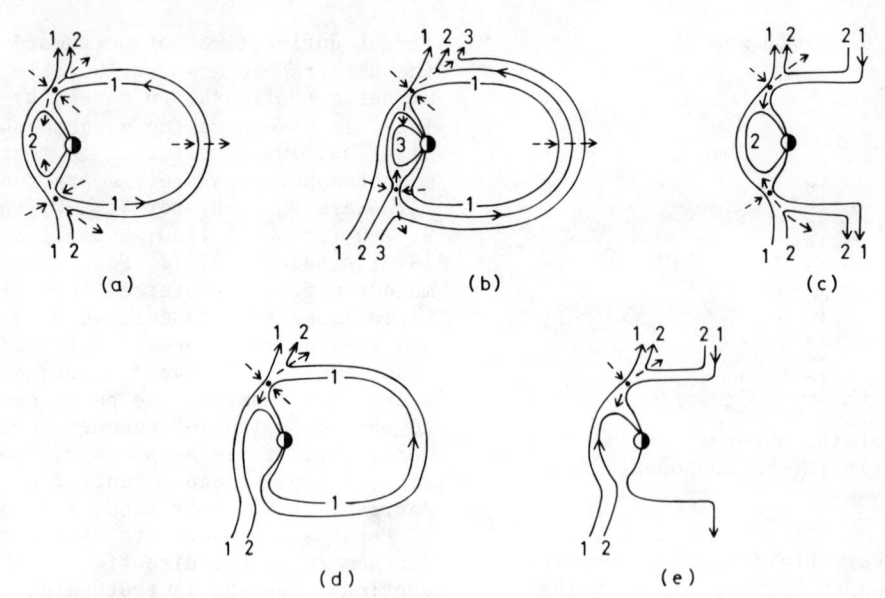

Fig. 5. Schematic sketches of merging between northward directed IMF lines and open and closed tail lobes. In (a) to (c) connection occurs to both lobes, although not simultaneously in (b); in (d) and (e) connection of a given IMF line occurs to only one lobe owing to IMF $B_y \neq 0$, although other IMF lines may similarly connect to the other lobe. The numbers indicate the field line sequence and the heavy dot the location of the x-line. The dashed lines indicate the flow in the vicinity of the latter. Case (a) is that discussed by Dungey [1963], case (d) by McDiarmid et al. [1980], and case (e) by Russell [1972], Maezawa [1976], and Crooker [1979] [from Cowley, 1982].

McDiarmid et al., 1980; Burke et al., 1979; Horwitz and Akasofu, 1979], and laboratory experiments [Dubinin and Potanin, 1980]. Figure 6 shows a schematic [from Reiff and Burch, 1984] plus two examples of a four-cell convection pattern expected for strongly northward IMF. Unfortunately, IMF data are not available for this time period. (Atmosphere Explorer data courtesy R. A. Heelis.) One can even discern a reversed convection "throat" near noon. The two sequential orbits show that the pattern is not merely a time variation.

Furthermore, there may well be atmospheric flywheel effects occuring, if the IMF has just recently turned northward. It is clear that ionospheric convection can set the upper atmosphere in motion [Killeen et al., 1982]; thus, if one has strong magnetospheric convection for several hours, a switchoff of magnetic merging on the dayside will stop the external driving currents, but the upper atmosphere may keep the ionosphere moving for several more hours. Evidence for this was found by Wygant et al. [1983] who found a several-hour decay of the polar-cap

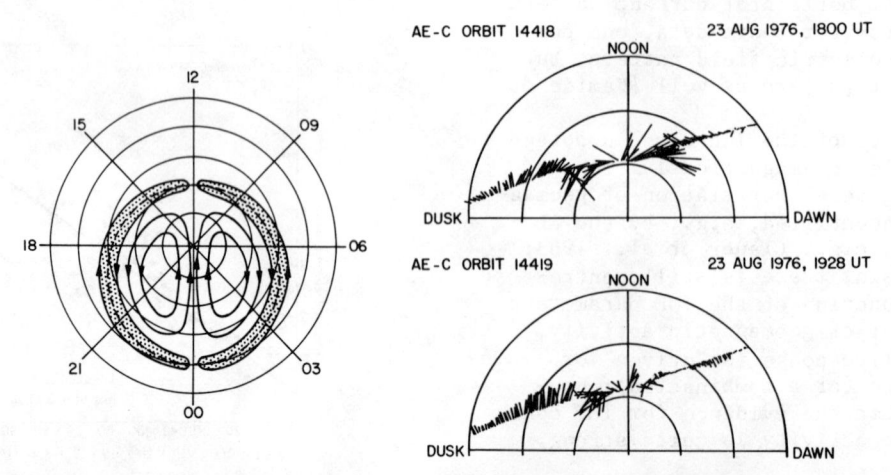

Fig. 6. (a) Schematic diagram of a four-cell reversed convection pattern expected for $B_z > 0$, $|B_y| < |B_z|$ [from Reiff and Burch, 1984]. (b) Two examples of this kind of convection from Atmosphere Explorer-C, showing a reverse throat (data courtesy of R. A. Heelis).

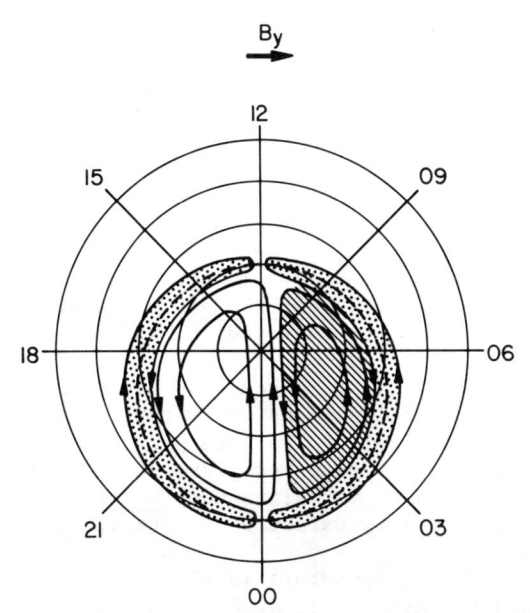

Fig. 7. Schematic convection pattern inferred to occur during theta arcs for cases with $B_z > 0$, $B_y < 0$. Note difference in flow direction from Figure 6a. The diffuse bar is seen in the sunward flowing region in the central polar cap; a discrete aurora adjoins it at the upward sheet current just dawnward of the bar [from Reiff and Burch, 1984].

potential drop after a northward turning, eventually approaching a value near 20 kV. Rezhenov [1981] also showed convective flow patterns consistent with a four cell Burke-type reversed convection pattern added to a decaying standard convection pattern.

The theta arc phenomenon, a broad arc stretching across the polar cap from the dayside to the nightside [Frank et al., 1982] may also be explainable by magnetic merging. Sunward convection in the polar cap for cases with a large $|B_y|$ may not take the form of counterrotating flow cells, as in Figure 6a, but may be consistent with two flow cells rotating the same direction (Figure 7, from Reiff and Burch [1984]). The double theta bar is interpreted as a combination of a diffuse auroral signature on closed sunward-flowing field lines in the central polar cap and a discrete arc along a sheet of upward current separating the two polar cap flow cells.

Evidence from Particle Distributions

Near the site of magnetic merging, the detailed particle distributions and electric and magnetic fields can be compared to ascertain whether the plasma shows evidence of the acceleration (or deceleration) expected from magnetic merging [Crooker, 1979]. At mid- and low-altitudes, the observer is several earth radii down the field line from the merging point, and the particles require a finite time to reach that location. The convection intrinsic to the merging process implies that, in that finite time, the particles have also drifted poleward (and eastward or westward, depending on the sign of B_y). Thus at a given altitude, the particles with lower initial parallel energy will be observed poleward of the particles with higher parallel energy. This effect was first observed in the plasma mantle region [Rosenbauer et al., 1975] and later in the cusp [Reiff et al., 1977; Shelley et al., 1976].

A similar signature caused by time-of-flight, rather than convective, dispersion was observed in the cusp by Torbert and Carlson [1976]. This may be the low altitude signature of intrinsically time-dependent merging, i.e., flux transfer events [Russell and Elphic, 1979]. Merging need not only be impulsive, however; the Burch et al. [1983] study shows merging dispersions lasting twenty minutes or more; therefore, quasi-steady-state reconnection must also occur. Variability in the intensity of the injected ions occurs during these extended cusp ion observations, but no significant periodicities have been determined [Frahm, 1983].

The magnetic mirror force (or, equivalently, conservation of magnetic moment) increases the pitch angle of the particles as they approach the earth. Thus at low altitudes (300-3000 km), all particles (even those observed at 90° pitch angle) started at roughly the same pitch angle (near 0°) and required roughly the same amount of time to traverse the field line. At mid-altitudes (2-6 R_E), however, particles with different observed pitch angles have spent varying amounts of time traversing the field line; thus, the characteristic ion energy at a given location is a function of pitch angle, and the energy-time distribution shows a series of V's, much like rick-rack [Burch et al., 1982]. From the width of the individual V's, one can determine the field-aligned distance to the acceleration region in one spacecraft spin period (6 seconds). In all the cases analyzed so far, the energy versus pitch angle dispersion is consistent with an X-point at an altitude of roughly 7-9 R_E, suggesting a cusp merging entry point [Burch et al., 1982]. These time-of-flight calculations were performed using a dipole field model; a more realistic model might yield consistency with injection near the nose.

It should be emphasized, however, that magnetic merging is not the only mechanism that will produce this kind of energy-latitude-pitch angle dispersion. Any source which is narrow in both the direction along the flow and the direction along the field line will produce a similar signature, and in fact a similar signature has been seen on the nightside, which could be either a time-of-flight dispersion [Klumpar et al., 1983] or a similar convective dispersion [Winningham et al., 1984].

Because the dispersion spreads out the initial distribution over several degrees of invariant latitude, it is not trivial to reconstruct the initial plasma characteristics (density, temperature, flow velocity) in order to ascertain whether or not the plasma was accelerated by merging. One can either average the distribution over the spatial extent of the dispersion [Shelley et al., 1976], or reconstruct the source distribution by unfolding the dispersion [Hill and Reiff, 1977]. In either case, the analysis is hampered by the fact that most particles mirror before reaching low altitudes, so that one can only infer the field-aligned portion of the source distribution. Nevertheless, the distribution is consistent with ion energization of roughly a kilovolt, which is about what is expected from merging [Hill and Reiff, 1977]. As yet, no simultaneous comparisons have been made between magnetosheath (source) distributions and those observed at low altitudes, but it may be possible to do so as a joint study between, say, ISEE-1 or 2 and Dynamics Explorer, or as part of the International Solar-Terrestrial Research Program. Recent work by Candidi and Meng [1984] does show an enhancement in cusp electron flux with increasing solar wind density. In addition, they show that cusp electron fluxes are stronger if the IMF is southward, also evidence for merging control.

For cases of northward IMF, when the convection is sunward, one might expect to observe dispersions of the reverse type, with higher energies more poleward and lower energies equatorward. Such signatures have been observed both from Atmosphere Explorer [Burch et al., 1980] and from Aureol-3 [Bosqued et al., 1983]. Thus the particle distributions are consistent with merging for both northward and southward IMF.

Conclusions

This paper has attempted to highlight some of the lines of evidence that have convinced many scientists that magnetic merging occurs frequently on the dayside when the IMF is southward. The evidence is by necessity indirect and much of it can be explained by other mechanisms. However, to my knowledge, no closed-model process has, for example, explained the correlation of east-west flows on the dayside with the y-component of the IMF, which is readily explained by the merging model.

There are several interesting studies that remain to be done, including, for example, comparison of the magnetosheath and low altitude plasma distributions and detailed multisatellite studies of magnetic topologies and flow patterns for both northward and southward IMF. However, the effect of these studies, I believe, will be to refine our understanding of the merging process rather than to diminish its role in theories of the solar-wind magnetosphere interaction.

Acknowledgments. The author thanks J. L. Burch and T. W. Hill for helpful comments. Atmosphere Explorer convection data courtesy R. A. Heelis. This work was supported by the Atmospheric Sciences Section of the National Science Foundation under grant ATM-8306772 and by the National Aeronautics and Space Administration under the Solar-Terrestrial Theory Program, grant NAGW-482.

References

Akasofu, S.-I., What is a magnetospheric substorm?, in Dynamics of the Magnetosphere, ed. S.-I. Akasofu, pp. 447-460, D. Reidel, Boston, 1980.

Akasofu, S.-I., D. N. Covey, and C.-I. Meng, Dependence of the geometry of the region of open field lines on the interplanetary magnetic field, Planet. Space Sci., 29, 803-807, 1981.

Alfvén, H, and C. G. Fälthammar, Cosmical Electrodynamics, Oxford Univ. Press, London, 1963.

Arnoldy, R. L., Signature in the interplanetary medium for substorms, J. Geophys. Res., 76, 5189-5201, 1971.

Axford, W. I., and C. O. Hines, A unifying theory of high-latitude geophysical phenomena and geomagnetic storms, Can. J. Phys., 39, 1433-1464, 1961.

Banks, P. M., J.-P. St. Maurice, R. A. Heelis, and W. B. Hanson, Electric fields and electrostatic potentials in the high latitude ionosphere, in Exploration of the Polar Upper Atmosphere, ed. J. Holtet and C. S. Deehr, D. Reidel, Dordrecht, 1981.

Bosqued, J. M., J. A. Sauvaud, H. Rème, D. Roux, Y. I. Galperin, R. A. Kovrazhkin, and V. A. Gladyshev, Aureol-3 charged particle measurements in the polar cusp (abstract), IAGA Bull., 48, 296, 1983.

Burch, J. L., Precipitation of low energy electrons at high latitudes: Effects of interplanetary magnetic field and dipole tilt angle, J. Geophys. Res., 77, 6696-6707, 1972.

Burch, J. L., Effects of the interplanetary magnetic field on the auroral oval and plasmapause, Space Sci. Rev., 23, 449-464, 1979.

Burch, J. L., Energy transfer in the quiet and disturbed magnetosphere, Rev. Geophys. Space Phys., 21, 463-473, 1983.

Burch, J. L., and R. A. Heelis, IMF changes and polar cap electric fields and currents, in Dynamics of the Magnetosphere, ed. S.-I. Akasofu, pp. 47-62, D. Reidel, Boston, MA, 1979.

Burch, J. L., P. H. Reiff, R. A. Heelis, R. W. Spiro, and S. A. Fields, Cusp region particle precipitation and ion convection for northward Interplanetary Magnetic Field, Geophys. Res. Lett., 7, 393-396, 1980.

Burch, J. L., P. H. Reiff, R. A. Heelis, J. D. Winningham, W. B. Hanson, C. Gurgiola, J. D. Menietti, R. A. Hoffman, and J. N. Barfield, Plasma injection and transport in the mid-altitude polar cusp, Geophys. Res. Lett., 9, 921-924, 1982.

Burch, J. L., P. H. Reiff, J. D. Menietti, J. D.

Winningham, S. D. Shawhan, R. A. Heelis, W. B. Hanson, E. G. Shelley, and M. Sugiura, B_y-dependent plasma flow and Birkeland currents in the dayside magnetosphere: 1. Dynamics Explorer observations, J. Geophys. Res., (submitted), 1984.

Burke, W. J., M. C. Kelley, R. C. Sagalyn, M. Smiddy, and S. T. Lai, Polar cap electric field structures with a northward interplanetary magnetic field, Geophys. Res. Lett., 6, 21-24, 1979.

Candidi, M., and C.-I. Meng, The relation of the cusp precipitating electron flux to the solar wind and interplanetary magnetic field, J. Geophys. Res., (submitted), 1984.

Clauer, C. R., R. L. McPherron, and C. Searls, Solar wind controls of the low-latitude asymmetric magnetic-disturbance field, J. Geophys. Res., 88, 2123-2130, 1983.

Cowley, S. W. H., A quantitative study of the reconnection between the earth's magnetic field and an interplanetary field of arbitrary orientation, Radio Sci., 8, 903, 1973.

Cowley, S. W. H., Magnetospheric asymmetries associated with the y-component of the IMF, Planet. Space Sci., 29, 79-96, 1981a.

Cowley, S. W. H., Asymmetry effects associated with the x-component of the IMF in a magnetically open magnetosphere, Planet. Space Sci., 29, 809-818, 1981b.

Cowley, S. W. H., Magnetospheric and ionospheric flow and the interplanetary magnetic field, in The Physical Basis of the Ionosphere in the Solar-Terrestrial System, pp. 4-1--4-14, AGARD, Neuilly sur Seine, France, 1982.

Crooker, N. U., Dayside merging and cusp geometry, J. Geophys. Res., 84, 951-959, 1979.

Crooker, N. U., The half-wave rectifier response of the magnetosphere and antiparallel merging, J. Geophys. Res., 85, 575-578, 1980.

Doyle, M. A., and W. J. Burke, S3-2 measurements of the polar cap potential, J. Geophys. Res., 88, 9125-9134, 1983.

Dubinin, E. M., and Yu. N. Potanin, The magnetosphere boundary phenomena and the polar cap convection, Moscow Academy of Sciences publication D-293, Moscow, 1980.

Dungey, J. W., Interplanetary magnetic field and the auroral zones, Phys. Rev. Lett., 6, 47, 1961.

Dungey, J. W., The structure of the ionosphere, or adventures in velocity space, in Geophysics: The Earth's Environment, ed. C. DeWitt, J. Hiebolt, and A. Lebeau, pp. 526-536, Gordon and Breach, New York, 1963.

Frahm, R. A., Cusp particle detection and ion injection source oscillations, M.S. thesis, Rice University, Houston, TX, 1983.

Frank, L. A., J. D. Craven, J. L. Burch, and J. D. Winningham, Polar views of the Earth's aurora with Dynamics Explorer, Geophys. Res. Lett., 9, 1001-1004, 1982.

Gonzales, W. D., and F. S. Mozer, A quantitative model for the potential resulting from reconnection with an arbitrary interplanetary magnetic field, J. Geophys. Res., 79, 4186-4194, 1974.

Gussenhoven, M. S., D. A. Hardy, and W. J. Burke, DMSP/F2 electron observations of equatorward auroral boundaries and their relationship to magnetospheric electric fields, J. Geophys. Res., 86, 768-778, 1981.

Haerendel, G., G. Paschmann, N. Sckopke, H. Rosenbauer, and P. C. Hedgecock, The frontside boundary layer of the magnetosphere and the problem of reconnection, J. Geophys. Res., 83, 3195-3216, 1978.

Heelis, R. A., The effects of interplanetary magnetic field orientation on dayside high latitude ionospheric convection, J. Geophys. Res., 89, in press, 1984.

Heelis, R. A., W. B. Hanson, and J. L. Burch, Ion convection velocity reversals in the dayside cleft, J. Geophys. Res., 81, 3803-3809, 1976.

Heelis, R. A., J. D. Winningham, W. B. Hanson, and J. L. Burch, The relationships between high-latitude convection reversals and the energetic particle morphology observed by Atmospheric Explorer, J. Geophys. Res., 85, 3315-3324, 1980.

Heppner, J. P., Polar cap electric field distributions related to the interplanetary magnetic field direction, J. Geophys. Res., 77, 4877-4887, 1972.

Heppner, J. P., Empirical models of high-latitude electric fields, J. Geophys. Res., 82, 1115-1125, 1977.

Hill, T. W., Magnetic merging in a collisionless plasma, J. Geophys. Res., 80, 4689-4699, 1975.

Hill, T. W., Solar-wind magnetosphere coupling, in Solar-Terrestrial Physics, ed. R. L. Carovillano and J. M. Forbes, pp. 261-302, D. Reidel, Hingham, MA, 1983.

Hill, T. W., and P. H. Reiff, Evidence of magnetospheric cusp proton acceleration by magnetic merging at the dayside magnetopause, J. Geophys. Res., 82, 3623-3628, 1977.

Hill, T. W., A. J. Dessler, and R. A. Wolf, Mercury and Mars: The role of ionospheric conductivity in the acceleration of magnetospheric particles, Geophys. Res. Lett., 3, 429-432, 1976.

Horwitz, J. L., and S.-I. Akasofu, On the relationship of the polar cap current system to the North-South component of the Interplanetary Magnetic Field, J. Geophys. Res., 84, 2567-2572, 1979.

Iijima, T., and T. A. Potemra, The relationship between interplanetary quantities and Birkeland current densities, Geophys. Res. Lett., 9, 442-445, 1982.

Kamide, Y., A. D. Richmond, and S. Matsushita, Estimation of ionospheric electric fields, ionospheric currents, and field aligned currents from ground magnetic records, J. Geophys. Res., 86, 801, 1981.

Killeen, T. L., P. B. Hays, N. W. Spencer, and L. E. Wharton, Neutral winds in the polar thermosphere as measured from Dynamics Explorer, Geophys. Res. Lett., 9, 957-960, 1982.

Klumpar, D. M., J. R. Burrows, and M. D. Wilson, Simultaneous observations of field-aligned currents and particle fluxes in the post-midnight sector, Geophys. Res. Lett., 3, 395-398, 1976.

Klumpar, D. M., W. K. Peterson, and E. G. Shelley, L:ocalized magnetospheric ion injection outside the cusp (abstract), EOS Trans. AGU, 64, 296, 1983.

Lei, W., R. Gendrin, B. Higel, and J. Berchem, Relationships between the solar wind electric field and the magnetospheric convection electric field, Geophys. Res. Lett., 8, 1099-1102, 1981.

Longenecker, D., and J. G. Roederer, Polar cap electric field dependence on solar wind and magnetotail parameters, Geophys. Res. Lett., 8, 1261-1264, 1981.

Luhmann, J. G., R. J. Walker, C. T. Russell, N. U. Crooker, J. R. Spreiter, and S. S. Stahara, Patterns of potential magnetic field merging sites on the dayside magnetopause, J. Geophys. Res., 89, in press, 1984.

Maezawa, K., Magnetic convection induced by the positive and negative z-components of the interplanetary magnetic field: Quantitative analysis using polar cap magnetic records, J. Geophys. Res., 81, 2289-2303, 1976.

Matsushita, S., and D. E. Trotter, IMF sector behavior deduced from geomagnetic data, J. Geophys. Res., 85, 2354-2365, 1980.

McDiarmid, I. B., J. R. Burrows, and M. D. Wilson, Magnetic field perturbations in the dayside cleft and their relationship to the IMF, J. Geophys. Res., 83, 5753-5756, 1978.

McDiarmid, I. B., J. R. Burrows, and M. D. Wilson, Comparison of magnetic field perturbations and solar electron profiles in the polar cap, J. Geophys. Res., 85, 1163-1170, 1980.

Meng, C.-I., Polar cap variations and the interplanetary magnetic field, in Dynamics of the Magnetosphere, ed. S.-I. Akasofu, pp. 23-46, D. Reidel, Dordrecht, Holland, 1979.

Meng, C.-I., S.-I. Akasofu, and K. A. Anderson, Dawn-dusk gradient of the precipitation of low-energy electrons over the polar cap and its relation to the interplanetary magnetic field, J. Geophys. Res., 82, 5271-5275, 1977.

Nishida, A., IMF control on the earth's magnetosphere, Space Sci. Rev., (in press), 1983.

Oberc, P., Magnetospheric tail dynamics and a concept of combined action of viscous drag and magnetic merging, Planet. Space Sci., 27, 1087-1093, 1979.

Reiff, P. H., Low altitude signatures of the boundary layers, in Magnetospheric Boundary Layers, ed. B. Battrick, ESA SP-148, Noordwijk, the Netherlands, pp. 167-173, 1979.

Reiff, P. H., Sunward convection in both polar caps, J. Geophys. Res., 87, 5976-5980, 1982.

Reiff, P. H., and J. L. Burch, B_y-dependent plasma flow and Birkeland currents in the dayside magnetosphere: 2. A global model for northward and southward IMF, J. Geophys. Res., (submitted), 1984.

Reiff, P. H., T. W. Hill, and J. L. Burch, Solar wind plasma injection at the dayside magnetospheric cusp, J. Geophys. Res., 82, 479-491, 1977.

Reiff, P. H., J. L. Burch, and R. W. Spiro, Cusp proton signatures and the Interplanetary Magnetic Field, J. Geophys. Res., 85, 5997-6005, 1980.

Reiff, P. H., R. W. Spiro, and T. W. Hill, Dependence of polar cap potential drop on interplanetary parameters, J. Geophys. Res., 86, 7639-7648, 1981.

Rezhenov, B. V., Convection at high latitudes when the interplanetary magnetic field is northward, Planet. Space Sci., 29, 687-693, 1981.

Rosenbauer, H., H. Brunwaldt, M. D. Montgomery, G. Paschmann, and N. Sckopke, Heos 2 plasma observations in the distant polar cusp near local noon: The plasma mantle, J. Geophys. Res., 80, 2723-2737, 1975.

Russell, C. T., The configuration of the magnetosphere, in Critical Problems of Magnetospheric Physics, ed. E. R. Dyer, Jr., pp. 1-16, National Academy of Sciences, Washington, D.C., 1972.

Russell, C. T., and G. Atkinson, Comments on a paper by J. P. Heppner, 'Polar cap electric field distributions related to interplanetary magnetic field direction,' J. Geophys. Res., 78, 4001-4002, 1973.

Russell, C. T., and R. C. Elphic, ISEE observations of flux-transfer events at the magnetopause, Geophys. Res. Lett., 6, 33-36, 1979.

Russell, C. T., and R. L. McPherron, Semiannual variation of geomagnetic activity, J. Geophys. Res., 78, 92-108, 1973.

Saflekos, N. A., and T. A. Potemra, The orientation of Birkeland current sheets in the dayside polar region and its relationship to the IMF, J. Geophys. Res., 85, 1987-1994, 1980.

Sergeev, V. A., and B. M. Kuznetsov, Quantitative dependence of the polar cap electric field on the IMF B_z-component and solar wind velocity, Planet. Space Sci., 29, 205-213, 1981.

Shelley, E. G., R. D. Sharp, and R. G. Johnson, He^+ flux measurements in the dayside cusp, J. Geophys. Res., 81, 2363-2370, 1976.

Sonnerup, B. U. O., Magnetopause reconnection rate, J. Geophys. Res., 79, 1546-1549, 1974.

Stern, D. P., A study of the electric field in an open magnetospheric model, J. Geophys. Res., 78, 7292, 1973.

Torbert, R. B., and C. W. Carlson, Impulsive ion injection into the polar cusp, in Magnetospheric Particles and Fields, ed. B. M. McCormac, pp. 47-54, D. Reidel, Hingham, MA, 1976.

Vasyliunas, V. M., Theoretical models of magnetic field-line merging, 1, Rev. Geophys. Space Phys., 13, 303-336, 1975.

Vasyliunas, V. M., An overview of magnetospheric dynamics, in Magnetospheric Particles and

Fields, ed. B. M. McCormac, pp. 99-110, D. Reidel, Hingham, MA, 1976.

Winningham, J. D., J. L. Burch, and R. A. Frahm, Bands of ions and angular V's: A conjugate manifestation of ionospheric ion acceleration, J. Geophys. Res., (submitted), 1984.

Wygant, J. R., R. B. Torbert, and F. S. Mozer, Comparison of S3-3 polar cap potential drops with the interplanetary magnetic field and models of magnetopause reconnection, J. Geophys. Res., 85, 5727-5735, 1983.

Questions and Answers

Lui: When you inferred the source region for the dispersion features observed by low altitude satellites, the result indicates the source to be about 8 to 12 R_e above the satellite, i.e. near the cusp. How is this result compatible with the reconnection taking place near the equator on the dayside?

Reiff: I agree, this is puzzling. Our fits to the energy-pitch angle V's are done with a simple dipole model and perhaps a more realistic magnetic field may allow us to infer sources nearer the equator. However, injection from merging in the other hemisphere injects a plasma which is not only farther away, it is also moving with a smaller earthward velocity. Thus it may not reach the earth at all before being swept by field line motion into the distant tail.

Moore: I didn't follow your remark about θ aurora. Are you suggesting that you have evidence that they are driven by dayside reconnection?

Reiff: The convection data suggest to me that theta aurorae are found at sheets of upward Birkeland currents consistent with a four-cell convection pattern that includes viscous cells at the flanks and two adjacent counterclockwise cells driven by magnetic merging with an IMF with strong $+B_z$ and $-B_y$ components. The sheet current separates the two counterclockwise cells. I have put together a simple model that is driven by merging of tail field lines behind the cusp, reconnecting in the magnetotail.

Russell: In the mid 1970s Carlson and Torbert reported dispersive ion events in the polar cusp. I have always thought these were low altitude manifestations of FTEs. Do your DE observations support this interpretation? Do you have any comment on Paschmann et al.'s (1983) alternative explanation in terms of a moving acceleration region at low altitudes? Perhaps G. Paschmann should comment too.

Reiff: Rocket-based measurements and high-altitude DE measurements are ideal to see FTEs, since the FTE is more of a time variation than steady-state reconnection, and one would expect to see time-of-flight dispersions rather than convective dispersions. I think that the Torbert and Carlson results are probably low-altitude signature of FTEs. So far, I have not observed any FTEs from DE yet; however, the coverage of DE observations just at noon was smaller than at 9-10 MLT where our best cusp dispersions were seen.

Paschmann: (Comment) Consistent latitudinal sense of velocity dispersion would require the low-altitude source to always move in the same direction.

Vasyliunas: (Comment) It is remarkable that you observe ion dispersion (which implies a narrow localized source) for long periods, up to 20 minutes; these periods are comparable to the flow time of magnetosheath plasma from the subsolar region to the cusps. The observations thus suggest that the source location may be relatively fixed rather than moving with the magnetosheath plasma.

Reiff: I agree; this implies that merging is (at least at times) quasi-steady-state with a (relatively) fixed neutral line location.

Questioner: When IMF is northward, has there been any observational verification of the entrainment of magnetosheath plasma on newly closed dayside field lines (i.e., in the boundary layer and cusp ionosphere?).

Reiff: This is just being looked at in detail. In my CDAW-2 paper [Reiff, JGR, 1982], we saw ~1 keV ions in the polar cap that looked trapped and may be evidence. Meng's data from DMSP and McDiarmid et al.'s ISIS data for northward IMF may be interpreted also in that fashion. And finally, the theta arc plasma data also suggest trapped ~1 keV ions in the polar cap. However, all of these measurements can be interpreted as merely plasma sheet intrusions, i.e., motion of the polar cap boundary toward the pole only on one side. Clearly more work needs to be done to sort this out.

Patricia H. Reiff, Department of Space Physics and Astronomy, Rice University, P.O. Box 1892, Houston, TX 77251.

PLASMA AND PARTICLE OBSERVATIONS AT THE MAGNETOPAUSE: IMPLICATIONS FOR RECONNECTION

Götz Paschmann

Max-Planck-Institut für Physik und Astrophysik, Institut für extraterrestrische Physik
8046 Garching b. München, West Germany

Abstract. The paper discusses the evidence for magnetopause reconnection obtained from *in situ* plasma and energetic particle measurements onboard the International Sun-Earth Explorer satellites. The strongest evidence is provided by the test for tangential momentum balance between the plasma transmitted through the magnetopause and the sharply curved magnetic field. The test of the energy balance suffers from larger uncertainties, but still is consistent with momentum balance results. More indirect support is provided by tracing the magnetic field topology at the magnetopause with energetic particles. The solar wind plasma entering the magnetosphere forms a boundary layer which extends all along the magnetopause. Boundary layer observations indicate that reconnection is probably not the only interaction process operating at the magnetopause.

Introduction

Reconnection at the Earth's magnetopause implies a number of unique signatures in plasma and energetic particle properties (see overview paper by Sonnerup, this volume). Adequate measurements of these properties have, however, only become available with the recent International Sun-Earth Explorer (ISEE) mission. It is the primary purpose of this paper to present the observational evidence for reconnection in the quasi-steady configuration originally proposed by Petschek (Section 2). The paper will not deal with the observations pertaining to the so-called flux-transfer-events which appear to be a manifestation of small-scale, impulsive reconnection. These events are discussed in detail elsewhere in this volume. In order to evaluate the role of reconnection in relation to other interaction processes at the magnetopause, the relevant features of the magnetospheric boundary layer are reviewed in Section 3.

Quasi-Steady Reconnection at the Magnetopause

The question of the *in situ* evidence for reconnection at the magnetopause can be approached in a number of ways: by measuring the momentum which particles gain from (or lose to) the magnetic field when they are transmitted through or reflected by the magnetopause (momentum balance); by relating the various forms of particle energies to the electromagnetic energy converted in reconnection (energy balance); or by trying to establish the characteristic magnetic field topology. In the following the application of these three approaches will be discussed.

Momentum Balance

The most definitive *in situ* evidence for the occurrence of reconnection at the magnetopause has been provided by considering what happens to the flow velocity of the plasma as it crosses the magnetopause (cf. Fig. 1). A combination of tangential momentum balance between plasma and magnetic field and the continuity of the tangential electric field causes a change in the flow velocity across a rotational discontinuity. For an isotropic plasma, this change is given by the expression [e.g., Landau and Lifshitz, 1960]

$$\underline{v}_2 - \underline{v}_1 = \pm (\underline{B}_2 - \underline{B}_1)/\sqrt{\mu_0 \rho} \qquad (1)$$

where $\underline{v}$ is the velocity, $\underline{B}$ the magnetic field, and ρ the mass density. The indices 1 and 2 refer to the magnetosheath and magnetosphere sides, respec-

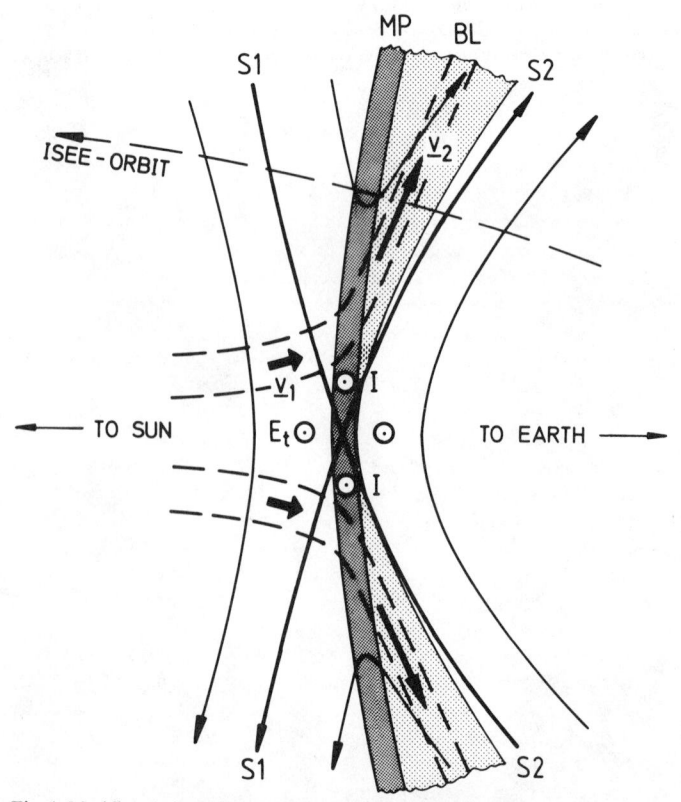

Fig. 1. Meridional view of magnetopause reconnection configuration for anti-parallel external and internal tangential magnetic fields. The magnetopause (MP) is shown as a current layer of finite thickness, with an adjoining boundary layer (BL). The field lines denoted by S1 and S2 mark the outer and inner separatrix, respectively. The reconnection electric field E_t is aligned with the magnetopause current I. Dashed lines are stream lines [from Sonnerup et al., 1981].

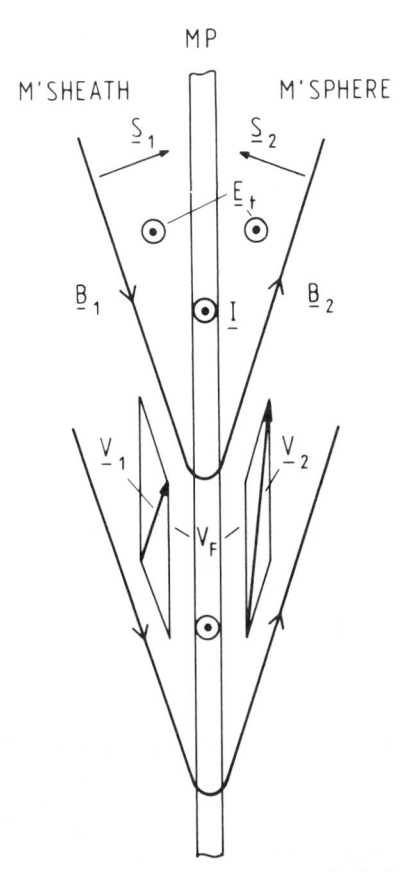

Fig. 2. Geometric construction of plasma flow acceleration in the de Hoffmann-Teller frame which moves along the magnetopause with speed v_F.

tively. The plus and minus signs refer to $B_n < 0$ and $B_n > 0$, respectively; i.e., to the situation north or south of the separator line. Viscous stresses have been ignored in the derivation of equation (1). It is conceivable that they might not be negligible, in particular near the center of the magnetopause current layer. The effects of plasma pressure anisotropy can easily be incorporated into equation (1) [Hudson 1970] but have been omitted for convenience.

The velocity changes implied by equation (1) are large, of the order of hundreds of km/sec, unless the field rotation angle becomes small, i.e., much less than 90°. Equation (1) does, however, not necessarily imply that the plasma is speeded up. If $(\underline{B}_2 - \underline{B}_1)/(\mu_0 \rho)^{1/2}$ and $\underline{v}_1$ are oppositely directed and $\underline{v}_1$ is sufficiently large the plasma speed can be reduced. This is precisely the situation which characterizes the tail magnetopause [e.g., Lee and Roederer, 1982].

An equivalent way of looking at the flow acceleration is illustrated in Fig. 2. If one considers the situation in a coordinate system sliding along the magnetopause at such a speed that the plasma flows along the magnetic field, no acceleration can occur and the plasma retains the same speed when crossing the magnetopause. The proper transformation velocity is obtained by decomposing the incident velocity, $\underline{v}_1$, into a component parallel to $\underline{B}_1$ and a component v_F parallel to the magnetopause. Transformation back to the frame at rest then yields the velocity $\underline{v}_2$ in the magnetosphere. This procedure is analogous to that frequently used in the analysis of shocks or shock-reflected particles [de Hoffman and Teller, 1950; Sonnerup, 1969]. Neglecting magnetopause motions, the transformation velocity is given by $v_F = E_t/B_n$ where E_t is the tangential electric field. v_F can be thought of as the "field-line speed" [Cowley, 1982]. The field-aligned velocity is the local Alfvén velocity, $\underline{B}/(\mu_0 \rho)^{1/2}$ (for isotropic pressure). The velocity in the frame at rest can thus be written as

$$\underline{v} = \underline{v}_F \pm \underline{B}/\sqrt{\mu_0 \rho} \qquad (2)$$

Equation (2) predicts a linear relationship between the three components of $\underline{v}$ and $\underline{B}$. Figure 3 shows a test (for one component) of the relationship for the 8 September 1978 magnetopause crossing [Paschmann et al., 1979]. The agreement of the measurements with equation (2) is very good, permitting the conclusion that that particular magnetopause was indeed a rotational discontinuity.

Eleven other magnetopause crossings have been identified as rotational discontinuities in much the same way, i.e., by the comparison of measured plasma velocities in high-speed flows with those predicted by equation (1), or, rather, the more general version which includes pressure anisotropy effects [Sonnerup et al., 1981; Gosling et al., 1982].

The expected acceleration of particles at the magnetopause may also be understood in a simple single-particle picture [Cowley, 1982]. This is illustrated in Fig. 4. For an arbitrary distribution function of incident ions, the spectrum of transmitted ions is just the mirror image of the incident spectrum about the field line speed v_F if one assumes that the pitch-angle is preserved in the process. If a fraction of the incident ions is reflected at the magnetopause, their spectrum is obtained the same way. When reflection occurs, the intensities of transmitted and reflected ions are, of course, scaled down from those of the incident ions. Note that the incident ions need not only be the magnetosheath ions discussed in the previous section, but could also be magnetospheric energetic ions. If both particle populations are partially transmitted and reflected, an observer on the magnetosheath side will record three types of particles, incident and reflected magnetosheath ions, and transmitted magnetospheric ions. On the magnetospheric side of the magnetopause the equivalent situation will be observed, the spectrum now comprising transmitted magnetosheath ions, and incident and reflected magnetospheric ions.

For the magnetosheath, such a situation has been observed during the 8 September 1978 magnetopause crossing [Sonnerup et al., 1981]. Recently, Scholer and Ipavich [1983] have reported a case of reflected energetic ions on the

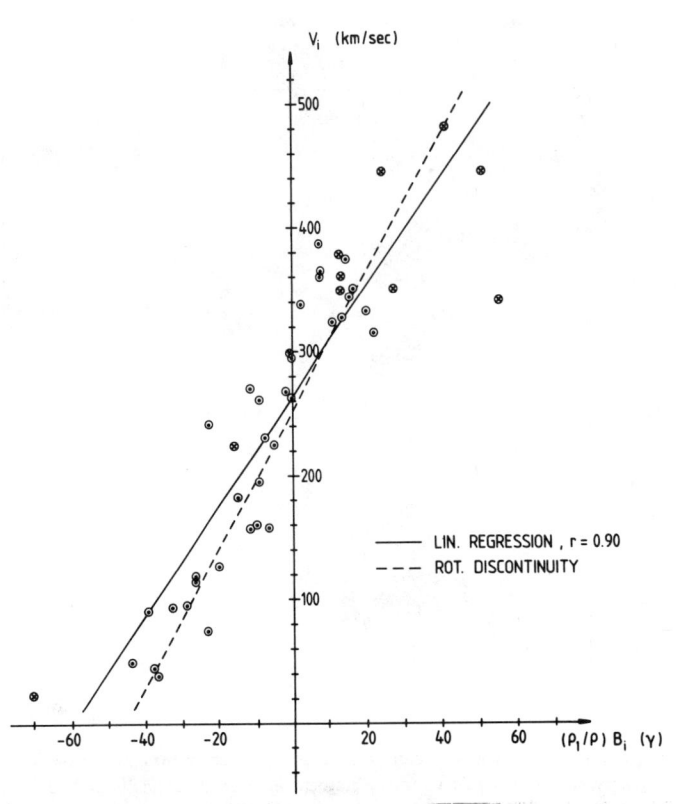

Fig. 3. Test of the linear relation between v_i and B_i predicted by equation (2). As the test actually included effects of plasma anisotropy, magnetic field is weighted by the density variation [from Paschmann et al., 1979].

Fig. 4. Spectra of particles transmitted (subscript tr) or reflected (re) at the magnetopause can be obtained for arbitrary incident (in) spectra of magnetosheath (MS) or ring current (RC) particles by mirroring the incident spectra around the field line speed v_F and adjusting the intensity for transmission/reflection coefficient. Part (a) shows the superposition of the spectra which would be observed in the magnetosphere in the direction along B; part (b) the same for the magnetosheath side of the magnetopause [after Cowley, 1982].

magnetospheric side of a magnetopause, which from the plasma analysis of Sonnerup et al. [1981] was already known to be a rotational discontinuity. The important aspects of the measurements, shown in Fig. 5, are the enhanced intensities of the ions returning from the magnetopause and the agreement between the magnitudes of the spectral slopes of incident and reflected particles. The latter property ensures that the reflected spectrum can be constructed as a mirror image of the incident spectrum; the former implies that the reflected particles have been accelerated. Of course, a wide range of combinations of field-line speeds and reflection coefficients can be chosen to match the data (just two are shown in Fig. 5). It is interesting to note that Sonnerup et al. [1981] have determined a field-line speed of only 130 km/sec for this case, which would require ~100% reflection. As Scholer and Ipavich point out, the relevant v_F for their comparison is that at the location where the field line which the particles are on crosses the magnetopause, while Sonnerup et al. determine v_F where the satellite crosses the magnetopause. The former location is much closer to the separator line where v_F could be larger.

Energy Balance

Another *in situ* check of reconnection is the investigation of the energy budget. The question is, whether the electromagnetic energy, converted per unit area of the magnetopause, can be accounted for in terms of the increased kinetic energy in the high-speed flows, discussed in the previous section; in terms of increased enthalpy of the plasma; or in terms of heat flowing away from the magnetopause. For a steady-state, one-dimensional current sheet the energy balance can be written as [e.g., Paschmann et al., 1984]

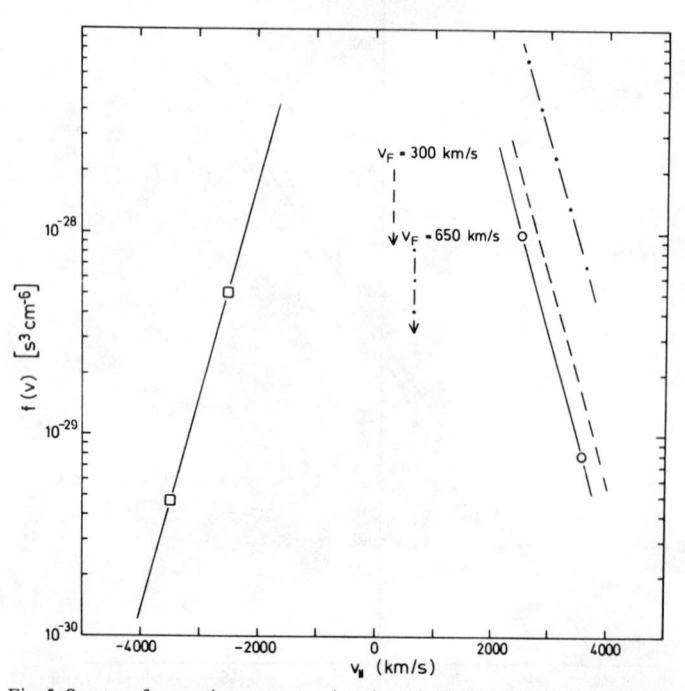

$$\rho(\underline{v}\cdot\underline{n})\Delta\left\{\frac{v^2}{2}+U\right\}+\Delta\{\underline{v}\cdot\underline{\underline{p}}\cdot\underline{n}\}+\Delta\{\underline{q}\cdot\underline{n}\}+\Delta\{(\underline{E}\times\underline{B})\cdot\underline{n}/\mu_0\}=0 \quad (3)$$

where U is the internal energy per unit mass of the plasma, $\underline{\underline{p}}$ its pressure tensor, $\underline{q}$ the heat flow, $\underline{E}\times\underline{B}/\mu_0$ is the Poynting flux (denoted by $\underline{S}$ in Fig. 2).

In terms of more directly observable quantities the energy balance can be rewritten as

$$\Delta\left\{\frac{v^2}{2}\right\}+\Delta\left\{\frac{5}{2}\frac{p_\perp}{\rho}\right\}+Q=-\Delta\left\{\frac{B^2}{\mu_0\rho}\right\}\pm\Delta\left\{\frac{\underline{v}\cdot\underline{B}}{\sqrt{\rho/\mu_0}}\right\} \quad (4)$$

$$(1) \qquad\qquad (2) \qquad\qquad (3) \qquad\qquad (4)$$

Fig. 5. Spectra of energetic protons moving along the field in the direction towards (squares) and away (circles) from the magnetopause. Dashed and dot-dashed lines are obtained by mirroring the spectrum at the left around $v_F = 300$ and 600 km/sec, respectively. Both curves can be made to fit the observed spectrum on the right by introducing appropriate reflection coefficients [from Scholer and Ipavich, 1983].

Terms (1) and (2) represent the changes in kinetic energy and enthalpy per unit mass, while terms (3) and (4) are the electromagnetic energy input obtained by expansion of the vector triple product in the last term of equation (3). For simplicity, the effects of pressure anisotropy are ignored in equation (4), as are resistive parts of the Poynting flux and the work by viscous stresses. Use has also been made of the relation between $\underline{v}\cdot\underline{n}$ and $\underline{B}\cdot\underline{n}$ for a rotational

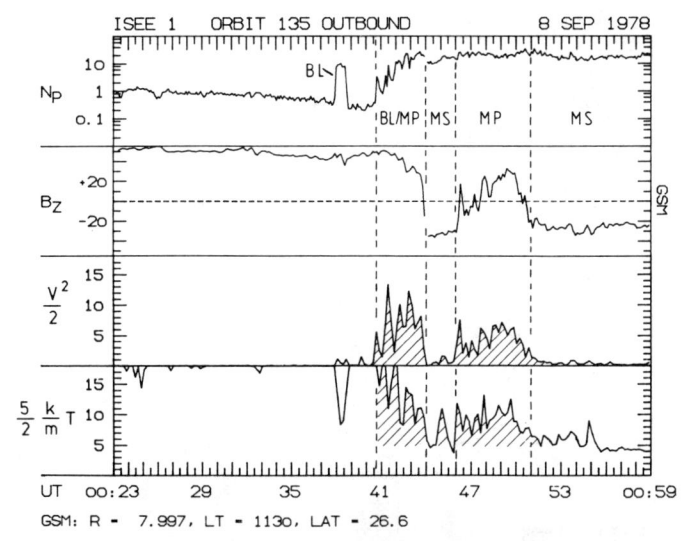

Fig. 6. Profiles of density N(cm^{-3}), north-south magnetic field B_z(nT), $v^2/2$ and $5/2$ kT/m (both in 10^{14} cm^2/sec^2) for the ISEE-1 magnetopause crossing on 8 September 1978. The latter two quantities represent terms (1) and (2) in the energy balance equation. The shading indicates the increase of both quantities in the magnetopause and boundary layer.

discontinuity. The plus and minus signs refer to cases with $B_n < 0$ and $B_n > 0$, respectively. Q is the heat flux per unit mass flux.

For an ideal rotational discontinuity, i.e., a standing Alfvén wave, there should be no change in enthalpy (term 2). In Fig. 6 we show the behavior of terms 1 and 2 across the magnetopause as measured during the 8 September 1978 ISEE-1 crossing. It is evident that the enthalpy actually increases substantially when entering the current sheet. Such a dissipative nature has been found for many of the magnetopause crossings examined.

Testing the entire energy balance (equation 4) is a more difficult task than the test of the momentum balance, because higher moments of the plasma distribution function (i.e., pressure and heat flux) enter, and because errors in the velocity vector will strongly affect term 4. Nevertheless, we have tested equation (4) (including pressure anisotropy effects, but ignoring the heat flux), for two pronounced magnetopause reconnection cases [Paschmann et al., 1984]. The result for the 8 September case is shown in Fig. 7. Many of the data points clearly do not obey perfect energy balance. In addition to experimental errors, a portion of the discrepancies could be due to the neglect of some contributions in equation (4). For example, the group of points lying well above the diagonal represents measurements taken near the center of the current sheet, where viscous effects are expected to maximize. It can be shown that qualitatively these effects would move the points closer to the diagonal. Nevertheless, it is clear from Fig. 7 that one would not want to rely solely on the energy balance result for the identification of this crossing as a reconnection case. It is, however, clearly consistent with this interpretation.

In the above discussion of the energy balance we were dealing with a situation where $\underline{I} \cdot \underline{E}$ is positive, such as shown in Figs. 1 and 2. In this situation, normally expected for the dayside magnetopause, the magnetopause is a "load," i.e., electromagnetic energy is converted into mechanical energy. As already mentioned earlier, the opposite situation must develop further tailward, for example past the polar cusps, where work is done on the field, and the plasma accordingly has to slow down. In this case, the magnetopause becomes a "generator" ($\underline{I} \cdot \underline{E} < 0$). The "generator" situation has been investigated theoretically be Lee and Roederer [1982] and by Swift and Lee [1982].

In principle, a generator situation could also occur on the dayside, when the separator line is shifted to fairly high latitudes. Such a case has recently been reported by Aggson et al. [1983]. It appears, however, that such a situation, which requires a reversal of the tangential electric field, could not be achieved in a steady state.

Magnetic Field Topology

At a first glance it would appear that the simplest method to establish the occurrence of reconnection would have been the determination of the normal magnetic field component B_n implied by the reconnection topology (Fig. 1). This is, however, not so, because B_n is usually too small to be reliably distinguished from zero (the same argument applies even more to the plasma flow velocity normal to the magnetopause). The field topology does, however, have another, more indirect consequence, namely that it should facilitate the loss of energetic particles from the magnetosphere. This hypothesis has been checked for some of the reconnection cases, with variable success. While Scholer et al. [1981] and Sonnerup et al. [1981] found that energetic ions in the magnetosheath were streaming along field lines in the expected sense, the inspection of anisotropies earthward of the 8 September 1978 magnetopause provided some conflicting evidence. Eastman and Frank [1982] reported that ≥ 40 keV electrons retained a "trapped" angular distribution right up to the magnetopause, which they interpreted as indicating closed field lines. Figure 8 combines the data showing the field- aligned ion and trapped electron distributions. In a subsequent paper, Scholer et al. [1982b] investigated the ion anisotropies earthward of the magnetopause and established that the ions were already exhibiting some streaming in the expected sense. Daly and Fritz [1982] finally showed that the electron angular distribution could have been the result of trapping in the minimum field region observed right at the magnetopause on 8 September 1978. Still, it needs to be explained why the electrons, contrary to the protons, show a pronounced intensity cutoff right at the magnetopause. Interestingly enough, similar intensity reductions of energetic electrons are also observed inside the magnetosheath portion of flux tubes which have been reconnected in the so-called magnetic flux transfer events [Daly et al., 1981; Scholer et al., 1982a].

Another question concerning the behavior of energetic particles escaping across the magnetopause has to do with their resupply. As illustrated by Fig. 9,

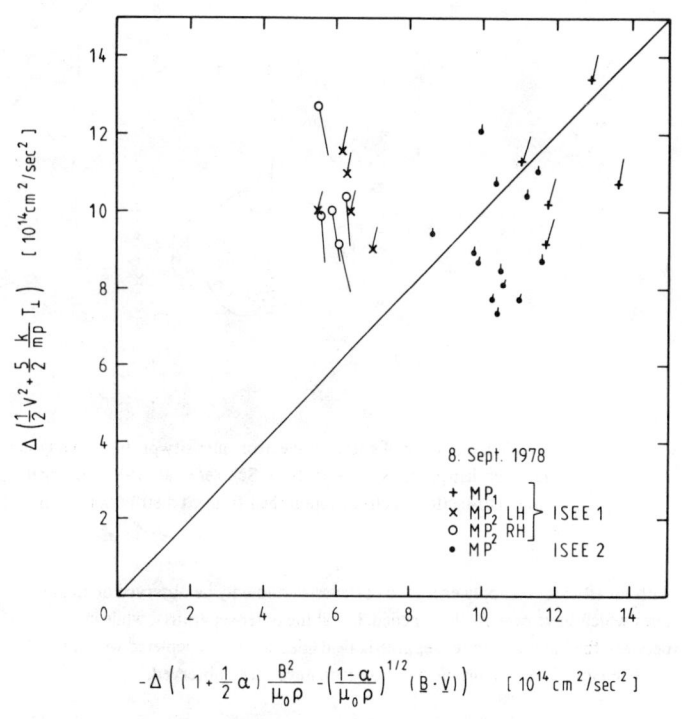

Fig. 7. Test of energy balance for the 8 September magnetopause crossing. The ordinate shows the sum of terms (1) and (2) in equation 4, the abscissa the sum of terms (3) and (4) (the terms are modified to include pressure anisotropy effects). The largest discrepancies are for those measurements (marked by the x and o symbols) which were taken near the center of the magnetopause current sheet, where viscous effects, ignored in equation 4, might become important.

ENERGETIC ELECTRONS (≥ 45 keV)
AVERAGE INTENSITIES AND ANGULAR DISTRIBUTIONS

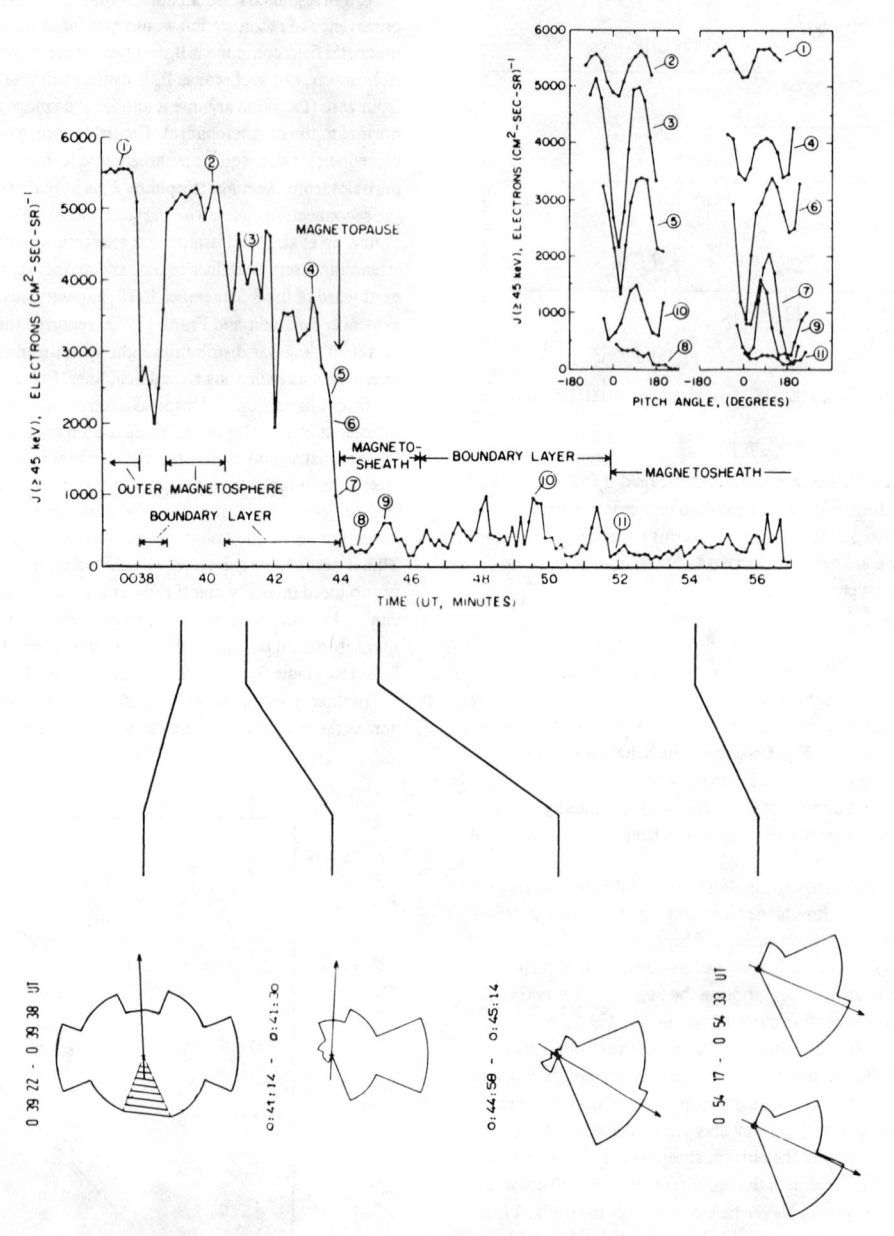

Fig. 8. Composite of energetic electron intensity profile and angular distributions [top, from Eastman and Frank, 1982] and energetic proton angular distributions [bottom, from Scholer et al., 1981]. Energetic protons show expected strong field-aligned streaming in the magnetosheath. Electrons retain their trapped distributions right to the magnetopause.

without efficient resupply energetic particles should only be observed on field lines which have been freshly opened, i.e., at the outer separatrix, while in between the inner and outer separatrix field lines should be depleted within a bounce period. Such a profile has, however, never been observed.

Occurrence

Figure 10 shows the location of ISEE magnetopause crossings which have been identified as reconnection cases. Of the total of 17 cases, 11 are from Sonnerup et al. [1981], 1 from Gosling et al. [1982], and the remaining 5 have been discovered subsequently. It is evident from Fig. 10 that (a) the observed reconnection cases cover a wide range of longitudes, and (b) that their division into cases with $B_n > 0$ and $B_n < 0$ is consistent with the separator line going through the subsolar point, particularly if one considers the expected tilt of the separator as a function of the B_y component. There is only one case where the separator probably was located at high latitudes.

The duration of the events shown in Fig. 10 ranged from a few minutes up to ~10 minutes. Noting that the time it takes the magnetosheath flow to traverse the entire frontside magnetopause is about 10 minutes, one would conclude that reconnection in these cases could be described at least as quasi-stationary. In one case, the magnetopause was crossed several times over a 5-hour period and each time showed the reconnection behavior.

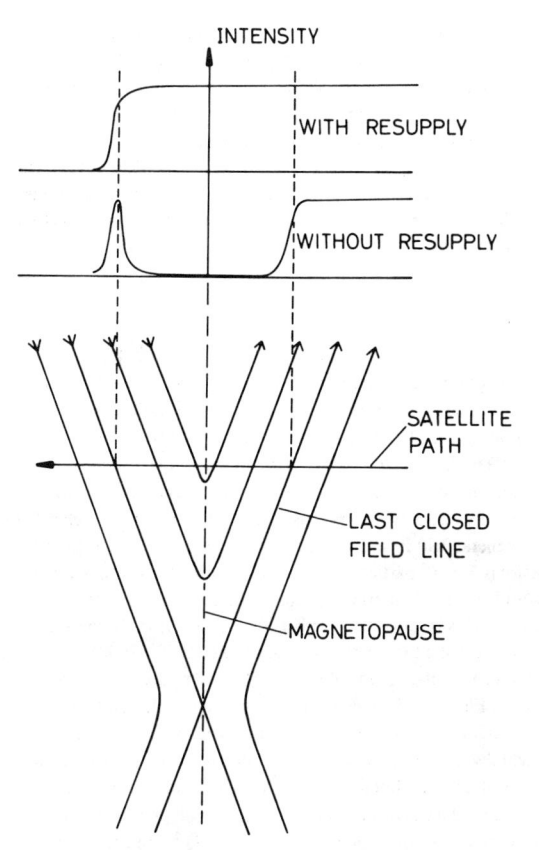

Fig. 9. Schematic diagram showing the energetic particle intensity profile which should be observed with (top) and without (center) efficient resupply.

rotated by ~ 170° but the changes in magnetic field through the current layer were not accompanied by the changes in plasma flow velocities predicted by equation (1). The change in B_y, for example, should have caused a change in v_y by up to 250 km/sec.

In view of the large number of ISEE magnetopause crossings, the number of cases positively identified as reconnection cases seems unduly small. The problem is that most crossings cannot be reliably classified, neither as rotational nor as tangential discontinuities, because they occur so rapidly that not enough, or no accurate flow measurements are obtained. The number of cases where brief occurrences of high-speed flows are observed is actually much larger than Fig. 10 indicates.

The Magnetopause Boundary Layer

Overview

Observations by a multitude of spacecraft have shown that a boundary layer of magnetosheath-like plasma is commonly found within the magnetopause. This boundary layer carries much of the momentum and energy transferred by the solar wind/magnetosphere interaction. The question is whether that interaction is solely through magnetic reconnection or whether other processes have to be considered.

Figure 12 illustrates the general morphology of the boundary layer. The various regions identified in Fig. 13 are distinguished by different spatial/temporal properties and plasma flow patterns. (For a review of the boundary layer properties see Paschmann, 1979; Eastman and Hones, 1979.)

The low-latitude boundary layer on the dayside is characterized by small thickness and highly variable flow directions. Towards the flanks the flows become more ordered and are directed tailwards, with a large cross-field component. The entry layer is not just the high-latitude extension of the LLBL. It owes its singular role to its proximity to the exterior cusp region where flows are very turbulent (cf. Figure 13). The entry layer usually is very thick, and flows both parallel and antiparallel to B occur.

The plasma mantle can be looked upon in two ways. One, it is the outflow region for the entry layer under the influence of tailward convection of magnetic field lines. This picture accurately describes the tailward field-aligned flow in the mantle as well as the characteristic variation of density, velocity, and temperature with distance from the magnetopause ("velocity filter" effect). On the other hand, there should be continued entry along the tail magnetopause into the mantle if the

The events shown in Fig. 10 had magnetic field rotation angles at the magnetopause between ~ 80° and ~ 170°. Large magnetic field angles are, however, not sufficient for the onset of reconnection, certainly not over the entire dayside magnetopause. Figure 11 shows a magnetopause crossing where the field

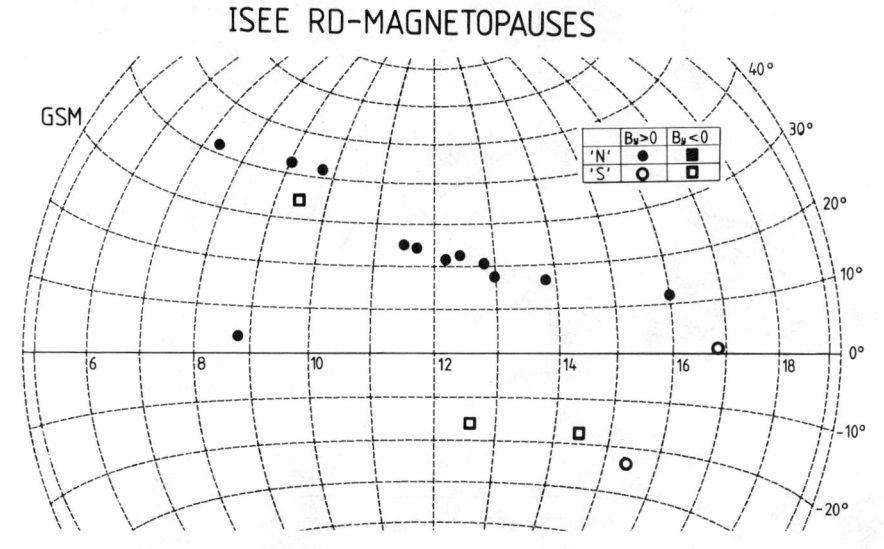

Fig. 10. Location of the 17 magnetopause crossings identified as rotational discontinuities from tangential momentum balance. Open and closed symbols refer to cases south and north of the separator.

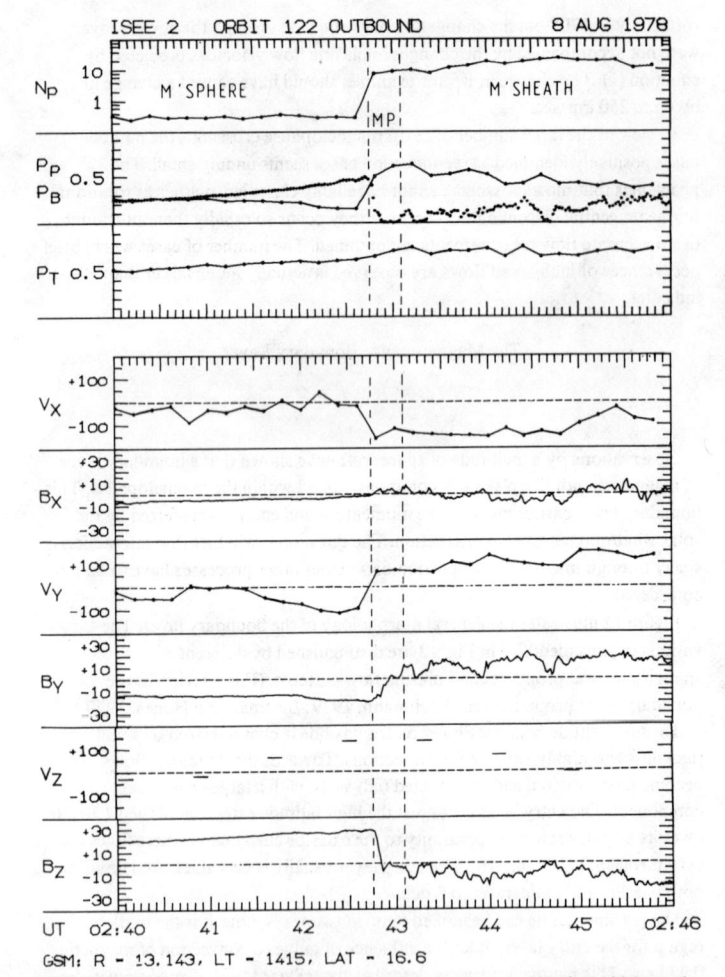

Fig. 11. Plasma and magnetic field data for a magnetopause crossing which can be identified as a tangential discontinuity in spite of the fact that magnetic fields were nearly antiparallel on the two sides. Note also the lack of a boundary layer [from Papamastorakis et al., 1984].

magnetopause is open (i.e., a rotational discontinuity). In this framework the mantle appears as an expansion fan [e.g., Swift and Lee, 1982].

It is interesting to note that of all the boundary layer regions, only the plasma mantle shows clear correlations in occurrence with the IMF.

The field lines threading the boundary layers connect to the earth, thus transverse momentum is coupled to the ionosphere via field-aligned currents [see Vasyliunas, 1979; Sonnerup, 1980; Hones, 1984]. This coupling is also evidenced by the often substantial contribution of ionospheric ions to the boundary plasma.

Entry Processes

Reconnection produces a boundary layer by direct fluid flow across the entire open portion of the magnetopause as discussed in Section 2. If no diffusion broadens the layer, the reconnection boundary layer is located outward from the inner separatrix (Fig. 1), i.e., on open field lines. As discussed in the previous section, there is clear *in situ* evidence from much of the dayside magnetopause that reconnection occurs. There is, however, no direct measurement of the normal mass flux density in these cases, nor of the size of the region. Using a normal magnetic field $B_n \approx 2$ nT and a density $n \approx 20$ cm^{-3}, the normal number flux density is 2×10^7 particles/cm^2 sec. Estimating the width of the open region at 3 R_E, and the length at 10 R_E, the total influx on the dayside is 2×10^{26} particles/sec. This is about the number of particles typically streaming downtail in the plasma mantle [Sckopke and Paschmann, 1978]. Similar fluxes could also result from flux transfer events if reasonable assumptions about the occurrence rate is made. Flux transfer events might also explain the "blobbiness" of the low-latitude boundary layer [Cowley, 1982], although Kelvin-Helmholtz instability of an originally laminar boundary layer is another plausible candidate [Sonnerup, 1980]. Thus it appears that under conditions when reconnection occurs it could sustain the boundary layer. It is also consistent with present observations that quasi-steady reconnection and flux transfer events alone account for the typical cross-magnetosphere electric potentials [Cowley, 1982].

There is evidence, however, that reconnection might not be the only process producing plasma entry into the boundary layer, but that some other process must operate. The strongest evidence is that the dayside boundary layers are observed even when the IMF is directed such that dayside reconnection should not occur with any significant rate [Haerendel et al., 1978]. On the basis of energetic electron angular distributions, it has also been suggested that the low-latitude boundary layer is located on closed field lines, at least in many cases

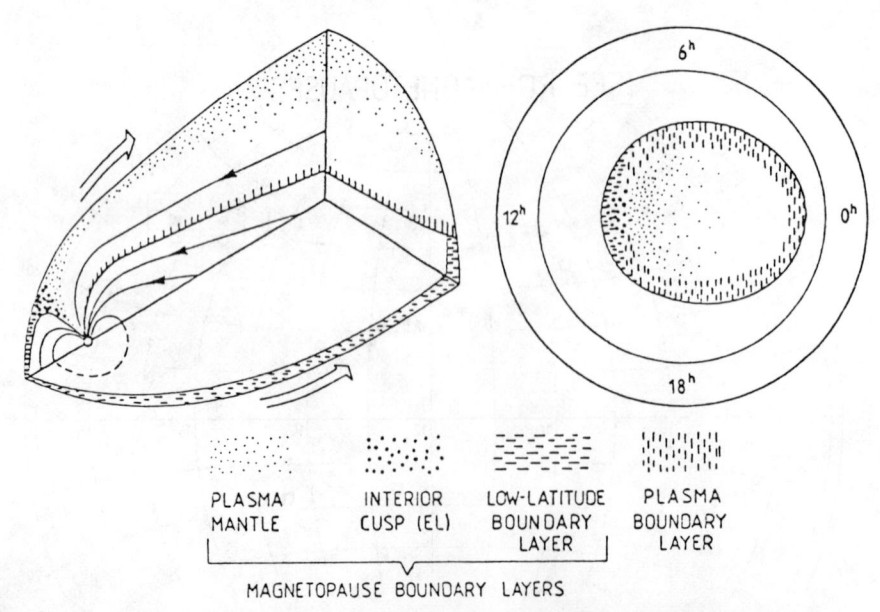

Fig. 12. Schematic diagram showing (a) the morphology of the magnetospheric boundary layers and (b) the ionospheric location of the magnetic field lines threading the boundary layers [after Vasyliunas, 1979].

Entry Regions

The fact that boundary layers are found essentially everywhere along the magnetopause does not necessarily mean that entry must also occur everywhere. Transport processes within the boundary layer will tend to distribute the plasma along the boundary (except for regions of strong earthward convection) and can produce a boundary layer at locations where no or little actual entry occurs.

From the evidence presented in Section 2 it is clear that entry definitely occurs at low latitudes on the dayside magnetopause (and since the open flux will be carried along the magnetopause into the tail, all along the path of the open flux tubes). There is evidence, however, suggesting that entry at high latitudes, i.e., the polar cusps, may be of primary importance too.

The polar cusps have long been expected to facilitate plasma entry because of their unique geometry. Originally, that property was attributed to the magnetic neutral points which in "closed" magnetospheric models identify the cusps [e.g., Willis, 1969]. More recently, attention has been drawn to the turbulent nature of the exterior cusp (cf. Fig. 13) as a source for driving plasma entry [Haerendel, 1978]. Evidence that entry actually does occur in the cusps is provided by the noted similarity between plasma densities in the exterior cusp and the adjacent boundary layer, as well as the large thickness of that layer. Cusp entry has also been inferred from velocity dispersions observed at the low-altitude extension of the cusps [see review by Reiff, 1984].

A novel way to distinguish high-latitude from low-latitude entry has been used by Hones et al. [1982]. Figure 15 illustrates how dramatically different field orientations at low latitudes can occur depending on where the interaction is located. Hones et al. have reported a case of "reverse" draping of field lines which according to Fig. 15 is evidence for a high-latitude interaction.

Summary

In situ plasma and energetic particle measurements have confirmed the occurrence of quasi-steady reconnection at the Earth's magnetopause. The most definitive evidence has been provided by the quantitative test of tangential momentum balance across the magnetopause. In the few cases where the energy balance could also be tested, the result was consistent with reconnection. The degree of agreement is, however, much less than in the case of momentum

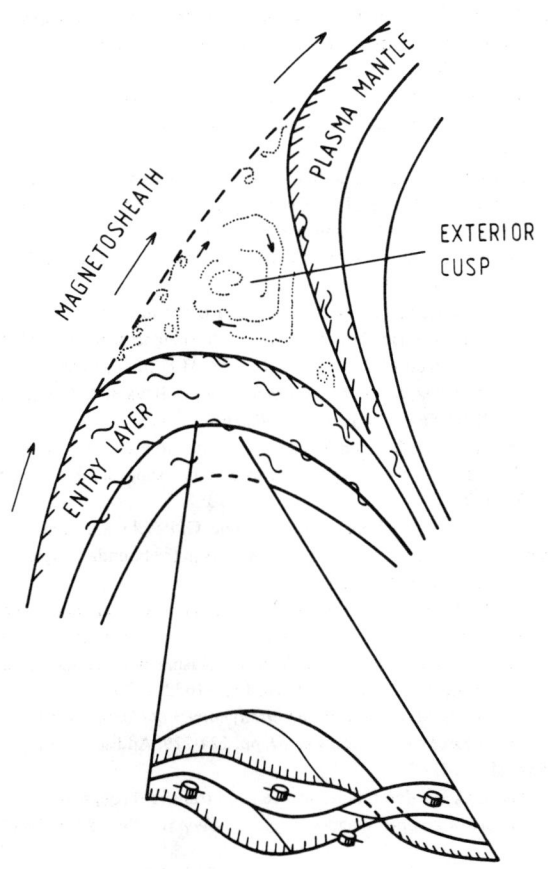

Fig. 13. Meridional cut through the northern polar cusp region illustrating the relationship between boundary layers and the turbulent eddy flows in the exterior cusp [from Haerendel et al., 1982].

[Eastman and Hones, 1979]. Recall that pure reconnection populates open field lines only.

Sckopke et al. [1981] have calculated the diffusion coefficient needed to populate the low-latitude boundary layer near the flanks of the magnetopause and obtained $D \approx 10^9$ m^2/sec if the diffusion layer is 400 km thick, i.e., of the order of the magnetopause thickness. Tsurutani and Thorne [1983] have recently shown that such a diffusion coefficient is not inconsistent with the observed low frequency electrostatic waves at the magnetopause.

Diffusion could also play the role of transporting plasma further across magnetic field lines, regardless of how the plasma actually crossed the magnetopause. Many of the density profiles of the boundary layer, notably those in the entry layer, are, however, very flat and thus inconsistent with diffusion. Flat density profiles, on the other hand, could be produced by the complete mixing of the boundary layer resulting from eddy convection, as proposed by Haerendel [1978].

Cowley et al. [1983] have recently compared the signatures in magnetic field deflections and plasma flow directions expected for reconnection and diffusion, and identified two sectors on the dayside magnetopause where these signatures would be very different. This point is illustrated in Fig. 14. For diffusive entry the boundary flows are expected to be directed essentially parallel to the external flow, i.e., radially away from the stagnation point. For reconnection with an IMF which has a substantial B_y component, magnetic tensions are such that the flow pattern is no longer symmetric around the noon meridian and thus is greatly different in certain regions.

It is interesting to note that at certain times and locations essentially no boundary layer is observed at all, and thus none of the entry or transport mechanisms could have been very effective. Figure 11 shows such a case.

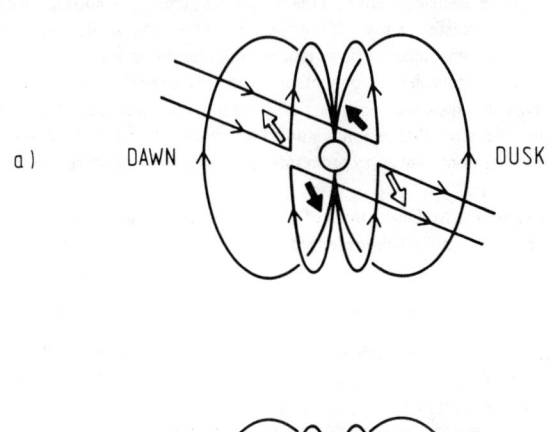

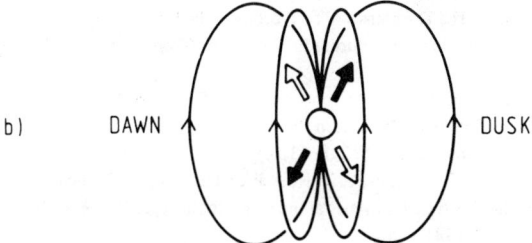

Fig. 14. Boundary layer flow patterns in the subsolar region for (a) reconnection and (b) diffusion.

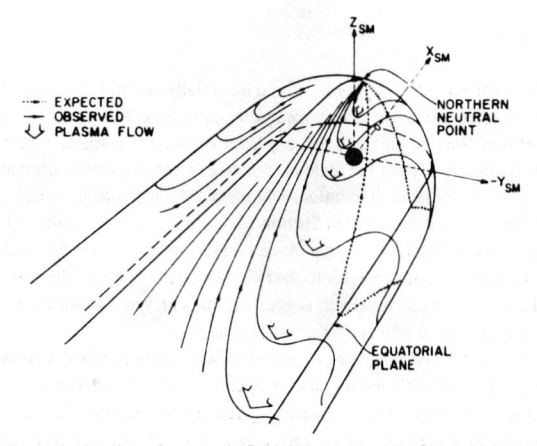

Fig. 15. Reverse draping of magnetic field lines as a result of high-latitude interaction (solid lines). The dotted lines show the draping expected for low-latitude interaction [after Hones et al., 1982].

balance. An important future extension of this kind of quantitative analysis of momentum and energy balance would have to include the effects of viscous stresses. This is, however, beyond the capabilities of the instrumentation flown so far.

Tracing the magnetic field topology at the magnetopause with energetic particles of magnetospheric origin has generally provided consistency checks with the results obtained from momentum balance. There are, however, some features in these data which are yet unexplained.

The results so far described to a large extent leave open the rate and scale size of the reconnection process as well as the precise onset conditions. Further progress in this area requires multi-spacecraft measurements with separation distances up to several earth radii [e.g., Haerendel et al., 1982]. Presently available observations also do not allow to judge the effectiveness of other entry or interaction processes, as well as to decide where the dominant entry sites are located. Within present constraints it is entirely possible that for southward IMF reconnection in the subsolar region is capable of providing the mass influx required to maintain the boundary layer as well as setting up the cross-magnetosphere electric potential. This is particular true if one adds the contribution from flux transfer events. (Of course, if it can be demonstrated that even under these circumstances the low- latitude boundary layer is on closed field lines, processes other than reconnection would have to be invoked.) As to the alternative entry sites, there is strong evidence that the polar cusp might play an important role. Until that region is studied with adequate instrumentation, the question of the dominant entry processes and sites will probably remain open.

Acknowledgement. Discussions of the paper with G. Haerendel and B. U. O. Sonnerup are gratefully acknowledged.

References

Aggson, T. L., P. J. Gambardella, N. C. Maynard, K. W. Ogilvie, and J. D. Scudder, Observation of plasma deceleration at a rotational magnetopause discontinuity, subm. to *Geophys. Res. Lett.*, 1983.

Cowley, S. W. H., The causes of convection in the Earth's magnetosphere: A review of developments during the IMS, *Rev. Geophys. Space Phys., 20*, 531-565, 1982.

Cowley, S. W. H., D. J. Southwood, and M. A. Saunders, Interpretation of magnetic field perturbations in the Earth's magnetopause boundary layers, subm. to *Planet. Space Sci.*, 1983.

Daly, P. W., D. J. Williams, C. T. Russell, and E. Keppler, Particle signature of magnetic flux transfer events at the magnetopause, *J. Geophys. Res., 86*, 1628-1632, 1981.

Daly, P. W. and T. A. Fritz, Trapped electron distributions on open magnetic field lines, *J. Geophys. Res., 87*, 6081-6088, 1982.

Eastman, T. E. and E. W. Hones, Jr., Characteristics of the magnetospheric boundary layer and magnetopause layer as observed by Imp 6, *J. Geophys. Res., 84*, 2019-2028, 1979.

Eastman, T. E. and L. A. Frank, Observations of high-speed plasma flow near the earth's magnetopause: Evidence for reconnection?, *J. Geophys. Res., 87*, 2187-2201, 1982.

Gosling, J. T., J. R. Asbridge, S. J. Bame, W. C. Feldman, G. Paschmann, N. Sckopke, and C. T. Russell, Evidence for quasi-stationary reconnection at the dayside magnetopause, *J. Geophys. Res., 87*, 2147-2158, 1982.

Haerendel, G., Microscopic plasma processes related to reconnection, *J. Atmos. Terr. Phys., 40*, 343-353, 1978.

Haerendel, G., G. Paschmann, N. Sckopke, H. Rosenbauer, and P. C. Hedgecock, The frontside boundary layer of the magnetosphere and the problem of reconnection, *J. Geophys. Res., 83*, 3195-3216, 1978.

Haerendel, G., A. Roux, M. Blanc, G. Paschmann, D. Bryant, A. Korth, and B. Hultqvist, *CLUSTER: Study in three dimensions of plasma turbulence and small-scale structure*, Mission Proposal submitted to ESA, 15 Nov. 1982.

Hoffman, F. de and E. Teller, Magneto-hydrodynamic shocks, *Phys. Rev., 80*, 692-703, 1950.

Hones, E. W., Jr., B. U. O:. Sonnerup, S. J. Bame, G. Paschmann, and C. T. Russell, Reverse draping of magnetic field lines in the boundary layer, *Geophys. Res. Lett., 9*, 523-526, 1982.

Hones, E. W., Jr., Field-aligned currents near the magnetosphere boundary, in: *Magnetospheric Currents* pp. 171-179, (ed. T. A. Potemra), AGU, 1984.

Hudson, P. D., Discontinuities in an anisotropic plasma and their identification in the solar wind, *Planet. Space Sci., 18*, 1611-1622, 1970.

Landau, L. D. and E. M. Lifshitz, ln: *Electrodynamics of Continuous Media (Course of Theoretical Physics, Vol. 8)*, pp. 224-229, Addison-Wesley, Reading, Mass., 1960.

Lee, L. C. and J. G. Roederer, Solar wind energy transfer through the magnetopause of an open magnetosphere, *J. Geophys. Res., 87*, 1439-1444, 1982.

Papamastorakis, I., G. Paschmann, N. Sckopke, S. J. Bame, and J. Berchem, The magnetopause as a tangential discontinuity for large field rotation angles, *J. Geophys. Res., 89*, 127-136, 1984.

Paschmann, G., Plasma structure of the magnetopause and boundary layer, in: *Magnetospheric Boundary Layers* (B. Battrick, ed.), pp. 25-36, Paris: ESA, SP-148, 1979.

Paschmann, G., B. U. Ö. Sonnerup, I. Papamastorakis, N. Sckopke, G. Haerendel, S. J. Bame, J. R. Asbridge, J. T. Gosling, C. T. Russell, and R. C. Elphic, Plasma acceleration at the earth's magnetopause: Evidence for reconnection, *Nature, 282*, 243-246, 1979.

Paschmann, G., I. Papamastorakis, B. U. Ö. Sonnerup, N. Sckopke, C. T. Russell, ISEE observations of magnetopause reconnection: The energy balance, to be submitted to *J. Geophys. Res.*, 1984.

Reiff, P., Ground-based and polar satellite observations, 1984 (this volume).

Scholer, M., F. M. Ipavich, G. Gloeckler, D. Hovestadt, and B. Klecker, Leakage of magnetospheric ions into the magnetosheath along reconnected field lines at the dayside magnetopause, *J. Geophys. Res., 86*, 1299-1304, 1981.

Scholer, M., D. Hovestadt, F. M. Ipavich, and G. Gloeckler, Energetic protons, alpha particles, and electrons in magnetic flux transfer events, *J. Geophys. Res., 87*, 2169-2175, 1982a.

Scholer, M., P. W. Daly, G. Paschmann, and T. A. Fritz, Field line topology determined by energetic particles during a possible magnetopause reconnection event, *J. Geophys. Res., 87*, 6073-6080, 1982b.

Scholer, M. and F. M. Ipavich, Interaction of ring current ions with the magnetopause, *J. Geophys. Res., 88*, 6937-6943, 1983.

Sckopke, N. and G. Paschmann, The plasma mantle: A survey of boundary layer observations, *J. Atmos. Terr. Phys., 40*, 261-278, 1978.

Sckopke, N., G. Paschmann, G. Haerendel, B. U. Ö. Sonnerup, S. J. Bame, T. G. Forbes, E. W. Hones, Jr., and C. T. Russell, Structure of the low latitude boundary layer, *J. Geophys. Res., 86*, 2099-2110, 1981.

Sonnerup, B. U. Ö., Acceleration of particles reflected at a shock front, *J. Geophys. Res., 74*, 1301-1304, 1969.

Sonnerup, B. U. Ö., Theory of the low-latitude boundary layer, *J. Geophys. Res.,*

85, 2017-2026, 1980.

Sonnerup, B. U. O., G. Paschmann, I. Papamastorakis, N. Sckopke, G. Haerendel, S. J. Bame, J. R. Asbridge, J. T. Gosling, and C. T. Russell, Evidence for magnetic field reconnection at the earth's magnetopause, *J. Geophys. Res., 86*, 10049-10067, 1981.

Swift, D. W. and L. C. Lee, The magnetotail boundary and energy transfer processes, *Geophys. Res. Lett., 9*, 527-530, 1982.

Tsurutani, B. T., and R. M. Thorne, Diffusion processes in the magnetopause boundary layer, *Geophys. Res. Lett., 9*, 1247-1250, 1982.

Vasyliunas, V. M., Interaction between the magnetospheric boundary layers and the ionosphere, in: *Magnetospheric Boundary Layers* (B. Battrick, ed.), pp. 387-393, Paris: ESA, SP-148, 1979.

Willis, D. M., The influx of charged particles at the magnetic cusps on the boundary of the magnetosphere, *Planet. Space Sci., 17*, 339-348, 1969.

Questions and Answers

Lui: How often do you see magnetopause crossings in which there is no indication of reconnection signatures and boundary layer?

Paschmann: Only on rare occasions is neither of these two signatures observed.

Vasyliunas: One should not assume that plasma transported across the magnetopause by diffusion necessarily retains its tangential momentum; whether it does or not depends entirely on the properties of whatever stochastic process is responsible for the diffusion. Simple collisional diffusion would lead to a broadening of the magnetopause current layer, which is not observed; the process, if it occurs, must therefore be more complex, involving waves or magnetic structures which in principle could take up some tangential momentum.

Paschmann: I did not mean the plasma to retain its tangential momentum, but that its velocity would tend to be aligned with that of the external flow.

RECONNECTION AT THE EARTH'S MAGNETOPAUSE: MAGNETIC FIELD OBSERVATIONS AND FLUX TRANSFER EVENTS

C. T. Russell

Department of Earth and Space Sciences and Institute of Geophysics and
Planetary Physics, University of California, Los Angeles, California 90024

Abstract. The concept of the magnetopause is over 50 years old and the first observation over 20 years old but there is still much to learn about this boundary. It was clear to some that the earliest observations of the magnetosphere demanded the reconnection of the interplanetary magnetic field with the terrestrial magnetic field across the dayside magnetopause, but early observations of the structure of the dayside magnetopause were unable to prove that this mechanism did, in fact, accelerate plasma as proposed. It was not until the advent of the ISEE-1 and -2 measurements beginning in 1977 that the three-dimensional high resolution plasma measurements were available to demonstrate that the magnetopause responded to southward interplanetary magnetic fields as predicted. Nevertheless, the ISEE measurements had their surprises. They revealed that reconnection was seldom a steady state process. Rather, the magnetopause was often subjected to patchy reconnection producing a feature called the flux transfer event. Flux transfer events are tubes of twisted magnetic field about 1 Re across, connecting the magnetosheath to the magnetosphere. Their transport to the tail transfers enough magnetic flux to be significant for magnetospheric dynamics. Despite all this progress, we still do not understand why we sometimes observe flux transfer events and sometimes steady-state reconnection. Nor do we know where reconnection is first initiated. There is still much to be learned about the reconnecting magnetopause.

Early History

The concept of a magnetopause, the boundary between the flowing solar wind, and the terrestrial magnetic field is now just over fifty years old. Chapman and Ferraro (1931a; 1931b; 1933; 1940) proposed that a neutral stream of ionized gas was emitted by the sun at geomagnetically disturbed times. When this stream met the geomagnetic field, it was deflected at the magnetopause, compressing the geomagnetic field and causing the initial increase in the field sensed on the surface of the earth. Later some of this plasma entered the field, inflating it and forming the ring current. Thus, the formation of the magnetopause was an integral part of the conventional wisdom concerning the cause of geomagnetic storms which could be sensed on the surface of the earth. Hence, when the magnetopause was finally detected at the dawn of the space age 30 years later (Heppner et al., 1963; Cahill and Amazeen, 1963), it came as no surprise.

The year 1961 was a very important year for the magnetopause both observationally and theoretically. On March 25, 1961, Explorer 10 was launched on a very eccentric orbit with apogee at 47 Re, roughly antisunwards. Only 52 hours of calibrated operation were obtained on the battery powered mission, providing magnetic field measurements out to 43 Re. On six principal occasions the spacecraft left the geomagnetic tail and entered what we know now as the magnetosheath. Figure 1 shows the magnetic records during the first

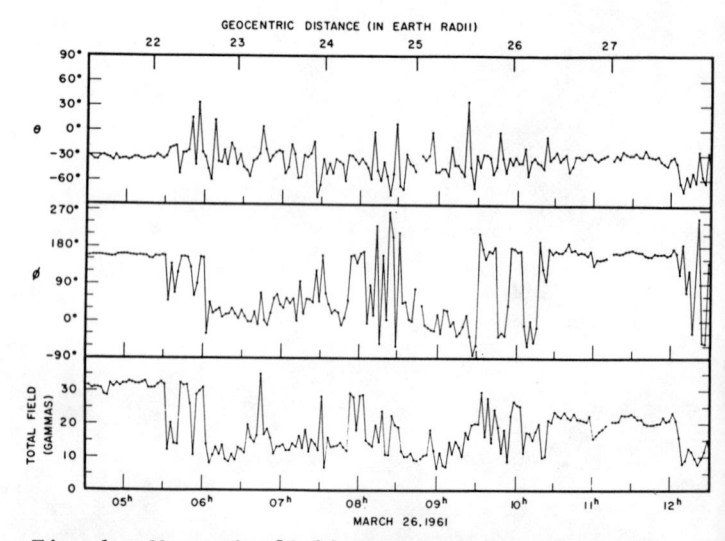

Fig. 1. Magnetic field measurements made by Explorer 10 on 3/26/1961, 15 Re behind the earth. These are the first measurements taken across the magnetopause and the first measurements of the geomagnetic tail (Heppner et al., 1963).

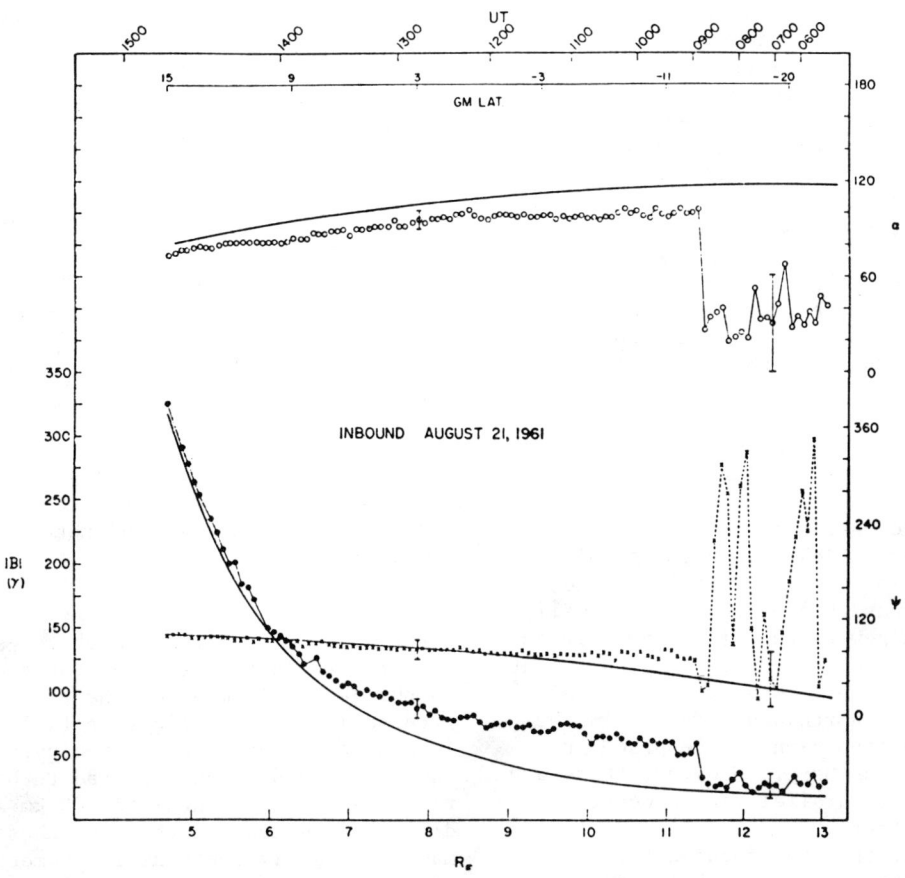

Fig. 2. Magnetic field measurements made by Explorer 12 on 8/21/1961. These are the earliest published measurements taken across the subsolar magnetopause (Cahill and Amazeen, 1963; Cahill and Patel, 1967).

magnetopause crossing (Heppner et al., 1963).

Later this same year, on August 16, 1961, Explorer 12 was launched into an elliptical orbit with an apogee of 14 Re, initially near local noon. Apogee moved around the sun and when transmission ceased almost four months later on December 6, 1961, Explorer 12 had sampled the entire dawnside near equatorial magnetopause (Cahill and Amazeen, 1963; Cahill and Patel, 1967). Figure 2 shows the earliest published magnetopause crossing from this mission. The subsequent analysis of these first data on the dayside magnetopause provided the foundation of the magnetopause paradigm for the next decade and a half despite the launch of Explorer 14, OGO-1, -3 and -5 and innumerable IMP's. Fairfield and Cahill (1966) provided the first real evidence for the reconnecting magnetopause when they pointed out that, when the magnetosheath magnetic field was southward as seen by Explorer 12, ground-based auroral zone magnetometers recorded substorms and, when the magnetosheath field turned northward, ground level disturbances ceased. The same year Patel and Dessler (1966) showed that the distance to the magnetopause was smaller for high A_p. Unfortunately, because of the large scatter they dismissed the correlation.

Properly interpreted this was evidence for what was later to be called erosion of the magnetopause by reconnection (Aubry et al., 1970). The next year Patel et al. (1967) plotted the correlation between geomagnetic activity and the angle between the magnetosheath field and the magnetospheric field at magnetopause crossings. In accord with the results of Fairfield and Cahill (1966), antiparallel fields, i.e., southward directed magnetosheath fields, were associated with the largest activity. However, again because of the scatter they dismissed the correlation.

Another important contribution to the magnetopause paradigm was provided in a series of papers by Sonnerup and Cahill (1967; 1968) and Sonnerup (1971). Sonnerup and Cahill pioneered the use of minimum variance analysis to determine the normal to the magnetopause and the magnetic field along the normal. Two different types of magnetopause crossings were distinguished: those resembling rotational discontinuities and those resembling tangential discontinuities. Finally, the sense of rotation of the magnetic field through the boundary was found to follow the direction of electron rotation about the normal component which in the reconnection model points inward north of the merging line and outward south of the merging line. Of the 12 rotational magnetopause crossings examined, 11 followed the relation expected if the magnetopause had "electron" polarization. Simple magnetohydrodynamic theory does not show any preference for the electron mode (Landau and Liftshitz, 1960). Thus, Su and Sonnerup (1968) worked out the first-order orbit theory of a large-amplitude rotational-discontinuity in a collision-free gyrotropic plasma. As we discuss in a following section, this work was unnecessary as the correlation was merely a coincidence.

Finally, the year 1961 was important theoretically because it was the year that Dungey's classic model of the reconnection of interplanetary magnetic field lines with the magnetosphere appeared (Dungey, 1961). As shown in Figure 3 in the top panel, southward field lines, convected along by the solar wind, break in half and join partners with magnetospheric field lines. The interplanetary magnetic field now has one foot on the ground and the other off at infinity. The field line convects past the earth but eventually drifts down, joins a new partner (the same partner in 2-D) and convects back to earth, only to repeat the process sometime later. When the interplanetary magnetic field is northward as shown in the bottom panel reconnection cannot take place at the nose of the magnetosphere. However, there are other places where antiparallel fields occur and it might take place there.

Dungey pointed out that the reconnection rate at the nose and in the tail had to balance on the average. Thus if there is reconnection on the dayside magnetopause you must eventually have reconnection in the tail. However, these rates do not have to balance instantaneously. If they did balance instantaneously we would not have magnetospheric substorms (Russell and McPherron, 1973a). The fact that these rates can get out of balance makes the magnetosphere interesting, but understanding it very difficult.

Although reconnection should have taken the field by storm, there was much reluctance to embrace Dungey's ideas. One of the clearest pieces of evidence supporting Dungey's picture is the erosion of the magnetosphere. Yet Aubry et al. (1970) chose not to reference Dungey, in order to facilitate the acceptance of their paper. Similarly substorm models based on reconnection were not initially widely acclaimed (McPherron et al., 1973; Russell and McPherron, 1973). It is only since the appearance of the ISEE-1 and -2 observations of the magnetopause that the reconnecting magnetosphere has become the conventional wisdom and that oposition has shrunk to a small, but vocal, minority.

Thickness of the Magnetopause Current Layer

The simplest models of the magnetopause give as the thickness of the current sheet, the distance required for the incident particles to be turned around by the magnetic field, i.e., a

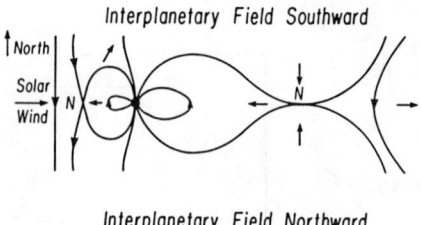

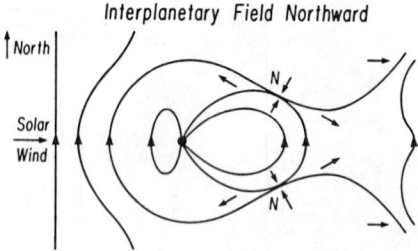

Fig. 3. The magnetic topology of the reconnecting magnetosphere for southward (top) and northward (bottom) interplanetary magnetic fields (Dungey, 1963).

gyroradius. If there is a charge-separation electric field then this distance can be as small as the geometric mean of the electron and proton gyroradii. If this electric field is shorted out, it can be as large as the ion gyroradius. Thus, the magnetopause was expected to be a thin current layer from about 1 to 100 km across. It is difficult to measure this thickness because the magnetopause is continually in motion. Most early measurements used data from multiple magnetopause crossings, assuming that the boundary was oscillating radially in sinusoidal motion. Speeds of about 10 km/sec were determined leading to a thickness of about 100-1000 km for the magnetopause current layer (Cahill and Amazeen, 1963; Holzer et al., 1966; Ogilvie et al., 1971).

There is a steep radial gradient in energetic protons at the magnetopause. By examining the flux of protons at the same pitch angle but with different guiding centers, one can remotely probe this gradient and gauge the velocity of the magnetopause. Kaufman and Konradi (1973) used the energetic ion experiment on Explorer 12 to do this and found that the magnetopause usually moved with a velocity of 20 km/sec or less and had a thickness of the order of 10 times the gyroradius of a 1 keV proton or about 1000 km. They noted that some current sheets may be substantially thinner, but their technique was inadequate to study thinner magnetopause crossings. Despite these measurements, the magnetopause current was usually considered to be of gyroradius scale size, perhaps because of the existence of a simple gyrocurrent model (cf. Neugebauer et al., 1974).

The launch in October 1977 of the dual satellites ISEE-1 and -2, co-orbiting in a single highly elliptic orbit, provided the first major advance in studying the structure of the magnetopause since the initial Explorer 12 observations. The two satellites had variable separation, ranging from less than 100 to several thousand

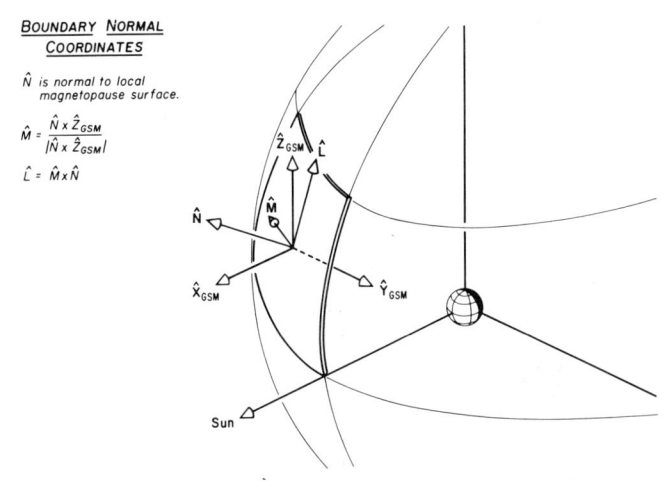

Fig. 4. The definition of boundary normal coordinates (Elphic and Russell, 1979).

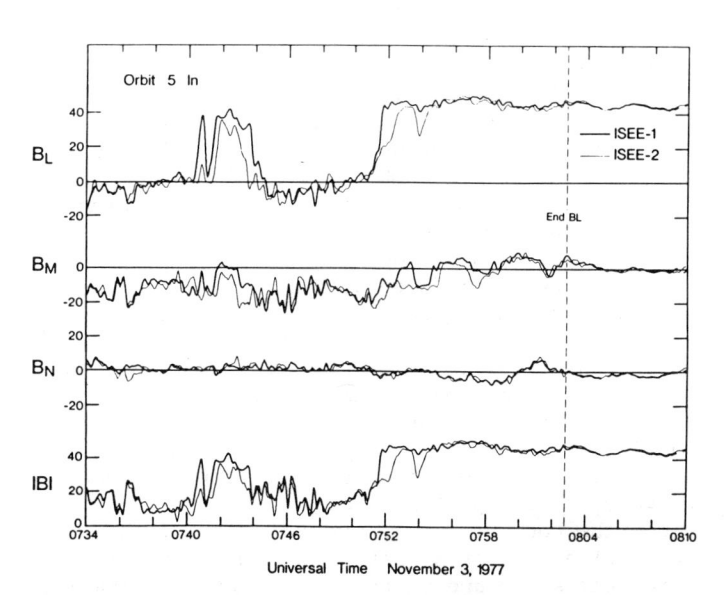

Fig. 5. ISEE-1 and -2 magnetic field measurements across the magnetopause on November 3, 1977 (Russell and Elphic, 1978).

Km at the magnetopause. Observations of the same feature at two different times at the two locations or simultaneous measurements at different locations in the same gradient permitted estimates of the velocity and thickness of the current layer. Since a knowledge of the thickness and motion of the magnetopause is crucial to understanding the physics of the magnetopause current layer, we will take some time to review how these are calculated and the results of these calculations.

In order to understand the behavior of the magnetopause magnetic field it is important to display the data in a geophysically meaningful coordinate system. If a variation of the magnetic field occurs in only two dimensions, the underlying physical cause may be simple to divine. However, if care is not taken in the choice of the coordinate system, one might not realize the field behavior is two-dimensional or simple. We have chosen to use boundary normal coordinates in which the N direction is along the boundary normal as is shown in Figure 4. The L and M directions are in the plane of the boundary. We have chosen the L direction to be along the projection of the Z GSM direction on the plane of the boundary. Alternatively, one could choose the Z direction to be the projection of the magnetospheric field in the plane of the boundary. There are also several ways for determining the direction of the boundary normal N. One could use the average shape of the magnetopause and the known location of the spacecraft to determine a geometric normal. One could assume that the magnetopause was a tangential discontinuity and use the cross product of the magnetic fields on either side. Finally, one can use the minimum variance direction as the direction of the normal. For most purposes the precise direction of the boundary normal coordinate directions is not crucial. However, for determining the normal component of the magnetic field, it is critical. Unfortunately, the errors inherent in determining the magnetopause normal are so large that the error in the component of the magnetic field along the normal is usually greater than the expected size of the normal component. Existing error estimates ignore both the physics of the magnetopause current layer and the source of the errors in the orientation of the normal so that caution should be used when discussing the significance of the normal component for individual crossings.

Figure 5 shows magnetic field measurements in boundary normal coordinates obtained from both

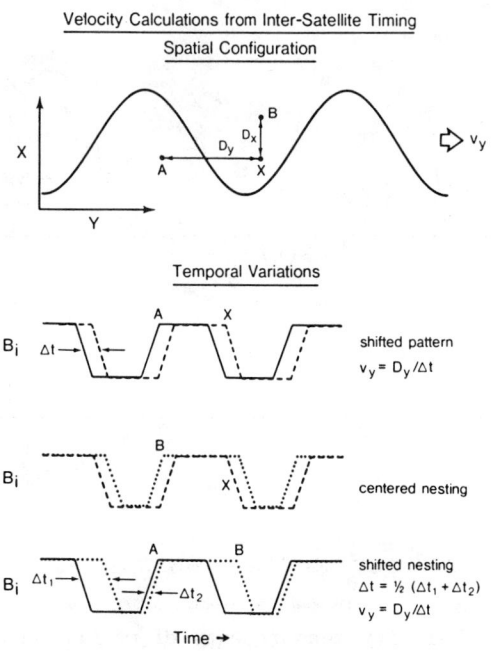

Fig. 6. Velocity calculations from intersatellite timing.

ISEE-1 and -2 (Russell and Elphic, 1978). These data exhibit a characteristic nesting pattern caused by multiple oscillations of the magnetopause across the two spacecraft which are at different radial distances from the earth. The use of the time delays between the boundary traversals by the two spacecraft is illustrated in Figure 6. Here we illustrate the patterns expected due to a periodic surface wave propagating along the magnetopause at three locations: at spacecraft A, spacecraft B, and at position X which is radially aligned with spacecraft B at the same radial distance as spacecraft A. An observer at position X would see the same disturbance pattern as at spacecraft A but shifted in time. The observer at position X would see a symmetrically nested disturbance pattern when compared to the pattern seen at spacecraft B, since the boundary would always cross X first and arrive back at X last. These antisymmetric time delays are due to the radial motion. Finally, the comparison of the observations at both A and B are shown in the bottom panel of Figure 6. The patterns are both shifted and nested.

We can solve for both the velocity of the radial displacement and the velocity of the surface wave if these velocities are constant between observations and if the inward and outward velocities are equal and opposite. In general we would need four spacecraft to measure these velocities: two baselines for the two components of velocity of the surface wave and one for the velocity of the normal displacement. If the two spacecraft are very nearly aligned along the normal or if the surface wave velocity is very high

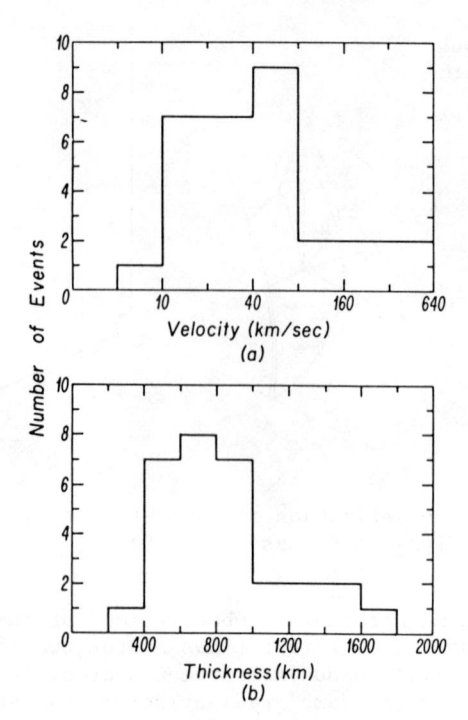

Fig. 8. Velocity and thickness of magnetopause (Berchem and Russell, 1982a).

compared to the velocity along the normal, then the contribution to the intersatellite time delay by the surface wave is small and we can use the delays between the two satellites as measures of the velocity along the normal. This we have done to measure the velocity and thickness of the magnetopause using ISEE-1 and -2 (Berchem and Russell, 1982a).

Figure 7 shows an expansion of the top trace of Figure 5. The horizontal lines on the top trace show the delay between ISEE-1 and -2 at fixed distances through the current sheet. This time delay divided into the separation along the normal gives the magnetopause velocity along the normal and is shown by vertical lines in the bottom panel. It is important to note that the velocity is large and variable and far from sinusoidal. In fact, the velocity can vary markedly during a single boundary traversal. Thus, we must be very careful in measuring the magnetopause velocity and thickness, and use only observations taken when the separation of the two spacecraft is comparable to or less than the thickness of the current sheet.

Figure 8 shows the velocity and thickness for thirty magnetopause crossings. The thickness estimate refers to the distance over which 80% of the change in the B_L component occurs. This avoids including boundary layer currents as part of the magnetopause. About 70% of the magnetopause thicknesses lie between 400 and 1000 km. Thus, the magnetopause current layer is generally thick as measured in terms of the ion gyro-radius. This is an important result in understanding the

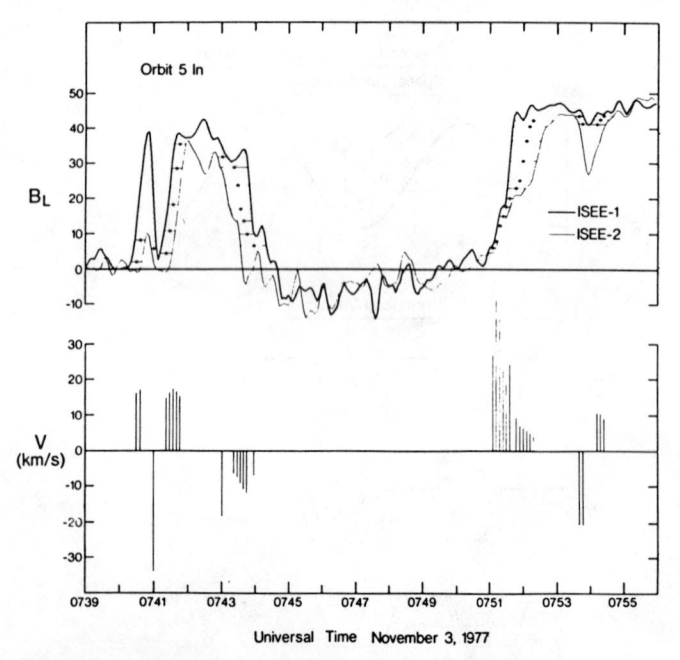

Fig. 7. Velocity measurements of magnetopause motion on November 3, 1977 (Russell and Elphic, 1978).

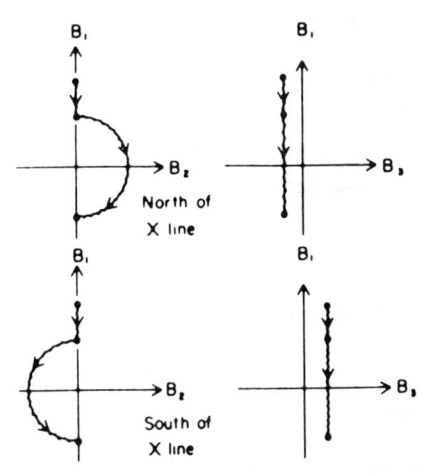

Fig. 9. Hodograms of the magnetic field through the magnetopause according to the classical reconnection paradigm (Sonnerup and Ledley, 1979).

physics of the magnetopause. First, it shows that the current layer is not simply due to the gyro-reflection of the magnetosheath flow. Many of the current carrying particles are trapped particles executing complete orbits in the current layer. Second, it bears on the suggestion of Su and Sonnerup (1968) that the apparent preference of the magnetopause rotation for electron polarization is due to the thinness of the boundary. They, in fact, state that proton polarization may occur only for layers thicker than about 4000 km. Since the thickest magnetopause crossings that we observed were less than half this thick, the electron polarization should have been exclusively observed if their theory was applicable to these current layers. As we discuss in the next section, electron and proton polarizations are seen with equal frequency in the ISEE data.

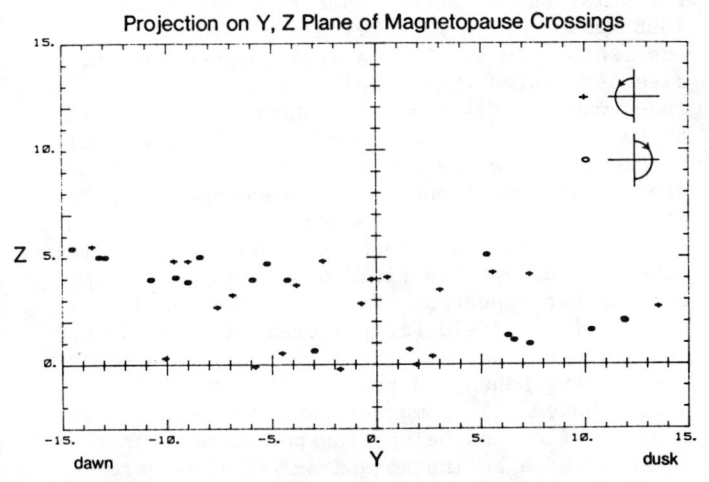

Fig. 10. Polarity of rotational forms through the magnetopause as a function of position (Berchem and Russell, 1982b).

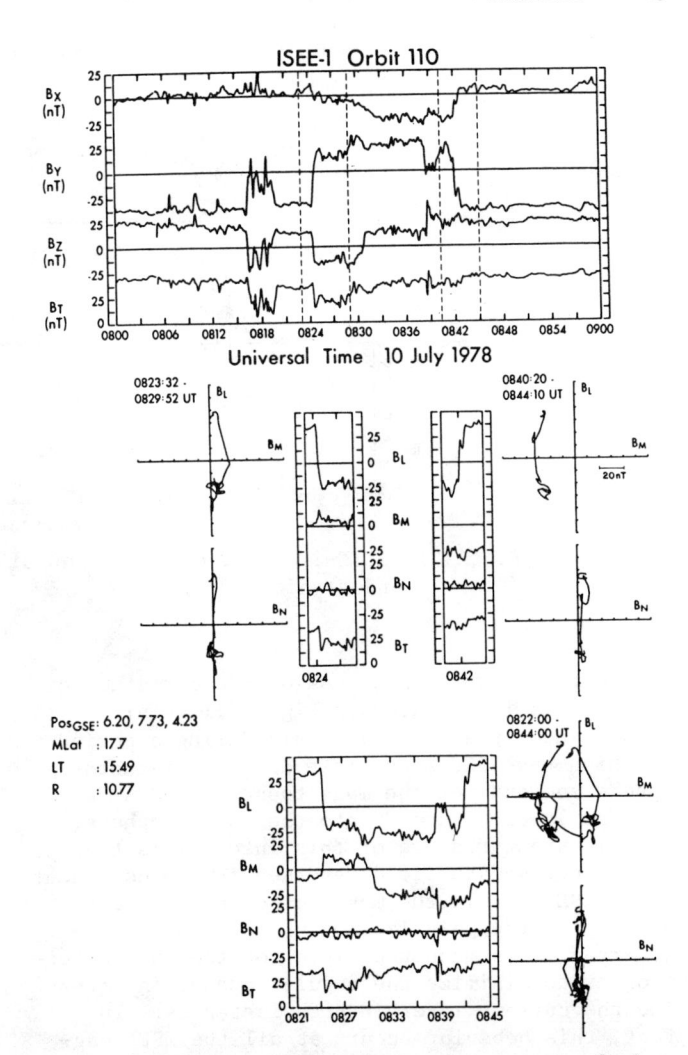

Fig. 11. Example of magnetosheath control of the polarity of the rotation through the magnetopause current sheet (Berchem and Russell, 1982b).

The Polarization of the Rotation in the Current Layer

In the "classical" theory of reconnection (Sonnerup and Ledley, 1979) the magnetic field component along the magnetopause normal is inward in the northern hemisphere and outward in the southern hemisphere. If the magnetopause is thin (as it is observed to be) the direction of rotation of magnetic field is expected to follow that of the rotation of an electron about the normal component (Su and Sonnerup, 1968). This pattern is shown in Figure 9. The ISEE-1 and -2 measurements do not support this picture. Figure 10 shows the direction of rotation of magnetopause crossings with clear rotational discontinuities. The pattern is random. Both rotation directions occur independent of location (Berchem and Russell, 1982b). This independence of position is not caused by motion of the merging line. The direction of rotation is also independent of the normal component when it can be measured.

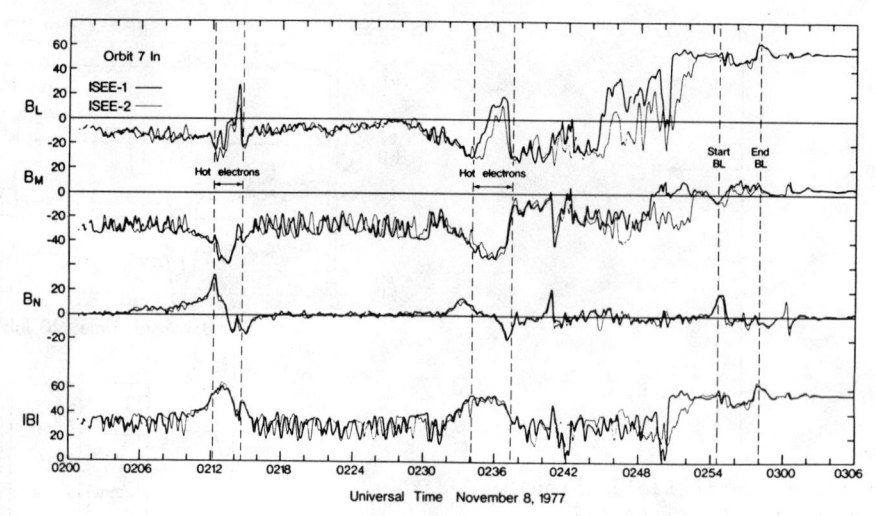

Fig. 12. ISEE-1 and -2 observations of flux transfer events on November 8, 1977 (Russell and Elphic, 1978).

The reason for randomness of the direction of rotation is illustrated in Figure 11. This figure shows the magnetic field during a period in which the ISEE spacecraft left the magnetosphere and entered the magnetosheath and a few minutes later returned into the magnetosphere. The important feature of this interval is that the magnetosheath field reversed its B_M component while ISEE was in the magnetosheath. Correspondingly, the polarization of the crossing reversed. The magnetic field rotated in a direction as to minimize the angular change in crossing the current layer in both instances. In fact, this behavior occurs at all the ISEE magnetopause crossings. Whenever there is a clear rotation of the magnetic field through the magnetopause, it rotates less than 180°. The direction of the magnetosheath magnetic field controls the direction of rotation of the field through the boundary.

While these observations are in contradiction to the interpretation of Sonnerup and Cahill (1968) and the theory of Su and Sonnerup (1968) they are not in contradiction to the Explorer 12 data. All of the Explorer 12 rotations were less than 180°. In fact, Sonnerup and Cahill (1968) conclude that for this to occur and be consistent with their rotational polarizations the magnetosheath field had to be biased in their data set so that the magnetosheath field was predominantly southeast for their observations below the equator and southwest above it. It is clear they considered that the fact that 12 of the 12 rotations were less than 180° was accidental but that the fact that 11 of the 12 crossings obeyed their polarization rule proved their theory. In fact, the reverse was true.

Finally, we emphasize that the magnetopause current layer often does not produce a simple rotational form. This was true in both the Sonnerup and Cahill (1967; 1968) studies and the Berchem and Russell (1983) study. In fact, examination of the combined plasma and field data on ISEE show that often the magnetopause is a tangential discontinuity even when the interplanetary magnetic field was directed strongly southward and the fields on the two sides of the magnetopause were very nearly antiparallel (Papamastorakis et al., 1983).

Flux Transfer Events

In 1978 two pieces of evidence appeared supporting the concept of patchy reconnection. The first came from the field and plasma measurements of Heos 2 in the high latitude boundary layer (Haerendel et al., 1978). In this region short duration increases, or spikes, in the magnetospheric field strength were observed. Haerendel et al. (1978) attributed these increases to temporally and spatially limited reconnection in the polar cusp. At the same time the initial ISEE measurements were being analyzed and a curious feature in the low latitude magnetosheath, also associated with a spike in the field magnitude, was found. Figure 12 shows the first of these events that were found. In the center of each event were energetic electrons, as well as protons streaming out of the magnetosphere. Yet, at the same time at low energies the plasma was clearly flowing magnetosheath plasma. The direction of the magnetic field also changed in a curious but repeatable fashion. The normal component of the field first increased outwards and then turned inwards. The azimuthal or B_M component strengthened and the vertical or B_L component behaved like a magnetopause crossing. Since magnetic flux was being transported and the phenomenon had a beginning and an end they were termed flux transfer events. The presence of energetic magnetospheric electrons and ions in these flux tubes indicated their connection to

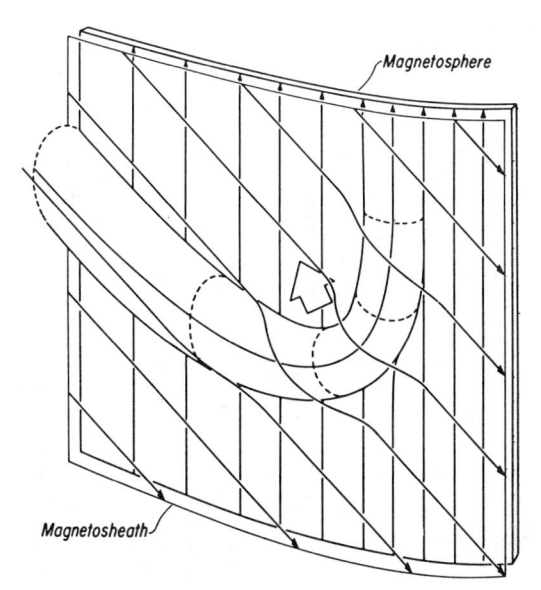

Fig. 13. Schematic of a flux transfer event (Russell and Elphic, 1978).

the magnetosphere and the streaming of the ions indicated that one end of the flux tube was open to the magnetosheath. Taken together the simplest explanation of the phenomenon was that spatially and temporally limited reconnection had taken place in the near equatorial region and the connected flux tube had been carried over the spacecraft (Russell and Elphic, 1978; 1979). The flux transport rate in these events is about 2×10^4 Webers/sec which is significant for geomagnetic processes. The total rate on the magnetopause could be somewhat greater because ISEE might not be in the proper position to detect all the events occurring at one particular time. For example, if flux transfer events may be initiated on both sides of noon the FTE's associated with only one of these sites would normally be detected by a spacecraft. Thus, the voltage applied across the polar cap by the FTE process would be on the average about 40 kilovolts. This value is comparable to the normal potential drop across the polar cap and is much larger than the 6 kV thought to be associated with viscous processes.

Podgorny et al. (1980) have proposed that reconnection takes place near the polar cusps and through the tearing instability forms magnetic islands on the dayside magnetopause. In order to explain the observed behavior of the normal component of the field the islands would have to drift towards the equator from the cusp against the magnetosheath flow. Further, we see no evidence in the particle anisotropies for field lines that close on themselves surrounding an 'O-type' neutral point. Finally, minimum variance of flux transfer events and the observed draping signature in the magnetosheath surrounding the events suggest that FTE's are intrinsically three-dimensional structures with nearly cylindrical cross-section. Thus, flux transfer events seem to be more than just simple tearing of the magnetopause.

Our interpretation of these events is shown in Figure 13. Reconnection takes place in a limited region for a short period of time and the resulting tube of reconnected flux is carried away by the magnetosheath flow. Surrounding field lines in the 'sheath drape over the tube and tension in the field lines attempts to straighten the bend in the tube accelerating the flow in this region. The duration of these events together with the plasma flow velocity gives dimensions along the boundary of up to several Re. The size of the variation of the normal component suggests that the dimensions normal to the boundary are similar.

Sometimes flux transfer events show an evolution in their signature as the spacecraft approaches the magnetopause. The bottom panel of Figure 14 shows such a case. The field perturbation far from the magnetopause is mainly along B_L and B_N. Closer to the magnetopause it is along B_M and B_N. Adjacent to the magnetopause the B_M perturbation is even stronger (Elphic and

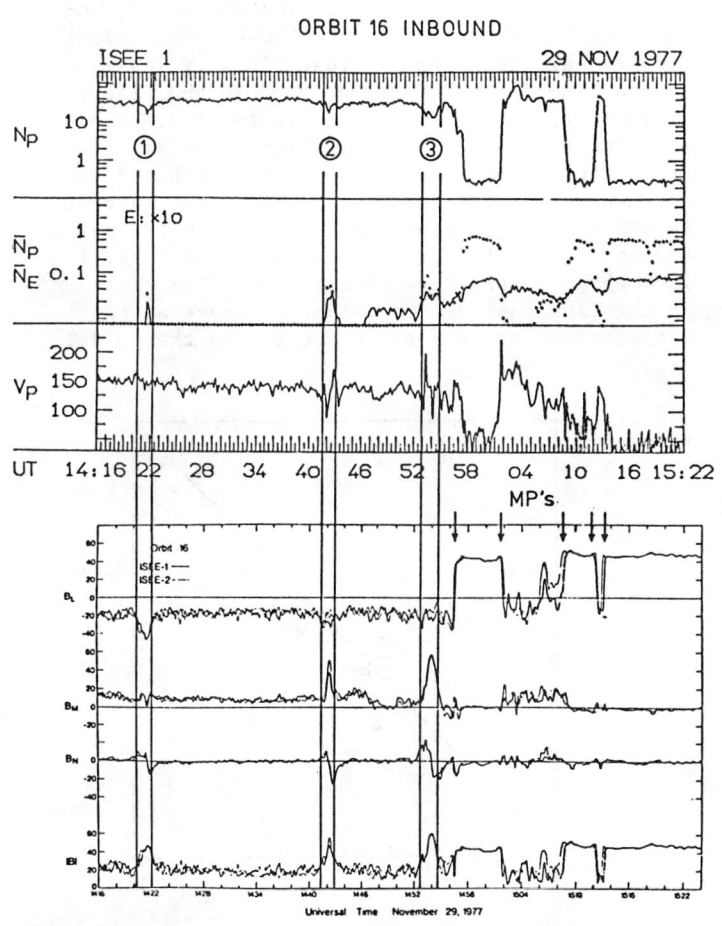

Fig. 14. Plasma and magnetic field data across flux transfer events shown in Fig. 15 (Pashmann et al., 1982). N_p is proton number density; V_p is 2-D proton velocity; $\bar{N}_p$ is number of energetic protons; $\bar{N}_E$ is number of energetic electrons.

Russell, 1979). The top panel of Figure 14 shows the plasma and field data during this event (Paschmann et al., 1982). There are some but very minor density decreases in the FTE. There are some velocity perturbations increasing as the magnetopause is reached and finally there are energetic particles increasing in flux as the magnetopause is reached. Thus, there is internal structure to flux transfer events. They are not just the simple tube of flux that Figure 14 might suggest.

One bit of evidence of what this structure might be comes from the combined plasma and field data (Paschmann et al., 1982). Figure 15 shows the change in magnetic pressure plus thermal pressure upon entry into an FTE versus a measure of the Maxwell stress of the magnetic field as exhibited by the size of the normal component and the tangential components of the field. The balance shown here suggests that the magnetic field wraps around the flux tube to contain the over-pressure, i.e., the magnetic field in the tube is twisted.

Saunders et al. (1983) have used dual satellite measurements of FTE's when ISEE-1 and ISEE-2 were well separated both to measure the dimensions of FTE's and to measure the twistedness of flux ropes. They confirm that FTE's typically have a diameter of the order of 1 Re and that the field in flux transfer events is twisted. The observed twist is usually such as to be associated with an inward current in the north and an outward current in the south. Thus, the current is not simply a current associated with the predominant loss of charge carriers of one sign. It is however in the direction expected for twisting of the flux tube due to velocity shears at the magnetopause as they are convected to the poles.

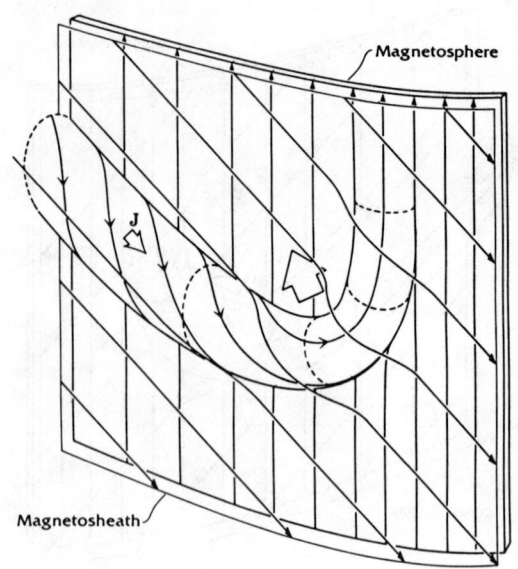

Fig. 16. Flux transfer event with twisted field lines.

Figure 16 shows our modified FTE sketch indicating the predominant twist in the northern hemisphere. The twist in the field continues into the magnetosphere as shown in the boundary layer events studied by Sckopke et al. (1981) and Cowley et al. (1983).

Flux transfer events perturb the region just inside the magnetopause as well as just outside. Figure 17 shows how the magnetospheric field as well as the magnetosheath field is disturbed by an FTE (Cowley, 1982). Figure 18 shows a sequence of flux transfer events going from inside the magnetosphere on the left (unshaded) into the magnetosheath on the left (shaded).

The initial observations with the ISEE spacecraft were all in the northern hemisphere where we would expect the perturbations in the normal component both due to draping outside the FTE and twisting inside the FTE to be first positive (outwards) and then negative (inwards) as an FTE convected away from the equatorial regions. If the observation point is below the equator one would expect the opposite signature as shown in Figure 19 (Cowley, 1982). This signature has been observed (Rijnbeek et al., 1982). Figure 20 shows a series of these reverse FTE's (Berchem and Russell, 1983).

ISEE-1 and -2 have now surveyed the dayside magnetopause from about +5 Re to -5 Re Z solar ecliptic distance (Berchem and Russell, 1983; Rijnbeek et al., 1983). Flux transfer events are seen throughout this region. Figure 21 shows the magnetic latitude and local time of the flux transfer events seen of both standard and reverse types. Standard flux transfer events are mainly a northern hemisphere phenomenon and reverse are mainly a southern hemisphere phenomenon. However, there are exceptions to this simple picture. Perhaps these exceptions result from non-

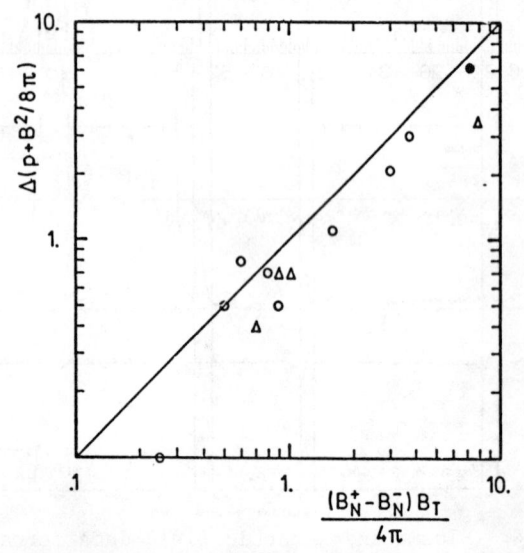

Fig. 15. Overpressure in flux transfer event versus Maxwell stress calculated from normal and tangential fields at an FTE (Paschmann et al., 1982).

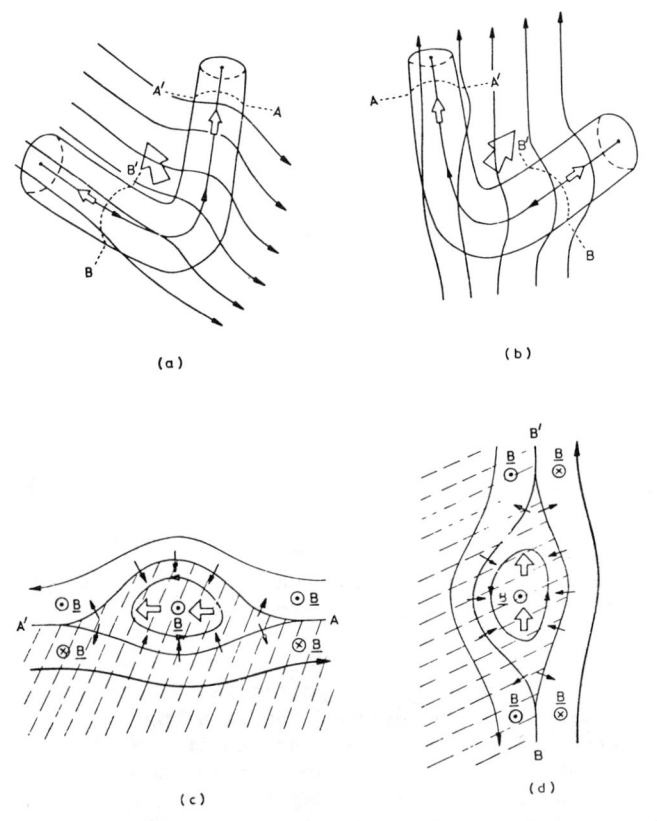

Fig. 17. Front and back view and cross-sections of a northward moving flux transfer event (Cowley, 1982).

slightly northward. As shown by Berchem and Russell (this volume) in their plot of the rate of FTE occurrence as a function of the direction of the interplanetary magnetic field as measured by IMP-8 and ISEE-3 with appropriate delays, FTE's almost never occur for northward IMF. The magnetopause truly looks like a half-wave rectifier (cf. Burton et al., 1975). If FTE's transfer magnetic flux to the magnetotail, we would expect to see some evidence for them at the tail magnetopause. IMP-8 data, in fact, clearly show FTE's at high latitude tail magnetopause boundaries (Sibeck, 1982). This is apparent both in the filamentary nature of the tail boundary region with short sections of unidirectional field adjacent to others of a different direction. It also appears as the varying direction of the magnetotail magnetopause which appears to be fluted with the flute axis parallel to the solar wind flow.

Finally, we note that flux transfer events are not solely a terrestrial phenomenon. Figure 22 shows an FTE at the Mercury magnetopause (Russell, 1983) and Figure 23 one at the Jovian magnetopause (Walker and Russell, 1983).

Discussion and Conclusions

There is no longer any doubt that reconnection is a significant process for the energization of the magnetosphere. The control of magnetospheric dynamics demands it and the observed behavior of the magnetosphere is consistent with it. Nevertheless, there are many unanswered questions. We have not addressed steady-state reconnection in this review. It clearly is observed (Paschmann et al., 1979; Sonnerup et al., 1981). It is not yet clear what is the relationship of steady-state reconnection to patchy reconnection. It is an either-or relationship or do both occur simultaneously in different regions of the magnetopause? If they are mutually exclusive phenomena, what controls their occurrence?

equatorial reconnection. If so, higher latitude measurements are highly desirable.

The occurrence of flux transfer events is controlled by the direction of the interplanetary magnetic field. Rijnbeek et al. (1983) have compared the occurrence of FTE's with the direction of the magnetosheath field. FTE's occur rarely when the magnetosheath field is even

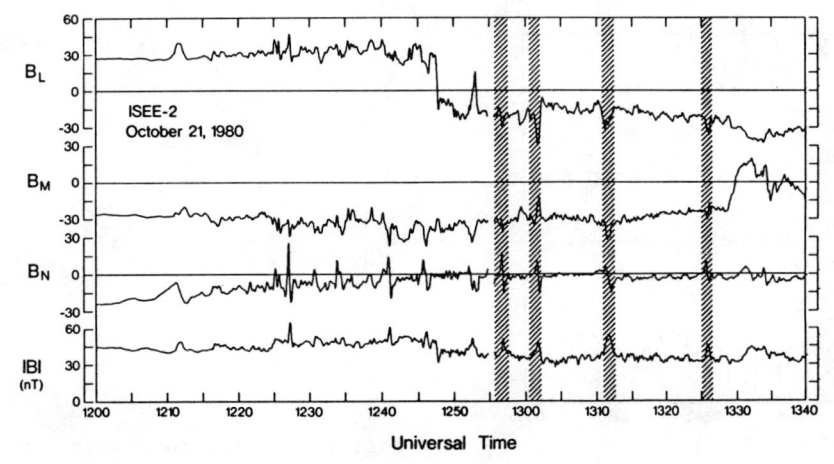

Fig. 18. Flux transfer events inside and outside the magnetosphere (Berchem and Russell, 1983).

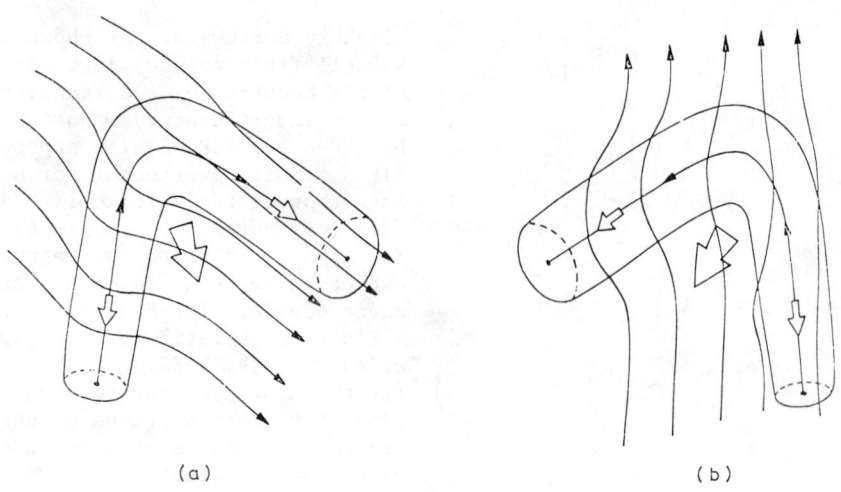

Fig. 19. Southward moving flux transfer event (Cowley, 1982).

Flux transfer events have been seen (Rijnbeek et al., 1983) on the same passes as steady-state merging was reported by Sonnerup et al. (1982). However, since these were observed at different times it is not clear whether the different types of reconnection were ever simultaneously present. To settle this question will require simultaneously widely separated probing of the magnetopause.

The existence of flux transfer events, their scale size, occurrence rates, and flux transfer rates are all well-defined but other questions about flux transfer events are not. What initiates and what terminates an FTE? What causes the twist in FTE's? Where are FTE's initiated? Do they begin in the subsolar region only? Are they initiated in the cusp? Or do they arise along movable merging lines whose locations are sensitive functions of the IMF direction as proposed by Crooker (1979)?

The question about the location of the separator, the line along which reconnection is initiated, where the field lines actually change partners, is an old one. Most would agree that when the interplanetary magnetic field is exactly southward the reconnection line is along the magnetospheric equator or close to it. When the IMF rotates to an angle away from due south, does the reconnection line tilt as Sonnerup (1974) has proposed, or does it break into two parts which move separately to the north and south away from the equator as the IMF becomes more horizontal as Crooker (1979) has proposed? The Sonnerup (1974) proposal seems reasonable on intuitive grounds. One would not expect the magnetosphere to behave much differently for fields that were moderately southward than those that were exactly southward. On the other hand, Sonnerup's model has been criticized by Cowley (1976) who pointed out that the orientation of the reconnection line was arbitrary. Crooker's proposal, on the other hand, is based on the supposition that reconnection takes place at the points on the magnetopause

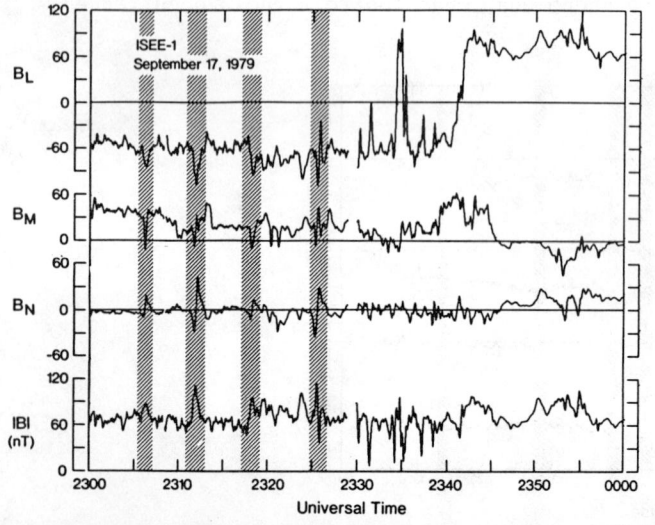

Fig. 20. Southward moving flux transfer events on 9/17/79 (Berchem and Russell, 1983).

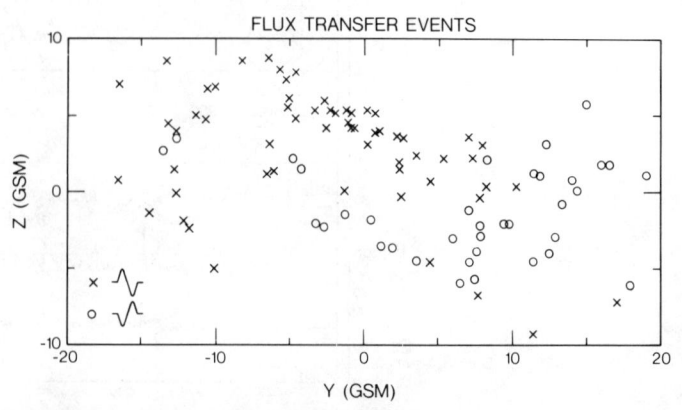

Fig. 21. Location of two types of flux transfer events on the magnetopause (Berchem and Russell, 1983).

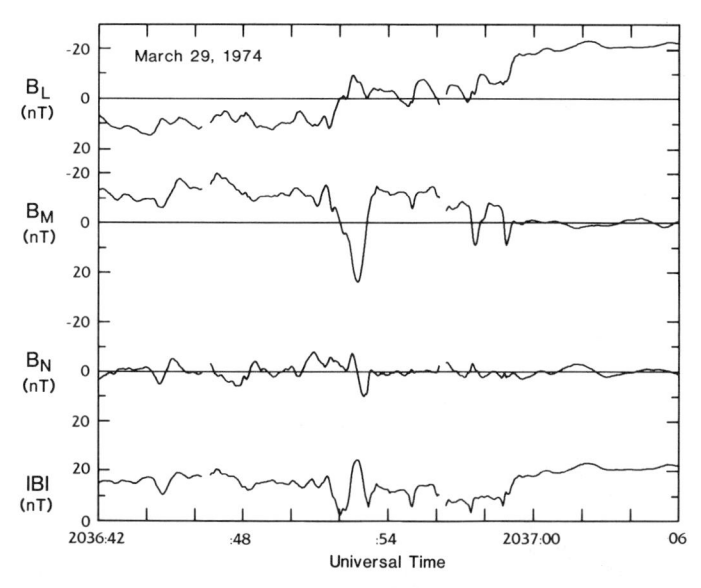

Fig. 22. A flux transfer event at the mercury magnetopause (Russell, 1983).

where the magnetosheath and magnetospheric fields are very nearly antiparallel. It has the pleasing feature that it produces a magnetosphere which has a half-wave rectification for flux transfer to the tail (Crooker, 1980). Such half-wave rectification is necessary to produce the semi-annual variation of geomagnetic activity (Russell and McPherron, 1973b) and the functional dependence of ring current injection on the IMF orientation (Burton et al., 1975).

Two recent studies bear on this question, one theoretical and one experimental. Luhmann et al. (1983) have used the Spreiter and Stahara (1980)

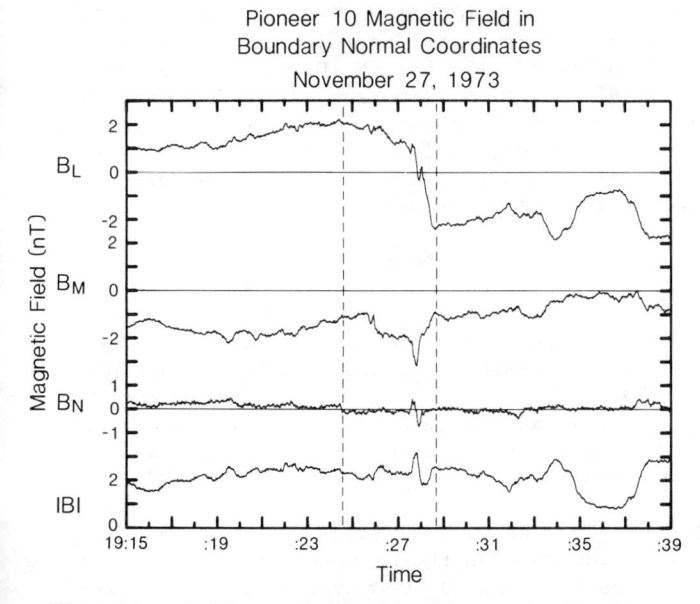

Fig. 23. A flux transfer event at the Jovian magnetopause (Walker and Russell, 1983).

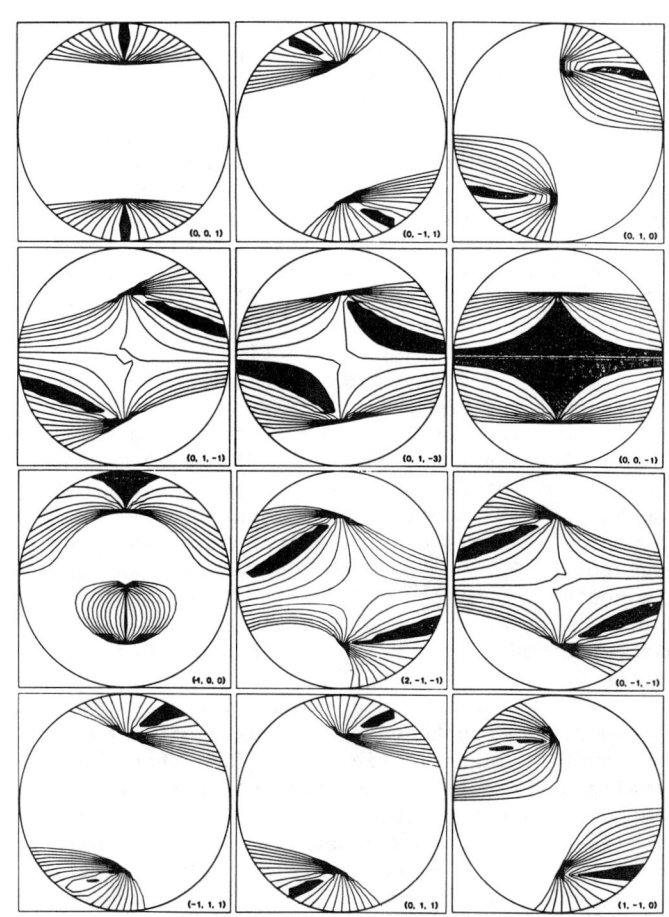

Fig. 24. Location of reconnection regions as a function of the direction of the IMF according to the Crooker hypothesis (Luhmann et al., 1983). View is of magnetopause as seen from the sun. North is at the top of each panel. The vector in parentheses is parallel to the IMF direction.

gasdynamic code to calculate the magnetic field in the magnetosheath as a function of the direction of the IMF. Then they have calculated the angle between this field and a model magnetospheric field to locate the antiparallel field regions. Their results are shown in Figure 24. The vectors in the lower right-hand corner of each panel give the direction of the IMF used in the calculation. Since it is a gasdynamic calculation the magnitude of the magnetic field is arbitrary. These panels show that, if the Crooker conjecture is right, merging will occur on open or tail field lines whenever the IMF is northward and on closed field lines whenever the IMF is southward. When the IMF is southward and $B_M < 0$, i.e., $B_Y > 0$, the reconnection line is above the equatorial plane on the afternoon side and below the equatorial plane on the morning side. For $B_M > 0$, i.e., $B_Y < 0$, the reconnection line is below the equatorial plane on the afternoon side and above on the morning side

The experimental data are the observations of

ion anisotropies in flux transfer events by Daly et al. (1983). These are shown in Figure 25. The open circles are flux transfer events with ions flowing antiparallel to the magnetic field. These flux transfer events are thus connected to the northern hemisphere. The solid points are those flux transfer events with ions flowing parallel to the magnetic field. These flux transfer events are thus connected to the southern hemisphere. The vertical lines indicate the direction of supposed motion based on whether the normal component has a +/- or -/+ signature. Usually the magnetosheath flow field determines the direction of motion but sometimes field line tension can overcome the magnetosheath flow and pull the FTE against the flow. We will not concern ourselves further with this point. The important aspect of Figure 25 is the location of points connected to the north and the south. The IMF controls this connectivity. On the dawn side there is connection to the northern hemisphere for $B_M < 0$ above the GSM equator and to the southern hemisphere in the same region for $B_M > 0$. The afternoon side does not show such a marked dependence on the IMF. In part this is due to the fewer FTE's in this region.

If we examine the proposed merging sites in Figure 24 for southward and duskward oriented interplanetary magnetic fields, we find that the merging site is on the equator only for directly southward magnetic fields. Otherwise the merging site is split into a southern latitude dawn site and a northern latitude dusk site. This splitting could explain the appearance of southern connectivity at northern sites in the energetic ion data of Daly et al. (1983). However, it leaves a gap in FTE occurrence near noon local time, a gap that does not seem to be observed as illustrated in Figure 21. Thus, we do not believe that the Crooker hypothesized merging sites play a role in FTE production.

One can, in fact, explain the Daly et al. result entirely in terms of stagnation point reconnection if one remembers that the separator between the two halves of the reconnected tube will be tilted roughly at right angles to the flux tubes and thus the equator is not necessarily the dividing line between the north and south sections of an FTE. We note that the observed north-south asymmetry in the +/- and -/+ signature in the normal component is consistent with the twist in the flux ropes as they are carried away from noon, rolling along the magnetopause surface and fixed at the elbow of the tube.

In conclusion we have learned a great deal about the magnetopause and about reconnection, beginning with Explorer 12 and continuing to the recent measurements with ISEE. There is still much we do not know. However, the questions we are asking now are much better posed and much more sophisticated than in the past. Research has become quantitative rather than qualitative and theoreticians have been presented with clearly defined problems and well determined boundary conditions. We anticipate much progress in solving the remaining problems over the next few years.

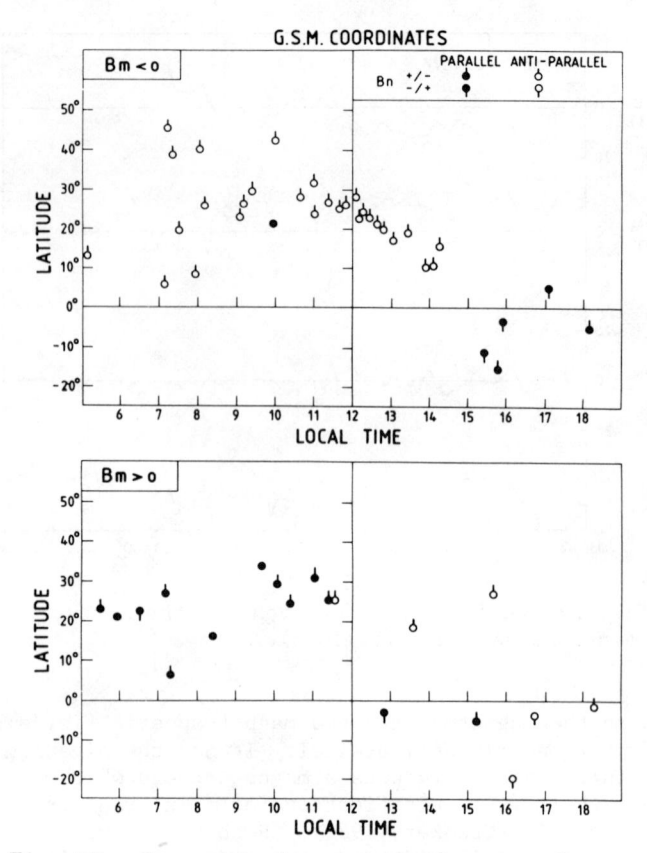

Fig. 25. Energetic ion asymmetries as a function of position and magnetosheath field direction (Daly et al., 1983).

Acknowledgments. In the course of his studies of the magnetopause the author has worked with and benefited from interactions with many individuals. The author is particularly indebted to M. Aubry, J. Berchem, N. U. Crooker, P. Daly, R. C. Elphic, M. G. Kivelson, J. G. Luhmann, M. Neugebauer, G. Paschmann, R. Rijnbeek, M. Saunders and B. Sonnerup. This work was supported by the National Aeronautics and Space Administration under contract NAS5-25772.

References

Aubry, M. P., C. T. Russell and M. G. Kivelson, On the inward motion of the magnetopause preceding a substorm, J. Geophys. Res., 75, 7018, 1970.

Berchem, J. and C. T. Russell, The thickness of the magnetopause current layer: ISEE-1 and -2 observations, J. Geophys. Res., 87, 2108-2114, 1982a.

Berchem, J. and C. T. Russell, Magnetic field rotation through the magnetopause: ISEE-1 and -2 observations, J. Geophys. Res., 87, 8139-8148, 1982b.

Berchem, J. and C. T. Russell, Flux transfer

events on the dayside magnetopause: Spatial distribution and controlling factors, J. Geophys. Res., in press, 1983.

Burton, R. K., R. L. McPherron and C. T. Russell, The terrestrial magnetosphere: A half-wave rectifier of the interplanetary electric field, Science, 189, 717, 1975.

Cahill, L. J., Jr. and P. G. Amazeen, The boundary of the geomagnetic field, J. Geophys. Res., 68, 1835-1843, 1963.

Cahill, L. J. and V. L. Patel, The boundary of the geomagnetic field, August to November 1961, Planet. and Space Sci., 15, 997-1033, 1967.

Chapman, S. and V. C. A. Ferraro, A new theory of magnetic storms, Terr. Mag., 36, 77-97, 1931a.

Chapman, S. and V. C. A. Ferraro, A new theory of magnetic storms, Terr. Mag., 36, 171-186, 1931b.

Chapman, S. and V. C. A. Ferraro, A new theory of magnetic storms, Terr. Mag., 37, 147-156, 1932.

Chapman, S. and V. C. A. Ferraro, A new theory of magnetic storms, II, The main phase, Terr. Mag., 38, 79, 1933.

Chapman, S. and V. C. A. Ferraro, The theory of the first phase of the geomagnetic storm, Terr. Mag., 45, 245, 1940.

Cowley, S. W. H., Comments on the merging of non-antiparallel magnetic fields, J. Geophys. Res., 81, 3455-3458, 1976.

Cowley, S. W. H., The causes of convection in the Earth's magnetosphere: A review of developments during the IMS, Rev. Geophys. Space Phys., 20, 531-565, 1982.

Cowley, S. W. H., D. J. Southwood and M. A. Saunders, Interpretation of magnetic field perturbations in the earth's magnetopause boundary layers, Planet. Space Sci., 31, 1237-1258, 1983.

Crooker, N. U., Dayside merging and cusp geometry, J. Geophys. Res., 84, 951-959, 1979.

Crooker, N. U., The half-wave rectifier response of the magnetosphere and antiparallel merging, J. Geophys. Res., 85, 575-578, 1980.

Daly, P. W., M. A. Saunders, R. P. Rijnbeek, N. Sckopke and C. T. Russell, The distribution of reconnection geometry in flux transfer events using energetic ion, plasma and magnetic field data, J. Geophys. Res., 88, submitted, 1983.

Dungey, J. W., Interplanetary magnetic field and the auroral zones, Phys. Rev. Lett., 6, 47-48, 1961.

Dungey, J. W., The structure of the exosphere or adventures in velocity space, in Geophysics, The Earth's Environment, C. DeWitt, J. Hieblot, A. Lebeau, Eds., Gordon and Breach, New York, 1963.

Elphic, R. C. and C. T. Russell, ISEE-1 and -2 magnetometer observations of the magnetopause, in Magnetospheric Boundary Layers, (B. Battrick, ed.), pp. 43-50, ESA SP-148, Paris, 1979.

Fairfield, D. H. and L. J. Cahill, Jr., Transition region magnetic field and polar magnetic disturbances, J. Geophys. Res., 71, 155-169, 1966.

Haerendel, G., G. Paschmann, N. Sckopke, H. Rosenbauer and P. C. Hedgecock, The front side boundary layer of the magnetosphere and the problem of reconnection, J. Geophys. Res., 83, 3195, 1978.

Heppner, J. P., N. F. Ness, C. S. Scearce and T. L. Skillman, Explorer 10 magnetic field measurements, J. Geophys. Res., 68, 1, 1963.

Holzer, R. E., M. G. McLeod and E. J. Smith, Preliminary results from the OGO-1 search coil magnetometer: Boundary positions and magnetic noise spectrum, J. Geophys. Res., 71, 1481, 1966.

Kaufmann, R. L. and A. Konradi, Speed and thickness of the magnetopause, J. Geophys. Res., 78(28), 6549-6569, 1973.

Landau, L. D. and E. M. Lifshitz, Electrodynamics of Continuous Media, Addison-Wesley Publishing Co., Reading, MA, 1960.

Luhmann, J. G., R. J. Walker, C. T. Russell, N. U. Crooker, J. R. Spreiter and S. S. Stahara, Patterns of magnetic field merging sites on the magnetopause, Geophys. Res. Lett., submitted, 1983.

McPherron, R. L., C. T. Russell and M. P. Aubry, Satellite studies of magnetospheric substorms on August 15, 1968. 9. Phenomenological model for substorms, J. Geophys. Res., 78(16), 3131-3149, 1973.

Neugebauer, M., C. T. Russell and E. J. Smith, Observations of the internal structure of the magnetopause, J. Geophys. Res., 79, 499-510, 1974.

Ogilvie, K. W., J. D. Scudder and M. Sugiura Magnetic field and electron observations near the dawn magnetopause, J. Geophys. Res., 76, 3574, 1971.

Papamastorakis, I., G. Paschmann, N. Sckopke, S. J. Bame and J. Berchem, The magnetopause as a tangential discontinuity for large field rotation angles, J. Geophys. Res., submitted, 1983.

Paschmann, G., B. U. O. Sonnerup, I. Papamastorakis, N. Sckopke, G. Haerendel, S. J. Bame, J. R. Asbridge, J. T. Gosling, C. T. Russell and R. C. Elphic, Plasma acceleration at the earth's magnetopause: Evidence for reconnection, Nature, 282, 243-246, 1979.

Paschmann, G., G. Haerendel, I. Papamastorakis, N. Sckopke, S. J. Bame, J. T. Gosling and C. T. Russell, Plasma and magnetic field characteristics of magnetic flux transfer events, J. Geophys. Res., 87, 2159-2168, 1982.

Patel, V. L., L. J. Cahill, Jr. and A. J. Dessler, Magnetosheath, field, geomagnetic index ap and stability of magnetopause, J. Geophys. Res., 72, 426-430, 1967.

Patel, V. L. and A. J. Dessler, Geomagnetic activity and size of magnetospheric cavity, J. Geophys. Res., 71, 1940-1942, 1966.

Podgorny, I. M., E. M. Dubinin and Yu. N. Potanin, On the magnetic curl in front of the magnetosphere boundary, Geophys. Res. Lett., 7, 247-250, 1980.

Rijnbeek, R. P., S. W. H. Cowley, D. J. Southwood

and C. T. Russell, Observations of reverse polarity flux transfer events at the earth's dayside magnetopause, Nature, 300, 23-26, 1982.

Rijnbeek, R. P., S. W. H. Cowley, D. J. Southwood and C. T. Russell, A survey of dayside flux transfer events, observed by the ISEE-1 and -2 magnetometers, J. Geophys. Res., in press, 1983.

Russell, Patchy reconnection and magnetic ropes in astrophysical plasmas, IAU Proceedings, Symposium 107, in press, 1983.

Russell, C. T. and R. C. Elphic, Initial ISEE magnetometer results: Magnetopause observations, Space Sci. Rev., 22, 681, 1978.

Russell, C. T. and R. C. Elphic, ISEE observations of flux transfer events at the dayside magnetopause, Geophys. Res. Lett., 6, 33-36, 1979.

Russell, C. T. and R. L. McPherron, The magnetotail and substorms, Space Sci. Rev., 15, 205, 1973a.

Russell, C. T. and R. L. McPherron, Semiannual variation of geomagnetic activity, J. Geophys. Res., 78, 92-108, 1973b.

Saunders, M. A., C. T. Russell and N. Sckopke, Flux transfer events: Scale size and interior structure, Geophys. Res. Lett., submitted, 1983.

Sckopke, N., G. Paschmann, G. Haerendel, B. U. O. Sonnerup, S. J. Bame, T. G. Forbes, E. W. Hones, Jr. and C. T. Russell, Structure of the low latitude boundary layer, J. Geophys. Res., 86, 2099, 1981.

Sibeck, D. G., An explanation of multiple boundary crossings near the magnetotail magnetopause, M.S. Thesis, Department of Atmospheric Science, University of California, Los Angeles, 1982.

Sonnerup, B. U. O., Magnetopause reconnection rate, J. Geophys. Res., 79, 1546, 1974.

Sonnerup, B. U. O., Magnetopause structure during the magnetic storm of September 24, 1961, J. Geophys. Res., 76, 6717, 1971.

Sonnerup, B. U. O. and L. J. Cahill, Jr., Magnetopause structure and attitude from Explorer 12 observations, J. Geophys. Res., 72, 171, 1967.

Sonnerup, B. U. O. and L. J. Cahill, Jr., Explorer 12 observations of the magnetopause current layer, J. Geophys. Res., 73, 1757, 1968.

Sonnerup, B. U. O. and B. G. Ledley, OGO-5 magnetopause structure and classical reconnection, J. Geophys. Res., 84, 399-405, 1979a.

Sonnerup, B. U. O., G. Paschmann, I Papmastorakis, N. Sckopke, G. Haerendel, S. J. Bame, J. R. Asbridge, J. T. Gosling and C. T. Russell, Evidence for magnetic field reconnection at the Earth's magnetopause, J. Geophys. Res., 86, 10,049-10,067, 1981.

Spreiter, J. R. and S. S. Stahara, A new predictive model for determining solar wind terrestrial planet interactions, J. Geophys. Res., 85, 6769, 1980.

Su, S-Y. and B. U. O. Sonnerup, First-order orbit theory of the rotational discontinuity, Phys. Fluids, 11, 851, 1968.

Walker, R. J. and C. T. Russell, Paper P.A. 20, IAGA Symposium, Hamburg, August, 1983.

Questions and Answers

Priest: The structure and stability of flux tubes have been studied in great detail by solar and laboratory theorists. What are the sizes of the plasma beta and twist in your tubes and what do you think causes the twist?

Russell: In the magnetosheath adjacent to the flux transfer events, the beta of the plasma is close to unity. However, internal to the flux transfer event the magnetic field strength is often much larger. Hence, since the plasma in the FTE is very sheath-like, the beta is often much less than unity. The wavelength of the twist in an FTE is about 4 to 6 R_E. This corresponds to a current of 1 to 2×10^5 Amps distributed over a diameter of about 1 R_E.

Speiser: Could you comment on any systematics to the direction of twist in the twisted FTEs (i.e., the direction of current?) Have attempts been made to identify the particles causing this current?

Russell: Figure 24 effectively addresses the question of twist. Mainly the twist corresponds to a current into the magnetosphere in the north and out of the magnetosphere in the south (see Figure 19 also). There are, of course, exceptions to this rule. Estimates by J. D. Scudder suggest that the low energy electrons carry much of the required current.

RECENT INVESTIGATIONS OF FLUX TRANSFER EVENTS OBSERVED AT THE DAYSIDE MAGNETOPAUSE

R. P. Rijnbeek, S. W. H. Cowley, D. J. Southwood

Blackett Laboratory, Imperial College,
London. SW 7 2BZ. United Kingdom

C. T. Russell

Institute of Geophysics and Planetary Physics, University of California,
Los Angeles, California CA 90024. U. S. A.

Abstract. We summarise the results of our survey of dayside flux transfer events (FTEs). FTEs are a common feature of the dayside magnetopause covered by the ISEE 1 and 2 spacecraft. They occur essentially whenever the magnetosheath magnetic field has a southward component. The properties of FTEs observed in the magnetosheath and in the magnetosphere are similar. FTEs contribute significantly to the flux erosion process at the dayside boundary when the external field has a southward component. Intervals of quasi-steady reconnection occur in close association with observations of FTE signatures. Finally, we outline some remaining questions relating to FTEs.

Introduction.

The ISEE 1 and 2 plasma and magnetic field data have provided convincing experimental evidence for at least the occasional occurrence of quasi-steady reconnection at the dayside magnetopause [Paschmann et al.,1979; Sonnerup et al.,1981; Gosling et al.,1982; Sonnerup,1984; Paschmann,1984]. More frequently, however, reconnection manifests itself in a more impulsive form, first identified in the ISEE magnetosheath magnetic field data by Russell and Elphic [1978] and termed the flux transfer event (FTE). Previously, Haerendel et al. [1978] and Haerendel and Paschmann [1982] had also interpreted certain magnetic field perturbations which they found inside the magnetopause in HEOS 2 data in terms of small-scale, transient flux erosion. A large number of publications on FTEs have now appeared, confirming for the most part their interpretation in terms of impulsive reconnection (see reviews by Cowley [1982] and Saunders [1983] and references therein). Using medium resolution ISEE 1 and 2 magnetometer data we have made a survey of FTE magnetic signatures observed both inside and outside the magnetopause region [Rijnbeek et al.,1984]. Other complementary statistical studies of FTEs have also appeared recently. Daly et al. [1984a,b] have made additional use of plasma and energetic particle data in order to determine whether individual FTE flux tubes are connected to the Northern or to the Southern Hemisphere. Berchem and Russell [1984a,b] have used ISEE magnetic field data from 1977-1981 in order to investigate properties of FTEs. They, however, chose stricter criteria for the identification of FTE magnetic signatures. Also, the latter two studies only treat FTEs observed in the magnetosheath in their statistics. In this paper we summarise the results of our survey of FTEs observed during the first two years (1977 and 1978) of the ISEE mission. In addition, we go on to discuss some questions concerning FTEs which remain topics for future investigation.

Observations and Results.

To illustrate the process of identifying FTE magnetic signatures, we show in Figure 1 magnetic field data from an inbound pass on November 12,1977. The four panels show the magnetic field components in the boundary normal (LMN) coordinate system [Russell and Elphic, 1978] and the total field strength B. The L and the M components point roughly northward and westward along the local magnetopause surface while the N component is along the outward normal. The dashed vertical lines indicate the magnetic signatures identified as FTEs. The easiest way to recognise them is from their characteristic bipolar signature in B_N. We refer to FTEs with positive-negative B_N perturbations (such as shown in Figure 1) as standard FTEs, and those with reverse, negative-positive B_N perturbations as reverse FTEs. Occasionally, the B_N signature is not strictly bipolar (e.g. at 22:25 UT in Figure 1) in which case the FTE

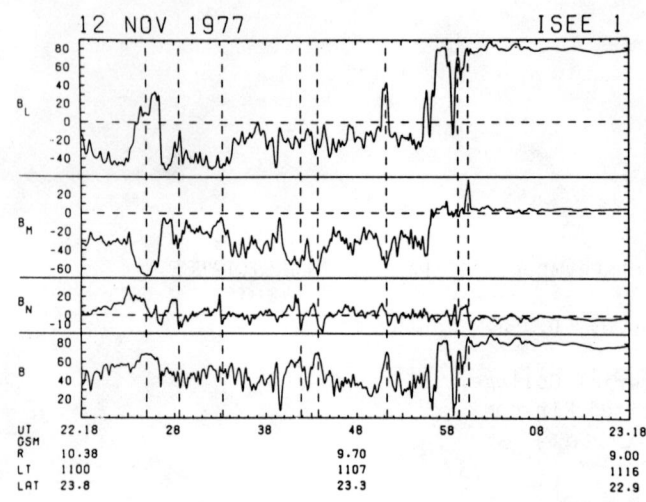

Fig. 1. Medium resolution (12 s averages plotted every 4 s) ISEE 1 magnetic field data for the inbound pass on November 12, 1977. The data are plotted in the boundary normal coordinate system. The normal direction in GSM components (0.948, -0.143, 0.284) was determined using the Fairfield magnetopause model. From top to bottom the four panels show respectively the L, M and N components and the total magnetic field strength B in nanoteslas. The horizontal axis shows the Universal Time in hours and min. The spacecraft location is given at the bottom and shows the radial distance (R_E), GSM local time and GSM latitude. FTEs are indicated by the dashed vertical lines. Standard FTEs occur at 22:28, 22:33, 22:44 and 22:51 UT the magnetosheath and at 22:59 and 23:00 UT in the magnetopause and boundary layer regions. One irregular FTE occurs at 22:25 UT in the magnetosheath.

is classed as irregular. FTEs also show field deflections in the L and M components tangential to the magnetopause and the FTE at 22:51 UT in Figure 1 is a good example showing this. Most of the events are also accompanied by increases in field strength, as reported by Paschmann et al. [1982]. However, we have found cases where the field strength increases are not very clear (e.g. 22:28 and 22:33 UT in Figure 1) and on occasion the field strength inside a FTE may actually show a decrease (Figure 2 in Rijnbeek et al., 1984). In addition to classifying FTEs in terms of their B_N signatures, we also make a distinction in our survey between FTEs observed in the magnetosheath and those observed inside the magnetosphere.

Figure 2 shows the spatial distribution of magnetosheath FTE signatures using the classification procedure described above. In this plot we have only considered well-defined FTEs with peak-to-peak B_N amplitudes greater than 10 nT and duration longer than one min. If at least one magnetosheath FTE was observed during a particular inbound or outbound pass then a symbol is plotted at the Y-Z (GSM) location of the corresponding magnetopause crossing. These locations do not, of course, differ substantially from those of the FTEs themselves. The different symbols indicate the appropiate FTF signatures, as explained in the caption. The diagonally banded distribution of the symbols results from the limited ISEE coverage of the dayside magnetopause during 1977 and 1978. A plot of crossings without FTEs would in fact show a similar distribution. The main feature to note here is that standard FTEs occur mostly at northerly latitudes, while reverse FTEs are seen more often at southerly or low latitudes. It is important to note that "mixed" crossings are rare, i.e. standard and reverse FTEs are very rarely seen together on the same pass. If we simplistically assume that for the most part the standard FTEs and reverse FTEs are northerly moving tubes connected to the Northern Hemisphere and southerly moving tubes connected to the Southern Hemisphere, respectively, then the spatial distribution argues that the source region of the FTEs is near the magnetic equator. The general absence of "mixed" signatures then also indicates that the location of the regions of intermittent reconnection is relatively stable, and does not move widely over the magnetopause surface at any one time. The use of additional plasma and energetic particle data to identify the hemisphere connection of FTEs has, however, enabled Daly et al. [1984a,b] to give a more accurate appraisal of the situation.

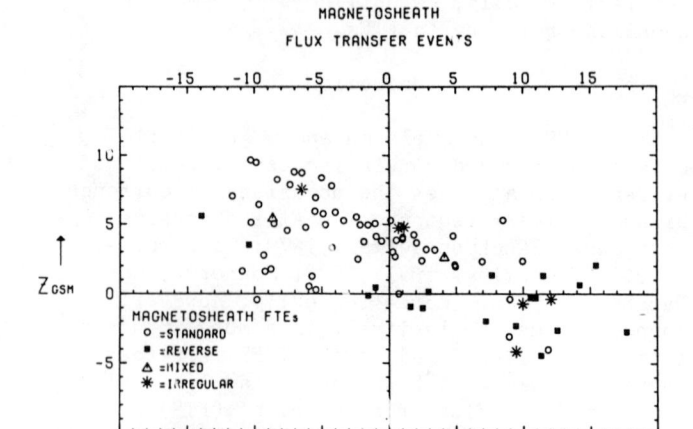

Fig. 2. Magnetopause crossing locations associated with observations of magnetosheath FTEs plotted in the Y-Z GSM plane as seen from the Sun. Different symbols denote the type of signature displayed by the FTEs in the B_N component. Open circles indicate standard FTEs, solid squares reverse FTEs, open triangles both standard and reverse FTEs, i.e. "mixed" crossings and stars irregular FTEs. In the case of both standard (reverse) and irregular FTEs being observed, only the standard (reverse) FTE symbol has been plotted.

They show that the direction of motion of FTE flux tubes, which is assumed to determine the sense of the B_N signature, is dependent on both the velocity of contraction of the tube and the ambient magnetosheath plasma velocity, the former dominating in the sub-alfvenic region and the latter dominating in the super-alfvenic region. Although there is some evidence that the reconnection line may tilt in response to the IMF B_y component, Daly et al. [1984a,b] show that their results are still consistent with a near-equatorial source region for FTEs. A more extensive coverage of the magnetopause region is needed, however, to determine whether the possibility can be ruled out of FTE signatures resulting from reconnection occuring at high latitudes, as suggested e.g. by Haerendel et al. [1978] and Crooker [1979] (see also Luhmann et al., 1984).

We have also determined whether the orientation of the local magnetosheath magnetic field has any influence on FTE occurrence. In particular, we tried to find whether FTEs occur more often during periods of southward magnetosheath field, as would be expected if FTEs are reconnection phenomena. Figure 3 shows that this is indeed the case. In Figure 3a we show the locations of magnetopause crossings with magnetosheath FTEs, while Figure 3b is the corresponding plot for the crossings without magnetosheath FTEs. For each crossing, the sign of the magnetosheath B_L component was determined from the half hour magnetosheath interval adjacent to the magnetopause. Comparison of the two plots shows that there is a clear correlation between FTE occurrence and the sign of magnetosheath B_L, with FTEs occurring predominantly when the field is southward (negative B_L) and not when it is northward (positive B_L). In their separate study of magnetosheath FTEs, Berchem and Russell [1984a,b] report a similar dependence of magnetosheath FTE occurrence on the IMF direction determined upstream in the solar wind. There is as yet no indication in any study of any dependence of FTE occurrence on the sign of IMF B_y.

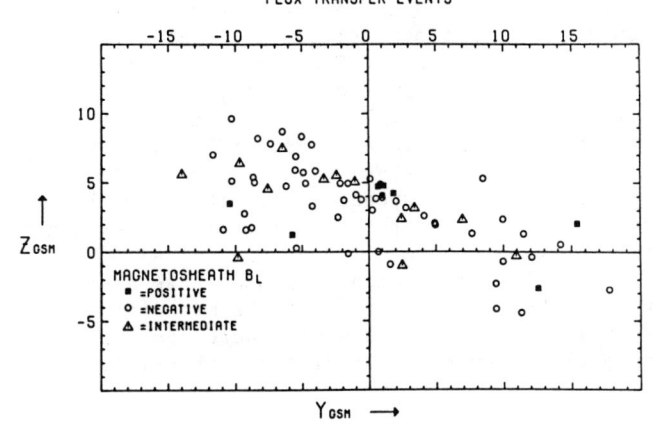

A similar analysis to that shown in Figures 3a,b for magnetosheath FTEs has also been performed for magnetospheric (boundary layer) FTEs, and with similar results [Rijnbeek et al.,1984]. During periods of southward magnetosheath field we find that the average number (~5) and recurrence time (~8 min.) of magnetosheath FTEs is similar to the average number (~4) and recurrence time (~7 min.) of magnetosphere FTEs observed per crossing. These results strongly suggest that magnetospheric and magnetosheath FTEs are part of the same reconnection-associated physical phenomenon.

Using values of 1 R_E and 2 R_E for the average scale sizes of FTEs normal to and along the magnetopause, respectively (see Saunders et al., 1984a,b), together with a field strength of ~50 nT inside an FTE flux tube, we calculate a value of ~4 x 10^6 Wb. for the magnetic flux transported by a typical FTE. Dividing this number by the above recurrence time (~8 min.) then gives a lower limit of ~10 kV for the typical voltage associated with FTEs during periods when the magnetosheath magnetic field has a southward component. However, because of the limited local time extent of FTEs and because we have used only single spacecraft observations, we cannot expect to see every FTE occurring at any particular time as indicated in Figure 4. Since FTEs are observed by ISEE at all local times at

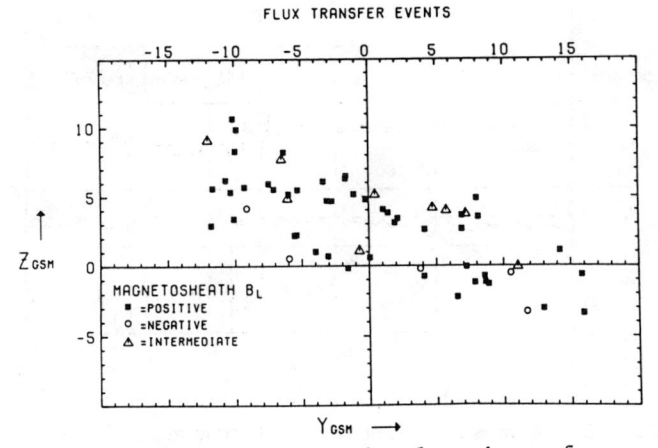

Fig. 3. Magnetopause crossing locations of passes with magnetosheath FTEs (Figure 3a) and passes without magnetosheath FTEs (Figure 3b). Different symbols show the polarity of the magnetosheath B_L component determined in the half hour interval before (after) the inward (outward) bound magnetopause crossing. In both plots solid squares denote crossings with B_L positive, open circles those with B_L negative, and open triangles those with intermediate B_L, i.e. B_L was either less than ~5 nT in magnitude or not predominantly positive or negative during the half hour magnetosheath interval.

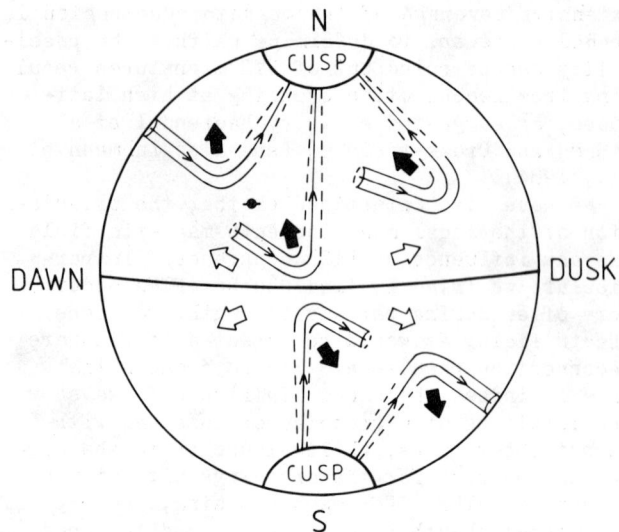

Fig. 4. Qualitative sketch of the dayside magnetopause viewed from the Sun. Three Northern and two Southern Hemisphere connected open flux tubes are shown propagating along the magnetopause surface. The solid and the dashed parts of the tubes show the magnetosheath and the magnetosphere parts overlying and underlying the magnetopause surface, respectively. The central arrowed lines show the magnetic field direction along the axis of the tubes. Broad closed arrows show the direction of contraction velocity of the tubes and broad open arrows show the direction of the ambient magnetosheath plasma bulk velocity. The motion of the flux tubes is a resultant of both these velocities. The symbol ⊷ shows a typical spacecraft location to indicate that not all flux tubes will be detected.

the dayside magnetopause virtually whenever the magnetosheath field is southward (Figure 3), this suggests that the ~10 kV value represents only a small fraction of the actual FTE voltage. The estimate of ~2 R_E for the scale size of FTFs along the magnetopause corresponds to the closest local time spacing of FTFs we can expect. Therefore, the maximum number of FTEs we can fit side by side across the equatorial dayside magnetopause is ~15 FTFs, thus giving an upper limit of ~150 kV for the FTE flux removal rate. The total FTE-associated voltage may therefore be within range of typically observed (50-100 kV) cross-magnetosphere voltages. FTEs are thus certainly responsible for a significant level of flux erosion and could be the dominant process needed to drive geomagnetic disturbances when the IMF is southward.

Quasi-steady Reconnection and FTEs.

So far, data from twelve dayside magnetopause passes made by ISEE 1 and 2 have been published containing evidence for the occurrence of quasi-steady reconnection, i.e. intervals of reconnection lasting for time-scales of five min. or longer [Sonnerup et al.,1981; Gosling et al.,1982]. Additional events have also been reported by Paschmann [1984]. During the course of our survey we have looked at the magnetic field data from these twelve passes and we find that in all cases FTF signatures occur in close association with the quasi-steady flows at the magnetopause. A few examples showing this close association have already been published [Paschmann et al.,1982; Rijnbeek et al.,1982; Saunders,1983; Rijnbeek et al.,1984], and in Figure 5 we show an additional example. The magnetic field data are from ISEE 1 during the outbound pass on September 3,1978. Plasma data for part of the interval can be found in Sonnerup et al. [1981] and they indicate a region of high speed boundary layer flows between 7:10-7:12 and 7:17-7:21 UT. Clear FTE signatures in Figure 5 can be identified in the boundary layer and magnetopause regions at 7:00, 7:11 and 7:19 UT, and in the magnetosheath at 7:23, 7:25, 7:32 and 7:34 UT. Several FTEs occur during the intervals of high flow speeds inside the outer boundary of the magnetopause as may be seen by comparison with Figure 5 in Sonnerup et al. [1981]. Paschmann et al. [1982] have also reported that FTFs observed at the magnetopause show large plasma flow speed increases. A possible interpretation of these observations is

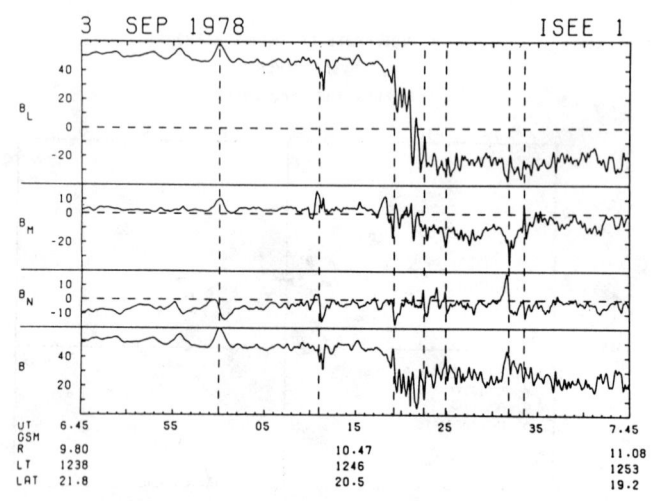

Fig. 5. A one hour interval of magnetic field data from ISEE 1 during the outbound pass on September 3, 1978. The format is the same as in Figure 1. The normal direction in GSM components (0.904, 0.018, 0.428) was determined by minimum variance analysis of the magnetopause crossing. Regions of northerly directed high speed plasma flows occur in the intervals 7:10-7:12 and 7:17-7:21 UT, as shown in Figure 5 of Sonnerup et al. (1981). Magnetosphere FTEs occur at 7:00, 7:11 and 7:19 UT and magnetosheath FTEs occur at 7:23, 7:25, 7:32 and 7:34 UT. Note the "flute-like" appearance of the B_L trace and the small-scale field variations near the magnetopause.

that between the localized bursts of rapid reconnection (FTEs) the reconnection rate at the magnetopause does not drop to zero in these cases but remains at a relatively steady value. It may be noted in this connection that the typical spatial dimension of the layer of open field lines perpendicular to the magnetopause is much larger in FTEs (~6000 km) than in the quasi-steady reconnection examples (less than ~1500 km).

Another feature to note in Figure 5 is the appearance of short time-scale field perturbations at and near the magnetopause, in particular the "flute-like" appearance of the P_L component. These features often have the appearance of small amplitude, small time-scale FTEs, and we find that their occurrence is common during periods of southward magnetosheath field. Russell [1984] has suggested that these features could result from small-scale tearing-mode structures.

Conclusions and Some Remaining Questions.

Our results give added suppport to the interpretation of FTEs in terms of impulsive reconnection. In particular, we find that FTEs occur predominantly during periods when the magnetosheath field has a southward component and that magnetosheath and magnetospheric FTEs have similar occurrence statistics. The latter result suggests that FTEs observed in the two regions are part of the same physical phenomenon, i.e. that they are reconnected open flux tubes observed on either side of the magnetopause. Also, it appears likely that FTEs play a significant role in the flux erosion process at the dayside magnetopause when the external field has a southward component. A further significant observation is that intervals of quasi-steady reconnection occur in close association with FTEs. This suggests that the two types of reconnection are closely related. The two phenomena could represent aspects of the same physical process. For instance, FTEs could indicate the peeling away of flux filaments from the edge of a region of steady reconnection or the two distinct signatures could reflect no more than variations in reconnection rate.

There still remain many topics relating to FTEs requiring further investigation, some of which we detail below.
a. Clarification is required of the precise form of the relationship between the quasisteady and more impulsive, localised FTE type of reconnection. Detailed examination is therefore necessary of the plasma and magnetic field data during the passes discussed by Sonnerup et al. [1981] and Gosling et al. [1982], together with the quasi-steady events identified subsequently and reported by Paschmann [1984].
b. Previous work [Saunders et al., 1984a,b] has shown that the spatial structure inside a FTE is complicated involving a bending of the open field lines as well as a twisting of the field around the tube axis. The twisting of the field is related to vortex motion of the plasma inside the FTE flux tube. The full 3-D spatial structure of FTEs is not fully understood, however, and further investigation is needed.
c. The evolution of the signatures of FTEs as they move over the dayside magnetopause and into the tail needs to be established. In the ISFE data we have found evidence that FTE signatures are more often seen at higher latitudes [Rijnbeek et al.,1984] and an examination of Figure 4 will show that this feature is not inconsistent with an equatorial source region for FTEs since they cover a larger area of the dayside at higher latitudes. It would be worthwhile to reexamine earlier satellite data sets (e.g. HEOS 2) which cover different areas of the magnetopause to ISFE 1 and 2. This possibility is being investigated.
d. The relationship between magnetospheric FTEs and the appearance of boundary layer plasma inside the magnetosphere also needs to be clearly established. ISFE plasma data show that the boundary layer inside the magnetopause often appears in the form of isolated pulses [e.g. Paschmann et al., 1978;1982 and Sckopke et al.,1981] and data have been published from a few selected magnetopause crossings showing that magnetosphere FTEs occur in close association with the boundary layer pulses [e.g. Paschmann et al.,1982; Daly and Keppler,1982; Cowley,1982; Saunders,1983]. A larger study is needed, however, to determine the precise relationship of FTEs and the appearance of the low-latitude boundary layer plasma.
e. Finally, an important future topic is the search for FTE signatures at ionospheric altitudes. It is important to note that the spatial scale of FTEs at these altitudes, which can be estimated by conservation of flux from the magnetic flux they contain at the magnetopause, is ~200 km and hence quite large. It is straightforward to give a scenario of what should take place in the ionosphere. FTEs should be observable at the boundary between the auroral zone and the polar cap which should represent the boundary between open and closed field lines. The impulsive nature of FTEs will result in that boundary making rapid periodic "jumps" equatorward in restricted local time zones, giving the boundary a corrugated appearance on spatial scales of a few hundreds of kilometres. Assuming reconnection switches off completely between FTEs the dayside boundary will move poleward with the plasma flow during these intervals. The average location of the boundary may remain fixed for steady external conditions, however, as a result of the subsequent rapid equatorward "jumps" in the open-closed field line boundary associated with the formation of FTEs. Following a "jump", the large-scale flow in the ionosphere region connected to the FTE will respond to the stresses

imposed by the magnetosheath plasma flow and magnetic field. In addition, vortical motions may occur in these regions associated with the field twisting observed near the magnetopause inside the FTE flux tubes [Saunders et al., 1984a,b].

Acknowledgments. RPR was supported by a UK SERC studentship. Work at UCLA was supported by NASA under research contract NAS-5-25772.

References.

Berchem, J. and C. T. Russell, Flux transfer events on the dayside magnetopause: Spatial distribution and controlling factors, in review, J. Geophys. Res., 1984a.

Berchem, J. and C. T. Russell, these proceedings, 1984b.

Cowley, S. W. H., The causes of convection in the Earth's magnetosphere - a review of developments during the IMS, Revs. Geophys. Space Phys., 20, 531, 1982.

Crooker, N. U., Dayside merging and cusp geometry, J. Geophys. Res., 84, 951, 1979.

Daly, P. W. and E. Keppler, Observation of a flux transfer event on the earthward side of the magnetopause, Planet. Space Sci., 30, 331, 1982.

Daly, P. W., M. A. Saunders, R. P. Rijnbeek, N. Sckopke and C. T. Russell, The distribution of reconnection geometry in flux transfer events, using energetic ion, plasma and magnetic data, submitted to J. Geophys. Res., 1984a.

Daly, P. W., M. A. Saunders, R. P. Rijnbeek, N. Sckopke and E. Keppler, these proceedings, 1984b.

Gosling, J. T., J. R. Asbridge, S. J. Bame, W. C. Feldman, G. Paschmann, N. Sckopke and C. T. Russell, Evidence for quasi-stationary reconnection at the dayside magnetopause, J. Geophys. Res., 87, 2147, 1982.

Haerendel, G., G. Paschmann, N. Sckopke, H. Rosenbauer and P. C. Hedgecock, The frontside boundary layer of the magnetosphere and the problem of reconnection, J. Geophys. Res., 83, 3195, 1978.

Haerendel, G. and G. Paschmann, Interaction of the solar wind with the dayside magnetosphere, in Magnetospheric Plasma Physics, edited by A. Nishida, p.118-122, D. Reidel Publishing Company, Dordrecht, 1982.

Luhmann, J. G., R. J. Walker, C. T. Russell, N. U. Crooker, J. R. Spreiter and S. S. Stahara, these proceedings, 1984.

Paschmann, G., these proceedings, 1984.

Paschmann, G., N. Sckopke, G. Haerendel, J. Papamastorakis, S. J. Bame, J. R. Asbridge, J. T. Gosling, E. W. Hones Jr. and E. R. Tech, ISEE plasma observations near the subsolar magnetopause, Space Sci. Rev., 22, 717, 1978.

Paschmann, G., B. U. O. Sonnerup, I. Papamastorakis, N. Sckopke, G. Haerendel, S. J. Bame, J. R. Asbridge, J. T. Gosling, C. T. Russell and R. C. Elphic, Plasma acceleration at the Earth's magnetopause: evidence for reconnection, Nature, 282, 243, 1979.

Paschmann, G., G. Haerendel, I. Papamastorakis, N. Sckopke, S. J. Bame, J. T. Gosling and C. T. Russell, Plasma and magnetic field characteristics of magnetic flux transfer events, J. Geophys. Res., 87, 2159, 1982.

Rijnbeek, R. P., S. W. H. Cowley, D. J. Southwood and C. T. Russell, Observations of "reverse polarity" flux transfer events at the Earth's dayside magnetopause, Nature, 300, 23, 1982.

Rijnbeek, R. P., S. W. H. Cowley, D. J. Southwood and C. T. Russell, A survey of dayside flux transfer events observed by ISEE 1 and 2 magnetometers, to be published in J. Geophys. Res., 1984.

Russell, C. T., these proceedings, 1984.

Russell, C. T. and R. C. Elphic, Initial ISEE magnetometer results: Magnetopause observations, Space Sci. Rev., 22, 681, 1978.

Saunders, M. A., Recent ISEE observations of the magnetopause and low latitude boundary layer: a review, J. Geophys., 52, 190, 1983.

Saunders, M. A., C. T. Russell and N. Sckopke, Flux transfer events: scale size and interior structure, to be published in Geophys. Res. Lett., 1984a.

Saunders, M. A., C. T. Russell and N. Sckopke, these proceedings, 1984b.

Sckopke, N., G. Paschmann, G. Haerendel, B. U. O. Sonnerup, S. J. Bame, T. G. Forbes, E. W. Hones Jr. and C. T. Russell, Structure of the low-latitude boundary layer, J. Geophys. Res., 86, 2099, 1981.

B. U. O. Sonnerup, these proceedings, 1984.

Sonnerup, B. U. O., G. Paschmann, I. Papamastorakis, N. Sckopke, G. Haerendel, S. J. Bame, J. R. Asbridge, J. T. Gosling and C. T. Russell, Evidence for magnetic field reconnection at the Earth's magnetopause, J. Geophys. Res., 86, 10049, 1981.

Questions and Answers

Nishida: The connected FTE field lines inside the magnetosphere are expected to interact with the closed field lines which surround them. Have you found any signatures of such an interaction?

Rijnbeek: In the course of doing our survey we have not specifically looked for any signatures which might indicate an interaction between the open flux tube inside the magnetosphere (i.e., the magnetosphere FTE) and the surrounding unreconnected magnetospheric field lines although we would expect there to be such signatures. If the orientation of the open flux tube is not aligned with the ambient magnetospheric field lines, then the motion of the flux tube could cause deflections in the field outside the tube in the direction tangent to the magnetopause in addition to the draping effect normal to the magnetopause. Also, the boundary of the open flux tube may not coincide sharply with the outer edges of a boundary layer pulse. If, as we expect, diffusive processes occur at the boundary region of the tubes associated with the observed waves, then some magnetosheath plasma could escape onto closed field lines. This may be related to the halo regions reported by Sckopke et al. (1981). A detailed investigation of magnetic field and plasma data from the boundary layer region is therefore needed to determine the nature and signature of the interaction between FTEs and the surrounding closed field lines.

A DUAL-SATELLITE STUDY OF THE SPATIAL PROPERTIES OF FTEs

M. A. Saunders*

Blackett Laboratory, Imperial College, London SW7 2BZ, England

C. T. Russell

Institute of Geophysics and Planetary Physics, University of California,
Los Angeles, California 90024

N. Sckopke

Max-Planck-Institut für Physik und Astrophysik, Institut für
Extraterrestrische Physik, 8046 Garching, Federal Republic of Germany

Abstract. Reconnection at the Earth's dayside magnetopause may manifest itself primarily as a localised and transient process called a flux transfer event (FTE). We have investigated directly the spatial properties of FTEs by examining data from the ISEE satellite pair when the satellites were separated by more than 1000 km in the vicinity of the magnetopause. We show examples of magnetosheath and boundary layer FTEs, each having a dimension normal to the magnetopause of order an Earth radius, R_E, and we substantiate statistically this scale size result for magnetosheath FTEs. When combined with other information, a 1 R_E normal dimension implies that the voltage associated with the FTE process at one magnetopause location is at least 10 kV. Our study strengthens the view that the magnetic field comprising an FTE is twisted, this twisting appearing to be continuous in sense across the magnetopause and corresponding to a core field-aligned current of magnitude a few x 10^5A. We observe changes in plasma flow speed and direction associated with FTEs. The transverse field and flow perturbations accompanying the three magnetosheath FTEs studied here satisfy approximately the Walén relation, the relation which describes a propagating Alfvén wave.

Introduction

Flux transfer events (FTEs) are evidence for magnetic reconnection occurring on short space and time scales at the dayside boundary of planetary magnetospheres (see the reviews by Cowley (1982), Saunders (1983), and Russell (1984)). The term

*Presently at Institute of Geophysics and Planetary Physics, University of California, Los Angeles, CA 90024

'flux transfer event' was introduced by Russell and Elphic (1978) to describe features in the initial ISEE magnetometer and particle observations from the Earth's dayside magnetosheath near the magnetopause. Earlier, Haerendel et al. (1978), using HEOS 2 magnetometer and plasma data also from the dayside, but in the boundary layer Earthward of the magnetopause and at higher latitudes than sampled by ISEE, had also suggested that magnetic reconnection occurs as a localised and transient process. Haerendel et al. (1978) termed their observations 'flux erosion events'. The question as to whether FTEs and flux erosion events represented the same physical process has recently been resolved in the affirmative by Rijnbeek and Cowley (1984), who reexamined the Haerendel et al. (1978) events using the boundary normal (LMN) coordinate system which allows FTEs to be easily identified (e.g., Russell and Elphic, 1978).

FTEs have been detected by other Earth-orbiting satellites, for example at synchronous orbit, by SCATHA (D. Croley, personal communication, 1983) and by GEOS (G. Wrenn, personal communication, 1983). FTEs have also been observed elsewhere in the solar system, at the magnetopauses of Mercury and Jupiter (Walker and Russell, 1983; Russell, 1984).

The expanding catalog of FTE identifications listed above, together with the recent statistical studies of FTE characteristics and occurrence based on data from a single ISEE satellite (Rijnbeek et al., 1984a,b; Berchem and Russell, 1984; Daly et al., 1984), leave little doubt that FTEs are a common feature of the magnetopause region when the magnetosheath magnetic field is directed southward. To assess the importance of FTEs in relation to other processes, for transporting magnetic flux and driving magnetospheric

146 SPATIAL PROPERTIES OF FTEs

while also permitting the spatial structure of an FTE to be probed.

Saunders et al. (1984) reported the first investigation exploiting the ISEE dual-satellite capability to study the spatial properties of FTEs. Here we substantiate and extend the Saunders et al. results. We present the dual-ISEE observations in two sections, the first of which describes magnetosheath FTEs, and the second discussing boundary layer FTEs. The next section considers the significance of our findings, especially with regard to the voltage associated with the FTE process. This section also includes a fresh interpretation for the origin of the bipolar magnetic signal which characterises FTEs. Conclusions comprise the final section.

Magnetometer and plasma data from both ISEE-1 and -2 are presented; the former measurements come from the UCLA fluxgate magnetometers (Russell, 1978), while the latter are data from the LANL/MPE fast plasma analysers (Bame et al., 1978).

Magnetosheath FTE Observations

Data

The data in Figure 1 illustrate the scale size which magnetosheath FTEs can attain in the magnetopause normal direction. The display shows 42 minutes of ISEE-1 (darker trace) and ISEE-2 (lighter trace) magnetometer and plasma data from the outbound orbit on October 23, 1978. The satellites were in the dawn magnetosheath at a GSM local time and latitude of 1000 hours and 40°N, and were separated by 5700 km with ISEE-2 leading in orbit. The plasma parameters are plotted in the top five panels and comprise moments of two-dimensional distributions (e.g. Paschmann et al., 1978) obtained at 6s resolution (ISEE-1) and 12s resolution (ISEE-2). Three-dimensional plasma data are not presented because their time resolution of 48s was too long to be useful. The three densities (cm^{-3}) plotted are N_P the total plasma density, $\bar{N}_P$ the density of energetic (9-40 keV) ions, and $\bar{N}_E$ the density of energetic (2-20 keV) electrons. Cool magnetosheath plasma dominates N_P, while hot particles of magnetospheric origin dominate the energetic particle densities. The third to fifth panels show the magnitude V (km s^{-1}) and components (V_X, V_Y) of the equatorial plasma bulk flow, with the latter displayed in satellite (approximately GSE) coordinates.

The magnetometer measurements in Fig. 1, as in subsequent Figures, are 12s averages plotted every 4s. They are presented here in boundary normal (LMN) coordinates calculated using the Fairfield (1971) model magnetopause normal. The two panels nearest the bottom show the field magnitude B, and the field angle α_{LM} in the LM plane (tangential to the magnetopause) defined such that $\alpha_{LM} = 0°$ is directed along L (northward) and $\alpha_{LM} = 90°$ points towards M (westward).

At 13.26 (hr. min.) UT ISEE-2 was 5540 km further from the magnetopause than ISEE-1. The

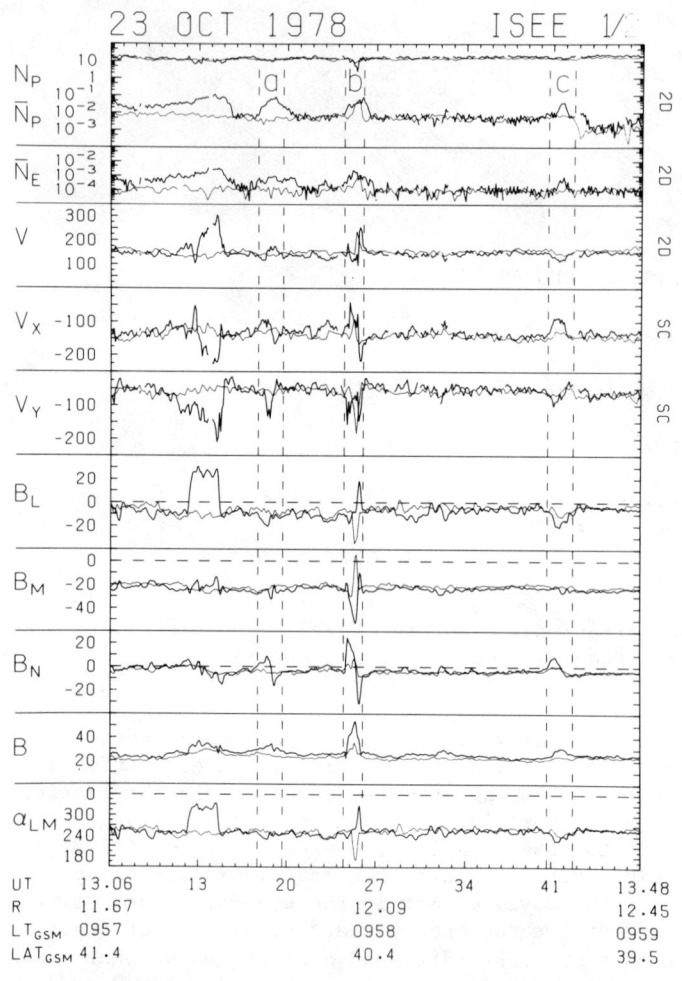

Fig. 1. ISEE-1 (heavy trace) and ISEE-2 (light trace) magnetic field and plasma measurements from the outbound magnetosheath passage on October 23, 1978. The plasma data are shown in the top five panels, and consist of various 2D density and flow parameters which are described fully in the text. The magnetic field recordings are displayed in boundary normal coordinates where B_N is outward along the boundary normal, B_L is along the projection of the GSM Z-axis in the magnetopause plane, and B_M completes the orthogonal triad and points westward. The normal direction had GSM-components (0.794, -0.300, 0.529) and was calculated using the Fairfield (1971) model. The bottom panel shows the field angle in the LM plane, defined by $\alpha_{LM} = \tan^{-1}(B_M/B_L)$. ISEE-1's position is given at the base of the Figure in terms of geocentric radial distance (R) in Earth radii, and GSM local time (LT$_{GSM}$) and latitude (LAT$_{GSM}$). At 13.26 UT the satellites were separated by 5540 km along the model normal.

motion, it is necessary to use multi-satellite measurements to determine the scale size of individual FTE flux tubes. The variable dual-ISEE satellite separation allows the FTE dimension normal to the magnetopause to be examined directly, satellite separation in LMN coordinates was (890, 1100, -5540) km measured from ISEE-2 to ISEE-1.

Pairs of dashed vertical guidelines, labelled (a), (b) and (c), bracket three magnetosheath FTEs. These FTEs show the positive-negative signature in B_N characterising a standard (rather than a reverse) polarity event (Rijnbeek et al., 1982). To study the FTE scale size let us consider the magnetic field and density observations. Further aspects of the Fig. 1 data will be considered later.

Event (b) at 13.25 UT is seen by both satellites. This is clear from the B_N signals, the field strength increases and the order of magnitude rises in the fluxes of energetic ring current ions and electrons. Indeed these latter approach levels observed during an apparent magnetopause encounter by ISEE-1 between 13.12 and 13.15 UT. As noted previously for magnetosheath FTEs (Paschmann et al., 1982), this behaviour of the energetic particles coupled with the slight decrease in the total plasma density are features expected for plasma mixing along open field lines.

FTE (a) at 13.18 UT illustrates a case where only the satellite closest to the magnetopause, ISEE-1, sees the event. The third standard polarity FTE, (c), at 13.41 UT is also an instance where only ISEE-1 observes the event. Although the magnetometer data suggest that ISEE-2 just grazed this latter event, the absence of an increase in the flux of energetic particles indicates that this satellite did not enter the open flux tube.

Scale Size in N Direction

The Fig. 1 data show that magnetosheath FTEs can have a scale size L_N normal to the magnetopause of at least 5500 km. To investigate further the magnetosheath FTE scale size L_N we have examined dual-ISEE satellite magnetometer data for a total interval exceeding six months when the satellites were well separated (separation > 1000 km) at the dayside magnetopause. A value for L_N has been determined statistically by finding the probability, as a function of satellite separation in the N direction, that both ISEE-1 and -2 see the same FTE.

The intervals examined covered the periods between September 1, 1978 and January 21, 1979, and from October 4, 1979 to November 23, 1979. Apart from a further interval in late 1981, these periods include all the large separation dayside observations recorded by the mission to date. Separations were directed mainly normal to the boundary, and varied from between ∼ 1000 km to ∼ 20000 km. However, the difficulty in tracking simultaneously two satellites at large separation meant that few intervals existed with overlapping data for separations above ∼ 7000 km. This limitation, coupled with our event selection criteria which are described below, restricted our data set (comprising magnetosheath FTEs seen at one or both satellites when the satellite separation parallel to N exceeded 1000 km), to 78 events. Both satellites had to be in the magnetosheath before and after the FTE encounter for an event to be included in our catalog.

In most cases we did not possess energetic particle data to support our magnetosheath FTE identifications, so extra care was taken to select only those events whose characteristics in the magnetometer data were fairly unambiguous. These characteristics have been documented elsewhere (e.g. Russell and Elphic, 1978, 1979; Elphic and Russell, 1979; Paschmann et al., 1982; Rijnbeek et al., 1984a,b; Berchem and Russell, 1984), and consist typically of a bipolar oscillation in the field component B_N normal to the magnetopause, together with a deflection in the tangential field direction and an increase in field strength. The data were analysed in boundary normal coordinates with normals defined both by the Fairfield (1971) model and by the tangential discontinuity technique. To be acceptable an FTE usually had to be identified using both normals. We restricted our data set to prominent events (B_N magnitude $\geq$ 10 nT peak-to-peak, and time scale $\geq$ 1 minute). The 78 events were observed at GSM local times between 0600 and 1500 hours, and at GSM latitudes between 40°N (outbound orbits) and 15°S (inbound orbits). The 78 events comprised 68 with standard polarity signatures, and 10 with reverse polarity signals.

The results of our statistical scale size study are shown in Figure 2. Satellite separations along the boundary normal were calculated using the Fairfield (1971) model. Probabilities were calculated by averaging the observations within bins of 1000 km width, and the error bars were obtained by using Poisson statistics. There were no instances of simultaneous ISEE-1 and -2 coverage with the satellites separated by between 7000 and 8000 km along N and with at least one satellite seeing an FTE. The straight line indicates the expected probability fit for an L_N of 1 R_E (R_E = Earth radius). This fit follows by considering the FTE cross-sections shown in Figure 4, and the motion of these past the satellites with the latter separated by different fractions of an L_N, where L_N is the maximum dimension of the FTE tube normal to the magnetopause. For a stationary magnetopause the fit depends only on this L_N value; for example, if the satellites are separated along N by $L_N/2$, there is a 50% probability that when one satellite sees an FTE the other satellite will also observe the same FTE as the open tube sweeps past. Despite the scatter, the observations in Fig. 2 suggest an L_N value close to 1 R_E for the prominent magnetosheath FTEs. This dimension is similar to the 0.7 R_E value for L_N obtained less directly by Rijnbeek et al. (1984a).

Field and Flow Perturbations

A significant feature of the data in Fig. 1 is the oppositely directed field tilting in the magnetopause plane associated with FTE (b). While

148   SPATIAL PROPERTIES OF FTEs

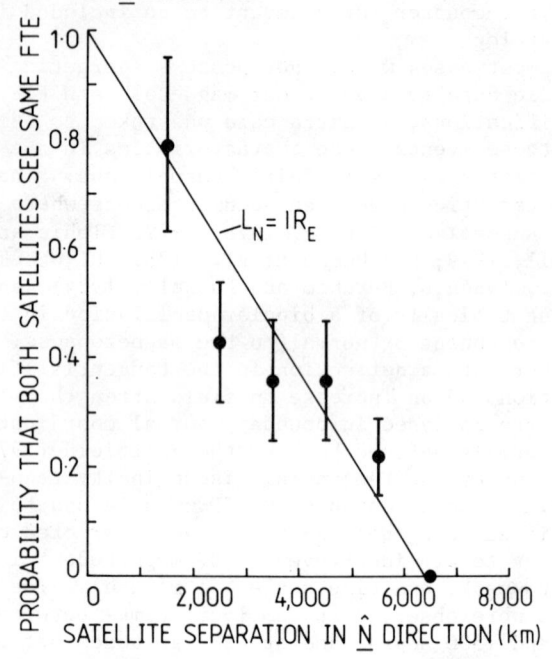

Fig. 2. Statistical result for the magnetosheath FTE scale size in the boundary normal direction based on ISEE dual-satellite magnetometer recordings. The data set consisted of 78 events.

the field at ISEE-1 rotates 70° towards the magnetospheric field direction, at ISEE-2 the field simultaneously rotates 70° in the opposite direction. We have seen differently or oppositely directed field tilting associated with FTEs on several occasions when the satellites have been well separated (separation > 10^3 km), with event (b) the most prominent example found in the data intervals examined. This feature substantiates previous suggestions (Paschmann et al., 1982; Cowley, 1982) based on the characteristic bipolar form of the FTE magnetic signal, that the magnetic field comprising an FTE appears to be twisted. We shall illustrate schematically the field twisting later, but first let us consider the plasma flow behaviour during the FTEs in Fig. 1 as this will help clarify the reason why the magnetic field appears twisted.

Changes in plasma flow speed and direction accompany the three October 23, 1978 events. The differently directed flows at the two satellites in FTE (b) indicate the presence of plasma vorticity which could be related to the field twisting (also see Paschmann et al., 1982). Closer inspection suggests a basic relationship between the flow and the field perturbations. The X and Y satellite coordinate directions correspond respectively within 37° of the N and -M directions in boundary normal coordinates. Perturbations in V_X and B_N appear in phase during the three FTEs, as also do the perturbations in V_Y and $-B_M$ if the change in B_M associated with the change in field strength is eliminated.

To check these relationships in greater detail we replotted the field measurements in GSE (approximately satellite) coordinates so that the field and flow data were in a similar coordinate system. The field and flow components perpendicular to the background magnetosheath field, $\underline{B}_o$, were then obtained by removing the components parallel to $\underline{B}_o$; a procedure possible for the flow recordings because the ambient field lay in the ecliptic plane.

In all events the X and Y GSE component field and flow perturbations transverse to B_o satisfy approximately the Walén relation which describes a propagating Alfvén wave (Walén, 1944; also see Ferraro and Plumpton, 1966); namely $\underline{b}_\perp = \pm B_o \underline{v}_\perp /A$, where $\underline{b}_\perp$ and $\underline{v}_\perp$ are the transverse field and flow perturbations, B_o is the background field strength, A is the Alfvén speed and the sign indicates whether $\underline{b}_\perp$ and $\underline{v}_\perp$ are in phase or in antiphase. The fact that $\underline{b}_\perp$ and $\underline{v}_\perp$ are in phase for the standard polarity events here implies that the wave was propagating antiparallel to the background field, that is, away from the equator. As an Alfvén wave carries both a field-aligned current density ($\underline{\nabla}_\perp \times \underline{b}_\perp /\mu_o$) and a parallel vorticity ($\underline{\nabla}_\perp \times \underline{v}_\perp$), it is clear that such a wave propagating along a tube could produce the field twisting and vortex motion which appear associated with the FTEs in Fig. 1.

Boundary Layer FTE Observations

Figure 3 displays 42 minutes of ISEE-1 and -2 magnetometer data for the outbound boundary layer and magnetopause crossing on October 24, 1979. The satellite location was close to that on October 23, 1978, due to the ISEE orbit remaining stationary in space as the Earth rotates about the Sun. The magnetosheath field had a similar orientation on the two days. The data in Fig. 3 are presented in a format identical to that employed for the magnetometer data in Fig. 1. ISEE-1 is again the trailing satellite as shown by its later exit from the magnetosphere at 5.07:45 (hr. min: sec) UT. At 5.03 UT the satellite separation in LMN coordinates, calculated using the Fairfield (1971) model and measured from ISEE-2 to ISEE-1 was (1360, 680, -5860) km. Thus ISEE-2 was 5860 km further outward along the boundary normal than ISEE-1.

Three prominent standard polarity FTEs, labelled (d), (e) and (f), are marked by the dashed vertical guidelines. Measurements by the fast plasma analyser (FPA) experiments (not presented) have been examined to support our FTE identifications based on the magnetometer data, although degradation of the ISEE-1 FPA instrument at low energies precluded the use of thermal plasma data from this satellite. The first event, (d), is encountered at 4.34 UT when both satellites were inside the magnetosphere. The particle measurements indicate that this event is seen only by ISEE-2, the satellite closest to the magnetopause, though it is interesting to note that a weaker but similar signal in the normal magnetic field occurs at ISEE-1 slightly later. The α_{LM} plot shows that at ISEE-2 the magnetic field in the magnetopause plane associated with this event tilts 'away' from the magnetosheath field direction ($\sim 260°$) by $\sim 20°$.

A second clear boundary layer FTE, labelled (f) occurs at 5.03 UT. This event is seen by both satellites (the energetic proton and electron fluxes at ISEE-1 showing similar drop-outs to those recorded by ISEE-2 for event (d)). The amplitude of the B_N signal and the field deflection tangential to the magnetopause associated with this FTE are different at the two locations. ISEE-2, the satellite closest to the magnetopause, sees the larger amplitude B_N signal. While ISEE-1 sees field tilting 'away' from the magnetosheath field direction in the same sense as ISEE-2 saw for event (d), ISEE-2 now sees field tilting 'towards' the magnetosheath field direction.

The third prominent FTE, (e), marked in Fig. 3 occurs at 4.58 UT and is seen only by ISEE-2. This event occurs adjacent to the magnetopause and we identify it as a magnetosheath FTE in which the field in the LM plane exhibits slight 'towards' tilting.

The Fig. 3 data complement the observations in Fig. 1 in two respects. Firstly, as expected from magnetic flux conservation, these data show that boundary layer FTEs can have a normal scale size comparable to the magnetosheath FTE L_N value of $\sim 1 R_E$. Secondly, the Fig. 3 data confirm that as for magnetosheath FTEs the magnetic field comprising boundary layer FTEs also appears to be twisted. The sense of twisting about the FTE tube axis is the same in both the boundary layer and in the magnetosheath. The next section illustrates these points and discusses their consequences.

Interpretation and Discussion

Our direct measurement of the FTE scale size allows the flux transfer effected by an FTE to be estimated. By combining the typical FTE time duration of 1-2 minutes with the FTE speed along the magnetopause of 100-200 km.s^{-1} (assumed to be equal to the average plasma bulk flow) one obtains a value of about 2 R_E for the FTE dimension tangential to the magnetopause. This value is similar to that obtained by Daly and Keppler (1983) from using energetic particles to remote sense an FTE. For an FTE normal dimension of 1 R_E and a field strength of 50 nT, the magnetic flux comprising an FTE is then about 4×10^6 Wb. Since FTEs recur at a particular location about every 8 minutes (Rijnbeek et al., 1984a), the voltage associated with the process is at least ~ 10 kV.

This is a lower limit for the total cross-magnetosphere voltage attributable to FTEs (e.g. Rijnbeek et al., 1984a). Firstly, FTEs are seen over the entire dayside boundary region when the magnetosheath field is directed southward (Rijnbeek et al., 1984a,b; Berchem and Russell, 1984). As there is no reason to believe that any given event

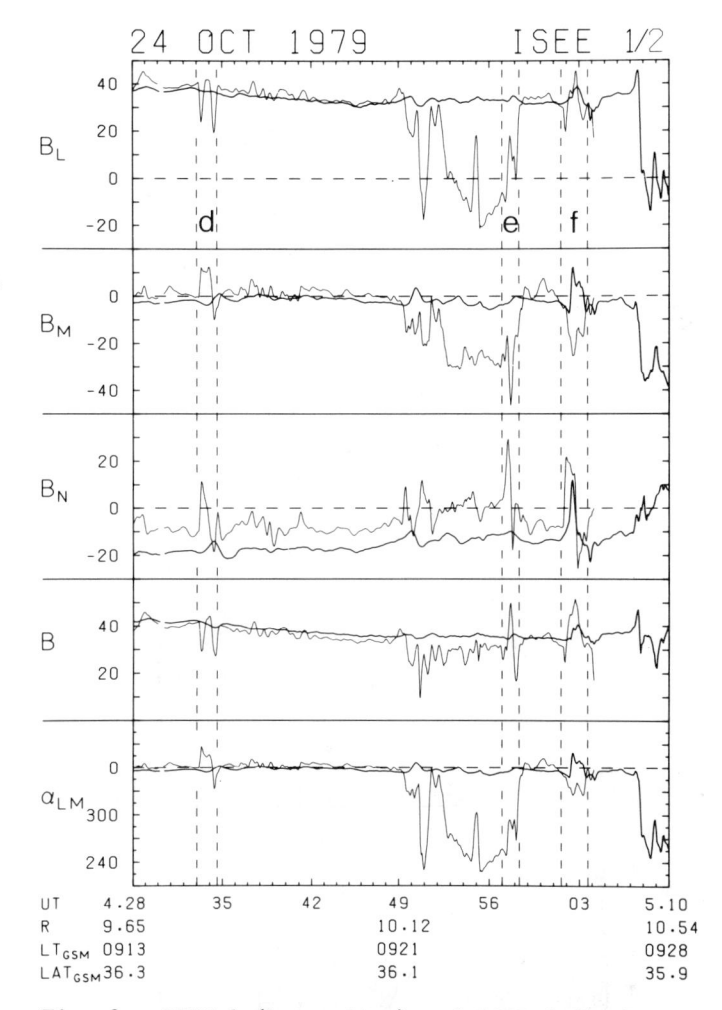

Fig. 3. ISEE-1 (heavy line) and ISEE-2 (light line) magnetometer measurements for the outbound boundary layer crossing on October 24, 1979. The display format is the same as in Fig. 1. The boundary normal is based on the Fairfield (1971) model and has GSM-components (0.818, -0.343, 0.462). The satellite position information is given for ISEE-1, which at 5.03 UT trailed ISEE-2 by 5860 km in the model normal direction.

will move in such a way that it can be detected at all magnetopause latitudes and longitudes, a satellite near the magnetopause is unlikely to see every event. Secondly, we have considered only the prominent cases, while there is evidence that a spectrum of smaller sized FTE signals may exist at the magnetopause (e.g. see Figs. 1 and 3; Russell and Elphic, 1978; Rijnbeek et al., 1984a,b; Berchem and Russell, 1984). Thus it seems fair to conclude that FTEs can make a significant contribution to the typical cross-magnetosphere voltage of 50-100 kV.

The observations reported here strengthen the view that the magnetic field comprising both magnetosheath and boundary layer FTEs is twisted. If no parallel $\nabla \times \underline{b}$ existed within the events, the field tension force acting on open field lines

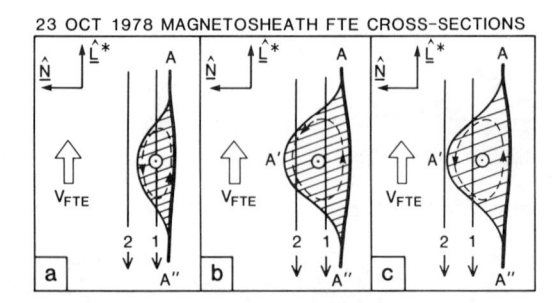

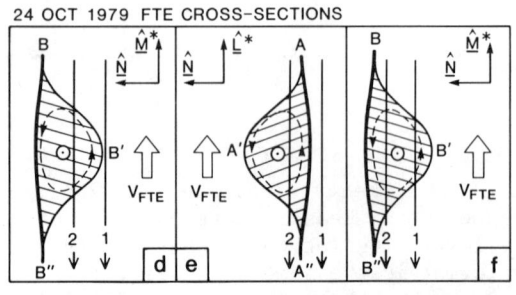

Fig. 4. Sketches illustrating the field twisting associated with a northward moving FTE tube and an easterly (-M) directed magnetosheath field as implied by the observations in Figs. 1 and 3. The upper drawing shows the twisting in three dimensions viewed from outside the magnetosphere. The panels below show the view looking along the axis of each of the six FTE tubes (a to f) marked in Figs. 1 and 3 as each tube sweeps past the two ISEE satellites.

would ensure that the FTE magnetic field shows generally 'towards' tilting with respect to the ambient field direction on the opposite side of the magnetopause (see Cowley et al. (1983) for a general discussion of field perturbations in the magnetopause vicinity in the absence of field twisting). Thus the frequently reported instances of 'away' field tilting associated with FTEs in the magnetosheath and in the boundary layer (examples may be found in Elphic and Russell, 1979; Paschmann et al., 1982; Rijnbeek et al., 1984a; Berchem and Russell, 1984; Rijnbeek and Cowley, 1984, as well as in Figs. 1 and 3 here) provides further evidence, albeit indirect, for field twisting. Similarly, Saunders (1983) pointed out that field twisting in FTEs could explain the 'away' tilting of the magnetic perturbations accompanying the pulses of boundary layer seen furthest from the magnetopause on the much-studied November 6, 1977 boundary layer encounter (also see Sckopke et al., 1981; Paschmann et al., 1982; Cowley, 1982).

Figure 4 illustrates schematically the field twisting indicated by the FTE observations in Figs. 1 and 3. The upper diagram shows how one might visualise part of a northward moving FTE tube viewed from the magnetosheath. The reconnected tube is shown speckled and crosses the magnetopause near the 'hairpin' bend. The 'ends' of the open tube connected to the magnetosheath and to the magnetosphere are indicated in the sketch; a satellite Earthward of the magnetopause would see the latter as a boundary layer FTE. Field tension causes the reconnected tube to contract northward and westward along the magnetopause in the direction of the open arrow.

The field line spiralling about the FTE tube axis indicates the observed sense of field twisting. The twisting in the magnetosheath and boundary layer segments has the same sense about the tube axis, suggesting that the twisting is continuous across the magnetopause, as illustrated in Fig. 4. By taking average FTE field values and assuming that the field pitch is uniform about the FTE tube axis one might argue that near the boundary of the open tube the twisting has a pitch length of $\sim 6\ R_E$ (= $2\pi r\ B_z/B_\Theta$, where r is the tube radius (3000 km), B_Θ is the azimuthal field component (15 nT) and B_z is the field component along the tube axis (30 nT)). The twisting in Fig. 4 is associated with a field-aligned current along the center of the flux tube directed towards the ionosphere. There may also be present an oppositely directed field-aligned current near the perimeter of the open tube. For a 1 R_E diameter flux tube associated with a B_N signal amplitude of 15 nT, Ampere's Law shows that the magnitude of the FTE Birkeland current(s) is a few x 10^5 Amps.

A satellite in the magnetosheath (magnetosphere) encountering the tube near the location AA'A" (BB'B") sees a standard polarity (+/-) B_N signal due to the field twisting about the flux tube axis. The direction of field tilting seen in the LM plane depends on whether the satellite bisects the FTE Earthward or beyond the open tube center as the FTE sweeps past. For the field twist geometry in Fig. 4 one would expect to observe associated with both the magnetosheath and boundary layer FTE tube segments, towards (away) field tilting with respect to the ambient field direction on the opposite side of the boundary in the section of the tube closest to (furthest from) the magnetopause.

The spatial variation which we envisage for the FTE magnetic field is illustrated more clearly in the two panels in the lower part of Fig. 4. In the upper of these panels, cross-sections through the plane AA'A" of the open flux tube (see upper sketch) are shown for the three FTE events (a, b and c) in Fig. 1. Similar cross-sections are shown in the lower panel for the events (d, e and f) in Fig. 3, though two of these are for the plane BB'B". The sketches are drawn looking antiparallel to the FTE axial field (dotted circle) and illustrate either the NL* or NM* planes where L*(M*) point northward in directions nearly perpendicular (parallel) to the ambient magnetosheath magnetic field. The open tubes are shown hatched with the magnetopause marked by a thicker line. FTE motion is indicated by the open arrows labelled V_{FTE}, and the trajectories of the ISEE satellites through each event are shown by the lines marked 1 (ISEE-1) and 2 (ISEE-2). These lines are drawn straight for simplicity, thus ignoring magnetopause motion during the encounter. The sketches are drawn to the same scale with the satellites separated in the N direction by 5500 km in the upper panel, and by 5900 km in the lower panel. The anticlockwise arrowed dashed lines indicate the sense of field twisting about the FTE tube axis. The observed FTE magnetic signals can be explained qualitatively by this sense of twisting together with the indicated location of the FTE encounter crossings.

In event (a) ISEE-1 passes through the center of the structure and thus sees a clear standard polarity B_N signal and no field tilting in the plane tangential to the magnetopause. In event (b), which has a larger scale size than (a), ISEE-1 encounters the tube slightly Earthward of its center and observes a substantial B_N signal together with slight 'towards' (northward) field tilting which increases sharply later in the event possibly due to outward magnetopause motion. Meanwhile ISEE-2, 5500 km further from the magnetopause, sees a smaller amplitude B_N signal and large 'away' (southward) tilting of the field. In event (c) ISEE-1 passes through the tube beyond its center and observes a clear B_N signal and away field tilting.

In event (d), a boundary layer FTE, ISEE-2 bisects the FTE Earthward of its center and observes 'away' (westward) field tilting. In the later boundary layer event, (f), ISEE-1 encounters the tube close to where ISEE-2 passed through event (d), while ISEE-2 5900 km closer to the magnetopause, crosses the FTE tube beyond its center and observes a larger amplitude B_N signal and 'towards' (eastward) tangential tilting. In the magnetosheath FTE, event (e), ISEE-2 passes through the FTE tube slightly Earthward of its center and sees a substantial B_N signal together with slight towards (northward) field tilting. ISEE-1 is inside the magnetosphere and does not observe this event.

Conclusions

The value of multi-satellite and multi-instrument data for studying the dimension and structure of FTEs has been demonstrated. Prominent FTEs (B_N magnitude $\geq$ 10 nT peak-to-peak, and time scale $\geq$ 1 minute) are large scale phenomena with a dimension normal to the magnetopause of order an Earth radius. This scale size indicates that FTEs contribute significantly to the flux erosion process which drives magnetospheric convection. The evidence that the magnetic field comprising both magnetosheath and boundary layer FTEs is twisted is convincing. The twisting appears to have the same sense in both FTE segments. The relationship between the field and flow perturbations associated with three magnetosheath FTEs studied here indicates that the twisting propagates along the reconnected tube as an Alfvén wave. The FTE magnetic signal in the normal component therefore is not simply due to the draping of exterior field lines around the open tube.

Further examples of the field and flow perturbations comprising FTEs should be examined to establish the generality of the result found here and to determine where the field twisting appears to be imposed. The higher time resolution flow information available on ISEE-1 from the GSFC low energy electron spectrometer, and from the UCB and GSFC D. C. electric field instruments, may be useful in this regard. The key remaining question concerns the mechansim which gives rise to the twisting.

Acknowledgments. Most of this work was performed while M. A. S. was supported by a UK SERC post-doctoral research award. C. T. R. was supported by NASA under contract NAS-5-25772, while Max-Planck-Institut portions of this work were supported by Bundesministerium für Forschung und Technologie. S. W. H. Cowley, D. J. Southwood and R. P. Rijnbeek are thanked for several helpful and generous discussions. It is a pleasure also to thank S. J. Bame and G. Paschmann for kindly making their data available.

References

Berchem, J. and C. T. Russell, Flux transfer events on the dayside magnetopause: spatial distribution and controlling factors, J. Geophys. Res., in press, 1984.

Bame, S. J., J. R. Asbridge, H. E. Felthauser, J. P. Glore, G. Paschmann, P. Hemmerich, K. Lehmann and H. Rosenauer, ISEE-1 and ISEE-2 fast plasma experiment and the ISEE-1 solar wind experiment, IEEE Trans. Geosci. Electron., GE-16, 216-220, 1978.

Cowley, S. W. H., The causes of convection in the Earth's magnetosphere: a review of developments during the IMS, Rev. Geophys. Space Phys., 20, 531-565, 1982.

Cowley, S. W. H., D. J. Southwood and M. A.

Saunders, Interpretation of magnetic field perturbations in the Earth's magnetopause boundary layers, Planet. Space Sci., 31, 1237-1258, 1983.

Daly, P. W. and E. Keppler, Remote sensing of a flux transfer event with energetic particles, J. Geophys. Res., 88, 3971-3980, 1983.

Daly, P. W., M. A. Saunders, R. P. Rijnbeek, N. Sckopke and C. T. Russell, The distribution of reconnection geometry in flux transfer events using energetic ion, plasma and magnetic data, J. Geophys. Res., in press, 1984.

Elphic, R. C. and C. T. Russell, ISEE-1 and -2 observations of the magnetopause, in Magnetospheric Boundary Layers, edited by B. Battrick, Rep. ESA SP-148, pp. 51-65, Noordwijk, Netherlands, 1979.

Ferraro, V. C. A. and C. Plumpton, An Introduction to Magneto-fluid Mechanics, 254 pp., Clarendon Press, Oxford, 1966.

Fairfield, D. H., Average and unusual locations of the Earth's magnetopause and bow shock, J. Geophys. Res., 76, 6700-6716, 1971.

Haerendel, G., G. Paschmann, N. Sckopke, H. Rosenbauer and P. C. Hedgecock, The frontside boundary layer of the magnetosphere and the problem of reconnection, J. Geophys. Res., 83, 3195-3216, 1978.

Paschmann, G., N. Sckopke, G. Haerendel, I. Papamastorakis, S. J. Bame, J. R. Asbridge, J. T. Gosling, E. W. Hones, Jr. and E. R. Tech, ISEE plasma observations near the subsolar magnetopause, Space Sci. Rev., 22, 717-737, 1978.

Paschmann, G., G. Haerendel, I. Papamastorakis, N. Sckopke, S. J. Bame, J. T. Gosling and C. T. Russell, Plasma and magnetic field characteristics of magnetic flux transfer events, J. Geophys. Res., 87, 2159-2168, 1982.

Rijnbeek, R. P. and S. W. H. Cowley, Flux erosion events are flux transfer events, Nature, in press, 1984.

Rijnbeek, R. P., S. W. H. Cowley, D. J. Southwood and C. T. Russell, Observations of reverse polarity flux transfer events at the Earth's dayside magnetopause, Nature, 300, 23-26, 1982.

Rijnbeek, R. P., S. W. H. Cowley, D. J. Southwood and C. T. Russell, A survey of dayside flux transfer events observed by the ISEE-1 and -2 magnetometers, J. Geophys. Res., 89, 786-800, 1984a.

Rijnbeek, R. P., S. W. H. Cowley, D. J. Southwood and C. T. Russell, Recent investigations of flux transfer events observed at the dayside magnetopause, these proceedings, 1984b.

Russell, C. T., The ISEE-1 and -2 fluxgate magnetometers, IEEE Trans. Geosci. Electron., GE-16, 239-242, 1978.

Russell, C. T., Reconnection at the Earth's magnetopause: magnetic field observations and FTEs, these proceedings, 1984.

Russell, C. T. and R. C. Elphic, Initial ISEE magnetometer results: magnetopause observations, Space Sci. Rev., 22, 681-715, 1978.

Russell, C. T. and R. C. Elphic, ISEE observations of flux transfer events at the dayside magnetopause, Geophys. Res. Lett., 6, 33-36, 1979.

Saunders, M. A., Recent ISEE observations of the magnetopause and low latitude boundary layer: a review, J. Geophys., 52, 190-198, 1983.

Saunders, M. A., C. T. Russell and N. Sckopke, Flux transfer events: scale size and interior structure, Geophys. Res. Lett., 11, 131-134, 1984.

Sckopke, N., G. Paschmann, G. Haerendel, B. U. O. Sonnerup, S. J. Bame, T. G. Forbes, E. W. Hones, Jr. and C. T. Russell, Structure of the low latitude boundary layer, J. Geophys. Res., 86, 2099-2110, 1981.

Walén, C., On the theory of sunspots, Arkiv för Matematik, Astronomi och Fysik, 30A, Paper 15, 1-87, 1944.

Walker, R. J. and C. T. Russell, Dayside reconnection at Jupiter and the Earth, Paper PA.20 (abstract), XVIII IAGA General Assembly, Hamburg, August, 1983.

Questions and Answers

Kling: 1. Do FTEs in the magnetospheric boundary later have a scale size normal to the magnetopause comparable to the 1 R_E which you reported for magnetosheath FTEs? 2. Is there evidence for, or would you expect, the FTE tube radius to change with time?

Saunders: 1. Although we have not made a statistical study of boundary layer FTE scale sizes, we have identified a case when the ISEE satellites were separated by ~6000 km in the magnetopause normal direction in the boundary layer, where both satellites saw clear FTE signatures in the magnetic field and particle data. 2. The limited number of FTE events observed during intervals of large ISEE satellite separation makes such a study difficult. However, there are reasons for expecting the FTE tube cross-sectional shape to evolve with time. In particular, field curvature forces associated with the overlying field would exert normal stresses causing a spherical or elliptical shape tube to become elongated into a long thin slab.

THE RELATION OF FLUX TRANSFER EVENTS TO MAGNETIC RECONNECTION

J. D. Scudder and K. W. Ogilvie

NASA/Goddard Space Flight Center, LEP
Greenbelt, MD 20771

C. T. Russell

University of California Los Angeles, IGPP
Los Angeles, CA 90024

Disssipation at the separator is a theoretically necessary requirement of any reconnection model. It is very difficult to produce this dissipation in a localized region without simultaneously providing a local source of heat for electrons. The magnetic tubes of force which are just inside the separatrices act as conduits for heat emanating from the diffusion region. Those field lines whose extremities are controlled by dissipative MHD should delineate layers of heat flow enhanced over the ambient level directed away from the diffusion region as shown in Figure 1.

Electron data from 31 previously identified flux transfer events have been examined to look for this heat flow signature. Most, but not all FTE's are observed to be accompanied by the heat flow layer, which we have used as an operational method of defining separatrix traversal. The interval between heat flow peaks (the flux transfer interval FTI) is distinguishable from the FTE whose duration is dominated by the draping signatures. Within the FTI surface, strong acceleration or deceleration, large parallel pressure anisotropies, and temperatures and densities intermediate between magnetosphere and magnetosheath are usually observed. Some events show evidence of field aligned ionospheric electrons. Usually the velocity inside of the separatrix shears to become more nearly orthogonal to the local magnetic field than externally. This has led to the suggestion of a more detailed model of FTE's as shown in Figure 2. All of the essential features of FTI's will be illustrated for a previously documented rotational magnetopause crossing. FTI's and rotational magnetopause crossings are nearly indistinguishable except by the conditions of their observation as indicated in Figure 3. From the timing of the arrival of conduction and acoustic fronts within the FTI at the spacecraft the clearing of the reconnected flux tubes from the diffusion region can be sensed. In this way the angular half widths of the reconnected tubes have been estimated to be ~20-40°. The observed thickness of the heat layers imply scale lengths in the diffusion region. Examples of FTE traversals below and above the separator (as defined by magnetic records) will be shown to illustrate the observed reversal of the local heat flow q on the separatrix with respect to B implied by Figure 1. The model suggests ways that imprints of transient reconnection can be seen with or without magnetic signatures. However, when both magnetic B, bulk flow and heat flux Q signatures occur for the same event reconnection is still in progress. "Q but not B" signatures indicate a transient start up phase and "B but not Q" a transient cool down phase. The magnetic flux transfer rate implied by the occurrence rate and duration of FTI's can replenish, between substorms, the magnetic flux lost

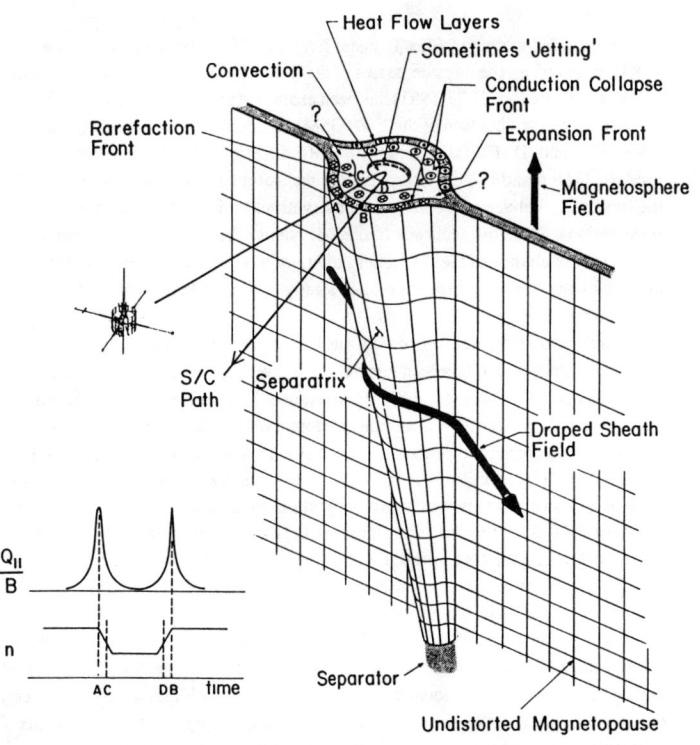

Fig. 1

Fig. 2

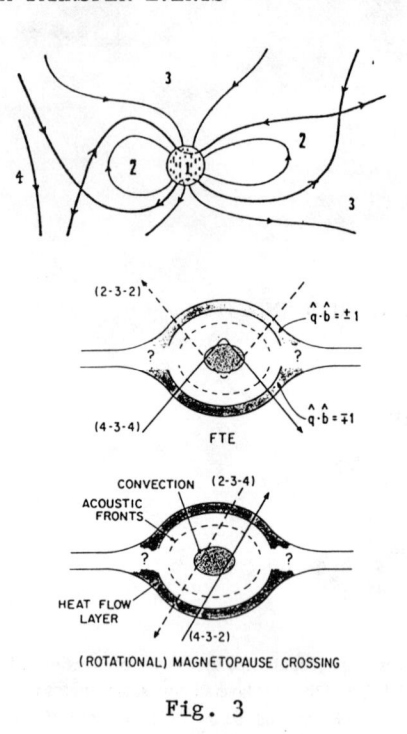

(2-3-2)

$\hat{q}\cdot\hat{b} = \pm 1$

(4-3-4)

$\hat{q}\cdot\hat{b} = \mp 1$

FTE

CONVECTION (2-3-4)
ACOUSTIC
FRONTS

HEAT FLOW
LAYER
(4-3-2)

(ROTATIONAL) MAGNETOPAUSE CROSSING

Fig. 3

during a substorm. Dynamical tests $\triangle \underline{V} = \alpha \triangle (B/\sqrt{\rho})$ have been performed on the largest and best resolved FTE to illustrate that it is consistent with a one-sided passage into and out of a rotational shear layer, RSL. The fit constant of proportionality is consistent with the theoretical value for an RSL in the presence of anisotropic pressure. Unlike the apparent situation in the quasi-steady reconnection phenomena reported by Paschmann and Sonnerup the RSL cannot be described as standing in the magnetosheath flow and its direction of propagation is not aligned with the local boundary normal direction. Field aligned currents have directly been measured to be present in FTE's but only occupy a central core of the FTI.

A full report of these results will appear in a paper to be submitted to the *Journal of Geophysical Research*.

Questions and Answers

Vasyliunas: 1) Is the electron heat flux vector you calculate defined with respect to the bulk flow of the plasma or of the electrons? 2) It is unlikely that the field-aligned (Birkeland) current density you infer near the magnetopause will map unmodified to the ionosphere: there are large gradients of pressure and flow, associated with the relatively small scale of the structures transverse to the field, and thus one may anticipate significant divergence of the perpendicular (and hence also of the parallel) current.

Scudder: 1) *Q reported is defined with respect to the electron bulk velocity. Note, however that the $\bar{q}$ layers so determined are not colocated with the zone of the suggested currents, but places where $\bar{U}_e \sim \bar{U}_i$ as reflected in the Alfvén test reported.* 2) *The spirit of the comparison was to give a relative measure of the current densities suggested by the data analysis—and that, while large, is not unreasonable.*

FLUX TRANSFER EVENTS AND INTERPLANETARY MAGNETIC FIELD CONDITIONS

J. Berchem and C. T. Russell

Institute of Geophysics and Planetary Physics
University of California, Los Angeles, CA 90024

The extended survey of Flux Transfer Events (FTEs) [Berchem and Russell, 1983] observed on the dayside passes of the ISEE spacecraft during the first five years of the mission (1977-1981) has been examined in order to relate the FTE occurrence to the direction of the Y-Z (GSM) component of the Interplanetary Magnetic Field (IMF). To determine the IMF direction for each FTE, we have used the IMP 8 and ISEE 3 interplanetary measurements after having estimated the time delay between the solar wind observations and the magnetopause observations. Since the accuracy of the IMF values determined depends greatly on the rate of change of the IMF orientation, we have omitted determinations obtained when the field does not remain steady enough in the neighborhood of the expected time delays.

Figure 1 shows the number of FTEs as a function of the IMF angle in the Y-Z GSM plane which has been normalized to the number of magnetopause crossings occurring for the same IMF conditions. This normalization avoids a bias in the FTE distribution due to the predominant duskward orientation of the IMF observed during the periods surveyed. As is readily seen in this figure, only a few of the FTEs identified occur during slightly northward IMF conditions and none of them for strongly northward orientation. Most of the FTEs are observed when the IMF is southward. Since there are only a small number of FTEs when we can determine the IMF without ambiguity, we do not attribute too much significance to the decrease of the rate of FTE occurrence with the increase in magnitude of the IMF B_y component.

Another point is that there is not a noticeable dependence of the rate of FTE occurrence on the sign of the IMF B_y component. However, the direction of the IMF B_y component might locally order the location of the FTE and the polarities observed. By polarity we mean the sense of the magnetic signature when the flux tube passes the spacecraft. When the tube is moving toward the north the perturbation in the magnetic component normal to the magnetopause is first positive and then negative (direct FTE) or vice versa (reverse FTE) when the tube is moving toward the south [Cowley, 1982; Rijnbeek et al., 1982].

Figure 2 shows the sense of the IMF for each FTE observation. At the projection on the Y-Z GSM plane (viewed from the sun) of the location of the

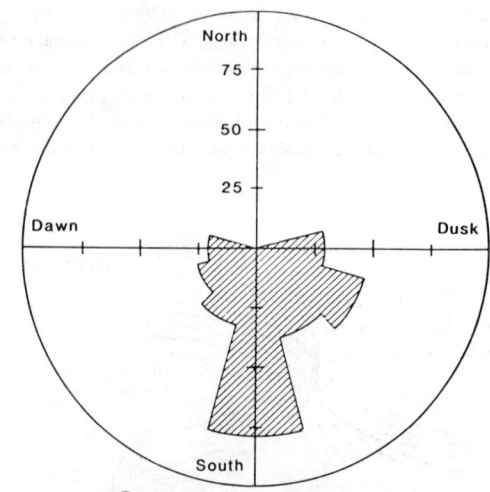

Percent of occurrence of FTE's as a function of the IMF angle in the Y-Z GSM plane

Figure 1

FLUX TRANSFER EVENTS

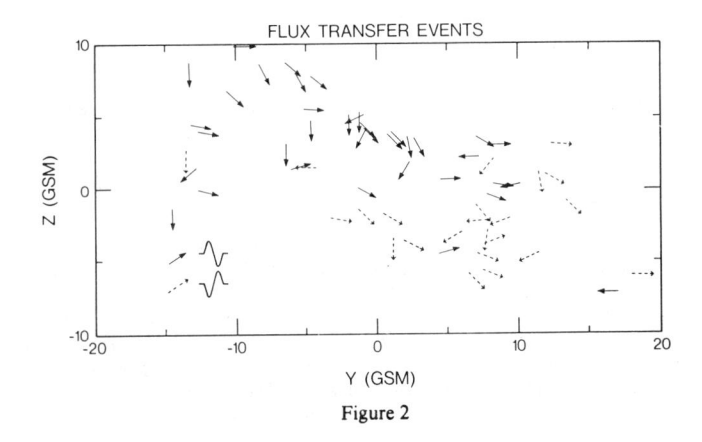

Figure 2

FTEs, we show with an arrow the orientation of the Y-Z GSM component of the IMF. We have indicated the polarity of the magnetic signature: solid arrows denote direct FTEs, while dashed arrows refer to reverse FTEs. When looking at the spatial distribution of the FTEs and the polarity, one may have expected to observe direct FTEs in the northern hemisphere and reverse in the southern hemisphere, making the assumption of an equatorial merging region. That is basically what we see here, but numerous cases do not agree with this simple order. The picture becomes more coherent when considering the IMF B_y component of the deviant cases. Indeed, for most of them, especially those observed near the flanks, their "abnormal" polarities become consistent with their locations if the merging line tilts with changes in the IMF B_y direction, as it has been postulated for the quasi-steady reconnection observations [Sonnerup et al., 1981]. However, a few discrepancies remain and further studies should resolve whether nonsteady IMF conditions can explain these deviations or if more sophisticated merging patterns have to be considered or if we must radically change the global picture that we have.

In conclusion, FTEs predominantly occur when the IMF is southward oriented. The tilt of the merging line with the B_y component of the IMF might explain the observation of unexpected FTE polarities at certain locations.

References

Berchem J. and C. T. Russell, Flux transfer events on the dayside magnetopause: spatial distribution and controlling factors, *J. Geophys. Res.*, submitted, 1983.

Cowley, S. W. H., The causes of convection in the earth's magnetosphere: a review of developments during the IMS, *Rev. Geophys. Space Phys.*, 20, 531, 1982.

Rijnbeek, R. P., S. W. H. Cowley, D. J. Southwood and C. T. Russell, Observations of reverse polarity flux transfer events at the earth's dayside magnetopause, *Nature*, 300, 23, 1982.

Sonnerup, B. U. Ö, G. Paschmann, I. Papamastorakis, N. Sckopke, G. Haerendel, S. J. Bame, J. R. Asbridge, J. T. Gosling, and C. T. Russell, Evidence for magnetic field reconnection at the earth's magnetopause, *J. Geophys. Res.*, 86, 10049, 1981.

SURVEY OF ION DISTRIBUTIONS IN FLUX TRANSFER EVENTS

P. W. Daly

Space Science Department of ESA, ESTEC
Noordwijk, The Netherlands

M. A. Saunders, R. P. Rijnbeek

Blackett Laboratory, Imperial College,
London SW7 2AZ, England

N. Sckopke

Max-Planck-Institut für Physik und Astrophysik
8046 Garching, Fed. Republic of Germany

E. Keppler

Max-Planck-Institut für Aeronomie
3411 Katlenburg-Lindau 3, Fed. Republic of Germany

The direction of energetic ion streaming relative to the magnetic field indicates whether the field lines in the flux transfer event (FTE) are connected to the northern or southern hemisphere. We have used the medium energy particle spectrometer on board ISEE-2 to determine this streaming from FTEs in the dayside magnetosheath from latitude -10 to 40. The results are plotted in Fig. 1. On the morningside a very strong correlation between ion distribution and the sign of the east-west component of magnetic field is seen. On the afternoonside, the correlation is with latitude. Using magnetic signatures and plasma data it is possible to produce a consistent picture of an equatorial origin of FTEs. There is some evidence that, whereas quasi-steady reconnection occurs at the geocentric solar magnetic equator, the FTEs may be controlled more by the geomagnetic equator. A full report of these results is given in Daly et al., 1983.

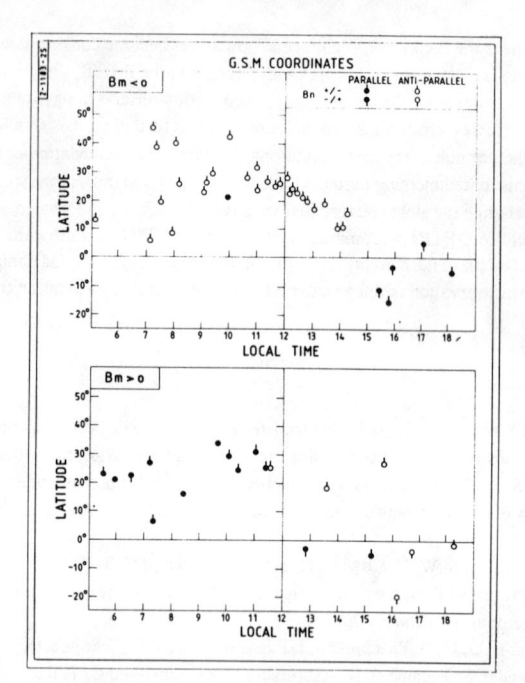

Figure 1. The distribution of FTEs about the dayside magnetopause, in geocentric solar magnetic (GSM) coordinates. The upper panel is for $B_M < 0$ (field westward), the lower for $B_M > 0$ (eastward). Closed and open circles indicate that ions were observed streaming parallel and antiparallel, respectively, to the magnetic field. A tick on the upper or lower side shows whether the B_N signature is +/− or −/+, respectively.

Reference

Daly, P. W., M. A. Saunders, R. P. Rijnbeek, N. Sckopke and C. T. Russell, The distribution of reconnection geometry in flux transfer events using energetic ion, plasma and magnetic field data, *J. Geophys. Res.*, submitted, 1983.

Questions and Answers

Paschmann: At higher latitudes on the dawnward side of the magnetopause, your distribution of events shows only "south-connected" cases (which move northward), but not their corresponding "north-connected" partners?

Daly: If FTE formation occurs primarily near noon, then when $B_M > 0$ (as in these cases), the "south-connected" flux tube lies mainly on the dawnside and the "north-connected" flux tube on the duskside. This would decrease the probability of observing the "north-connected" FTE's on the dawnside when $B_M > 0$. These events would also be moving very fast and might be overlooked for this reason.

PATTERNS OF MAGNETIC FIELD MERGING SITES ON THE MAGNETOPAUSE

J. G. Luhmann, R. J. Walker, C. T. Russell

Institute of Geophysics and Planetary Physics, University of California
Los Angeles, CA 90024

N. U. Crooker

Department of Atmospheric Sciences, University of California
Los Angeles, CA 90024

J. R. Spreiter

Department of Applied Mechanics, Stanford University
Stanford, CA 94305

S. S. Stahara

Neilsen Engineering and Research, Inc., Mountain View, CA 94043

Several years ago, Crooker [1979] presented a qualitative picture of the merging sites on the magnetopause defined as the points where the magnetospheric and magnetosheath fields are antiparallel. However, Cowley [1976] pointed out that merging can also occur where these fields are not exactly antiparallel, but merely have antiparallel components. Using realistic models of the magnetosphere and magnetosheath magnetic fields, the angles between the fields at the magnetopause boundary were determined for different interplanetary field orientations, including radial field and Parker spiral field. The results are summarized in Figure 1, which displays contours on the dayside magnetopause (viewed from the sun, i.e., the GSE Y-Z plane projection) of equal value of the cosine of the angle between the magnetospheric and magnetosheath model fields. Only contours with negative values, implying some antiparallel component, are shown. Values at the contours, starting with the contour filled with shading, are −.98, −.95, −.9, −.8, −.7, −.6, −.5, −.4, −.3, −.2, −.1, −.0. The interplanetary field orientations are indicated in vector notation in the lower right corners. In particular, the pattern for a 45° cone angle is shown in the fourth diagram in the right hand column and patterns for northward and southward fields occupy the first position in the first column and the second position in the right hand column. These results can be used for comparisons with observed distributions of flux transfer events and for studies of magnetospheric particle leakage. A full report can be found in Luhmann et al. [1984].

References

Cowley, S. W. H., Comments on the merging of nonantiparallel magnetic fields, *J. Geophys. Res. 81*, 3455, 1976.

Crooker, N. U., Dayside merging and cusp geometry, *J. Geophys. Res. 84*, 951, 1979.

Luhmann, J. G., R. J. Walker, C. T. Russell, N. U. Crooker, J. R. Spreiter and S. S. Stahara, Patterns of potential magnetic field merging sites on the dayside magnetopause, *J. Geophys. Res.*, in press, 1984.

Questions and Answers

Vasyliunas: The direction of the magnetic field just outside the magnetopause predicted by the Spreiter et al. model is independent of B_x, the radial component (or more precisely, the component parallel to the solar wind flow) of the interplanetary magnetic field. The reason is that in the model the magnetic field at any point within the magnetosheath is a linear function of the three interplanetary field components, and the coefficients of B_y and B_z grow without limit as the magnetopause is approached while the coefficient of B_x remains finite. Any effect of B_x on the field direction just outside the magnetopause can therefore result only if there are departures from the idealized Spreiter model; these may be entirely reasonable, but they should be explicitly identified and examined.

Luhmann: *The magnetic field in the gasdynamic model does become infinite on the stagnation streamline for the reason Dr. Vasyliunas states. However, we assume that this is a point of zero measure and that in nature other boundary effects such as the magnetic field's influence on the flow and viscosity make the present gasdynamic model invalid at the magnetopause in any case. The best we can do with the gasdynamic magnetosheath model is to predict the outer boundary conditions for the rotational discontinuities over the magnetopause. This is what we have done using the direction of the tangential (to the magnetopause) component of the magnetosheath field a few tenths of an earth radius outside of the stagnation streamline in the gasdynamic model.*

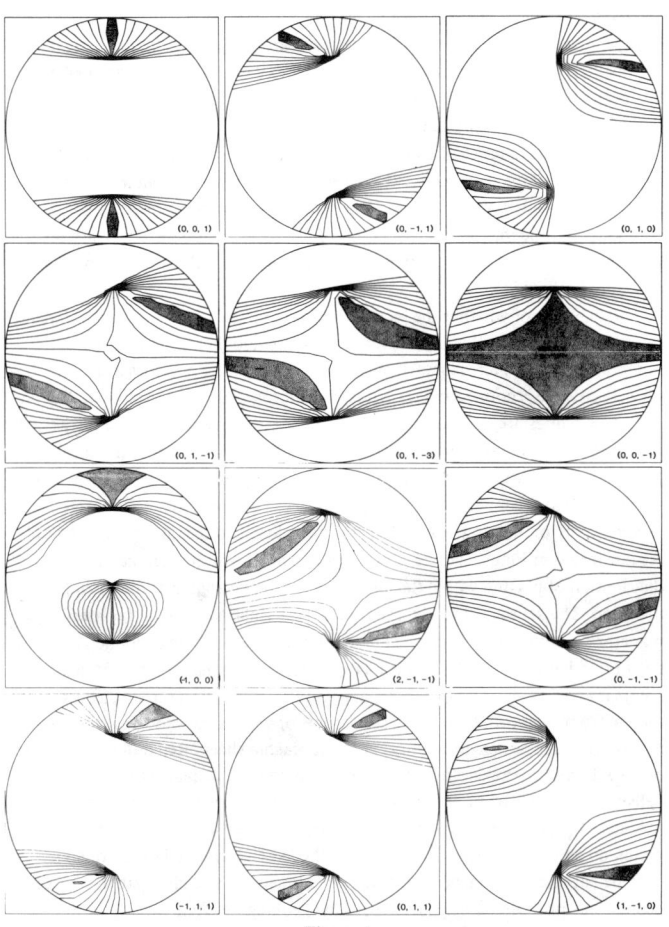

Figure 1

ISEE-3 PLASMA MEASUREMENTS IN THE LOBES OF THE DISTANT GEOMAGNETIC TAIL: INFERENCES CONCERNING RECONNECTION AT THE DAYSIDE MAGNETOPAUSE

J. T. Gosling

Los Alamos National Laboratory
Los Alamos, NM 87545

The lobes of the geomagnetic tail are the regions located north and south of the plasma sheet and extending to the magnetopause. Beyond ~10 R_E tailward of the earth they are identified as strong field regions in which the magnetic field points nearly parallel or anti-parallel to the sun-earth line. It is commonly believed that lobe field lines are "open," one end of a lobe field line being rooted in the ionosphere and the other end being connected to the interplanetary magnetic field. This picture necessarily assumes that reconnection at least occasionally occurs at the dayside magnetopause, lobe field lines being reconnected field lines which have been dragged back into the tail by the flow of the solar wind.

Inside about 20 R_E the lobe plasma is very tenuous everywhere except where it abuts the magnetopause, presumably being populated by particles of ionospheric origin at densities between ~10^{-3}–10^{-2} cm^{-3}. Adjacent to but inside the tail magnetopause a denser boundary layer of flowing plasma, commonly called the mantle, is often observed. The mantle is usually thought to be composed primarily of magnetosheath plasma which enters on the dayside at the polar clefts and is subsequently mirrored in the polar regions and directed down the tail. Ionospheric plasma is believed to be a minor constituent of the mantle. The

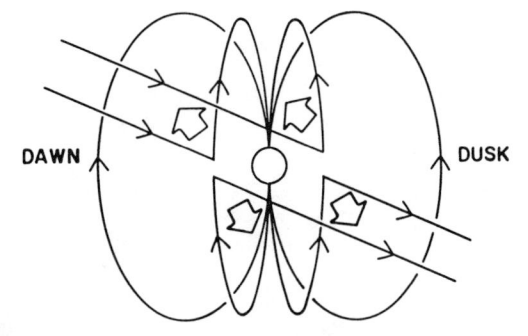

Figure 1. A schematic drawing of dayside reconnection as viewed from the sun when the interplanetary magnetic field has a positive y-component. The particular reconnected field lines drawn can be thought of as flux transfer events or as representative of reconnected field lines over the entire dayside magnetopause.

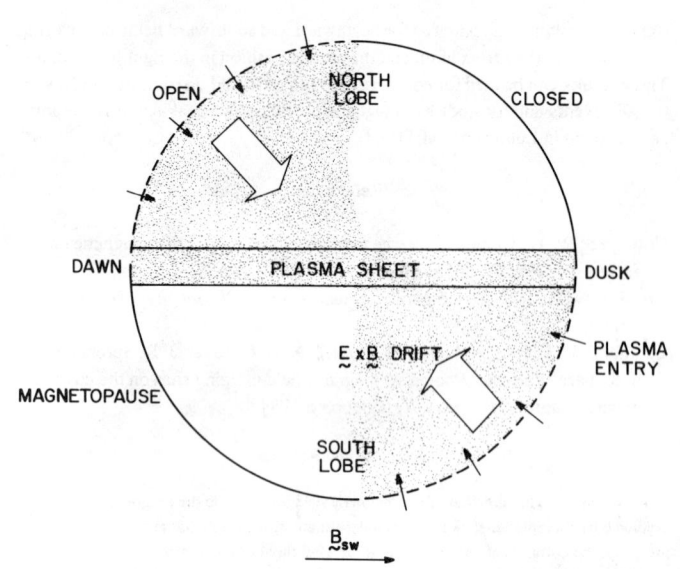

Figure 2. A schematic drawing of a cross section of the distant geomagnetic tail illustrating plasma entry into the tail for a positive y-component of the interplanetary magnetic field. The view is from the earth looking tailward and has been drawn assuming reconnection over the entire dayside magnetopause. The pattern is essentially the same, but patchier, if reconnection occurs preferentially in association with flux transfer events.

cross tail (dawn-to-dusk) electric field causes mantle plasma to drift toward the plasma sheet so that at lunar orbit (~ 60 R_E) mantle plasma is often found adjacent to or near the plasma sheet. If the cleft were the only region where magnetosheath plasma gains entry to lobe field lines, then far down the tail we would not expect to observe a dense lobe plasma adjacent to the magnetopause.

By way of contrast, the Los Alamos plasma analyzer on ISEE-3 often (but not always) observes a dense (~ 0.1-1 cm^{-3}), tailward flowing (~ 200 km s^{-1}) plasma on lobe field lines adjacent to the magnetopause. This boundary layer-like plasma has been observed over the entire range of ISEE3's orbital sampling of the distant tail (-235 $R_E <$ X_{SE} < 0; -23 R_E $<$ Z_{SE} $<$ $+23$ R_E). Thus it appears that plasma from the magnetosheath often enters the tail lobes along much of the length of the tail magnetopause. Where such entry occurs the tail magnetopause is probably locally "open," i.e. there exists a finite normal magnetic field component at the magnetopause. On the other hand, where the boundary layer-like plasma is absent, local entry is prevented and the local tail magnetopause is probably "closed."

The question of whether the local tail magnetopause is open or closed depends upon reconnection at the dayside magnetopause and the recent polarity of the y-component of the interplanetary magnetic field. Figure 1 provides a schematic view of dayside reconnection for a positive interplanetary y-component. Reconnected field lines are pulled preferentially toward the dawn side of the tail across the northern polar cap, and toward the dusk side of the tail across the southern polar cap. Thus when the y-component of the interplanetary field is positive, one expects to find the magnetopause adjacent to the northern lobe to be open preferentially on the dawn side of the tail and closed on the dusk side of the tail. As illustrated in Figure 2 there should result a north-south asymmetry in the distribution of plasma within the lobes which reverses sign near the midnight meridian. The sense of this asymmetry reverses when the sign of the interplanetary y-component changes. Direct evidence for such an asymmetry has been provided by the ISEE-3 plasma measurements in the deep tail. Measurements made during rapid traversals across the plasma sheet on the dawn side of the tail from the north lobe to the south lobe, or vice versa, show that the north and south tail lobe densities adjacent to the plasma sheet often differ by a factor of ~ 10. In all traversals studied to date the southern lobe plasma has the higher density when the earth is within a toward interplanetary magnetic sector (negative B_y), and the northern lobe has the higher density when the earth is within an away sector (positive B_y). Thus the ISEE-3 deep tail measurements provide substantial evidence for dayside reconnection and illustrate that reconnection has profound consequences for the geomagnetic tail.

RECONNECTION IN EARTH'S MAGNETOTAIL: AN OVERVIEW

A. Nishida

Institute of Space and Astronautical Science, Komaba, Meguro, Tokyo 153, Japan

Abstract. The reported evidences and counter-evidences of reconnection in Earth's magnetotail are critically reviewed. We conclude that convincing observational bases for the tail reconnection exist, but emphasize that large-scale reconnection in the near-earth tail is not the only driving agent of the plasma dynamics in the tail. Confusion was created when it was believed that the reconnection should explain every observable feature. Physical mechanism of the tail reconnection is not yet fully understood, however, and we discuss important issues which remain to be tackled by future observations.

Introduction

The idea that the magnetic field energy stored in the magnetotail is occasionally released and converted into kinetic energies through the reconnection process has been with us for more than a decade. Following early theoretical suggestions and pioneering analyses, in 1973 several papers were published which claimed that particle and field observations during substorms can be interpreted adequately by the reconnection model (see Nishida and Nagayama (1973) and Nishida et al. (1981) for references). However, it was also evident that the reconnection model cannot explain every one of the dynamical features observed in the magnetotail, and this led some to cast doubt on the validity of the idea.

This overview is intended to be a critical assessment of the observational basis of the reconnection model. First we shall examine whether or not the signatures which have been taken as the basic evidence of reconnection are liable to alternative interpretations. Second we shall pick up the observations which have been taken to represent serious difficulties of the reconnection model and discuss if these difficulties are indeed real. Third we shall enumerate several key questions that remain to be answered. Since theoretical aspects of reconnection are dealt with in other papers, this paper concentrates on the observational aspects of the subject.

The Reason Why We Believe That Reconnection Occurs in the Near-Earth Magnetotail

The basic observational evidence of reconnection in the magnetotail is that the southward polarity of the magnetic field and the tailward polarity of the plasma flow occur combined. This signature is expected to be observable when the spacecraft is located on the tailward side of the X-type neutral line where reconnection of field lines takes place (Figure 1a). However, reconnection is not the only conceivable mechanism which can give rise to this combination. Several other configurations of the field line and the plasma flow can be envisaged which also involve the southward field polarity and the tailward polarity of the plasma flow (Figure 1b through 1e). In order to be sure that reconnection indeed occurs, observations have to be compared in greater detail with each of these alternative interpretations. (Hereafter we shall sometimes omit the term "polarity" for brevity and use the terms "southward field" and "tailward flow" while in fact it is the polarity of the field or flow rather than their entire vector that is southward or tailward.)

In Figure 1b, the tailward flow is aligned to magnetic field lines which are tilted in the same way as the field lines in the tail lobe. If this is what is actually happening, $\mathbf{E} = -\mathbf{u} \times \mathbf{B}$ should be zero, or equivalently, the drift velocity $\mathbf{u}_d = \mathbf{E} \times \mathbf{B}/B^2$ should be zero. In Figure 2 we show the electric and magnetic field data for an event where the northward component B_z of the magnetic field (panel c) became negative for about 15 minutes (shaded).

Clearly the duskward component E_y of the electric field (panel b) increased when B_z turned negative, and the drift velocity was non-zero and its earthward component was negative, namely directed tailward (panel h) during this event which was observed near the 01 LT meridian at the distance of $19R_E$ from the earth. The model of Figure 1b is therefore not consistent with this and similar cases reported in Nishida et al. (1983) and Cattell and Mozer (1984).

We note in passing that in the present event the northward component of the drift velocity was positive at the spacecraft which was located on the southern side of the neutral sheet (see panel f). This flow direction is consistent with the progress of the plasma sheet thinning which has long been counted among the signatures of the near-earth reconnection (Hones et al., 1973). Recently Forbes et al. (1981) have measured the velocity of the thinning by two methods, namely by direct flow measurements and by comparison of the boundary crossing times at two spacecraft, and obtained the values of about 20 km/s.

In Figure 1c, the southward polarity of the magnetic field is caused by the wavy structure of the neutral sheet. A characteristic feature of this model is that the polarity of B_z changes across the neutral sheet, so that chances of observing the northward B_z should be at least as high as that of observing the southward B_z. In the top panel of Figure 3 we show the occurrence frequency of the inclination angle θ, which the magnetic field vector makes with respect to the solar magnetospheric equatorial plane, for the intervals where the plasma flows tailward with speeds greater than 300 km/s (Hayakawa et al., 1982). Clearly the observed polarity is southward much more often than it is northward. Hence the wavy neutral sheet model is not a valid alternative.

Note in passing in the bottom panel of Figure 4 that the magnetic polarity is predominantly northward when the flow is fast but earthward. Predominance of the northward polarity is even more pronounced when the earthward flow is slow (<300 km/s) (Hayakawa et al., 1982).

In Figure 1d, field lines have a spiral structure and southward B_z appears locally at certain phases of the spiral. This structure can be produced when strong currents flow along the magnetic field. Similar

160   RECONNECTION IN EARTH'S MAGNETOTAIL

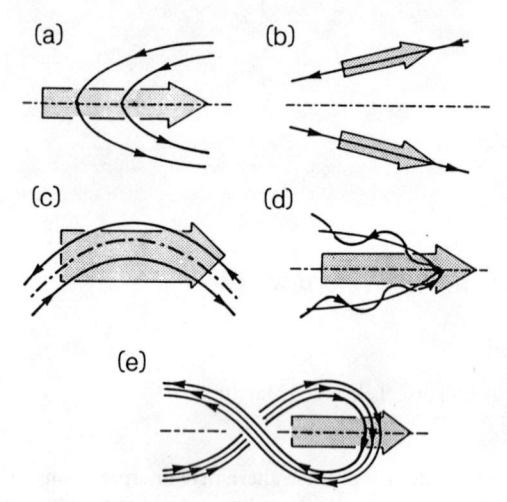

Fig. 1. A variety of field line configurations which locally have the southward polarity. The earth is on the left hand side of each diagram.

but two-dimensional undulations of field lines can also produce localized regions of southward B_z. Such undulations would accompany a wavy structure of the plasma-sheet boundary. In these models northward and southward B_z polarities appear alternatively along a given field line, so that their detection probabilities are expected to be equal. However, as we have already seen in Figure 3, southward B_z accompanies the fast tailward flow much more often than northward B_z does.

Figure 1e is a model of the twisted magnetic tubes of force. Because of the twist the polarity of the X-component (i.e. sunward component) of the magnetic field is negative on the northern side of the neutral sheet and positive on the southern side, namely opposite to the B_x polarity under normal conditions. In order to check this point observationally, it is necessary to know whether the neutral sheet moved northward or southward across the spacecraft position in individual instances. This check is possible in principle when the component of the plasma flow velocity perpendicular to the neutral sheet is measured, but to my knowledge such an analysis has not yet been reported. Hence the argument against the twisted tube model has to be conceptual at present; it is difficult to understand why the tip of the twisted tubes of force moves rapidly away from the earth. Since field lines are closed in this model, the Lorentz force is directed in an opposite direction, namely toward the earth. To drive the tubes of force rapidly tailward, pressure should be very high in the region encompassed by the twisted tube of force, or plasma from the magnetopause boundary layer should penetrate as deep into the tail as to the midnight region even at radial distances of less than $30R_E$. So far we have not seen indications that plasma pressure becomes very high in the near-earth tail region during disturbed times, and the penetration of the boundary layer plasma into such deep interior of the magnetotail has not been proven. Moreover, we have not seen signatures of the earthward contraction of the southward oriented field lines of the twisted tube. For these reasons we consider it unlikely that Figure 1e is a realistic alternative.

Thus we are left with the reconnected field configuration of Figure 1a, where the tailward flow can be understood as a natural consequence of the anti-earthward Lorentz force which results from the coupling of the southward B_z and the dawn-to-dusk current across the tail. Needless to say, we are not denying that some of the field configurations shown in Figure 1b through 1e may sometimes occur. Our point is that they cannot provide an alternative interpretation to the observations that have been taken as the basic evidence of reconnection.

Reconnection in the magnetotail is associated with thinning of the plasma sheet, burst-like increase of energetic electron flux, and substorm signatures observable on the ground. Figure 4 shows these signatures in a summary form. The figure is constructed by superposing data of the flow velocity, the plasma number density, northward component of the magnetic field, and the AE index of the geomagnetic activity (Bieber et al., 1982). The fiducial mark is the first observation of intense streaming of energetic (> 200 keV) electrons, and 16 events identified by Imp-8 are superposed. The observed sequence of events can be interpreted as follows on the basis of the reconnection model. First, the tailward plasma flow builds up from −15 to −10 min. This is

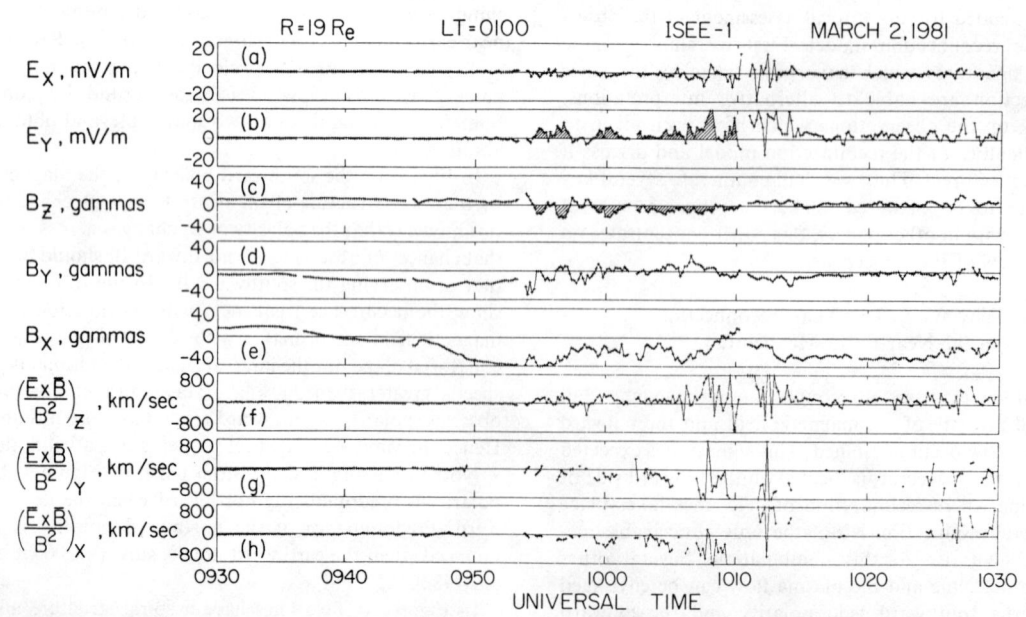

Fig. 2. Electric and magnetic field observations during a reconnection event. The components refer to solar ecliptic coordinates. E_z is derived from the assumption of $\mathbf{E} \perp \mathbf{B}$, only when $|B_x/B_z|$ are not too large (<5). (Nishida et al., 1983).

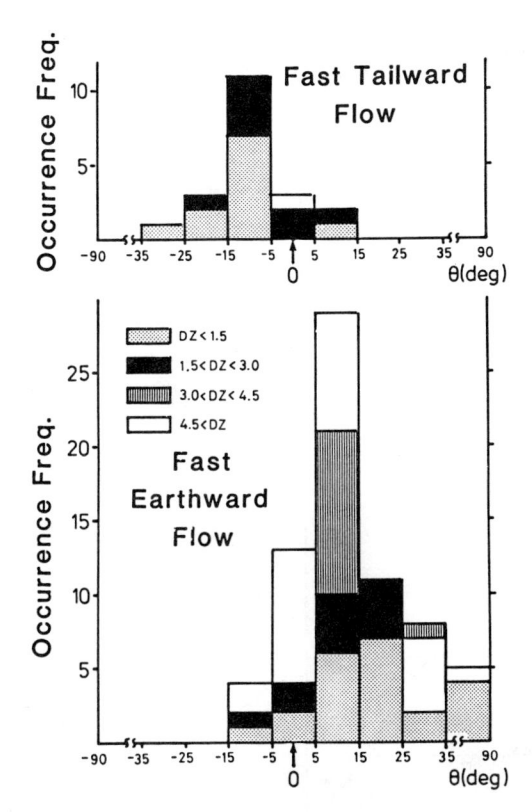

Fig. 3. Distribution of the inclination angle of the magnetic field with respect to the solar magnetospheric equatorial plane during the intervals when the X_{SM} component of the speed exceeded 300 km/s. Top: for tailward flow events, and bottom: for earthward flow events (Hayakawa et al., 1982).

unobtainable because of low density of the plasma, but still it has to be said that there was no indication of the tailward flow just before the southward turning of B_z. Instead, strong earthward flows were observed at such times. These observations were presented by Lui (1980) as counter-evidences of the reconnection model.

As is well known, however, reconnection is not the only cause of the southward B_z polarity in the magnetotail. In the high latitude lobes of the tail the magnetic field is always inclined southward. When substorms occur, B_z in the lobe tends to become more negative first, and then at the onset of the expansion phase it starts to recover while $|B_x|$ begins to decrease. In contrast, in low latitudes around the neutral sheet the entry of B_z into the southward range occurs at the expansion phase onset. In low-latitudes B_x either varies irregularly or its absolute magnitude increases around the onset (Nishida and Nagayama, 1973). When we compare the B_x–B_z relation of Figure 5 with this result of a synoptic study, we note that the last event around 00 UT had the character of an event detected in the lobe. The first two events were recorded in the boundary region between the lobe and the plasma sheet as judged from the magnitude of the distance dZ_{SM}. Strong earthward flows have been detected frequently at the high-latitude boundary region of the plasma sheet during disturbed times (e.g., DeCoster and

because the tailward flow from the reconnection region drives the plasma on the downstream side. Second, B_z turns southward as the reconnected field lines reach the observing site. In some instances electron temperature increases by a factor of 2 to 5 during this period. This can be interpreted to be a consequence of the energy dissipation in the reconnection region. Third, at 0 min plasma density drops and streaming of energetic electrons is detected. This probably means that the spacecraft exited from the thinned plasma sheet into the region of open field lines, where energetic electrons escape freely by streaming along field lines. The AE index increases sharply in approximate coincidence with the start of the tailward flow. This suggests that part of the magnetic field energy released from the magnetotail is deposited in the polar ionosphere.

Probable Reasons Why Signatures of Reconnection Were Not Evident in Some Earlier Analyses

While we have seen, on one hand, a number of instances where observations in particles and fields agree nicely with the reconnection model, in the literature many other events have also been reported which are apparently inconsistent with expectations by this model. For example, Lui (1980) reported some cases where B_z was southward but the flow direction was earthward, and Frank et al. (1976) noted that the B_z often remained northward when the flow direction was tailward. In this section we shall discuss the nature and the implications of the discordant events.

Figure 5, taken from Lui (1980), shows at least three intervals where B_z was southward. During these intervals the flow velocity was mostly

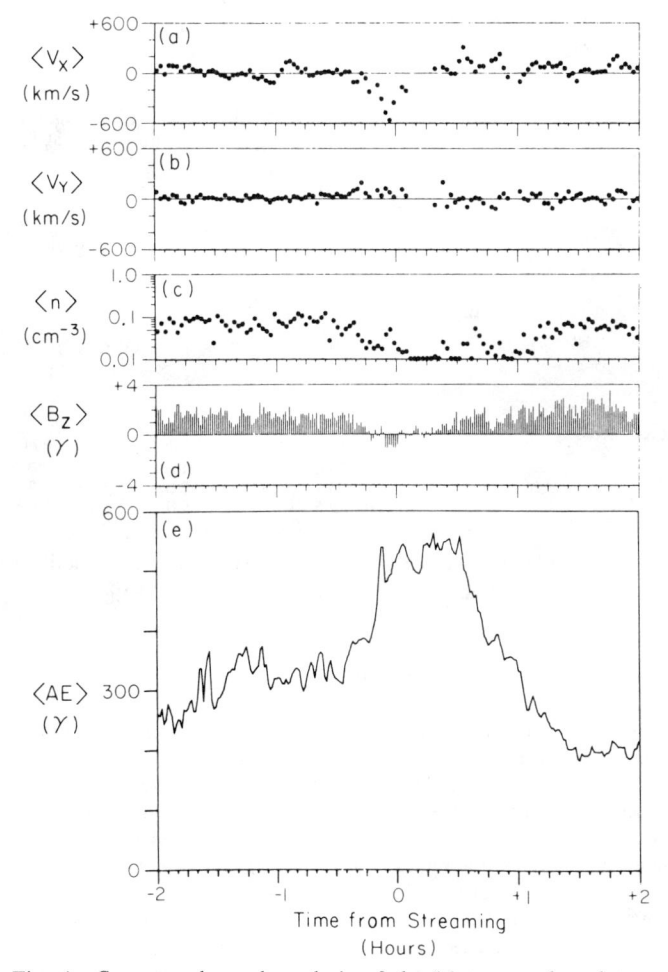

Fig. 4. Superposed epoch analysis of the 16 events where intense tailward streaming of >200 keV electrons were observed in the tail at radial distances of about 40 R_E. (a) and (b): solar ecliptic components of the flow velocity, (c): plasma density, (d): solar magnetospheric northward component of the magnetic field, and (e): AE index (Bieber et al., 1982).

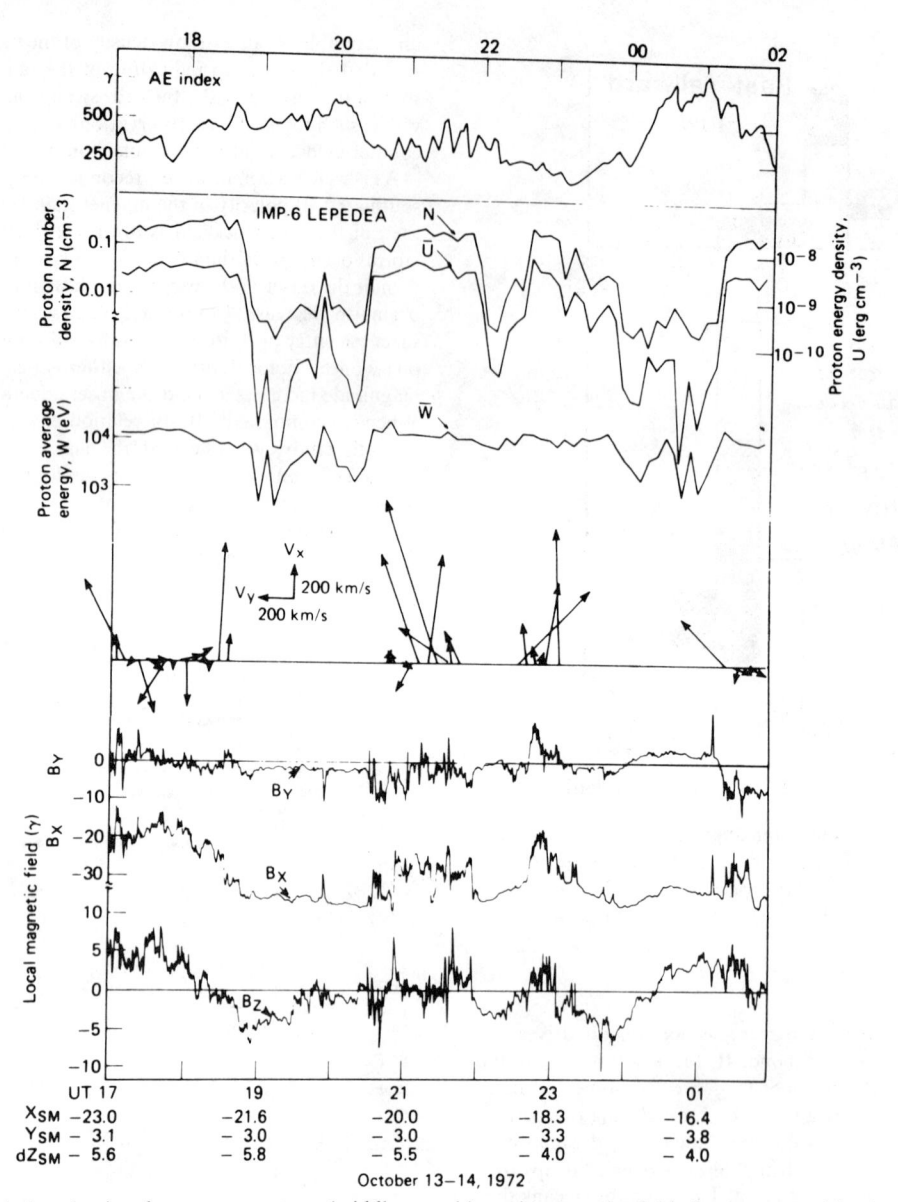

Fig. 5. AE index (top), plasma parameters (middle panels) and magnetic field (bottom panels) observed in the magnetotail during an active period (Lui, 1980).

Frank (1979)) and the earthward flows noted in Figure 5 seem to belong to this category.

In order to avoid the confusion that can arise if observations in the lobe or the plasma-sheet boundary region are mixed with observations in the neighbourhood of the neutral sheet, we have selected from Imp 6 observations the intervals of the southward B_z by the following criterion. Among the 15 successive 15.36 s averages of the solar magnetospheric latitudinal angle θ of the magnetic field, (1) at least 10 satisfied $\theta < 0°$, (2) at least 5 satisfied $\theta < -10°$, and (3) the difference between maximum and minimum values of θ was greater than $5°$. The last condition is imposed to discriminate against the observations in the tail lobe where changes in θ are relatively slow. For each event selected by these criteria, we have recorded the position dZ_{SM} of the observation from the Russell-Brody neutral sheet model and the direction of the associated plasma flow, and the result is presented in Figure 6. We note first that majority of the cases were obtained at $|dZ_{SM}| < 3\,R_E$. This confirms that these measurements were made indeed in the neigh-

bourhood of the neutral sheet. We can also see that for these cases the flow direction was predominantly tailward (Nishida et al., 1981). Thus if proper care is taken to concentrate on the observations in the neighbourhood of the neutral sheet, the association of the southward B_z and the tailward flow can be recognized clearly.

The above conclusion, however, refers to the southward B_z which lasts longer than certain length of time which is about 4 min in the above analysis. Inspection of the spacecraft data reveals numerous other occasions where B_z turns southward but recovers to northward within about one minute. In such cases the flow is often weak and does not show preference for the tailward direction. Before discussing them, we would like to examine the cases where the B_z polarity remains northward in spite of the tailward polarity of the associated flow.

Figure 7 shows the occurrence frequency of B_z (15.36 sec averages) observed during the intervals of the tailward plasma flow. The cases where the tailward component of the flow speed was faster than certain threshold values (200 km/s or 300 km/s) are hatched. It is seen that

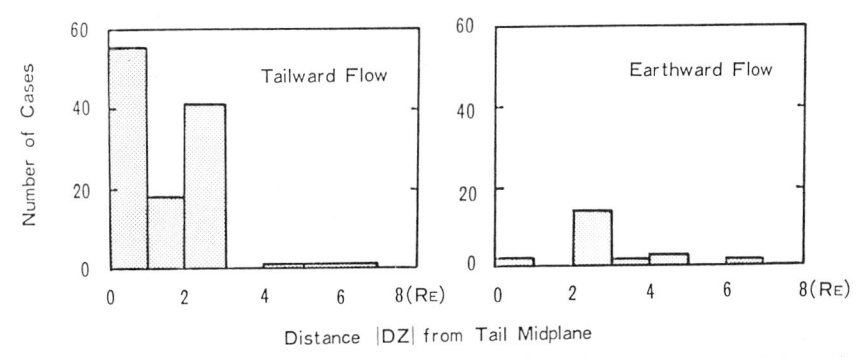

Fig. 6. Occurrence frequency of tailward (left) and earthward (right) flow components versus distance from the neutral sheet during intervals of the southward field polarity. The selection criteria of these intervals are given in the text (Nishida et al., 1981).

when the tailward flow velocity exceeded 300 km/s, the occurrence frequency of the southward B_z was substantially higher than that of the northward B_z; this is exactly what we saw in Figure 3. However, the dominance of the southward B_z becomes much less marked when the threshold is reduced to 200 km/s, and furthermore the probability of B_z being northward or southward becomes almost equal to each other when all cases where $|V_x|$ was greater than 100 km/s are taken. This means that B_z is much more likely to be northward when the flow is tailward but slow, namely less than about 300 km/s (Hayakawa et al., 1982). As an example of such instances, we reproduce one of Frank et al. (1976)'s figures in Figure 8. Although the flow direction was often tailward its speed was slower than about 300 km/s in most cases (bottom panel), and the latitudinal angle of the magnetic field stayed northward except for a few short intervals.

Thus the relation of the tailward plasma flow to the polarity of B_z can be divided statistically into (at least) two types by using the flow speed as a separation parameter. It would then be natural to consider that more than one process cause the tailward flow in the magnetotail. Reconnection in the near-earth tail region is just one of these processes, and the occurrence of the events which cannot be explained by the reconnection model simply means that such events were not due to the reconnection; it has no unique implication regarding to the validity of the reconnection model.

At this point some caution is warranted concerning the flow speed that is used throughout this paper. The concept of the flow speed is valid only when the velocity distribution is reasonably symmetric around a peak. Recently Paschmann et al. (1983) have looked at the velocity distribution of the "slow tailward flow" events and found that the peak of the distribution was not substantially different from zero but the non-zero average speed was produced due to asymmetry in the high-energy tail of the distribution function. This observation is extremely important for identifying the nature of the "slow flow" and we look forward to seeing more cases analyzed. They also found that "fast tailward flow" indeed represented the shift of the peak of the distribution from zero.

The presence in Figure 7 of some exceptional cases where the tailward flow speed exceeded 300 km/s but the field polarity was northward may probably reflect the fact that this threshold speed is a statistically derived value and not a sharply defined parameter. However, occasional associations of the strong tailward flow with the northward B_z can be more intimately related to reconnection process. In fact, the occurrence of this association is one of the essential check points of the reconnection model. As illustrated in Figure 9, the plasma driven tailward from the neutral line is associated with the southward B_z in the neighbourhood of the neutral line. This plasma pushes and drives the plasma on the downstream side by transmitting the compressional wave, so that the tailward flow is expected to be produced in the region of the northward B_z (e.g., Hones, 1980). This expectation is entirely consistent with observations presented in Figure 2, where prior to the enhancements of the dawnward E_y (second panel) and the southward B_z around 0954, there occurred a few minutes' interval of the duskward E_y and northward B_z. Another possible cause of the association of the tailward flow with the northward B_z is formation of magnetic loops by instabilities which may be facilitated by the presence of the tailward flow. Mutually correlated oscillations of E_y and B_z in Figure 2 (hatched) demonstrate that magnetic loops were indeed being transported tailward. The loop structure is also illustrated schematically in Figure 9.

We have mentioned earlier that the inspection of the magnetic field data obtained in the neutral sheet region of tail often reveals short southward excursions lasting less than a minute. These short excursions have not been included in the correlational analysis between polarities of magnetic field and flow direction which we have presented so far. Not much has been found about the nature of these events

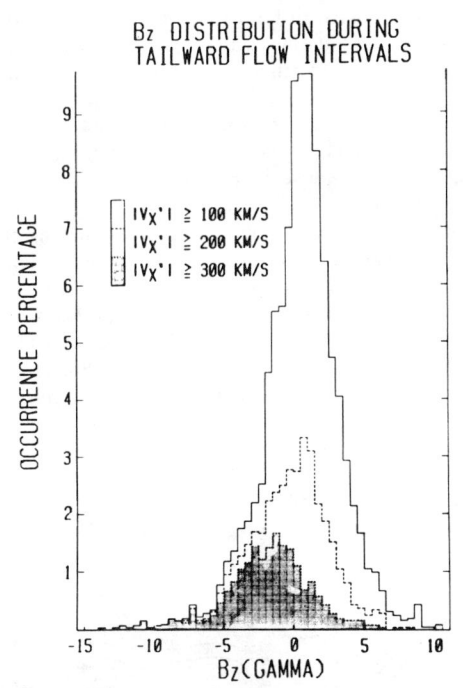

Fig. 7. Distribution of the northward component B_z of the magnetic field during the intervals of the tailward flow. Data are categorized by the flow speed (Hayakawa et al., 1982).

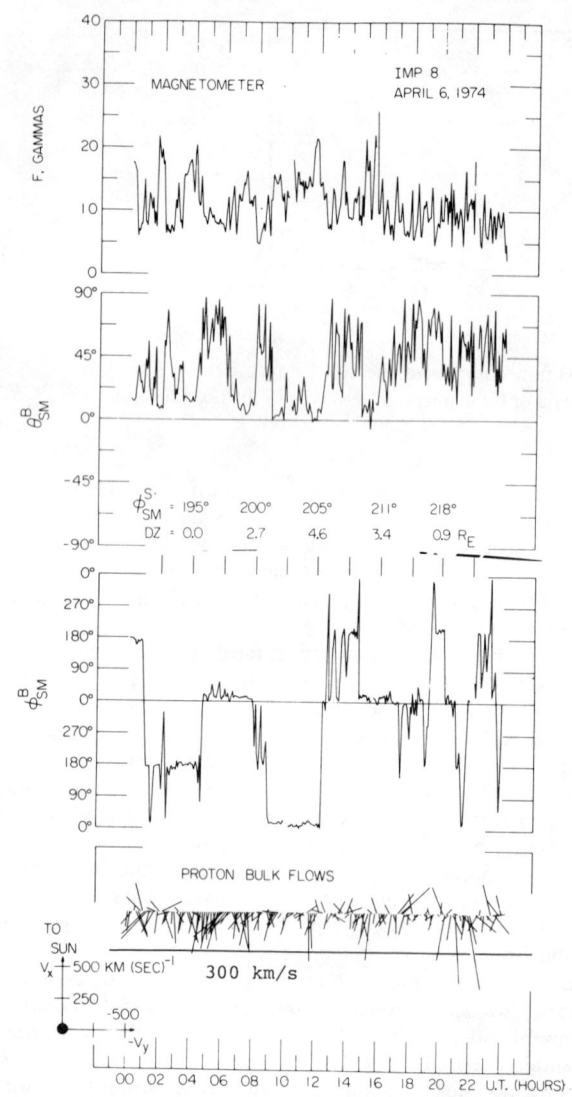

Fig. 8. Simultaneous observations of magnetic field and proton bulk flows within the plasma sheet. A line has been drawn in the bottom panel to indicate the 300 km/s level (Frank et al., 1976).

because until quite recently the derivation of one flow velocity vector took more than 80 s which is often longer than the duration of these southward excursions. However, when we studied the structure of the tail when the flow speed is very low—less than 200 km/s for extended periods, we obtained some information that pertains to them. In this analysis the data set was not divided by direction of the flow.

We have found that among the neutral sheet crossings that occurred during such periods, the magnetic field showed the southward polarity in about 20% of the cases. The distribution of occurrence probabilities of the southward polarity in 5-R_E bins with respect to the solar magnetospheric x and y coordinates are shown in Figure 10. Southward polarity in the neutral sheet is observed beyond $x \sim -15 R_E$ and is more frequent in the midnight region than at earlier local times. (This observation was limited to the positive range of y.) Since the number of the neutral sheet crossings by Imp 6 at $|x| \leq 15 R_E$ was very small, the statistics of Figure 10 is not strong enough to allow one to state that the B_z polarity was always northward inside 15 R_E. However, the increase in the probability of the southward B_z with an increasing distance from the earth seems consistent with the expectation that it would become easier to reverse the B_z polarity at greater distances where the control of the intrinsic dipole field is weaker. Similarly, the decrease in this probability with increase in y can probably be attributed to the fact that statistically B_z is larger at larger $|y|$ (Fairfield, 1979) and hence more difficult to reverse. There is no unique preference in the direction of the weak flow during the 80 s intervals that encompass the neutral sheet crossings where B_z is southward. The southward polarity lasted less than 1 min in about 80% of the above cases. This suggests that either the southward polarity is confined to a narrow region around the neutral sheet or its life time is short, or both.

Thus, in addition to large-scale reconnection events which last for several minutes or more and produce tailward flows faster than about 300 km/s, there seem to be minor events which have much smaller spatial/temporal scale and whose effect on the flow speed is not recognizable with the resolution of the available Imp 6 data. It is tempting to identify the "minor" reconnection with the tearing mode instability at the neutral sheet. While the large-scale reconnection has been known to be closely associated with the expansion phase of substorms (e.g., Nishida and Hones, 1974), the minor reconnection shows no obvious relation to the geomagnetic activity. This can be seen by comparing the top and the bottom panels of Figure 10 and recognizing that most of these cases occurred during intervals of low AE ($\lesssim 100$ nT). Lui (1980) has once pointed out that AE often stays low when the field polarity at the neutral sheet is southward. His data set seems to involve cases of the minor reconnections as defined above.

We have seen in Figure 10 that the occurrence probability of the southward B_z decreases as one goes toward the tail magnetopause and suggested that this may be related to the higher mean value of B_z at shorter distances from the tail surface. Similar tendency has been noted for the occurrence frequency of the southward B_z and the tailward flow ascribable to the large-scale reconnection. Observations of these signatures are concentrated to the inner tail of $|y| \lesssim 10 R_E$ (Hones, 1980; Nishida and Nagayama, 1973). Hence it is not surprising that neither ISEE-1 nor Imp-8 detected the signatures of reconnection at $|y| \gtrsim 10 R_E$ during a geomagnetically disturbed period of 19 May, 1978 (Huang et al., 1983). For the same reason, Kirsch et al. (1981)'s observation at $y_{SM} = 18 R_E$, that (1) bursts of energetic particles above 0.2 MeV were associated with both positive and negative B_z, and (2) negative B_z values were not always associated with observable particle bursts, might be relevant but cannot readily be compared with reported signatures of reconnection. Frank et al. (1976) once proposed that reconnection takes place in the close neighbourhood of the magnetopause in the region which appeared to be identifiable as the boundary layer, but this view does not seem to have been substantiated by later studies (Lui et al., 1984).

Observations of energetic protons and electrons having energies of 50 keV to several MeV in the magnetotail have been reviewed extensively by Krimigis and Sarris (1980). Their review emphasized the complexity of the behaviors of these energetic particles. Sometimes 50 keV and 500 keV protons counterstream, and significant difference can exist between flux variations and/or anisotropy directions observed at spacecraft which were only about 1 R_E apart. The impact of these observations on the near-earth reconnection model is difficult to assess exactly, since in most instances the plasma and field parameters which signify the progress of reconnection have not been examined. Because the duration of each episode of the near-earth reconnection seems to be several minutes, it is possible that the data collected randomly during

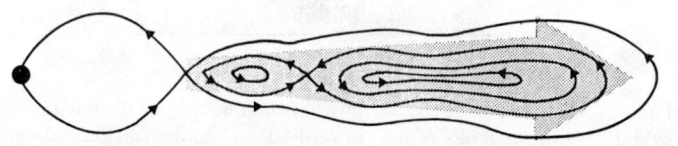

Fig. 9. Schematic illustration of the field lines and flow direction when reconnection is in progress.

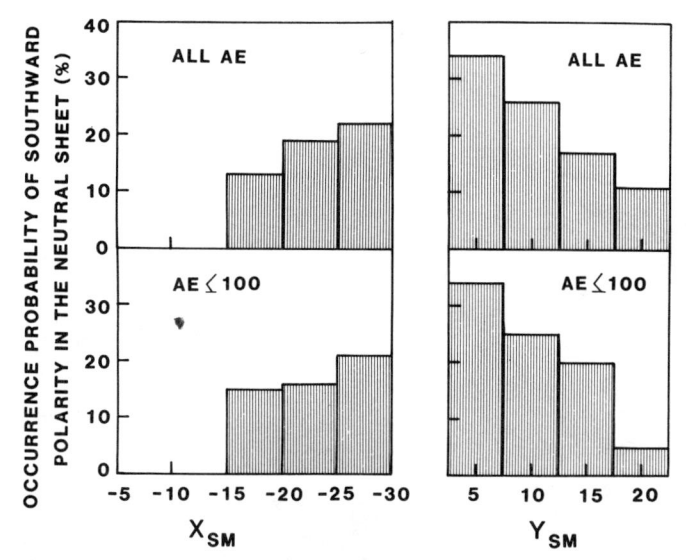

Fig. 10. Occurrence probability of the southward polarity in the neutral sheet in the quiet magnetotail with respect to x and y coordinates. Top: all cases, and bottom: cases observed when AE was less than 100 nT.

high magnetic activity are often unrelated directly to the near-earth reconnection. The need to separate a brief burst that occurs at the expansion phase onset from a more prolonged rapid-rise slow-decay type enhancement at the substorm recovery has indeed been widely recognized (e.g., Belian et al., 1981).

However, a detailed analysis of energetic particles conducted at the CDAW 6 (Fritz and Baker, 1983) indeed seems to reveal that the behavior of energetic particles is more complex than that of the low-energy plasma. A readily conceivable reason for the complexity is the high mobility of energetic particles. They have high speeds, long free paths, and ability to drift swiftly across field lines. The high mobility allows these particles, depending upon the occasion, to flow away along the open field lines, to be reflected from the mirror point or from the closed end of a magnetic loop, and to drift onto field lines on which reconnection has not taken place. As compared to low-energy plasma, energetic particles reflect the structure and the process over a more extensive domain of the magnetotail. Another reason for the complexity is the presence of energetic particles streaming earthward in the high-latitude boundary region of the plasma sheet and subsequently reflected at the mirror point (Williams, 1981). In addition, the possibility that energetic trapped particles in the inner magnetosphere somehow can find their way to the magnetotail without significant reduction in their energy has not been eliminated. It is clear that the near-earth reconnection is not the only possible cause of the intensification and streaming of energetic particles in the magnetotail, and it is important to differentiate the features belonging to different causes.

Basic Information Which Remains to Be Found or Expanded

(i) *Precursor and Triggering of the Near-Earth Reconnection*

Large-scale reconnection in the near-earth magnetotail occurs when the magnetosphere is in the disturbed state. The X-type neutral line where magnetic field lines are reconnected is produced transiently when part of the cross-tail current is diverted from the near-earth tail region to the ionosphere. This diversion of the current reduces the magnetic field energy contained in the magnetotail and allows it to be converted to kinetic energies of the resident particles through the reconnection process and also through the formation of the field-aligned electric field.

A fundamental question which we have to answer is, what causes the onset of the current diversion and reconnection? One naturally expects that the onset is preceded by gradual variations in parameters characterizing the structure of the near-earth magnetotail and that it is triggered when some of them cross a threshold value. Observations at the geosynchronous altitude on the nightside have demonstrated that magnetic field lines are stretched further for about one hour preceding the onset of the expansion phase of substorm (e.g., Baker et al. 1982; Nagai, 1982). This means that the cross-tail current is intensified above the geosynchronous altitude, and its disruption by instability can be suggested as a possible cause of the current diversion and reconnection. Unfortunately this idea cannot yet be tested experimentally because the low-latitude magnetotail between the geosynchronous altitude and about $20R_E$ has been covered only infrequently by observations. Most of the spacecraft which explored the magnetotail in the past had too high inclinations to spend enough time in this important region and sample transient disturbance phenomena. It is very important to fill this gap.

Figure 2, which shows observations made at the radial distance of $19R_E$, does not reveal any precursor feature that can be readily identified prior to the onset of the tailward plasma drift. It is striking that the electric field is absolutely quiet until then. On the other hand, Fairfield et al. (1981) found indications that the plasma β was sometimes reduced prior to the onset of the expansion phase of substorms. This decrease in β was due to the field increase and temperature decrease. Their observations may suggest that the plasma sheet thinning precedes the onset of reconnection even at distances beyond $15R_E$. The thinning can escape attention of the available electric-field or flow measurements if the associated flow is field-aligned or slow as compared to the background velocity level.

The frequent presence of magnetic loops which we have called products of "minor" reconnection may play an important role in the generation of the "large-scale" reconnection. The presence of these loops suggests that reconnection of field lines across the neutral sheet can occur sporadically and spontaneously. In normal, quiet times the growth of these loops is apparently restrained, but when certain conditions are fulfilled the loops can grow to the scale of the magnetotail and the large-scale reconnection happens. If this picture is correct, our task is observationally to identify the condition which governs the development of the tearing mode instability which is probably responsible for the formation of the loops.

In the above discussion we have assumed that the reconnection is initiated or accelerated by a process which takes place in the near-earth tail. But a completely different model may be suggested in which the current diversion originates from a process taking place in low altitudes, namely in the ionosphere or in the field-aligned acceleration region. The idea may be that a demand arises for a stronger flow of current along field lines and the diversion of the cross-tail current is caused as a result. An observational check of this idea is to measure the time difference between the disturbances in low altitudes and in the tail so as to find out where the disturbance occurred first. Such an analysis should be performed when observations in the near-earth tail region become available. However, it should be noted that it is not obvious that the cross-tail current really responds to the demand as supposed here. The current across the tail is intimately related to the spatial distribution of the plasma pressure, so that the diversion of the current requires redistribution of plasma. Whether or not this can be realized remains to be examined theoretically.

(ii) *Local Time Extent of the Reconnection Region*

As we have mentioned earlier the signature of the (large-scale) reconnection has been detected in the central part of the tail at $|y| \lesssim 10R_E$. This implies that the longitudinal extent of the reconnection region can be as long as about $20R_E$. Unfortunately there have

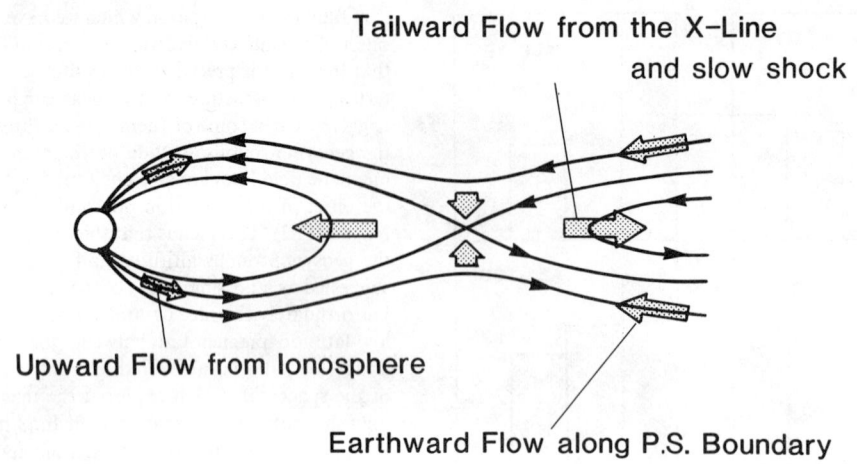

Fig. 11. A suggested flow pattern in active periods.

been few opportunities to estimate this extent by multiple spacecraft observations of reconnection signatures for individual events.

An indirect way for estimating the longitudinal extent of the reconnection region is to use as an intermediary the substorm observations on the ground. When signatures of the reconnection are observed in the tail, signatures of the substorm expansion phase are observed on the ground (e.g., Hones, 1980; Nishida et al., 1981). Hones and Schindler (1979) studied this relation from an opposite direction by searching for signatures of reconnection in the tail when substorms were observed on the ground. If the reconnection region indeed covers $y = -10R_E$ to $10R_E$, reconnection signatures should be observable whenever spacecraft is located within $|y| \lesssim 10R_E$ at the substorm expansion phase. Their analysis revealed, however, that the signatures of reconnection were detectable only in about 50% of such cases. Roughly speaking this means that the longitudinal extent of the reconnection region is about one half of $20R_E$. The reconnection region is not necessarily a single entity and may be divided into multiple longitudinal segments of smaller extents.

It is very important to prove this conjecture by simultaneous observations by multiple spacecraft. A basic assumption which underlies the above estimate is that an expansion phase of a substorm is associated necessarily with the reconnection process that operates in some local-time sector in the near-earth tail. This assumption cannot be proved until multiple spacecraft survey is accomplished. With the limited material available at present we cannot completely deny the possibility that there are substorms where the current diversion is not extensive enough to create the X-type neutral line in the near-earth tail. Substorms with and without the near-earth reconnection are expected to show a characteristic difference in the energy spectrum of the associated energetic particle phenomena, because suprathermal particles produced by reconnection would be missing in substorms of the latter category, if indeed such cases exist.

(iii) Particle Acceleration at the X-type Neutral Line

The intensity and spectrum of the accelerated particles are determined by two factors. The first is the strength of the electric field and the second is the mobility of particles in the direction of the electric force.

At the onset of reconnection a significant transformation occurs in the field line configuration. This reconfiguration would accompany an intensive inductive electric field. *In situ* measurements of $\partial \mathbf{B}/\partial t$ and the electric field at the X-type neutral line and its close vicinity are of prime importance for evaluating the acceleration effect of the reconnection. It is equally important to be able to measure the features that govern the mobility of particles. The energy which an individual particle obtains by reconnection depends upon the time it spends in the region of the rapid magnetic field variation as well as the distance it travels in the direction of the electric force. These resident times and travel distances can be influenced by turbulence.

Turbulence around the X-type neutral line is often attributed to a lower hybrid drift instability, but observations have not revealed intense LHR waves in the neutral sheet. In fact, the particle mobility may be governed essentially by the inertia alone. When Sato et al. (1982) numerically calculated the particle acceleration by tracing motions of test particles near the X-type neutral line they disregarded scattering. On the other hand, Terasawa (private communication) has recently pointed out that the Hall effect can play a vitally important role in the energy conversion process, when the normal component of the magnetic field across the neutral sheet is small but non-zero prior to the onset of reconnection as it actually is in the magnetotail.

It should be of great importance to compare *in situ* observations of intensity, spectrum and anisotropy of the accelerated particles with theoretical estimates based upon various models and identify the microscopic energy conversion mechanism which dominates in the reconnection region.

(iv) Mechanism of the Plasma Supply

Operation of reconnection in the near-earth tail requires supply of plasma. The plasmas which flow tailward or earthward away from the X-type neutral line have to be supplied from high-latitude sides.

In the earliest stage of the reconnection process the plasma residing in the high-latitude regions of the plasma sheet clearly serves as the source. If particles that originally occupy a box which is $20R_E$ long in the earth-antiearth direction and $10R_E$ wide perpendicular to the neutral sheet leak out of a slit of $4R_E$ wide with a speed of 400 km/s, it takes 800 s before all the particles are lost. Durations of fast tailward flows are less than this in great majority of the cases, but very infrequently there are occasions when the fast flow lasts longer than this. In these cases the source of plasma should be found elsewhere too. Tenuous and cold plasma from the tail lobe may be suggested as a candidate, but potentially a more affluent source would be the earthward streaming plasma which is often found in the high-latitude boundary region of the plasma sheet. Since fast tailward flows are observed only at $|dZ_{SM}| < 3R_E$ (see Figure 3), it can be conjectured that the earthward plasma flow is present outside this range even when the tailward flow is present in the neighbourhood of the neutral sheet. This source mechanism, illustrated in Figure 11, would be particularly effective during the disturbed times where a new expansion phase begins during the recovery phase of an earlier substorm. In this model the near-earth reconnection reaccelerates plasma which has been accelerated at another neutral line that has moved to a distant tail.

Another possible source of plasma is the upward flow of the ionospheric plasma. Flows of heavy ions from the ionosphere have now been well documented, and their contribution to the particle budget of the plasma sheet may sometimes become significant. It should be interesting to look at the ionic composition of the tailward flowing plasma.

References

Baker, D. N., E. W. Hones, Jr., R. D. Belian, P. R. Higbie, R. P. Lepping, and P. Stauning, Multi-spacecraft and correlated riometer study of magnetospheric substorm phenomena, *J. Geophys. Res., 87,* 6121, 1982.

Belian, R. D., D. N. Baker, E. W. Hones, Jr., P. R. Higbie, S. J. Bame, and J. R. Asbridge, Timing of energetic proton enhancements relative to magnetospheric substorm activity and its implication for substorm studies, *J. Geophys. Res., 86,* 1415, 1981.

Bieber, J. W., E. C. Stone, E. W. Hones, Jr., D. N. Baker, and S. J. Bame, Plasma behavior during energetic electron streaming events: Further evidence for substorm-associated magnetic reconnection, *Geophys. Res. Lett., 9,* 664 1982.

Cattell, C. A., and F. S. Mozer, Substorm electric fields in the earth's magnetotail, 1984.

DeCoster, R. J., and L. A. Frank, Observations pertaining to the dynamics of the plasma sheet, *J. Geophys. Res., 84,* 5099, 1979.

Fairfield, D. H., On the average configuration of the geomagnetic tail, *J. Geophys. Res., 84,* 1950, 1979.

Fairfield, D. H., R. P. Lepping, E. W. Hones, Jr., S. J. Bame, and J. R. Asbridge, Simultaneous measurements of magnetotail dynamics by Imp spacecraft, *J. Geophys. Res., 86,* 1396, 1981.

Forbes, T. G., E. W. Hones, Jr., S. J. Bame, J. R. Asbridge, G. Paschmann, N. Sckopke, and C. T. Russell, Substorm-related plasma sheet motions as determined from differential timing of plasma changes at the Isee satellites, *J. Geophys. Res., 86,* 3459, 1981.

Frank, L. A., K. L. Ackerson, and R. P. Lepping, On hot tenuous plasmas, fireballs, and boundary layers in the earth's magnetotail, *J. Geophys. Res., 81,* 5859, 1976.

Fritz, T. A., and D. N. Baker, Energetic particle observations during the CDAW 6 substorm event of March 22, 1979, *IAGA Bulletin No. 48,* A 13, 1983.

Hayakawa, H., A. Nishida, E. W. Hones, Jr., and S. J. Bame, Statistical characteristics of plasma flows in the magnetotail, *J. Geophys. Res., 87,* 277, 1982.

Hones, E. W., J. R. Asbridge, S. J. Bame, and S. Singer, Substorm variations of the magnetotail plasma sheet from $X_{SM} \approx -6R_E$ to $X_{SM} \approx -60R_E$, *J. Geophys. Res., 78,* 109, 1973.

Hones, E. W., Jr., and K. Shindler, Magnetotail plasma flow during substorms: A survey with Imp 6 and Imp 8 satellites, *J. Geophys. Res., 84,* 7155, 1979.

Hones, E. W., Jr., Plasma flow in the magnetotail and its implications for substorm theories, in S.-I. Akasofu (ed.), *Dynamics of the Magnetosphere,* 545, 1980.

Huang, C. Y., T. E. Eastman, L. A. Frank, and D. J. Williams, Periodic substorm activity in the geomagnetic tail, 1983.

Kirsch, E., S. M. Krimigis, E. T. Sarris, and R. P. Leppig, Detailed study on acceleration and propagation of energetic protons and electrons in the magnetotail during substorm activity, *J. Geophys. Res., 86,* 6727, 1981.

Krimigis, S. M., and E. T. Sarris, Energetic particle bursts in the earth's magnetotail, in S.-I. Akasofu (ed.), *Dynamics of the Magnetosphere,* 599, 1980.

Lui, A. T. Y., Observations on plasma sheet dynamics during magnetospheric substorms, in S.-I. Akasofu (ed.), *Dynamics of the Magnetosphere,* 563, D. Reidel Pub. Co., 1980.

Lui, A. T. Y., D. J. Williams, T. E. Eastman, L. A. Frank, and S.-I. Akasofu, Streaming reversal of energetic particles in the magnetotail during a substorm, *J. Geophys. Res., 89,* in press, 1984.

Nagai, T., Observed magnetic substorm signatures at synchronous altitude, *J. Geophys. Res., 87,* 4405, 1982.

Nishida, A., and N. Nagayama, Synoptic survey for the neutral line in the magnetotail during the substorm expansion phase, *J. Geophys. Res., 78,* 3782, 1973.

Nishida, A., and E. W. Hones, Jr., Association of plasma sheet thinning with neutral line formation in the magnetotail, *J. Geophys. Res., 79,* 535, 1974.

Nishida, A., H. Hayakawa, and E. W. Hones, Jr., Observed signatures of reconnection in the magnetotail, *J. Geophys. Res., 86,* 1422, 1981.

Nishida, A., Y. K. Tulunay, F. S. Mozer, C. A. Cattell, E. W. Hones, Jr., and J. Birn, Electric field evidence for tailward flow at substorm onset, *J. Geophys. Res., 88,* 9109, 1983.

Paschmann, G., D. H. Fairfield, E. W. Hones, Jr., C. Huang, W. Lennartsson, and S. Orsini, Plasma signatures of substorm activity in the magnetotail: ISEE-1 and -2 observations on 22 March 1979, *IAGA Bulletin No. 48,* A 12, 1983.

Sato, T., H. Matsumoto, and K. Nagai, Particle acceleration in time-developing magnetic reconnection process, *J. Geophys. Res., 87,* 6089, 1982.

Williams, D. J., Energetic ion beams at the edge of the plasma sheet: ISEE 1 observations plus a simple explanatory model, *J. Geophys. Res., 86,* 5507, 1981.

Questions and Answers

Lui: In discussing the acceleration associated with reconnection, you mention the possibility of $E_\parallel$ as a possible means to accelerate particles. However, in the first example supporting the tail reconnection, you have used the assumption of $E \cdot B = 0$ in your data presentation. Would you like to comment on these two seemingly contradictory expectations of reconnection in the tail?

Nishida: I included $E_\parallel$ as a possible agent of acceleration mainly for the sake of generality since some people favor the idea. I personally am not convinced that tens of kV or more can be generated in the field-aligned direction. Even if $E_\parallel$ might exist, it is likely to be localized and the assumption of $\bar{E} \cdot \bar{B} = 0$ would be valid in most circumstances.

Schindler: In your presentation you distinguished between "global reconnection" and tearing. If we assume that both processes occur spontaneously, it is likely that both phenomena are tearing modes of different length scales and amplitudes. This view is supported by large scale computer models. Did your distinction refer to such morphological differences or do you envisage a more basic physical distinction?

Nishida: I did not mean that there should be a basic physical distinction between the tearing mode and the global reconnection. The energy dissipation and conversion mechanisms involved in these two phenomena would probably be the same. The essential distinction is that sometimes the tearing mode is saturated at a very localized scale while in other times the disturbance grows into a global scale. The important question to answer is what causes this distinction.

MAGNETOTAIL ENERGY STORAGE AND THE VARIABILITY OF THE
MAGNETOTAIL CURRENT SHEET

D. H. Fairfield

Code 695, Goddard Space Flight Center
Greenbelt, Maryland 20771

Abstract. The anti-parallel field configuration in the earth's magnetotail is a prime location to search for the important astrophysical process known as magnetic reconnection. The magnetic evidence that reconnection occurs in the earth's magnetotail is (1) the energy of the tail invariably decreases at the time of global substorm onsets, and (2) an increased number of closed field lines are seen in the near earth magnetotail after substorms. This increased northward flux is often convecting earthward from the expected site of the X line after the substorm onset while southward flux is often convecting tailward. Prior to a substorm, energy accumulates in the tail as the field assumes a more tail-like configuration with a smaller field component across the equatorial plane. The high β plasma sheet becomes thin and has been observed to be at least as thin as a few thousand km near substorm onset times. The small B_n component and thin plasma sheet are the conditions which favor the onset of the tearing mode instability which is thought to lead to neutral line formation and the reconnection associated with substorm onset. Intervals of weak and highly variable fields in the plasma sheet provide evidence for the tearing mode instability.

Introduction

One of the clearest and most fundamental results of experimental space plasma physics has been the demonstration that when a magnetized body is immersed in a flowing plasma containing a magnetic field, an extended magnetotail is formed on the downstream side of the body (e.g., Schindler and Birn, 1978; Ness et al., 1975; Behannon et al., 1981a; 1981b). Field lines emanating from the north and south poles of a planetary dipole are distended on the night side forming two anti-parallel bundles of magnetic flux. These bundles or tail lobes are separated by a diamagnetic plasma sheet of reduced field intensity which contains a limited number of field lines connecting the northern and southern hemispheres. This anti-parallel field configuration in the earth's magnetotail has long been viewed as a likely site of magnetic reconnection. Furthermore its accessibility makes it a prime location for in situ exploration of this important astrophysical process.

Although reconnection is often approached theoretically as a steady state process, the evidence in the earth's magnetotail suggests that reconnection occurs sporadically in association with episodes of global energy dissipation known as magnetospheric substorms. Assuming that reconnection does occur, the question then is what causes its onset. A favored answer to this question has been that the tearing mode instability leads to the formation of a neutral line in the closed field line region of the magnetotail which allows reconnection to proceed. Perhaps steady state reconnection can even be viewed as the nonlinear saturation of the tearing mode instability [Von Hoven, 1976].

Considerable work has been done to determine when the tearing mode might go unstable. Schindler [1974] investigated the behavior of an initally stable magnetotail configuration that was changed in accord with pre-substorm observations; the plasma sheet was thinned and the field component normal to the current sheet, B_n, was diminished. Using linear theory Schindler found that the configuration went tearing mode unstable for a small enough B_n and a thin enough plasma sheet, and he suggested that this was the cause of substorm onsets. Further discussions have focused on how efficiently the B_n component stabilizes this configuration [Galeev and Zeleni, 1976] and to what extent phenomena such as high frequency noise might inhibit this stabilizing influence of B_n [Coroniti, 1980]. The potential importance of the tearing mode instability has been further confirmed by nonlinear studies [Galeev et al., 1978; Terasawa, 1981] that suggest that the tearing mode grows explosively in time and saturates when the width of the magnetic islands become comparable to the thickness of the current sheet. Other numerical simulations that add resistivity to an otherwise stable tail configuration have also found resistive tearing and

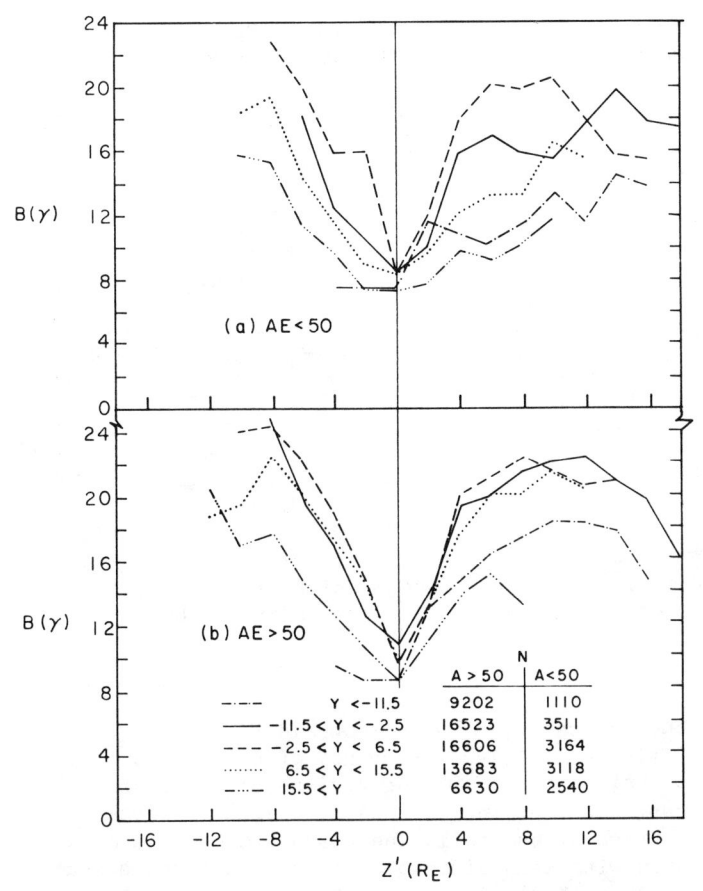

Fig. 1 Average field strength as a function of north-south distance from the tail midplane for AE < 50 and AE > 50 nT. The five curves in each panel correspond to five east-west sectors across tail. N represents the number of 2.5 min averages used in the study.

the creation of a near earth neutral line with many of the accompanying substorm-like features [Birn, 1980; Birn and Hones, 1981]. All these theoretical studies indicate that the magnitude of B_n and the thickness of the plasma sheet and current sheet are critical parameters in controlling tail instabilities.

Experimental investigations of the magnetotail are hampered by ambiguities in distinguishing between space and time variations. Since the tail often changes on time scales that are comparable to those for which the spacecraft traverses spatial boundaries, it is often difficult to know if an observed change reflects a real global change in the tail configuration or if it is due to relative motion between the spacecraft and a time independent spatial feature. In the latter case the velocity of the feature is often large compared to the spacecraft velocity and it is generally not clear if a thin structure is slowly crossing the spacecraft or a thicker structure is crossing at a faster rate. Two methods can be used to circumvent these experimental problems. One alternative is to use multiple spacecraft measurements in an attempt to determine if real configuration changes are occurring simultaneously at two or more points or if the changes occur sequentially in the manner that might be expected if features were moving past the spacecraft. A second alternative is to use large quantities of data and statistical analyses in an attempt to repeatedly observe certain features at similar spatial locations or particular times relative to other events.

This paper will review the experimental magnetic field data as it relates to the reconnection process. We will first review the general configuration of the magnetotail as it is revealed by statistical studies. We will then consider variations in the configuration and how these changes are relevant to reconnection and tearing mode theory. Finally we will consider the state of the current sheet during magnetospheric substorms when reconnection and tearing are thought to be occurring.

The Average Magnetotail

The earth's magnetotail is confined to an approximately cylindrical volume whose diameter is about 50 R_e at a distance 60 R_e behind the earth [Behannon, 1970; Howe and Binsack 1972] and 45 R_e at a distance 25 R_e behind the earth [Scarf et al., 1977]. The field lines near the boundaries make an angle with the tail axis which is about $15°$ at 25 R_e [Fairfield, 1979] and decreases gradually to values near $0°$ beyond approximately 100 R_e [Slavin et al.,1983]. At the nearer earth locations where the angle is appreciable, there is a component of the solar wind kinetic pressure normal to the boundary which compresses the tail field to typical values of 20 nT at 25 R_e [e.g. Fairfield, 1979]. As the angle decreases, the pressure and tail field magnitude decrease until beyond 100 R_e only the magnetosheath field and thermal plasma contribute to the external pressure [Slavin et al.,1983]. Tail field strength is thus related to both the solar wind kinetic pressure and the flaring angle of the field [Holzer and Slavin, 1978]. This angle is determined by how much magnetic flux is contained in the tail and this in turn is controlled by the north-south component of the interplanetary field [e.g. Holzer and Slavin,1979]. The geomagnetic tail has been observed as far from the earth as 500 and 1000 R_e [e.g. Walker et al., 1975].

An important and interesting region of the magnetotail is the volume of weak field and hot diamagnetic plasma surrounding the equatorial plane which is known as the plasma sheet. A dawn-to-dusk cross-tail current flows in the plasma sheet and it is this current that produces the anti-parallel magnetotail fields in the two halves of the tail. In Figure 1 [Fairfield, 1979] a statistical summary of measurements of the average field strength are shown as a function of distance relative to the equatorial field reversal region, z'. This distance z' is only an

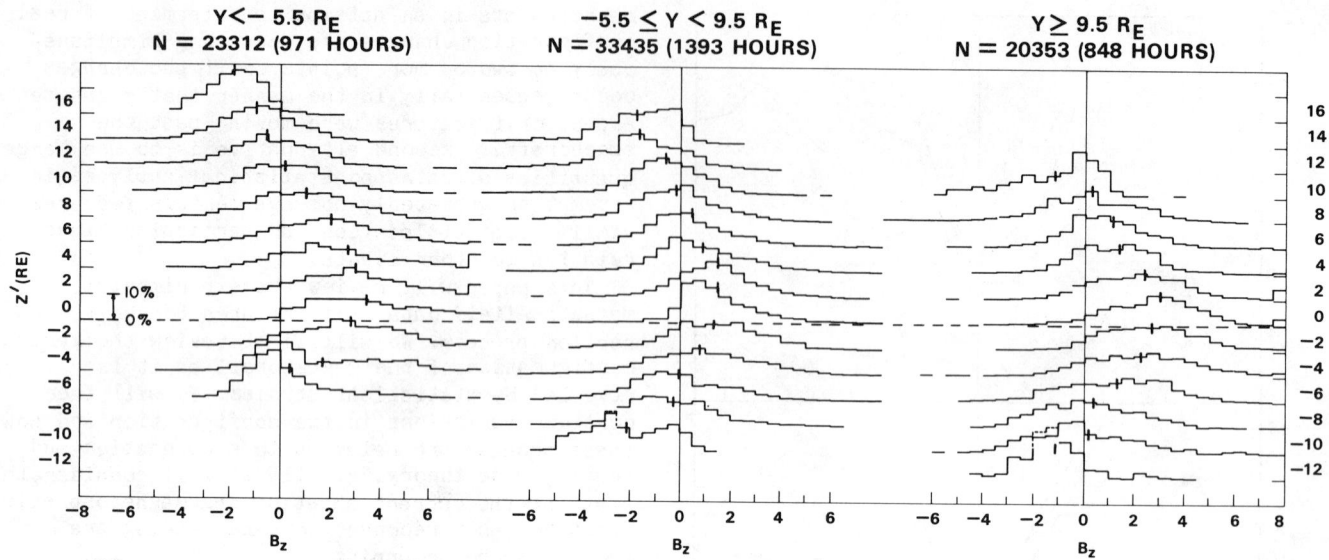

Fig. 2 Histograms of B_z for 2 R_e intervals of z' in dawn, dusk, and midnight sectors of the tail between 20 and 33 R_e. The vertical bars through each histogram indicate the average B_z for that z' interval.

estimated distance from the expected center of the current sheet [Fairfield, 1980] because this reversal region is continually moving and has an unpredictable component to its motion. The exact location is known only at those instants when the spacecraft detects the reversal. Data in Figure 1 are taken by the Imp 6 spacecraft between 20 and 33 R_e down the tail. The five traces correspond to 5 spatial intervals across the tail (negative y towards dawn, positive y towards dusk). The top and bottom panels correspond to geomagnetically quiet and disturbed times respectively and N indicates the number of 2.5 minute data averages used in the analysis. Field values near z'=0 are roughly half the lobe values which are near 20 nT for the more disturbed data set and somewhat less during quiet times. In the midnight sector the field strengths increase with increasing z' and reach their lobe values at about +6 R_e. This fact implies that the thickness of the plasma sheet is hardly ever thicker than 12 R_e in the midnight region and undoubtedly has a typical value more like the 4-6 R_e that was determined from plasma measurements by Bame et al. [1967]. The fields increase more slowly with z' in the dawn and dusk sectors. This slower increase again supports plasma measurements that indicate a plasma sheet thicker by about a factor of 2 in these regions. It should be emphasized that the 10 nT average field at z'=0 is considerably larger than the typical few nT minimum field at the current reversal [e.g. Speiser and Forbes, 1981] because of inaccuracies in z' and the fact that measurements in this central region were made throughout +2R_e intervals rather than at the current reversal.

Information about another important tail parameter - the field component normal to the average current sheet, B_z - is shown in Figure 2 [Fairfield, 1979]. Frequency of occurrence distributions of B_z for various z' distance ranges are shown for dawn, midnight and dusk sectors of the tail. The heavy tic mark associated with each distribution indicates the average B_z at that z'. In each sector the average B_z near z'=0 is positive indicating flux crossing the equatorial plane connecting northern and southern hemispheres. At large z', B_z becomes negative indicating lobe field lines that are flaring away from the equatorial plane as the diameter of the tail increases. The average B_z at z'=0 in the dawn and dusk regions is over 3 nT which is about twice the average B_z in the midnight sector. This result suggests that if perturbing currents are to reduce the B_z component and create a neutral line, this as apt to be done most easily in the midnight sector. The three dimensional simulations of Birn and Hones [1981] do indeed show that reconnection effects are most prominent near the center of the tail for this reason. This y dependence also suggests that if a neutral line were formed and if B_z decreases with distance down the tail, the line might tend to be located nearest the earth near midnight and at greater distances on the flanks. The occasional occurrence of negative B_z values near z'=0 in Figure 2 is suggestive of a spacecraft being downstream of a neutral line. Note that frequent, or extended intervals of large southward B_z are not expected to be produced by reconnection; if a 2nT southward field were being carried tailward at 700 km/s all across the tail, 10% of the tail flux would be lost in 2.5 minutes - the duration of the averaging interval for the points used [Fairfield, 1979].

Knowing the thickness of the plasma sheet is

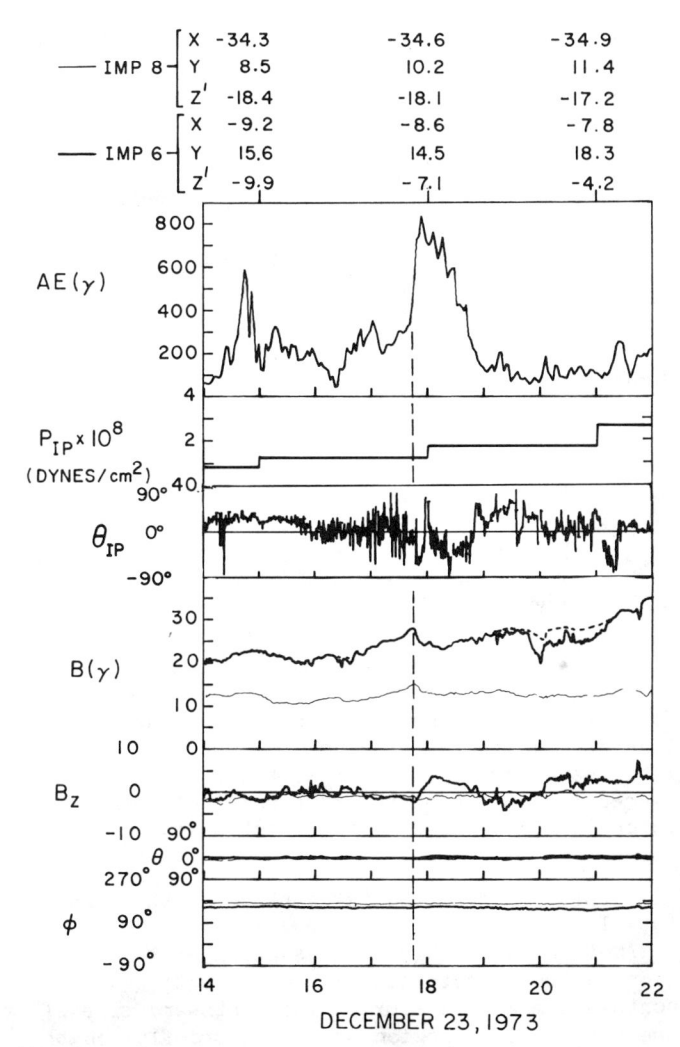

Fig. 3 Magnetic field data from two spacecraft in the tail lobe separated by a large X distance. Data indicate how tail energy typically increases and decreases at both spacecraft before and after a large AE increase.

field is proportional to the spatial variation through the current sheet with the constant of proportionality the velocity of the sheet ($dB_x/dt = v_z \, dB_x/dz$). After statistically analyzing 244 hours of Imp 4 data taken near the current sheet on 15 different days, they argued that most of the current was flowing in a region 2.5 R_e thick and the entire structure was moving back and forth past the spacecraft with a typical velocity of 90 km/s and an amplitude of +2 R_e. This amplitude estimate was based on the distribution of observed current sheet crossings but they did not consider the diurnal up and down motion caused by the rotation of the dipole and it would seem that they may have overestimated the thickness and velocity. We will suggest below that their assumption of constant structure is frequently not valid and that much thinner current sheets are present at the time of magnetospheric substorms. Speiser and Forbes [1981] also argue that the current tends to be concentrated near the central region.

Even if the current distribution within the plasma sheet is unknown, it is possible to monitor the total current in the tail circuit by monitoring the tail lobe field strength. If we model the tail current as either an infinite current sheet across the tail or as infinite solenoids producing each tail lobe, the field is related to the current as B=629J where B is in nT and J in amps/m. Since the field is proportional to the current strength, we can monitor the current by measuring the field in the tail lobe. For a tail field strength of 20 nT the current is .032 amps/m for each meter of x distance. If we assume this current is distributed across a sheet of thickness 2.5 R_e as determined by Bowling and Wolf [1974] we obtain a current density of 2×10^{-9} amps/m^2. In the next section we review how monitoring this lobe field has led to the concept of tail energy storage and dissipation during substorms.

not the same as knowing the thickness of the primary current carrying region since the currents could be concentrated in a thin equatorial region rather than distributed more equally over the thickness of the plasma sheet. Bowling and Wolf [1974] carried out an analysis to determine this current sheet thickness but they were forced to make a fundamental assumption whose validity can be questioned. They assumed that the structure of the current sheet was constant in time with B_x increasing monotonically with distance from the current sheet. Changes in B_x were attributed exclusively to motion of the entire structure past the spacecraft. Under this assumption there is a one-to-one relation between z, the distance from the center of the current sheet, and B_x. Ordering the data according to B_x is equivalent to ordering it according to z and a measurement of the time rate of change of the

Tail Energy Storage

For many years spacecraft have been monitoring the tail lobe field strength during substorms and it has become clear that the tail invariably gains energy prior to substorms and looses energy at the time of global substorm onsets [e.g. Fairfield and Ness, 1970; Caan et al., 1973; 1975]. An example of such tail energy loss is shown in Figure 3 [Fairfield et al., 1981a]. The Imp 6 and 8 spacecraft are measuring the total energy density at locations 9 and 35 R_e down the tail in the dusk sector. The field magnitude B (containing virtually all this energy density) is shown along with the north-south field component B_z, the latitude and longitude angles θ and φ, the ground activity index AE, interplanetary ram pressure P_{Ip}, and north-south field angle $θ_{Ip}$. A substorm onset at 1743 UT (vertical dashed line) separates an interval of increasing total energy density prior to the substorm from an interval of decreasing energy density after the substorm. The simultaneous observation of such decreases at two widely separated points suggests that the whole magnetotail loses energy. Using these measured decreases from 28 to 23 nT at Imp 6 and 15 to 13 nT at Imp 8 and interpolating to other locations, Fairfield et al. [1981a] estimate that the tail lost 4×10^{21} ergs from the region between x=-15 and x=-35 R_e. Distributing this loss over a time of 30 minutes gives a rate of energy loss for the tail of 2×10^{18} ergs/s which is at least as much as an estimate of the energy dissipated during the substorm.

Akasofu [1978] has questioned the importance of tail energy storage during substorms and suggested that the solar wind directly drives the substorm current system. Indeed there do seem to be directly driven currents both before the sudden onsets [Fairfield and Cahill, 1966; Burch, 1973; Kisabeth and Rostoker, 1974] and sometimes for extended intervals during disturbed periods [Hones et al. 1976; Caan et al., 1973; Pytte et al., 1978a], but the fact remains that large intensifications of auroral zone currents invariably correspond to tail energy loss.

Rostoker [1983] has further confirmed the role of the tail in substorms by investigating the relationship between the interplanetary magnetic field during three isolated substorms that were monitored by an extensive network of ground observatories. He found evidence for weak currents driven by a southward interplanetary field prior to the main onset followed by the intense substorm currents that began when the interplanetary field turned northward. He concluded that the substorm energy could not have come directly from the solar wind and must have come from the tail under these circumstances but he presented no tail data. For one of these events ISEE 1 was outbound in the high latitude, post-midnight magnetotail, and in Figure 4 we present the magnitude and B_z magnetic field measurements of C. T. Russell (private communication). Rostoker showed that the interplanetary magnetic field shifted suddenly southward at 0548 and then shifted suddenly northward at 0718 with the later time corresponding to a large substorm expansive phase. These times are marked by vertical lines in Figure 4. It is clear that the tail field increased during the southward interval, especially if the magnitude is calculated relative to the decreasing dipole field associated with outward motion of the spacecraft. The field then decreased dramatically beginning at the time of the 0718 UT onset and 40% of the local tail energy density had disappeared by 0815 UT. ISEE 1 low energy electron data of K. W. Ogilvie (private communication) confirm that plasma was not contributing any significant energy density until the plasma sheet recovery at 0815. Since conversion of magnetic energy to other forms of energy is one of the primary characteristic of reconnection, tail energy loss at the time of substorms provides support for the

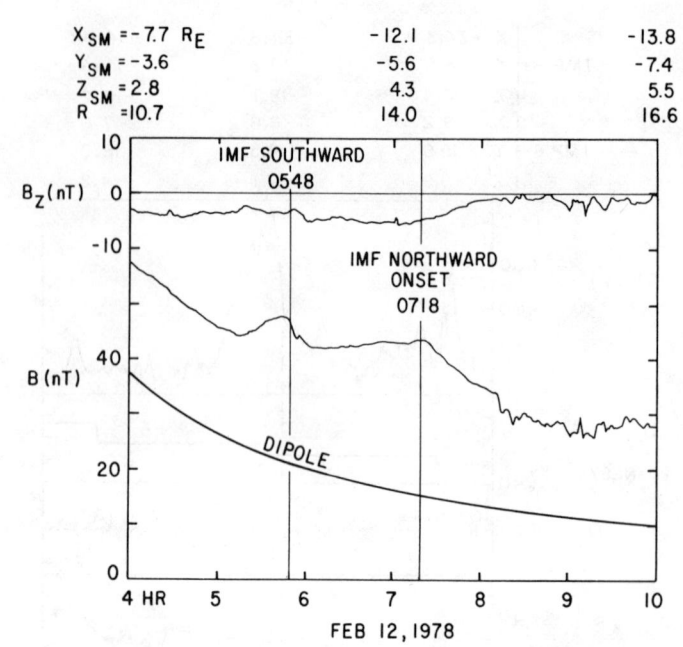

Fig. 4 Tail magnetic field magnitude and the B_z component for an isolated substorm studied by Rostoker [1983]. The tail field magnitude increases during a period of southward interplanetary field prior to the 0718 onset and decreases after a northward turning of the interplanetary field at the onset time.

view that reconnection takes place in the magnetotail.

Another observable manifestation of reconnection in the magnetotail is the production of northward magnetic flux on the earthward side of the reconnection region and southward flux on the tailward side. A number of case studies have documented the presence of the southward fields and tailward flowing plasmas at substorm times and these are covered in the review by Hones [this volume]. A great deal of additional evidence suggests that an increased amount of northward flux is generated earthward of 15 R_e which is suspected to be the typical site of the reconnection region. Inside this distance the field invariably exhibits a sharp increase near the time of onset [Russell and McPherron, 1973]. The effect is particularly clear and well documented at the geosynchronous orbit [see review by Baker, this volume]. This increase is invariably clear and relatively sudden inside 15 R_e, and the effect is increasingly smaller and more gradual at larger distances where an onset time cannot be determined to better than a few tens of minutes. Russell and McPherron [1973] have shown that the increase near earth rapidly propagates inward creating what has been called an "injection front" [Moore et al., 1981]. This propagating structure can be considered a dawn to dusk current sheet which is responsible for the stronger fields behind (tailward of) the front and weaker fields ahead (earthward) of the front. After the

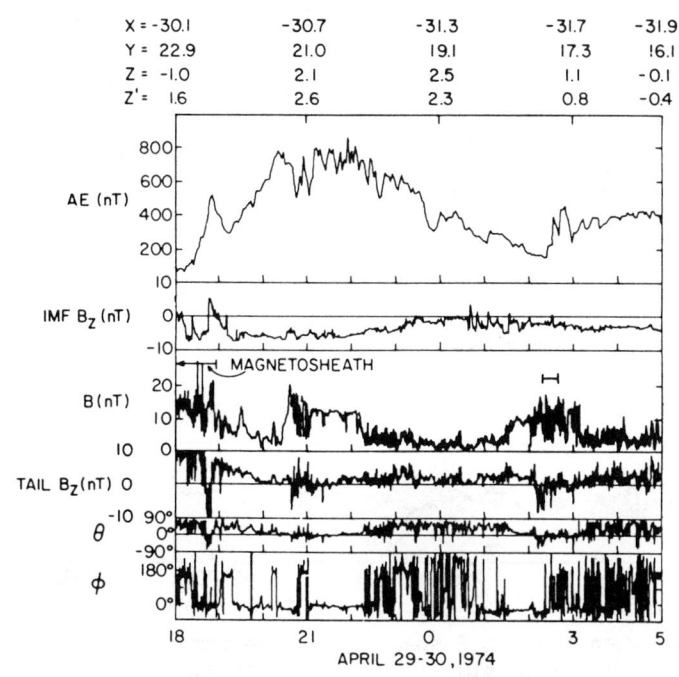

Fig. 5 The AE index and interplanetary B_z magnetic field component are shown along with the tail magnetic field data. A bar during a strong field region near a substorm onset indicates an interval whose detailed data are shown in fig 6.

substorm the tail is left in a more dipolar state with more flux crossing the equatorial plane. This more dipolar field is also evidence that the dawn-dusk current is reduced, with the "missing" current diverted through the high latitude ionosphere where it contributes to the auroral electrojet [McPherron, 1973]. An example of this tail increase in B_z inside 15 R_e can be seen in figure 3 where the Imp 6 spacecraft nearer the earth detects the increase in B_z at substorm onset. The effect is perceptible at 35 R_e in Figure 3 only as a slight increase in B_z between 1800 and 2000. This increase may indicate the tailward passage of the neutral line during substorm recovery.

The consistent and repeatable nature of the evidence for for both tail energy loss and increased northward magnetic flux at the time of substorms provides substantial evidence that reconnection occurs in the earth's magnetotail. Case studies of the field, particle, and plasma measurements on the downstream side of the suspected reconnection region further strengthen this case. Still, some workers choose to emphasize only the driven component of substorms [Akasofu, 1980] or interpret the data without invoking reconnection [e.g. Rostoker et al., 1983]. Although it is difficult to refute these alternate explanations, the existing data seem to be quite consistent with the fundamental predictions of reconnection theory. In the next section we will consider the magnetotail current sheet at the time of substorms.

Thin Plasma Sheets

To understand the reconnection process it is of interest to investigate measurements in the center of the plasma sheet at the time of substorm onsets. Anyone who searches for such measurements quickly learns about the phenomenon known as plasma sheet thinning [Hones et al., 1971]; the plasma sheet invariably becomes thin near the substorm onset time and observable plasma usually disappears at the spacecraft location unless the spacecraft is located quite near the midplane of the plasma sheet. The thinning is associated with an increase in the field magnitude and a decrease in the B_z component which leads to a more tail-like field configuration [e.g. Fairfield and Ness, 1970]. Inside of 15 R_e this thinning takes place before the substorm and the associated energetic particle behavior at synchronous orbit can actually be used as an accurate predictor of an impending substorm [Baker et al, 1978]. Beyond 15 R_e the thinning occurs nearer the time of the onset. [Nishida and Fujii, 1976].

An example of plasma sheet thinning is shown in Figure 5 [Fairfield et al., 1981b] where the AE geomagnetic activity index and the interplanetary magnetic field component B_z are plotted above 15-s averages of the Imp 8 magnetotail field. The spacecraft entered the dusk magnetotail at 1900 UT as geomagnetic activity was increasing. Even though the plasma sheet is normally very thick at this location the plasma disappeared and the spacecraft detected strong tail lobe fields as AE approached a maximum near 2030. As AE decreased the plasma sheet reappeared at 2210, as it normally does during substorm recovery, and the field remained weak and northward until 0130. Prior to another AE intensification near 0230 the plasma again disappeared, B_z and the θ angle decreased indicating a more tail-like configuration, and the field achieved values characteristic of the lobe even though the spacecraft was very near the expected location of the current reversal. Indeed, sudden changes of the ϕ angle indicate current reversals within the strong field regions near both 0240 and 2040 although they cannot be resolved on this time scale.

Fairfield et al. [1981b] investigated .32-s average field data during such strong field regions and the period 0220-0243 marked by a bar over the field magnitude in Figure 5 is shown in Figure 6. The total energy density (actually the magnetic field equivalent of the plasma plus field energy density) is indicated by the crosses plotted above the B trace. The dashed line is a $\beta=1$ line; field values above this line indicate a $\beta<1$ plasma where the field strength is nearly equivalent to the tail lobe field strength. The rectangles above the B trace indicate the lobe as defined as a $\beta<1$ region. From the field polarity we can deduce whether the northern ($\phi=0°$) or southern ($\phi=180°$) tail is being measured and this

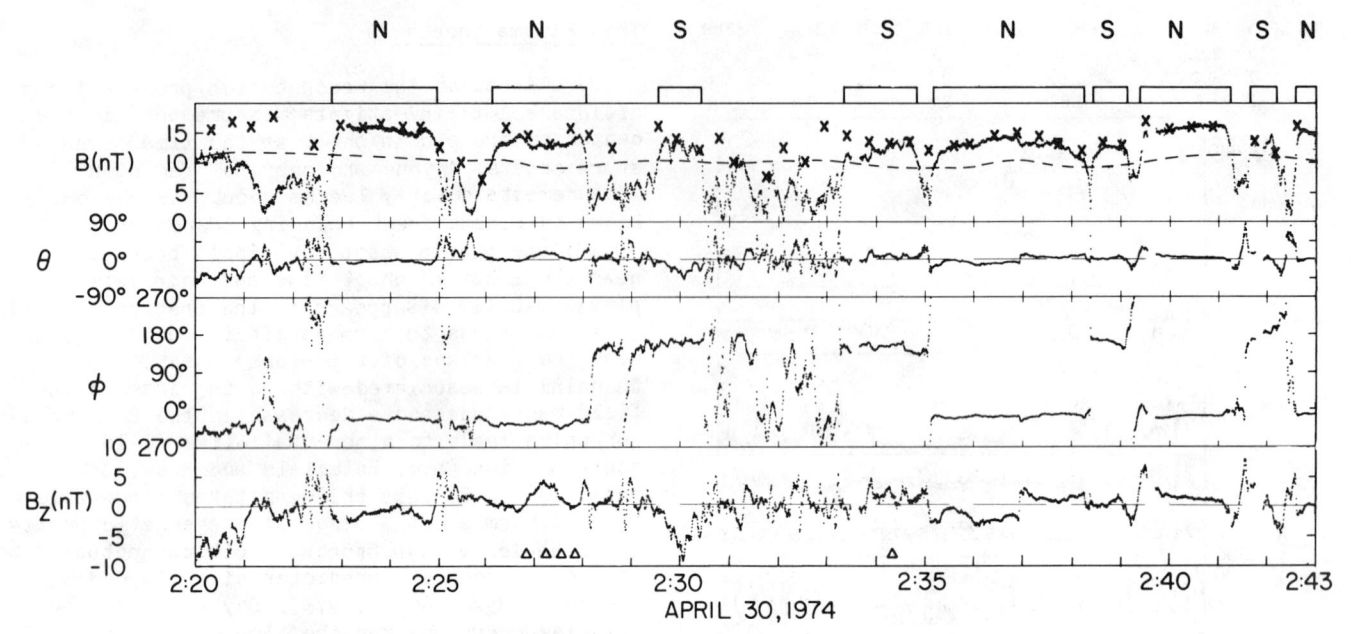

Fig. 6 Tail magnetic field data for an interval when the Imp 8 spacecraft made numerous traversals between the northern and southern tail lobes in times as short as 10 s.

is indicated by the letter above the rectangle. All significant currents are equatorward of a spacecraft in the β<1 region since the field is essentially uniform throughout the higher latitude region. Changes in the φ angle between 0° and 180° indicate crossings from one lobe of the tail to the other. Several of these traversals such as 023505, 023825, and 023915 illustrate complete crossings of the magnetotail current sheet and the entire high-β plasma sheet in times as short as 10 s. Such data indicate the presence of a thin current sheet-plasma sheet whose thickness, Fairfield et al. argue, is probably less than a few thousand km. Furthermore this current sheet is oscillating back and forth over the spacecraft during this substorm period [see also Toichi and Miyazaki, 1976].

Another example of a thin current sheet is contained in the well studied data set obtained on March 22, 1979. A large substorm onset was observed on the ground at 1052 and within 3 minutes the usual effects were seen at synchronous orbit. The ISEE 1 spacecraft was 15.4 R_e from the earth at solar magnetospheric coordinates (-13.6,-7.2,-0.8 R_e). ISEE 2 was 1.5 R_e closer to the earth with a solar magnetospheric z coordinate .3 R_e south of ISEE 1. The solar wind happened to be blowing from the north so that the current sheet was tilted away from its expected location near the z=0 plane and into the southern hemisphere. Magnetic field data from these two spacecraft are shown in Figure 7 in a coordinate system that has been rotated 8.5° about the y axis to account for this tilt. In this coordinate system the spacecraft were separated by ~ 0.6 R_e in z. ISEE 2 (the lighter trace in Figure 7) crossed the current sheet from the northern to southern hemisphere at 105730 UT and ISEE 1 did the same at 110645 UT. For 9 minutes in between these times the two spacecraft were observing high β lobe-like magnetic fields of opposite hemispheres. The current sheet and high β plasma sheet was located between the spacecraft and must have been considerably thinner than the 3700 km spacecraft separation. If we assume that the current sheet traversed the 3700 km distance in 9.25 minutes at an average velocity of 6.6 km/s, and then crossed each spacecraft in 45 s, we obtain a thickness for the high β plasma sheet of ~ 300 km. This thickness may be unreasonably small since a 10 kev thermal proton in a relatively small 10 nT field has a 1400 km gyroradius. During this period B_z remained small (.3 and 1.1 nT averages for ISEE 1 and 2 respectively for the interval shown) and not until about 1115 UT did B_z begin to increase (average solar magnetospheric B_z was 15.1 nT and 20.8 nT at the two spacecraft for the interval 1118-1128 UT.) Plasma and particle measurements show predominantly tailward and earthward flow prior to and after 1115 as if the spacecraft were tailward of and then earthward of the neutral line, but there are brief intervals of exception that do not agree with this simple model.

Figures 6 and 7 also illustrate the futility of trying to precisely answer the question of how small B_n may become at substorm times. Clearly the current sheet is often oscillating and if there is a coordinate system in which it is appropriate to measure B_n it is undoubtedly moving and is not easily determined from the field data. The small average value of B_n for these examples does however suggest that B_n probably is not more than a few percent of B.

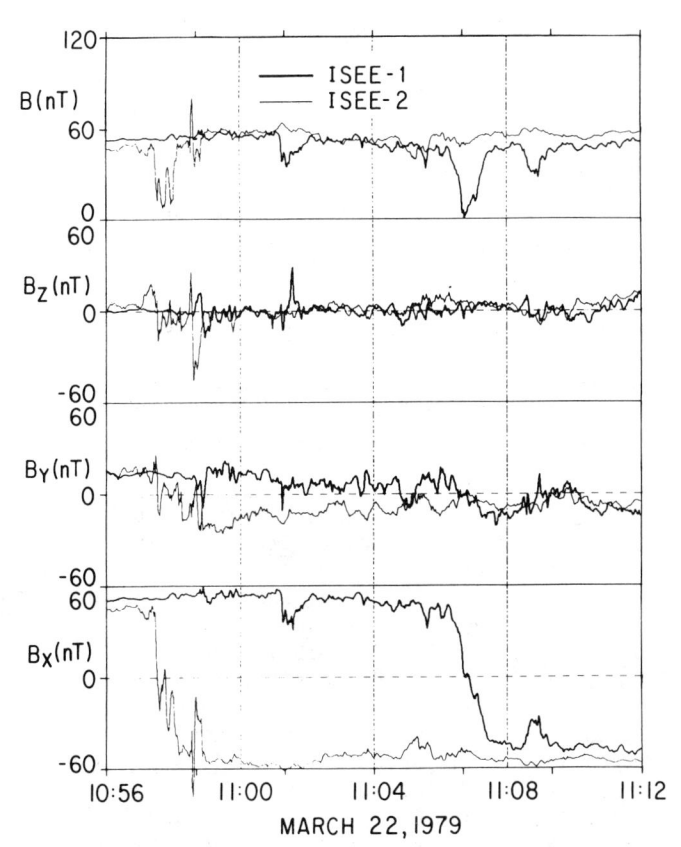

Fig. 7 ISEE 1 and 2 tail magnetic field data for a 16 min interval near 14 R_e. The two spacecraft observed strong fields of northern and southern hemispheres from 1058 to 1107 even thought their solar magnetospheric z separation was only 0.3 R_e.

During some intervals such as 0231-0233 UT in figure 6, a weak and highly variable field is detected near the current sheet where brief intervals of relatively strong northward and southward B_z occur. Such intervals have been cited as evidence for the current loops or magnetic islands which would be produced by the tearing mode instability [e.g. Coroniti et al., 1977; 1980; Speiser and Schindler, 1981]. Schindler and Ness [1972] statistically analyzed 12 hours of such data which occurred for longer intervals on 3 rather quiet days. They concluded that such variations could not have been produced simply by tilting the fields in a more uniform current sheet and hence they were more likely associated with loops produced by the tearing mode. Coroniti et al., [1980] and Lui and Meng [1979] suggest that the tearing mode may be common in the quiet magnetotail. Perhaps only during substorms does enough reconnection occur such that higher latitude open field lines reconnect and allow the downstream release of the plasma sheet and the further tapping of the stored tail lobe energy [Coroniti et al.,1980].

Summary and Conclusions

Magnetic field measurements in the earth's magnetotail lend considerable support to the idea that magnetic reconnection occurs at the time of magnetospheric substorms. The tail invariably loses energy at the time of large global substorm onsets and increased northward magnetic flux is invariably seen on the earthward side of the presumed reconnection region. Also southward flux is frequently seen tailward of the reconnection region in association with the tailward plasma flow that is expected to result from reconnection.

Although it can be argued that some of the evidence for reconnection is circumstantial (e.g. there is no proof that the decreased energy of the magnetotail is that dissipated by the substorm) it is difficult to see how the magnetic field evidence could be much more compelling. The global changes such as tail energy loss and field dipolarization - those phenomena that are more or less independent of spacecraft location - are seen with considerable clarity and repeatability. This is especially true of large events and less true of smaller substorms that may have more localized effects. It is true that "classical" examples of local phenomena such as 90° southward fields in tailward flowing plasmas are more difficult to find, but this is not unreasonable in view of (1) the limited region over which such phenomena are apt to occur relative to the probability that a spacecraft will be suitably located to see them, and (2) the likelihood that these events are more complex and time dependent than theories can easily describe.

Given that reconnection occurs in the geomagnetic tail, it is then of interest to inquire into what magnetic field data can tell us about the details of this interesting process. Here local measurements are necessary and there are the accompanying difficulties of having a spacecraft in an appropriate location and knowing the location relative to suspected theoretical structures.

Measurements indicate that the high β plasma sheet and current sheet can be very thin and have a small B_z component at the time of substorms. These facts are consistent with conditions conducive to the onset of the tearing mode instability. These small thicknesses are of the order of the ion gyroradius, but it is difficult to be very quantitative about the magnitude of the small B_n component because the entire structure seems to oscillate on a scale of minutes. This oscillation means that the coordinate system in which B_n should be measured is variable and cannot be determined. It may well be, however, that B_n is not more than a few percent of the total field of 10 to 20 nT at these times. The current sheet oscillations themselves are suggestive of an instability and the occurence of positive and negative B_z's are consistent with the occurrence of magnetic structures such as would be produced by the tearing mode instability.

In spite of experimental limitations, high resolution, multi-experiment multi-spacecraft studies in the earth's magnetotail hold considerable promise for determining further details of the reconnection process and the conditions leading to its occurrence.

References

Akasofu, S.-I., What is a magnetospheric substorm? in Dynamics of the Magnetosphere, editor S.-I. Akasofu, D. Reidel Pub. Co., Dordrecht, Holland, 447-460, 1978.

Akasofu, S.-I., The solar wind-magnetosphere energy coupling and magnetospheric disturbances, Planet. Space Sci., 28, 495, 1980.

Baker, D. N., P. R. Higbie, E. W. Hones, Jr., and R. D. Belian, High resolution energetic particle measurements at 6.6 R_E 3. low energy electron anistropies and short-term substorm predictions, J. Geophys. Res., 85, 4863-4868, 1978.

Bame, S. J., J. R. Asbridge, H. E. Felthauser, E. W. Hones, and I. B. Strong, Characteristics of the plasma sheet in the Earth's magnetotail, J. Geophys. Res., 72, 113-129, 1967.

Behannon, Kenneth W., Geometry of the geomagnetic tail, J. Geophys. Res., 75, 743-753, 1970.

Behannon, Kenneth W., J. E. P. Connerney, and N. F. Ness, Saturn's magnetic tail: Structure and dynamics, Nature, 292, 753, 1981a.

Behannon, K. W., L. F. Burlaga, and N. F. Ness, The Jovian magnetotail and its current sheet, J. Geophys. Res., 86, 8385, 1981b.

Birn, J., Computer studies of the dynamic evolution of the geomagnetic tail, J. Geophys. Res., 85, 1214-1222, 1980.

Birn, J., and E. W. Hones, Jr., Three-dimensional computer modeling of dynamic reconnection in the geomagnetic tail, J. Geophys. Res., 86, 6802-6808, 1981.

Bowling, S. D., and R. A. Wolf, The motion and magnetic structure of the plasma sheet near 30 R_E, Planet Space Sci., 22, 673, 1974.

Burch, J. L., Rate of erosion of day side magnetic flux based on quantitative study of the dependence of polar cusp latitude on the interplanentary magnetic field, Radio Science, 8, 955-962, 1973.

Caan, M. N., R. L. McPherron, and C. T. Russell, Solar wind and substorm- related changes in the lobes of the geomagnetic tail, J. Geophys. Res., 78, 8087-8096, 1973.

Caan, M. N., R. L. McPherron and C. T. Russell, Substorm and interplanetary magnetic field effects on the geomagnetic tail lobes, J. Geophys. Res., 80, 191-194, 1975.

Coroniti, F. V., F. L. Scarf, L. A. Frank and R. P. Lepping, Microstructure of a magnetotail fireball, Geophys. Res. Lett., 4, 219-222, 1977.

Coroniti, F. V., L. A. Frank, D. J. Williams, R. P. Lepping, F. L. Scarf, S. M. Krimigis, and G. Gloeckler, Variability of plasma sheet dynamics, J. Geophys. Res., 85, 2957-2977, 1980.

Coroniti, F. V., On the tearing mode in quasi-neutral sheets, J. Geophys. Res., 85, 6719-6728, 1980.

Fairfield, D. H., On the average configuration of the geomagnetic tail, J. Geophys. Res., 84, 1950-1958, 1979.

Fairfield, D. H., A statistical determination of the shape and position of the geomagnetic neutral sheet, J. Geophys. Res., 85, 775-780, 1980.

Fairfield, D. H., and L. J. Cahill, Jr., Transition region magnetic field and polar magnetic disturbances, J. Geophys. Res., 71, 155-169,1 1966.

Fairfield, D. H., and N. F. Ness, Configuration of the geomagnetic tail during substorms, J. Geophys. Res., 75, 7032-7047, 1970.

Fairfield, D. H., R. P. Lepping, E. W. Hones, Jr., S. J. Bame, and J. R. Asbridge, Simultaneous measurements of magnetotail dynamics by IMP spacecraft, J. Geophys. Res., 86, 1396-1414, 1981a.

Fairfield, D. H., E. W. Hones, Jr., and C. I. Meng, Multiple crossings of a very thin plasma sheet in the Earth's magnetotail, J. Geophys. Res., 86, 11,189-11,200, 1981b.

Galeev, A. A., and L. M. Zeleni, Tearing instability in plasma configurations, Sov. Phys. JETP, 43, 1113, 1976.

Galeev, A. A., F. V. Coroniti, and M. Ashour-Abdalla, Explosive tearing mode reconnection in the magnetospheric tail, Geophys. Res. Lett., 5, 707-710, 1978.

Holzer, R. W., and J. A. Slavin, Magnetic flux transfer associated with expansions and contractions of the dayside magnetosphere, J. Geophys. Res., 83, 3831-3839, 1978.

Holzer, R. E. and J. A. Slavin, A correlative study of magnetic flux transfer in the magnetosphere, J. Geophys. Res., 84, 2573-2578, 1979.

Hones, E. W., Jr., J. R. Asbridge and S. J. Bame, Time variations of the magnetotail plasma sheet at 18 R_E determined from concurrent observations by a pair of Vela satellites, J. Geophys. Res., 76, 4402-4419, 1971.

Hones, E. W., Jr., S. J. Bame, and J. R. Asbridge, Proton flow measurements in the magnetotail plasma sheet made with IMP 6, J. Geophys. Res., 81, 2270-234, 1976.

Howe, H. C., Jr., J. H. Binsack, Explorer 33 and 35 plasma observations of magnetosheath flow, J. Geophys. Res., 77, 3334-3344, 1972.

Kisabeth, J. L., and G. Rostoker, The expansive phase of magnetospheric substorms 1. Development of the auroral electrojet and auroral arc configuration during a substorm, J. Geophys. Res., 79, 972-984, 1974.

Lui, A. T. Y. and C.-I. Meng, Relevence of southward magnetic fields in the neutral sheet to anistropic distribution of energetic electrons and substorm activity, J. Geophys. Res., 84, 5817, 1979.

McPherron, R. L., C. T. Russell and M. P. Aubrey, Satellite studies of magnetospheric substorms on August 15, 1968, 9. Phenomenological model for substorms, J. Geophys. Res., 78, 3131-3149, 1973.

Moore, T. E., R. L. Arnoldy, J. Feynman, and D. A. Hardy, Propagating substorm injection fronts, J. Geophys. Res., 86, 6713-6726, 1981.

Ness, N. F., K. W. Behannon, R. P. Lepping, and Y. C. Whang, The magnetic field of Mercury: Part I, J. Geophys. Res., 80, 2708, 1975.

Nishida, A., and K. Fujii, Thinning of the near-earth ($10 \sim 15$ R_E) plasma sheet preceeding the substorm expansion phase, Planet Space Sci., 24, 849-853, 1976.

Pytte, T., R. L. McPherron, E. W. Hones, Jr., and H. I. West, Jr., Multiple - satellite studies of magnetospheric substorms: distinction between polar magnetic substorms and convection driven magnetic bays, J. Geophys. Res., 83, 663-679, 1978.

Rostoker, G., Triggering of expansive phase intensifications of magnetospheric substorms by northward turnings of the interplanetary magnetic field, J. Geophys. Res., 88, 6981-6993, 1983.

Russell, C. T. and R. L. McPherron, The magnetotail and substorms, Space Sci. Rev., 11, 111-222, 1973.

Scarf, F. L., L. A. Frank, and R. P. Lepping, Magnetospheric boundary observations along the IMP 7 orbit 1. Boundary locations and wave level variations, J. Geophys. Res., 82, 5171-5180, 1977.

Schindler, K. and J. Birn, Magnetospheric physics, Physics Reports, 47, No. 2, North-Holland Pub. Co., 109-165, 1978.

Schindler, K., and N. F. Ness, Internal structure of the geomagnetic neutral sheet, J. Geophys. Res., 77, 91-100, 1972.

Schindler, K., A theory of the substorm mechanism, J. Geophys. Res., 79, 2803-2810, 1974.

Slavin, J. A., B. T. Tsurutani, E. J. Smith, D. E. Jones, and D. G. Sibeck, Average configuration of the distant (< 220 R_E) magnetotail: initial ISEE-3 magnetic field results. Geophys. Res. Lett., 10, 9793-976, 1983.

Speiser, T. W., and K. Schindler, Magnetospheric substorm models: Comparison with neutral sheet magnetic field observations, Astrophys. and Space Science, 77, 443-453, 1981.

Speiser, T. W., and T. G. Forbes, Explorer 34 magnetic field measurements near the tail current sheet and auroral activity, Astrophys. and Space Science, 77, 409-442, 1981.

Terasawa, T. Numerical study of explosive tearing mode instability in one-component plasma, J. Geophys. Res., 86, 9007-9019, 1981.

Toichi, T. and T. Miyazaki, Flapping motions of the tail plasma sheet induced by the interplanetary magnetic field variations, Planet. Space Sci., 24, 147-159, 1976.

Von Hoven, G., Solar flares and plasma instabilities: Observations, mechanisms and experiments, Solar Phys., 49, 1976.

Walker, R. C., V. Villante, and A. J. Lazarus, Pioneer 7 observations of plasma flow and field reversal regions in the distant geomagnetic tail, J. Geophys. Res., 80, 1238-1244, 1975.

Questions and Answers

Fritz: What is the limitation on the thickness of the magnetic tail current sheet? Why?

Fairfield: *The magnetic field data suggest only that the magnetotail current sheet sometimes becomes quite thin—probably as thin as the ion gyroradius. Theory suggests that ions carry the cross tail current and therefore the current sheet would not be expected to be thinner than the gyroradius of these particles.*

Priest: How do the thickness of the sheet and the size of the normal field component compare qualitatively with the criteria for collisionless tearing mode onset?

Fairfield: *The thickness of the current sheet and the strength of the normal field component probably can become small enough to cause onset of the collisionless tearing mode instability but a quantitative comparison with theory is not possible because the stabilizing effect of adiabatic electrons is not known.*

Vasyliunas: The magnetic energy stored in the magnetotail depends not only on the field magnitude B_T but also on the volume of the magnetotail; in addition, if the boundary of the magnetotail changes, the work done by the external pressure on the boundary contributes to the energy budget. Deducing the changes of magnetic energy content from observations of B_T alone should therefore be viewed with some caution.

Fairfield: *Work by Maezawa (J. Geophys. Res., 80, 3543, 1975) shows that the radius of the tail increases as the field strength increases and decreases as the field decreases. This means that the tail energy storage and release is actually larger than what one would calculate by neglecting this effect.*

PLASMA SHEET BEHAVIOR DURING SUBSTORMS

Edward W. Hones, Jr.

University of California, Los Alamos National Laboratory
Los Alamos, NM 87545

Abstract. Auroral or magnetic substorms are periods of enhanced auroral and geomagnetic activity lasting one to a few hours that signify increased dissipation of energy from the magnetosphere to the earth. Data acquired during the past decade from satellites in the near-earth sector of the magnetotail have suggested that during a substorm part of the plasma sheet is severed from earth by magnetic reconnection, forming a "plasmoid," i.e., a body of plasma and closed magnetic loops, that flows out of the tail into the solar wind, thus returning plasma and energy that have earlier been accumulated from the solar wind. Very recently this picture has been dramatically confirmed by observations, with the ISEE 3 spacecraft in the magnetotail 220 R_E from earth, of plasmoids passing that location in clear delayed response to substorms. It now appears that plasmoid release is a fundamental process whereby the magnetosphere gives up excess stored energy and plasma, much like comets are seen to do, and that the phenomena of the substorm seen at earth are a by-product of that fundamental process.

Introduction

The solar wind carries energy away from the sun continually in the form of rapidly flowing, fully ionized hydrogen plasma. This is threaded with "frozen in" solar magnetic field which is stretched out by the flowing plasma and which is referred to as the interplanetary magnetic field (IMF). Typical values of solar wind parameters at the orbit of the earth are: density $\approx 10/cm^3$; flow speed ≈ 450 km/sec; magnetic field strength ≈ 10 nanoteslas. These constitute a particle flux of $\sim 4.5 \times 10^8/cm^2$-sec, a kinetic energy flux of roughly 1 erg/cm^2-sec, and a magnetic energy flux of $\sim 10^{-2}$ erg/cm^2-sec. All large-scale objects orbiting the sun encounter this continual flow of matter and energy and, because of the cohesive influence of the IMF, they can restrain, locally, the solar wind's flow and thereby store solar wind plasma and energy in their "magnetospheres." The magnetospheres of those planets having substantial intrinsic magnetic fields (e.g., Mercury, Earth, Jupiter, Saturn) are formed by tangential stresses that the solar wind applies as it gives up momentum to the outer regions of those fields, stretching them downstream to form long comet-like "magnetotails." The magnetospheres of comets and of planets that lack a substantial intrinsic field (e.g., Venus) are formed by the slowing down or "hanging-up" of the IMF by mass loading with cometary or planetary ions or by interaction with their conducting ionospheres.

Storage of energy cannot continue indefinitely and the magnetospheres must, either continually or intermittently, release stored energy and plasma to their parent body or back to the solar wind. In the case of comets, large releases of energy to the solar wind occur intermittently, simply by the comet's discarding part or all of its plasma tail. This process, illustrated in Figure 1, is thought to occur by magnetic reconnection (see the paper by M. B. Niedner, Jr., in this volume). In Figure 1, even as the detached tail of comet Morehouse drifts downstream in the solar wind, a new tail is seen to be forming as lines of the IMF fold around the comet's head. It is reasonable to ask whether the magnetosphere of the earth may also rid itself, intermittently, of excess energy and plasma in this way; and, indeed, there is very good evidence that it does. This evidence, gained during the past decade, largely from satellite measurements of particles, plasmas, and fields in the outer magnetosphere, suggests very strongly that during "substorms" the earth's magnetosphere spontaneously divests itself of a substantial portion of the "plasma sheet" that extends across the midplane of its magnetotail. In fact it now appears that this is the basic underlying physical process in a substorm, i.e., the magnetotail getting rid of stored plasma and energy that it can no longer restrain.

This paper will describe the evidence mentioned above, discussing first that obtained during the past decade from satellites orbiting relatively close to the earth and concluding with a discussion of some remarkable observations, made very recently with the ISEE-3 satellite far out in the magnetotail, of plasmoids, i.e., detached portions of the plasma sheet, flowing out of the magnetotail into the downstream solar wind.

Auroral and Magnetic Substorms

The auroras that regularly light the night skies of the northern and southern polar regions of the earth have been an object of amazement and great interest to man throughout recorded history. Over the centuries, many men of science, famous for their contributions to philosophy, mathematics, chemistry, and physics, have puzzled, also, about the aurora. For example the name, aurora borealis (northern dawn), is said to have been first used by Galileo in 1616. Edmund Halley, in 1716, was the first to suggest some relationship between auroras and the earth's magnetic field. But modern understanding of the auroras began in the late nineteenth and early twentieth centuries with extensive observational and theoretical work such as that of Birkeland and Störmer that showed the auroras to be caused by charged particles, guided by the earth's magnetic field, bombarding the upper atmosphere. (See Eather, 1980, for an excellent description of auroras and the development of our understanding of them from antiquity up to modern times.)

The International Geophysical Year (IGY) of 1957-1958 brought globally-organized research efforts, along with modern rocket and satellite technologies, to bear on geophysics problems, with a strong focus on auroral and magnetospheric research. Widespread networks of all-sky cameras and magnetometers were established to study the large-scale space and time variations of auroras and their associated magnetic signatures. Using data from these, Akasofu [1964] identified a sequence of systematic and characteristic auroral displays which he called an "auroral substorm." Quiet auroras which lie along the auroral oval are intermittently activated. It is at these times that bright, active, and spectacular auroral displays are seen. This activation generally originates near local midnight and rapidly spreads to earlier and later local times and to higher latitudes. The substorm reaches its peak in a rather short time (10-20 minutes) and gradually subsides. The auroral activation is accompanied by substantial (as great as 1 to 2 percent) variations of the local surface geomagnetic field, called "magnetic bays" and thus the name "magnetic substorm" is sometimes applied to these intervals of auroral and magnetic activation. On some days (called "quiet days") there

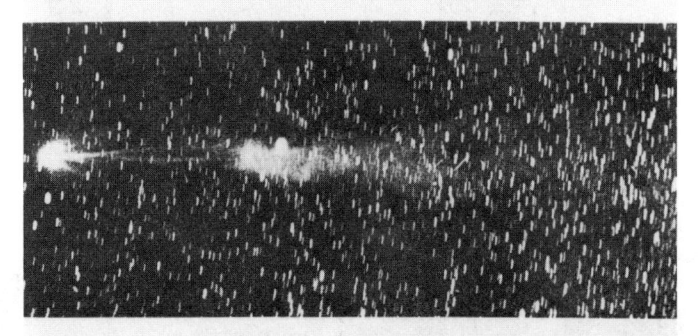

Fig. 1. Comet Morehouse (1908 III) on September 30, October 1, and October 2, 1908. This sequence shows the disconnection and drifting away of the plasma tail (Yerkes Observatory photograph).

may be only a very small number of substorms (or none) identified. But "active days" are more usual, and then substorms occur every few hours.

The Reconnection Model of Substorms

The advent of scientific spacecraft brought observations of plasmas, particles, and fields in distant space around the earth without which our understanding of substorms could not have gone much beyond simply their identification. The existence and general character of the earth's magnetotail were first fully recognized by Ness [1965] in magnetic data returned by the earth satellite Imp 1. The plasma sheet that carries the cross-tail electric current was discovered by Bame et al. [1967] with a plasma probe on the earth satellite Vela 2B. It was found, soon after these basic structural features were ascertained, that certain variations of the magnetotail and of its plasma and particle populations characteristically accompany substorms. As a result of these latter discoveries, the substorm quickly assumed a central role in magnetospheric physics, as a phenomenon that ordered a wide variety of outer magnetospheric observations and that thus seemed a fundamental part of the magnetosphere's functioning and required a physical explanation. Unfortunately, but perhaps not surprisingly, the complexity of such a global phenomenon and the relatively incomplete observational coverage of it that can be gained from limited numbers of ground stations sometimes make the substorm's identification imprecise. Nevertheless, the studies of magnetospheric dynamics over the past 16 years, with the substorm as a focus, have achieved a quite satisfactory understanding of the macroscopic aspects of the solar wind-magnetosphere interaction in terms of magnetic reconnection, including a reconnection model of substorms.

A full discussion of the reconnection model of substorms and of its evolution would take much more space than is available here. For such a discussion the reader is referred to Russell and McPherron [1973] and references to be found there. Here we shall treat only the plasma sheet and its substorm variations. We shall do so by first examining plasma sheet phenomena observed with Imp 8 satellite during a substorm on October 8, 1974, and then offering an explanation for these phenomena in terms of the reconnection model. The geomagnetic signature of the substorm was the 800 nT magnetic bay shown in the magnetometer record from the near-midnight Russian station, Dixon (Figure 2). Its onset at 18:27 UT was very sharp. It began to recover after its peak at ~19:05 UT and was rapidly recovering by 19:30 UT. This variation of the geomagnetic field was caused by a sudden "turning-on" of an intense westward current (called a westward electrojet) in the ionosphere above Dixon. Examination of records from other near-midnight stations revealed that the electrojet started nearly simultaneously over a several-hour extent in local time along the auroral oval. The electrojet resulted from diversion of part of the magnetotail's dawn-to-dusk cross-tail current along magnetic field lines into the ionosphere [McPherron et al., 1973].

At the time of this substorm Imp 8 satellite was in the magnetotail about 32 R_E from earth, very near local midnight and about 2 R_E below (southward of) the estimated location of the magnetic midplane of the tail. Data from the satellite are shown in Figure 3 where the onset time of the substorm (18:27 UT) is marked by a dashed vertical line. Coincident with the bay onset the plasma began to flow tailward and the energetic electron flux rose suddenly above background. The magnetic field latitude first became northward and then, at 18:28:45 UT it turned steeply southward. At 18:29:20 UT a dropout of plasma started (indicated by the rapid reduction of plasma electron energy density, U). A particularly important aspect of these data is that the energetic electrons were essentially isotropic when they first appeared at 18:27 UT and for about 4 minutes longer. Then, at 18:31 UT they suddenly began to display a pronounced tailward-streaming distribution, indicating that the satellite became enveloped at that instant by magnetic field lines that opened tailward [Baker and Stone, 1976].

From 18:29 UT until 19:06 UT the field latitude was mostly southward, becoming approximately zero at times when Imp 8 was largely outside the

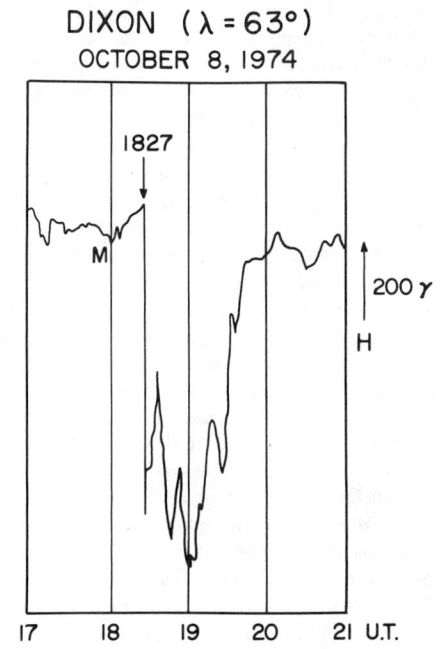

Fig. 2. Record of the horizontal component (H) of the geomagnetic field measured at Dixon, 17:00-21:00 UT, October 8, 1974. The letter M marks the time when Dixon is at local magnetic midnight. The arrow at right indicates the direction and magnitude of a positive change of H of 200γ, or 200 nanoteslas.

plasma sheet, as indicated by low values of U (e.g., 18:31-18:35 UT and 18:57-19:06 UT). Also, the plasma flow continued to be tailward and the energetic electron flux was low but streaming tailward when measurable. At 19:06 UT, when the bay at Dixon had begun showing rapid recovery, the field latitude became primarily northward, plasma flow turned earthward, and the energetic electron flux became more intense and isotropic.

The interpretation of these data is depicted schematically in Figure 4 where midnight meridian plane cross-sections of the plasma sheet are shown at several sequential times. (A dot represents the Imp 8 location at ~32 R_E from earth.) The processes depicted there do not prevail over the whole width of the plasma sheet but over perhaps the central one-half of its width. A "substorm neutral line" or X-line, N', forms in the near-earth sector of the plasma sheet at substorm onset (panel 2). Magnetic reconnection occurring there causes fast jetting of plasma earthward and tailward. The earthward jetting plasma flows along field lines to the ionosphere where, by intermediate processes that are still not fully understood, it creates the auroras. The tailward jetting plasma constitutes the tailward flow sensed by Imp 8 starting at 18:27 UT. The reconnection, continuing in panels 3 and 4, creates a structure of closed magnetic loops, and in panel 5 the last closed field line of the pre-substorm plasma sheet is pinched off by reconnection, leaving the plasma sheet tailward of N' magnetically detached from earth. Panels 6 and 7 show this detached plasma sheet, now a free "plasmoid," accelerating tailward under the influence of pre-existing plasma pressure gradients and of lobe field lines (lines 6 and 7) which reconnected after its detachment and are contracting tailward. In panels 6 and 7 the center of the plasmoid, the magnetic O-line, approaches and passes Imp 8 and this causes the change from positive to negative magnetic field latitude seen at 18:28:45 UT. In panels 7 and 8 the trailing edge of the plasmoid approaches and passes Imp 8 and this causes the plasma dropout that begins at 18:29:20 UT and that soon thereafter results in Imp 8's exit from the closed loops of the plasmoid to the later-reconnected field lines that envelop it. The exit from the plasmoid is indicated at 18:31 UT by the appearance of a tailward-streaming population of energetic electrons.

In panel 9 the satellite is shown near a very thin plasma sheet, downstream from the substorm X-line, which has remained at the near-earth location of its initial formation. In this thin downstream plasma sheet one expects to find plasma flowing tailward from the X-line, threaded with open field lines having southward latitude where they cross the midplane. And indeed, these are the conditions found in the Imp 8 data from ~18:35 UT to ~18:56 UT, where frequent neutral sheet crossings are indicated by the fluctuations of field longitude, the field latitude is predominantly southward, tailward plasma flow prevails, and the energetic electrons stream tailward, indicating open field lines.

Panel 10 shows the X-line, N', at a new distant location N", and the plasma sheet of closed field lines thickening over the satellite earthward of N". These are the conditions that were seen in the Imp 8 data starting at ~19:05 UT after an ~11 minute plasma dropout. The field latitude is predominantly northward, the plasma flow is earthward, the energetic electrons become isotropic, indicating closed field lines, and their intensity builds up (because they are now on earth-tied closed field lines and can no longer escape so readily).

The picture that evolves from the above discussion of data from the 18:27 UT substorm on October 8, 1974 is that, starting precisely at the substorm's onset, as observed at earth, the process of magnetic reconnection began at ~15 R_E from earth in the magnetotail. This led, within ~5 minutes, to the severance of a substantial longitudinal sector of the pre-substorm plasma sheet which then flowed rapidly tailward (at speeds approaching 1000 km/sec), presumably to join the solar wind far downstream. After the plasmoid's departure the substorm X-line remained near earth for about a half hour and then, in conjunction with recovery of the auroral zone negative bay, it moved suddenly and rapidly tailward beyond Imp 8, causing the plasma sheet to once more thicken and lengthen, filling with plasma jetting earthward from the retreating neutral line.

Many examples of this sequence were found [e.g., Hones, 1977; Bieber et al., 1982; Bieber, this volume], and the occurrence of plasmoid formation and release and its role in energy dissipation from the magnetosphere were quite well established by these relatively near-earth satellite observations. Just within the past year, data have been obtained from the satellite ISEE 3, as it traversed the magnetotail at ~220 R_E from earth, that provide remarkably detailed confirma-

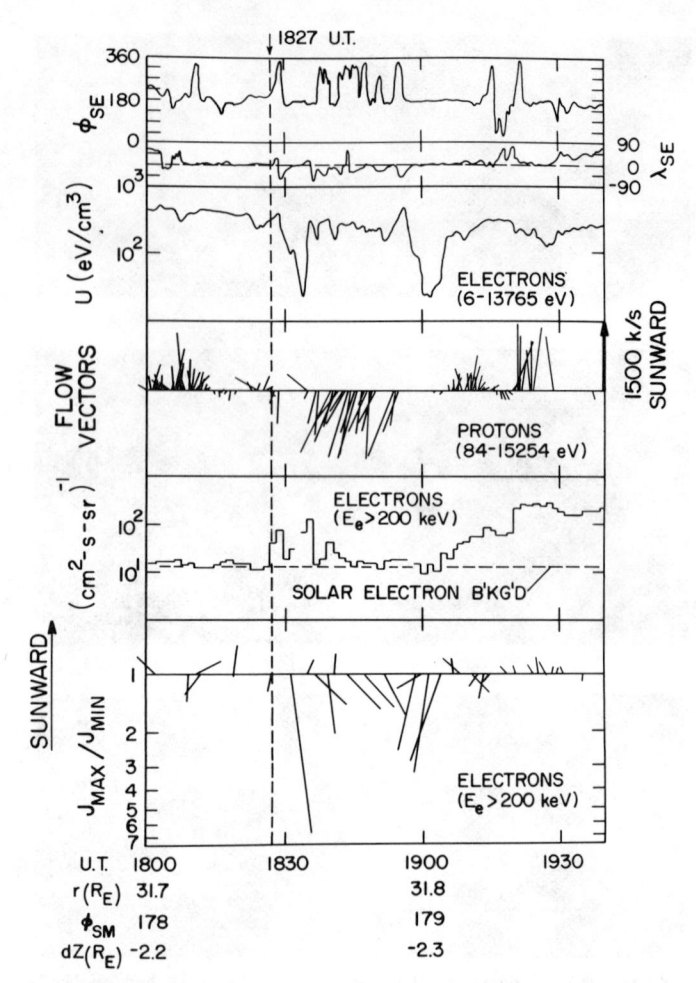

Fig. 3. Various data sets from Imp 8, 18:00-19:30 UT, October 8, 1974. First (top) panel: Longitude (ϕ_{SE}) of the magnetic field. Zero degrees is sunward, 180° is tailward, 90° is duskward. Second panel: Latitude (λ_{SE}) of the magnetic field. Plus 90° is due northward, −90° is due southward. Third panel: Energy density (U) of plasma electrons. Fourth panel: Vectors indicating the direction and magnitude of the bulk flow of plasma, derived from the measured distribution function of plasma ions. Fifth panel: Directional flux of energetic electrons. Sixth panel: Anisotropy parameter, J_{MAX}/J_{MIN}, of energetic electrons. Downward indicates tailward streaming, upward indicates earthward streaming. Numbers at the bottom are the universal time (UT), the satellite geocentric radial distance (r), its longitude (ϕ_{SM}), and its distance from the estimated position of the magnetotail neutral sheet (dZ). The units of r and dZ are earth radii (R_E) and 1 R_E = 6370 km. ϕ_{SM} = 180° is in the anti-sunward direction (from Hones, 1979a).

tion of these near-earth results. Briefly, ISEE 3 found that the plasma sheet at 220 R_E becomes thick, filled with hot plasma flowing rapidly tailward, about 30 minutes after the onset of a substorm at earth. The hot plasma is found to contain closed magnetic loops in a pattern consistent with a large plasmoid. These observations are the subject of the next section.

ISEE 3 Observations of Plasmoids 220 R_E from Earth

From December 1982 to March 1983, ISEE 3 traveled outward along the magnetotail to $X_{SE} \approx -220$ R_E and back to earth. (See Bame et al., 1983, for the configuration of the orbit.) Hones et al. [1984] have reported results of a study of data acquired during a 16-day interval in January-February 1983 when the satellite was quite near the center of the magnetotail about 220 R_E from earth. Figure 5 shows geomagnetic records from six auroral zone stations and one mid-latitude station (Rapid City, South Dakota) for January 25, 1983. Periods of

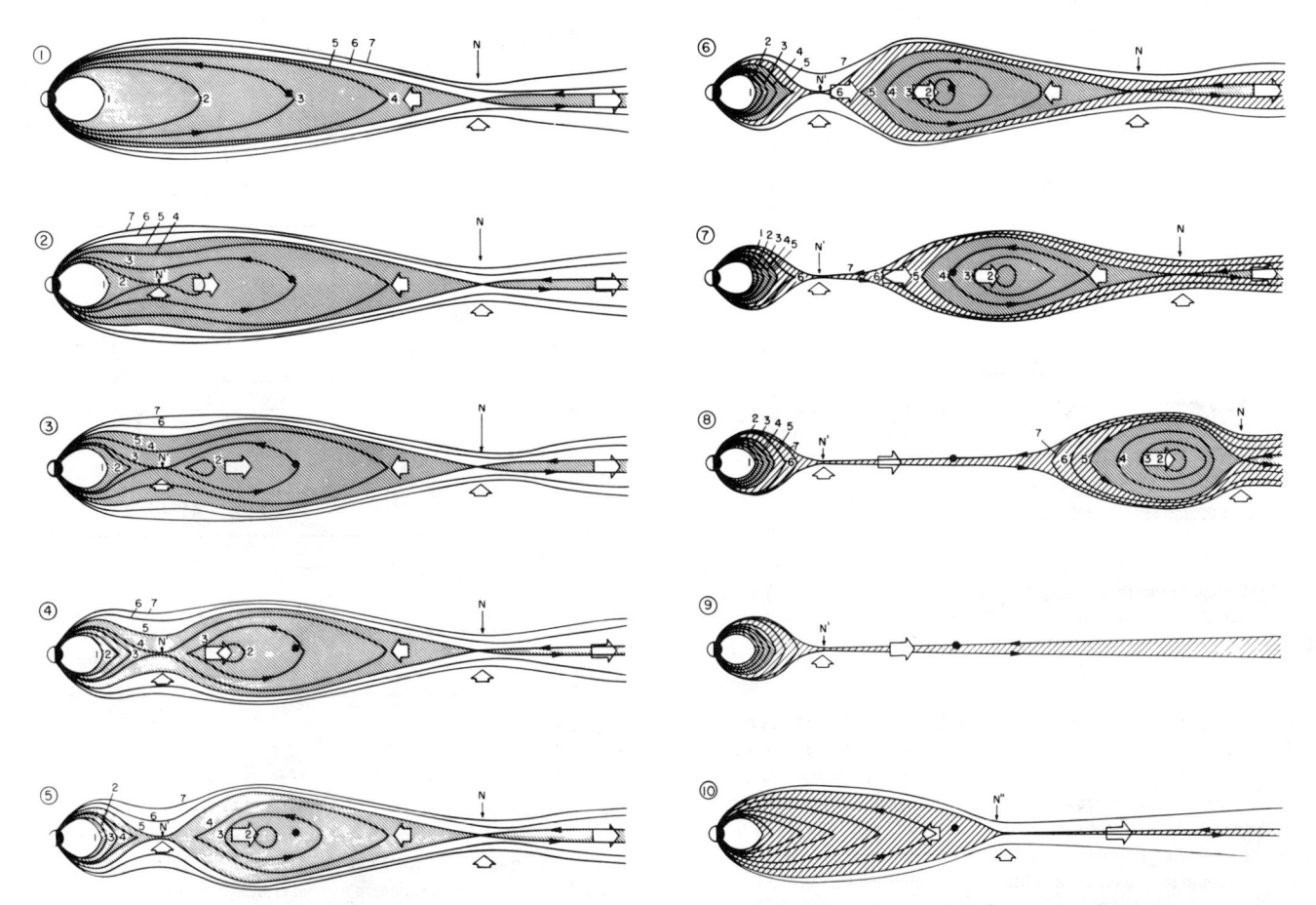

Fig. 4. Schematic representation of changes of the magnetotail plasma sheet that are thought to occur during substorms. These are cuts along the midnight meridian plane of the tail. Earth is at left and a dot near the center of each picture represents a satellite at $r \approx 35$ R_E and $dZ \approx +1$ R_E. Black lines are magnetic field lines and white arrows indicate plasma flow. A distant neutral line, N, is shown at $r \approx 60$ R_E and is thought to be a quasi-permanent feature of the magnetotail though its distance is not really known and is probably quite variable. The fine hatching indicates the plasma sheet, which contains closed field lines 1, 2, 3, 4 and is bounded by the "last closed field line," 5. Field lines 6 and 7 are in the lobe, outside the plasma sheet (from Hones, 1979b).

enhanced activity are seen at near-midnight stations at ~04:30-06:45 UT (NA, GW, FC, RPC), ~10:00-16:30 UT (RPC, ME, CO), and ~18:00-22:00 UT (AB). Plasma electron data acquired during the same day with a solar wind plasma instrument on ISEE 3 [Bame et al., 1978] are shown in Figure 6. Careful examination of extended periods of data such as these [Bame et al., 1983] and of accompanying magnetic data from the vector helium magnetometer on the satellite [Frandsen et al., 1978; Slavin et al., 1983] permit these authors to identify, with high confidence, intervals when ISEE 3 is in the magnetosheath, the tail lobes, and the plasma sheet. Briefly, the magnetosheath is characterized by the highest densities and lowest temperatures of electrons; the lobes are identified primarily by a nearly constant magnetic field with latitude ~0° and longitude ~0° or ~180°; the plasma sheet is identified primarily by the highest temperatures and, usually, the highest flow speeds (almost always tailward). Using such means, Hones et al. [1984] reached the identifications shown under the Temperature graph of Figure 6. At ~05:00 UT, ~30 minutes after the first interval of enhanced geomagnetic activity began (Figure 5), ISEE 3 crossed from the magnetosheath into the tail lobe, and then into the plasma sheet. About an hour later it went back through the lobe and into the magnetosheath. Again, at ~10:30 UT, about 30 minutes after the onset of the second interval of enhanced geomagnetic activity, ISEE 3 crossed from the magnetosheath into the tail and plasma sheet once more, remaining in the plasma sheet or lobe, this time, throughout essentially that whole extended period of geomagnetic activity. Finally, ISEE 3 was in the plasma sheet again from ~18:30 to ~19:30.

The point to be made from these data is that the magnetotail at 220 R_E appears to swell and envelop ISEE 3 about 30 minutes after the onset of intense geomagnetic activity. ISEE 3 usually encounters the plasma sheet shortly after such tail entries and the implication is that the plasma sheet thickens greatly and is responsible for the tail's swelling.

Figure 7 shows magnetic field data from ISEE 3 for the first tail encounter, ~05:00-07:00 UT. ISEE 3 presence in the plasma sheet is indicated by the reductions of field strength, B, during the interval ~05:25 UT to ~06:12 UT. (The other reductions signify magnetosheath.) The important point here is that the latitude of the field in the plasma sheet is first steeply northward and then steeply and more enduringly southward (top panel). This is the magnetic signature anticipated for the passage of the severed plasma sheet and the body of later reconnecting field lines that follow it.

Figure 8 shows plots of the magnetic field latitude during 14 instances when ISEE 3 made passages from the lobe into the plasma sheet. The beginning of each graph is the time of onset of the corresponding geomagnetic activity enhancement at earth. The time delays (Δt) between substorm onset at earth and entry of ISEE 3 into the plasma sheet are remarkably uniform and average about 30 minutes.

182 PLASMA SHEET BEHAVIOR

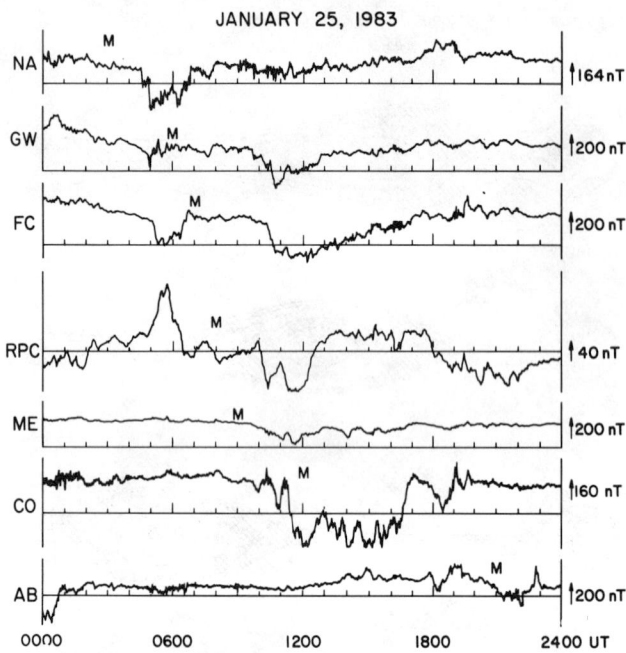

Fig. 5. Magnetograms for January 25, 1983 from Narssarssuaq (NA), Great Whale River (GW), Fort Churchill (FC), Rapid City (RPC), Meanook (ME), College (CO), and Absiko (AB). Magnetic midnight is indicated by M.

Figure 9 shows the interpretation given these ISEE 3 results by Hones et al. [1984]. The plasma sheet is severed by magnetic reconnection near the earth at T = 0 minutes. By T = 30 minutes the plasmoid reaches and envelopes ISEE 3, having expanded laterally because of reduced lobe magnetic field pressure at greater distances. By T = 50 minutes the plasmoid is well past ISEE 3 which remains for some time, however, in the body of later reconnected field lines contracting behind the plasmoid (and helping to accelerate it tailward).

We have found remarkably clear verification of important features of this model in reported measurements of energetic electrons made on ISEE 3 (Scholer et al., 1983; Scholer, this volume). Note that the model predicts that as the plasmoid approaches ISEE 3 the satellite should first be enveloped by open field lines that extend from near the substorm X-line and that were reconnected after the plasmoid's departure. Only after that will ISEE 3 enter the closed-loop structure of the plasmoid itself. Scholer et al. measure electrons (E_e = 75-115 keV) and find that the flux of these is enhanced during substorm-related plasma sheet encounters such as we have discussed above. And, remarkably, they find that the first-appearing electrons arrive before ISEE 3 is enveloped by the plasma sheet and they stream tailward, indicating open field lines. A few minutes later, as ISEE 3 becomes enveloped by the plasma sheet, the electrons become isotropic, indicating a closed magnetic structure. An example of this is shown in Figure 10. A half-hour enhancement of electrons began at ~03:05 UT on 16 February, 1983, some 20 minutes after onset of a very intense substorm at earth. The magnetic field remained lobe-like (i.e., ISEE 3 was not enveloped by high density plasma) and the electrons streamed tailward (open field lines) for about 12 minutes. Then at ~03:17 UT ISEE 3 entered the plasma sheet and the electron flux became isotropic, indicating a closed magnetic configuration. As noted above, such observations as these are in remarkable agreement with the model in Figure 9.

The size of the plasmoids can be estimated only roughly from the ISEE 3 measurements. Their dimension in the direction of motion must be something like 50-100 R_E because the northward and initial southward swing of the magnetic field takes about 10-15 minutes and the flow speed is ~500-1000 km/sec (~5 to 10 R_E/minute). Their north-south dimensions must be comparable to the diameter of the unperturbed tail (i.e., probably somewhat bigger than illustrated in Figure 9) because transitions from magnetosheath, through the lobe, to the plasma sheet are often seen (e.g., the ~0500 UT event in Figure 6). Thus that

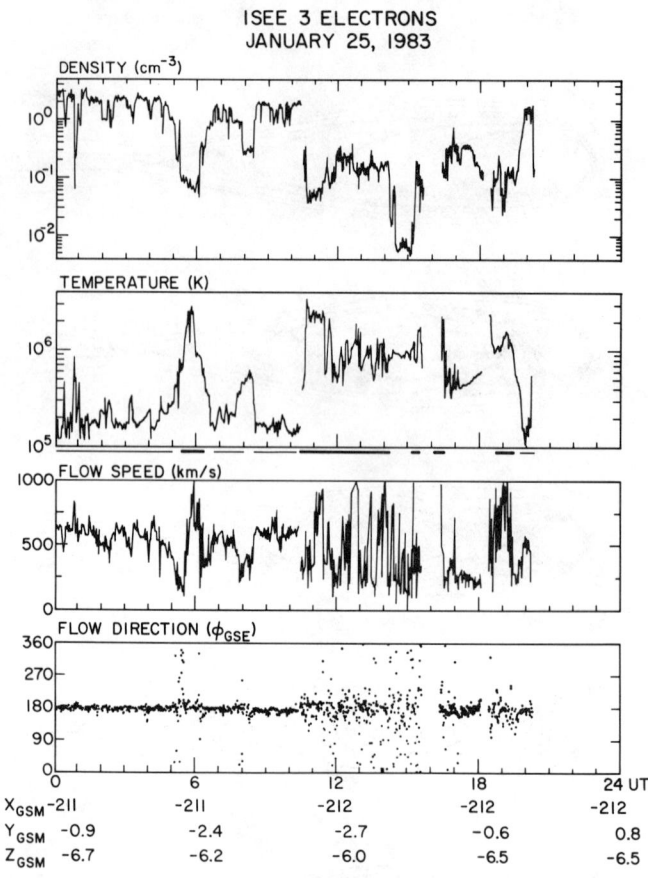

Fig. 6. Plasma electron parameters. A flow direction (ϕ_{GSE}) of 180° is tailward. ISEE 3 presence in the plasma sheet is indicated by heavy lines under the temperature graph. Magnetosheath intervals are indicated by light lines. The unmarked periods are lobe intervals (from Hones et al., 1984).

dimension may be ~20-30 R_E. Their east-west dimension also must be comparable to the tail's diameter because they are seen with high probability after substorms throughout the ISEE 3 traversal across the distant tail. Thus a plasmoid is a large structure and its passage causes a substantial deformation of the distant magnetotail.

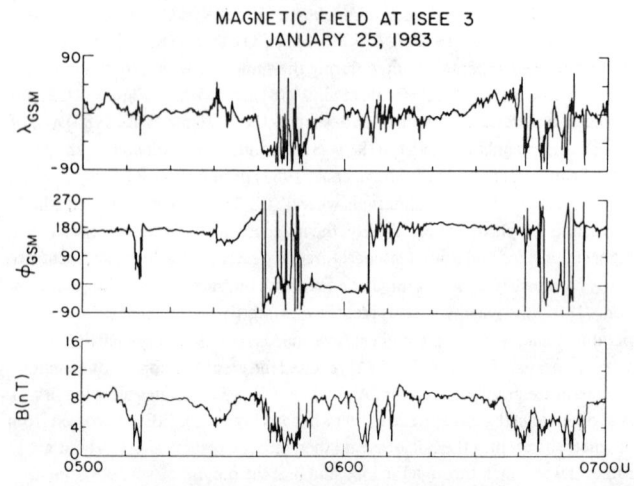

Fig. 7. Magnetic field parameters measured by ISEE 3, January 25, 1983. In the latitude (λ_{GSE}) graph 90° is due north, -90° is due south. In the longitude (ϕ_{GSE}) graph 0° is sunward, 180° is tailward. The magnitude (B) is given in nanoteslas.

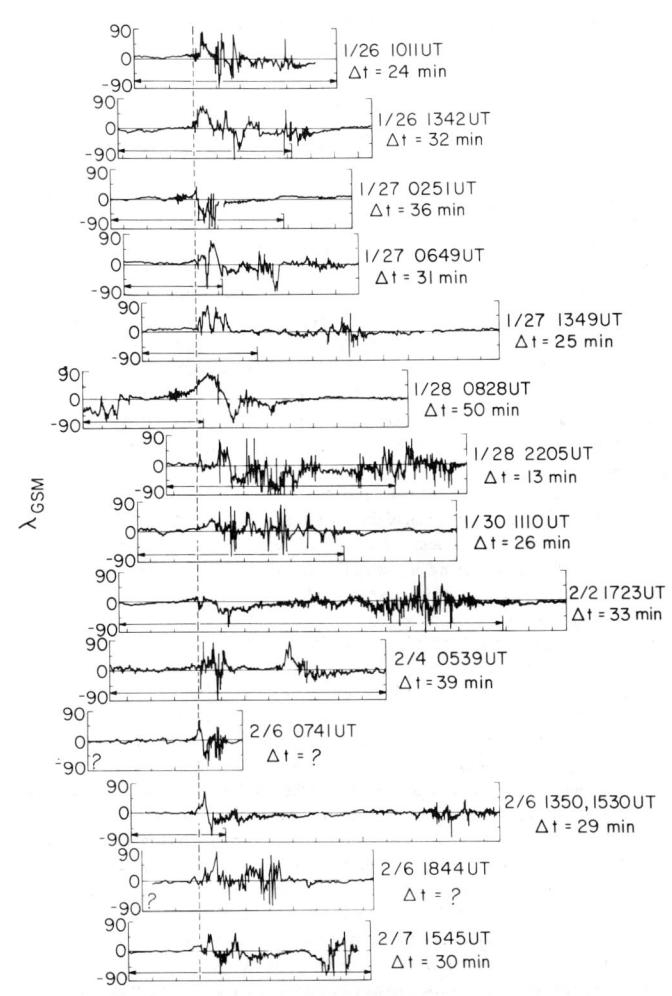

Fig. 8. Magnetic field latitude measured by ISEE 3 during plasma sheet encounters. The vertical dashed line marks the time of entry into the plasma sheet, and this time is given at the right of each curve. Horizontal lines with arrowheads mark the durations of substorms at earth. Values of Δt at the right are the number of minutes from substorm onset to plasma sheet entry. Question marks indicate lack of clear substorm onset times. Tic marks on base lines are 10 minutes apart (from Hones et al., 1984).

Conclusions

Large bodies orbiting the sun continually accumulate energy and plasma from the ever-flowing magnetized solar wind. This energy cannot build endlessly but must be dissipated somehow. We can see at least one form of this dissipation in comets, where the energy stored in their plasma tails is intermittently returned to the solar wind when the plasma tails are severed, probably by magnetic reconnection.

The auroral or magnetic substorm has been, for nearly two decades, an object of intense scientific research and has been regarded as a process by which energy from earth's magnetosphere is suddenly and rapidly dissipated primarily into the ionosphere. But several years ago data from satellites orbiting in the magnetotail within ~35 R_E of earth were interpreted to mean that during a substorm a sector of the plasma sheet is severed from earth by magnetic reconnection to form a free configuration of magnetized plasma, a plasmoid, that flows out of the magnetotail into the solar wind. Now, ISEE 3 satellite, in the magnetotail 220 R_E from earth, has observed these plasmoids flowing tailward past that location. The discreteness of these plasma releases through the magnetotail and their delayed association, at 220 R_E, with substorm onsets at earth suggest that they are consequences of spontaneous release, probably by magnetic reconnection, of

energy and plasma earlier stored in the magnetotail. It is likely that this spontaneous release of energy and plasma to the solar wind, analogous to that observed in comets, is the underlying physical process in substorms, and that the concurrent energy release to the earth, creating the auroral substorm, is simply a by-product of this underlying process.

It is reasonable to believe that other planetary magnetospheres (e.g., those of Mercury, Jupiter, and Saturn) intermittently give up energy through plasmoid formation and release. Indeed, evidence for magnetic reconnection in the near tail of Jupiter has been found (see papers by Behannon and by Nishida in this volume). Furthermore, plasma wave measurements from Voyager 2, ~7000 R_J (~3 AU) downstream from Jupiter, have shown variations that can, perhaps, be interpreted as due to the passage of plasmoids through (or from) the Jovian magnetotail [Kurth et al., 1982]. It is interesting to ask how long these planetary plasmoids maintain their identity as they are carried by the solar wind toward the edge of the heliosphere.

Finally we note that the plasmoids created in the magnetotail are somewhat analogous to those made in "compact toroid" experiments [e.g., "Spheromaks" and "Field Reversed Configurations (FRC's)] in laboratory fusion research (see papers by Hammer and by Milroy in this volume). A particularly interesting such analogy is found between the "translation" of FRC plasmoids that has been

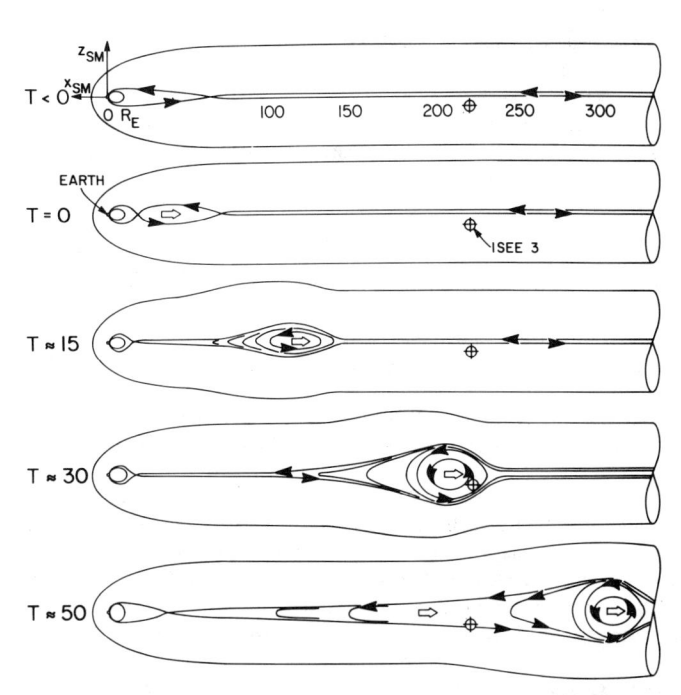

Fig. 9. Model depicting the severance of (a longitudinal sector of) the plasma sheet at substorm onset (T = 0) and its departure along the tail as a closed magnetic structure, a plasmoid. Black arrows indicate the magnetic field direction and white arrows indicate plasma flow (from Hones et al., 1984).

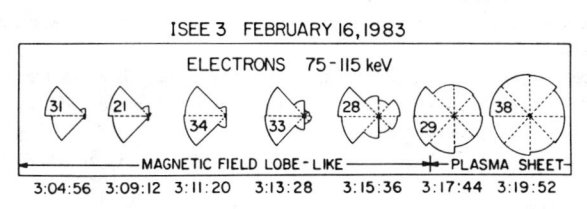

Fig. 10. Angular distributions of energetic electrons measured by ISEE 3 on February 16, 1983. The azimuthal distribution is measured in eight sectors, and the count rate in the most intense sector is written in that sector. The line at the bottom indicates that the magnetic field was lobe-like (ISEE 3 was not yet embedded in the plasma sheet) until ~03:17 UT (adapted from Scholer et al., 1983).

achieved in the laboratory [Rej et al., 1983] and the transport of plasmoids through the magnetotail.

Acknowledgments. I am grateful to E. J. Smith and co-workers, of the Jet Propulsion Laboratory, for the ISEE 3 magnetic field date used in this report. This work was performed under the auspices of the U.S. Department of Energy.

References

Akasofu, S.-I., The development of the auroral substorm, *Planet. Space Sci., 12*, 273, 1964.

Baker, D. N. and E. C. Stone, Energetic electron anisotropies in the magnetotail: Identification of open and closed field lines, *Geophys. Res. Lett. 3*, 557, 1976.

Bame, S. J., J. R. Asbridge, H. E. Felthauser, J. P. Glore, H. L. Hawk, and J. Chavez, ISEE-C solar wind experiment, *IEEE Transactions on Geoscience Electronics, GE-16*, 160, 1978.

Bame, S. J., R. C. Anderson, J. R. Asbridge, D. N. Baker, W. C. Feldman, J. T. Gosling, E. W. Hones, Jr., D. J. McComas, and R. D. Zwickl, Plasma regimes in the deep geomagnetic tail: ISEE-3, *Geophys. Res. Lett. 10*, 912, 1983.

Bieber, J. W., E. C. Stone, E. W. Hones, Jr., D. N. Baker, and S. J. Bame, Plasma behavior during energetic electron streaming events: Further evidence for substorm-associated magnetic reconnection, *Geophys. Res. Lett. 9*, 664, 1982.

Eather, R. H., *Majestic Lights*, American Geophysical Union, 1980.

Frandsen, A. M. A., B. V. Connor, J. Van Amersfoort, and E. J. Smith, The ISEE-C vector helium magnetometer, *IEEE Transactions on Geoscience Electronics, GE-16*, 195, 1978.

Hones, E. W., Jr., Substorm processes in the magnetotail: Comments on 'On hot tenuous plasmas, fireballs, and boundary layers in the earth's magnetotail' by L. A. Frank, K. L. Ackerson, and R. P. Lepping, *J. Geophys. Res. 82*, 5633, 1977.

Hones, E. W., Jr., Transient phenomena in the magnetotail and their relation to substorms, *Space Sci. Rev. 23*, 393, 1979a.

Hones, E. W., Jr., Plasma flow in the magnetotail and its implications for substorm theories, in *Dynamics of the Magnetosphere* (S. I. Akasofu, ed.), p. 545, D. Reidel Publ. Co., 1979b.

Hones, E. W., Jr., D. N. Baker, S. J. Bame, W. C. Feldman, J. T. Gosling, D. J. McComas, R. D. Zwickl, J. A. Slavin, E. J. Smith, and B. T. Tsurutani, Structure of the magnetotail at 220 R_E and its response to geomagnetic activity, *Geophys. Res. Lett., 11*, 5-7, 1984.

Kurth, W. S., J. D. Sullivan, D. A. Gurnett, F. L. Scarf, H. S. Bridge, and E. C. Sittler, Jr., Observations of Jupiter's distant magnetotail and wake, *J. Geophys. Res. 87*, 10373, 1982.

McPherron, R. L., C. T. Russell, and M. P. Aubry, Satellite studies of magnetospheric substorms on August 15, 1968. 9. Phenomenological model for substorms, *J. Geophys. Res. 78*, 3131, 1973.

Ness, N. F., The Earth's Magnetic Tail, *J. Geophys. Res. 70*, 2989, 1965.

Rej, D., T. Armstrong, R. Bartsch, T. Carroll, R. Chrien, J. Cochrane, R. Gribble, P. Klingner, P. Linford, K. McKenna, E. Sherwood, R. Siemon, M. Tuszewski and E. Yavornik, Axial transactions of FRC's in FRX-C, *Bull. Am. Phys. Soc. 28*, 1242, 1983.

Russell, C. T. and R. L. McPherron, The magnetotail and substorms, *Space Sci. Rev. 15*, 205, 1973.

Scholer, M., G. Gloeckler, B. Klecker, F. M. Ipavich, and D. Hovestadt, Fast moving plasma structures in the distant magnetotail, *J. Geophys. Res.*, submitted, September 1983.

Slavin, J. A., B. T. Tsurutani, E. J. Smith, D. E. Jones, and D. G. Sibeck, Average configuration of the distant ($\leq$220 R_E) magnetotail: Initial ISEE-3 magnetic field results, *Geophys. Res. Lett.*, in press, 1983.

Questions and Answers

Moore: Do you have an estimate of the total energy output down the tail in a substorm?

Hones: The plasma density in the plasmoids is ~0.3/cm³ We don't measure protons but their flow kinetic energy (at 500-1000 km/sec) is ~4 keV, and let's assume their thermal energy is comparable. Then taking a plasmoid length of 60 R_E and diameter of 20 R_E, we find the particle thermal and flow energy total ~1.5 × 10²² ergs. But the passage of the plasmoid (the closed-loop structure) takes only about 10 minutes and is often followed for several times this interval by the later-reconnected lobe field lines. Thus, I estimate that the total energy output down the tail is several times 10²² ergs.

Vasyliunas: There are two possible topological configurations for plasmoids. Earthward of the large-scale X line associated with the open magnetosphere, they have an X line on the earthward side joined to an O line in the middle; tailward of the large-scale X line they have an X line on the tailward side joined to an O line in the middle (see figures and associated discussion in Vasyliunas, this volume). It would be interesting to determine which topology is found in the plasmoids observed by ISEE 3, as this would tell us whether the large-scale X ring of the open magnetosphere does or does not usually extend beyond the ISEE 3 location. The signature of B_z is the same for both types of plasmoids, but the relative timing of plasma (indicative of closed field lines within the plasmoid) and of energetic electron streaming (indicative of open separatrix field lines) might provide some clues.

Hones: It seems to me that the best indication of whether or not ISEE 3 is tailward of the X ring is whether it observes fast tailward plasma flow whenever it crosses the plasma sheet. It does so at x ≃ -220 R_E, so I would say the X ring is almost always earthward of that distance. It's not clear to me how any better clues could be derived from the energetic electrons other than to use them to determine whether the field lines in the rapidly flowing plasma are open or closed.

Birn (Comment): When the plasmoid crosses the location of the distant X line, which is a part of a ring like X-line around the earth, the near-earth X-line becomes connected with the sunward part of the original X-line ring and the plasmoid neutral line becomes completely detached. The energetic particle data seem to indicate that the plasmoid observed at 220 R_E is surrounded by open field lines extending into interplanetary space and has thus passed this stage. The fact that the plasmoid is obviously tailward of the newly reformed X-line ring around the earth, however, does not allow to conclude that it is also tailward of the average position of a more or less stationary X-line ring.

Birn (Comment): Tim Eastman asked you whether the plasma flow in the presented events could possibly be boundary layer flow. I would like to comment on that question. The main difference between plasma sheet boundary layer flow and the signatures observed in the distant tail by ISEE 3 is the orientation of the flow relative to the magnetic field. The magnetic field orientation during the fast flow events observed by ISEE 3 in the distant plasma sheet was usually first strongly northward and then strongly southward, so that indeed a significant part of the flow was perpendicular to the field. This fact makes these events different from boundary layer flow which is basically field aligned.

Lui: In your distant tail streaming observation from ISEE 3, do you see the velocity filtering effect as you enter the plasma sheet as predicted by Sarris and Axford?

Hones: Sarris and Axford (Nature, 77, 460, 1979) discussed the "inverse velocity dispersion" of high energy protons in the impulsive (~30-40 second) bursts, observed by a satellite (at $X_{SM} \approx$ -30 R_E) at substorm expansive phase onset as the boundary of the thinning plasma sheet passed by, leaving the satellite in the tail lobe. Their interpretation was that the protons originated at the X-line and that their spectrum softened due to preferential loss of the faster protons along the field lines as the reconnected lines moved away from the reconnection point. The boundary's motion over the satellite thus presented spectra of increasing energy as the separatrix was approached from within the plasma sheet. ISEE 3 sees essentially the reverse of this scenario as it crosses the "separatrix layer" from lobe to plasma sheet as the plasmoid approaches and envelopes it. Scholer et al. report seeing, first, streaming energetic electrons (which effectively define the separatrix because of their very high speed) and, later, streaming high energy, then lower energy, protons. Whether they clearly see the reverse of this, as the plasmoid departs (analogous to the Sarris-Axford observations) has yet to be determined.

STREAMING ENERGETIC ELECTRONS IN RECONNECTION EVENTS

John W. Bieber

Bartol Research Foundation of The Franklin Institute
University of Delaware, Newark, Delaware 19716

Abstract. Energetic electrons can be used to probe the large-scale topology of magnetic fields in Earth's magnetotail. In the plasma sheet region near the tail's midplane, these particles normally exhibit the trapped or isotropic angular distributions characteristic of closed magnetic field lines, but brief intervals of intense tailward streaming, indicative of open field lines, are occasionally observed. Such streaming events occur preferentially near the time of substorm onset as the observing spacecraft exits the thinning plasma sheet, and they are usually preceded by a 5-10 minute interval of fast tailward plasma flow and southward magnetic field. These correlated phenomena have been interpreted as evidence for magnetic reconnection at a transient magnetic X-line located ~ 15 R_E tailward of Earth. Recent studies of energetic electron streaming events report novel reconnection-related phenomena, including heating of plasma electrons, bump-in-tail electron velocity distributions, and possible rotational and tangential magnetic discontinuities.

Introduction

The large-scale topology of magnetic field lines is an issue of central importance in studies of magnetic reconnection. In Earth's magnetosphere, the only currently available means of probing field line topology is through measurements of the angular distribution of energetic electrons. These particles are uniquely suited to this purpose due to their high speed (~ 30 R_E/s at 200 keV), which causes them to respond very quickly to changes in magnetic field conditions at locations remote from the observation point, and due to their small gyroradius (~ 0.02 R_E), which makes them less susceptible than ions to cross-field drifts and other effects that complicate interpretation of anisotropies.

Energetic electrons are a common feature of Earth's plasma sheet. They normally exhibit trapped (i.e., symmetric about a plane perpendicular to the magnetic field) or isotropic angular distributions (Baker and Stone, 1977). As indicated in Figure 1a, magnetic field lines in the normal plasma sheet are connected to Earth at both ends. Energetic electrons on these closed field lines mirror in the strong near-Earth fields at either end of the field line, leading to equal fluxes in the Earthward and tailward hemispheres of the angular distribution -- i.e., a trapped distribution. Note that even on the highly stretched field lines of the plasma sheet, the bounce period of a 200 keV electron is only seconds.

In contrast with the plasma sheet, strong intensities of energetic electrons are usually absent from the magnetotail lobes. This feature is attributed to the open topology of magnetic field lines in this region. As shown in Figure 1a, the lobe field lines connect to Earth at one end and to the interplanetary magnetic field at the other, so that any electrons that are somehow injected onto these field lines quickly escape into the solar wind. In the absence of a source to replenish escaped particles, the lobes would be void of energetic electrons, as is usually observed.

Note, however, that if such a source did temporarily exist, then a spacecraft located tailward of the source region would observe a large unidirectional anisotropy, with essentially all of the flux contained in the tailward hemisphere of the angular distribution and only background fluxes in the Earthward hemisphere. Such pronounced unidirectional anisotropies are in fact observed for brief intervals and have been taken as evidence that the spacecraft has moved into a region of open magnetic field topology (Baker and Stone, 1976).

Examination of concurrent plasma data reveals that energetic electron streaming events occur preferentially at times of substorm-related plasma dropouts (or plasma sheet thinnings). Further, it is found that streaming events are usually preceded by a 5-10 minute interval of fast tailward plasma flow and southward magnetic field (e.g., Hones, 1977; Bieber et al., 1982).

These correlated phenomena have been interpreted as evidence that, near the time of substorm expansion phase onset, a transient magnetic X-line forms across a portion of the near-Earth

186 STREAMING ENERGETIC ELECTRONS

(a) Normal Plasma Sheet

(b) Plasmoid Formation

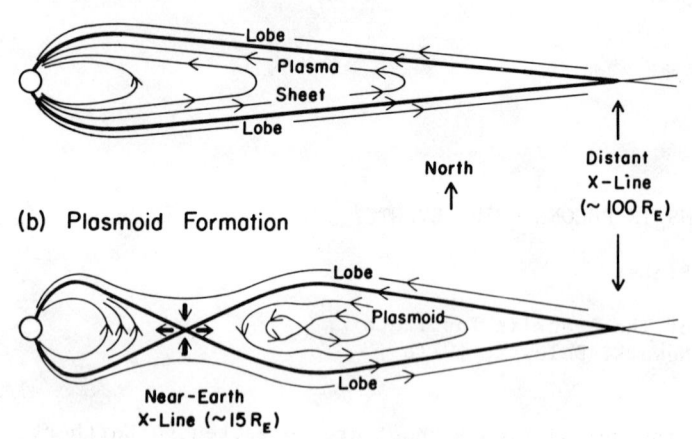

Fig. 1. (a) Schematic of magnetic field topology in the normal plasma sheet. The last closed field line (heavy line) lies at or near the interface of the plasma sheet with the magnetotail lobes and connects to a distant X-line believed to exist ~ 100 R_E tailward. In the plasma sheet, closed field lines connect to Earth at both ends. In the magnetotail lobes, open field lines connect to Earth at one end, and to the interplanetary magnetic field at the other (off right side of figure). (b) Schematic of magnetic field topology around a near-Earth X-line at the moment the last closed field line reconnects. Tailward of the near-Earth X-line, plasma sheet field lines now close in a magnetic island topology. This portion of the plasma sheet (the plasmoid) is no longer magnetically connected to Earth and is free to flow tailward into the solar wind. Thick arrows indicate plasma flow pattern in the vicinity of the near-Earth X-line. A second X-line is shown embedded in the plasmoid to suggest the complicated magnetic field behavior that would result from the formation of multiple X- and O-line pairs. Note that the near-Earth X-line would not necessarily extend across the full longitudinal width of the plasma sheet.

(~ 15 R_E) plasma sheet, causing this portion of the plasma sheet to undergo a large-scale change from the configuration shown in Figure 1a to the configuration shown in Figure 1b. If reconnection proceeds until the last closed field line (at the boundary of the plasma sheet) reconnects, then the portion of the plasma sheet tailward of the X-line is no longer magnetically tied to Earth and is ejected tailward into the solar wind (Russell and McPherron, 1973; Hones, 1977).

Energetic Electron Streaming Events

An example of an energetic electron streaming event observed on IMP-8 is shown in Figure 2. The > 200 keV electrons exhibit trapped/isotropic angular distributions for the first 3 samples shown (bottom panel), but at 15:37 UT there is a sudden transition to intense tailward streaming. The concurrent plasma data show that this streaming transition occurs precisely at the time the plasma density drops to very low levels (top panel). The second and fourth panels in Figure 2 show that the streaming transition is preceded by a 7-minute interval of fast (500-1000 km/s) tailward plasma flow and a progressively greater incidence of steeply inclined southward magnetic field. The beginning of this tailward flow interval coincides with a sharp substorm onset evident in the Cape Chelyuskin magnetogram, shown in Figure 3.

The interpretation of this event in terms of magnetic reconnection can be understood with reference to Figure 1b. A transient X-line has formed ~ 15 R_E tailward of Earth, resulting in the formation of a plasmoid with magnetic island

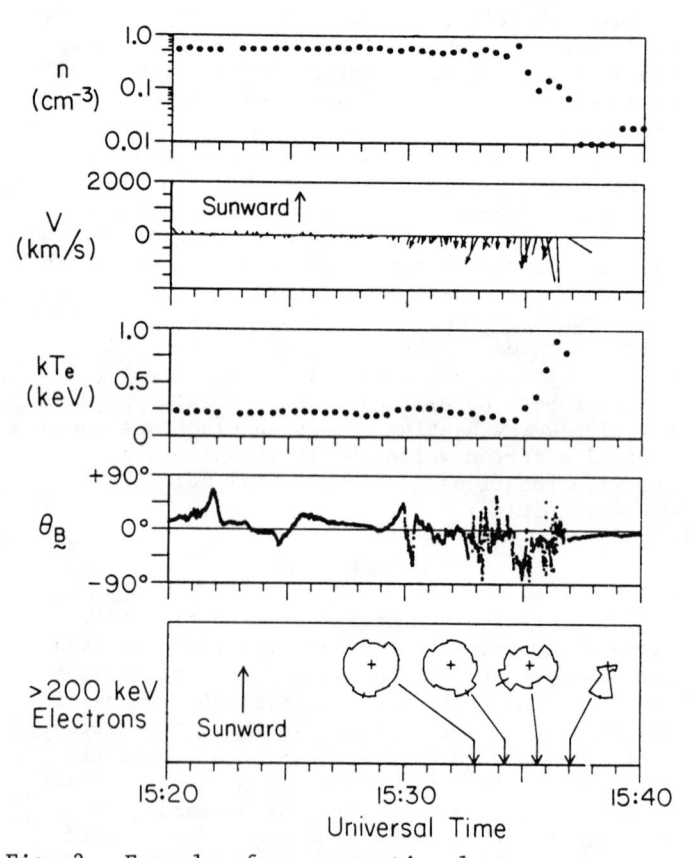

Fig. 2. Example of an energetic electron streaming event observed on IMP-8, located in the central plasma sheet 32 R_E tailward of Earth, on 14 November 1973. From top to bottom are shown plasma density n, bulk flow velocity V as calculated both from ion data (arrow-tipped vectors) and electron data (untipped vectors), electron temperature kT_e, elevation angle of magnetic field θ_B (GSM coordinates), and polar plots of angular distribution of > 200 keV electrons (normalized to a constant size of peak sector). From Bieber et al. (1983).

(or magnetic O-line) topology. A spacecraft located tailward of this X-line will observe tailward plasma flow due to the reconnection process and, eventually, due to tailward motion of the plasmoid. (For the event shown in Figure 2, IMP-8 was 32 R_E tailward of Earth.)

In the simplest model, the magnetic field lines in the plasmoid would form a single magnetic loop, and southward magnetic fields would thus be confined to the Earthward part of the loop. This is consistent with the observation that the greatest incidence of southward magnetic field is often observed during the latter part of the tailward flow interval. Certain reconnection theories, however, predict that multiple loops might form (Schindler, 1974; Galeev et al., 1978), and this could be partially responsible for the complicated magnetic field behavior observed in some events (Coroniti et al., 1977; Bieber et al., 1983).

Finally, after the last closed field line reconnects, the plasmoid is no longer magnetically connected to Earth, and is ejected tailward into the solar wind. As the spacecraft exits the plasmoid, a plasma dropout is observed, and, since the spacecraft is now on open lobe field lines, the angular distribution of energetic electrons changes from trapped/isotropic to intensely tailward streaming.

A point of considerable importance is that during the interval of tailward plasma flow, the energetic electrons continue to exhibit trapped/isotropic distributions, in support of the interpretation that the field lines in the plasmoid close in a magnetic island configuration. This magnetic island topology might actually be more complex than suggested in Figure 1b due to the presence of a finite B_y component (out of the plane of the figure -- see Hones et al., 1982). Nevertheless, plasmoid field lines are not open to the solar wind, so particle trapping would still occur.

Source of Energetic Electrons

The appearance of tailward-streaming energetic electrons at the moment the spacecraft exits the plasma sheet could indicate that processes at the near-Earth X-line accelerate particles to relativistic energies. The induction electric fields predicted by time-dependent reconnection theories appear capable of accelerating particles to the hundreds of keV energy range (Galeev et al., 1978). If the accelerated electrons were injected onto closed as well as open magnetic field lines, then this source could account for the trapped population of the normal plasma sheet as well.

Limited evidence in favor of this acceleration mechanism does exist. Terasawa and Nishida (1976) reported that the intensity of energetic electrons inside the plasmoid shows definite enhancements over normal levels. Meng et al. (1981) showed that energetic electron intensities in the

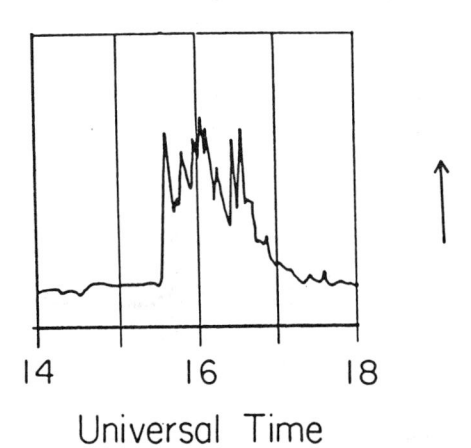

Fig. 3. Cape Chelyuskin magnetogram for a 4-hour interval which includes the energetic electron streaming event in Figure 2. The arrow to the right of the trace shows the magnitude and direction of a 200 nT negative perturbation. From Bieber et al. (1983).

plasma sheet are positively correlated with geomagnetic activity, as measured by the Kp index.

This evidence in favor of direct X-line acceleration of the streaming electrons is, however, rather indirect. An alternate possibility would be that plasma sheet electrons are accelerated by as yet unknown processes, and that the streaming occurs because reconnection-related changes in magnetic field topology allow the release of a pre-existing trapped population of energetic electrons onto open field lines.

It might be thought that perhaps tailward-streaming electrons are a more or less permanent feature of the plasma sheet boundary, and that the association of streaming events with substorm onset occurs simply because it is at this time that a spacecraft is likely to encounter the plasma sheet boundary, due to plasma sheet thinning. In this interpretation, it would be expected that streaming electrons would also be encountered when the spacecraft re-enters the plasma sheet from the lobes. However, statistical studies (see below) clearly show that intense streaming events occur preferentially at times of plasma dropout in temporal association with substorm expansion phase. Thus, the streaming electrons are not simply a spatial feature of the plasma sheet boundary.

Statistical Results

Statistical studies of energetic electron streaming events confirm that these events occur persistently in association with plasma/magnetic behavior indicative of magnetic reconnection. Bieber and Stone (1980) searched ~2 years of IMP-8 data and ~3 years of IMP-7 data for occur-

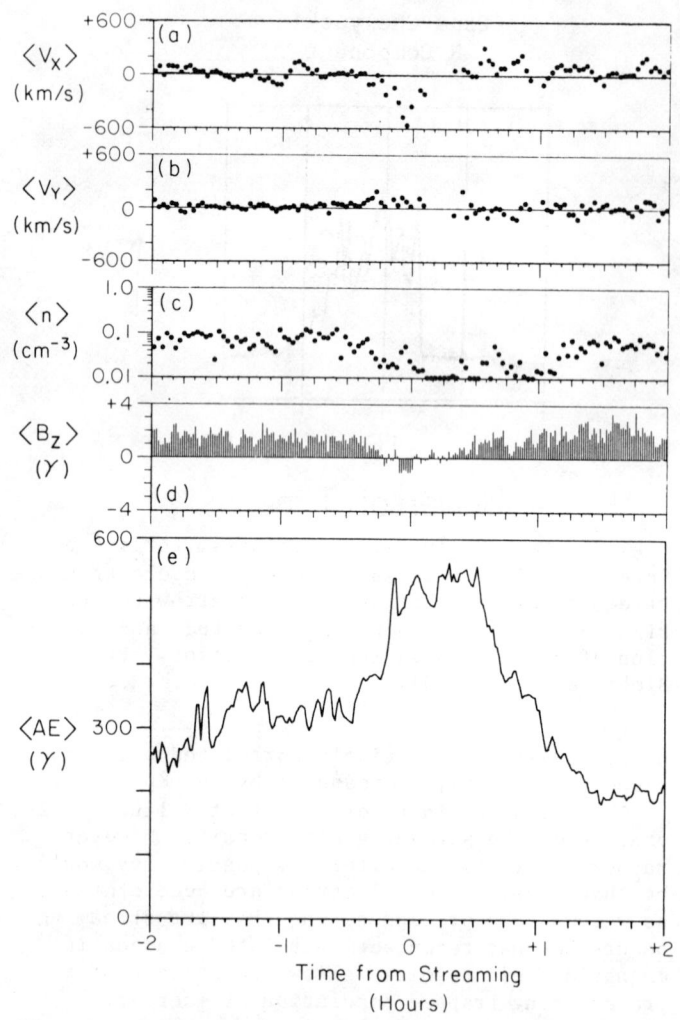

Fig. 4. Superposed epoch analysis of:
(a) X_{GSE}-component of plasma flow velocity;
(b) Y_{GSE}-component of plasma flow velocity;
(c) plasma density (plasma parameters are 2-minute averages for protons of energy 84 eV to 15 keV); (d) Z_{GSM}-component of magnetic field (1-minute averages); (e) AE index (1-minute values). Plotted values are the median over 16 events. Zero epoch time is determined by the first observation of intense streaming of > 200 keV electrons. From Bieber et al. (1982).

rences of pronounced unidirectional anisotropy in the angular distribution of > 200 keV magnetotail electrons. Specifically, the anisotropy amplitude (see Bieber and Stone, 1980, for definition) was required to be larger than unity. This stringent criterion essentially limits the events to those for which nearly all of the electron flux is contained in one hemisphere of the angular distribution. A total of 46 such events were found and were shown to occur preferentially during substorms in association with steep southward magnetic fields.

Although no directional requirement was imposed in the search for streaming events, it was found that the direction of streaming was tailward in all 46 events. This fact supports the idea that streaming energetic electrons can be used as an identifier of open magnetic field lines. If these observations were made on closed field lines, and the lack of Earthward fluxes were due, for example, to drift processes near the neutral sheet that moved the Earthward electrons onto a different flux tube, then it would be expected that the spacecraft would sometimes be located in the flux tube where pronounced Earthward anisotropies occur. The statistical results, however, show that pronounced Earthward anisotropies, if they occur at all, are quite rare compared to pronounced tailward anisotropies. This strongly supports the interpretation that intense streaming events reflect the tailward escape of electrons along open field lines into the solar wind.

The statistical study of energetic electron streaming was subsequently extended to include plasma data for the IMP-8 events (Bieber et al., 1982). In Figure 4 is shown a superposed epoch analysis of these 16 events, where zero epoch time represents the first observation of intense energetic electron streaming. The plotted points are the median over 16 events for various plasma/magnetic parameters and for the AE index. (Plasma flow velocities could not always be reliably determined for all events. For example, during some events the spacecraft was periodically in the magnetotail lobes, where the plasma density was too low to permit a calculation of the flow velocity. Thus, many of the points in the first 2 panels of Figure 4 are based on fewer than 16 events. Nevertheless, the plasma flow behavior indicated in Figure 4 is, without question, statistically significant. See discussion relating to Figure 5 below.)

Throughout most of the 4-hour interval shown in Figure 4, the average plasma flow velocity is fairly small and the magnetic field is clearly northward. However, during the ∼10 minutes prior to the energetic electron streaming transition, tailward plasma flow up to 600 km/s (median value) and southward magnetic fields are observed. Following the streaming transition, the plasma density drops to a level characteristic of the magnetotail lobe and remains there for ∼30 minutes. This shows that the sequence of plasma/magnetic behavior exhibited by the individual event in Figure 2 is a characteristic feature of most energetic electron streaming events.

Figure 4 also demonstrates the statistical association between energetic electron streaming events and geomagnetic activity. The median value of AE shows a clear substorm-like enhancement that begins ∼15 minutes before the streaming transition and lasts for ∼90 minutes. A random sample of AE values would yield a median value ∼150 nT (see Akasofu et al., 1983; the 150 nT value applies to 1-hour averages of AE, but Akasofu et al. state that the distribution of

1-minute values is quite similar). However, Figure 4 shows that at the time of the streaming transition the median value of AE is > 500 nT.

Further confirmation of the association of streaming events with tailward plasma flow is given in Figure 5. Figure 5a shows that the plasma flow speed during the pre-substorm interval is generally < 200 km/s, with tailward flow speeds in excess of 400 km/s obtained for only 4.3% of measurable samples. By contrast, during the interval from -4 to -2 minutes (the most negative median value appearing in Figure 4), tailward flows in excess of 400 km/s are observed in 8 of the 10 events for which the velocity could be reliably determined at this time.

A similar comparison of occurrence histograms is made for B_Z in Figure 6. Figure 6a indicates that only ∼10% of 5-minute averages of B_Z are southward (negative) at pre-substorm times. By contrast, Figure 6b indicates that B_Z averaged over the 5 minutes prior to electron streaming is southward ∼80% of the time.

It should be emphasized that the events included in Figures 5 and 6 were selected purely on the basis of energetic electron data, with no screening of events according to plasma or magnetic field behavior. These figures demonstrate

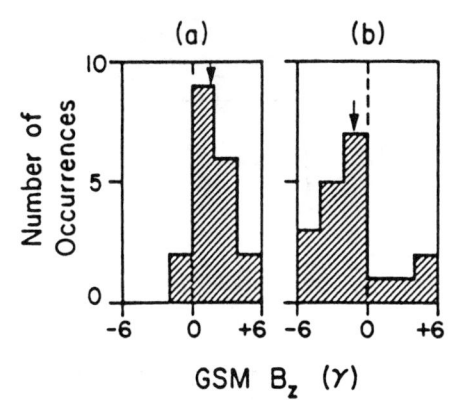

Fig. 6. Occurrence histograms of 5-minute averages of B_Z over intervals (a) from -60 to -55 minutes; (b) from -5 to 0 minutes. Zero time is the onset of intense streaming of energetic electrons. Arrows above histograms indicate mean values. These histograms are based on the 16 events included in Figure 4, plus 3 additional IMP-8 events not included in Figure 4 due to the lack of plasma data. From Bieber and Stone (1980).

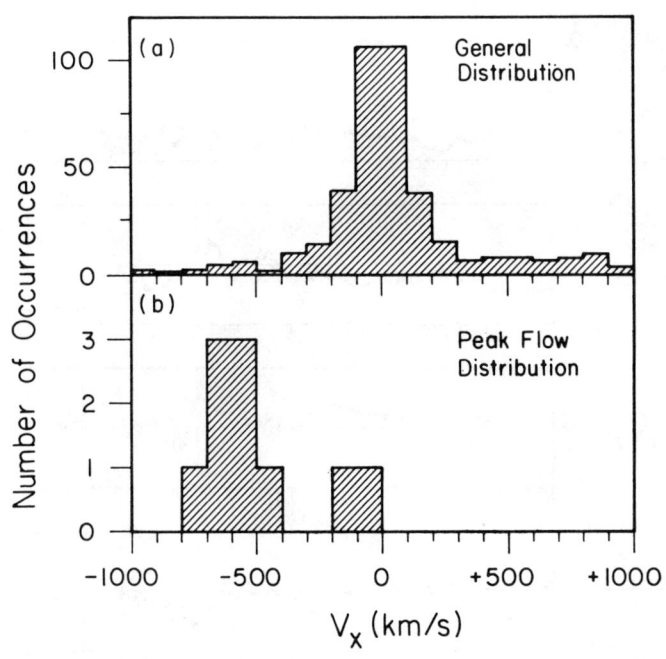

Fig. 5. (a) Distribution of 2-minute averages of V_X for the period -120 to -30 minutes. Of 720 possible samples, 392 appear in the histogram, 5 had V_X > +1000 km/s, and 323 were unmeasurable. (b) Distribution of 2-minute averages of V_X for the interval -4 to -2 minutes (the time of peak tailward flow in Figure 4). Of 16 possible samples, 10 appear in the histogram and 6 were unmeasurable. Zero time is the onset of intense streaming of energetic electrons. From Bieber et al. (1982).

that, while plasma/magnetic signatures of reconnection are not a universal feature of every electron streaming event, such signatures are observed in association with ≳ 50% of streaming events, a frequency far greater than would be expected on the basis of chance. This shows that the tailward plasma flow and southward magnetic fields evidenced in Figure 4 are definitely real and are not a statistical fluctuation.

The statistical results described above place severe constraints on the types of models that can be invoked to explain the substorm behavior of plasmas, magnetic fields, and energetic particles in the magnetotail. A realistic model must explain not only why fast plasma flows would arise in the substorm plasma sheet, but also why such flows would be preferentially tailward during the 10 minutes before streaming energetic electrons are observed. It must explain why the magnetic field 30 R_E tailward of Earth is persistently and preferentially southward during the 5 minutes before streaming electrons are observed. It must explain not only the occurrence of pronounced tailward anisotropy of > 200 keV electrons, but also the complete lack of return (i.e., Earthward) fluxes in a statistical survey of 46 events. All of these observational constraints are well satisfied by a model involving magnetic reconnection at a transient near-Earth X-line.

Heated-Electron Events

In the course of conducting the statistical study described above, it was noted that in 5 of the 16 events studied, the thermal energy of plasma sheet electrons rose sharply near the end of the tailward plasma flow interval, shortly be-

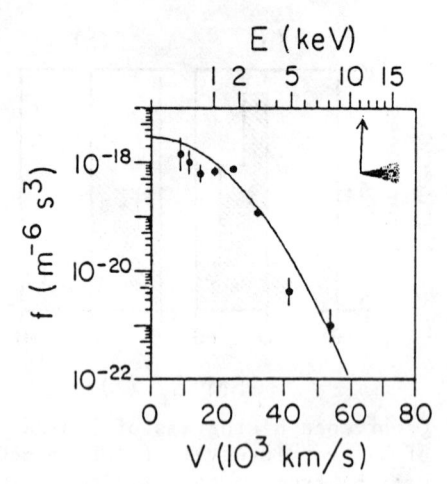

Fig. 7. Example of a bump-in-tail electron velocity distribution observed at the time of peak electron heating for the event shown in Figure 2. The curve is a fitted Maxwellian. Error bars on the points are due to counting statistics. At upper right, the arrow shows the azimuth of the magnetic field, and the shaded fan shows the azimuthal range of electron velocity vectors included in the distribution, where Earthward is up and dawnward is to the right. The distribution required ~0.25 s to accumulate. From Bieber et al. (1983).

fore onset of > 200 keV electron streaming (Bieber et al., 1982). The event displayed in Figure 2 is typical of these heated-electron events, with the heating pulse readily apparent in the third panel. (Note that these heated plasma electrons are distinct from the > 200 keV electrons, which have ~100 times higher energy, and which were present -- though not streaming -- prior to the heating pulse.)

New insights into the nature of magnetic reconnection in Earth's magnetotail were recently provided by a detailed study of heated-electron events (Bieber et al., 1983). Among the results reported in this study were:
(1) Measurements of the electron velocity distribution show that the energization of plasma electrons is a true heating process in that the velocity distribution changes from a narrower to a broader Maxwellian. The increase in electron temperature is accompanied by a decrease in density, so that the electron pressure (and energy density) remains roughly constant.
(2) Bump-in-tail electron velocity distributions are observed during each heating pulse. An example appears in Figure 7. These distributions suggest that the beam-plasma instability may play a role in the dynamics of reconnection.
(3) Very strong (< -10 nT) southward magnetic fields appear near the beginning of each heating pulse. Episodes of very strong southward field are often quite brief (~30 s), suggesting that magnetic reconnection in the magnetotail can be bursty or impulsive in nature. The magnetic field behavior during the event shown in Figure 2 can be seen at high time resolution in Figure 8. The pulse of very strong southward field is evident at 15:34:45 UT.
(4) Very strong (~10 nT) B_Y are observed at times (see Figure 8), emphasizing the importance of constructing fully 3-dimensional reconnection models. The signs of the observed B_Y are generally consistent with the "draping" mechanism proposed by Hones et al. (1982).
(5) Very rapid variations of the magnetic field occur during these events. Part of this variation could be due to tearing-mode magnetic turbulence (Schindler, 1974; Coroniti et al., 1977; Galeev et al., 1978), but minimum variance analysis (Sonnerup and Cahill, 1967) suggests that some of the variations reflect the passage of rotational or tangential magnetic discontinuities past the spacecraft. In Figure 9 is shown a possible rotational discontinuity observed near the beginning of an electron heating pulse.

In some respects, these novel magnetotail observations are similar to recent laboratory observations which emphasize the complex phenomenology of time-dependent magnetic reconnection in 3 dimensions (Stenzel et al., 1982; 1983). A detailed theoretical understanding of this phenomenology does not presently exist.

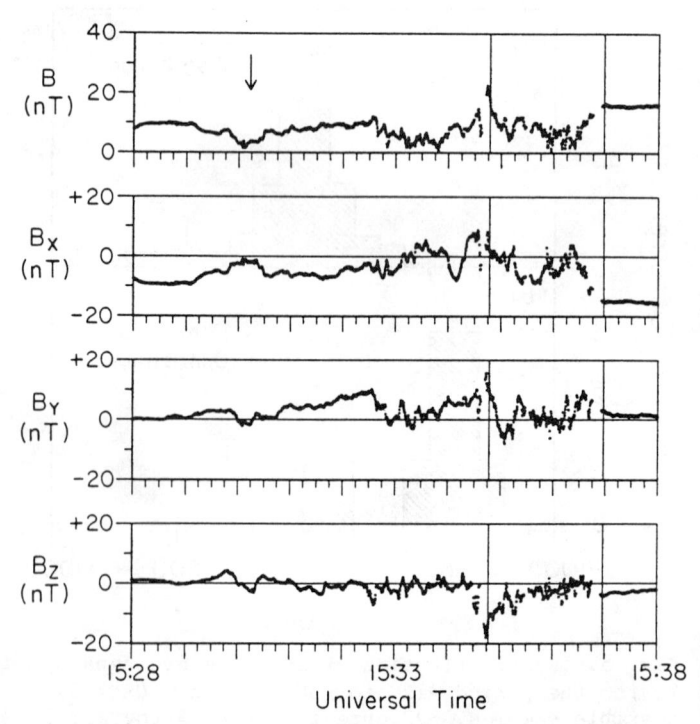

Fig. 8. Magnetic field behavior during the event shown in Figure 2, based upon 0.32 s vector samples. From top to bottom are shown magnetic field magnitude B and the three GSM components of the field. The arrow in the top panel indicates the onset of fast tailward plasma flow. Vertical bars indicate the interval of electron heating. From Bieber et al. (1983).

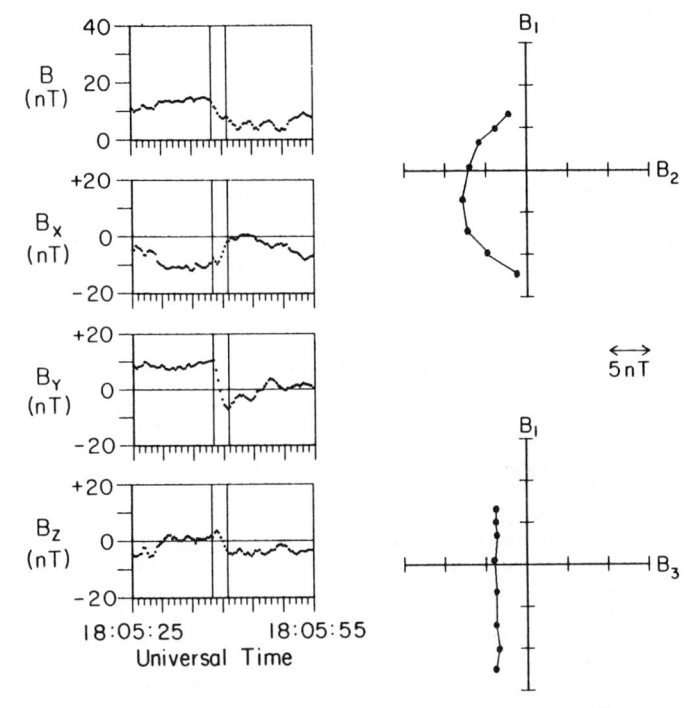

Fig. 9. A possible rotational magnetic discontinuity observed during a heated-electron event on 14 November 1973 (not the event shown in Figure 2, however). To the left are shown magnetic field magnitude and the three GSM components for a 30 s interval in which the possible discontinuity, indicated by vertical bars, was observed. To the right are shown hodograms of the magnetic field in minimum variance coordinates, where B_1, B_2, and B_3 are the magnetic field components along the maximum variance, median variance, and minimum variance directions, respectively. The three eigenvalues are 51, 5.8, and 0.02 nT^2. The minimum variance eigenvector has GSM components (X, Y, Z) = (0.68, 0.03, 0.73). From Bieber et al. (1983).

Summary

The past 10 years have seen the development of considerable observational support for the idea that magnetic reconnection at a near-Earth X-line plays a central role in the substorm dynamics of Earth's magnetotail. Energetic electron measurements have contributed to this development in several ways. Their use as tracers of magnetic field line topology confirms the topology implied by the reconnection theory. Statistical studies based on energetic electron streaming show the persistent occurrence of plasma and magnetic field behavior indicative of reconnection processes. Recent studies of heated-electron events (originally identified through the occurrence of energetic electron streaming) provide new insights into the complex phenomenology of substorm-associated magnetic reconnection in Earth's magnetotail.

Acknowledgement. This work was supported by the National Science Foundation under grant DPP-8300544.

References

Akasofu, S.-I., B.-H. Ahn, Y. Kamide, and J. H. Allen, A note on the accuracy of the auroral electrojet indices, J. Geophys. Res., 88, 5769-5772, 1983.

Baker, D. N., and E. C. Stone, Energetic electron anisotropies in the magnetotail: Identification of open and closed field lines, Geophys. Res. Lett., 3, 557-560, 1976.

Baker, D. N., and E. C. Stone, Observations of energetic electrons (E $\gtrsim$ 200 keV) in the earth's magnetotail: Plasma sheet and fireball observations, J. Geophys. Res., 82, 1532-1546, 1977.

Bieber, J. W., and E. C. Stone, Streaming energetic electrons in Earth's magnetotail: Evidence for substorm-associated magnetic reconnection, Geophys. Res. Lett., 7, 945-948, 1980.

Bieber, J. W., E. C. Stone, E. W. Hones, Jr., D. N. Baker, and S. J. Bame, Plasma behavior during energetic electron streaming events: Further evidence for substorm-associated magnetic reconnection, Geophys. Res. Lett., 9, 664-667, 1982.

Bieber, J. W., E. C. Stone, E. W. Hones, Jr., D. N. Baker, S. J. Bame, and R. P. Lepping, Microstructure of magnetic reconnection in Earth's magnetotail, J. Geophys. Res., submitted, 1983.

Coroniti, F. V., F. L. Scarf, L. A. Frank, and R. P. Lepping, Microstructure of a magnetotail fireball, Geophys. Res. Lett., 4, 219-222, 1977.

Galeev, A. A., F. V. Coroniti, and M. Ashour-Abdalla, Explosive tearing mode reconnection in the magnetospheric tail, Geophys. Res. Lett., 5, 707, 1978.

Hones, E. W., Jr., Substorm processes in the magnetotail: Comments on 'On hot tenuous plasmas, fireballs, and boundary layers in the earth's magnetotail' by L. A. Frank, K. L. Ackerson, and R. P. Lepping, J. Geophys. Res., 82, 5633-5640, 1977.

Hones, E. W., Jr., J. Birn, S. J. Bame, G. Paschmann, and C. T. Russell, On the three-dimensional magnetic structure of the plasmoid created in the magnetotail at substorm onset, Geophys. Res. Lett., 9, 203-206, 1982.

Meng, C.-I., A. T. Y. Lui, S. M. Krimigis, S. Ismail, and D. J. Williams, Spatial distribution of energetic particles in the distant magnetotail, J. Geophys. Res., 86, 5682-5700, 1981.

Russell, C. T., and R. L. McPherron, The magnetotail and substorms, Space Sci. Rev., 15, 205, 1973.

Schindler, K., A theory of the substorm mechanism, J. Geophys. Res., 79, 2803-2810, 1974.

Sonnerup, B. U. Ö., and L. J. Cahill, Jr., Magnetopause structure and attitude from Explorer 12 observations, J. Geophys. Res., 72, 171-183, 1967.

Stenzel, R. L., W. Gekelman, and N. Wild, Magnetic field line reconnection experiments, 4. Resistivity, heating, and energy flow, J. Geophys. Res., 87, 111-117, 1982.

Stenzel, R. L., W. Gekelman, and N. Wild, Magnetic field line reconnection experiments, 5. Current disruptions and double layers, J. Geophys. Res., 88, 4793-4804, 1983.

Terasawa, T., and A. Nishida, Simultaneous observations of relativistic electron bursts and neutral-line signatures in the magnetotail, Planet. Space Sci., 24, 855-866, 1976.

Questions and Answers

Fritz: In your presentation the energetic electrons appear to arrive coincident with or after the plasma and plasma flow enhancements. Since 200 keV electrons are moving with a velocity close to c, why don't the electrons arrive first due to their shorter time-of-flight time from the X-line source you discussed?

Bieber: As discussed in my written report, it is not clear that >200 keV plasma sheet electrons are accelerated at the near-Earth X-line. Further research is needed to establish the acceleration mechanism and to determine whether any time-of-flight effects exist. As for the streaming electrons, their appearance after the interval of fast tailward plasma flow is explained by their location on open lobe field lines just outside the plasma sheet. Energetic electrons inside the tailwind-moving plasmoid typically show trapped or isotropic angular distributions because the field lines in the plasmoid are closed in a magnetic island configuration.

Lui (Comment): I disagree with your statement that tailward streaming of energetic electrons necessarily implies an open magnetic field line. Since the electron bursts occur at the edge of the plasma sheet, the magnetic field line, if closed, would cross the neutral sheet far downstream, possibly ~100 R_E, where the magnetic field normal to the neutral sheet may be small, say 0.5 nT. The gyroradius of 1 MeV electrons in this field is ~1 R_E. Therefore, the dawn-dusk displacement of the electrons at the neutral sheet crossing would be significant and may move the electron beam outside the flux tube sampled by the spacecraft even if the electron beam comes back. This idea is documented in Lui et al. (JGR, accepted for publication in 1983).

Bieber: This proposed mechanism for tailward streaming on closed field lines seems inconsistent with observational results, at least for the very anisotropic events I reported on. First of all, a search for highly anisotropic >200 keV electron streaming found 46 cases of tailward streaming and no cases of earthward streaming. Based on the mechanism you describe, it would be expected that earthward streaming should be detected at times, contrary to observation. Secondly, the association of streaming events with plasma dropouts supports the idea that the streaming occurs on open lobe field lines just outside the plasma sheet.

Rostoker: The small number of energetic electron beams (16 in all) agreeing with the small numbers reported by Scholer - seem to indicate that acceleration of such beams is rare indeed. Since you suggest an X-type neutral point is the reason acceleration takes place, and substorms are very common, should you not see more events if near earth neutral points are actually associated with substorm expansive phases on a one-to-one basis?

Bieber: When IMP-8 is monitoring magnetospheric electrons in the central magnetotail region, it observes streaming electron events at an average rate of about 2 per day. The apparent scarcity of events in a survey of ~2 years of data is due, in large part, to the rather large inclination of the IMP-8 orbit, which carries the spacecraft through a Z_{GSE} range of about −20 to +20 R_E. On many orbits, IMP-8 does not encounter the plasma sheet at all. See Bieber and Stone [1980] for further information on the occurrence frequency of streaming events. Please note that I have not claimed that the streaming electrons are directly accelerated at the X-line (see written report).

Schindler (Comment): I should like to make a comment that might be useful to identify the underlying physics. The tearing instability can produce both the global plasmoid and the smaller islands that you showed in your schematic picture. Tearing is a configurational instability with a tendency to go to the largest wave length possible. Details depend on availability of dissipation and on its variability with time.

Speiser: You described the rise in kT_e as a heating pulse, which seems to imply a temporal nature. Would you comment on the temporal versus spatial nature?

Bieber: The term "pulse" was chosen to describe the characteristic shape of the electron temperature profile (see Figure 2) and does not preclude the possibility that this temporal variation is due to a spatial variation sweeping past the spacecraft. In fact, the heating pulse seems to be partly spatial and partly temporal in nature. It is spatial in that reversals in the sign of B_x were observed during each of the 5 heating pulses studied, suggesting that the heated electrons are localized near the neutral sheet. It is temporal in that the pulse always occurs near the end of the tailward plasma flow interval, even though the spacecraft encountered the neutral sheet prior to the observation of heating in some events.

Vasyliunas: In your discussion of the electron heating pulse, you seem to assume that it is a local heating of the plasma observed prior to the pulse. It is, however, just as likely that the spacecraft is entering a distinct spatial region of hot plasma, especially as the pulse is short-lived and its end must certainly correspond to the spacecraft exiting from the hot plasma region - there can be no collisionless cooling (any collisionless heating corresponds to phase-space mixing and hence a broadening of the distribution function, and an "unmixing" is of course impossible).

Bieber: A temporal variation is what is observed from the perspective of IMP-8, but this variation could, of course, be due to spatial variations moving past the spacecraft. An exact delineation of temporal versus spatial effects is not possible based on single-spacecraft observations, but it seems likely that the heating is both spatially and temporally localized.

Vasyliunas: In your superposed-epoch plot, AE, before the electron events, is about twice the background value, while after the event it is about equal to the background. This might seem to suggest that the events tend to be associated with the last substorm that ends a period of activity. It would be interesting to see if the individual events support such an interpretation.

D. N. Baker: I would like to reply to this question by Vasyliunas. The AE index can be enhanced by a significant amount prior to substorm expansion phase onset due to "growth phase" effects. Such effects are of variable duration (0.5 to 2.0 hours) and are indicative of enhanced global convective dissipation in the polar ionosphere. Thus, at least some of the high value in the AE superposed epoch analysis could be due to the pre-expansion loading of the magnetotail.

PARTICLE AND FIELD SIGNATURES OF SUBSTORMS IN THE NEAR MAGNETOTAIL

D. N. Baker

Los Alamos National Laboratory, Los Alamos, NM 87545

Abstract. The near-earth magnetotail ($10 \lesssim r \lesssim 20\ R_E$) portion of the terrestrial magnetosphere is very likely the region in which magnetospheric substorms are initiated. An observational advantage compared to other astrophysical regions is that the near magnetotail can be nearly continuously monitored by spacecraft that are relatively fixed in location. Based on numerous case studies, it is found that southward IMF is clearly related to increased energy input to the magnetosphere which manifests itself as an increase of magnetotail currents and field strength, while northward IMF "turns off" energy input and rapidly stops the progress of energy storage in the tail field. In studying hundreds of substorm events near local midnight with geostationary spacecraft instrumentation, we have found that most substorm injection events are preceded by cigar (growth) phase features (i.e., $j_\parallel > j_\perp$ for energetic electron distributions). In one particular study, for more than 100 cases of detected cigar phases, in 97 cases the cigar phase was terminated by a substorm injection event. In only 4 cases did a cigar phase occur with no identifiable substorm onset. Such results suggest that substorm expansion onsets occur if, and only if, stored magnetotail energy is increased above a quiet time level. We conclude that observations in the near-earth magnetotail show some of the clearest and most repeatable signatures available in support of the concept of loading and unloading of magnetic energy in association with substorms. The data illustrate that magnetic energy is accumulated and stored for $0.5 \sim 2.0$ hours in the tail lobes and then is rapidly dissipated at substorm expansion onset. The dissipation is manifested by the acceleration and rapid transport of hot plasma and energetic particle populations within the near-tail region. These energized plasmas provide an excellent tracer capability which allows a relatively clear determination of where, when, and how magnetic energy is converted to other forms during substorms. When near-tail data are considered in a global context of deep-tail measurements, numerical models, ground-based data, etc., they provide very strong evidence for the neutral line substorm model and, thus, for the regular occurrence of magnetic reconnection in the near-earth magnetotail.

Introduction

Magnetic merging is thought to be an important form of energy conversion in a variety of cosmic settings including planetary magnetospheres and solar flare sites. One particularly clear illustration of this is in the neutral line model of substorms [Russell and McPherron, 1973], wherein the earth's magnetotail plays a fundamental role in the occurrence and character of magnetospheric dynamics. This model focusses on magnetospheric substorms as the basic dynamical entity, and it describes substorms in terms of growth, expansion, and recovery phases. In the growth phase, energy from the solar wind is added to the earth's magnetosphere by means of a magnetic merging between the interplanetary magnetic field (IMF) and the terrestrial field. By virtue of the flow of the solar wind past the magnetosphere, interconnected field lines are dragged from the dayside to the nightside of the earth and, in effect, magnetic flux is added to the magnetotail. The result is that the magnetosphere enters an elevated energy state, with free energy stored in the form of enhanced magnetic energy density in the magnetotail lobes. The aspect of sudden release (and conversion) of this stored energy by magnetic reconnection at substorm expansion onset into hot jetting plasmas and energetic particle populations is termed the unloading process. In the unloading process, large-scale changes of magnetospheric configuration take place (see Figure 1).

Unlike many supposed sites of reconnection which are totally inaccessible to direct in situ observation (e.g., the sun), or else are only occasionally sampled by fast-moving spacecraft (e.g., the magnetopause), the near magnetotail can be nearly continuously monitored by spacecraft that are relatively fixed in location. In particular, spacecraft at, or near, geostationary orbit remain approximately constant in radial distance and move slowly through the nightside region near the place where substorm reconnection normally begins.

Using observations from instrumentation aboard numerous geostationary satellites one can show that magnetic fields and plasma distribution functions undergo a very regular and predictable sequence of variation in association with substorms. Following enhanced coupling between the solar wind and the magnetosphere (associated with southward interplanetary field), magnetic fields in the near-earth region exhibit a progressive development toward a more taillike configuration in the midnight sector. This taillike field is indicative of enhanced cross-tail currents and, thus, of increased storage of magnetic energy in the tail lobes. This available free energy in the magnetotail is the energy which is eventually dissipated during substorms in episodes of magnetic reconnection.

The development of a stressed, taillike magnetic field in the vicinity of geostationary orbit (where the field configuration is ordinarily nearly dipolar) leads to clear signatures in the distribution functions of energetic plasma particles. In particular, it is regularly observed in association with taillike field development that tens of keV electrons exhibit a progressive transition away from a trapped distribution character ($j_\perp > j_\parallel$) to a field-aligned distribution character ($j_\parallel > j_\perp$). This effect can be readily understood in terms of azimuthal particle drifts in the distorted, taillike magnetic field and occurs for approximately 0.5-2.0 hours prior to substorm expansive phase onsets.

Following substorm expansion onset (within a minute or so) there is a rapid relaxation of the stressed magnetic field configuration of the near-magnetotail around local midnight. Along with this dipolarization there is invariably the associated injection of hot plasma and energetic particles into the region of geostationary orbit. These plasma populations appear to be directly related to the rapid conversion of stored magnetic energy at reconnection sites in a limited segment of the magnetotail plasma sheet. The higher energy (hundreds of keV) particle population, further, appears to be accelerated very impulsively, probably due to intense induced electric fields in the magnetic merging region.

Once the substorm-generated hot plasma and energetic particle populations are injected into the inner magnetosphere, they tend to drift adiabatically in magnetically trapped orbits. The energetic particles, once produced, can provide a variety of tracer functions to determine characteristics of acceleration location and extent. These tracer aspects also include energy-dependent ion drift characteristics that allow identification of the principal region of overall substorm disturbance onset. Furthermore, ion gradient anisotropy information can allow

194 PARTICLE AND FIELD SIGNATURES

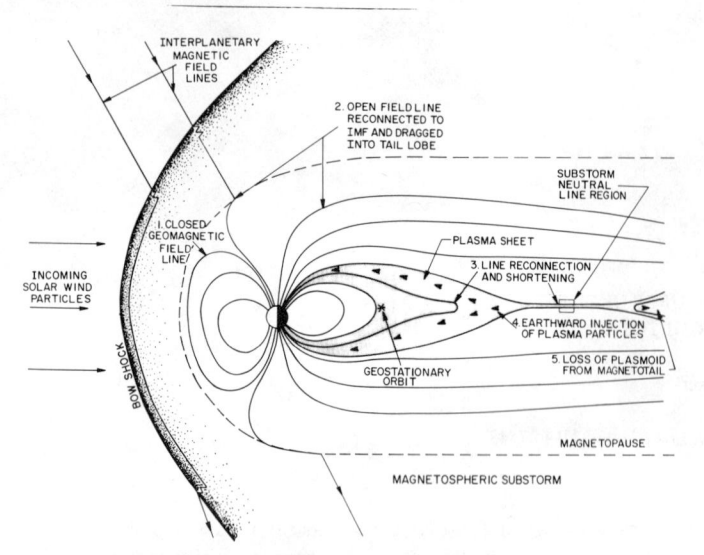

Fig. 1. A model of the sequence of events occurring in a magnetospheric substorm.

the remote sensing of moving density-gradient boundaries in the vicinity of observing spacecraft. In a very recent discovery, charge-state-dependent ion drift characteristics are found that permit identification and discrimination between solar wind and ionospheric sources for the accelerated plasma which forms the energetic particle population (and eventually constitutes the terrestrial ring current).

In this paper we review many of the observations, and interpretations of such observations, which are supportive of the near-earth reconnection model of substorm energy dissipation.

Loading the Magnetospheric System

Numerous empirical studies of coupling between the solar wind and the magnetosphere [e.g., Nishida, 1983 and references therein] have shown the important role of the interplanetary magnetic field (IMF) orientation in determining the occurrence of geomagnetic (substorm) activity. When the IMF turns southward, strong substorm activity ordinarily follows after about one hour of persistently southward IMF. Figure 2, taken from a study by Baker et al. [1983a] for the Coordinated Data Analysis Workshop (CDAW) 6, shows the relationship between a measure of solar wind energy input to the magnetosphere ($V \cdot B_z$) and substorm onset.

The solar wind speed (V) and IMF north-south component (B_z) observed at IMP-8 immediately upstream of the earth's magnetosphere showed a very clear example of a "southward turning" of B_z at 1010 UT on 22 March 1979. In Fig. 2 we plot $-V \cdot B_z$ (i.e., the east-west component of the interplanetary electric field) as an indicator of solar wind-magnetosphere coupling. Approximately 10 min after $-V \cdot B_z$ went positive we observed at 6.6 R_E that a very taillike magnetic field orientation began to occur. The field line inclination (0° would be ∼ dipolar, while 90° would be ∼ parallel to the ecliptic plane) measured at GOES-3 at geostationary orbit (∼ 135°W) is shown as the solid line. It is seen that between ∼1025 UT and ∼1055 UT the GOES-3 field line inclination reached ≳60° indicative of a very stressed, taillike field structure at 6.6 R_E. This is one of the classic signatures of the substorm growth phase [McPherron, 1970, 1972; Baker et al., 1978] and is the internal magnetosphere manifestation of the storage of energy in the near-earth magnetotail.

The expected effect of the kind of highly distorted, nondipolar magnetic field shown in Figure 2 would be to greatly distort trapped magnetospheric particle drift paths [Roederer, 1970]. Model calculations of azimuthal drift effects in a distorted magnetosphere are shown, for example, in Fig. 3 [Paulikas and Blake, 1979]. As seen in this illustration, particles with equatorial pitch angles (α) near 0° tend to drift nearer the earth at local noon and drift farther from the earth at local midnight. Conversely, $\alpha \approx 90°$ particles drift very much closer to the earth at local midnight than at local noon. Since relatively strong inward radial gradients exist (higher flux for lesser geocentric radial distances) the effect of a distorted, taillike magnetic field structure near local midnight is to produce a local particle distribution with enhanced fluxes near $\alpha \approx 0°$ and $\alpha \approx 180°$ and a depletion of fluxes near $\alpha \approx 90°$ [e.g., West, 1979]. Such "cigarlike" or "butterfly" bidirectional anisotropies are readily detected with present-day instrumentation.

An example of such effects is shown in Figure 4. Energetic electron data from two geostationary spacecraft 1977-007 and 1976-059, are shown for the period 0400-1000 UT on 8 Spetember 1977. A substorm expansion onset occurred at 0720 UT on this day and this was observed at geostationary orbit as an "injection" of 30 to ≳200 keV electrons. The injection of freshly accelerated particles was seen both premidnight (S/C 1977-007) and postmidnight (S/C 1976-059).

As is particularly clear in the premidnight data in the upper panels of Fig. 4, an extended development of taillike magnetic field occurred at the 77-007 position. The field line inclination (θ_B) went from ≲20° to ∼50° between 0520 and 0720 UT. Accompanying this local field development, the energetic electron distributions became more and more cigarlike (as shown by the small inset panels of counts vs cosα). The parameter C_2 [Baker et al., 1978] is a measure of the amplitude and direction of the second-order electron anisotropy: $C_2 > 0$ indicates a field-aligned (cigar) distribution, while $C_2 < 0$ indicates a $j_\perp > j_\parallel$ (pancake) distribution. A strong and progressive development of cigarlike distributions occurred in the >30 keV electrons in accompaniment with the taillike stretching of the local magnetic field. A weak cigar phase was also seen, as indicated in Fig. 4, postmidnight at the 76-059 position.

Figure 5 shows the relationship of the C_2 parameter for 8 September (Fig. 4) to the concurrently measured IMF (sheath) orientation. As discussed by Baker et al. [1982b], it is observed that dC_2/dt is positive for southward IMF ($\lambda_B < 0$), whereas dC_2/dt is approximately 0 for northward IMF ($\lambda_B > 0$). Thus, southward IMF is clearly related to increased energy input to the magnetosphere which manifests itself as an increase of magnetotail currents and field strength, while northward IMF "turns off" energy input and rapidly stops the progress of energy storage in the tail field.

In studying hundreds of substorm events near local midnight with geostationary spacecraft instrumentation, we have found that most substorm injection events are preceded by cigar (growth) phase features of the type discussed above.

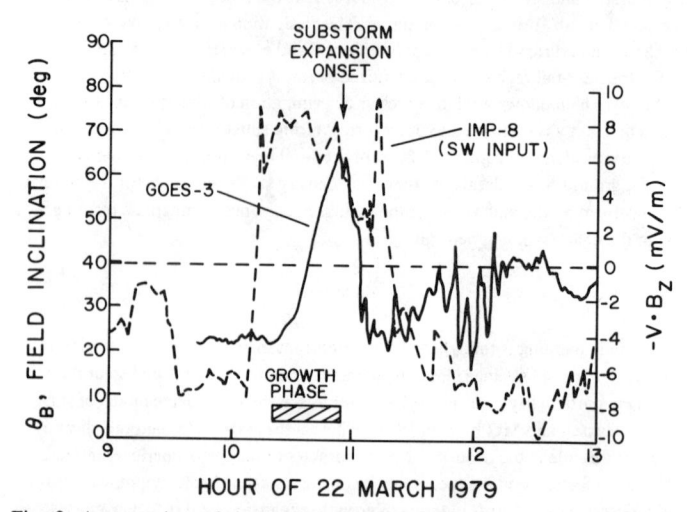

Fig. 2. A comparison of a solar wind energy input function ($-VB_z$, dashed line) as compared with the GOES-3 magnetic field line inclination (θ_B, solid line) measured at synchronous orbit for a portion of 22 March 1979. As discussed in the text, a substorm growth phase was observed for ∼1/2 hour prior to substorm expansion onset (at ∼1055 UT) and this growth phase was manifested by an extreme taillike field development near local midnight at geostationary orbit [from Baker et al., 1983a].

Table 1 summarizes our findings concerning the occurrence of cigar phases prior to substorm expansion onsets. For more than 100 cases of detected cigar phases of 0.5 ~3.0 hour duration, in 97 cases the cigar phase was terminated by a substorm injection event. In only 4 cases did a cigar phase occur with no identifiable substorm onset. Conversely, when no cigar phase occurred when the geostationary spacecraft was in the nighttime sector we saw no substorms at all on 15 of those occasions and we saw some substorm activity on only 2 occasions. Table 1 is highly diagonal and suggests that substorm expansion onsets occur if, and only if, stored magnetotail energy is increased above a quiet time level.

Direct evidence is found in the more distant tail for the storage of magnetic energy as cigar phases develop near geostationary orbit. Figure 6 shows data from S/C 1976-059 at 6.6 R_E for portions of 28 and 29 December 1976. At 0100 UT a sharp, intense 800 nT negative magnetic bay occurred at ground stations near local midnight on 29 December 1976 [Baker et al., 1981] and intense substorm activity followed. The injection of energetic protons and electrons at 6.6 R_E at the substorm expansion onset (0100) UT is well-illustrated in Figure 6 as is the very strong cigar phase evident in the lower panels of the figure. Between ~2330 UT and 0100 UT the field at geostationary orbit reached an inclination of nearly 80° ($\theta_B \sim 80°$) and C_2 became very large and positive ($C_2 \sim 3.0$). The period, 2330-0100 UT, thus appears to have been an interval of strong magnetotail energy storage.

Concurrent data for this time from IMP-8 high in the southern tail lobe at ~35 R_E geocentric radial distance (and near local midnight) is shown in Figure 7. Ancillary data available [Baker et al., 1981] show that IMP-8 stayed near the magnetopause boundary for this entire interval and it is seen from the magnetic records of Figure 7 that $|\vec{B}|$ increased progressively from ~25 nT to ~40 nT between 2330 and 0100 UT. Thus, precisely during the geostationary orbit cigar phase, magnetotail energy densities greatly increased. Furthermore, right at the

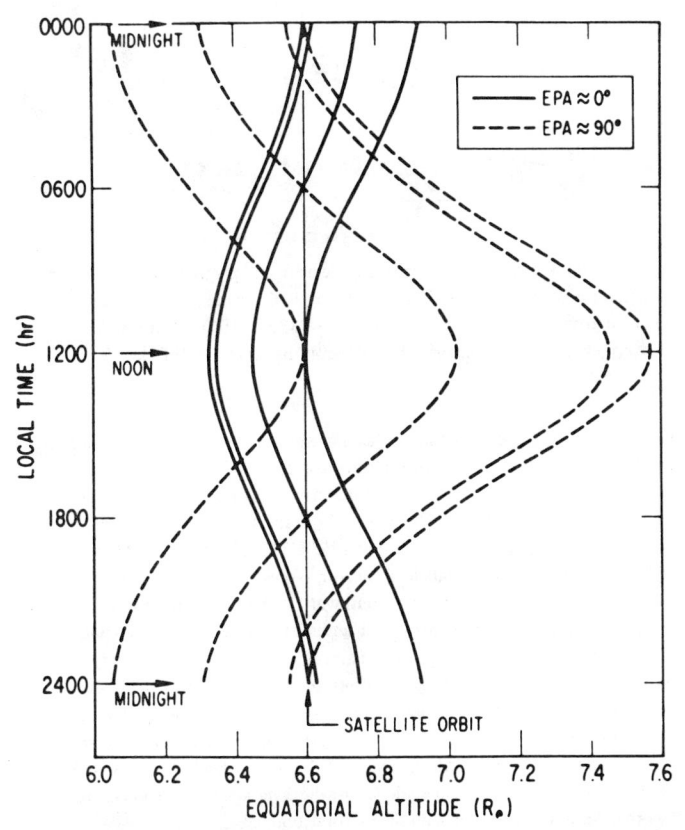

Fig. 3. Representative electron drift paths (equatorial crossing altitudes) for those particles mirroring near the equator (EPA = 90°) and those mirroring at high latitudes (EPA = 0°) [from Paulikas and Blake, 1979].

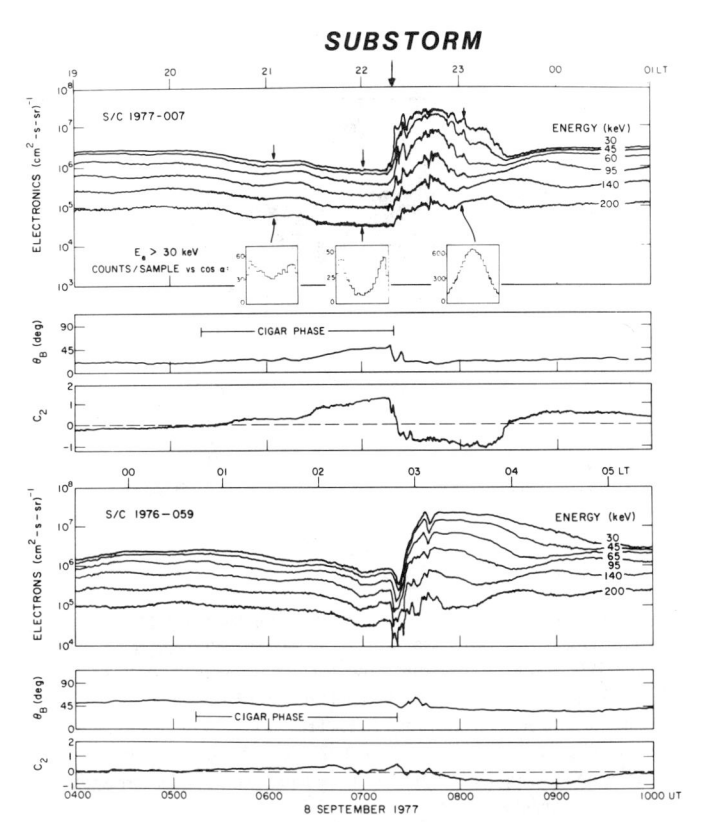

Fig. 4. A detailed plot of the geostationary orbit spin-averaged energetic electron fluxes, local magnetic field line tilt angle (θ_B), and second-order anisotropy amplitude (C_2) on September 8, 1977. All electron channels (energies as labeled) have a common upper cutoff energy of 300 keV. The upper panels show data for spacecraft 1977-007, while the lower panels show data for spacecraft 1976-059. Universal time is shown along the bottom of the figure, while geographical local time is shown for each satellite. A substorm injection event is seen at ~0720 UT, preceded by a substorm growth (cigar) phase of ~2-hour duration. [From Baker et al., 1982b].

time of substorm expansion phase onset (and particle injections, Fig. 6) the magnetic energy density in the tail at ~35 R_E rapidly decreased [e.g., Fairfield and Ness, 1970]. Figure 7 illustrates this feature very clearly since $|\vec{B}|$ decreased strongly and rapidly from 40 nT back to ~30 nT between 0100 and 0130 UT. Thus, the stored magnetic energy in the magnetotail was rapidly dissipated at substorm onset in this case.

Using ISEE-3 data in the very deep tail (80-220 R_E) we have now observed many examples of magnetotail diametrical expansion in association with cigar phases at 6.6 R_E. A good example, shown in Figure 8, is that of 26 January 1983. ISEE near the Y = 0 region of the aberrated tail at 213 R_E radial distance was located in the magnetosheath between 0800 and 0910 UT. A cigar phase began at 0850 at 6.6 R_E and culminated in a substorm expansion with sharp, dispersionless particle injection at 6.6 R_E at 0950 UT. At 1011 UT ISEE-3 instruments saw the plasma sheet suddenly envelop the spacecraft with plasma bulk flow velocities in excess of 1000 km/s at times (Hones et al., 1983).

The sequence of events in Fig. 8, borne out by many other examples (Baker et al., 1983b), is that energy storage near earth gives rise to the cigar phase signature. With 20 minute delay, the distant ($r \sim 220$ R_E) tail increases substantially in diameter and ISEE-3 will then go progressively from the magnetosheath into the tail lobe. After a typical cigar phase development time (1 hour) a substorm expansion phase onset with particle injection, etc., in the near-earth region occurs. Again with a delay of 20-30 minutes from the time of near-tail onset phenomena, the hot, jetting plasmas of a reconnection-produced plasmoid (Hones et al., 1983) reach ISEE-3, having been released from the near-

196 PARTICLE AND FIELD SIGNATURES

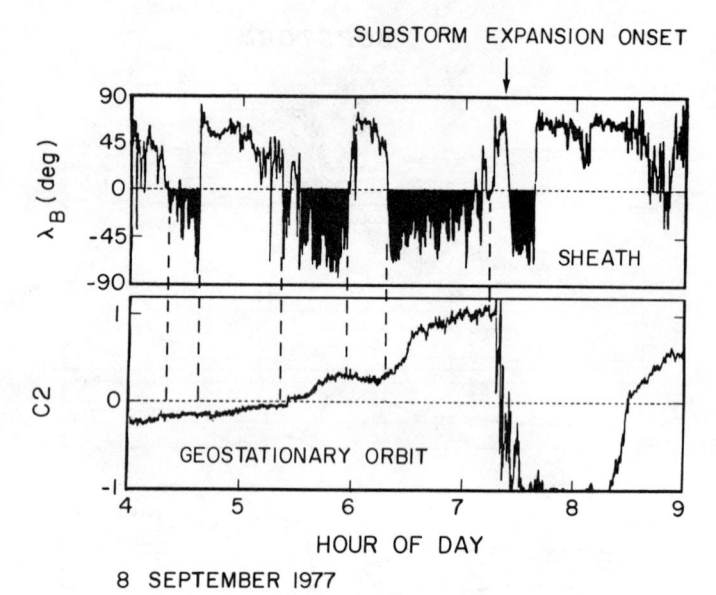

Fig. 5. A detailed comparison of the concurrently measured magnetic field inclination (λ_B) at IMP 8 and the second-order electron anisotropy amplitude (C_2) at spacecraft 1977-007. The figure shows data for September 8, 1977 (compare Fig. 4). Periods of southward sheath or interplanetary fields ($\lambda_B < 0$) have been emphasized by black shading, while times of strong northward or southward rotation of λ_B are shown by the vertical dashed lines. Periods of positive growth of C_2 are seen to correspond to $\lambda_B < 0$, while periods of constant C_2 correspond to $\lambda_B > 0$. [From Baker et al., 1982b].

earth plasma sheet at substorm onset. We have seen many tens of these correlated events with near-midnight spacecraft at 6.6 R_E and ISEE-3 at 80-220 R_E. These results demonstrate quite clearly that the entire tail participates in the storage and sudden release of magnetic energy during substorms.

Unloading the Magnetospheric System

As the examples from the data presented above demonstrate, one cannot discuss the loading of the magnetosphere (growth phase) without discussing the expansive phase (unloading). These two parts of the substorm are intimately related and, indeed, during quite disturbed times the loading and unloading processes are often proceeding concurrently. As we strive to understand geomagnetic activity, however, it is normally very useful to begin by trying to comprehend simpler, less complicated events rather than immediately trying to untangle very complex, very disturbed geomagnetic patterns. As long as one recognizes that highly disturbed periods may represent a nonlinear superposition of effects seen during moderate substorm events, one often can get a clearer picture of substorm processes by examining details of relatively isolated events.

Figure 2 above demonstrated many of the gross temporal relationships that exist as one follows the flow of energy from the solar wind through the magnetosphere to its eventual dissipation in substorm processes. As noted previously, 10 min after the southward turning of the IMF the growth phase of

TABLE 1. Precursory Cigar-Phase Association with Substorm

	No Substorm	Substorm Observed
No Cigar-Phase Observed	15	2
Cigar-Phase Observed	4	97

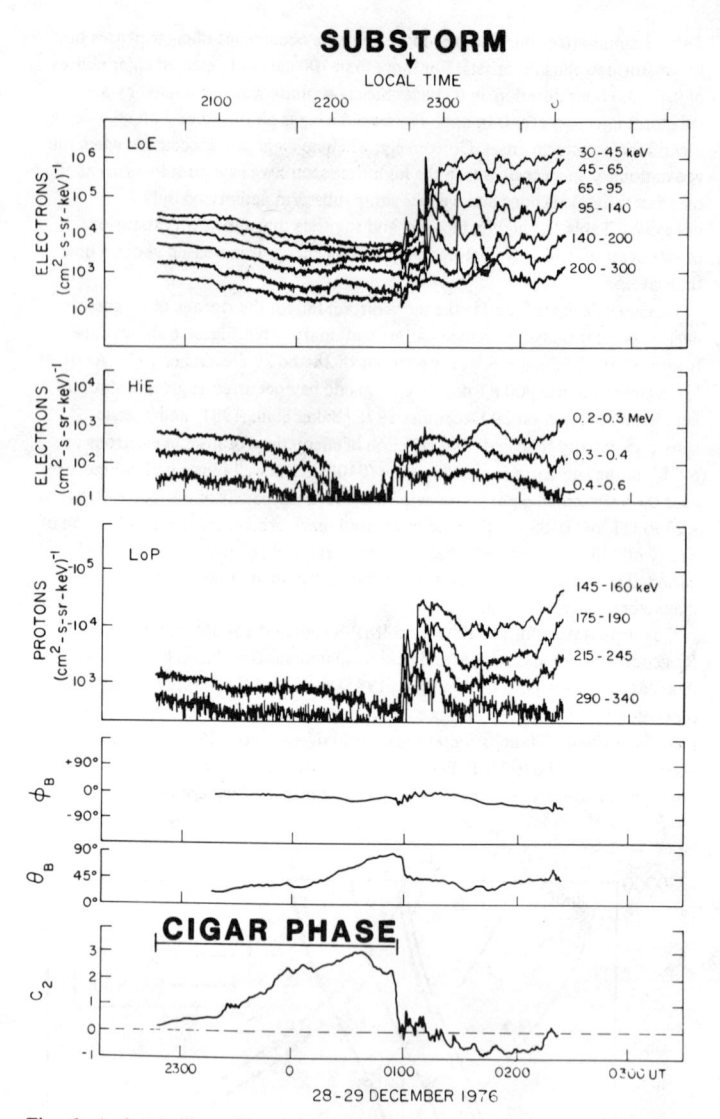

Fig. 6. A plot similar to Fig. 4 showing electron and proton differential fluxes as labelled for a portion of 28 and 29 December 1976. As described in the text, a substorm growth phase was observed from ~2330 to 0100 UT at which time a substorm expansion phase commenced. [From Baker et al., 1981].

substorm activity began with taillike field development, etc. However, in keeping with the general statistical results of Bargatze et al. [1983] and many other researchers [cf., Nishida, 1983], the substorm expansion onset was delayed by a much longer time. In the case of 22 March 1979 shown in Fig. 2, the substorm onset was very well identified to occur at 1055 UT [McPherron and Manka, 1983]. This initiation time is labelled "substorm onset" in Fig. 2 and is indicated by the vertical arrow. It followed the southward IMF turning by 45 min.

Note that precisely at the substorm onset time in Fig. 2, the field inclination at 6.6 R_E went from $\theta_B \sim 65°$ to $\theta_B \sim 20°$. This rapid "dipolarization" is very characteristic of substorm expansion onsets and is generally seen in a region of several hours local-time width near midnight at geostationary orbit. (The second panel from the top of Fig. 4 and the second-to-bottom panel of Fig. 6 show precisely the same effect; cf. McPherron [1972] and Fairfield et al. [1981].) This dipolarization is taken as direct evidence for the diversion of a portion of the cross-magnetotail current through the ionosphere.

In general, observations show that a significant part of the energy stored in the magnetotail is dissipated through the ionospheric part of the so-called substorm current wedge. This substorm current wedge is set up by the sudden disruption of

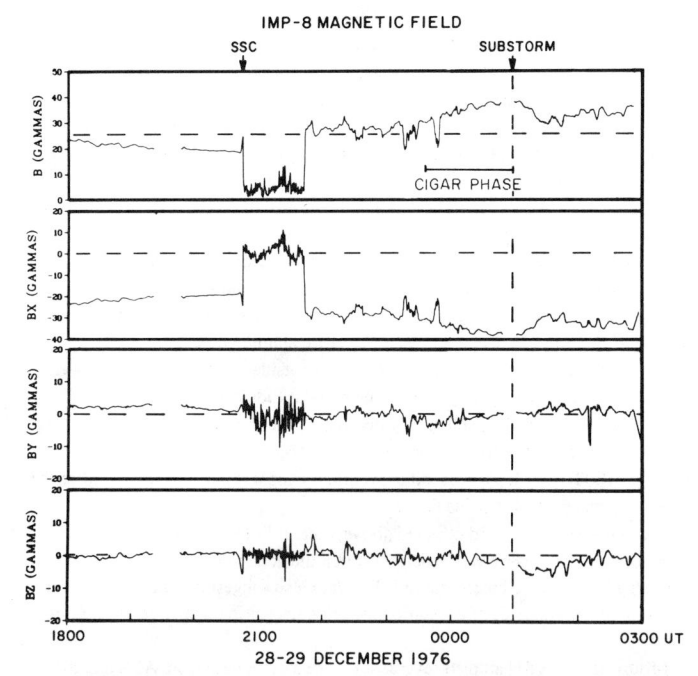

Fig. 7. IMP 8 magnetic field data showing 15.36-s field averages from 1800 UT on December 28 to 0300 UT on December 29, 1976. The upper panel shows the total field B, while the succeeding lower panels show the X, Y, and Z vector components of B in solar magnetospheric coordinates. IMP 8 was located in the high southern tail region at this time at ~ local midnight. At 1800 UT the GSM coordinates (in R_E) were X = -30.7, Y = 0.1, Z = -16.8, while at 0300 UT the spacecraft coordinates were X = -32.5, Y = -5.9, Z = -13.9.

an azimuthally confined section of the enhanced cross-tail current and its diversion to the auroral ionosphere via field-aligned currents [see McPherron et al. (1973) and Bostrom (1974)]. The existence of the current wedge has been known for a long time from ground-based data and recent results have clarified the physical mechanism leading to its formation. Strong support for the neutral line model of substorms was obtained when three-dimensional MHD simulations of magnetotail reconnection [Birn and Hones, 1981, and Sato et al., 1983] showed that a pair of oppositely directed field-aligned currents are an inherent part of the neutral line model.

Geostationary satellites in the near-magnetotail show that the magnetospheric part of the substorm current wedge is azimuthally confined and expands eastward and (especially) westward during the course of the substorm expansion phase [see Nagai, 1982 and Nagai et al., 1983]. The near-earth part of the current wedge can also be studied by ground-based observations. Present results suggest that the substorm current wedge and the rapid dipolarization of magnetic fields near midnight at geostationary orbit are the direct results of the onset of magnetic reconnection in the near-tail region (10-20 R_E).

As demonstrated by the data of Figs. 4 and 6, the rapid collapse of the magnetic field at 6.6 R_E at substorm onset is accompanied in precise time coincidence with the sudden appearance of hot plasma [DeForest and McIlwain, 1971] and energetic particles. Note in Fig. 6, for example, that both electrons and ions up to >100 keV kinetic energy suddenly appear. The injection events become larger in flux amplitude as one goes to lower energy (Fig. 4, Fig. 6) and, in fact, the injection events are very prominent down into the plasma energy (< 1 keV) regime [DeForest and McIlwain, 1971]. Recall also from Fig. 7 that the appearance of these freshly injected particles occurs precisely during the time that magnetic energy density in the deep tail rapidly decreases. The evidence in many, many instances therefore points to a model in which stored magnetotail energy is rapidly converted to hot plasma and strong field-aligned currents which are resistively dissipated in the ionosphere.

Examination of the details of plasma and energetic particle properties during expansion phase onsets can reveal further important characteristics of such events. For example, it would be possible that particle flux variations at substorm onsets do not represent true flux variations, but rather are adiabatic changes associated with magnetic field increases and decreases. To test this, we have used combined plasma, energetic particle, and magnetic field data in another Coordinated Data Analysis Workshop (CDAW-2). Taking the phase space density as $f = j/2m_o \mu B$ (where m_o is the particle mass, μ is the magnetic moment of the particle, and B is the local field strength), we have calculated the variations of the distribution function of electrons and ions, at constant μ, for CDAW-2 time intervals [Baker et al., 1982a].

As is evident, the advantage of studying the phase space density at constant μ is that magnetic field variations are removed. Thus true particle density increases are revealed, and sources (or sinks) of particles can be identified.

The CDAW-2 analysis concentrated on a substorm onset which occurred at 1200 UT on 29 July 1977. Figure 9 shows examples of the phase space densities for electrons at μ = 1, 10, and 100 MeV/G (1-300 keV in kinetic energy). The most evident features in the upper panel (0300 LT grouping) were the following: (1) Even with removal of adiabatic effects, the flux dropout persists; (2) The phase space densities at constant μ were identical before the dropout (1130 UT) and after the dropout (1155 UT); and (3) True phase space density increases were observed for all magnetic moments (energies) after 1200 UT.

Thus, by examining geostationary orbit flux and phase space density variations (particularly near local midnight), it is established that fresh particles (up to several hundred MeV/G) appear at synchronous orbit during substorms. A remaining question about such particles is where the particles come from. The best available tool for examining the question of the general source region for the injected hot plasma and energetic particles is provided by ion gradient measurements. Because of their large gyroradii, 10-1000 keV ions can provide good

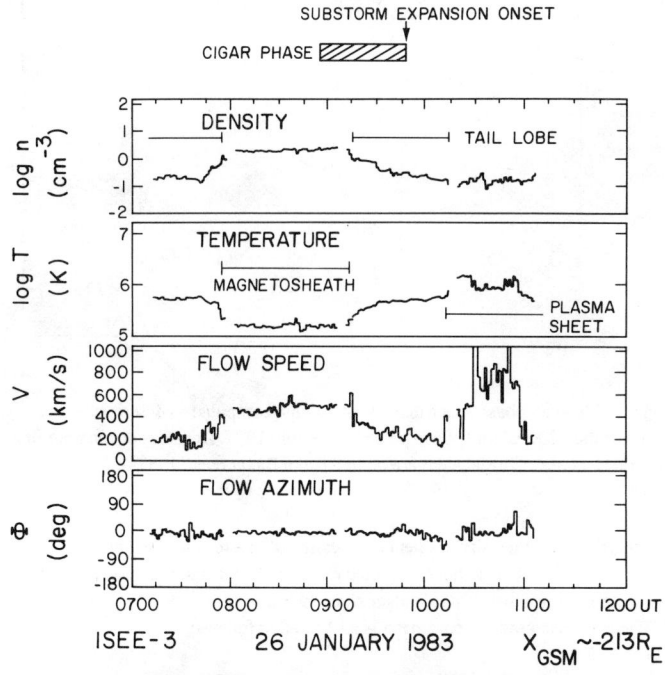

Fig. 8. A comparison of ISEE-3 electron distribution function moments at ~220 R_E in the center of the distant magnetotail with measurements from spacecraft 1981-025 at geostationary orbit (~135°W) near local midnight. The substorm growth, or cigar, phase at 6.6 R_E occurred as indicated between ~0850 UT and ~0950 UT at which time a sharp, intense substorm expansion phase onset occurred. With delays of 20-30 minutes, ISEE-3 saw closely related events such that at ~0910 UT it went from the sheath into the tail lobe and at ~1011 UT it went from the lobe into the plasma sheet where very high tailward plasma flows were seen. These data show that the very distant tail expands during growth phases as energy is added to the tail lobes [from Baker et al., 1983b].

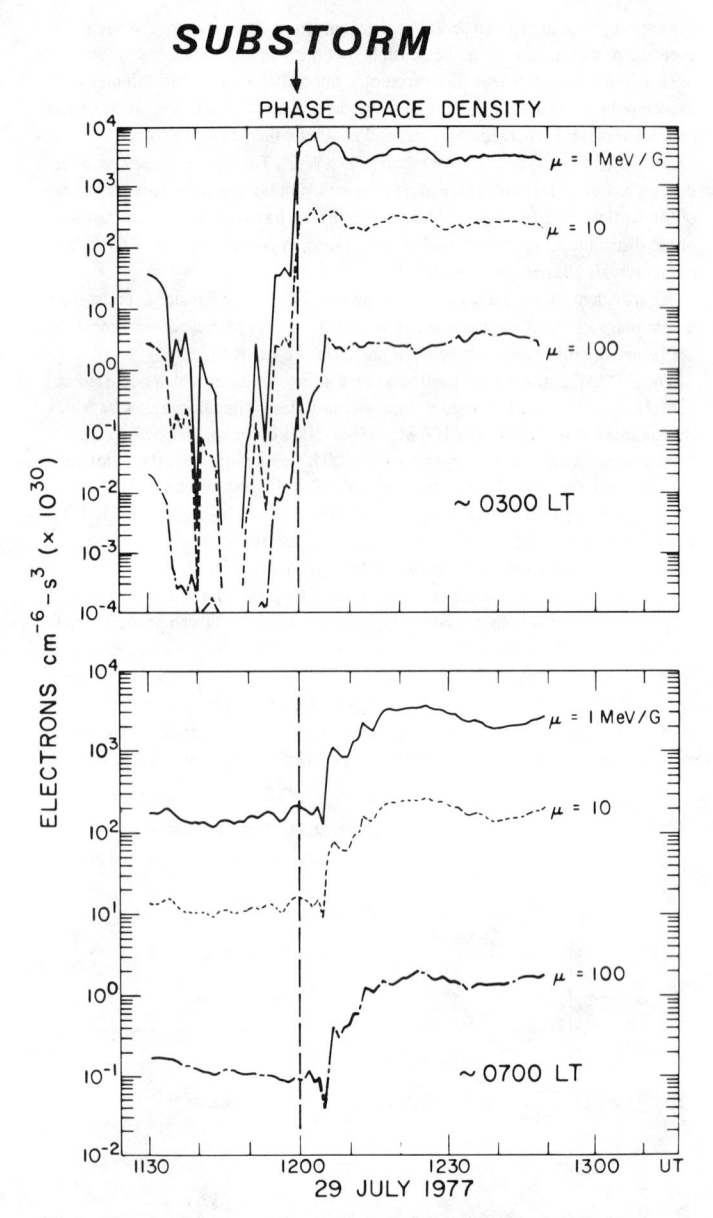

Fig. 9. Electron phase space density variations (computed as described in the text) for the 1200 UT substorm period on 29 July 1977. Densities at constant first invariant values (μ, as labeled) are plotted [from Baker et al., 1982a].

information about density gradients that exist within a region of strong radial intensity variations or within an injected cloud of plasma and energetic particles [Fritz and Fahnenstiel, 1982; Walker et al., 1976].

The east-west gradient parameter is computed as follows:

$A_{EW} = (E - W)/(E + W)$ where E is the proton flux ($E_p > 145$ keV)

measured in the sector with the detector looking eastward, and W is the proton flux measured looking westward. Given the direction of the normal magnetic field in the vicinity of the geostationary orbit satellites, and using the sense of gyration of protons, $A_{EW} > 0$ generally implies a higher particle density at altitudes greater than that of the spacecraft. For a stretched (taillike) magnetic field orientation (as distinguished from a completely dipole field), one also obtains some secondary information from A_{EW}.

Figure 10 shows the A_{EW} (dashed line) values calculated for 29 July 1977 from the 77-007 energetic proton data (E > 145 keV). The solid line shows the measured >145 keV proton flux for the same interval. Looking at A_{EW} and intensity variations together, the following sequence of events is seen. Between 1155 and 1200, i.e., during the recovery from the flux dropout, A_{EW} was strongly positive. This suggests that the higher particle density was inside the spacecraft. Since concurrent data showed the field to be very taillike during this period, the suggestion of a boundary motion during the dropout, with the high flux region moving earthward and equatorward, is fully borne out. As the fluxes recover, the spacecraft was enveloped from inside and from below.

At 1200 UT, A_{EW} went strongly negative. This period corresponded precisely to the first energetic particle and hot plasma injection into synchronous orbit. The character of A_{EW} showed that the injected particles came from outside the spacecraft location. The conclusion is therefore unambiguous in this case, viz., the injected particles arrive at 6.6 R_E from the outside and from above. This very likely means that these particles filled the high-latitude plasma sheet and that these filled field lines then collapsed inward over the spacecraft. After the leading edge of the particle injection passed over the spacecraft, A_{EW} went strongly positive (1202-1205 UT). This indicates that the highest particle density, after the injection, was generally inside 6.6 R_E.

From studies of the kind outlined above, there is little doubt that substorms produce freshly accelerated particles through the action of conversion of magnetic energy into particle kinetic energy. The data also suggest that this conversion generally occurs outside of geostationary orbit, deeper in the magnetotail plasma sheet.

Moore et al. [1981] studied several substorms using data from ATS 6 and SCATHA (P78-2) with particular attention to the abrupt and dispersionless nature of the leading edges of many events. They found that events which were abrupt at both spacecraft and travelled 1 R_E in as little as one minute, implying average speeds up to 100 km/s and boundary thickness 0.1 R_E. They further argued, on the basis of electron energy spectral changes that the moving plasma boundary which they referred to as the "injection front," was the precipitation-flow boundary described by Kennel [1969]. The plasma increases studied by Moore et al. [1981] were closely associated with large (factor 2-3) equally abrupt increases of local magnetic field strength and with rotations to a more dipolar orientation. On the basis of this fact, the agent of injection was identified as the

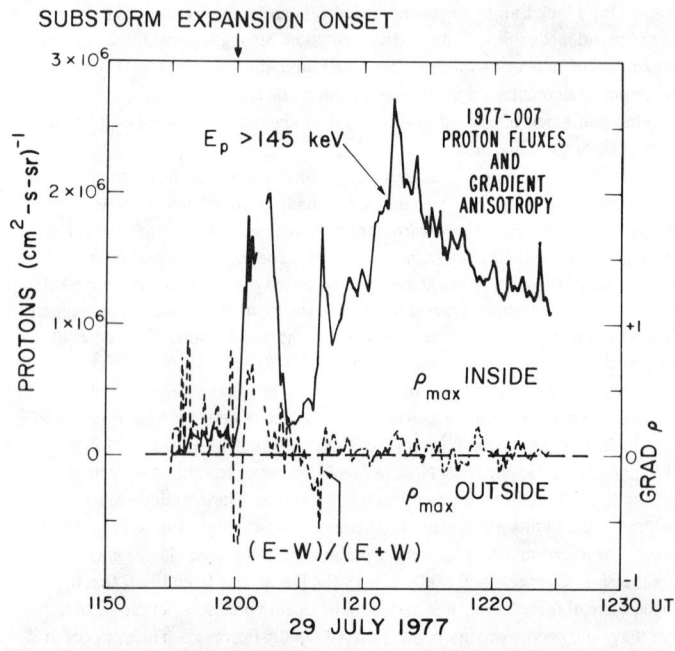

Fig. 10. A comparison of the >145-keV proton flux (solid line) and the associated east-west gradient anisotropy (dotted line) for the period illustrated in Fig. 9. Strong gradient anisotropies occur as new energetic particles are injected near synchronous orbit [from Baker et al., 1982a].

earthward propagating compression wave previously observed by Russell and McPherron [1973].

It should be noted that the compression wave observed by Russell and McPherron [1973], and hypothesized as an agent of injection, contains a dawn-to-dusk directed current sheet and has an associated dawn-to-dusk directed induced electric field (10 mV/m) which is consistent with recent electric field observations [Pedersen et al., 1978; Aggson et al., 1983]. The motion of such a wave corresponds directly to the collapse of stretched magnetic field lines to a less stretched configuration [cf. Baker et al., 1982a], the collapse being communicated to more earthward locations by the propagating wave. An interesting aspect of such a wave is that it propagates into a region of decreasing phase speed. In such circumstances, the wave will steepen and may break, forming a shock, or be reflected. In either case, it should be unable to propagate very deeply into the plasmasphere, and there should be a well defined earthward limit of injection effects.

The energization mechanisms described above do not appear capable of producing the high energy (>100 keV) component of substorm-related particle enhancements. Such particles could, however, be rapidly accelerated in the parallel electric field which exists along a near-earth neutral line. Note that induction effects related to dynamic reconnection can raise the total potential drop along the neutral line far above its expected steady-state value [Baker et al., 1982a]. Moreover, acceleration processes at the neutral line could explain the electron heating pulse observed during some reconnection events [Bieber et al., 1982].

As has been shown above, enhancements of the fluxes of particles having energies of several hundred keV are commonly observed at synchronous orbit and in the magnetotail during geomagnetically active periods. In nearly all cases these flux enhancements are closely associated with individual magnetospheric substorms. As we have demonstrated here using energetic particle data from synchronous orbiting satellites and from satellites in the magnetotail, it is found that many features of the timing of particle enhancements relative to substorm onsets and recoveries (derived from ground magnetic records) and relative to plasma sheet thinnings and recoveries (measured with plasma probes on earth-orbiting satellites) can be understood in terms of the neutral line model of substorms in which the particles are impulsively accelerated during a brief period at substorm onset near a site of magnetic reconnection.

Particularly striking are very high energy proton (ion) phenomena associated with substorms. In approximately 10-20% of substorms, >0.3 MeV ions appear throughout the magnetosphere and its environs in close association with expansion phase onset [Belian et al., 1978]. Oftentimes, ion bursts may be identifiable as distinct particle bunches ("drift echoes") which drift azimuthally around the earth through several (as many as 5) circulations (cf. Fig. 11). A comprehensive model for the morphology of energetic ion enhancements is illustrated in Figure 12 [Baker et al., 1979]. This model suggests that after acceleration at the substorm X-line in the plasma sheet, the ions stream both sunward and tailward. Those reaching the synchronous orbit region are transported westward around the earth via curvature and grad-B drifts.

The tailward-streaming ions produced at the same time as drift-echo ions appear as "impulsive bursts". The inverse velocity dispersion (i.e., observation of slower particles before faster ones) exhibited by these bursts is supportive of their hypothesized origin at a magnetic X-line [Sarris and Axford, 1979]. As suggested by the inset at the bottom of Figure 12, a spacecraft in the thinning plasma sheet successively samples field lines that have reconnected more and more recently at the X-line. These field lines contain ion distributions that are less depleted at the high energy end of the spectrum by escape of the faster particles. Finally, just as the spacecraft enters the lobe, it samples preferentially the fastest ions streaming along field lines connected directly to the X-line source. Concurrent plasma observations confirm that impulsive bursts do indeed occur right at times of plasma dropouts [Belian et al., 1981].

As a final component of this picture, the more commonly observed non-impulsive (rapid-rise, slow-decay) plasma sheet ion enhancements are attributed in this model to envelopment of the observing satellite by the recovering (i.e., expanding) plasma sheet, into which have leaked ions previously injected into the outer radiation zone. The subsequent decay is explained by a combination of

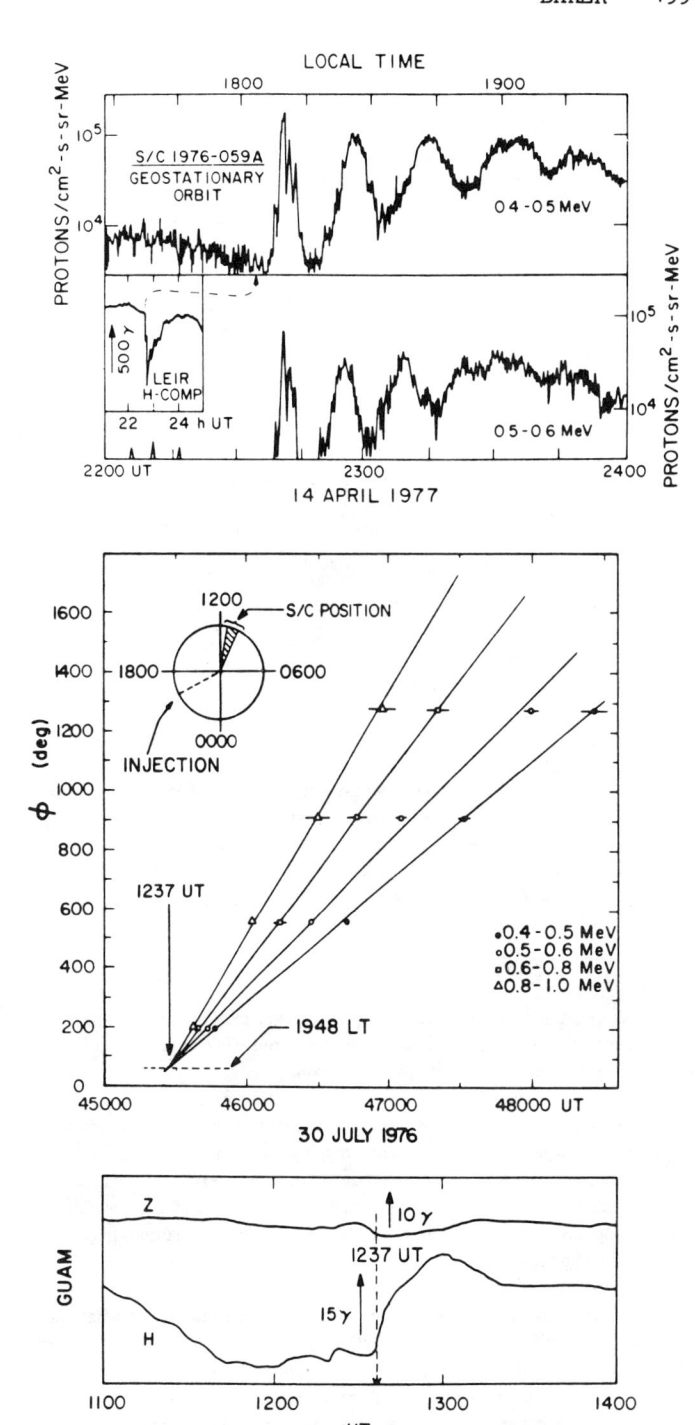

Fig. 11. The upper panel is a representative illustration of an energetic ion drift echo event as observed by spacecraft 1976-059 on 14 April 1977. Two energy channels are shown (0.4-0.5 MeV and 0.5-0.6 MeV) and the left-hand inset illustrates that the first of the 5 drift echoes seen occurred at precisely the time of a substorm expansion onset seen at Leirvogur, Iceland. The middle panel is for a similar event on 30 July 1976 and shows a plot of azimuthal position of occurrence of proton drift echo pulses versus the UT of their observation at the spacecraft. Several energy ranges, as labeled, are included in the analysis, and the intersection of the several lines indicates the time and location of the proton injection. The lower panel shows the Guam magnetogram for 1100-1400 UT of July 30, 1976 and shows a substorm expansion onset at precisely 1237 UT [from Belian et al., 1978, 1981].

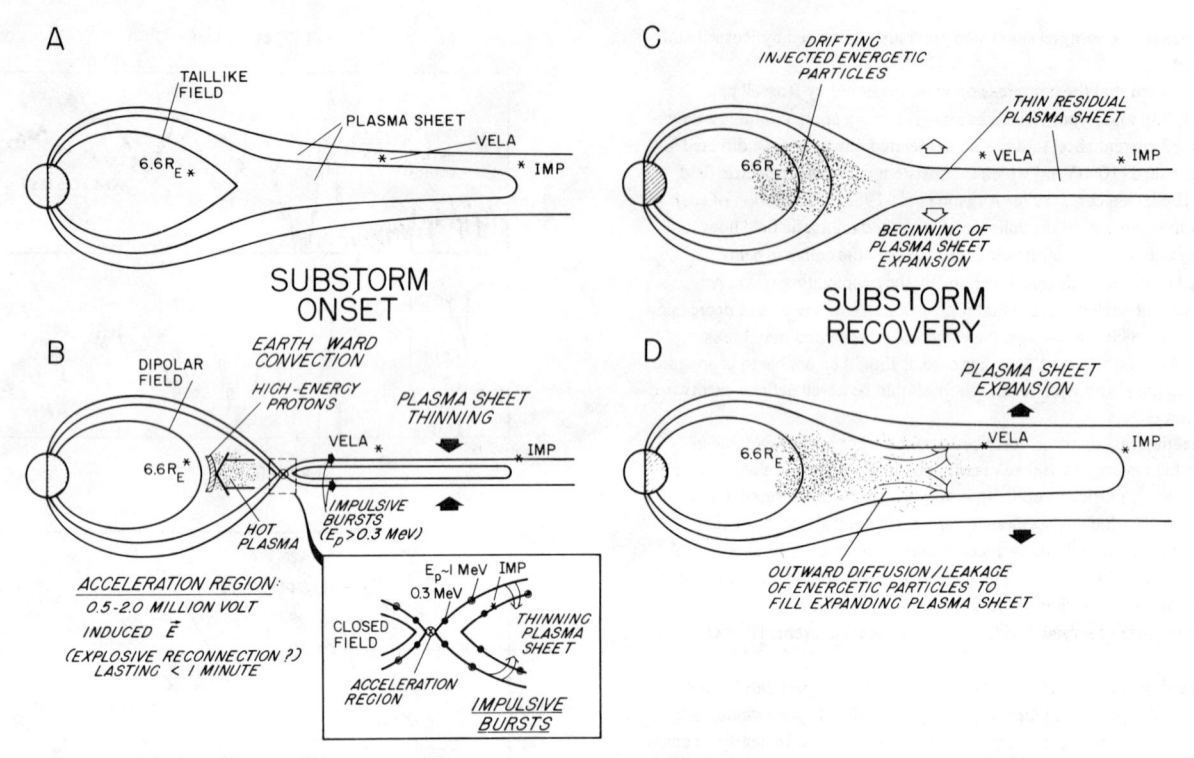

Fig. 12. Schematic depicting the sequence of energetic particle events predicted by the model of Baker et al. [1979]. (a) The inner magnetosphere just prior to substorm onset showing the buildup of stress evidenced by the taillike field. (b) The magnetosphere just after onset showing a dipolar field configuration and the accelerated ion bunches streaming sunward toward the trapped radiation zones and antisunward along the thinning plasma sheet. (c) Conditions just prior to substorm recovery and the beginning of the plasma sheet expansion. (d) Expansion of the plasma sheet and the subsequent filling of the expanding sheet with energetic protons diffusing out of the trapped region.

plasma sheet expansion, adiabatic cooling, and escape mechanisms. Escape from the magnetotail could in turn account for the appearance of energetic proton bursts in the magnetosheath and upstream region.

Although the number densities of energetic particles are relatively small compared to plasma number densities, the energy density in this component can be reasonably large throughout the outer trapping region and the magnetotail. Furthermore, once such particles are produced, they make excellent diagnostic tools for establishing substorm timing [e.g., Belian et al., 1981], for examining field line topology, and for remotely probing plasma boundary motions [e.g., Baker et al., 1982a].

An interesting application of the observation of drift-echo ion events comes about from the close inspection of the temporal and compositional characteristics of such events. Drift echo peaks frequently are observed to have a complex shape and do not always exhibit a simple rise and fall. One sees several peaks very close together. These "structured peaks" could result from several causes including: an injection multiple in time or longitude; there being more than one relatively abundant charge state of a given ion; multiple ion species being detected in a single sensor; and charge exchange causing the charge state of an individual ion to vary significantly on the drift time scale. Observation of the evolution of a structured peak with longitudinal drift can help to distinguish between some of these possibilities.

The angular velocity associated with the azimuthal drift of an energetic ion is given by the relationship [cf, Blake et al., 1983.]

$$\Omega_3 \propto f(\alpha_o)LE/q$$

where E is the kinetic energy of the ion, q is the charge state and α_o the equatorial pitch angle. Here it is assumed (as is the case for drift-echo events) that the energy of the ion is sufficiently high ($E \gtrsim 100$ keV) that the effect of magnetospheric electric fields on drift is not significant. The important feature to note for present purposes is that the drift velocity of an ion at a given L value and pitch angle depends upon the ratio of its energy to the effective charge of the ion. Thus a measurement of the drift speed and the energy of an ion determines the charge state.

The mean charge states of heavy ions trapped in the earth's magnetosphere, in particular those of the abundant heavy ions He, C, and O, are important observational parameters predicted by various theories (e.g., Spjeldvik and Fritz, 1978) of the origin and evolution of the magnetospheric plasma. Up to the present time satellite instrumentation capable of determining the charge state of heavy ions with energies above ~ 50 keV has not been flown. Consequently there is motivation to develop indirect methods of determining the charge state of energetic magnetospheric ions, even if such methods do not have universal applicability. One such method is to measure the drift speed of an ion in the magnetic field of the earth.

A major difficulty is to tag an ion in some way in order to be able to measure its drift speed. The experimental procedure employed by Blake et al. (1983) utilizes observations of the transient, highly peaked enhancements in the ion fluxes represented by ion drift echo events. An ion drift echo event seen in several proton channels and two helium channels on board the SCATHA spacecraft between 2100 and 2200 UT on 25 February 1979 is shown in Figure 13.

The data of Fig. 13 are for pitch angles of 90° ±30°. The drift speed of an ion is a function of pitch angle; however, a dipolar calculation predicts only a 5% difference between the extremes of 60° and 90°. The dispersion in arrival time as a function of ion energy can be seen clearly in Figure 13 although, because of the averaging of the data that was done to generate the figure, it cannot be used for quantitative timing purposes. Note that the peak in the 363-717 keV proton channel occurs prior to either of the helium peaks. In other data (Blake et al., 1983) not shown here it was seen that there was only one CNO count in the five

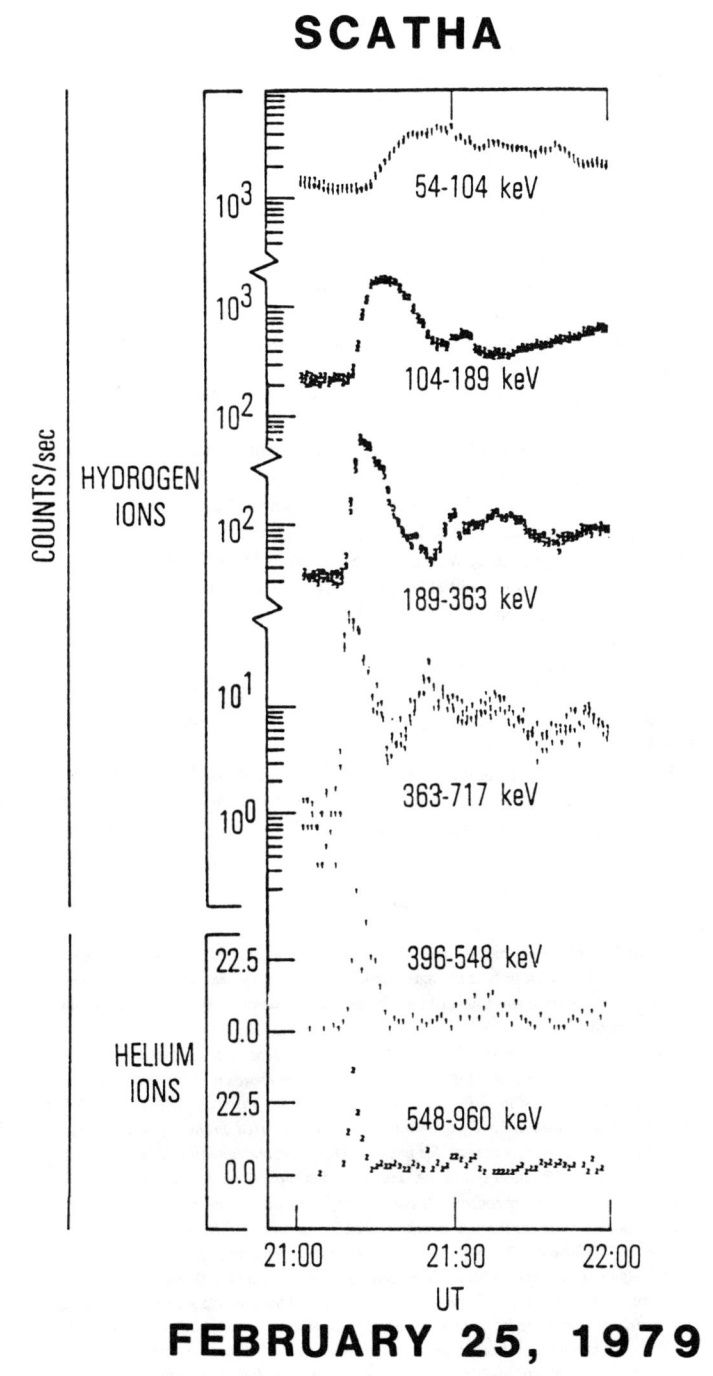

Fig. 13. The temporal history of proton and helium ion count rates for a drift echo event measured by SCATHA instruments on 25 February 1979. As described in the text, examination of the relative timing of H and He peaks allows a charge state determination of the heavier ions [from Blake et al., 1983].

region for a long time before acceleration. If it had been, then charge exchange would have transformed the stripped, or nearly stripped, solar wind ions (Spjeldvik and Fritz, 1978) to a lower charge state than observed. A model in which plasma sheet ions, originally from the solar wind, are brought in from the tail and accelerated would fit the observations.

As discussed by Blake et al. (1983), studies have shown that the energetic heavy-ion fluxes in the synchronous altitude region are highly time variable and that, above a few hundred keV, CNO ions are the most abundant. These results suggest that the most-energetic ions in the synchronous altitude region result from injections of plasma sheet ions accelerated by strong electric fields. If the plasma had a steep energy spectrum prior to acceleration, the several-fold increase in the CNO energy relative to that of protons, because of their high charge state, could make them most abundant in the energized plasma population.

Summary

Observations in the near-earth magnetotail show some of the clearest and most repeatable signatures available in support of the concept of loading and unloading of magnetic energy in association with substorms. The data illustrate that magnetic energy is accumulated and stored for 0.5 ~2.0 hours in the tail lobes and then is rapidly dissipated at substorm expansion onset. The dissipation is manifested by the acceleration and rapid transport of hot plasma and energetic particle populations within the near-tail region. These energized plasmas provide an excellent tracer capability which allows a relatively clear determination of where, when, and how magnetic energy is converted to other forms during substorms.

When near-tail data are considered in a global context of deep-tail measurements, numerical models, ground-based data, etc., they provide very strong evidence for the neutral line substorm model and, thus, for the regular occurrence of magnetic reconnection in the near-earth magnetotail.

Acknowledgments. The author would like to thank his many colleagues who have contributed significantly to the work reviewed in this paper. In particular, thanks are extended to E. Hones, R. Belian, P. Higbie, T. Fritz, J. Gosling, S. Bame, and R. Zwickl of Los Alamos, R. McPherron of UCLA, and J. Blake and J. Fennell of Aerospace Corporation. Sincere appreciation is also extended to M. Halbig, R. Robinson, E. Tech, and R. Anderson for data analysis support. Very useful discussions with K. Schindler and A. Nishida are gratefully acknowledged.

References

Aggson, T. L., J. P. Hepper, and N. C. Maynard, Observations of large magnetospheric electric fields during the onset phase of a substorm, *J. Geophys. Res.*, 88, 3981, 1983.

Baker, D. N., P. R. Higbie, E. W. Hones, Jr., and R. D. Belian, High-resolution energetic particle measurements at 6.6 R_e, 3, Low-energy electron anisotropies and short-term substorm predictions, *J. Geophys. Res.*, 83, 4863, 1978.

Baker, D. N., R. D. Belian, P. R. Higbie, and E. W. Hones, Jr., High-energy magnetospheric protons and their dependence on geomagnetic and interplanetary conditions, *J. Geophys. Res.*, 84, 7183, 1979.

Baker, D. N., E. W. Hones, Jr., P. R. Higbie, R. D. Belian, and P. Stauning, Global properties of the magnetosphere during a substorm growth phase: A case study, *J. Geophys. Res.*, 86, 8941, 1981.

Baker, D. N., T. A. Fritz, B. Wilken, P. R. Higbie, S. M. Kaye, M. G. Kivelson, T. E. Moore, W. Studemann, A. J. Masley, P. H. Smith, and A. L. Vampola, Observation and modeling of energetic particles at synchronous orbit on July 29, 1977, *J. Geophys. Res.*, 87, 5917, 1982a.

Baker, D. N., E. W. Hones, Jr., R. D. Belian, P. R. Higbie, R. P. Lepping, and P. Stauning, Multiple-spacecraft and correlated riometer study of magnetospheric substorm phenomena, *J. Geophys. Res.*, 87, 6121, 1982b.

Baker, D. N., et al., Evidence for magnetotail energy storage and sudden release during substorms of the CDAW-6 intervals, *J. Geophys. Res.*, to be submitted, 1983a.

hours preceding the event, and it was in the lowest energy channel; the many observed CNO counts seen after the substorm onset could, therefore, also be associated with the drift echo event with confidence.

The results of the Blake et al. study show that the helium ions were fully stripped (charge state +2) and that the CNO ions observed by SCATHA were probably of charge state 5 or higher, and definitely not charge state 1 or 2. These results indicate that the source of the accelerated plasma was not the ionosphere. Furthermore the plasma could not have been resident in the synchronous altitude

Baker, D. N., S. J. Bame, R. D. Belian, W. C. Feldman, J. T. Gosling, P. R. Higbie, E. W. Hones, Jr., D. J. McComas, and R. D. Zwickl, Correlated dynamical changes in the near-earth and distant magnetotail regions: ISEE-3, *J. Geophys. Res.*, in press, 1983b.

Bargatze, L. F., D. N. Baker, R. L. McPherron, and E. W. Hones, Jr., Magnetospheric response for many levels of geomagnetic activity, *J. Geophys. Res.*, in press, 1983.

Belian, R. D., D. N. Baker, P. R. Higbie, and E. W. Hones, Jr., High-resolution energetic particle measurements at 6.6 R_E, 2, High-energy proton drift echoes, *J. Geophys. Res., 83*, 4857, 1978.

Belian, R. D., D. N. Baker, E. W. Hones, Jr., P. R. Higbie, S. J. Bame, and J. R. Asbridge, Timing of energetic proton enhancements relative to magnetospheric substorm activity and its implication for substorm theories, *J. Geophys. Res., 86*, 1415, 1981.

Bieber, J. W., E. C. Stone, E. W. Hones, Jr., D. N. Baker, and S. J. Bame, Plasma behavior during energetic electron streaming events: further evidence for substorm-associated magnetic reconnection, *Geophys. Res. Lett., 9*, 664, 1982.

Birn, J. and E. W. Hones, Jr., Three-dimensional computer modeling of dynamic reconnection in the geomagnetic tail, *J. Geophys. Res., 86*, 6802, 1981.

Blake, J. B., J. F. Fennell, D. N. Baker, R. D. Belian, and P. R. Higbie, A determination of the charge state of energetic magnetospheric ions by the observation of drift echoes, *Geophys. Res. Letters*, in press, 1983.

Bostrom, R., Ionosphere-magnetosphere coupling, in: *Magnetospheric Physics*, B. M. McCormac (ed.), pp. 45-59, Dordrecht, D. Reidel, 1974.

DeForest, S. E., and C. E. McIlwain, Plasma clouds in the magnetosphere, *J. Geophys. Res., 76*, 3587, 1971.

Fairfield, D. H., and N. F. Ness, Configuration of the geomagnetic tail during substorms, *J. Geophys. Res., 75*, 7032, 1970.

Fairfield, D. H., R. P. Lepping, E. W. Hones, Jr., S. J. Bame, and J. R. Asbridge, Simultaneous measurements of magnetotail dynamics by IMP spacecraft, *J. Geophys. Res., 86*, 1396, 1981.

Fritz, T. A., and S. C. Fahnenstiel, High temporal resolution energetic particle soundings at the magnetopause on November 8, 1977, using ISEE-2, *J. Geophys. Res., 87*, 2125, 1982.

Hones, E. W., Jr., D. N. Baker, S. J. Bame, W. C. Feldman, J. T. Gosling, D. J. McComas, R. D. Zwickl, J. Slavin, E. J. Smith, B. T. Tsurutani, Structure of the magnetotail at 220 R_E and its response to geomagnetic activity, *Geophys. Res. Letters*, in press, 1983.

Kennel, C. F., Consequences of a magnetospheric plasma, *Rev. Geophys. Sp. Phys., 7*, 379, 1969.

McPherron, R. L., Growth phase of magnetospheric substorms, *J. Geophys. Res., 28*, 5592, 1970.

McPherron, R. L., Substorm related changes in the geomagnetic tail: The growth phase, *Planet. Space Sci., 20*, 1521, 1972.

McPherron, R. L., and R. H. Manka, Dynamics of the March 22, 1979 substorm event: CDAW-6, *J. Geophys. Res.*, in preparation, 1983.

McPherron, R. L., C. T. Russell, and M. P. Aubry, Satellite studies of magnetospheric substorms on August 15, 1968, 9. Phenomenological model for substorms, *J. Geophys. Res., 78*, 3131, 1973.

Moore, T. E., R. L. Arnoldy, J. Feynman, and D. A. Hardy, Propagating substorm injection fronts, *J. Geophys. Res., 86*, 6713, 1981.

Nagai, T., Local-time dependence of electron flux changes during substorms derived from multi-satellite observation at synchronous orbit, *J. Geophys. Res., 87*, 3456, 1982.

Nagai, T., D. N. Baker, and P. R. Higbie, Development of substorm activity in multiple onset substorms at synchronous orbit, *J. Geophys. Res., 88*, 6994, 1983.

Nishida, A., IMF control of the earth's magnetosphere, *Space Sci. Rev., 34*, 185, 1983.

Paulikas, G. A., and J. B. Blake, Effects of the solar wind on magnetospheric dynamics: energetic electrons at the synchronous orbit, in *Quantitative Modeling of Magnetospheric Processes*, W. P. Olson, Editor, American Geophys. Union, Washington, D.C., 1979.

Pedersen, A., R. Grard, K. Knott, D. Jones, and A. Gonfalone, Measurements of quasi-static electric fields between 3 and 7 earth radii on Geos-1, *Space Sci. Rev., 22*, 333, 1978.

Roederer, J. G., *Dynamics of Geomagnetically Trapped Radiation*, Springer Verlag, New York, 1970.

Russell, C. T. and R. L. McPherron, The magnetotail and substorms, *Space Sci. Rev., 15*, 205, 1973.

Sarris, E. T. and W. I. Axford, Energetic protons near the plasma sheet boundary, *Nature, 77*, 460, 1979.

Sato, T., T. Hayashi, R. J. Walker, and M. Ashour-Abdalla, Neutral sheet current interruption and field-aligned current generation by three dimensional driven reconnection, *Geophys. Res. Letters, 10*, 221, 1983.

Spjeldvik, W. N., and T. A. Fritz, Theory for charge states of energetic oxygen ions in the earth's radiation belts, *J. Geophys. Res., 83*, 1583, 1978.

Walker, R. J., K. N. Erickson, R. L. Swanson, and J. R. Winckler, Substorm-associated particle boundary motion at synchronous orbit, *J. Geophys. Res., 81*, 5541, 1976.

West, H. I., Jr., "The Signatures of the Various Regions of the Outer Magnetosphere in the Pitch Angle Distribution of Energetic Particles," *Quantitative Modelling of Magnetospheric Processes*, (ed. W. P. Olson), AGU, 1979.

Questions and Answers

Russell: How close can the neutral point get to synchronous orbit near midnight? Can the large increase in field strength accompanying a very near earth neutral point formation and subsequent tailward retreat cause the observed very high energizations sometimes observed?

D. N. Baker: We believe that the substorm neutral line sometimes approaches as close as perhaps 1-2 R_E of the geostationary orbit. This means that the neutral line would be formed at X_{GSM} ~-8 to -9 R_E. We reach this conclusion based on the extreme tail-like field orientation seen at 6.6 R_E on many occasions and based on the apparently very thin plasma sheet that often exists at this position. Our energetic ion data allow reasonably good estimates of the position, shape, and velocity of plasma boundaries in the midnight sector; and, thus, we are reasonably confident of these inferences. With regard to the second part of your question, it is quite probable that the very high-energy events that we observe (~10-20% of all substorms) are those that correspond to neutral line formation relatively near the earth. This picture would also be consistent with, and would tend to explain, the nondispersive character of these events. If the particle source were very far tailward, we would expect much more energy dispersion (due to gradient and curvature drifts) than is observed. In fact, as you suggest, it is possible that the formation of a very near-earth neutral line could lead to very small magnetic field strengths and then, subsequently, to very large field increases. This large field strength change could, in principle, produce significant particle accelerations of the kind observed.

IMPLICATIONS OF THE 1100 UT MARCH 22, 1979 CDAW 6 SUBSTORM EVENT FOR THE ROLE OF MAGNETIC RECONNECTION IN THE GEOMAGNETIC TAIL

T. A. Fritz and D. N. Baker

Earth and Space Sciences Division, Los Alamos National Laboratory,
Los Alamos, NM 87545

R. L. McPherron

Institute of Geophysics and Planetary Physics, University of California,
Los Angeles, CA 90024

W. Lennartsson

Lockheed Palo Alto Research Laboratory,
3251 Hanover Street, Palo Alto, CA 94304

Abstract. The event of March 22, 1979 has been the object of a concentrated study effort as a part of the Coordinated Data Analysis Workshop activity designated CDAW-6. Energetic electron and magnetic field measurements from a set of four satellites aligned from 6.6 to 13 R_E at the 0200 LT meridian at the time of the magnetospheric substorm event of 1100 UT are presented. These data are used to show that a magnetic X-line formed spontaneously in the region of 7 to 10 R_E in response to a steady build-up of magnetic stress in the geomagnetic tail.

Introduction

Since the introduction of the concept of an open magnetosphere by Dungey in 1961 with the requirement for magnetic field merging regions near the subsolar magnetopause and in the geomagnetic tail, the role of reconnection in the magnetospheric substorm process has been a topic of debate. Dungey [1961] argued that if there is reconnection at the subsolar magnetopause there must be reconnection in the magnetic tail and the rate at which reconnection occurs in these two regions must be equal on the average but need not balance instantaneously. When these rates fail to balance there is a net transport of magnetic flux into the geomagnetic tail [Russell and McPherron, 1973]. This in turn leads to a build-up of magnetic stresses in the tail. Much evidence now exists which points to the formation of X- and O-type neutral lines that form as a result of magnetic reconnection. This process can occur spontaneously from a collisionless tearing mode instability under certain conditions [Galeev, 1982; Schindler, 1983, Cowley, 1984]. Three-dimensional MHD modeling [Birn and Hones, 1981] has demonstrated that this newly formed reconnection region can form well earthward of the preexisting tail reconnection site required in Dungey's open model of the magnetosphere.

Although much evidence exists for the formation of a new region of reconnection in association with the magnetospheric substorm, a significant point to note is that little concensus exists as to the location of this new site; as evidenced by this conference and these proceedings its location is discussed as occurring anywhere from a value of $X(GSE) = -15$ R_E to beyond $X(GSE) \approx -200$ R_E. We present here data obtained by a set of four satellites in the near earth magnetotail during a well-studied substorm which indicate that the near earth neutral-line formed in the vicinity of the geostationary orbit ($r = 7$ to 10 R_E). We also demonstrate that the formation of the region this close to the earth was able to draw appreciable quantities of ions from the ionosphere. This result indicates that the formation and movement of the X-line is only one aspect of the magnetospheric substorm process. Specifically, the ensuing change in the ion composition in the plasma sheet is an effect which is demonstrated to persist for many hours after the near-earth X-line has disappeared.

Data Presentation

At 0826 UT on March 22, 1979 a propagating interplanetary shock front struck the magnetopause and subsequently there developed moderate-to-strong storm activity (Dst(max) = −70nT). Two discrete substorms occurred following the SSC at 0826 UT with expansion phase onsets at 1055 UT and 1436 UT. This period has been the object of a concentrated study effort as part of the Coordinated Data Analysis Workshop activity known as CDAW 6 [McPherron and Manka, 1984]. We report here on observations obtained from a set of four satellites aligned along the 0200 LT meridian at the time of a large magnetospheric substorm occurring near 1100 UT. The location of these satellites is presented in Figure 1.

Energetic electron measurements made by satellite 1977-007 at the geostationary orbit are presented in Figure 2 [c.f. Baker et al., 1982]. The fluxes of these electrons are constant prior to 1040 UT at which time they decrease sharply to near background levels. Approximately ten minutes later they increase sharply back to their pre-dropout levels and at 1104 UT undergo a second sharp increase. Although the satellite is located at the geographic equator it is at a magnetic latitude of ~5° and we interpret the electron flux dropout from 1042 to 1052 UT as an indication that the satellite moved through the trapping boundary into the high latitude tail lobe as a result of the geomagnetic field becoming highly stressed and tail-like due to the transfer of magnetic flux into the geomagnetic field.

The energetic electron measurements from spacecraft 1977-007 can be used to calculate the local magnetic field orientation in a self-consistent manner. Using a spherical harmonic analysis of the >30 keV electron distributions [cf., Baker et al., 1982] we have computed the field line inclination in a dipole meridional plane. This field inclination, which we call θ_B, is shown in the lower panel of Figure 2. A value of $\theta_B = 0°$ would correspond to a magnetic field parallel to the dipole axis, while a value of $\theta_B \sim 90°$ would correspond to a very taillike configuration with the magnetic field nearly parallel to the dipole equatorial plane. The data of

204 IMPLICATIONS OF THE 1100 UT

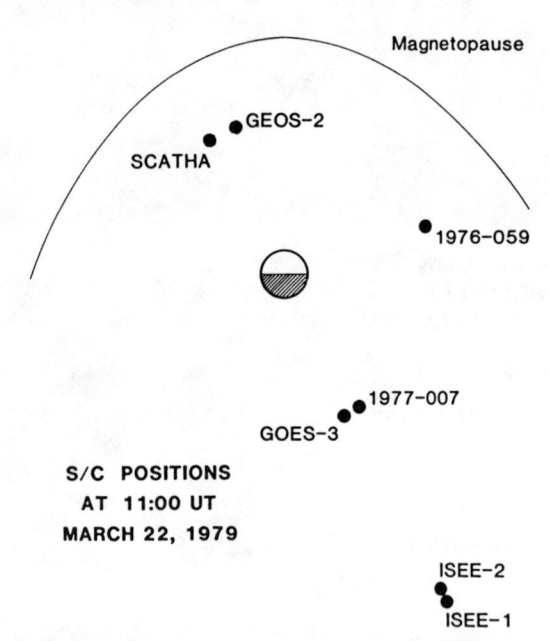

Figure 1. Location of seven spacecraft during the CDAW-6 substorm of 11 UT on March 22, 1979. Note the radial alignment of four satellites along the 0200 LT meridian with ISEE-1 at a GSE position of (–13.6 R_E, –7.2 R_E, 0.6 R_E) and ISEE-2 at (-12.1 R_E, -7.1 R_E, 0.3 R_E). The relative positions of these satellites changed very little during the event discussed here.

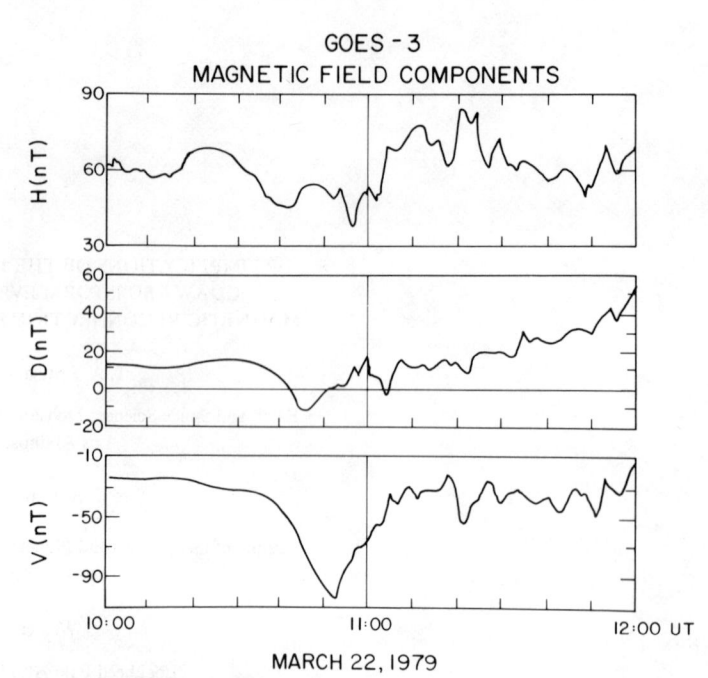

Figure 3. Magnetic field measurements made by the magnetometer on satellite GOES-3 during the CDAW-6 substorm of 11 UT on March 22, 1979. See text for definition of V, D, and H components.

Figure 2 show that prior to the substorm expansion onset θ_B (when calculable) reached values approaching 90° between ~1045 and ~1055 UT.

At the GOES-3 satellite, nearly colocated with satellite 1977-007 at synchronous orbit, onboard measurements of the magnetic field confirm this picture. In Figure 3 the measured V, D, and H components of the magnetic field at GOES-3 are presented where H is parallel to the earth's rotational axis, V is radially away from the earth, and D is positive eastward (V X D = H) completing the right-handed system [Fritz and Neeley, 1982]. Note the change in the V-component as it departs from its dipolar value of –25nT at ~1020 UT and steadily becomes more negative reaching a value of –110nT at 1052 UT. This again is direct indication of the tail-like development of field at the geostationary orbit due to the

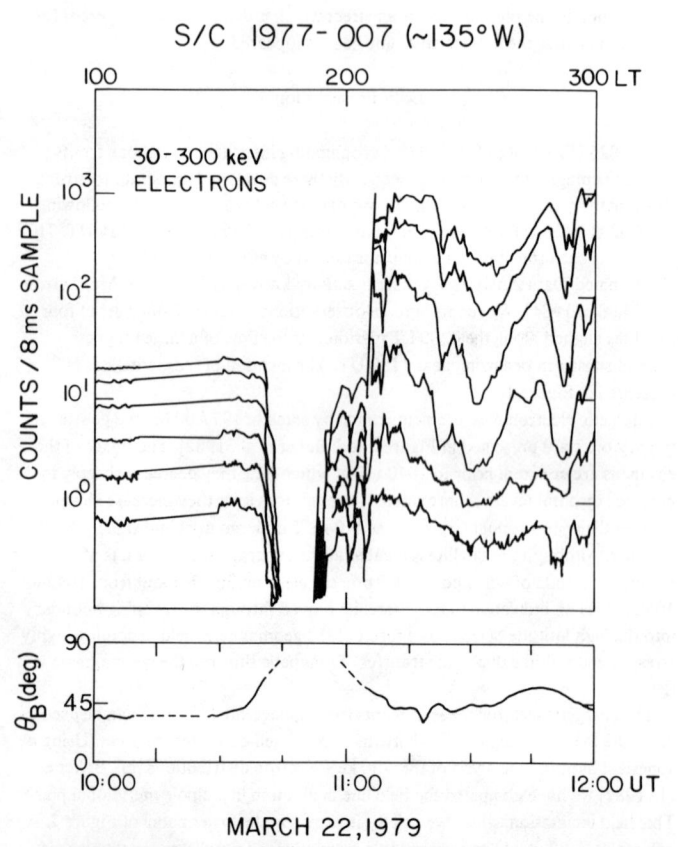

Figure 2. Energetic electron intensities measured by instruments on satellite 1977-007 during the CDAW-6 substorm of 11 UT on March 22, 1979. θ_B is described in the text.

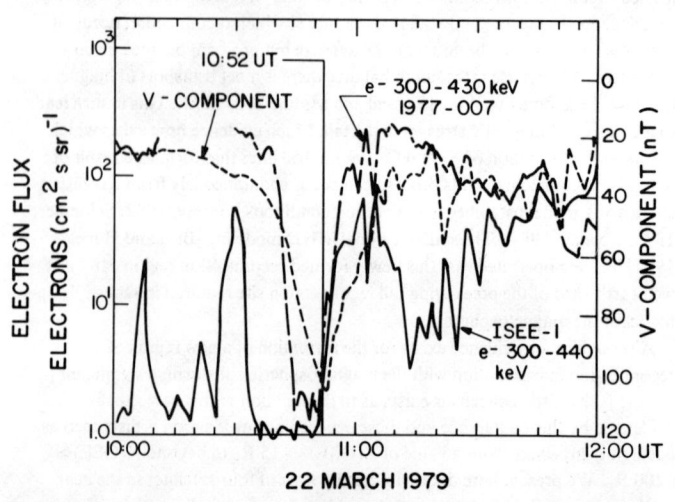

Figure 4. Composite plot of the GOES-3 magnetic field V-component and two similar electron energy channels on satellite 1977-007 at the geostationary orbit and satellite ISEE-1 at 13 R_E.

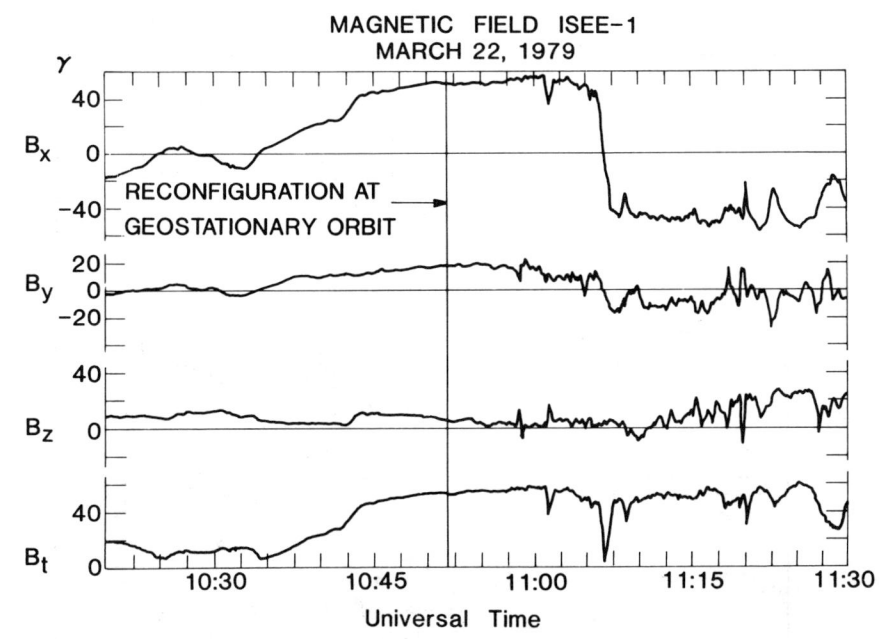

Figure 5. Magnetic field measurements in GSE coordinates made by the magnetometer on satellite ISEE-1 during the CDAW-6 substorm of 11 UT on March 22, 1979.

build-up of stresses during the period 1010 UT to 1052 UT - a phase described by McPherron [1972] as the substorm "growth" phase.

At 1052 UT (to an accuracy of $\sim \pm 1$ min) as determined both by the reappearance of the energetic electrons at satellite 1977-007 and the sharp change in the magnetic field V-component at GOES-3, the magnetic field suddenly relaxed back toward a more dipolar configuration. The electron intensity variation (300–430 keV) from S/C 1977-007 is plotted in Figure 4 along with the GOES-3 V-component to illustrate the simultaneous recovery seen in these two measurements. In addition, an identical electron energy passband on the ISEE-1 satellite (D. Hovestadt, Principal Investigator) is presented and, surprisingly, this channel shows the same recovery at 1052 UT. This behavior probably is not an *in situ* energization of these electrons. Rather, we interpret this as an indication that these electrons were excluded from reaching either 1977-007 at 6.6 R_E (5° magnetic latitude) or ISEE-1 at 13 R_E. This exclusion would most likely be due to the highly stressed tail-like magnetic field configuration prior to 1052 UT, while the energetic electrons were later able to drift to each satellite location after the field reconfiguration at 1052 UT.

The three GSE components of the magnetic field measured at ISEE-1 (C. T. Russell, Principal Investigator) for this time interval are presented in Figure 5. Note the steadily increasing value of the B_x component and total magnitude of

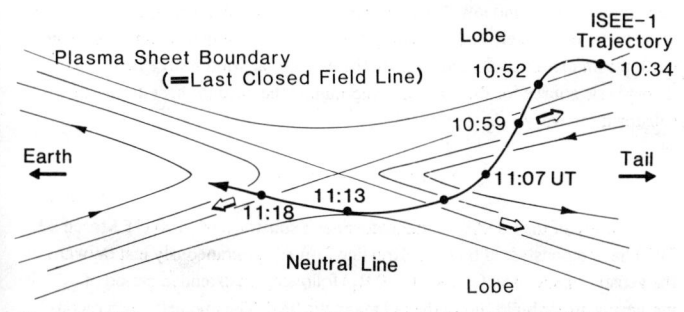

Figure 6. Diagram illustrating a possible interpretation of the ISEE-1 observations during the 10:55 UT substorm event of March 22, 1979.

the field from $\sim$1035 UT to beyond 1100 UT. As demonstrated by Fairfield [1984] this increase of the magnetic field is further evidence for the enhanced transport of magnetic flux into the geomagnetic tail and the resultant build-up of stresses there. The point to note in Figure 5 is that there is no evidence for a magnetic field reconfiguration at 1052 UT at ISEE-1 located at 13 R_E. In fact the first evidence of any particle acceleration at ISEE-1 associated with the magnetic field reconfiguration which occurred at GOES-3 and 1977-007 at 1052 UT was seven minutes later at 1059 UT [Fritz, et al., 1983; Paschmann et al., 1983].

It is beyond the scope of this paper to present detailed observations of plasma and energetic particle flow measurements. Nonetheless, it is useful to review the main points found by Fritz et al. and Paschmann et al. concerning the 1055 UT substorm:

- At 1059 UT, and for $\sim$20 minutes thereafter, there was moderate-to-strong tailward bulk flow of plasma recorded by ISEE 1,2.
- During this same interval, energetic ion (>25 keV) and energetic electron (>20 keV) measurements onboard the IEEE-1 spacecraft showed strong unidirectional (tailward) streaming.
- During this period of predominantly tailward plasma and particle flow, the local plasma sheet magnetic field was usually southward, often very strongly so.
- At $\sim$1118 UT the plasma flow at ISEE 1,2 became strongly earthward and energetic particle distributions became bidirectional and/or isotropic.

Our analysis of the magnetic field, energetic particle flux variations, and ion gradient anisotropies at geostationary orbit (c.f. Fig. 2-4 above) suggest very strongly that a particle acceleration region formed *outside* of 6.6R_E. The very taillike field at this location prior to 1055 UT is taken as being indicative of a very thin plasma sheet region very near the earth. Our interpretation is that a neutral line formed near, but outside of, geostationary orbit at the substorm expansive phase onset.

The data from ISEE 1,2 at the same local time meridian (but at $\sim$13 R_E geocentric distance) are also consistent in a general sense with near-earth neutral line formation. However, as outlined above, the ISEE data are indicative of a neutral line *earthward* of 13R_E. The strong tailward flow of plasma, the tailward jetting of energetic ions and electrons, and the presence of southward magnetic field in the plasma sheet at this location are three of the primary pieces of evidence. Figure 6 illustrates the geometry of this event and our concept of the near-earth neutral line model for this event. As a result of the steady build-up of

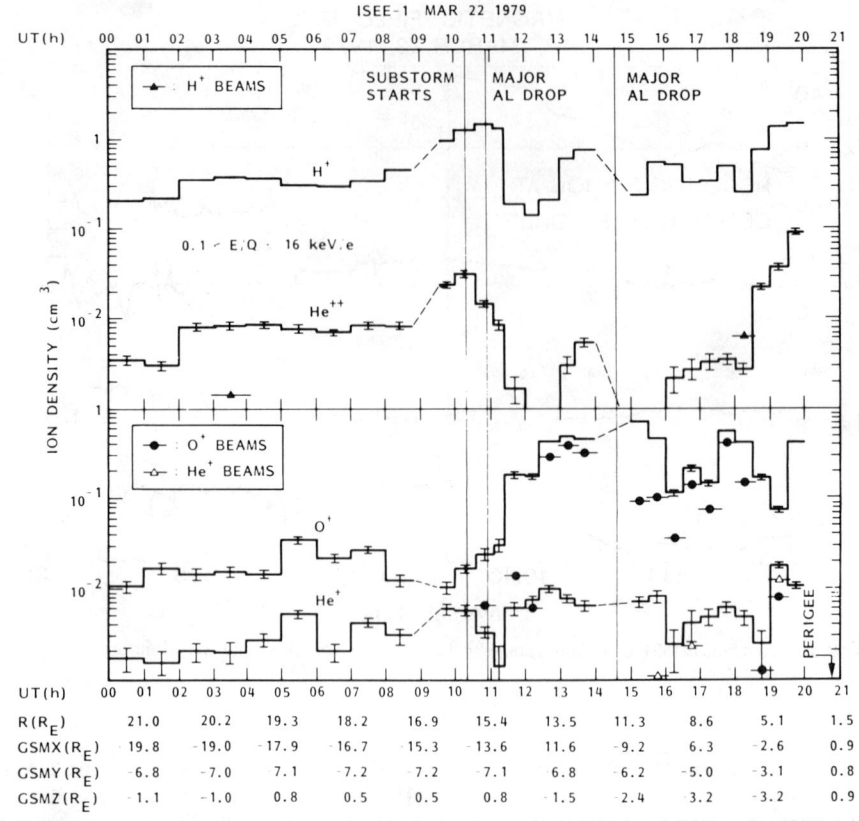

Figure 7. *In situ* measurements of the ionic compositional densities in the plasma sheet at the location of satellite ISEE-1 during the CDAW-6 event interval on March 22, 1979 [from Lennartsson et al., 1984]. The number densities are averaged over time intervals ranging from about 17 min to 60 min in length. Each interval corresponds to a horizontal section on the histograms. The circular and triangular symbols indicate what fraction of the density of each population is due to narrowly collimated beams. The error bars show the propagated uncertainty in the counting statistics ($\pm 1\sigma$). The vertical lines extending from top to bottom indicate the times of prominent changes in the auroral electrojet activity, as determined by groundbased magnetometers [cf. McPherron and Manka, 1984].

the stressed magnetic field at ISEE-1, we feel that this satellite crossed the boundary of the plasma sheet at 10:34 UT and entered the lobe. Inside the plasma sheet electrons are capable of executing their bounce motion on closed field lines. The field line associated with the plasma sheet boundary maps to the distant X-line required in the Dungey (1961) model. This is the last closed field line. The disappearance of the electron fluxes at ISEE-1 at 10:34 UT represents the thinning of the plasma sheet which caused the resultant exit of ISEE-1 onto open field lines and into a lobe-like environment. At 10:52 UT the magnetic field began to relax toward a more dipolar configuration at the geostationary orbit. Simultaneously the energetic electrons reappeared at ISEE-1, indicating this satellite reentered the region of closed field lines. We feel that the neutral line formed earthward of ISEE 1 and 2 at this time by merging oppositely directed magnetic field lines across the neutral sheet. These field lines were "closed" in the sense that they were joined across the neutral sheet at some greater distance down the tail. The resultant configuration formed an "O" geometry. As further field lines reconnected across the newly-formed near-earth neutral line, the region of the "O" geometry grew in size until those field lines connecting the ISEE-1 and ISEE-2 satellites to the earth were reconnected and the resultant energized particles and plasma flowing out along the field lines were observed at ISEE-1 at 10:59 UT, seven minutes after the initial neutral line formation. Such a model would permit any sequence of time-of-observation of respective flows at ISEE-1 and ISEE-2 depending on which field line was the first to connect the satellite to the neutral line region. Eastman and Frank [1984] in these proceedings have argued that the neutral line explanation is not valid because ISEE-1, located ~ 1 R_E further down the tail than ISEE-2, observed features in advance of ISEE-2. However, ISEE-1 was located ~ 0.3 R_E closer to the normal position of the plasma sheet mid-plane in GSM coordinates.

As the field lines continued to reconnect, the field line associated with the plasma sheet boundary finally was reached at $\sim 11:20$ UT. After 11:20 UT the data indicate that this near-earth neutral line moved rapidly tailward.

The *in situ* measured densities of various ion species at ISEE-1 are presented in Figure 7 [Lennartsson et al., 1983]. In the upper portion of the figure the measured density of ions associated with a solar wind source is plotted while in the lower panel ion densities associated with a putative ionospheric source are presented. Note that prior to the 1052 UT magnetospheric substorm onset (and the associated magnetic field reconfiguration at geostationary orbit) the plasma sheet at the ISEE-1 position was dominated by ions of solar wind origin (as indicated by the large density of He^{++} ions and low density of O^+ ions), whereas during the substorm recovery phase and later, the composition of the plasma sheet was dominated by ions of ionospheric origin (as indicated by the large density of O^+ ions and low density of He^{++} ions). This magnetospheric substorm was able to completely alter the composition of the plasma resident in the near-earth plasma sheet by switching from the usual solar wind source to an ionospheric source for the plasma to repopulate the plasma sheet during the substorm recovery.

Summary and Conclusions

The study of the CDAW-6 magnetospheric substorm of 1100 UT March 22, 1979 has demonstrated that a neutral line formed spontaneously just tailward of the geostationary orbit (e.g. ~ 7 to 10 R_E) following an extended period of increasing stress build-up in the tail magnetic field. The magnetic field reconfiguration associated with the formation of this reconnection region was not initially observed at ISEE-1 suggesting that the region was therefore localized well earthward of 13 R_E. The formation of the X-line and the eventual ejection of a plasmoid down the tail are apparently only the first manifestation of the magnetospheric substorm expansion process since appreciable fluxes of ions can be lifted out of the ionosphere, energized and used to reform the plasma sheet in association with the magnetospheric substorm process.

Acknowledgments. The CDAW-6 activity has extensively used the facilities of the National Space Science Data Center (NSSDC) at NASA/Goddard Space Flight Center and we thank Dr. James I. Vette and his organization for their efforts. Many other scientists have contributed their efforts to the interpretation of the CDAW-6 events and we acknowledge their collective contributions and the use of their data. We also wish to thank Dr. R. H. Manka for his extensive organizational efforts in making CDAW-6 the success it has become.

References

Baker, D. N., T. A. Fritz, B. Wilken, P. R. Higbie, S. M. Kaye, M. G. Kivelson, T. E. Moore, W. Stüdemann, A. J. Masley, P. H. Smith, and A. L. Vampola, Observation and modeling of energetic particles at synchronous orbit on July 29, 1977, *J. Geophys. Res., 87*, 5917, 1982.

Birn, J. and E. W. Hones, Jr., Three-dimensional computer modeling of dynamic reconnection in the geomagnetic tail, *J. Geophys. Res., 86*, 6802, 1981.

Cowley, S. W. H., The distant geomagnetic tail in theory and observation, this volume, 1984.

Dungey, J. W., Interplanetary magnetic fields and the auroral zones, *Phys. Rev. Lett., 6*, 47, 1961.

Eastman, T. E., and L. A. Frank, Boundary layers of the earth's outer magnetosphere, this volume, 1984.

Fairfield, D. H., Solar wind control of magnetospheric pressure, *J. Geophys. Res.* (CDAW-6 results, in preparation), 1984.

Fritz, T. A. and C. Arthur Neeley, Geostationary satellites ATS-6 and SMS/GOES: Description, position, and data availability during the IMS, in *The IMS Source Book* (C. Russell and D. Southwood, editors) AGU Press, Washington, D.C., 53, 1982.

Fritz, T. A. et al., An energetic particle perspective of the 11 UT March 22, 1979 substorm: CDAW-6, *J. Geophys. Res.* (CDAW-6 results, in preparation), 1984.

Galeev, A. A., Magnetospheric tail dynamics, in *Magnetospheric Plasma Physics* (A. Nishida, editor), D. Reidel, Dordrecht, 143, 1982.

Lennartsson, W., R. D. Sharp, and R. D. Zwickl, Substorm effects on the plasma sheet composition during the CDAW-6 event, *J. Geophys. Res.* (CDAW-6 results, in preparation), 1984.

McPherron, R. L. and R. H. Manka, Dynamics of the March 22, 1979 substorm event: CDAW-6, *J. Geophys. Res.* (CDAW-6 results, in preparation), 1984.

McPherron, R. L., Substorm related changes in the geomagnetic tail: The growth phase, *Planet. Space Sci., 20*, 1521, 1972.

Paschmann, G. et al., Plasma measurements during the 11 UT substorm of March 22, 1979: CDAW-6, *J. Geophys. Res.*, (CDAW-6 results, in preparation), 1984.

Russell, C. T. and R. L. McPherron, The magnetotail and substorms, *Space Sci. Rev., 15*, 205, 1973.

Schindler, K., Spontaneous Reconnection, this volume, 1984.

SUBSTORM ELECTRIC FIELDS IN THE EARTH'S MAGNETOTAIL

C. A. Cattell and F. S. Mozer*

Space Sciences Laboratory, University of California, Berkeley, California 94720

Abstract. A survey has been made of all the electric field data from the University of California, Berkeley, double probe experiment on ISEE-1 (apogee ~22 R_E) during 1980 when the satellite was in the magnetotail. This study was restricted to the 74 events where **E** × **B** flows could be calculated and were ⩾100 km/s. Substorm times were determined by examining the Ae index for peaks ⩾250 γ. In association with substorms, ~70% of the flows were earthward, and ~20% had a signature called "near satellite reconnection" (first described by Nishida et al., 1983) of tailward flow followed by earthward flow which can be interpreted in terms of a model where the x-line forms earthward of the satellite and subsequently propagates tailward past it. This type event occurred only tailward of $X_{GSM} = -21\ R_E$ and within $|Y_{GSM}| \leqslant 4.5\ R_E$. These data suggest that the near earth x-line usually forms tailward of $X_{GSM} \approx -20\ R_E$.

Introduction

The phenomenology of plasma flows, magnetic fields and energetic particles in the earth's magnetotail in association with substorms has been well described in the literature [Hones, 1977; 1979; 1980; Hones and Schindler, 1979; Nishida and Nagayama, 1973; Nishida et al., 1981; Frank et al., 1976; Lui, 1980; and references in these articles] and in other papers at this conference. However, the region of the tail which is well sampled by the ISEE satellite ($R \approx -10$ to $-23\ R_E$) has not been previously studied in detail. Since most of the studies referenced above suggest that a near-earth x-line forms in this distance range at substorm onset, it is interesting and important to see if the location can be pinpointed more exactly using data from ISEE.

In this paper, results of a survey of the electric field data from the University of California, Berkeley, double probe experiment (in particular, the (**E** × **B**)/B^2 velocities) will be described. The electric field is an important quantity in reconnection, since near the x-line, a component of the plasma flow must be **E** × **B**, and **E** · **j** > 0 for conversion of electromagnetic energy to particle energy.

In section II, the data and methodology of the study are described. Examples of the event types are presented in section III. The statistical results are given in section IV. The locations of the events and what information they yield about the location of the x-line are discussed in section V. Conclusions are described in section VI.

---
*Also Physics Department.

Instrumentation and Methodology

The data for this study are from the ISEE-1 satellite (apogee ~22.3 R_E) when it was in the earth's magnetotail from February through May, 1980.

The electric fields were obtained by the University of California, Berkeley, double probe experiment [Mozer et al., 1978], which measures the component of the electric field in the spin plane (which is within ~2° of the ecliptic plane) along the boom 8 (32) times per second in low (high) bit rate. The components in the duskward ($\hat{y}$) direction and sunward ($\hat{x}$) direction are determined by least-squares fitting of the data over one spin period (3 seconds). The absolute accuracy of the measurement is ~½ mV/m for E_y [Mozer et al., 1983; Pedersen et al., 1983]. The error in E_x is larger (the order of ~1 mV/m) as determined using the same techniques described in the two papers referenced above. Variations of a smaller magnitude can be observed in both components. Although E_z is not measured, for a subset of events when the magnetic field geometry is such that any errors in E_x and E_y are not multiplied by too large a factor (B_x/B_z and $B_y/B_z < 5$), the assumption that **E** · **B** = 0 provides a reasonable estimate of $E_z = -E_x B_x/B_z - E_y B_y/B_z$. For these events, all three components of the (**E** × **B**)/B^2 convective flow velocity could be calculated. (This restriction on B_y/B_z and B_x/B_z generally means that the satellite was in the plasmasheet or plasmasheet boundary rather than in the lobe). The value of E_z was examined for all of the data. Events in which E_z was the dominant component in the calculation of the sunward component of the convection velocity ($v_x = (E_y B_z - E_z B_y)/B^2$) were not included in the study in order to avoid possible errors in the direction of v_x due to errors in E_x. The consistent relationship observed during the flow events between E_y and B_z and the fact that, during any specific event, the flow direction remains constant even though the magnetic field ratios vary widely both provide strong evidence that the results are not influenced by errors in the E_x measurement. All the electric field data were examined for times when $E_y \geqslant 2$ mV/m (note that E_y is often equal to zero to within ½ mV/m for many hours). ~400 events were found.

The magnetic field data are from the University of California, Los Angeles, fluxgate magnetometers on ISEE-1 [Russell, 1978]. 3-second averages of the data in geocentric solar-ecliptic coordinates ($\hat{x}$, sunward; $\hat{y}$, duskward; $\hat{z}$, perpendicular to the ecliptic) are plotted herein.

The list of substorms was determined by examining Ae (from the World Data Center C2 for Geomagnetism Data Book No. 7) for all peaks ⩾ 250 γ separated by 1 hour or more. ~180 sub-

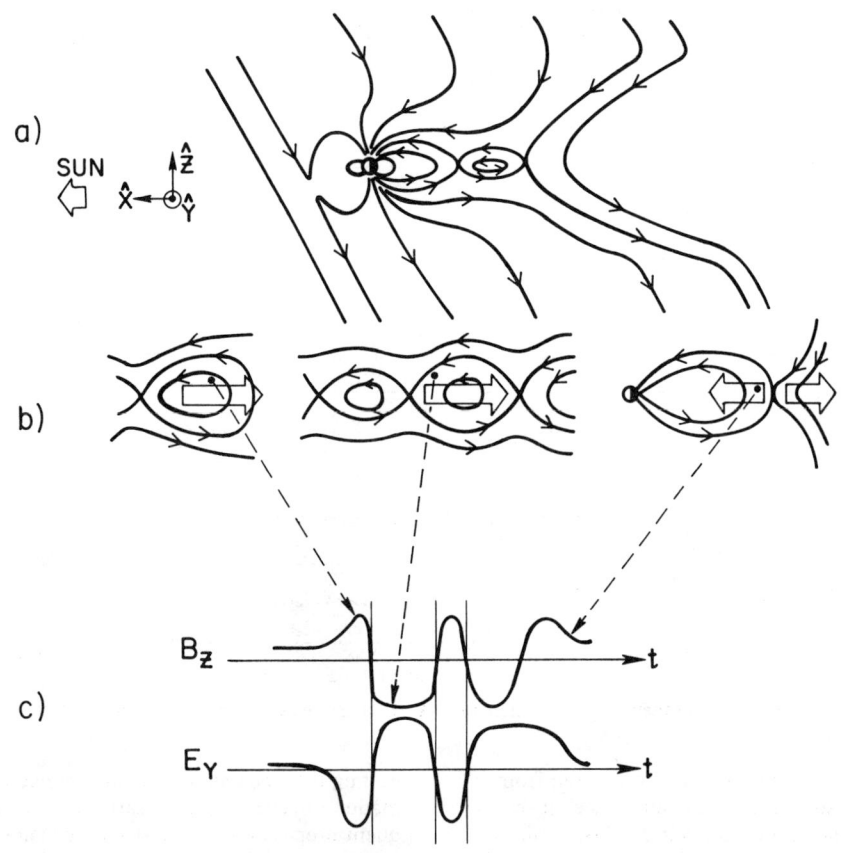

Fig. 1. (a) Schematic drawing of a possible magnetic field topology when reconnection is occurring both in the tail and at the frontside magnetopause, (b) Expanded drawings of part of the magnetotail showing the expected flow direction (large arrow), and (c) The electric and magnetic field signature which would be observed by a satellite located at the point marked by "·" in (b).

storm times were found (when the satellite was in the tail at ≥ 10 R_E and good electric field data were available).

For this study, only the subset of electric field events for which $\mathbf{E} \times \mathbf{B}$ flows could be calculated and the x-component of the flow was ≥ 100 k/s was examined. There were 74 such events. This precludes the possibility of determining how frequently any specific signature in $\mathbf{E}$ is seen in response to substorms compared to nonsubstorm times. Therefore, a future study will include all 400 electric field events.

Examples of Event Types

The types of electric field/$\mathbf{E} \times \mathbf{B}$ flow events that were observed in this study can be described as follows: (1) "near satellite reconnection" events (defined below); (2) earthward flows associated with enhancements in the duskward electric field; (3) earthward flows associated with variable electric fields; (4) tailward flows; and (5) examples with both tailward and earthward flows but not in the configuration of (1).

A schematic drawing of a possible magnetic field topology during reconnection is shown in Figure 1a. In Figure 1b, part of the magnetotail region is expanded to show the expected plasma flow and magnetic field direction. In the first picture, corresponding to the onset of near earth reconnection, the satellite (marked by a dot) is located at the tailward end of an island where B_z is northward and the flow is tailward. E_y is negative (dawnward) (Figure 1c). As time progresses, the magnetic structures move past the satellite so that it is located at the earthward edge of the island and, therefore, B_z is southward while the flow remains tailward. During this time E_y would be positive (duskward) (Figure 1c). If more than one island structure passed over the satellite, alternate regions of northward B_z with negative E_y, and southward B_z with positive E_y would be observed. When the satellite is finally located earthward of the most earthward x-line, B_z is northward and the flow is earthward. The corresponding E_y is positive. In summary, if an x-line (or multiple x-lines) formed earthward of the satellite and subsequently propagated tailward past the satellite, first anticorrelated E_y and B_z (tailward flow) and then correlated E_y and B_z with B_z northward (earthward flow) would be observed.

This type of behavior in E_y and B_z was first described by Nishida et al. [1983]. It is the most distinctive type of electric field/flow event found in the present survey and will be called "near satellite reconnection." An example [Nishida et al., 1983] is shown in Figure 2. The format of this and the following 5 figures is as follows: panels (a) - (c) contain the 3 components of the magnetic field in GSE; panel (d) contains the y-component of the electric field; and panels (e) - (f) contain the 3 components of the $\mathbf{E} \times \mathbf{B}$ drift velocity in GSE. The x- and y-components of the drift velocity are plotted only when the criteria, $B_x/B_z < 5$ and $B_y/B_z < 5$, are met. Gaps in all 7 panels are due to the operation of the active experiment on ISEE-1. The Ae index for the day of the event is also shown. At ~0230, there was a rapid enhancement in Ae. Before this time, E_y was zero to within ½ mV/m. At ~0228, B_z became more northward, E_y became negative (dawnward), and the x-component of

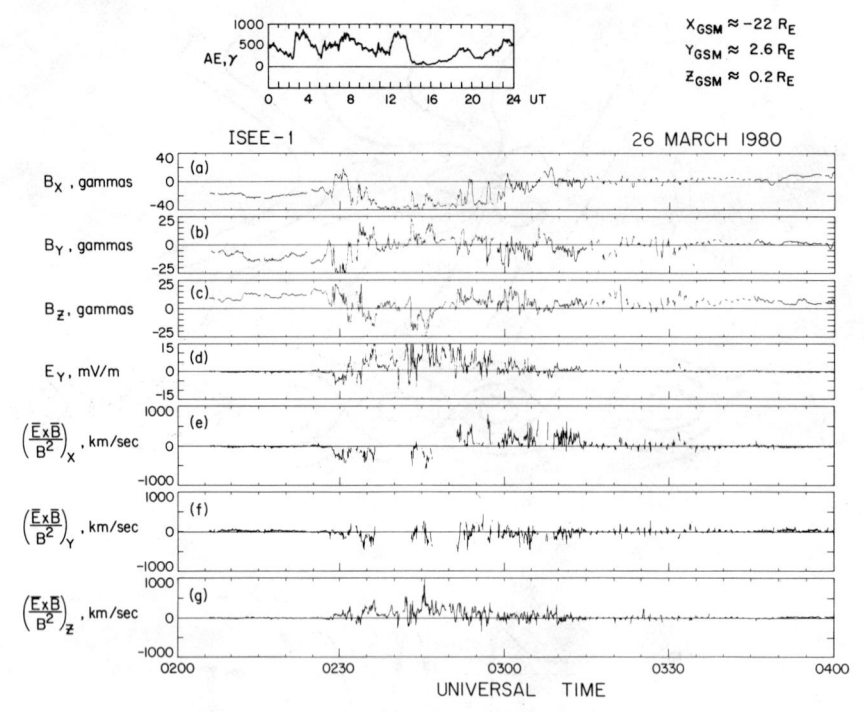

Fig. 2. The electric and magnetic fields and signature $(\mathbf{E} \times \mathbf{B})/B^2$ velocities for March 26, 1980, 0200 - 0400.

the drift velocity was ~500 k/s tailward. As can be seen from B_x (panel a), the satellite also crossed the neutral sheet at this time. Subsequently, B_z turned southward and E_y duskward. This anticorrelation of E_y and B_z (tailward flow) continued until ~0248; after that, B_z remained northward and E_y duskward, corresponding to earthward $\mathbf{E} \times \mathbf{B}$ flow (to ~700 k/s) until ~0315. This pattern is exactly the same as that in Figure 1, suggesting that the data can be interpreted as being due to the formation of several x-lines earthward of the satellite and subsequent propagation of the structures tailward past the satellite. Note that, on the average, the $\mathbf{E} \times \mathbf{B}$ velocity is towards the neutral sheet since v_z is usually positive (northward) and the satellite is generally in the southern lobe.

Another example of this event type is shown in Figure 3 and

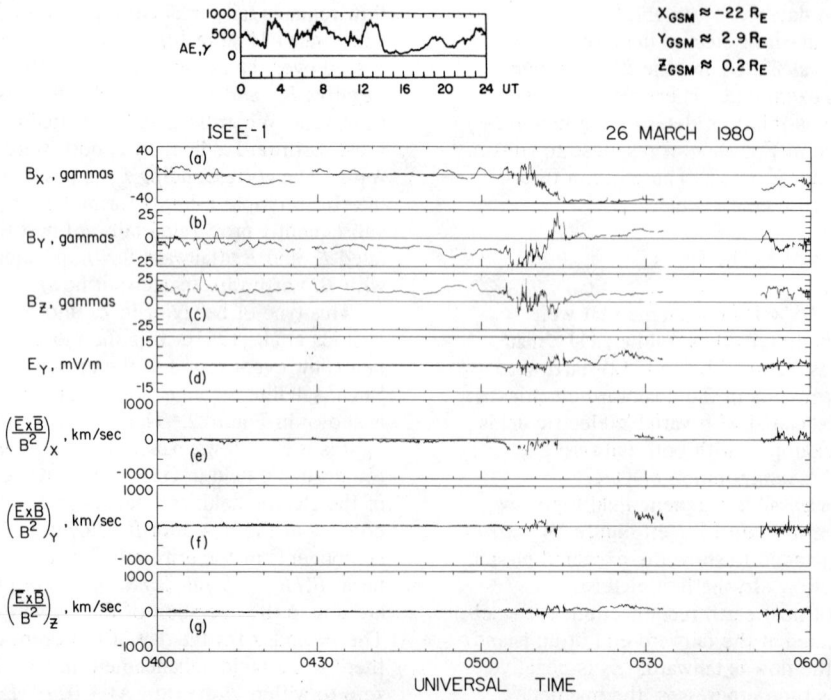

Fig. 3. The electric and magnetic fields and $(\mathbf{E} \times \mathbf{B})/B^2$ velocities for March 26, 1980, 0400 - 0600.

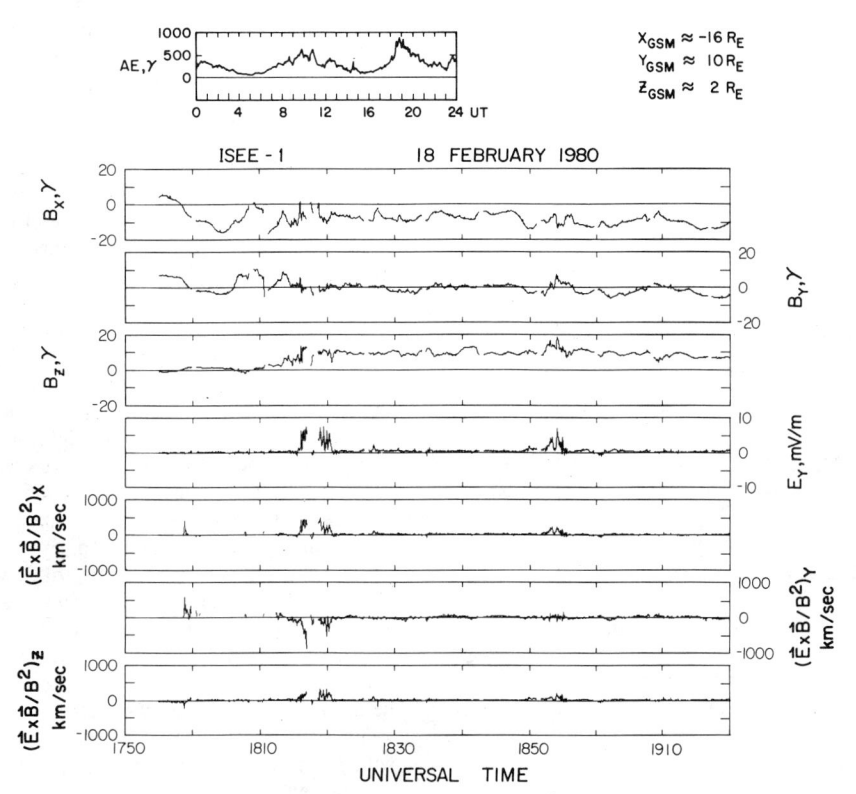

Fig. 4. The electric and magnetic fields and $(\mathbf{E} \times \mathbf{B})/B^2$ velocities for Feb. 18, 1980.

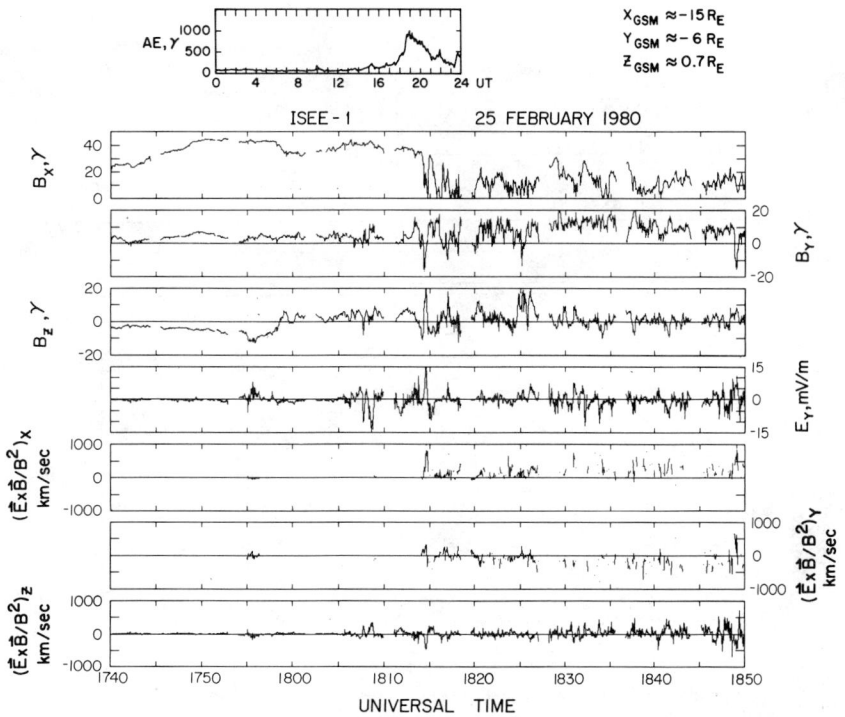

Fig. 5. The electric and magnetic fields and $(\mathbf{E} \times \mathbf{B})/B^2$ velocities for Feb. 25, 1980.

212 SUBSTORM ELECTRIC FIELDS

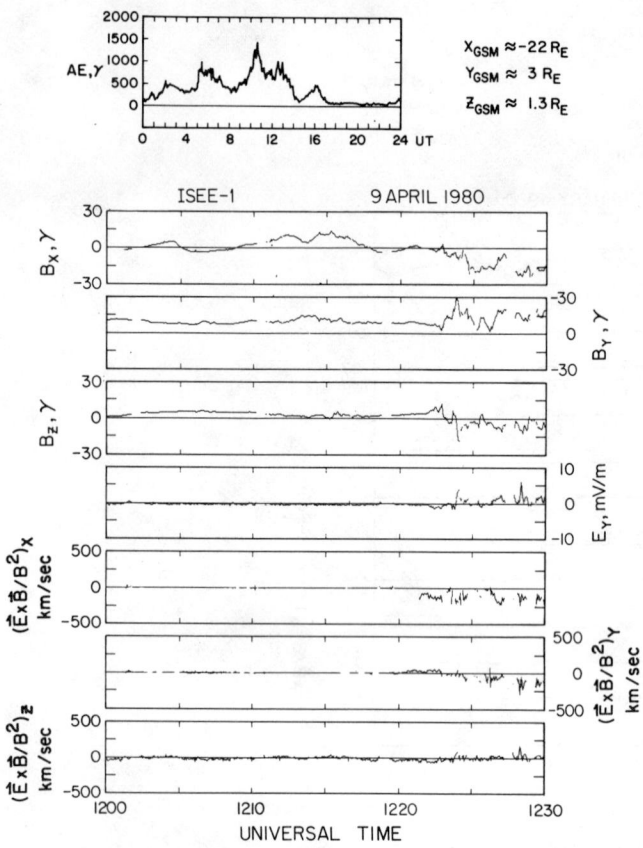

Fig. 6. The electric and magnetic fields and $(\mathbf{E} \times \mathbf{B})/B^2$ velocities for April 9, 1980.

occurred ~ 3 hours later than the event in Figure 2. This is characteristic of the "near satellite reconnection" events. When one event was observed, at least one more was observed within ~ 6 hours. There was a rapid enhancement in Ae at ~0510. At ~0505 B_z turned sharply northward, E_y became negative, and tailward drift velocities of ~300 k/s were observed. This anticorrelation of B_z and E_y continued until ~0515, when the fields became correlated. Once again, the satellite was near the neutral sheet at the beginning of the event.

An example of the second type of event found in this survey, earthward flow associated with enhancements in E_y, is presented in Figure 4. A rapid increase in Ae occurred after 1800. At ~1813, there was a strong enhancement in E_y to ~8 mV/m lasting for ~5 minutes. B_z was northward; and B_x and B_y were close to zero, indicating proximity to the neutral sheet. The $\mathbf{E} \times \mathbf{B}$ flow was ~500 k/s earthward. The flow was also toward the neutral sheet (v_z was northward and the satellite was in the southern lobe) and toward midnight, since the satellite was in the dusk half of the tail and the flow was dawnward. There was another strong enhancement at ~1853 associated with an earthward flow of ~250 k/s. This event could be interpreted in terms of the third picture in Figure 1b.

The third event type (earthward flow associated with variable electric field) is exemplified by Figure 5. Ae increased before 1800. The plasmasheet was probably very narrow, since both the plasmasheet boundary and the neutral sheet were observed. There were large spiky electric fields to ~10 mV/m. These fields are typical of the plasmasheet boundary [Cattell et al., 1982; Levin et al., 1983]. B_z was also variable and correlated with E_y, so the $\mathbf{E} \times \mathbf{B}$ flow was earthward (~100 - 800 k/s).

Figure 6 presents an example of the next event type, tailward flow. There was an increase in Ae at ~1215. Before this time, E_y was equal to zero within ½ mV/m. E_y became negative at

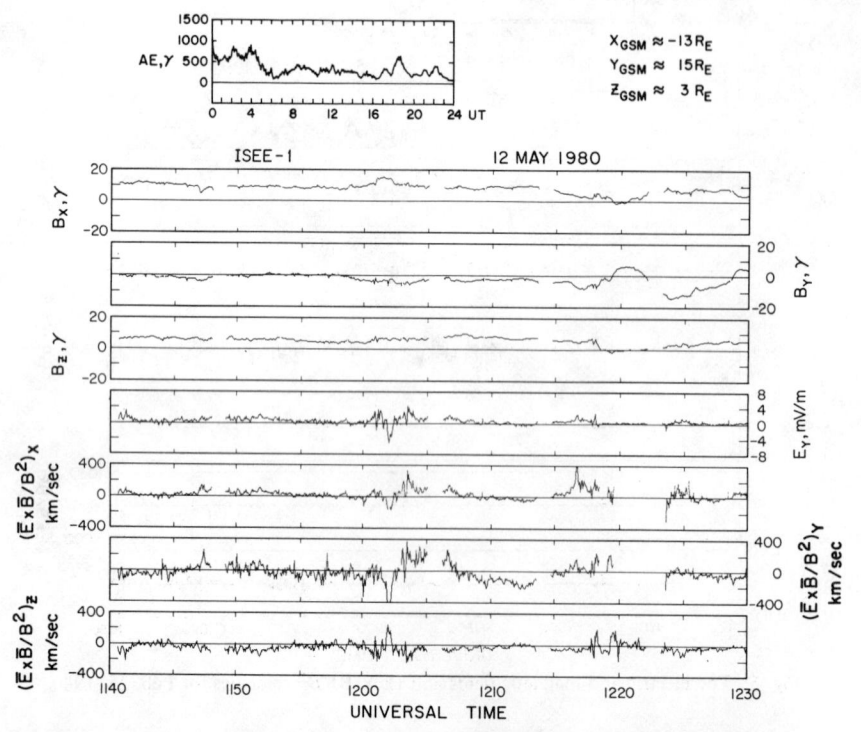

Fig. 7. The electric and magnetic fields and $(\mathbf{E} \times \mathbf{B})/B^2$ velocities for May 12, 1980.

TABLE 1

Flow	"Substorm" events ($Ae \geq 250\,\gamma$) (42)	"Nonsubstorm" Events ($Ae < 250\,\gamma$) (32)	w/$Ae > 100\,\gamma$
Earthward ≥ 250 k/s	16 (38%)	13 (41%)	11
Earthward < 250 k/s	13 (31%)	16 (50%)	9
Tailward	3 (7%)	2 (6%)	1
"Near Satellite Reconnection"	8 (19%)	0 (0%)	0
Both	2 (5%)	1 (3%)	1

~1222 with positive B_z, then E_y became positive and B_z negative. The $\mathbf{E} \times \mathbf{B}$ flow was tailward (100 - 250 k/s).

The final event type (shown in Figure 7) had both tailward and earthward flow, but not in the specific pattern described as "near satellite reconnection". B_z was northward throughout the event (except ~1220). E_y was variable. The $\mathbf{E} \times \mathbf{B}$ flow was generally earthward, but there were several regions of tailward flow also.

Statistics

The statistical results are summarized in Table 1. 42 events occurred within 30 minutes of a peak in $Ae \geq 250\,\gamma$ ("substorm"). Of these 29 (69%) were earthward $\mathbf{E} \times \mathbf{B}$ drifts, 3 (7%) were tailward, 8 (19%) had the signature described as "near satellite reconnection", and 2 (5%) had flow in both directions. There were 32 events that did not occur with 30 minutes of a peak in $Ae \geq 250\,\gamma$. (Note that 22 were within 30 min of a peak in $Ae > 100\,\gamma$ and thus were associated with ground magnetic activity and possibly smaller substorms). 29 (91%) were earthward and none showed the "near satellite reconnection" signature.

The earthward drift velocity events can be further characterized by the type and size of electric field with which they were associated. ~70% of the earthward flows were associated with enhancements in E_y in the duskward direction and northward B_z. Only one event was associated with an E_y enhancement in the dawnward direction and southward B_z. The other ~30% were due to variable E_y and, therefore, variable B_z. Variable electric fields were more common with the smaller earthward flows. For flows ≥ 250 k/s, the enhancements in E_y were usually ~ 3 - 15 mV/m (average ~6 mV/m) and lasted ~2 - 20 minutes. The variable electric fields were usually ~6 - 20 mV/m (average ~9 mV/m). For flows < 250 k/s, the enhancements were slightly smaller − 2 - 10 mV/m (average ~4 mV/m), and the variable E_y averaged ~6 mV/m.

In summary, most of the $\mathbf{E} \times \mathbf{B}$ flows in the tail at distances of 10 to 22 R_E associated with substorms were earthward. These flows usually occurred with enhancements of the duskward electric field and northward B_z. This is very different from the results obtained at greater radial distances (-28 to -35 R_E) by Hones and Schindler [1979]. At those distances, the plasma flow was usually tailward. These two facts suggest that the near-earth x-line usually forms tailward of the ISEE apogee (~22 R_E) and earthward of 27 R_E.

Location of the Events

Additional information can be gained by looking at the locations of the events. In particular, if the flows are assumed to be due to reconnection, constraints may be placed on the location of x-line formation. To do this, it is necessary to examine the satellite coverage to insure that the observed occurrences are not biased due to sampling. The number of minutes of good data in each 1 $R_E \times 1\,R_E$ box in the tail is plotted in Figure 8. Figure 8(a) shows the results for the $X_{GSM} - Y_{GSM}$ plane (summed over Z_{GSM}). There is good coverage tailward of ~15 R_E and

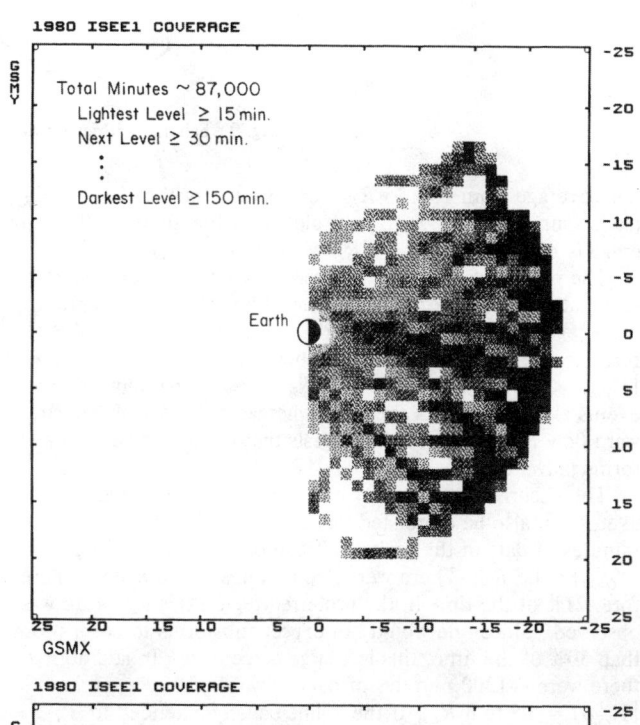

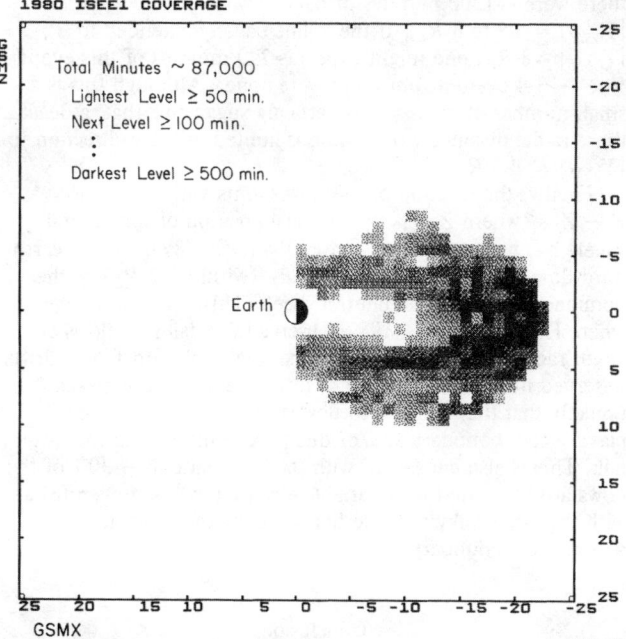

Fig. 8. (a) The number of minutes of data in 1 $R_E \times 1\,R_E$ boxes included in this study, plotted in the $X_{GSM} - Y_{GSM}$ plane. (b) Same as (a) for the $X_{GSM} - Z_{GSM}$ plane.

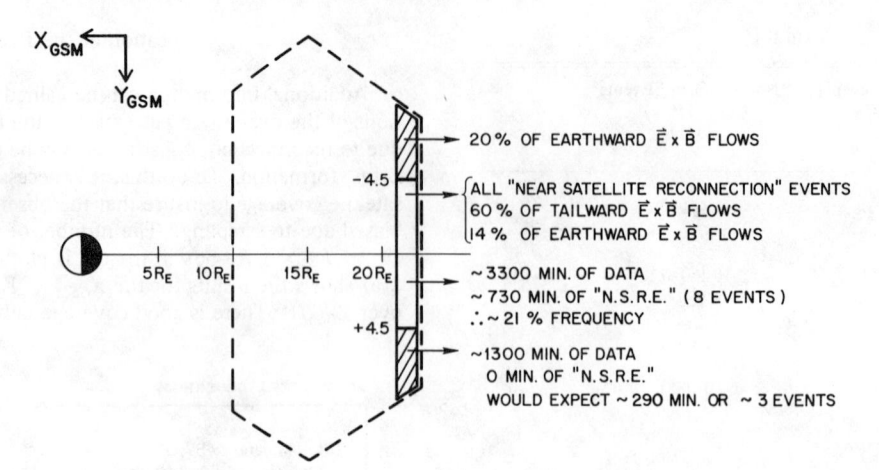

Fig. 9. Schematic drawing of event locations in the $X_{GSM} - Y_{GSM}$ plane.

fair coverage from 10 - 15 R_E. The results for the $X_{GSM} - Z_{GSM}$ plane (summed over Y_{GSM}) are plotted in Figure 8(b). The coverage is best for $|Z_{GSM}| \leq 5 R_E$ when $X_{GSM} \gtrsim 15 R_E$.

The results of the study of event locations are summarized schematically in Figure 9. The dashed line shows the approximate region covered by this study. All of the "near satellite reconnection" events occurred tailward of $X_{GSM} = -21 R_E$, with $|Y_{GSM}| < 4.5 R_E$ and $|Z_{GSM}| < 3 R_E$. 60% of the tailward flow events also occurred in this box, whereas only 14% of the earthward flow events did. This suggests that the x-line generally forms tailward of $X_{GSM} \sim 20 - 21 R_E$.

The occurrence frequency of "near satellite reconnection" events can also be calculated. The study included ~3300 minutes of data in the region tailward of 21 R_E and within $|Y_{GSM}| \leq 4.5 R_E$. There were 730 minutes of the events, therefore, 21% of the time in the preferred region this signature was observed. Since one would not expect substorms to occur more than 50% of the time, this is a large percentage. In addition, there were ~1300 minutes of data tailward of 21 R_E with $|Y_{GSM}| = 4.5$ to 8 R_E. If the x-line usually extended to $|Y_{GSM}| \approx 8 R_E$, one might expect ~290 minutes of this event type (~3-4 events), but there were none. Although this is a small number of events, it is certainly suggestive that, at least at these radial distances, the x-line is limited in the y-direction to $|Y_{GSM}| \lesssim 4.5 R_E$.

Finally, the location of the flow events with respect to $Z - Z_{NS}$ (where Z_{NS} is the nominal position of the neutral sheet) has been examined. More than 95% (85%) of the earthward flow events were within $\pm 3 R_E$ (within $\pm 2 R_E$) of the nominal position of the neutral sheet. This is the same region where Hayakawa et al. [1982] observed fast tailward flows at larger radial distances. The fact that the earthward $\mathbf{E} \times \mathbf{B}$ drifts observed in this study are close to the neutral sheet makes it unlikely that they are due to flows occurring only on the plasmasheet boundary and/or due to reconnection in the distant tail. This is also consistent with the fact that only ~30% of the flows are associated with variable electric fields and even fewer with the large spiky variable fields usually observed at the plasmasheet boundary.

Conclusions

At $X_{GSM} = -10$ to $-22 R_E$, the $\mathbf{E} \times \mathbf{B}$ flows observed in association with substorms (as indicated by Ae) were usually earthward, and were most often associated with enhancements in the duskward electric field and northward B_z and less often with variable E_y and mixed B_z. Almost all of the earthward flows occurred within 2 - 3 R_E of the nominal position of the neutral sheet. If these flows are interpreted as being due to reconnection, this suggests that the near-earth x-line usually forms tailward of the ISEE apogee (~22 R_E).

This conclusion can be refined by examining the location of the $\mathbf{E} \times \mathbf{B}$ flow events. "Near satellite reconnection" events (which can be explained by a model in which the x-line forms earthward of the satellite and then propagates tailward past it) were observed tailward of $X_{GSM} = -21 R_E$ and within $|Y_{GSM}| \leq 4.5 R_E$. 60% of the tailward flow events also occurred in that region. Since the velocity of the x-line (or if it remains stationary for some period) is not known, it is not possible to pinpoint exactly where the x-line forms. However, the fact that tailward flow was only rarely observed earthward of $X_{GSM} = -21 R_E$ and that the "near satellite reconnection" signatures were all tailward of that location strongly suggests that the x-line usually forms tailward of $X_{GSM} \approx -20$ to $-21 R_E$. It may also be significant that all but 2 of the "near satellite reconnection" events occurred on either the most disturbed or second most disturbed day of the month ($\sum K_p > 31$). It may be that the x-line forms this close to the earth only during very active periods. The data also suggest that (at least at these radial distances) the x-line is limited in the y-direction. Three-dimensional computer simulations [Birn and Hones, 1981] also show that the x-line does not extend across the full width of the tail.

The events were all of duration longer than ~2 minutes. It is possible that the conclusions would be modified if the reconnection structures (plasmoids) were usually smaller or if the plasma flow was either faster or more field-aligned at radial distances closer to the earth so that the duration of "near satellite reconnection" events or tailward flow events was systematically shorter.

Since this survey was restricted to periods when $(\mathbf{E} \times \mathbf{B})/B^2$ flows could be determined and were ≥ 100 km/s, it was not possible to determine how often any particular signature in the electric field was observed in association with substorms in comparison to nonsubstorm times. Examination of all the electric field events will clarify this point.

Acknowledgments. We would like to thank C. T. Russell for the magnetic field data from the UCLA fluxgate magnetometers, and Carolyn

Overhoff for the preparation of the manuscript. This work was performed under NASA Contract NAS5-25770.

References

Birn, J., and E. W. Hones, Jr., Three-dimensional computer modeling of dynamic reconnection in the geomagnetic tail, *J. Geophys. Res., 86*, 6802, 1981.

Cattell, C. A., M. Kim, R. P. Lin, and F. S. Mozer, Observations of large electric fields near the plasmasheet boundary by ISEE-1, *Geophys. Res. Lett., 9*, 539, 1982.

Cattell, C. A., and F. S. Mozer, Electric fields measured by ISEE-1 within and near the neutral sheet during quiet and active times, *Geophys. Res. Lett., 9*, 539, 1982.

Frank, L. A., K. L. Ackerson, and R. P. Lepping, On hot tenuous plasmas, fireballs and boundary layers in the earth's magnetotail, *J. Geophys. Res., 76*, 5859, 1976.

Hones, E. W., Jr., Substorm processes in the magnetotail: Comments on 'On hot tenuous plasmas, fireballs and boundary layers in the earth's magnetotail', by L. A. Frank, K. L. Ackerson, and R. P. Lepping, *J. Geophys. Res., 82*, 5633, 1977.

Hones, E. W., Jr., Transient phenomena in the magnetotail and their relation to substorms, *Space Sci. Rev., 23*, 393, 1979.

Hones, E. W., Jr., Plasma flow in the magnetotail and its implications for substorm theories, in *Dynamics of the Magnetosphere*, S.-I. Akasofu (ed.), 545, 1980.

Hones, E. W., Jr., and K. Schindler, Magnetotail plasma flow during substorms: A survey with IMP 6 and IMP 8 satellites, *J. Geophys. Res., 84*, 7155, 1979.

Levin, S., K. Whitley, and F. S. Mozer, A statistical study of large electric field events in the earth's magnetotail, *J. Geophys. Res., 88*, 7765, 1983.

Lui, A. T. Y., Observations of plasmasheet dynamics during magnetospheric substorms, in *Dynamics of the Magnetosphere*, S.-I. Akasofu (ed.), 545, 1980.

Mozer, F. S., R. B. Torbert, U. V. Fahleson, C.-G. Fälthammar, A. Gonfalone, and A. Pedersen, Measurements of quasistatic and low frequency electric fields with spherical double probes on the ISEE-1 spacecraft, *IEEE Trans. Geosci. Electron., GE-16*, 258, 1978.

Mozer, F. S., and E. W. Hones, Jr., Comparison of spherical double probe electric field measurements with plasma bulk flows in plasmas having densities less than 1 cm^{-3}, *Geophys. Res. Lett., 10*, 737, 1983.

Nishida, A., and N. Nagayama, Synoptic survey for the neutral line in the magnetotail during the substorm expansion phase, *J. Geophys. Res., 78*, 3782, 1973.

Nishida, A., H. Hayakawa, and E. W. Hones, Jr., Observed signatures of reconnection in the magnetotail, *J. Geophys. Res., 86*, 1422, 1981.

Nishida, A., Y. K. Tulunay, F. S. Mozer, C. A. Cattell, E. W. Hones, Jr., and J. Birn, Electric field evidence for tailward flow at substorm onset, *J. Geophys. Res., 88*, 9109, 1983.

Pedersen, A., C.-G. Fälthammar, V. Formisano, P.-A. Lindqvist, F. Mozer, and R. Torbert, Quasistatic electric field measurements with spherical double probes on GEOS and ISEE satellites, *Space Sci. Rev.*, in press, 1983.

Russell, C. T., The ISEE 1 and 2 fluxgate magnetometers, *IEEE Trans. Geosci. Electron., GE-16*, 239, 1978.

Questions and Answers

D. N. Baker: A comment in response to a remark here by Gordon Rostoker: I was asked by Chris Russell during my talk how close I believed the reconnection neutral line might sometimes approach geostationary orbit. My answer was that sometimes it might get as close as a couple of earth radii from synchronous altitude and, thus, it may reach geocentric distances of ~9 or 10 R_E. This is not necessarily a common location, however. A question: In your talk you said you might expect substorms ~50% of the time. On average one might, however, take a substorm duration of $\lesssim 2$ hours and also take ~4 substorms per day suggesting $\lesssim 8/24$ (or $\lesssim 30\%$) substorm probability. How would you estimate a number as high as 50%?

Cattell: In response to your comment, I would certainly not disagree that the neutral line might sometimes form quite close to the earth. My point is just that, statistically, it usually occurs tailward of ~20 R_E. The 50% number was just a guess at an upper limit. I just wanted to make the point that the occurrence frequency of "near satellite reconnection" events was very high.

Fritz: You have presented the electric field data in terms of $E \times B$. Have you done a detailed comparison of your determination of the plasma flow velocity and the in situ measurement of the same quantity by the Iowa and/or LANL experiments? If so, what is the quality of the agreement?

Cattell: Yes, comparisons between our electric field and the Los Alamos plasma measurements have been published (Mozer et al., GRL, 10, 737, 1983) for data in the plasma sheet and tail lobe. The agreement is generally within the statistical uncertainties in the measurements.

Lui: 1) We have examined the north-south component of electric field (E_z) in the plasma sheet (Lui and Akasofu, GRL, 1981) and we have found the magnitude to be large, comparable to the enhanced dawn-dusk component (E_y) you have reported here. Have you compared E_y and E_z for your events, and if so, what is the result? 2) In your substorm onset study, you have found that the majority of plasma flow is earthward. How is this result compared with the earlier result of Schindler and Hones (JGR, 1979) that almost all flows are tailward at substorm onset?

Cattell: 1) In many of the events in this study, the z-component of the electric field (deduced from $\underline{E} \cdot \underline{B} = 0$) is comparable to E_y. You can see that in this figure (another of the "near satellite reconnection" events) which has plots of all 3 components of $\underline{E}$. 2) The data are certainly consistent with the results of Hones and Schindler since all their events were at greater radial distances ($\gtrsim 27\ R_E$).

Vasyliunas: 1) The electric field you discuss is measured, of course, in the frame of reference of the spacecraft, and one should keep in mind that it may differ from the field discussed, for example, in theoretical models which is mostly referred to a frame where the magnetotail structures are stationary. (For instance, your measured electric field always has a curl associated with the time-varying magnetic field seen at the spacecraft.) 2) From the electric field one can deduce only the perpendicular components of the bulk velocity, whereas the physically significant quantity is the full vector. For example, one may have a locally uniform velocity but a nonuniform magnetic field; $\underline{V}$ is then constant but $\underline{V}_\perp$ may exhibit large variations which have no particular physical significance in this case (other than reflecting the variations in the direction of $\underline{B}$.

Cattell: 1) You are, of course, correct that we don't measure the electric field in the "theoretical frame." I question whether there is, in fact, an inertial frame where the magnetotail structures are stationary. Since we don't know and, therefore, can't transform our field measurements, I think it is quite reasonable to look at the data in an earth-fixed reference frame as we do. 2) Both the total flow and the $\underline{E} \times \underline{B}$ flow are physically significant. For reconnection to occur, there must be $\underline{E} \times \underline{B}$ flow of the plasma into and out of the separatrix region. Our analysis was based on the flow direction (not magnitude), and I think it is highly unlikely that the $\underline{E} \times \underline{B}$ flow would be earthward while the total flow was tailward.

ENERGETIC IONS AND ELECTRONS AND THEIR ACCELERATION PROCESSES IN THE MAGNETOTAIL

Manfred Scholer

Max-Planck-Institut für Physik und Astrophysik,
Institut für Extraterrestrische Physik, 8046 Garching, FRG

Abstract. Many years of observations of energetic particle fluxes in the geomagnetic tail have shown that these particles exhibit a bursty appearance on all time scales. However, often the bursty appearance is merely due to multiple entries and exits of the spacecraft into and out of the plasma sheet which always contains varying fluxes of energetic particles. Therefore these bursts should not in each case be immediately associated with reconnection. Nevertheless the fact that charged particles are accelerated to high energies within the magnetosphere has to be explained and reconnection may ultimately be a promising candidate. In addition to these entries and exits into and out of the plasma sheet there occur short term bursts well within the plasma sheet which may be the direct signature of reconnection. At the boundary of the recovering plasma sheet earthward directed beams of energetic ions have been observed which may be due to more steady state reconnection in the distant tail. During plasma sheet dropout at substorm onset short lived (~40 s) high energy particle bursts occur which are related to the newly created earthward neutral line. Recent results from the ISEE 3 deep tail mission have revealed the existence of fast tailward moving plasma structures which are preceded by energetic electron and ion beams. The observed velocity dispersions during the appearance of these beams allow a determination of the source location. Finally it is noted that the vast literature on energetic burst observations in the geomagnetic tail has to be contrasted with the existence of only a few theoretical papers which deal with particle acceleration to high energies during reconnection in a more quantitative way.

Introduction

The literature on energetic particle observations in the geomagnetic tail started almost two decades ago with the work of Anderson [1965] on energetic electron bursts and of Armstrong and Krimigis [1968] on energetic proton bursts. It was, however, not until the mid-seventies that a systematic study of the energetic particle phenomena in the geomagnetic tail was undertaken. This became possible due to improved instrumentation on the Vela, IMP 7, and IMP 8 spacecraft. Since then, the literature on energetic bursts in the tail has grown almost exponentially. Single bursts have been studied in great detail, statistical analyses on the burst occurrence pattern have been performed, and anisotropies and spectra have been studied on all time scales. However, it becomes increasingly difficult to bring all these detailed observations into some larger context. Instead of reviewing the vast amount of literature we will therefore try to differentiate between the various kinds of energetic particle phenomena and delineate their relationship to the occurrence of magnetospheric substorms.

Why is there an interest in high-energy phenomena in the geomagnetic tail? Historically such an interest was more or less enforced because in the early days of space exploration only measuring techniques for the higher energy ranges were available. With advanced techniques the interest shifted indeed to lower and lower energies, i.e. to the energy range of the thermal plasma responsible for the dynamics of the geomagnetic tail. Nevertheless, to use an expression of Pellinen

and Heikkila [1978], it is a hard fact of life that charged particles are accelerated within the magnetotail to energies of 1 MeV or more although the cross-tail electrostatic field during quiet times typically has a modest potential difference of only ~50 kV. Fluxes of high energy particles are not just an extension of the thermal plasma sheet population but are so copious that, e.g., the energy residing in ions above ~50 keV can, at times, account for ~25% of the depression in the local magnetic field [Kirsch et al., 1977]. To understand the acceleration processes occurring in the earth's magnetotail is the more important as they may also be occurring in other astrophysical situations. In addition, energetic particles can be used as probes to obtain information from distant acceleration sites or about the large scale topology of the magnetic field.

Energetic particles in the geomagnetic tail exhibit a bursty character on all time scales. Figure 1 taken from Fan et al. [1976], presents energetic ion and electron pulses observed in the magnetotail over a six day period. But the occurrence of a burst should not in each case immediately be connected with an ongoing acceleration process. Quite often the bursty nature is simply due to repeated entries and exits of the satellite into and out of the plasma sheet which always contains varying fluxes of energetic ions. In the first section we will discuss observations of the suprathermal and high energy component of the plasma sheet. We will then present observations of energetic particle bursts in the plasma sheet proper, which may be due to a locally ongoing acceleration process. In the third section we will discuss energetic particle phenomena occurring near the edge of the plasma sheet, either during thinning or during recovery, and

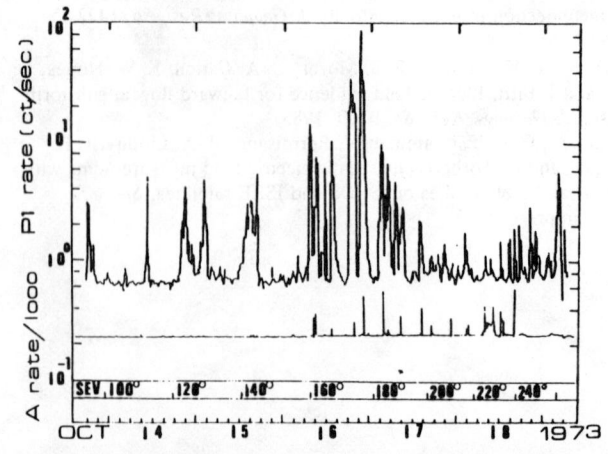

Fig. 1. 125-160 keV/Q ion and E ≳ 300 keV electron rates as measured by IMP 7 in the geomagnetic tail on October 14-18, 1973. SEV is the sun-earth-vehicle angle. [after Fan et al., 1976].

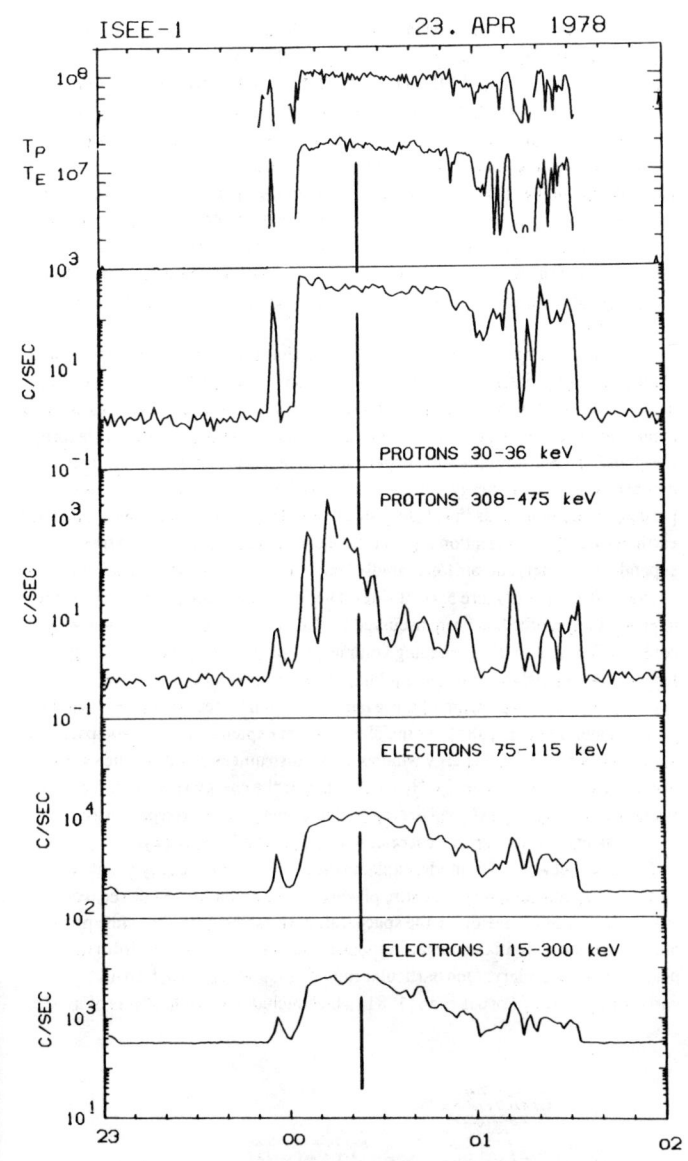

Fig. 2. From top to bottom: proton (upper trace) and electron (lower trace) plasma temperature, count rates of 30-36 keV protons, 308-475 keV protons, 75-115 keV electrons and 115-300 keV electrons for the 23 April 1978, plasma sheet recovery event. The vertical bar indicates the time of maximum energetic electron intensity. [Scholer et al., 1983a].

shorter time scales are often due to repeated entries and exits of the satellite into and out of the plasma sheet. An example of a typical plasma sheet recovery is shown in Figure 2 [taken from Scholer et al., 1983a]. The top panel shows the plasma sheet proton and electron temperatures. The high energy protons exhibit a typical fast rise - slow decay intensity time profile. The short time burst at the beginning and the end of the event are clearly correlated with fluctuations in the temperatures due to multiple entries and exits into and out of the plasma sheet. We note that the energetic electron profiles are different from the energetic proton profile, i.e. they do not show the fast rise followed by a slow decay. The electrons during this and many other recoveries reach their maximum intensity considerably after the appearance of the plasma sheet as determined by the thermal electrons. At this time the high energy proton intensity is already in its decaying phase. Many of the events reported by, e.g., Hones et al. [1976] and Sarris et al. [1976] show the rapid rise - slow decay profile. However, Hones et al. [1976] were first to point out the close relationship to the recovering plasma sheet. Carbary and Krimigis [1979] noted that event onsets may be understood in terms of the spacecraft's entering the plasma sheet while an event was in progress. But they considered a burst as an independent event and did apparently not consider the fact that the plasma sheet always contains a high-energy particle component.

Ipavich and Scholer [1983] have shown that the characteristic decay times after plasma sheet recovery are energy dependent with higher energy proton

finally we will present some recent results from the ISEE 3 deep tail mission bearing on energetic particle acceleration. In the last section we will briefly review the present status of the theory of particle acceleration within the magnetotail.

The High Energy Component of the Plasma Sheet

Many high-energy proton burst events are simply due to entry of the satellite into the plasma sheet during plasma sheet recovery. Baker et al. [1979] found that ~75% of the periods in which they observed energetic proton events were characterized first by a plasma sheet thinning followed by a plasma sheet expansion. As the expanding plasma sheet eventually envelopes the spacecraft the rapid rise in energetic proton intensity is observed since the recovering plasma sheet always contains varying fluxes of energetic particles. Following the rapid rise the energetic proton fluxes usually exhibit a more or less exponential decay over time scales ranging from ~30 min. up to several hours. Multiple bursts on

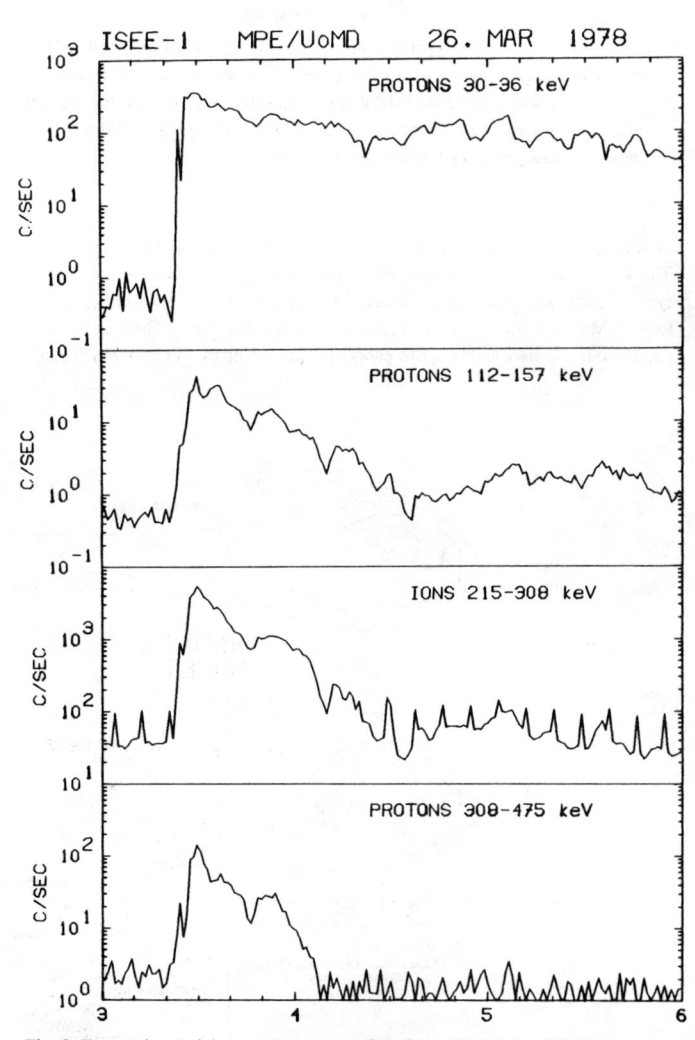

Fig. 3. Energetic particle counting rate profiles from ISEE 1 for 0300-0600 UT on March 26, 1978 during a plasma sheet recovery event. [Ipavich and Scholer, 1983].

218   ENERGETIC IONS AND ELECTRONS

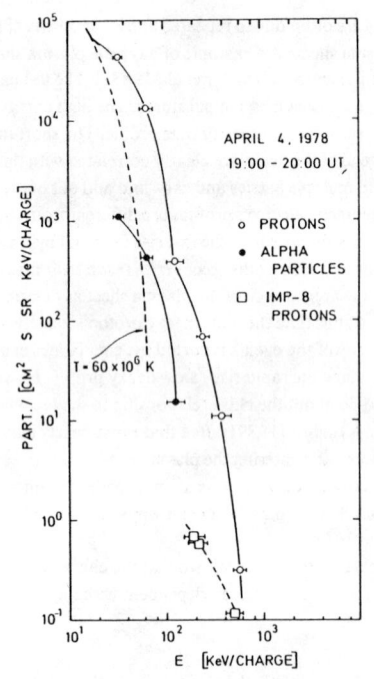

Fig. 4. Proton and alpha particle spectra for 1900-2000 UT on April 4, 1978. The dashed line represents a Maxwellian spectrum with T = 6 × 10^7 K and n = 0.35 cm^{-3}. Open squares are IMP 8 proton measurements outside the magnetosphere and represent a solar background. Differential intensity is plotted versus energy per charge [Ipavich and Scholer, 1983].

fluxes having shorter decay times. A typical example can be seen in Figure 3. This leads to a steepening of the particle spectra in the course of an event. A typical energy spectrum (one hour average) starting at ~12 keV and extending up to ~0.5 MeV is shown in Figure 4 [Ipavich and Scholer, 1983]. Three different populations are discernable: a low energy population up to ~18 keV which can be approximated by a Maxwellian with 6×10^7 K and a number density of 0.35 cm^{-3}, a suprathermal particle population which follows closely an exponential in energy up to 150 keV and a third high-energy population which also cannot be fit by a power law. It should be noted in this respect that measurements made by rate channels of instruments that respond to all ions in a given total energy range need to be carefully interpreted when the particle spectra are steep. If, for example, the alpha to proton ratio is 1% at all energy per nucleon values and if spectra are power laws with γ = −5, then the alpha to proton ratio is ~250% at the same total energy in a given rate channel. If spectra are exponentials in energy per charge with an e-folding energy of 15 keV/Q and the ratio is ~8% at all energy per charge values, then the alpha particle contribution dominates above ~80 keV.

Although we want to caution in this section that every energetic particle burst should not immediately be connected with an on-going acceleration process, the mere existence of high-energy particles in the plasma sheet has nevertheless to be explained. Baker et al. [1979] have proposed a model whereby during substorm onset and plasma sheet thinning explosive reconnection lasting for less than ~1 min accelerates high-energy particles up to ~1 MeV (see section on energetic particle phenomena near the plasma sheet boundary). These particles are injected earthward of the acceleration region onto closed field lines. As the plasma sheet expands these energetic protons rapidly leak out to fill the newly available volume of closed field lines (Figure 5, right side). The following exponential decay of the intensity time profile has been discussed by Baker et al. [1979] as an adiabatic cooling effect due to the expanding volume occupied by the plasma sheet. In terms of this model the different e-folding times at different energies is a consequence of the spectrum of these particles which is steeper than a power law [Ipavich and Scholer, 1983]. As the plasma sheet expands the energetic particles are cooled and the fixed energy window of the instrument counts particles that originally had a higher energy. Hence the steeper the energy spectrum, the faster the observed counting rate will decrease. Since the spectrum steepens with increasing energy, the highest energies will display the faster decay.

The Baker et al. [1979] model explains the rapid rise-slow decay profile with a temporal process affecting the entire plasma sheet population. Since the recovering plasma sheet is enveloping the spacecraft at the same time, this same profile may be related to spatial changes in spectra with varying distances from the plasma sheet boundary. One particular model that addresses such spatial variations is that of Forbes et al. [1981] which includes possible effects of a

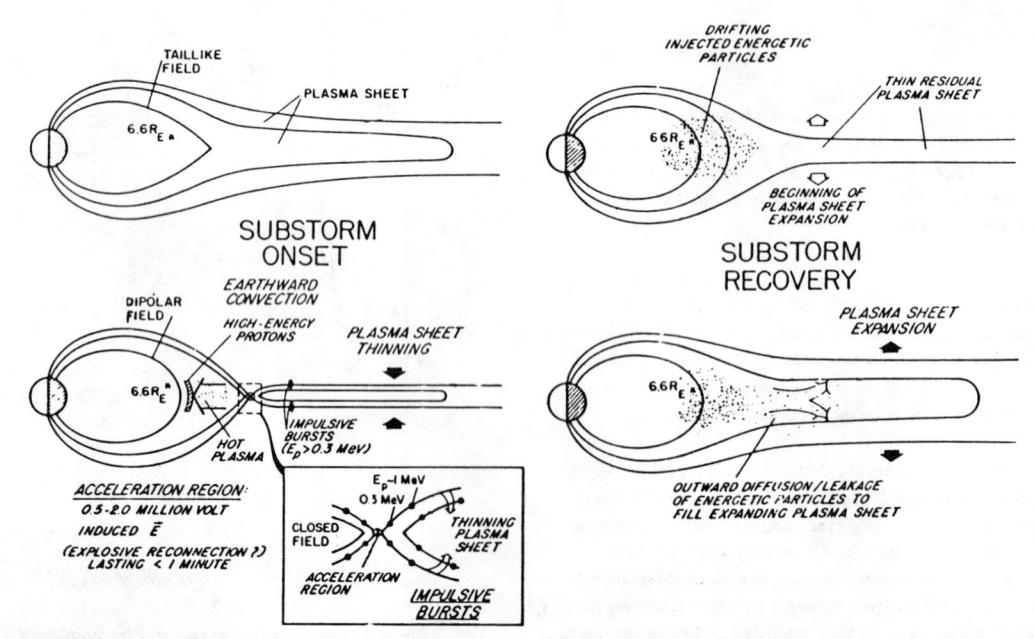

Fig. 5. A schematic illustration of proposed sequences of events which occur at substorm onset (left side) and at substorm recovery (right side) [Baker et al., 1979].

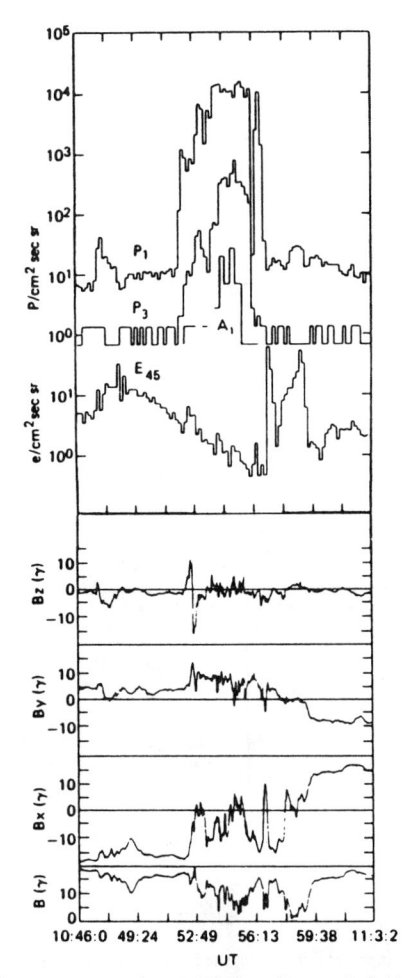

Fig. 6. particle measurements (upper panel) and magnetic field measurements (lower panels) during a high-intensity burst [after Kirsch et al., 1981].

distant neutral line which can add new field lines on to the existing plasma sheet. Forbes et al. [1981] have presented evidence that the recovering plasma sheet is due to continuous reconnection at some tailward retreating neutral line. The movement of the plasma sheet surface results then from the propagation of the energetic particle and plasma source, i.e. the reconnection region, onto new magnetic field lines which progressively map to higher polar latitudes. Parks et al. [1979], Möbius et al. [1980] and Williams [1981] have recently noted the importance of energetic particle effects near the boundary of the plasma sheet. The rapid rise-slow decay signature may at times be simply related to a localization of higher energy particles to the plasma sheet boundary (see below). Another process which can lead to the observed exponential decay is leakage of the energetic particles into the magnetosheath. Such a leakage may be due to cross-tail drift of energetic particles crossing the neutral sheet and is expected to depend on the gyroradius, i.e. is rigidity dependent. Leakage of energetic particles into the magnetosheath and even the upstream medium has indeed been established from simultaneous multi-spacecraft observations by, e.g., Sarris et al. [1978]. Kirsch et al. [1980] found for a specific particle burst in the magnetosheath a characteristic softening of the spectrum with time and suggested a rigidity dependent escape of particles from the plasma sheet into the magnetosheath as the reason. Such a rigidity dependent escape can explain the different decay time scales at different energies observed within the plasma sheet [Scholer et al., 1983a]. In order to prove or disprove the magnetosheath leakage model it should be investigated whether the softening of spectra observed in the magnetosheath is due to an increase of the fluxes at the lower energy end or due to a decrease of the fluxes at the high energy end of the spectrum with progressing time. In the case of an energy dependent leakage the first sequence of events should be observed in magnetosheath, the second sequence of events should be (and indeed is) observed in the plasma sheet. Finally, it should be noted that leakage may instead of being a time dependent process, be a spatial process resulting in a more rapid decrease of the highest energy ions with increasing distance from the plasma sheet boundary. As the satellite passes into the plasma sheet at recovery, this spatial profile is observed as a temporal profile in which also more energetic ions decrease more rapidly in intensity.

Energetic Particle Bursts Within the Plasma Sheet

In this section we will discuss the characteristics of energetic particle bursts which have been observed within the plasma sheet proper. Figure 6, taken from the work of Kirsch et al. [1981], shows particle and magnetic field data during a high intensity burst observed by IMP 8 at a distance of ~ 30 R_E near the dusk magnetopause. The burst had a duration of ~ 10 min and occurred during a neutral sheet crossing, as can be seen from the magnetic field B_x component. Alpha particles of 2.36-4.56 MeV total energy (A_1) were observed while no flux enhancements of protons at about the same total energy have been detected. From this the authors conclude that the acceleration process is charge-dependent. This, as well as many other bursts investigated by the authors, occurred in association with rapid variations of the magnetic field magnitude and direction. Both energetic protons and electrons during this burst are streaming tailward, which suggests that the source of the energetic particles is located earthward of the spacecraft, i.e. within ~ 25 R_E. Assuming that the magnetic field changes are temporal rather than due to convection of spatial structures Kirsch et al. [1981] calculated an inductive electric field of 40 mV/m.

Simultaneous observations of energetic proton bursts from two closely spaced spacecraft have been reported by Sarris et al. [1976a] and are shown in Figure 7. The location of the two spacecraft at the right lower corner shows that the two spacecraft were separated by ~ 1 R_E along the x-axis and have the same y and z coordinates. The panel at the upper right shows the intensity time profile of a burst at IMP 7 with the numbers indicating the time intervals when anisotropy vectors were measured. The anisotropy time history at both spacecraft is shown on the left hand side of Figure 8. First both satellites observed an earthward directed anisotropy with a strong dawn-dusk component. A few minutes later the anisotropy was directed earthward at the position of IMP 6 and tailward at position of IMP 7. The observations are consistent with the existence of a slowly earthward moving (~ 30 km/s) source of energetic particles. The dimension of the source along the x-axis was of the order of ~ 2000 km.

During some of the energetic particle bursts oppositely directed proton and electron anisotropies have been observed [Kirsch et al., 1977]. An example, taken from the paper by Krimigis and Sarris [1979] is shown in Figure 9. Shown are energetic proton and electron observations from IMP 8 when the spacecraft was located in the center of the magnetotail at a distance of ~ 37 R_E from the earth. Angular distributions (5 sec samples) of the various energetic particles at different stages of the burst are shown in the lower part of the figure. Both high and low energy protons are moving tailward and primarily from the solar direction. The 200 keV electrons, however, show at least during some of the 5 sec samples an earthward directed anisotropy. The authors have discussed the oppositely directed anisotropy of protons and electrons in terms of a field-aligned electric field. In order to produce spectral changes giving rise to the observed anisotropies a potential difference of ~ 30 kV is needed.

One of the deficiencies of many energetic burst observations in the geomagnetic tail is the lack of simultaneous magnetic field as well as plasma data presentations. In particular whether magnetic field changes are due to sudden global changes of the magnetic field configuration and indicative of large inductive electric fields or whether magnetic field configurations are simply convected over the satellite can only be determined by analyzing the plasma flow. Hones [1979] studied the energetic electron behaviour together with magnetic field and plasma data during several individual substorms. More recently, Bieber et al. [1982] investigated the plasma behaviour during energetic electron streaming events in the magnetotail. However, these papers were not so much concerned with the high energy particle acceleration process but they rather used

220 ENERGETIC IONS AND ELECTRONS

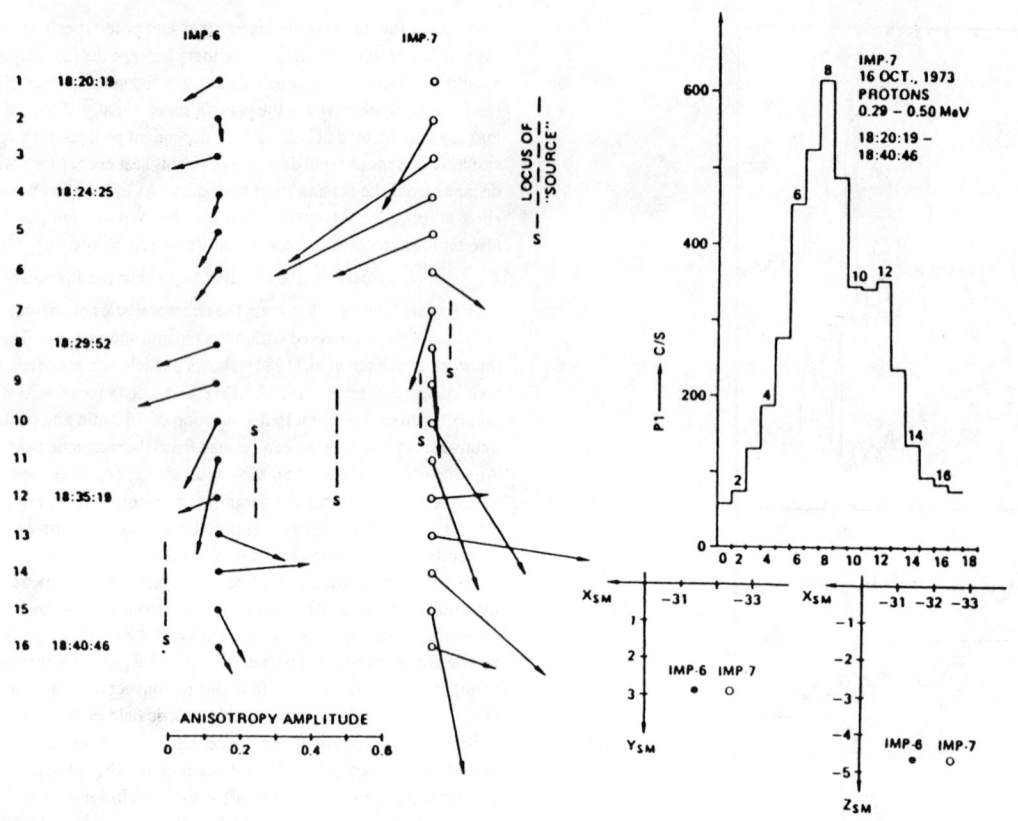

Fig. 7. Anisotropy measurements of a burst event simultaneously seen by IMP 7 and IMP 8. For details see text. [Sarris et al., 1976a].

the energetic electron angular distributions as a diagnostic tool in order to identify open (streaming electrons) or closed (isotropic electrons) field lines.

Recently Möbius et al. [1983] combined for two successive energetic particle burst events in the geomagnetic tail plasma, magnetic field, energetic proton and energetic electron data. In addition, they used plasma data from two closely spaced satellites (ISEE-1 and -2) in order to delineate spatial from temporal effects. The satellites were well within the plasma sheet at ~ 14 R_E from the earth and at ~ 6 R_E to the dawn side from the midnight meridian. Two energetic particle bursts (~ 400 keV) were observed within one hour and were superimposed on an exponentially decaying intensity time profile following an earlier plasma sheet recovery. Several minutes before the occurrence of both bursts the magnetic field first turns northward and then back again into the earthward direction. These changes are accompanied by changes in the absolute value and direction of the plasma velocity. The plasma starts flowing earthward shortly before the magnetic field returns in the earthward direction. This flow represents a convective motion of the plasma: after the magnetic field has returned to the earthward direction the plasma flow changes from an earthward to a tailward direction which is mostly field aligned.

By comparing the plasma flow at ISEE-1 and ISEE-2 (earthward of ISEE-1) it could be shown that the magnetic field lines with a northward component are convected past the spacecraft in the earthward direction. The flow signature is observed on ISEE-1, which is tailward of ISEE-2, ~ 12 seconds before it is observed on ISEE-2. This observation results in an earthward motion of the flow pattern with a velocity of ~ 100 km/s. Möbius et al. [1983] have suggested that the earthward motion of a neutral line to the dusk side of the satellite can explain this sequence of events: the satellite starts out in a region with earthward pointing magnetic field, then encounters northward pointing field within a region of earthward directed flow and measures finally an earthward pointing field in a region of tailward flow tailward of the neutral line.

The energetic protons exhibit, during the burst appearance, a dusk to dawn anisotropy almost perpendicular to the magnetic field. This anisotropy has been explained by Möbius et al. [1983] in terms of a grad B drift of the particles from a localized source, i.e. the neutral line limited to a region in the center of the tail. Although a grad B drift does not produce any anisotropy within a spatially homogeneous particle distribution, anisotropic distributions will occur near a localized source above and below a neutral sheet. Such a situation near the neutral sheet, seen from the tailward direction, is sketched in Figure 9, where the dotted region represents the acceleration region. Particle transport to the dusk side can, in principle, occur within two gyroradii above and below the satellite. The meandering particles will lead to a dawn to dusk anisotropy. To the dawn side the grad B drifting particles above and below a neutral sheet can produce a dusk to dawn anisotropy.

It should be noted that the region over which such an anisotropy can be observed may be a large fraction of the total plasma sheet extension in the z direction. The gyroradius of a 300 keV proton in a 10 γ field is ~ 1.2 R_E. Thus, anisotropy effects may be observed over a region of ~ 2.5 R_E above and below the neutral sheet, i.e. over 5 R_E, that has to be compared with an average plasma sheet thickness of ~ 6 R_E, at a distance of ~ 20 R_E.

The various observations of energetic particle bursts within the plasma sheet have provided significant evidence for the existence of spatially limited, tailward as well as earthward, moving acceleration regions although there are complications involving, e.g., near-earth particle mirroring [Williams, 1981] as will be discussed in the next section. Combined plasma, magnetic field and particle data have shown that the source is most probably near a magnetic neutral line, so that reconnection is ultimately responsible for the acceleration. However, parallel electric fields seem also to play an important role. Further progress is possible only by detailed analysis of magnetic field, plasma as well as energetic particle data, possibly from more than one spacecraft.

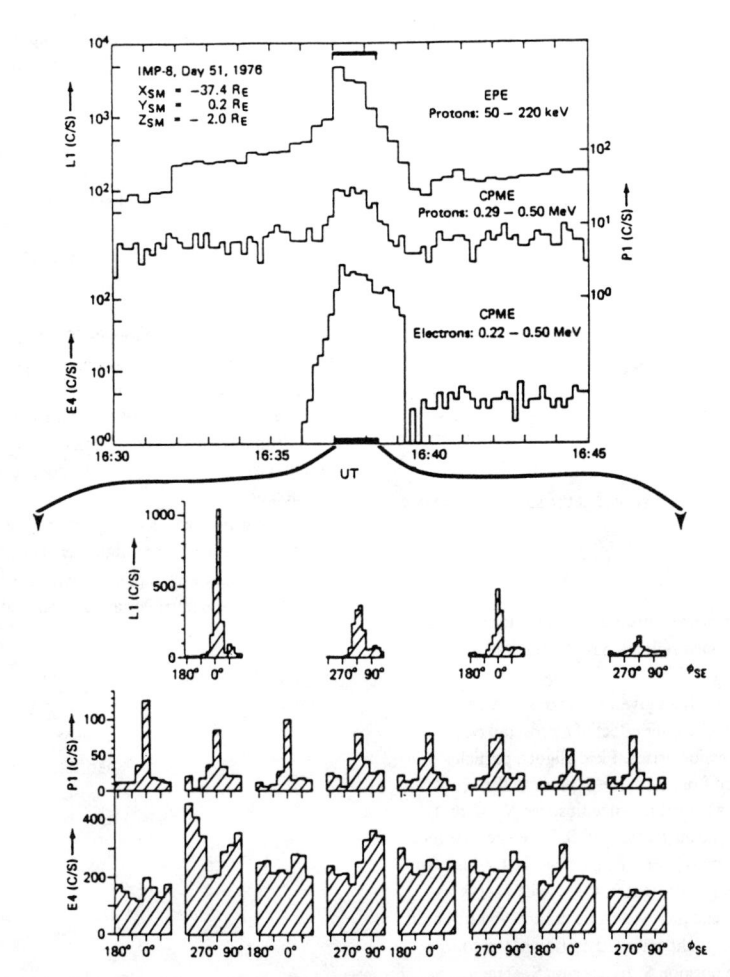

Fig. 8. An acceleration event in the center of the magnetotail (upper panel). The detailed angular distributions during the peak (lower panels) show anti-parallel flows of protons and electrons (second snapshot from left). The direction of the magnetic field longitude is $0° \pm 30°$ during this time interval. [Krimigis and Sarris, 1979].

Energetic Particle Phenomena Near the Plasma Sheet Boundary

Measurements by the ISEE satellites have revealed the existence of a thin layer of non-thermal particles streaming highly collimated along the tail field at the edge of the plasma sheet [Parks et al., 1979; Möbius et al., 1980; Williams, 1981; Andrews et al., 1981a, 1981b]. A detailed investigation of the energetic ions $\gtrsim 30$ keV at the edge of the plasma sheet has been performed by Williams [1981]. The three-dimensional measurements showed strong streaming and beam-like characteristics at each transition from low to high and high to low intensities in the energetic particle population. The azimuthal asymmetries indicate that the energetic ion streaming is located within ~2000 km of the plasma sheet edge. In general, when encountering the plasma sheet, ions are first observed at higher energies and are streaming in the earthward direction (spectrum turns over towards lower energies). Later in the event earthward flowing particles are also observed at lower energies. At the same time as low energy earthward ion jetting

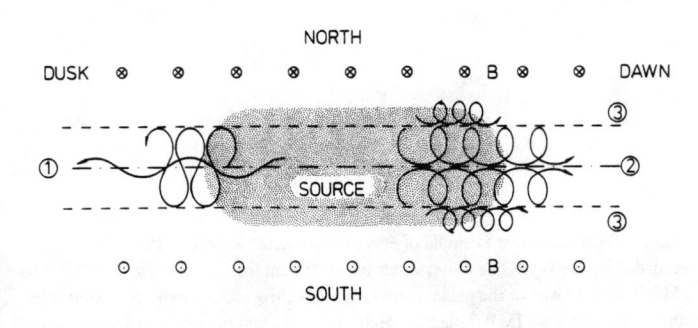

Fig. 9. Schematic view of a source of energetic particles in the neutral sheet seen from the tailward direction. [Möbius et al., 1983].

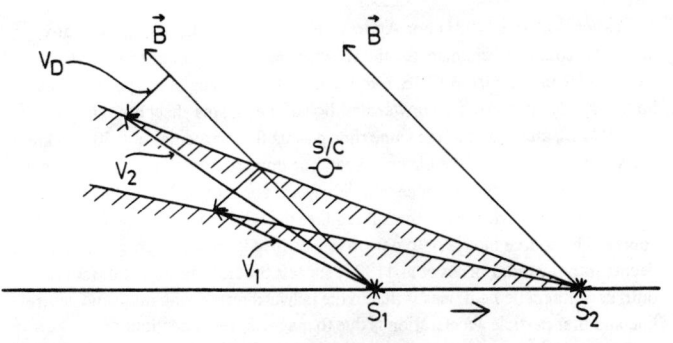

Fig. 10a. Diagram to illustrate how a tailward retreating source together with the ExB drift causes upward moving energy dependent boundaries.

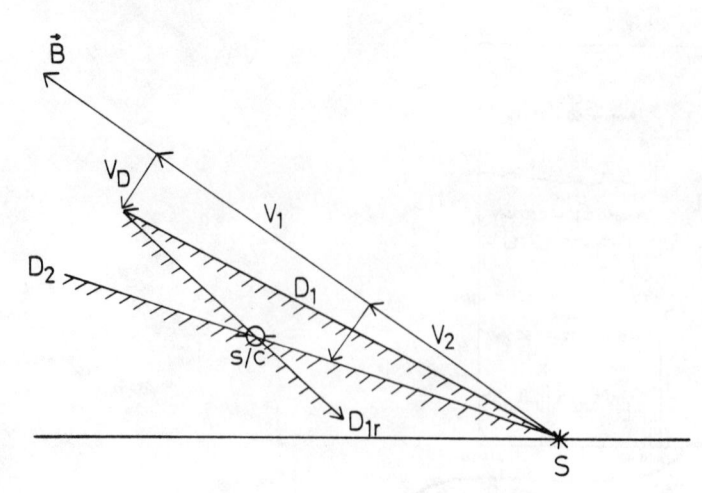

Fig. 10b. Trajectories of particles which can be observed at the satellite when near-earth reflection is included.

is seen, tailward directed fluxes of higher energy ions are detected. These tailward streaming beams are the result of mirror point reflection of the earthward moving ion population.

An explanation of the peaked spectra has been given by Andrews et al. [1981a,b] in terms of the well-known velocity filter effect of a crosstail electric field. Figure 10a illustrates how this dispersion arises. Field aligned particles of low and high speeds V_1 and V_2 are ejected from a source S_1 close to the neutral sheet. The ExB drift causes both particles to drift down with speed V_D so that their trajectories lie at different angles to the magnetic field B. Thus the energetic ions are spatially dispersed such that the most energetic particles lie furthest from the neutral sheet. A satellite moving towards the neutral sheet will observe successively particles at lower and lower energies. Andrews et al. [1981a,b] have combined this velocity dispersion effect with the tailward motion of the source. Figure 10b shows a source moving from position S_1 to position S_2. The connection of the endpoints of V_1 and V_2 with the source position S_2 then define upward moving fronts of slow and fast particles. Eventually the satellite will be crossed by these layers of particles of different energy. Figure 12b demonstrates how slow earthward moving particles and fast tailward moving particles can be observed simultaneously. For simplicity we neglect the motion of the source and consider only the ExB drift effect. Particles of slow velocity V_2 are ejected from a source at position S. Their velocity and the ExB drift velocity then defines a line D_2 where particles of the velocity V_2 can be found. Particles of higher velocity V_1 can be found at a line D_1 further away from the neutral sheet. Let us assume that these particles are adiabatically reflected in the near earth magnetic field. They will then move along a line D_{1r} which intersects with line D_2 at some location. A satellite at this location will observe simultaneously earthward jetting ions of velocity V_2 and tailward moving particles of velocity V_1. Due to tailward motion of the source, regions with different particle velocities will move across the satellite.

Neither Williams [1981] nor Andrews et al. [1981a,b] have suggested any particular source mechanism for the energetic particles. Forbes et al. [1981] studied plasma data from ISEE-1 and -2 during a crossing of the plasma sheet boundary. They inferred an upward motion of the plasma sheet boundary of 20 ± 10 km/s and found at the same time plasma flow velocities of $\sim 30 \pm 10$ km/s toward the midplane of the plasma sheet. The upward advance of the surface of the plasma sheet in the presence of a downward convective flow induced by a dawn to dusk electric field in the tail requires the tailward motion of the particle source. The source moves onto new magnetic field lines which progressively map deeper into the tail. Forbes et al. [1981] suggest that this motion of the source onto new magnetic field lines is due to the tailward retreat of a magnetic neutral line and that particle acceleration is due to magnetic reconnection.

Another energetic particle phenomenon in the geomagnetic tail are the very brief energetic proton enhancements termed impulsive bursts [Sarris et al. 1976].

These events are characterized by rapid onsets and rapid decays, with time constants of the order of ~ 10 seconds. The whole event typically lasts for less than one minute and the maximum intensity along the magnetic field reaches up to 10^5 (cm² sec sr MeV)$^{-1}$ at ~ 0.3 MeV. These bursts are observed only at the dusk side of the magnetotail and are not accompanied by energetic electron bursts. Baker et al. [1979] have proposed that these protons are energized during substorm onset by induced electric fields associated with the onset of magnetic field line merging at a magnetic neutral line earthward of the satellite, i.e. some 10 to 15 R_E behind the earth (Figure 5, left side). Belian et al. [1980] have studied more closely the relation of these bursts to plasma sheet thinning. Figure 11 shows in the upper panel two of the bursts as observed by Sarris et al. [1976b] in three different energy channels (P1: 0.29-0.5 MeV, P2: 0.5-0.97 MeV, P3: 0.97-1.85 MeV). The middle panel shows plasma densities for a period of three hours including the onset time of the two bursts [Belian et al., 1980]. It can be seen that the occurrence of the impulsive bursts are closely related to plasma sheet drop out. From the Kiruna magnetogram (bottom panel) it can be seen that the burst occurrence and the plasma sheet drop out are closely associated with the substorm onset.

The impulsive bursts exhibit an inverse velocity dispersion, with lower energy particles observed before higher energy ones (see Figure 11). Sarris and Axford [1979] have given an explanation for the inverse velocity dispersion. They assumed that the particles are accelerated simultaneously in a small region

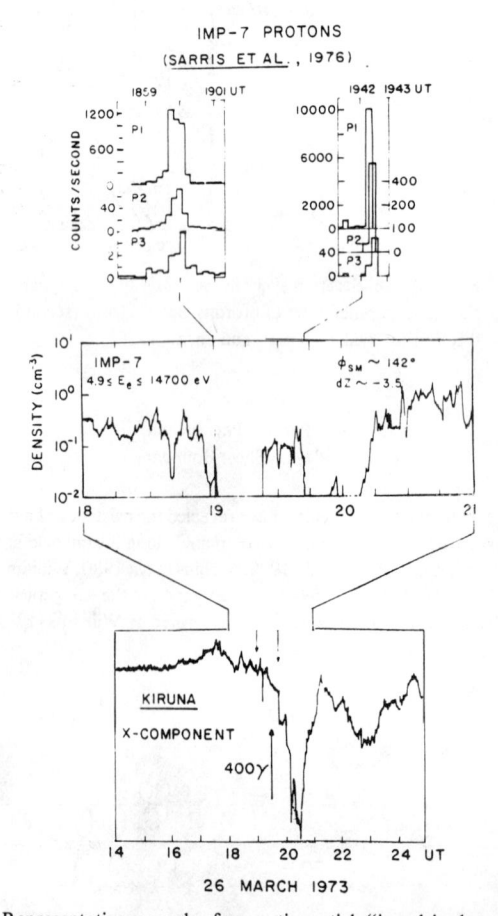

Fig. 11. Representative example of energetic particle "impulsive bursts" as recorded by the IMP-7 satellite, upper panels (taken from Sarris et al. (1976)). The IMP-7 satellite was in the plasma sheet, near the bow shock on the dusk side. The middle panel shows IMP-7 plasma electron density and the lower panel magnetogram X-component as recorded at Kiruna for the period of time bracketing the bursts. [Belian et al., 1980].

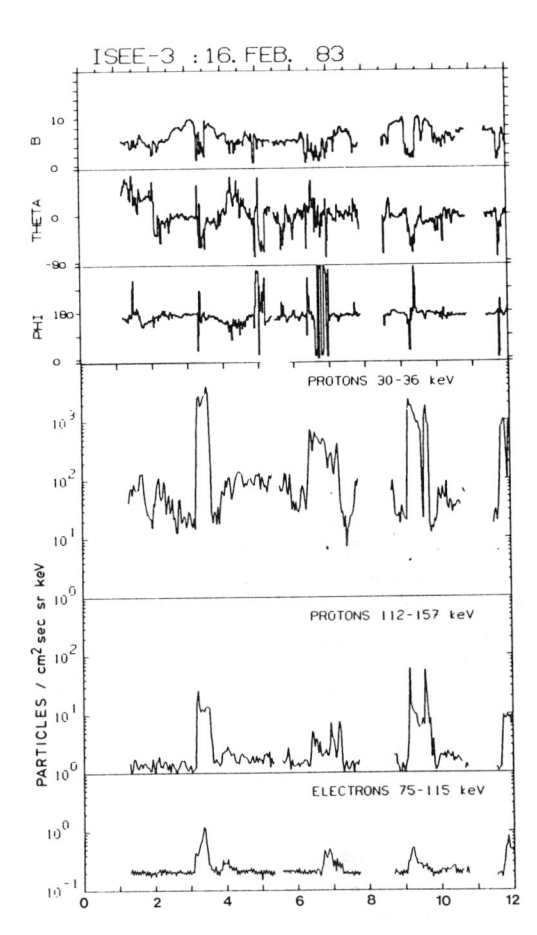

Fig. 12. From top to bottom: magnetic field magnitude in gamma, magnetic field θ – and φ-component (GSE), intensity of 30-36 keV protons, 112-157 keV protons and 75-115 keV electrons in the distant geomagnetic tail at 220 R_E on February 16, 1983. [Scholer et al., 1983b].

surrounding the neutral line. Following reconnection, the magnetic field is convected earthward and tailward away from the neutral line. The energetic particles disperse along the magnetic field line, so that during an outbound crossing lower energy particles are observed at the inner field lines and higher energy particles on outer field lines with respect to the plasma sheet. This description is equivalent to the velocity filter effect of a cross-tail electric field discussed in connection with the ion beams at the plasma sheet boundary.

Baker et al. [1979] have proposed that the earthward flowing particles during these impulsive bursts are injected onto closed field lines, leading to a relatively high density of quasi-trapped energetic particles in the outer radiation zone. As the plasma sheet expands during substorm recovery, the energetic ions of the outer radiation zone rapidly leak out into the newly expanded plasma sheet to fill the available volume of closed field lines. As already mentioned in the first chapter, the Baker et al. [1979] model may have a problem in explaining the observed fluxes following substorm recovery. let us assume that protons of velocity v are accelerated near a neutral line extending all across the tail. If there is no energy redistribution the ratio of the differential flux in the recovering plasma sheet to the differential flux in the impulsive burst is given by $j_r/j_i = \Delta t \cdot v \Delta L_z/(2L_zL_x)$, wherein Δt is the burst duration, ΔL_z the extension of the source in the z-direction, L_z and L_x the z- and x-dimension of the recovering plasma sheet. Assuming $\Delta L_z = 1\ R_E$, $L_z = 4\ R_E$, $L_x = 100\ R_E$ and $\Delta t = 40$ sec we obtain $j_r/j_i \approx 1/20$, i.e. the differential flux in the burst has to be, by more than an order of magnitude, larger than the observed flux in the expanding plasma sheet. Representative differential fluxes at 400 keV in the plasma sheet are ~10 p/(cm² s sr keV) [Ipavich and Scholer, 1983] and differential fluxes during the three burst events reported by Sarris et al. [1976] range from 3 to 30 p/(cm² s sr keV). Unless Δt is considerably larger than observed at the satellite position the predicted fluxes in the recovering plasma sheet are, by more than an order of magnitude, too small.

Lui and Krimigis [1983] have recently presented high time resolution measurements of energetic particles in the tail lobe during plasma sheet drop out. They found prominent peaks in the proton energy spectra at 0.1 to 0.7 MeV. By fitting the observed spectra to a drifting Kappa distribution a proton temperature of 15 to 45 keV and a velocity between 3500 an d 7000 km/s in the tailward direction was derived. Therefore, Lui and Krimigis (1983) interpreted their observations in terms of a hot beam in the tail lobe jetting away from the earth. So far, this has been the only example of such a possibly hot and fast beam in the literature.

Recent ISEE-3 Results in the Distant Tail

A new exploration of the distant geotail is now possible with data obtained from the ISEE-3 spacecraft. ISEE-3, originally positioned after its launch in August 1978 into an orbit around the sunward libration point L_1, was transferred in the summer of 1982 into an earth orbit which brings the satellite close to the tailward Lagrangian point L_2 at ~220 R_E. Also, three times the satellite probed the magnetotail close to the orbit of the moon.

Figure 12 shows a typical example of energetic particle measurements in the distant tail at ~220 R_E, taken from the work of Scholer et al. [1983b]. From top to bottom are shown 30-36 keV protons, 112-157 keV protons and 75-115 keV electrons. Several bursts can be recognized in the proton as well as in the electron data. Figure 13 shows angular distributions at the beginning of one of the burst events. The time of the beginning of each 128 sec averaging interval is shown at the bottom of the figure. The intensity is plotted linearly versus the instrument look direction, the sun is to the left of the figure. The intensity is normalized to the intensity in the sector with the maximum count rate. The event starts with highly collimated electrons streaming from the earth in the tailward direction. During the next 2 min readout no energetic protons are yet seen. Beam-like tailward streaming protons at the higher energy are first seen ~4 min following the energetic electron appearance. At this time, the 30-36 keV protons are still near background, they appear for the first time ~2 min after the appearance of the 112-157 keV protons. About 10 min after the first appearance of the energetic electrons, the distribution becomes more and more isotropic. The isotropic electron component is, however, not due to pitch angle scattering of the beam electrons, but is a different and independent distribution. From the timing of the first appearance of ~120 keV protons and ~32 kV protons Scholer et al. [1983b] derived a source location ~ 105 R_E earthward of the satellite, i.e. at ~115 R_E. The timing of the energetic electrons and the 32 keV protons leads to a considerably larger distance and Scholer et al. [1983b] suggested that this can be explained by assuming the electrons are generated or leave the source region considerably before the protons, i.e. ~120 s earlier.

The electron and proton data can be consistently explained in terms of an encounter with a plasmoid which was released near the earth following new creation of a neutral line [Hones et al., 1984]. The tailward streaming high energy electrons mark the encounter of the satellite with a field line which has just been reconnected at the near earth neutral line i.e. the separatrix. At later times, the satellite encounters particles with continually lower energies. They have also been accelerated at the time of the reconnection, but the time of flight effect leads to the observed energy dispersion. This assumes that the satellite does not cross field lines. Finally, the satellite encounters the plasmoid with closed magnetic field lines leading to isotropic electron distributions. At the time of the plasmoid encounter, the near earth neutral line has during this particular event already moved down the tail to a distance of ~100 R_E.

Further information on the acceleration process leading to these fast moving plasma structures can be obtained by studying the particle composition. Figure 14 shows proton and alpha particle differential spectra averaged over the event. It should be noted that both proton and alpha particle spectra cannot be approximated by a power law since the spectrum steepens considerably towards higher energies. Gloeckler et al. [1983] found that exponentials in energy per charge are in general better approximations to observed spectra in the distant tail.

224 ENERGETIC IONS AND ELECTRONS

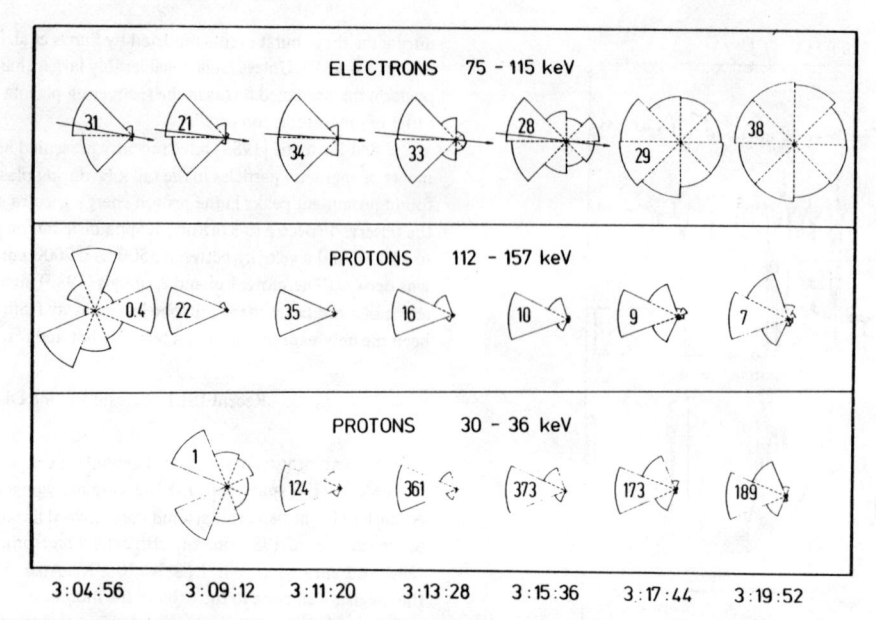

Fig. 13. 128 s averaged angular distributions of (from top to bottom) 75-115 keV electrons, 112-157 keV protons and 30-36 keV protons in the beginning of a particle burst on February 16, 1983 in the geomagnetic tail. The earthward direction is to the left of the Figure; the intensity is plotted linearly versus the instrument look direction. [Scholer et al., 1983b].

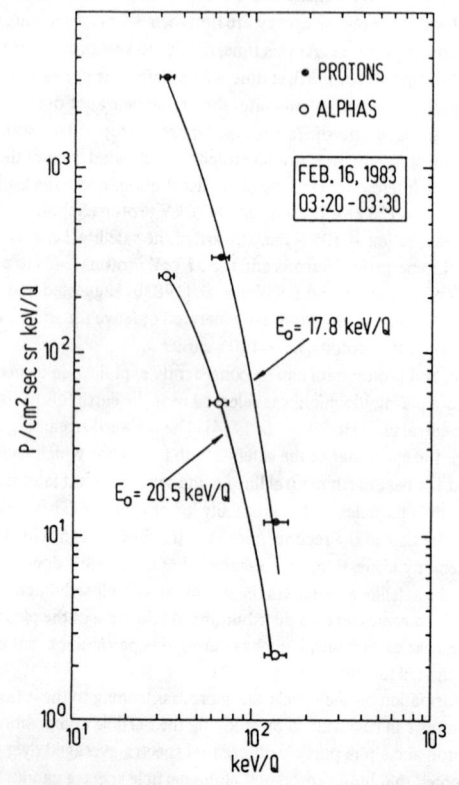

Fig. 14. Proton and alpha particle differential intensities in a log-log representation during a burst observed by ISEE 3 in the distant tail. Solid lines are exponential fits through the data.

During the event discussed above protons and alpha particle fluxes can be approximated by exponentials with e-folding energies of 17.7 keV and 20.5 keV/Q. The similarity of the spectra at the same energy/charge indicates the importance of electric field related acceleration processes.

Theory of Energetic Particle Acceleration in the Magnetotail

The vast amount of observational papers on energetic particles in the geomagnetic tail has to be contrasted with relatively few theoretical papers dealing with the acceleration in a more quantitative way. Several authors have studied particle trajectories in a stationary neutral sheet [e.g. Speiser, 1965, 1967; Sonnerup, 1971; Wagner et al., 1979]. Recently more detailed calculations of particle acceleration in current sheets have been performed by Cowley and Southwood [1980] and Lyons and Speiser [1982]. However, in all these calculations, the maximum energy a particle can gain is, at best, given by the cross-tail potential, which is of the order of ~50 kV.

Pellinen and Heikkila [1978] proposed that the rotational electric field induced by a time-dependent magnetic field accelerates particles to considerably higher energies. They did not compute the complicated time dependent magnetic and electric fields occurring during sudden reconnection in a self-consistent way, but rather assumed growing magnetic field disturbances in the neutral sheet current leading to X- and O-type magnetic neutral lines. This does not take into account the response of the plasma to the induced electric fields. The perturbations are such that they lead to the generation of X- and O-type neutral lines. Pellinen and Heikkila [1978] then performed trajectory calculations in the time dependent magnetic and electric field configuration. Figure 15 shows the path of protons and electrons starting near the O- (right side) and X-type (left side) neutral line. Protons starting at the X-type neutral line gain essentially the potential along the line by linear acceleration. Electrons and protons starting at the O-type neutral line and electrons starting at the X-type neutral line are accelerated in a two step process. They gain energy first by linear acceleration and then get deflected and trapped with a large magnetic moment in the magnetic field. After that they undergo betatron acceleration, keeping their first adiabatic invariant approximately constant. The problem with this approach is clearly that the magnetic and electric fields are not calculated in a self-consistent way.

Sato et al. [1982] have recently investigated particle trajectories in a develop-

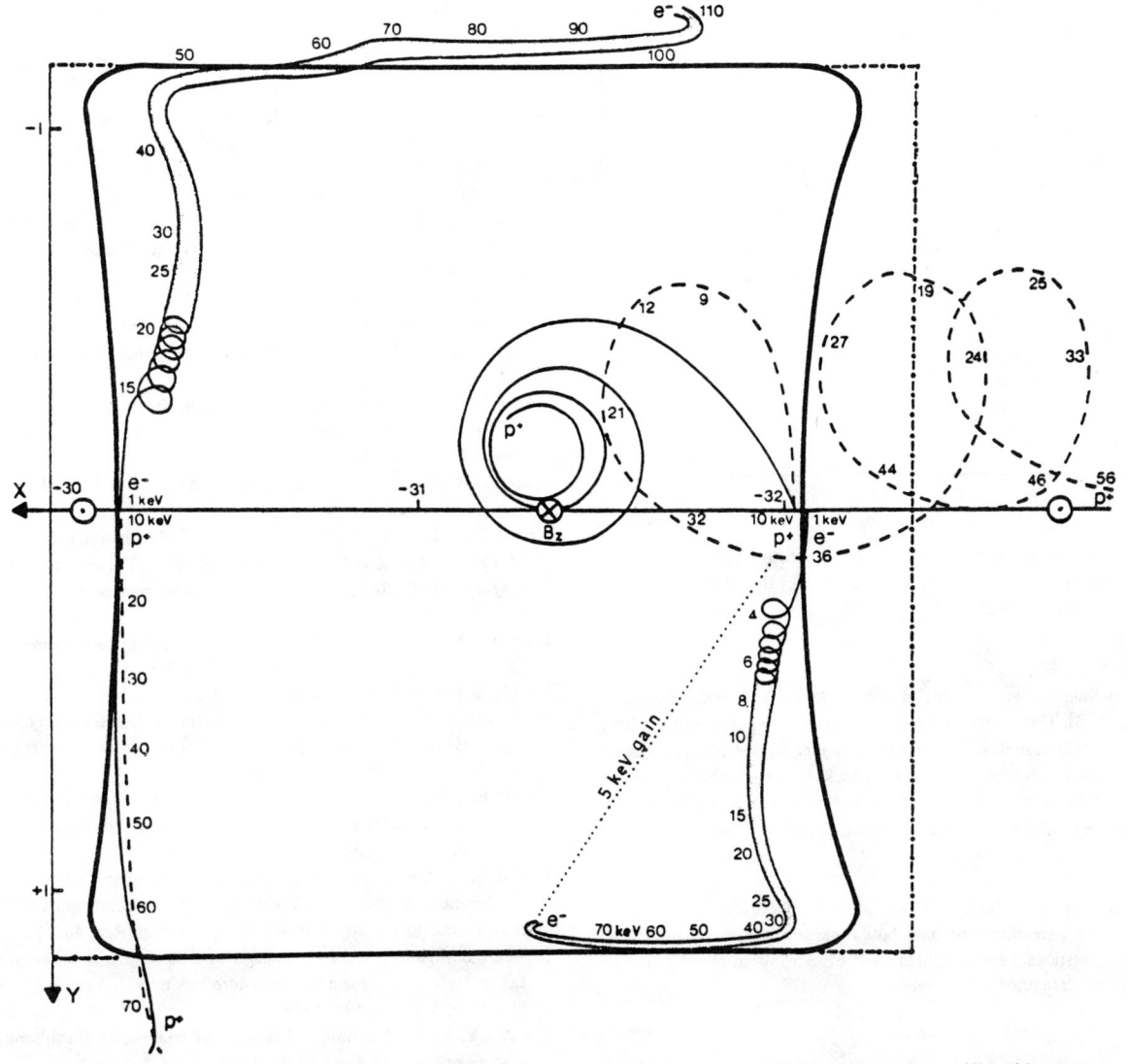

Fig. 15. Paths of two protons and two electrons, starting with typical plasma sheet energies near the O and X type neutral lines (shown as a continuous heavy line). The solid curves are normalized to the size of the growing disturbance. The straight line path (dotted line) for one electron would involve an energy gain of only 5 keV, as compared with the actual gain of 75 keV for the tortuous path; the difference is due to the electromotive force of the induced electric field. [Pellinen and Heikkila, 1979].

ing reconnection process. They used the time-varying electric and magnetic field model derived from a two-dimensional MHD simulation of externally driven magnetic reconnection and computed, numerically, orbits of test particles injected at various positions relative to the developing neutral line. Protons starting well away from the neutral line drift towards the neutral sheet and encounter the slow shock generated by the forced reconnection. These protons gain, due to the interaction with the slow shocks, an energy of ~10 keV. Larger energy gains are obtained by particles that end up close to the neutral line after their drift towards the neutral sheet (small off-distance). Electrons reaching the neutral line within a few gyroradii gain a maximum energy of about 10 keV no matter how small their off-distance is. The authors give no explanation for this result. However, this may be due to the fact that electrons with their small gyroradius are "tied" to the field lines: newly reconnected magnetic lines of force are convected away from the neutral line region and so are the electrons tied to these field lines. It is interesting to note that Pellinen and Heikkila [1978] also find that the electrons reach about 15 keV when they leave the X-type neutral line. In order to reach higher energies Pellinen and Heikkila have to assume, as the second stage, a betatron process. Electrons placed exactly at the neutral line can

in the Sato et al. [1981] model, in principle, be accelerated unlimitedly. Protons reaching the X-type neutral line gain energies of a few hundred keV. Figure 16 shows the energy gain of a proton versus the initial velocity parallel to the x-axis, which is proportional to the final off-distance from the neutral line. There is a sharp cut-off at a distance of the order of a few gyroradii. Protons reaching the neutral sheet at positions where the distance to the neutral line is smaller than a few gyroradii all reach, independent of this distance, about the same final energy, i.e. a few hundred keV.

Birn and Hones [1981] studied numerically the sudden occurrence of reconnection due to enhancement of the resistivity in a three-dimensional configuration. They found a limitation of the reconnection region in the y-direction. The region tailward of the X-line is characterized by low magnetic field B_z values and this region is bounded in the y-direction by regions of much stronger northward B_z values. Birn and Hones [1981] suggested that particles that move adiabatically in the y-direction would be turned back by the stronger B_z outside and could cross the nonadiabatic neutral sheet region several times, thereby reaching a considerably higher energy than from just one crossing.

Finally, it should not be forgotten that particles gain energy in the reconnec-

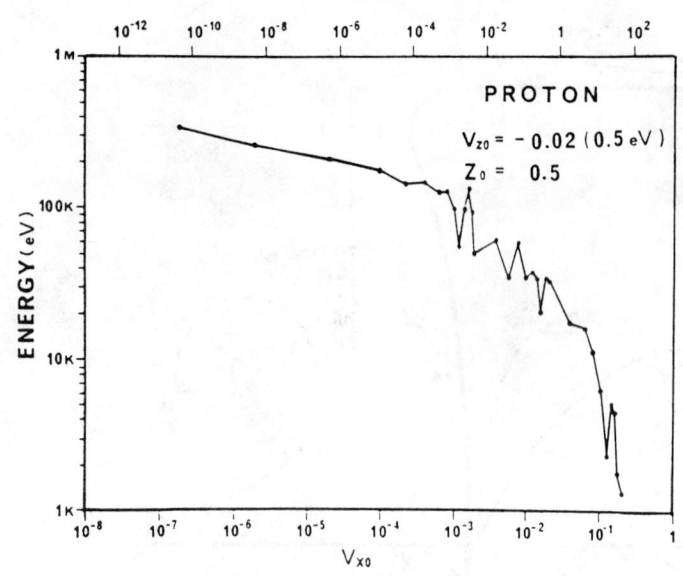

Fig. 16. Energy gain versus initial velocity (along x) for protons starting at x = 0 and z = 0.5. Note the parallel velocity represents essentially the off-distance from the neutral line [Sato et al., 1982].

tion region every time they are transmitted through one of the slow shocks [see, e.g., Sonnerup, 1973]. This energy gain is incorporated in the calculations of Sato et al. [1981]. If particle scattering in the reconnection region is strong, it may well be that diffusive shock acceleration at the slow shocks is an important acceleration process [Isenberg, private communication, 1983]. Particle scattering will also change the results of Pellinen and Heikkila [1978] and Sato et al. [1982].

Acknowledgements. I am grateful to E. W. Hones, Jr., for discussions of the ISEE 3 observations. I should like to thank both Referees for their constructive comments and suggestions. I am grateful to my ISEE-3 fellow co-investigators for using distant tail data prior to publication.

References

Anderson, K.A., Energetic electron fluxes in the tail of the geomagnetic field, *J. Geophys. Res., 70*, 4741, 1965.

Andrews, M.K., P.W. Daly, and E. Keppler, Ion jetting at the plasma sheet boundary: simultaneous observations of incident and reflected particles, *Geophys. Res. Lett., 8*, 987, 1981a.

Andrews, M.K., E. Keppler, and P.W. Daly, Plasma sheet motions inferred from medium-energy ion measurements, *J. Geophys. Res., 86*, 7543, 1981b.

Armstrong, T.P., and S.M. Krimigis, Observations of protons in the magnetosphere and magnetotail with Explorer 33, *J. Geophys. Res., 73*, 143, 1968.

Baker, D.N., R.D. Belian, P.R. Higbie, and E.W. Hones, Jr., High-energy magnetospheric protons and their dependence on geomagnetic and interplanetary conditions, *J. Geophys. Res., 84*, 7138, 1979.

Belian, R.D., D.N. Baker, E.W. Hones, Jr., P.R. Higbie, S.J. Bame, and J.R. Asbridge, Timing of energetic proton enhancements relative to magnetospheric substorm activity and its implication for substorm theories, *J. Geophys. Res., 86*, 1415, 1981.

Bieber, J.W., E.C. Stone, E.W. Hones, Jr., D.N. Baker, and S.J. Bame, Plasma behavior during energetic electron streaming events: further evidence for substorm-associated magnetic reconnection, *Geophys. Res. Lett., 9*, 664, 1982.

Birn, J., and E.W. Hones, Jr., Three-dimensional computer modeling of dynamic reconnection in the geomagnetic tail, *J. Geophys. Res., 86*, 6802, 1981.

Carbary, J.F., and S.M. Krimigis, Energetic particle activity at 5-min and 10-s time resolution in the magnetotail and its relation to auroral activity, *J. Geophys. Res., 84*, 7123, 1979.

Cowley, S.W.H., and D.J. Southwood, Some properties of a steady state geomagnetic tail, *Geophys. Res. Lett., 7*, 833, 1980.

Fan, C.Y., G. Gloeckler, and D. Hovestadt, Energy spectra and charge states of H, He and heavy ions in the earth's magnetosheath and magnetotail, *Phys. Rev. Lett., 34*, 495, 1975.

Forbes, T.G., E.W. Hones, Jr., S.J. Bame, J.R. Asbridge, G. Paschmann, N. Sckopke, and C.T. Russell, Evidence for the tailward retreat of a magnetic neutral line in the magnetotail during substorm recovery, *Geophys. Res. Lett., 8*, 261, 1981.

Gloeckler, G., F.M. Ipavich, D. Hovestadt, B. Klecker, M. Scholer, and C.Y. Fan, Suprathermal H$^+$ and He^{++} in the distant geomagnetic tail: ISEE-3 observations, 18th Intern. Cosmic Ray Conf., Bangalore, India, paper MG 2.2-3, 1983.

Hones, E.W., Jr., I.D. Palmer, and P.R. Higbie, Energetic protons of magnetospheric origin in the plasma sheet associated with substorms, *J. Geophys. Res., 81*, 3866, 1976.

Hones, E.W., Jr., Transient phenomena in the magnetotail and their relation to substorms, *Space Sci. Rev., 23*, 393, 1979.

Hones, E.W., Jr., D.N. Baker, S.J. Bame, W.C. Feldman, J.T. Gosling, D.J. McComas, R.D. Zwickl, J. Slavin, E.J. Smith, B.T. Tsurutani, Structure of the magnetotail at 220 R_E and its response to geomagnetic activity, *Geophys. Res. Lett., 11*, 5, 1984.

Ipavich, F.M., and M. Scholer, Thermal and suprathermal protons and alpha particles in the earth's plasma sheet, *J. Geophys. Res., 88*, 150, 1983.

Kirsch, E., S.M. Krimigis, E.T. Sarris, R.P. Lepping, and T.P. Armstrong, Possible evidence for large, transient electric fields in the magnetotail from oppositely directed anisotropies of energetic protons and electrons, *Geophys. Res. Lett., 4*, 137, 1977.

Kirsch, E., E.T. Sarris, and S.M. Krimigis, Two spacecraft observation of particle bursts at the distant magnetopause and in the magnetotail boundary layer, *Planet. Space Sci., 28*, 487, 1980.

Kirsch, E., S.M. Krimigis, E.T. Sarris, and R.P. Lepping, Detailed study on acceleration and propagation of energetic protons and electrons in the magnetotail during substorm activity, *J. Geophys. Res., 86*, 6727, 1981.

Krimigis, S.M., and E.T. Sarris, Energetic particle bursts in the earth's magnetotail, in *Dynamics of the Magnetosphere*, (ed. by S.-I. Akasofu) D. Riedel, Hingham, Mass., p. 599, 1979.

Lui, A.T.Y., and S.M. Krimigis, Energetic ion beam in the Earth's magnetotail lobe, *Geophys. Res. Lett., 10*, 13, 1983.

Lyons, L.R., and T.W. Speiser, Evidence for current-sheet acceleration in the geomagnetic tail, *J. Geophys. Res., 87*, 2276, 1982.

Möbius, E., F.M. Ipavich, M. Scholer, G. Gloeckler, D. Hovestadt, and B. Klecker, Observations of a non-thermal ion layer at the plasma sheet boundary during substorm recovery, *J. Geophys. Res., 85*, 5143, 1980.

Möbius, E., M. Scholer, D. Hovestadt, G. Paschmann, and G. Gloeckler, Energetic particles in the vicinity of a possible neutral line in the plasma sheet, *J. Geophys Res., 88*, 7742, 1983.

Parks, G.K., D. Leaver, C.S. Lin, K.A. Anderson, R.P. Lin, and H. Réme, ISEE 1/2 particle observations of outer plasma sheet boundary, *J. Geophys. Res., 84*, 6471, 1979.

Pellinen, R.J., and W.J. Heikkila, Energization of charged particles to high energies by an induced substorm electric field within the magnetotail, *J. Geophys. Res., 83*, 1544, 1978.

Sarris, E.T., S.M. Krimigis, T. Iijima, C.O. Bostrom, and T.P. Armstrong, Location of the source of magnetospheric energetic particle bursts by multispacecraft observations, *Geophys. Res. Lett, 3*, 437, 1976a.

Sarris, E.T., S.M. Krimigis, and T.P. Armstrong, Observations of magnetospheric bursts of high-energy protons and electrons at 35 R_E with IMP 7, *J. Geophys. Res., 81*, 2341, 1976b.

Sarris, E.T., S.M. Krimigis, C.O. Bostrom, and T.P. Armstrong, Simultaneous multispacecraft observations of energetic proton bursts inside and outside the magnetosphere, *J. Geophys. Res., 83*, 4289, 1978.

Sarris, E.T., and W.I. Axford, Energetic protons near the plasma sheet boundary, *Nature, 277,* 460, 1979.

Sato, T., Strong plasma acceleration by slow shocks resulting from magnetic reconnection, *J. Geophys. Res., 84,* 7177, 1979.

Sato, T., H. Matsumoto, and K. Nagai, Particle acceleration in time-developing magnetic reconnection process, *J. Geophys. Res., 87,* 6089, 1982.

Scholer, M., N. Sckopke, F.M. Ipavich, and D. Hovestadt, Relation between energetic electrons, protons, and the thermal plasma sheet population: plasma sheet recovery events, submitted to *J. Geophys. Res.,* 1983a.

Scholer, M., G. Gloeckler, B. Klecker, F.M. Ipavich, and D. Hovestadt, Fast moving plasma structures in the distant magnetotail, submitted to *J. Geophys. Res.,* 1983b.

Speiser, T.W., Particle trajectories in model current sheets, 1, Analytical solutions, *J. Geophys. Res., 70,* 4219, 1965.

Sonnerup, B.U.Ö., Adiabatic particle orbits in a magnetic null sheet, *J. Geophys. Res., 76,* 8211, 1971.

Sonnerup, B.U.Ö., Magnetic reconnection and particle acceleration, in *High Energy Phenomena on the Sun*, ed. R. Ramaty and R.G. Stone, GSFC-Greenbelt, Maryland, p. 357, 1973.

Wagner, J.S., J.R. Kan, and S.-I. Akasofu, Particle dynamics in the plasma sheet, *J. Geophys. Res., 84,* 891, 1979.

Williams, D.J., Energetic ion beams at the edge of the plasma sheet - ISEE 1 observations plus a simple explanatory model, *J. Geophys. Res., 86,* 5507, 1981.

THE DISTANT GEOMAGNETIC TAIL IN THEORY AND OBSERVATION

S.W.H. Cowley

Blackett Laboratory, Imperial College
London SW7 2BZ, United Kingdom

Abstract. Properties of the geomagnetic tail expected in the reconnection model of the magnetosphere are discussed and related to observations made in the distant tail by Pioneer 7 and 8, and by ISEE-3. Particular attention is given to the acceleration of tail lobe plasma in the current sheets which lie at the tail centre plane downstream from the neutral line, and to the structure of the resulting energized plasma populations. This study shows how a simple model in which reconnection occurs essentially continuously provides a consistent framework within which to interpret a wide range of data, both in the near and distant magnetotail.

1. Introduction

Theoretical discussion of the form and extent of the distant geomagnetic tail, and its dependence on the nature of the large-scale interaction between the Earth's magnetosphere and the solar wind, has been a subject largely in abeyance for the past twenty years [Dessler, 1964; Dungey 1965]. This has been due in part to the lack of detailed observations at various distances along the extended tail, although, in fact considerable indirect and *in situ* information has been available for much of this period.

The indirect information referred to above has resulted from low altitude spacecraft studies of energetic solar particle access to the Earth's polar caps. The data were found to be consistent with the access of essentially free particles along the field lines of an open magnetic tail, of length several hundred Earth radii (see *eg* the review by Paulikas [1974], and references therein). These results were therefore in agreement with the Dungey [1961] open magnetosphere, for which Dungey [1965] had estimated a typical tail length of ~ 1000 R_E on the basis of the cross-polar cap travel time of open field lines. The results did not favour the ~ 1 AU tail proposed by Dessler [1964], to which solar particles gain access only by slow diffusion inward from the magnetopause.

In addition to this indirect information, an important set of detailed field and plasma data were also obtained from the interplanetary probes Pioneer 7 and 8, which passed through the tail region at distances ~ 1000 R_E and ~ 500 R_E. Although these fly-bys took place in 1966 and 1968 respectively it was not until much later that a thorough appreciation and interpretation of the combined field and plasma data set became available [*eg* Bavassano et al., 1974; Walker et al., 1975].

In recent months studies of the distant tail have once more come to the fore with the ISEE-3 geotail mission to distances of ~ 220 R_E, results of which are now in the process of being published. It is therefore opportune to review in this paper previous theoretical and experimental work on this subject and its relationship to initial ISEE-3 results. The theoretical discussion will be based on a simple picture in which reconnection occurs essentially continuously in an open geomagnetic tail. The consequences of this picture and its relationship with Pioneer 7 and 8 and ISEE-3 data will then be discussed. In line with the title of this paper, and with the interests of the author, the theoretical picture will be presented first and will then be used as a framework for interpretation of the data. In practice, however, the development of these two aspects (*ie* theory and data interpretation) has been much more interdependent than might be indicated by this approach.

2. Particle Acceleration in the Geomagnetic Tail

In a magnetically open magnetosphere such as the Earth's whose properties are dominated by solar wind-driven internal convection (as opposed *eg* to corotation), considerable interest and importance attaches to the acceleration of plasma in the current sheets which lie down stream of reconnection regions. On the dayside of the magnetosphere, where reconnection results in the production of open field lines, the newly reconnected lines will in general initially contract over the magnetopause under the action of the magnetic tension forces, liberating magnetic energy to the magnetosheath plasma as they do so. The total power involved is $\sim 3 \times 10^{11}$ watts when IMF B_z is southward, sufficient to raise the energy of inflowing magnetosheath ions which interact with the current sheet by a few hundred eV, *ie* by an amount comparable with their initial energies [Cowley, 1980, 1983; Sonnerup et al, 1981]. In part, this acceleration represents a re-release to the magnetosheath plasma of energy previously extracted from it in compression and draping of the solar wind magnetic field over the magnetopause.

Following this initial period of liberation of magnetic energy to the plasma on the dayside, the open flux tubes begin to be stretched out on the nightside into the geomagnetic tail by the magnetosheath flow. The direction of the electric current at the magnetopause accordingly switches sign from dawn-to-dusk at the dayside, corresponding to $\underline{j} \cdot \underline{E}$ positive and plasma energization, to dusk-to-dawn over the tail magnetopause, corresponding to $\underline{j} \cdot \underline{E}$ negative and energy extraction from the streaming plasma [*eg* Axford, 1976]. This energy appears as the magnetic energy of the geomagnetic tail, and Poynting flux of electromagnetic energy flows continuously inward from the magnetosheath dynamo region ($\underline{j} \cdot \underline{E}$ negative) towards the centre plane of the tail. There, the electromagnetic energy is released back to the plasma as the open tail field lines reconnect and contract away from the tail neutral line toward and away from the Earth on either side. This reconnection may be continuous but is generally unsteady, being periodically greatly enhanced during substorms. The total power extracted from the magnetosheath plasma and fed into the tail depends on IMF B_z but is typically $\sim 5 \times 10^{12}$ watts [e.g. Cowley, 1980]. Much of this power is, however, directly returned to the magnetosheath in the form of accelerated plasma from the region tailward of the tail neutral line, while only a fraction $\sim 10^{12}$ watts appears subsequently in the form of the hot plasma sheet and ring current plasmas, ionospheric heating and auroral phenomena in the region Earthward of the tail neutral line. The latter estimate is based on the conclusion to be discussed further below that the tail neutral line lies within a distance ~ 100 R_E downtail from the Earth under usual circumstances.

Within this overall framework a primary focus of interest is on the plasma energization which takes place at the tail centre plane and the form of the accelerated plasma populations to which tail reconnection gives rise. The first quantitative investigations of this subject were undertaken by Speiser [1965,

1967, 1968] and Jaeger and Speiser [1974], following the qualitative description of particle behaviour in an X-type magnetic geometry given by Dungey [1953]. Speiser's studies centred on analytic and numerical investigations of individual particle motion in model current sheet electromagnetic fields representative of the geomagnetic tail, and showed that the process results in the formation of field-aligned accelerated particle beams flowing toward and away from the Earth on either side of the neutral line. The conditions imposed by the requirements of self-consistency and equilibrium were considered later, first by Alfvén [1968] and Cowley [1973] for the case of a strict magnetic neutral sheet system, and then by Rich et al. [1972], Eastwood [1972,1974], Bird and Beard [1972], Schindler [1972], Hill [1975] and Francfort and Pellat [1976], who investigated in various approximations the more general situation in which a weak field component (B_z in a magnetotail context) threads across the current sheet.

In the geomagnetic tail current sheet, three regimes of investigation may be identified. The first and most difficult is the region in the immediate vicinity of the neutral line itself, a region which remains relatively ill-understood theoretically. Complicating factors are the highly non-adiabatic nature of particle motions near the null, the possible importance of wave-particle interactions and anomalous resistance resulting from the unstable velocity distributions set up by these motions [eg Robertson et al., 1981], and the necessity of considering at least two spatial dimensions (ie either x - y and/or y - z) in modelling. Particle acceleration in this region is likely to be sensitively dependent on the initial conditions of the orbits which determine how far the particles travel in the current sheet in the direction of the magnetic separator (x-line) and cross-tail electric field before being ejected along the reconnected magnetic field lines away from the null by the Lorentz force associated with the B_z field [eg Speiser, 1965, 1967]. Particles having special orbits which remain close to the x-line for large distances will gain large energies, the gain being limited only by the total emf along the x-line. For continuous steady reconnection in the tail this emf will be just the total cross-magnetosphere potential ~60kV, but the emf is likely to be much higher (several hundred kV) along the near-Earth neutral line during substorms associated with the rapidly changing magnetic configuration and induced electric field. Indeed, Sonnerup [1979] has suggested that particles could make several traversals along the x-line in the latter case, returning along the 'O' line connected to it, it being noted that the emf along the 'O' line will generally be limited to small values [Vasyliunas, 1980]. If this occurs the particles would gain the few hundred kV emf along the x line each turn.

Although the region in the immediate vicinity of the x-line occupies a central position in the reconnection picture, it is, of course, a region of very limited spatial extent. Away from the x-type null in the quasi-steady state the B_z field in the current sheet may be expected to increase rapidly in magnitude to values determined by the plasma-field stress balance conditions, and thereafter will change only slowly with position in the flow downstream from the x-line in response to slowly varying inflow plasma conditions. The second regime of interest in the tail thus corresponds to the extended regions downstream from the x-line where a one-dimensional current sheet description should form a valid first approximation ie where the field may be taken to be $\underline{B} = (B_x(z), 0, B_z)$, with B_z nearly a constant over ion trajectories within the current layer. The field line tension is then balanced (in the field line "rest" frame) by the dynamic pressure of the inflowing tail lobe plasma, as will be described more fully below in relation to the details of particle acceleration in this case. Some previous authors have suggested that this stress balance condition as applied to the tail leads to inherent time dependence, $|B_z|$ decreasing and the current sheet thinning as time progresses [Eastwood, 1972; Hill and Reiff, 1980 a,b]. However, Cowley and Southwood [1980] have shown that this conclusion results from these authors having made arbitrary assumptions about the variation of B_z with distance in the current sheet, and that steady states may readily be obtained in the one-dimensional approximation by allowing B_z to slowly vary with position in response to slowly changing inflow plasma parameters, as may readily be envisaged. The Cowley-Southwood model will be described in greater detail in section (4) below.

With regard to particle motion in a one-dimensional current sheet, it should be noted that this may be either adiabatic or non-adiabatic depending upon the equilibrium thickness of the current layer. Numerical work suggests a thickness of order the ion larmour radius or less, with 'Speiser'-type non-adiabatic ion motion within the current layer [eg Eastwood, 1972, 1974]. However, adiabatic solutions with arbitrary scale length are also theoretically possible provided that a special magnetically trapped particle population exists within the current sheet [Francfort and Pellat, 1976; Cowley and Pellat, 1979].

The one-dimensional approximation should be valid away from the x-line at all points tailward of the neutral line (so long as a tail-like field geometry is maintained), and also for extended distances Earthward of the neutral line. As discussed further below, however, near the earth the approximation ceases to be valid due to the mirroring back into the current sheet of previously-accelerated ions. In this third region stress balance is then no longer determined in terms of tail lobe plasma newly flowing into the centre plane as in the two regimes just discussed, but is instead dominated by the pressure of the trapped hot plasma. This regime corresponds to the usual plasma sheet observed in the near-Earth tail, where the hot plasma is found to have been isotropized. In this case the field tension is balanced by the inward pressure gradient of the hot plasma. This pressure gradient, and plasma energization, is then in the simplest picture determined by adiabatic compression of the palsma on the Earthward-moving flux tubes. Isotropic pressure equilibria are thus at least two-dimensional in nature and in this case satisfy

$$\frac{B_x}{B_z} \approx \frac{L_x}{L_z}$$

where B_x is the tail lobe field, B_z the normal field threading the current sheet, L_z the sheet thickness and L_x the scale of variations along the current sheet. Detailed models of this nature have been presented by eg Bird and Beard [1972] and Birn et al. [1975, 1977], but since they apply only to the near-Earth tail, typically, say, within the lunar distance, they will not be discussed further here. Instead, we return to consider in more detail stress balance and plasma energization in one-dimensional current sheets applicable to the more distant tail.

3. Particle Acceleration in 1-D Current Sheets

Particle energization in a one-dimensional current sheet may be discussed in very simple and general terms due to the existence of a special frame transformation which removes the uniform electric fields in the plane of the current sheet [deHoffman and Teller, 1950; Speiser, 1965; Eastwood, 1972]. If the general 1-D current sheet fields are taken to be (in the usual tail coordinate system)

$$\underline{B} = (B_x(z), B_y(z), B_z) \text{ and } \underline{E} = (E_x, E_y, E_z(z))$$

where B_z is a constant (satisfying div $\underline{B} = 0$) and E_x, E_y are constants (satisfying curl $\underline{E} = \underline{0}$ then on moving to a frame having velocity $\underline{V}$ relative to the initial frame (the Earth's rest frame), where

$$\underline{V} = \left(\frac{E_y}{B_z}, -\frac{E_x}{B_z}, 0 \right),$$

the electromagnetic fields in the new frame are

$$\underline{B}' = \underline{B} \text{ and } \underline{E}' = (0,0,\underline{E} \cdot \underline{B}/B_z),$$

assuming the transformation is non-relativistic. Note that the transformation velocity above is just the $\underline{E} \wedge \underline{B}$ drift due to electric fields E_x, E_y in the uniform magnetic field B_z. The transformed frame will be referred to here as the field line "rest" frame (it is also known as the deHoffman-Teller frame), and has the special property that since the electric field has only a steady component $E'_z(z)$, the speed of individual particles depends only on z. If $E'_z = 0$ then the particle speed is constant on each trajectory since $\underline{E}' = \underline{0}$ at all points. In any case the particle energy remains unchanged in its interaction with the current sheet in this frame (assuming no net potential change across the current sheet for particles which are transmitted through it rather than being reflected), and the particle exits the current sheet into the uniform field region outside (where $\underline{E} \cdot \underline{B} = 0$) at the same speed at which it entered. The change in energy which occurs in the Earth's frame can then be simply explained with the aid of Figure 1, taken from Cowley and Shull [1983]. These are diagrams of velocity space appropriate to the region just outside (northward of) the central current sheet and tailward of the neutral line. The unprimed velocity coordinate system is the Earth's rest frame, with V_x pointing towards the Earth and V_z northward, normal to the current sheet. In this frame we for simplicity take the fields to be $\underline{B} = (B_x(z), 0, B_z)$ and $\underline{E} = (0, E_y, E_z(z))$, where B_z is negative and $E_z = 0$ outside the current layer. The primed frame is then the field line rest frame moving tailwards with the speed $V_x = -E_y/|B_z|$ as given by equation(1). The line at the angle θ to the V_x, V_x' axes and passing through the origin of the primed coorinate system O' is in the direction of the magnetic field outside the current sheet. A field-aligned particle which is moving slowly down-tail and convecting into the current sheet is then represented by point A in the upper diagram. In the Earth's frame its velocity is the sum of field-aligned speed V_L and $\underline{E} \wedge \underline{B}$ drift speed $\underline{V}_D$ perpendicular to $\underline{B}$. In the field line rest frame the particle simply moves along the field towards the current sheet with speed $v = |AO'|$. Let us now assume for definiteness that after interaction with the current sheet the particle remains field-aligned. It will then emerge from the current sheet at point B in velocity space, determined by the condition $|AO'| = |BO'|$ since particle speed is preserved in the field line rest frame. In the Earth's frame, however, the particle has gained considerable energy in moving from A to B in its interaction with the current sheet. From the geometry of the figure the field-aligned outflow speed V_B is given by

$$V_B = \frac{2E_y}{|B_z|} \cos\theta - V_L \approx \frac{2E_y}{|B_z|} - V_L \quad (2)$$

the approximate form being valid when $|B_x| \gg |B_z|$ outside the current sheet so that $\theta = \tan^{-1}(|B_z|/|B_x|)$ is small. This will generally be the case in systems of interest. The corresponding expression for the region Earthward of the neutral line (B_z positive) is easily shown to be

$$V_B \approx \frac{2E_y}{|B_z|} \cos\theta + V_L \approx \frac{2E_y}{|B_z|} + V_L \quad (3)$$

where V_L is again positive for tailward flows, but V_B is now positive if Earthward directed (see *eg* Cowley [1980]). Note that if V_L is small the particles emerge from the current sheet with a field-aligned speed nearly twice the field line speed $V_F = E_y/B_z$. Typically in magnetospheric tail applications this speed is of order hundreds of km s^{-1}, as will be discussed below. Ion energization in the current sheet is correspondingly of order keV. An electron with speed 10^3 km s^{-1}, however, has an energy of only 3eV, generally much less than electron thermal energies in the inflowing plasma. Electrons are therefore not much energized by this mechanism (at least not in one interaction with the current sheet), and consequently most of the energy liberated from the field to the plasma is fed (initially at least) into ions. The same conclusions can also be reached by considering the particle trajectories. The energy gained by each particle in the current sheet interaction is given by

$$W = qE_y \Delta Y$$

where q is the particle charge and ΔY the cross-system displacement in the current sheet. However ΔY is of order the particle larmour radius in the field B_z (computed using the particle speed in the field line "rest" frame) and is therefore generally much larger for ions than electrons.

The above discussion, and equations (2) and (3), were based on an assumption that a field-aligned ion beam would remain field-aligned after the current sheet interaction, or more generally that particle pitch angle is preserved in the field line "rest" frame (or is reflected about 90° pitch angle). Assuming similar inflow plasma properties on either side of the current sheet, the velocity distributon of outflowing ions would then simply be the image of the inflow distribution mirrored in the $V_{\parallel} = 0$ plane. In the lower diagram in Figure 1 the inflow distribution is indicated by three concentric circular contours of constant distribution function to the left of O', indicating a cold, isotropic tail lobe ion population having slow down-tail bulk speed V_L. If pitch angle is preserved in the

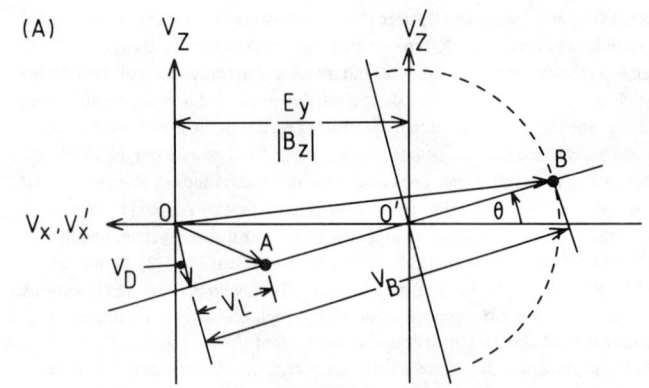

Fig. 1a. Velocity space diagram showing the relation between the velocities of a field-aligned particle before (point A) and after (point B) interaction with the current sheet. The diagram is appropriate to the region tailward of the neutral line and northward of the current layer.

field line rest frame then the outflow distribution would be similar circles centered on point B, having down-tail bulk speed V_B. The energy input to the ions would then appear solely as the kinetic energy of the resulting field-aligned ion jet. These considerations would certainly apply if the ion motion is adiabatic within the current layer (the particles would then emerge on the opposite side of the current sheet from which they entered, but this does not affect the energy argument given here). However, in the more probable non-adiabatic case some pitch angle scattering will occur, such that energy conservation in the field line "rest" frame will only locate the outflowing particle in the upper diagram of Figure 1 at some point on the dashed hemisphere radius $|AO'|$ centred on O'. For given E_y/B_z, the larger the exit pitch angle, the less the energy gain experienced by the particle in the Earth's frame (see also Cowley [1978]), equation (19). The result of pitch angle scattering due to non-adiabatic motion in the current sheet (or potentially, wave noise) on the outflow ion distribution is thus to spread the distributon along circles centred on O', as sketched in the lower diagram in Figure 1.

In the preceding discussion it was tacitly assumed that the field line speed in the Earth's frame $V_F = E_y/B_z$ is a known quantity. However, this of course is not actually the case, the speed must be determined self-consistently from the stress balance condition for the current sheet. Since by the 1-D assumption the field tension (x-z component of the stress tensor) is not balanced by a pressure gradient along the current sheet (as seems reasonable for the distant tail), it must instead be balanced by the dynamic pressure exerted (in the field line "rest" frame) by the inflowing tail lobe plasma. Assuming for simplicity that the plasma

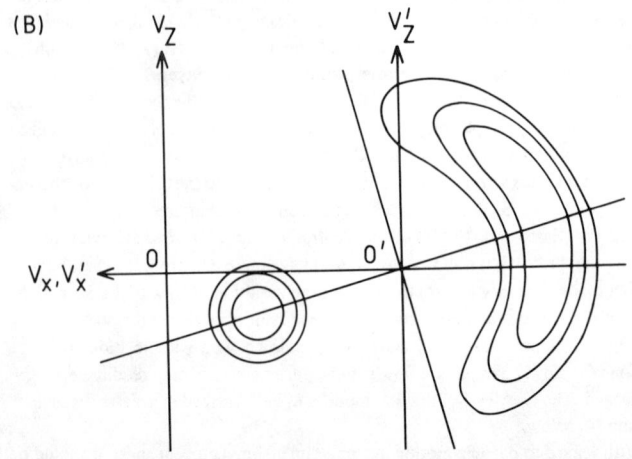

Fig. 1b. Sketch of inflowing and outflowing ion distribution functions based on the discussion of individual particle behaviour. See text for full description [After Cowley and Shull, 1983].

properties on either side of the current layer are identical, so that there is no net field-aligned flow in the field line "rest" frame, the stress exerted by the plasma can be expressed in terms of the stress due to anisotropic pressure and can be written as

$$P_\parallel - P_\perp = \frac{B^2}{\mu_0} \qquad (4)$$

where $P_\parallel$, $P_\perp$ are plasma pressures parallel and perpendicular to the field and B is the field strength in the region just outside the current sheet (eg Rich et al., [1972]; Eastwood [1972]). Generally the electrons will make little contribution to the l.h.s. of (4), in line with previous discussion, and if then the ions may be approximated as cold field-aligned beams of density n and speed v in the field line "rest" frame, then $P_\perp = 0$ and (4) becomes

$$2nm_i v^2 = \frac{B^2}{\mu_0} \text{ or } v = \frac{B}{\sqrt{2\mu_0 nm_i}} = V_A . \qquad (5)$$

For equilibrium, therefore, the ion speed in the field line "rest" frame must be just the Alfvén speed V_A outside the current sheet (where the total density of inflowing and outflowing beams combined is 2n). An opposite limiting approximation for the velocity distribution of the outflow ion beam is obtained by assuming the cold inflow beam is fully isotropized in pitch angle in the current sheet. The outflow distribution does not then contribute to the l.h.s. of (4), so that equations (5) become modified only by eliminating the factor 2 [Cowley and Shull, 1983]. Results are not, therefore, sensitively dependent on detailed assumptions.

Assuming that the field-aligned speed of the inflowing ions in the Earth's frame is known it is now possible to use (2), (3) and (5) to determine the field line speed $V_F = E_y/B_z$ required for equilibrium, and hence to determine the speed of the accelerated ions on either side of the neutral line. In using these equations we are effectively assuming cold ion beams, and this will be the only case treated here. We will also use only the small field angle approximations in (2) and (3). For the region Earthward (E) of the neutral line we have

$$V = V_A = \frac{E_y}{B_{zE}} + V_L$$

so that

$$V_{FE} = \frac{E_y}{B_{zE}} = V_A - V_L \text{ and } V_{BE} = 2V_A - V_L \qquad (6)$$

while for the region tailward of the neutral line we have

$$V = V_A = \frac{E_y}{|B_{zT}|} - V_L$$

so that

$$V_{FT} = \frac{E_y}{B_{zT}} = -(V_A + V_L) \text{ and } V_{BT} = -(2V_A + V_L). \qquad (7)$$

In (6) and (7) V_L is positive for tailward flows, while the signs of V_{FE}, V_{FT}, V_{BE} and V_{BT} have been written such that positive values correspond to Earthward flows and negative values to tailward flows.

In summarizing this and the preceding section it is first to be emphasized that the isotropic pressure equilibria used successfully to model the quiet-time near-Earth (cislunar) tail and plasma sheet are not applicable to the distant tail. Rather, the field line stress in this regime is balanced by the dynamic pressure of the inflowing tail lobe plasma such that the 1-D current sheet will form a useful approximation, at least away from the immediate vicinity of the neutral line. The energy input from the field to the plasma in the current sheet is then taken up mainly by ions, forming field-aligned jets moving away from the X-line on reconnected field lines on either side. For given inflow plasma density n and down-tail bulk speed V_L, the speed of these jets together with the field line speed in the current sheet $V_F = E_y/B_z$ may then be determined from the stress-balance condition (equations (6) and (7)). It should be noted that the analysis determines only the ratio E_y/B_z and not E_y and B_z individually. In other words stress balance determines the speed to which the inflowing ions are accelerated ($\sim 2V_A$), but not how many ions are accelerated to that speed per unit time. The latter is proportional to E_y, ie to the rate of reconnection, and, as found in other studies [eg Vasyliunas, 1975], this is not uniquely determined by local analysis. Once E_y is specified, however, eg by observation or by more global considerations, then B_z can be determined from V_F. In the next section these results will be applied in discussing the Cowley-Southwood model of the distant tail.

4. The Cowley-Southwood Model

In the Cowley and Southwood [1980] model the local 1-D equilibrium conditions (6) and (7) are used to determine the speed of the field lines and accelerated ions as a function of distance along the tail, for given down-tail variations of the field strength and tail lobe plasma bulk parameters. The latter variations are taken to be sufficiently slow that the 1-D approximation remains valid locally. Two solutions are obtained at each point, corresponding to whether the location is taken to be Earthward or tailward of the neutral line. The neutral line is then treated essentially as a discontinuity across which these two solutions are joined at some point downtail. The location of the neutral line is not, of course, determined by the analysis, although limitations are set on where it may occur in a steady state.

In any complete theory, the field and plasma parameters outside the current sheet required for equations (6) and (7) should first be determined from a model describing the tail lobe fields and particle motions therein. However, no model of this nature as yet exists. The most detailed calculation so far presented is the 3-D kinematic model of Cowley [1981], building on initial studies by Pilipp and Morfill [1976, 1978]. The Cowley [1981] model illustrates the essential features of ion motion in an open tail lobe, including the cross-tail asymmetries which are associated with IMF B_y which may be of relevance to ISEE-3 observations discussed in section (6). Here in Figure 2 we give a simpler presentation which illustrates the basic elements. We consider one particular (but quite general) open field line from the time (t = 0) when it maps into the dayside cusp a few minutes after dayside reconnection has occurred to the time 4 hours later when it maps to the tail neutral line and becomes closed. Distance along the field lines, S, is plotted on the vertical axis, while time is plotted on the horizontal axis. The distance S = 0 corresponds to the ionosphere, while the inclined solid line represents the location of the tail magnetopause as the field line is stretched out downtail at an assumed solar wind speed of 400 km s^{-1}. This line does not quite pass through the origin due to the finite ($\sim 10R_E$) distance of the magnetopause at the cusp. The dot-dash lines indicate schematically the tail centre-plane current sheets on either side of the tail neutral line following reconnection. The dashed lines then show the trajectories of field-aligned particles which arrive at an arbitrary point on the field line at an arbitrary time. The lines are marked with the corresponding particle field-aligned speed, again taken positive down-tail. It is seen that above a certain down-tail speed (~ 120km s^{-1} for the point chosen) particles arriving at 0 originate in the ionosphere, while for smaller and negative (Earthward) speeds the particles originate at various locations along the magnetopause. In order to determine the velocity distribution at 0, therefore, the source distributions must be known over wide regions of the magnetopause and ionospheric boundaries. It should be noted, however, that in the absence of pitch angle scattering ionospheric particles will be closely confined to the field direction at 0. At larger pitch angles essentially all particles will originate at the magnetopause, those arriving with large positive speeds doing so after mirroring near the Earth. It should also be noted that it is not entirely satisfactory to simply take a magnetosheath distribution as the source distribution at the magnetopause, since as pointed out above, the particles will change energy as they interact with and cross the magnetopause current sheet.

If the polar cap ionospheric ion source is confined to relatively low energies, as is probably generally the case then these particles will usually not form an important component of the lobe plasma in the distant tail, since these ions will convect into the tail centre plane relatively close to the Earth. More significant are tailward-flowing magnetosheath ions originating at the magnetopause. Consideration of Figure 2 then readily shows that such particles having a down-

232   DISTANT GEOMAGNETIC TAIL

across the neutral line at some $X > X_c$. The third panel shows the accelerated ion beam speeds V_{BE} and V_{BT}, together with V_L for comparison. Earthward of the neutral line the ions are accelerated to speeds of several 100 km s^{-1}, while tailward of the neutral line the outflow speed is nearly constant with distance at $\sim$600 km s^{-1}. the fourth and final panel shows B_{zE} and B_{zT}, computed from the field line speeds V_F by taking $E_y = 0.15$ mV m^{-1} ($\sim$50 kV across a $\sim$50 R_E diameter tail). B_{zE} becomes large near X_c such that the model then breaks down in this vicinity since (6) and (7) require $|B_z| \ll |B_x|$ outside the current sheet for their validity. Tailward of the neutral line B_{zT} is roughly constant at $\sim$–0.3 nT.

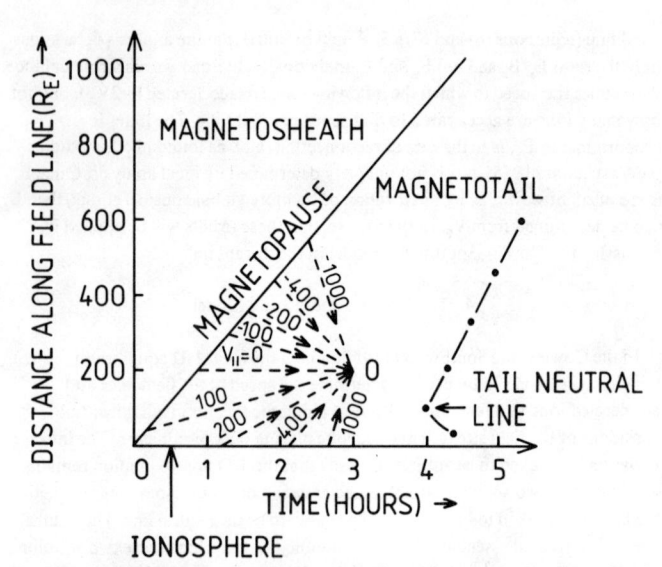

Fig. 2. Diagram demonstrating the "mapping" problem required to determine the magnetotail lobe ion distribution function. One particular (but quite general) open field line is considered as it convects over the polar cap. Distances along this field line from the ionosphere to the magnetopause are plotted vertically, against time horizontally. Time zero corresponds to the point when the newly opened dayside field line maps into the dayside cusp (when $s \approx 10 R_E$), and the line is followed until after it reconnects in the tail some 4 hours later. The solid inclined line represents the distance of the magnetopause along the field line, as the line is stretched out in the tail at an assumed steady solar wind speed of 400 km s^{-1}. The dot-dash lines indicate schematically the tail-centre current sheets following reconnection.

The paths of field-aligned particles arriving at arbitrary point 0 at some particular time are then shown by the dashed lines. They are marked with particle speed in km s^{-1}; positive speed indicates tailward motion. It is seen that, depending on the particle speed, the ions originate from wide regions of the tail magnetopause and ionosphere.

tail speed which is a fraction F of the external magnetosheath speed will fill the tail lobe and start entering the central current sheet at down-tail distance exceeding F times the total tail length. In Figure 2, for example, the tail length is 900 R_E (400 km s^{-1} times 4 hours). Particles starting at the magnetopause near the Earth with a downtail speed 200 km s^{-1} (half the sheath speed) will have travelled 450 R_E (half the tail length) when the field line is reconnected after 4 hours. Only beyond distances of $\sim$450 R_E, therefore, will 200 km s^{-1} magnetosheath ions fill the lobe and start entering the current sheet. Ions with speed 100 km s^{-1} will enter beyond distances of $\sim$225 R_E (one quarter of the total tail length) and so on. Clearly the result will be that the plasma entering the current sheet will increase in both density and down-tail bulk speed with increasing distance from Earth, until solar wind-like conditions prevail at extreme distances.

In the absence of a detailed model, the calculations presented by Cowley and Southwood [1980] used only the simplest assumed downtail variations of plasma bulk parameters in line with the above discussion, and with the observations to be outlined in subsequent sections. Both the inflow plasma density and the downtail bulk speed were taken to vary linearly with distance, from zero near the Earth to $n = 1$ cm^{-3} and $V_L = 400$ km s^{-1} respectively at 1000 R_E. The tail lobe field strength was taken to be uniform at 8nT. The results are shown in Figure 3. The top panel shows the assumed down-tail inflow plasma speed V_L and the Alfvén speed V_A defined by (5) plotted versus X. An immediate feature is that beyond a certain distance, $X_c \approx -460 R_E$ in this model, the lobe plasma speed exceeds the Alfvén speed. According to (6) therefore, V_{FE} is then negative, implying that closed flux tubes beyond this distance would be blown down tail by the dynamic pressure of the inflowing plasma. A similar feature appears in the work of Hill and Reiff [1980b]. In a steady-state, therefore, the neutral line must lie Earthward of X_c. The second panel shows V_{FE} and V_{FT} versus distance, and we imagine an almost discontinuous transition from the upper curve to lower curve

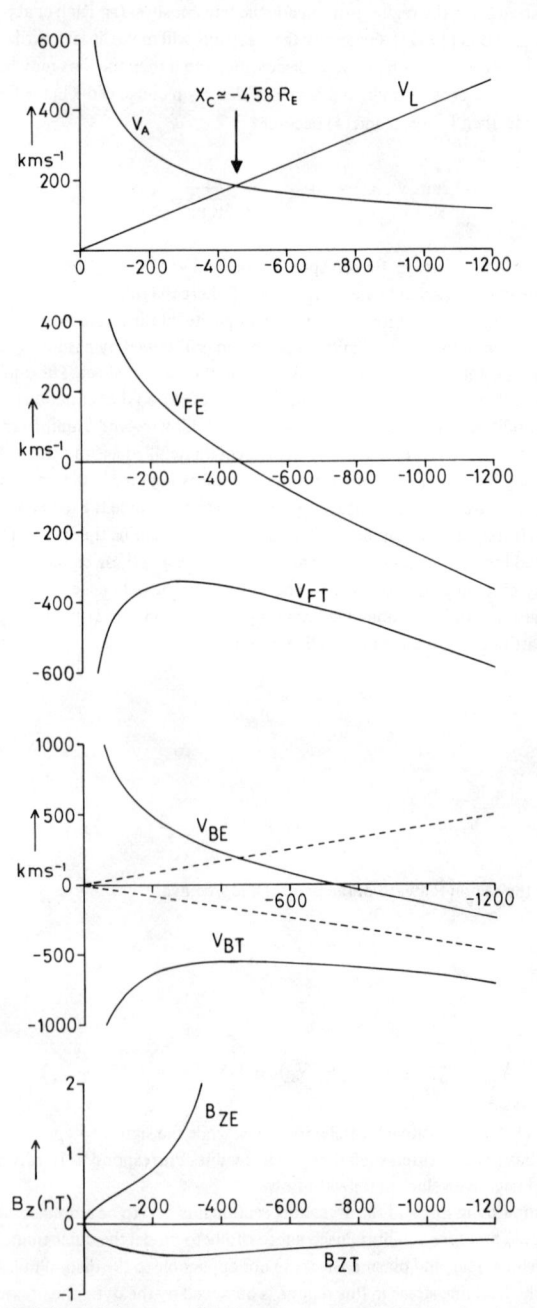

Fig. 3. Variation of plasma and field parameters in the Cowley-Southwood model of the distant tail. The figure has been adapted from Figure 1 of Cowley and Southwood [1980] to bring symbols into line with the present text. The top panel shows the assumed down-tail bulk speed V_L of the plasma convecting into central current sheet, together with the Alfvén speed V_A. Subsequent panels show the field line speed V_F, the accelerated beam speed V_B and the B_z field threading the current sheet for regions either Earthward (E) or Tailward (T) of the tail neutral line.

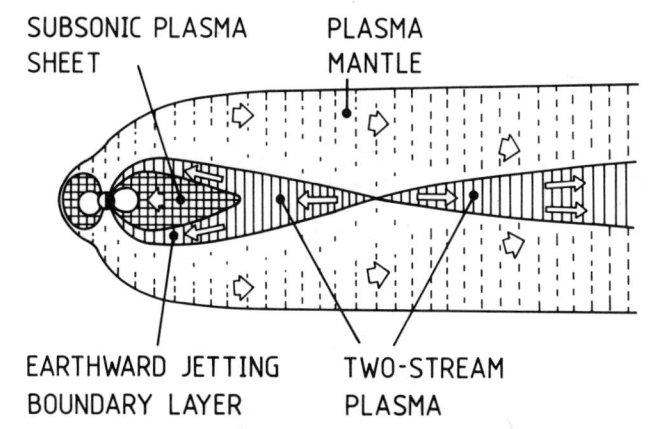

Fig. 4. Sketch of the form of the accelerated plasma populations downstream from the tail neutral line when steady reconnection is occurring.

5. Structure of the Accelerated Plasma Populations

In this section we will now discuss the spatial structure of the accelerated plasma populations to which models of the Cowley-Southwood type give rise on reconnected field lines downstream from the neutral line. The qualitative features are shown in Figure 4, following the discussion given by Cowley [1980] and Cowley and Southwood [1980].

In the region tailward of the neutral line the accelerated ion beam will form a wedge-shaped layer of fast tailward-flowing plasma across the tail centre plane. Outside the current layer these accelerated ions will flow through the slower incoming tail lobe plasma out of which it is formed, so that a "two stream" plasma will be produced. Exactly similar conditions will occur with Earthward flowing accelerated plasma and the tail lobe plasma for extended distances Earthward of the neutral line as shown in the figure, but this regime will be terminated in the near-Earth tail due to mirroring of the accelerated ions as outlined in section (2). The expected extent of the simple two-stream region Earthward of the neutral line may be illustrated eg by taking the neutral line to lie $\sim 250\, R_E$ down-tail in the Cowley and Southwood [1980] model illustrated in Figure 3. In this case mirrored accelerated ions first return to the current sheet only at the lunar distance $X \approx -60 R_E$, so that the simple situation in the absence of mirrored particles will prevail for a large distance downstream from the neutral line. Within the region where mirrored particles are present the plasma is observed to become rapidly isotropized to produce the usual near-Earth plasma sheet population [Forbes et al., 1981]. The Earthward-jetting ion beams will then form a layer on the outer surface of the isotropized region as illustrated in Figure 4. The thickness of this layer at a particular distance down tail may be estimated by combining the time of flight of the beam ions to the Earth and back with the north-south drift of the tail lobe field lines E_y/B_x. With $E_y = 0.15$ mV m^{-1} as before and $B_x \approx 15$ nT, say, the latter speed is ~ 10km s^{-1}. For beam ions of ~ 500 km s^{-1} the return time of flight from eg 25 R_E down-tail is ~ 10 min, leading to a layer thickness at this distance of $\sim 1 R_E$. Clearly the jetting ion layer thickness goes to zero near the Earth, while with increasing distance down-tail it widens at the expense of the isotropized plasma region until the latter disappears, usually within the near-Earth tail as shown in Figure 4.

Although the near-Earth tail is not the main concern of this paper it is worth pointing out here that Earthward-directed ion jets (together with tailward-moving lobe particles, corresponding to our "two-stream" plasma) are indeed observed on the outer surface of the expanding near-Earth plasma sheet during substorm recovery, and at other times as well [Lui et al., 1977, 1983; DeCoster and Frank, 1979; Forbes et al., 1981, Eastman et al., 1983a,b]. During the substorm expansion phase when the neutral line lies close to the Earth, however, fast tailward and Earthward flowing ion beams will lie at the centre plane in thin layers on either side of the neutral line, and the isotropic plasma sheet will be absent except very close to the Earth. We relate these current sheet-accelerated ion beams to the substorm-associated rapid flows reported eg by Hones et al. [1972], Hones [1973, 1977, 1980], Frank et al. [1976], Lui et al. [1978, 1979], Hones and Schindler [1979], Lui [1980], Coroniti et al. [1980], Fairfield et al. [1981], Nishida et al. [1981] and Hayakawa et al. [1982]. This association was first suggested by Jaeger and Speiser [1974]. The relationship between our simple model and observed flows in the distant tail will be discussed in more detail in the next section.

One point which has not perhaps been sufficiently emphasized theoretically in the past is that due to the finite speed of the accelerated ions along the field lines, the beam will not in fact fill the whole width of the region of reconnected field lines at any distance down stream from the neutral line [Hill, 1975]. This effect may also be viewed as being due to the small $\underline{E} \wedge \underline{B}$ drift of the accelerated ions which causes them to flow at a small angle relative to the reconnected field (see Figure 1).

For a given E_y the external field angle to the current sheet is given in terms of the field line speed V_F by

$$\theta \approx \frac{B_z}{B} = \frac{E_y}{|V_F|B}.$$

For the regions Earthward and tailward of the neutral line these become explicitly (using (6) and (7))

$$\theta_E \approx \frac{E_y}{(V_A - V_L)B} \quad \text{and} \quad \theta_T \approx \frac{E_y}{(V_A + V_L)B}. \qquad (8)$$

For a particle of field-aligned speed V in the Earth's frame the angle χ at which it moves relative to the field outside the current sheet due to the $\underline{E} \wedge \underline{B}$ drift E_y/B is then

$$\chi \approx \frac{E_y}{VB} \approx \frac{|V_F|}{V} \theta,$$

where we note that the $\underline{E} \wedge \underline{B}$ drift is directed towards the current sheet so that the flow is deflected in that direction away from the field (see Figure 5). The angle of flow relative to the current sheet is then

$$\phi = \theta - \chi \approx \left(1 - \frac{|V_F|}{V}\right) \theta. \qquad (9)$$

Note that angle $\phi = 0$ for $V = |V_F|$ and that $\phi \to \theta$ as V increases to become much larger than $|V_F|$, as expected. The particular cases of special interest are, of course, where V is taken to be the accelerated ion beam speeds given in (6) and (7). We then have

$$\phi_E \approx \frac{\theta}{\left(2 - \frac{V_L}{V_A}\right)} \quad \text{and} \quad \phi_T \approx \frac{\theta}{\left(2 + \frac{V_L}{V_A}\right)}. \qquad (10)$$

Similar expressions have previously been derived by Hill [1975] (his equations (38)-(41)). It should be noted that if V_L is small we have $\phi_E \approx \phi_T \approx \theta/2$ ie the beam will fill just half the region of reconnected field lines outside the current sheet. This is shown schematically in Figure 5a where the solid lines are magnetic field lines, while the hatched area indicates the region occupied by the ion beam in the case where $V_L = 0$. The thickness of the current layer is not represented for sake of clarity. Figure 5b illustrates the asymmetry introduced by tailward flow of the inflow plasma, specifically for the case $V_L = V_A/2$. The field angle, B_z, and the thickness of the wedge of reconnected field lines are then larger in the region Earthward of the neutral line (left of the figure) than tailward of it, while the speed of the accelerated plasma is larger tailward of the neutral line than Earthward of it (the difference being just $2V_L = V_A$ in the case illustrated). The ion beams occupy rather more than half the reconnected field wedge Earthward of the neutral line, and rather less than half in the tailward region (for $V_L = V_A/2$, $\phi_E \approx 2\theta_E/3$ and $\phi_T \approx 2\theta_T/5$).

In the above discussion it has effectively been assumed that the accelerated ion beam is essentially mono-energetic. This will, of course, not really be the case,

234 DISTANT GEOMAGNETIC TAIL

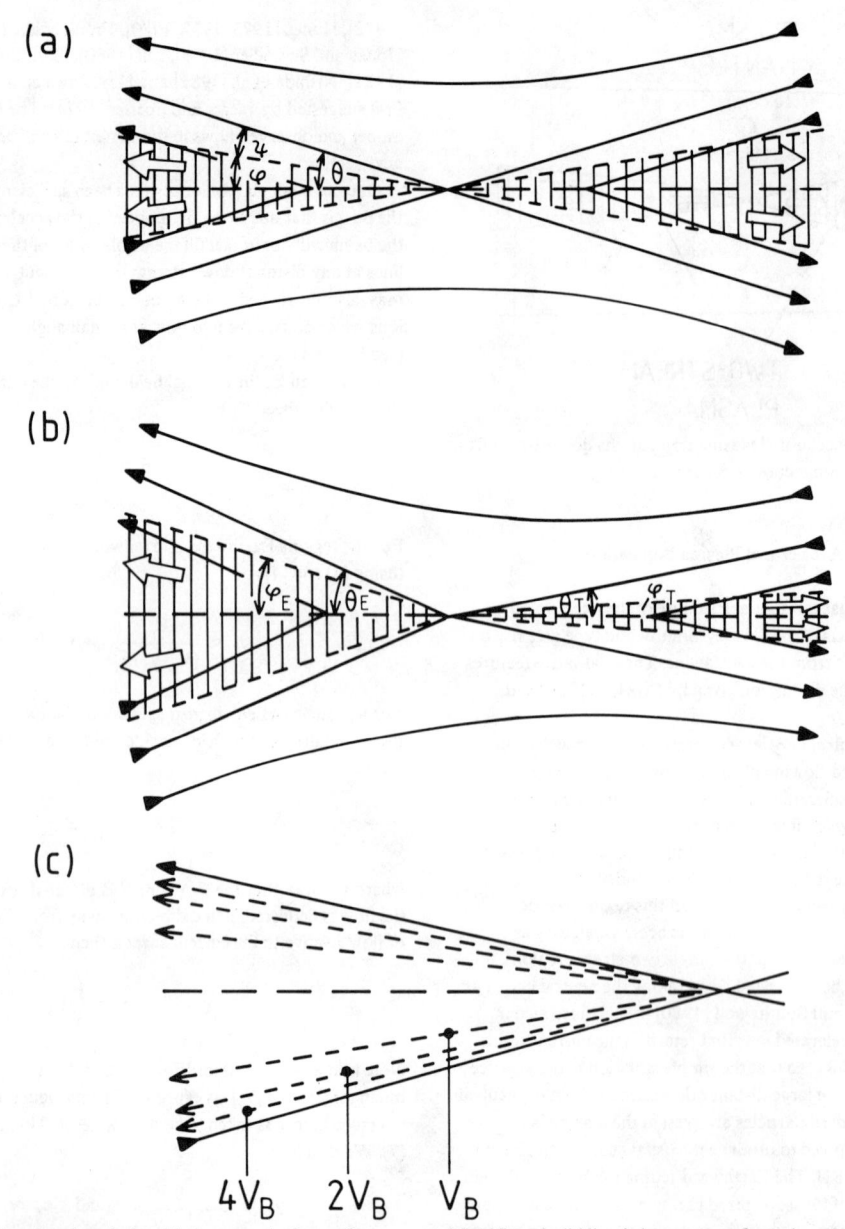

Fig. 5. Sketches showing the relationship between the layer of reconnected field lines (solid lines) downstream from the neutral line and the beam of accelerated tail lobe ions (shown hatched). Case (a) shows the symmetrical situation which obtains for zero down-tail bulk speed of the inflow plasma. The ion beam occupies just half the reconnected field wedge outside the current sheet. (See also Hill [1975], Figure 5). Case (b) shows the situation when the lobe plasma entering the current sheet flows tailward (left to right) with a speed $V_L = V_A/2$. The field angles are increased Earthward of the neutral line and reduced tailward of it, while the beam occupies more than half the reconnected field wedge in the former region and less than half in the latter. In sketch (c) we show for the case $V_L = 0$ the regions occupied by particles having speeds greater than the nominal beam speed, illustrating how energetic particles can form a layer extending outside the main ion beam, but within the reconnected field wedge. The field and beam angles in this figure have been exaggerated for sake of clarity.

and the spread of field-aligned ion speeds produced by the current sheet source will then be translated by the $\underline{E} \wedge \underline{B}$ drift into a spatial dispersion of velocities at the edge of the beam within the region of reconnected field lines. As this region is entered across the magnetic separatrix, high energy particles produced in the current sheet will be observed first, followed by progressively lower energy ions until the beam proper is reached roughly half way through the reconnected field layer. This is illustrated in Figure 5c where the regions occupied by ions of various field-aligned speeds are shown for the case $V_L = 0$. The speeds are given in terms of the beam speed given by (6) or (7) ie $V_B = 2V_A$. The main conclusion to be reached from these results is that a significant region containing high-energy particles produced in the current sheet can exist outside and adjacent to the region occupied by the main ion beams. In the near-Earth tail this conclusion relates directly to the energetic particle beams (tens of keV and above) observed at the outer surface of the plasma sheet eg by Mobius et al. [1980], Spjeldvik and Fritz [1981], Andrews et al. [1981] and Williams [1981].

Finally, we can use the above results to make estimates of the width of the region of reconnected field lines downstream from the neutral line, and the thickness of the ion jet. If we consider for simplicity the region where V_L is small compared with V_A, and take eg $E_y \approx 0.15$ mV m^{-1}, $B \approx 10$nT and $V_A \approx 250$ km s^{-1} (see eg Figure 3) then from (8) we find $\theta \approx 3.5$ degrees (ie $B_z \approx 0.6$ nT).

The half thickness of the reconnected field region is then ~0.6 R_E (~4000km) for every 10 R_E distance from the neutral line, while the thickness of the region of accelerated ions will be roughly half this value. The width of these regions, say, ~100 R_E from the neutral line will thus be of order a few Earth radii. The width of the central current sheet itself, however, is likely to be rather less than this, roughly an accelerated ion larmour diameter in the tail lobe field, ~500km.

6. Distant Tail Observations by Pioneers 7 and 8 and ISEE-3

We now turn to briefly review the *in situ* observations made in the distant tail, and their interpretation in terms of the theoretical picture given above. The majority of this discussion will concern the Pioneer 7 and 8 data obtained in 1966 and 1968 at down-tail distance ranges of 720-1050 R_E and 450-580 R_E respectively. Only a few preliminary results are avilable from ISEE-3 at the present time.

It is first worthwhile to briefly review the capabilities of the Pioneer instrumentation from which tail data have been published. Pioneer 7 carried two ion instruments intended mainly as solar wind detectors, which covered the energy range ~100 eV to 10 keV and gave detailed angular information, particularly in the solar direction. One instrument, provided by Ames Research Center (ARC), was an electrostatic analyser [Wolfe et al., 1967; Intriligator et al., 1969, 1972], while the other was a Faraday cup provided by MIT [Walker et al., 1975; Villante and Lazarus, 1975; Villante, 1974, a,b; 1975]. Both were capable of providing ion distributions with a time resolution of ~1 minute. Pioneer 7 also carried a GSFC monoaxial magnetometer providing vector information with 1.5 sec resolution [Ness et al., 1967; Fairfield, 1968; Ness 1969; Villante 1974 a, b; 1975; 1976]. Pioneer 8 similarly carried an ARC electrostatic analyser measuring ions from 150 eV to 15 keV with 1 minute resolution, [Intriligator et al., 1969; Wolf and Intriligator, 1970; Siscoe et al., 1970], a GSFC magnetometer having 1.3 sec time resolution [Mariani and Ness, 1969; Bavassano et al., 1974; Villante 1974b] and a rudimentary VLF electric field experiment measuring the broadband (100 Hz-100 kHz) electric field with 7.5 min time resolution and 400 Hz (± 15%) noise with 7 sec resolution [Scarf et al., 1970, Siscoe et al., 1970]. Both spacecraft therefore made detailed observations of the magnetic field and plasma ion environment in the distant tail, while only rudimentary VLF wave information was returned, and no information has been available regarding either the plasma electrons or energetic particles. With regard to the particle data it may be noted that precisely the converse situation applies to tail data from ISEE-3, in that detailed thermal electron and energetic particle data have been obtained, while the thermal ion detector was no longer operational at the time of the magnetotail passes. It may also be noted that the references given above represent, to the best of our knowledge, a complete bibliography of published Pioneer deep-tail observations.

The deep-tail data returned by the Pioneer probes contain many similarities and also many differences from that typical of the near-Earth (cislunar) region. A major similarity is that there exist regions, even at 1000 R_E, where the field has a similar character to that in the near-Earth tail lobes *ie* it is elevated in strength relative to the magnetosheath, and points in a steady direction either toward or away from the Earth. A major difference is that such tail-like fields at large distances are observed continuously only for relatively short intervals, typically ~15 min to ~4 hours at Pioneer 7, possibly for rather longer durations at Pioner 8 closer to the Earth. This led early investigators to suggest that the tail had become filamentary and turbulent at these distances [Ness et al., 1967; Wolfe and Intriligator, 1970], a conclusion strengthened by an apparent lack of correspondence between "in-tail" intervals deduced separately from magnetometer data and from the presence of "disturbed" ion spectra [Intriligator et al., 1972].

Later work showed, however, that two factors contribute significantly to the latter lack of correspondence. Firstly it was found that the tail lobes at these distances were nearly always filled with tailward streaming plasma which at times reaches magnetosheath-like intensities. More typically, however, the density is reduced compared with that in the magnetosheath (n ≈ 1 cm^{-3} as compared with ~5 cm^{-3} in the magnetosheath), and the down-tail bulk speed is also reduced, but only by ~70 km s^{-1} [Walker et al., 1975; Villante and Lazarus,

1975]. These results may now seem rather unsurprising, in terms *eg* of the discussion given in section (4), but it should be remembered that at the time of writing *eg* of the paper by Intriligator et al. [1972] the plasma mantle had yet to be discovered in the near-Earth tail [Rosenbauer et al., 1975].

The second factor is that clear field reversal regions separating northern and southern tail lobes were later identified, where the plasma data is unlike that in the magnetosheath, but where the field does not in general point toward or away from the Earth [Walker et al., 1975; Villante, 1976]. As a result of these and other studies (such as that by Bavassano et al., [1974] who investigated the systematics of magnetopause normal directions), later authors argued strongly that the Pioneer data is, in fact, consistent with a well-organized tail having a similar basic structure to that observed near the Earth and a diameter of 60-90 R_E [Bavassano et al., 1974; Villante, 1976; Walker et al., 1975]. Fairfield [1968] had earlier suggested that the rather fleeting nature of Pioneer tail encounters might be a result of changes in solar wind flow speed and/or direction, a suggestion later directly supported by the results obtained by Walker et al. [1975] for Pioneer 7 and Bavassano et al. [1974] for Pioneer 8.

We now turn to the crucial question of the nature of the plasma regime in the vicinity of the field reversal regions at these distances. Basically, it was found that these regions are characterized by the presence of rapid tailward flows, considerably exceeding the speed of flows in the adjacent magnetosheath. As a field reversal region is approached, a high-speed tailward-directed ion beam was generally observed, flowing through the slower plasma which pervades the tail lobes [Walker et al., 1975; Villante and Lazarus, 1975]. An interpretation in terms of our 'two stream' plasma region tailward of the neutral line discussed in section (5) and illustrated in Figure 4 is obvious, as first suggested by Cowley [1980] and Cowley and Southwood [1980]. An example is given in Figure 6, which shows field (top panel) and plasma data (lower two panels) from Pioneer 7 on September 25, 1966, and taken from Figure 3 of Walker et al. [1975]. When the two beams are present two velocities are indicated in the bottom panel, and also two densities in the middle panel, the density of the lower velocity peak being given by the dashed line. It should be noted that the speed of the fast stream in this data is typically 600 km s^{-1}, a value closely compatible with the results of the Cowley-Southwood model shown in Figure 3. This indicates that the observed acceleration is at least roughly consistent with the requirements of stress balance tailward of the neutral line.

It is also seen in Figure 6, however, that the double stream often does not extend fully within the field reversal region itself. Instead in the region of weak field, the two streams coalesce to produce a 'hot' tailward flowing plasma having a bulk speed comparable with, but sometimes rather less than that of the fast stream [see *eg* Villante and Lazarus, 1975]. These regions are accompanied by large (few nT) and variable B_z, suggestive of the presence of multiple field nulls and associated closed flux loops [Villante, 1975, 1976], possibly substorm-associated. These values of B_z may be compared with the 1-D current sheet Cowley-Southwood estimate of ~-0.3 nT for the region tailward of the neutral line and shown in Figure 3, which should represent, roughly, an average value of the B_z field threading the current sheet at these distances. However, in a detailed study of 14 Pioneer 7 field reversals Villante [1976] reported an average B_z field of 1.5 nT. This result is considerably at odds with expectations, and with the interpretation of the plasma data given above. The results of Villante's [1976] study appear to be severely compromised, however, by the inclusion of at least two clear magnetopause crossings in the set of field reversal regions. These are the current sheets observed at 1036 UT on 25 September (see Figure 6) and at 1733 UT on 26 September [see Ness 1969]. In both cases the magnetosheath field outside the magnetopause has a substantial positive B_z component such that these cases probably contribute importantly to the large average B_z field reported above.

We conclude this section with a few remarks on initial results obtained from the ISEE-3 deep-tail pass out to distances ~220 R_E. In this case plasma ion data are not available, but plasma densities and bulk speeds have been deduced from thermal electron data for two 12 hour intervals in January 1983 by Bame et al. [1983]. Detailed energetic particle magnetic field and wave data have also been obtained [Scholer et al., 1983; Tsurutani et al., 1983]. A principal result of the electron observations is that, as in the Pioneer observations, the "plasma sheet"

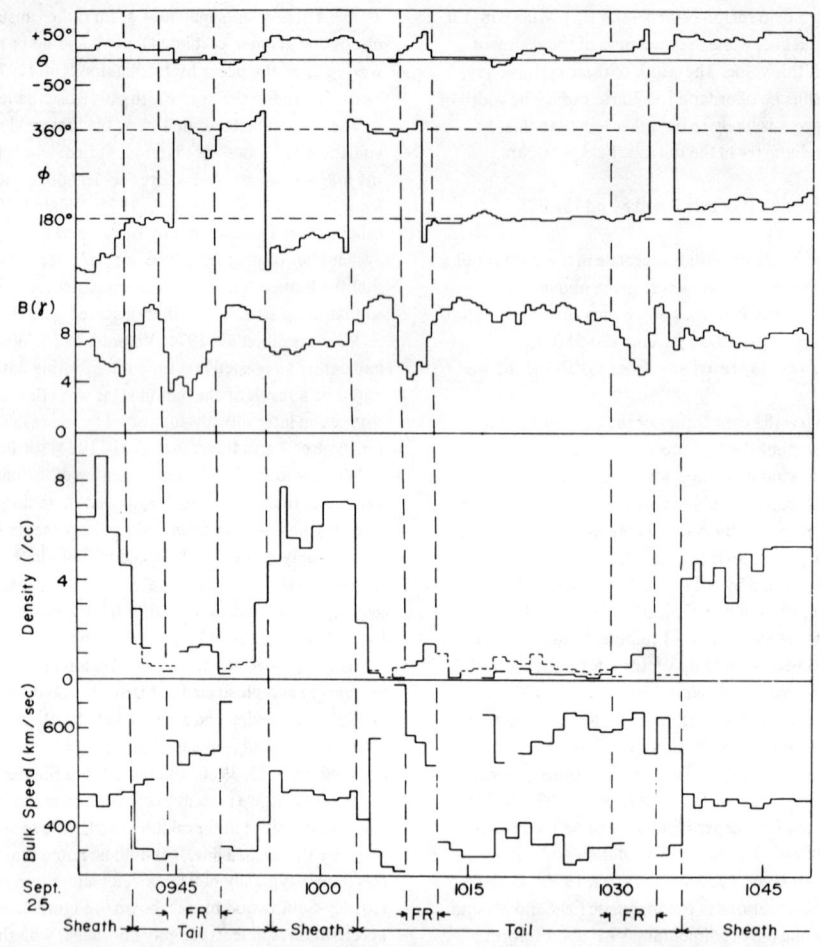

Fig. 6. Field and plasma data obtained by Pioneer 7 for tail crossings on September 25, 1966. Parameters from individual plasma spectra (taken every 70 sec) and 30 sec averages of the magnetic field are shown. θ and φ are the elevation and azimuth of the field respectively; θ is positive for north pointing fields, and φ = 180° and 360° for tailward and Earthward pointing fields respectively. Vertical lines divide the figure into tail and sheath regions, the former being further sub-divided to indicate observation of a field reversal (FR) region. When double-peaked velocity distributions are observed the two velocities are indicated, and the density corresponding to the lower velocity peak indicated by a dashed line. [After Walker et al., 1975].

population observed near field reversal regions generally streams rapidly tailwards. Intervals of consistent Earthward flow are observed at ~200 R_E, but are relatively rare. This result again indicates that the tail neutral line generally lies Earthward of ~200 R_E. This conclusion is supported in initial analysis of the magnetometer data [Tsurutani et al., 1983], which shows that the appearance of negative B_z within the plasma sheet region increases in relative frequency with increasing distance from Earth, such that at ~200 R_E the average B_z becomes negative. Outside the plasma sheet in the magnetic tail lobes tailward-streaming plasma of highly variable density is observed at these distances. During the two intervals presented by Bame et al. [1983] a large asymmetry in the plasma density was observed between northern and southern lobes, suggestive of the IMF B_y effect mentioned in section (4), and as previously reported at the lunar distance by Hardy et al. [1979].

Energetic (> 35 keV) ion data for the above two intervals discussed by Bame et al. [1983] have recently been presented by Cowley et al. [1984]. The flows seen in these particles are generally quite consistent with those deduced from the electron measurements. In particular, essentially unidirectional tailward ion streaming is observed within the tailward-streaming thermal "plasma sheet". Energetic fluxes, however, often extend well outside the region occupied by "plasma sheet" electrons, suggestive of the plasma sheet energetic particle boundary layer discussed in section (5) and illustrated in Figure 5. This layer is thus believed to be the counterpart tailward of the neutral line of that observed at the outer surface of the near-Earth plasma sheet as mentioned in section (5). In the magnetic tail lobes proper energetic ion fluxes are essentially identical in northern and southern lobes despite the extreme differences in thermal electron properties reported by Bame et al. [1983]. Fluxes in this regime are generally much lower than those observed in the "plasma sheet" and its environs, and are peaked perpendicular to the magnetic field.

7. Summary

Although the geomagnetic tail seems to retain a coherent structure out to distances of at least 1000 R_E, the plasma and field properties observed beyond ~100 R_E are considerably different from those usually observed in the near-Earth (cislunar) tail. The first difference is the considerable development with increasing distance of the density and bulk speed of the plasma in the magnetic tail lobes. This is due to inward convection of magnetosheath plasma entering at the magnetopause boundary towards the centre plane of the tail. At 1000 R_E this plasma fills essentially the entire tail lobe but remains rather less dense than that observed in the adjacent magnetosheath. At 200 R_E regions of the tail lobes remain in which the density is very low (< 0.1 cm^{-3}).

The second major difference concerns the energized plasma component observed at the centre plane of the tail. At 500 and 1000 R_E this plasma consistently flows rapidly tailwards, as is also usually the case at 200 R_E as observed by ISEE-3. These observations contrast sharply with the subsonic hot "plasma sheet" typical of the quiet-time near Earth tail, and suggest that at nearly all times the tail neutral line lies at distances of $\sim 100\ R_E$, or Earthward thereof (during substorms).

Theoretically, the reconnection model of the magnetosphere as applied to the geomagnetic tail provides a consistent framework within which much tail data can be interpreted in at least a qualitative and in part a quantitative fashion. Simple ideas consistently applied lead to natural descriptions of the formation of the hot subsonic plasma sheet in the near-Earth tail, the Earthward flow layer on its outer boundary, as well as the "two-stream" tailward-flowing plasma observed near field reversal regions in the distant tail. The relationship between the energized thermal plasma regimes and the exterior layers of energetic particles is also readily explained. At the same time, only the consequences of a very simple and essentially steady model have been addressed here. There is much scope remaining for introducing further complications, arising for example from time-dependent (*eg* substorm) processes, whose effects will certainly be present in observations.

References

Alfvén, H., Some properties of magnetospheric neutral surfaces, *J. Geophys. Res., 73*, 4379, 1968.

Andrews, M. K., P. W. Daly, and E. Keppler, Ion jetting at the plasma sheet boundary: simultaneous observations of incident and reflected particles, *Geophys. Res. Lett., 8*, 987, 1981.

Axford, W. I., Flow of mass and energy in the solar system, in *Physics of Solar Planetary Environments*, edited by D. J. Williams, p. 270, AGU, Washington, D.C., 1976.

Bame, S. J., R. C. Anderson, J. R. Asbridge, D. N. Baker, W. C. Feldman, J. T. Gosling, E. W. Hones, Jr., D. J. McComas, and R. D. Zwickl, Plasma regimes in the deep geomagnetic tail: ISEE-3, *Geophys. Res. Lett., 10*, 912, 1983.

Bavassano, B., F. Mariani, and U. Villante, Far magnetospheric field observations by pioneer 8: the distant bow shock and the extended tail, *Space Res., 14*, 403, 1974.

Bird, M. K., and D. B. Beard, The self-consistent geomagnetic tail under static conditions, *Planet. Space Sci., 20*, 2057, 1972.

Birn, J., R. R. Sommer, and K. Schindler, Open and closed magnetospheric tail configurations and their stability, *Astrophys. Space Sci., 35*, 389, 1975.

Birn, J., R. R. Sommer, and K. Schindler, Self-consistent theory of the quiet geomagnetic tail in three dimensions, *J. Geophys. Res., 82*, 147, 1977.

Coroniti, F. V., L. A. Frank, D. J. Williams, R. P. Lepping, F. L. Scarf, S. M. Krimigis, and G. Gloeckler, Variability of plasma sheet dynamics, *J. Geophys. Res., 85*, 2957, 1980.

Cowley, S. W. H., A self-consistent model of a simple magnetic neutral sheet system surrounded by a cold, collisionless plasma, *Cosmic Electrodyn., 3*, 338, 1973.

Cowley, S. W. H., A note on the motion of charged particles in one-dimensional magnetic current sheets, *Planet. Space Sci., 26*, 539, 1978.

Cowley, S. W. H., Plasma populations in a simple open model magnetosphere, *Space Sci. Rev., 25*, 217, 1980.

Cowley, S. W. H., Magnetospheric asymmetries associated with the y-component of the IMF, *Planet, Space Sci., 29*, 79, 1981.

Cowley, S. W. H., The causes of convection in the earth's magnetosphere: a review of developments during the IMS, *Rev. Geophys. Space Phys., 20*, 531, 1982.

Cowley, S. W. H., and R. Pellat, A note on adiabatic solutions of the one-dimensional current sheet problem, *Planet. Space Sci., 27*, 265, 1979.

Cowley, S. W. H., and D. J. Southwood, Some properties of a steady state geomagnetic tail, *Geophy. Res. Lett., 7*, 833, 1980.

Cowley, S. W. H., and P. Shull, Jr., Current sheet acceleration of ions in the geomagnetic tail and the properties of ion bursts observed at the lunar distance, *Planet Space Sci., 31*, 235, 1983.

Cowley, S. W. H., R. J. Hynds, I. G. Richardson, P. W. Daly, T. R. Sanderson, K. -P. Wenzel, J. A. Slavin and B. T. Tsurutani, Energetic ion regimes in the deep geomagnetic tail: ISEE-3, *Geophys. Res. Lett.*, in press, 1984.

DeCoster, R. J., and L. A. Frank, Observations pertaining to the dynamics of the plasma sheet, *J. Geophys. Res., 84*, 5099, 1979.

DeHoffman, F., and E. Teller, Magneto-hydrodynamic shocks, *Phys. Rev., 80*, 692, 1950.

Dessler, A. J., Length of magnetospheric tail, *J. Geophys. Res., 69*, 3913 1964.

Dungey, J. W., Conditions for the occurrence of electrical discharge in astrophysical systems, *Phil. Mag., 44*, 725 1953.

Dungey, J. W., Interplanetary magnetic field and the auroral zones, *Phys. Rev. Lett., 6*, 47, 1961.

Dungey, J. W., The length of the magnetospheric tail, *J. Geophys. Res., 70*, 1753, 1965.

Eastman, T. E., L. A. Frank, and C. Y. Huang, The boundary layers as the primary transport regions of the earth's magnetotail, submitted to *J. Geophys. Res.*, 1983a.

Eastman, T. E., L. A. Frank, W. K. Peterson, and W. Lennartsson, The plasma sheet boundary layer, submitted to *J. Geophys. Res.*, 1983b.

Eastwood, J. W., Consistency of fields and particle motion in the 'speiser' model of the current sheet, *Planet. Space Sci., 20*, 1555, 1972.

Eastwood, J. W., The warm current sheet model and its implications on the behaviour of the geomagnetic tail, *Planet. Space Sci., 22*, 1641, 1974.

Fairfield, D. H., Simultaneous measurements on three satellites and the observation of the geomagnetic tail at 1000 R_E, *J. Geophys. Res., 73*, 6179, 1968.

Fairfield, D. H., E. W. Hones, Jr., and C. -I. Meng, Multiple crossings of a very thin plasma sheet in the earth's magnetotail, *J. Geophys. Res., 86*, 11189, 1981.

Forbes, T. G., E. W. Hones, Jr., S. J. Bame, J. R. Asbridge, G. Paschmann, N. Sckopke and C. T. Russell, Evidence for the tailward retreat of a magnetic neutral line in the magnetotail during substorm recovery, *Geophys. Res. Lett., 8*, 261, 1981.

Francfort, P., and R. Pellat, Magnetic merging in collisionless plasmas, *Geophys. Res. Lett., 3*, 433, 1976.

Frank, L. A., K. L. Ackerson, and R. P. Lepping, On hot tenuous plasmas, fireballs, and boundary layers in the earth's magnetotail, *J. Geophys. Res., 81*, 5859, 1976.

Hardy, D. A., H. K. Hills, and J. W. Freeman, Occurrence of the lobe plasma at lunar distance, *J. Geophys. Res., 84*, 72, 1979.

Hayakawa, H., A Nishida, E. W. Hones, Jr., and S. J. Bame, Statistical characteristics of plasma flow in the magnetotail, *J. Geophys. Res., 87*, 277, 1982.

Hill, T. W., Magnetic merging in a collisionless plasma, *J. Geophys. Res., 80*, 4689, 1975.

Hill, T. W., and P. H. Reiff, On the cause of plasma sheet thinning during magnetospheric substorms, *Geophys. Res. Lett., 7*, 177, 1980a.

Hill, T. W., and P. H. Reiff, Plasma sheet dynamics and magnetospheric substorms, *Planet. Space Sci., 28*, 363, 1980b.

Hones, E. W., Jr., Plasma flow in the plasma sheet and its relation to substorms, *Radio Sci., 8*, 979, 1973.

Hones, E. W., Jr., Substorm processes in the magnetotail: comments on "on hot tenuous plasmas, fireballs, and boundary layers in the earth's magnetotail," by L. A. Frank, K. L. Ackerson and R. P. Lepping, *J. Geophys. Res., 82*, 5633, 1977.

Hones, E. W., Jr., Plasma flow in the magnetotail and its implications for substorm theories, in *"Dynamics of the Magnetosphere,"* edited by S. -I. Akasofu, p. 545, D. Reidel, Dordrecht, Holland, 1980.

Hones, E. W., Jr., J. R. Asbridge S. J. Bame, M. D. Montgomery, S. Singer, and S. -I. Akasofu, Measurements of magnetotail plasma flow with vela 4B, *J. Geophys. Res., 77*, 5503, 1972.

Hones, E. W., Jr., and K. Schindler, Magnetotail plasma flow during substorms: a survey with imp 6 and imp 8 satellites, *J. Geophys. Res., 84*, 7155, 1979.

Intriligator, D. S., J. H. Wolfe, D. D. McKibbin, and H. R. Collard, Preliminary comparison of solar wind plasma observations in the geomagnetospheric wake at 1000 and 500 earth radii, *Planet. Space Sci., 17*, 321, 1969.

Intriligator, D. S., J. H. Wolfe, and D. D. McKibbin, Simultaneous solar wind plasma and magnetic field measurements in the expected region of the extended geomagnetic tail, *J. Geophys. Res., 77*, 4645, 1972.

Jaeger, E. F., and T. W. Speiser, Energy and pitch angle distributions for auroral ions using the current sheet acceleration model, *Astrophys., Space Sci., 28,* 129, 1974.

Lui, A. T. Y., Observations of plasma sheet dynamics during magnetospheric substorms, in *"Dynamics of the Magnetosphere,"* edited by S. -I. Akasofu, p. 563, D. Reidel, Dordrecht, Holland, 1980.

Lui, A. T. Y., E. W. Hones, Jr., F. Yashuhara, S. -I. Akasofu, and S. J. Bame, Plasma flow during plasma sheet expansions: vela 5 and 6 and imp 6 observations, *J. Geophys. Res., 82,* 1235, 1977.

Lui, A. T. Y., L. A. Frank, K. L. Ackerson, C. -I. Meng, and S. -I. Akasofu, Plasma flows and magnetic field vectors in the plasma sheet during substorms, *J. Geophys. Res., 83,* 3849, 1978.

Lui, A. T. Y., L. A. Frank, K. L. Ackerson, C. -I. Meng, and S. -I. Akasofu, Correction, *J. Geophys. Res., 84,* 4471, 1979.

Lui, A. T. Y., T. E. Eastman, D. J. Williams, and L. A. Frank, Observations of ion streaming during substorms, *J. Geophys. Res., 88,* 7753, 1983.

Mariani, F., and N. F. Ness, Observations of the geomagnetic tail at 500 earth radii by pioneer 8, *J. Geophys. Res., 74,* 5633, 1969.

Mobius, E., F. M. Ipavich, M. Scholer, G. Gloeckler, D. Hovestadt, and B. Klecker, Observations of a nonthermal ion layer at the plasma sheet boundary during substorm recovery, *J. Geophys. Res., 85,* 5143, 1980.

Ness, N. F., The geomagnetic tail, *Rev. Geophys., 7,* 97, 1969.

Ness, N. F., C. S. Scearce, and S. C. Cantarano, Probable observations of the geomagnetic tail at 10^3 earth radii, *J. Geophys. Res., 72,* 3769, 1967.

Nishida, A., H. Hayakawa, and E. W. Hones, Jr., Observed signatures of reconnection in the magnetotail, *J. Geophys. Res., 86,* 1422, 1981.

Paulikas, G. A., Tracing of high latitude magnetic field lines by solar particles, *Rev. Geophys. Space Phys., 12,* 117, 1974.

Pilipp, W. G., and G. Morfill, The plasma mantle as the origin of the plasma sheet, in *"Magnetospheric Particles and Fields,"* edited by B. M. McCormac, D. Reidel, Dordrecht, Holland, p. 55, 1976.

Pilipp, W. G., and G. Morfill, The formation of the plasma sheet resulting from plasma mantle dynamics, *J. Geophys. Res., 83,* 5670, 1978.

Rich, F. J., V. M. Vasyliunas, and R. A. Wolf, On the balance of stresses in the plasma sheet, *J. Geophys. Res., 77,* 4670, 1972.

Robertson, C., S. W. H. Cowley, and J. W. Dungey, Wave-particle interactions in a magnetic neutral sheet, *Planet. Space. Sci., 29,* 399, 1981.

Rosenbauer, H., H. Grunwaldt, M. D. Montgomery, G. Paschmann, and N. Sckopke, Heos 2 plasma observations in the distant polar magnetosphere: the plasma mantle, *J. Geophys. Res., 80,* 2733, 1975.

Scarf, F. L., I. M. Green, G. L. Siscoe, D. S. Intriligator, D. D. McKibbin and J. H. Wolfe, Pioneer 8 electric field measurements in the distant geomagnetic tail, *J. Geophys. Res., 75,* 3167, 1970.

Schindler, K., A self-consistent theory of the tail of the magnetosphere, in *"Earth's Magnetospheric Processes,"* edited by B. M. McCormac, D. Reidel, Dordrecht, Holland, p. 200, 1972.

Scholer, M., G. Gloeckler, D. Hovestadt, F. M. Ipavich, B. Klecker, and C. Y. Fan, Anisotropies and flows of suprathermal particles in the distant magnetotail: ISEE-3 observations, *Geophys. Res. Lett., 10,* 1203, 1983.

Siscoe, G. L., F. L. Scarf, D. S. Intriligator, J. H. Wolfe, J. H. Binsack, H. S. Bridge and V. M. Vasyliunas, Evidence for a geomagnetic wake at 500 R_E, *J. Geophys. Res., 75,* 5319, 1970.

Sonnerup, B. U. Ö., Magnetic field reconnection, in *"Solar System Plasma Physics, Vol. III,"* edited by L. T. Lanzerotti, C. F. Kennel and E. N. Parker, North Holland, Amsterdam, p. 45, 1979.

Sonnerup, B. U. Ö., G. Paschmann, I. Papamastorakis, N. Sckopke, G. Haerendel, S. J. Bame, J. R. Asbridge, J. T. Gosling, and C. T. Russell, Evidence for magnetic field reconnection at the earth's magnetopause, *J. Geophys. Res., 86,* 10049, 1981.

Speiser, T. W., Particle trajectories in model current sheets, 1, analytical solutions, *J. Geophys. Res., 70,* 4219, 1965.

Speiser, T.W., Particle trajectories in model current sheets, 2, applications to auroras using a geomagnetic tail model, *J. Geophys. Res., 72,* 3919, 1967.

Speiser, T. W., On the uncoupling of parallel and perpendicular motion in a neutral sheet, *J. Geophys. Res., 73,* 1113, 1968.

Spjeldvik, W. N., and T. A. Fritz, Energetic ion and electron observations of the geomagnetic plasma sheet boundary layer: three-dimensional results from ISEE-1, *J. Geophys. Res., 86,* 2480, 1981.

Tsurutani, B. T., J. A. Slavin, E. J. Smith, R. Okida, and D. E. Jones, Magnetic structure of the distant geotail from −60 to −220 R_E: ISEE-3, Preprint JPL, Pasadena, Calif., USA, 1983.

Vasyliunas, V. M., Theoretical models of magnetic field line merging, 1, *Rev. Geophys. Space Phys., 13,* 303, 1975.

Vasyliunas, V. M., Upper limit on the electric field along a magnetic O line, *J. Geophys. Res., 85,* 4616, 1980.

Villante, U., Magnetopause observations at large geocentric distances, *Nuovo Cimento Letters, II,* 557, 1974a.

Villante, U., Physical properties of the distant geomagnetic tail, *Riv. Ital. Geofis, 23,* 27, 1974b.

Villante, U., Some remarks on the structure of the distant neutral sheet, *Planet. Space. Sci., 23,* 723, 1975.

Villante, U., Neutral sheet observations at 1000 R_E, *J. Geophys. Res., 81,* 212, 1976.

Villante, U., and A. J. Lazarus, Double streams of protons in the distant geomagnetic tail, *J. Geophys. Res., 80,* 1245, 1975.

Walker, R. C., U. Villante, and A. J. Lazarus, Pioneer 7 observations of plasma flow and field reversal regions in the distant geomagnetic tail, *J. Geophys. Res., 80,* 1238, 1975.

Williams, D. J., Energetic ion beams at the edge of the plasma sheet: ISEE-1 observations plus a simple explanatory model, *J. Geophys. Res., 86,* 5507, 1981.

Wolfe, J. H., R. W. Silva, D. D. McKibbin, and R. H. Mason, preliminary observations of a geomagnetospheric wake at 1000 earth radii, *J. Geophys. Res., 72,* 4577, 1967.

Wolfe, J. H., and D. S. Intriligator, The solar wind interaction with the geomagnetic field, *Space Sci. Rev., 10,* 511, 1970.

Questions and Answers

Fritz: Do you make any distinction in the physics associated with an X-line formed close to earth (<20 R_E), say by the tearing-mode instability discussed by Birn, and that of the deep-tail X-line (at 200 to 400 R_E) which is required to balance a day-side merging process. Specifically, would the presence of heavy ions affect your conclusions?

Cowley: *Cowley-Southwood-type models describe steady tail reconnection in a situation where the field stress is balanced by the dynamic pressure of the inflowing tail lobe plasma (in the de Hoffman-Teller or field-line rest frame). Under this condition the model is applicable both to the case of a near-Earth substorm-associated neutral line (after plasmoid ejection) and to a more distant neutral line during quiet times. The only restriction is that for a steady state the neutral line must lie earthward of the point where the tail lobe plasma flow into the current sheet becomes super-Alfvénic such that closed tubes would be blown down-tail at larger distances. The model does not describe the formation of a near-Earth X-line in the plasma sheet during substorms, nor is it directly applicable to the situation immediately after its formation, since reconnection then occurs in a β > 1 plasma (the plasma sheet), whereas β ≪ 1 in the tail lobe inflow plasma we have considered. However, the essential physics need not be different. Once the plasmoid has been ejected and the tail lobe is the source region immediately adjacent to the current sheet, then Cowley-Southwood-type models should become applicable as I have indicated. In the near-Earth tail this source plasma should be mainly ionospheric and may therefore be dominated by heavy ions particularly during solar maximum. It is trivial to elaborate the model to include different ion species. Changing a pure proton inflow plasma for pure oxygen of the same number density, for example, does not change the energy of the accelerated ions, which is essentially the magnetic energy per ion in the inflow region. However, if the inflow plasma is a mix of ions, then the current sheet mechanism results in ions emerging with essentially the same speed (determined from an Alfvén speed based on the mean ion mass), so that ion energy in this case is proportional to mass.*

Heikkila: In your model you emphasized the plasma mantle as a source of particles for the plasma sheet. Another source is the magnetopause boundary layer on closed field lines. Taking note of the massive BL flow (10^{27}/s), only a fraction of that is required for the plasma sheet (10^{26}/s); the rest of the boundary layer flow must be tailward, producing a topology such as I show in my paper (Fig. 2). How would a boundary layer source affect the reconnection model that you showed?

Cowley: *A variety of recent observations have shown that reconnection is usually the dominant process resulting in magnetospheric convection (see the review by Cowley, Rev.*

Geophys. Space. Phys., *20*, *531, 1982). Closed boundary layers driven, e.g., by a diffusive process appear to contribute at most 10-20% of the cross-magnetosphere potential. In the dominant reconnection-associated part of the flow the near-Earth plasma sheet is formed as closed field lines earthward of the neutral line. The source of plasma for these field lines must predominantly be the plasma on the pre-reconnected open field lines in the tail lobes, i.e., ionospheric plasma in the near-Earth tail and plasma mantle much beyond the lunar orbit. These sources are also quite adequate to account for plasma sheet formation. Particles from adjacent closed boundary layers on the flanks (or the magnetosheath in the absence of the latter), may leak "sideways" into the reconnection-associated current sheet, but this would give rise only to a narrow edge effect. With regard to the topology you show in your figure, it suffers from a major problem in that it does not properly account for the major source of magnetospheric convection, i.e., magnetosphere-IMF reconnection. Consequently, in my view, its predictions do not agree with either Pioneer 7 and 8 or ISEE-3 data. To be sure, the flow observed in the tail at large distances from Earth is nearly always tailward, but the plasma at the tail center plane flows* faster *than adjacent regions, indicating plasma acceleration tailward of the tail neutral line. For a closed boundary layer this flow should be* slower, *the difference being due to the opposite sign of the magnetic force acting on the plasma. I therefore believe that your picture of the tail is not correct.*

Reiff: The X-line configuration that you showed is stationary in the earth frame. The downstream flow of the incoming plasma implies that the X-line must be asymmetric. There are other configurations possible in which the X line moves downstream in the earth frame leading to a more symmetric X-line. Since the near-earth X line is observed to move tailward in substorm recovery phases, what theoretical considerations (e.g., back pressure of the earthward plasma sheet) affect the tailward speed? Once it starts moving, what would cause it to stop?

Cowley: *The location of the neutral line in the Cowley-Southwood model is arbitrary, provided it lies earthward of the point where the down-tail flow becomes super-Alfvénic. I think the model may also be valid for an arbitrarily moving neutral line, provided the motion is not too rapid. However, what actually determines the location of the neutral line and its motion is certainly not clear at the present time. It is possible to speculate about the effect of local conditions at the current sheet, etc., but more global considerations may also be significant.*

ISEE 3 MAGNETIC FIELD OBSERVATIONS IN THE MAGNETOTAIL: IMPLICATIONS FOR RECONNECTION

G. L. Siscoe and D. G. Sibeck

Department of Atmospheric Sciences, UCLA, Los Angeles, CA 90024

J. A. Slavin, E. J. Smith, B. T. Tsurutani

Jet Propulsion Laboratory, California Institute of Technology, Pasadena, CA 91109

D. E. Jones

Physics Department, Brigham Young University, Provo, UT 84600

Abstract. ISEE 3 explored the magnetotail to a distance of 220 R_E during its first two tail orbits. The characteristic cross-sectional structure of the tail near the earth was recognizable to apogee. The field strength in the lobes and the B_z component in the plasma sheet decreased systematically with distance. The frequency of occurrence of negative B_z increased with distance. Beyond 210 R_E, the average B_z was slightly negative. The observations suggest the existence of an x-type neutral line located on average near 200 R_E.

In addition to the systematic variations with distance, three types of transient events were recorded. The first type, which the spacecraft encountered on more than 30 occasions, is a discrete magnetic structure of 5 to 20 minutes duration distinguished by strong diamagnetism, a bipolar north-to-south B_z signature, hot plasma, high energy particles and increased wave activity. These events occur during intervals of geomagnetic activity. Hones has identified these structures as tailward moving plasmoids produced by reconnection within the tail.

The second type of discrete magnetic structures consists of solitary compressions of the lobe field. The compressions have amplitudes from 2 to 6 nT and last 5 to 30 minutes. These events are closely correlated with substorm onsets. They appear to be peripheral encounters with the plasmoids that make up the first class of events.

The third type of transient features are best described by how they differ from plasmoids. Whereas plasmoids cause a bipolar signature in B_z and have a weak field at their center, events of the new class have bipolar signatures in either B_z or B_y and have strong fields at their center. Whereas the axis of a plasmoid is y-aligned, the axis of the new kind of structures is x-aligned. They have the characteristics of magnetic flux ropes lying parallel to the tail.

Introduction

We present early results from our analysis of magnetometer measurements made during the first two tail passes of ISEE-3. Figure 1 shows the trajectories of the passes. The aberrated magnetopause of Howe and Binsack [1972] is shown for reference. The first pass cut nearly directly across the tail at an average geocentric distance of 75 R_E. The apogee of the deep tail pass at 220 R_E is near the center of the tail. The thick bands on the trajectories mark intervals where boundary crossings occurred. The most distant extensive tail measurements taken in the past were by Explorer 33 at 80 R_E. The results given here relate mainly to measurements taken beyond 80 R_E.

ISEE-3 boundary crossings in the remote tail conform reasonably well to the reference boundary of Howe and Binsack. The locations of the boundary crossings between 100 and 200 R_E are consistent with an average tail radius of approximately 30 R_E [Slavin et al., 1983]. The magnetic field signatures of the tail lobes, the plasma sheet and the magnetopause are recognizable to apogee [Slavin et al., 1983; Tsurutani et al., 1983]. We infer that the tail at apogee is a continuous extension of the well-described, cislunar tail. Nonetheless, the structure of the tail evolves with distance in important ways. We describe first the radial dependence of the average values of selected magnetic variables. Transient events are then discussed.

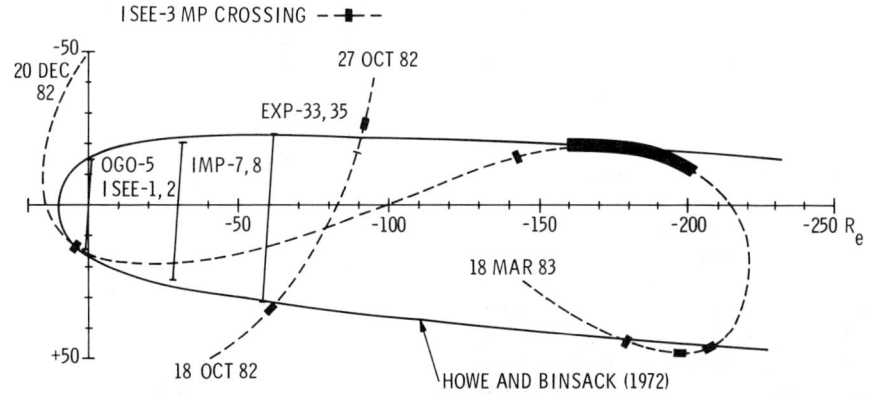

Fig. 1. The projection of the trajectories of the first two tail passes of ISEE-3 onto the GSE xy-plane. The regions of space reached by earlier spacecraft also indicated [see Slavin et al., 1983].

Systematic Variations with Distance

Figure 2 shows the distance dependence of selected tail parameters measured during the deep tail pass. As discussed by Slavin et al. and Tsurutani et al., the lobe regions were readily identified by field strength enhancements and solar/anti-solar orientations. However, ambiguities can arise in identifying the plasma sheet and magnetosheath regions with magnetic field data alone. ISEE-3 plasma electron data [Bame et al., 1983; J. T. Gosling, private communication, 1983] have been used to distinguish between the two regions in this study. The two top panels give the field strength and the GSM-z-component of the field in the plasma sheet. The bottom panel shows the field strength in the tail lobes. Gaps mark intervals where the spacecraft was not in the plasma sheet or lobe. The spacecraft was generally in the magnetosheath between $x = -170$ and $-180\ R_E$ which accounts for the data gap in all panels in that interval.

Each point in the figure is an average of the measurements made while ISEE-3 moved through consecutive 10 R_E segments parallel to the x-axis. Since the speed of the spacecraft parallel to the x-axis decreases markedly with distance, many more measurements were taken in the furthest 10 R_E interval than in the closest. For example, the spacecraft moved from $x = -60$ to $-70\ R_E$ in 17 hours, from $x = -100$ to $-110\ R_E$ in 28 hours, from $x = -150$ to $-160\ R_E$ in 52 hours, and from $x = -200$ to $-210\ R_E$ in 117 hours.

Look first at the bottom panel, which shows how the field strength in the tail lobe depends on distance. The small-scale irregularities almost certainly result from changes in solar wind conditions. Ignoring these, we see a curve that first decreases monotonically, then becomes constant around $x = -110\ R_E$. The power law which best fits the descending portion of the curve is $|\vec{B}| \propto |x|^{-0.6}$. [Slavin et al., 1983] This is stronger than the $|x|^{-0.3}$ dependence found by Behannon (1968) but weaker than the $|x|^{-0.74}$ dependence found by Mihalov et al. [1968]. Both earlier analyses used 1966 Explorer 33 data. The asymptotic constant value in all studies is approximately 8-9 nT, but that value was reached closer to earth in the Explorer 33 data than in the ISEE-3 data.

The curve of field strength in the plasma sheet shown in the top panel also has a descending part and a constant part. The constant part begins near $x = -60\ R_E$, considerably nearer Earth than for the lobe field strength. The average value of the plasma sheet field strength beyond 60 R_E is approximately 5 nT. The plasma-to-field pressure

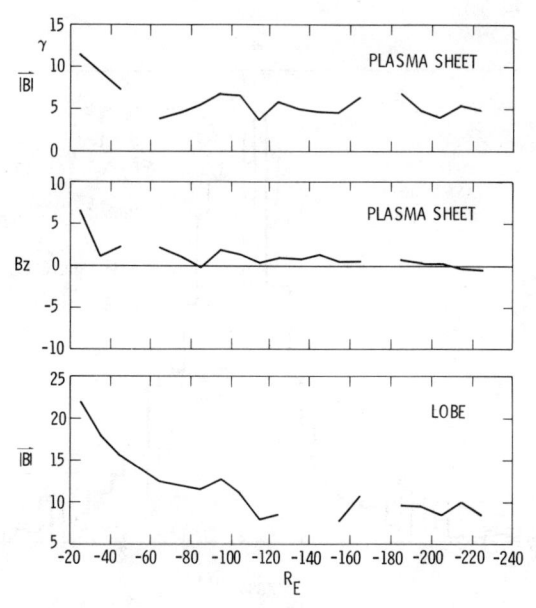

Fig. 2. Ten-R_E averages of (1) the field strength in the plasma sheet, 2) the solar-magnetospheric z-component of the field in the plasma sheet, and (3) the field strength in the tail lobes as measured during the first deep tail pass of ISEE-3.

ratio, β, in the plasma sheet can be estimated by comparing the lobe and plasma sheet field strengths under the assumption that the total pressure is the same in the two regions and that β in the lobes is negligible. The plasma sheet β decreases from a near-Earth value near three to about two by 110 R_E, then remains approximately constant to apogee.

The solar-magnetospheric z-component shown in the middle panel decreases markedly between the first pair of 10 R_E averages, then decreases slowly to apogee. The slope of the slow decrease is about 1 nT per 100 R_E. The average value of z appears to cross zero at 210 R_E. This is an important observation. The transition from positive to negative average B_z cannot be explained simply by the motion of the spacecraft in the yz-plane. The measurements shown in this panel were taken while ISEE-3 was in the plasma sheet. The field in the plasma sheet has a positive average B_z everywhere in the yz-plane, except beyond the hypothesized neutral line. A southward tilt of the tail also produces a negative B_z. But ISEE-3 was beyond 210 R_E for more than 10 days, which is a long time for a temporary tilt to be imposed by the solar wind. Furthermore, the sign of B_x showed that ISEE-3 was in the southern part of the plasma sheet. Since it was also never more than 5 R_E from the xy-equatorial plane, any southward tilt of the tail must have been small. Therefore, we conclude that for the first time an x-type neutral line (x-line) in the tail has revealed itself in data averaged over many transient-event times. The x-line of the time-average tail evidently lies near 210 R_E.

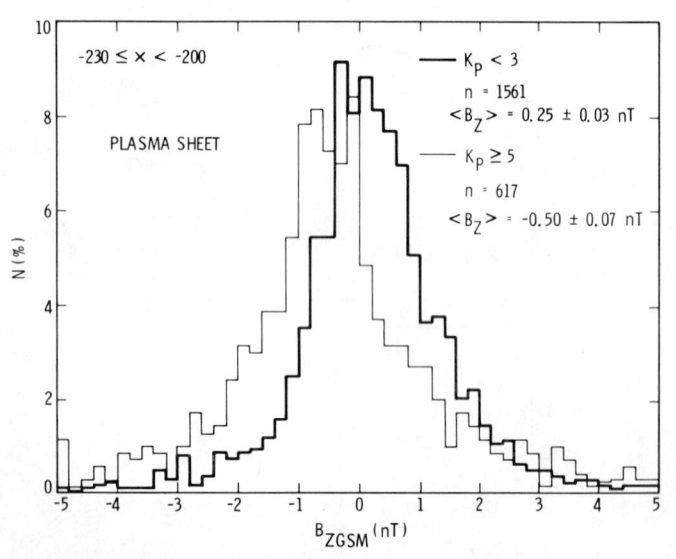

Fig. 3. Histograms of one-minutes averages of the solar-magnetospheric z-component of the field measured in the plasma sheet beyond 200 R_E by ISEE-3. The histograms show data measured during geomagnetically quiet times ($K_p < 3$, 1561 cases) and disturbed times ($K_p > 5$, 617 cases)

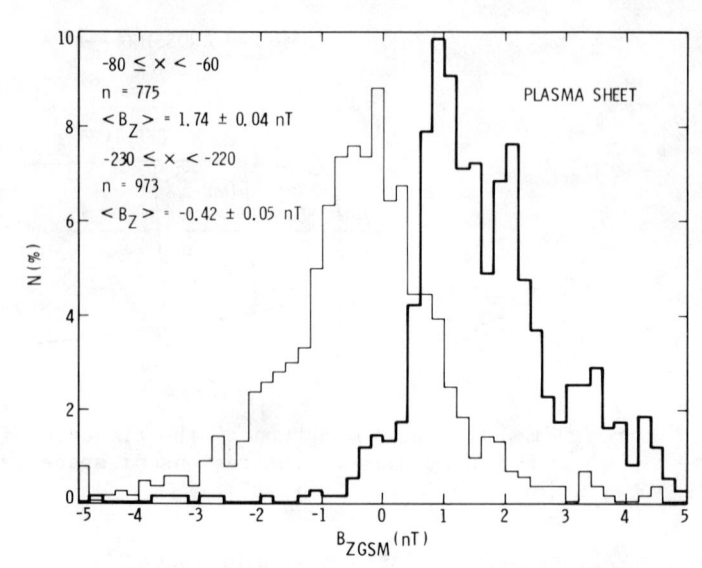

Fig. 4. Histograms of one-minute averages of the solar-magnetospheric z-component of the field measured in the plasma sheet by ISEE-3. The histograms show data acquired between 60 and 80 R_E (775 cases) and beyond 200 R_E (973 cases).

In contrast to the 10 R_E averages discussed above, one-minute averages of B_z measured between 200 R_E and apogee are generally positive and negative with comparable frequency, as shown in Figure 3. The two histograms in this figure give the distributions for quiet ($K_p < 3$) and disturbed ($K_p > 5$) geomagnetic conditions. The centroid of the histogram of quiet-time data lies to the right of zero, and the centroid of the other histogram lies to the left. The average of the quiet-time values is 0.25 ± 0.03 nT. The average of the disturbed-time values is -0.50 ± 0.07 nT. The range in values as characterized by the half-width-at-half-maximum of the histograms in both cases is approximately 1 nT. These results suggest the position of the x-line in the disturbed-time-averaged tail lies closer to Earth than 200 R_E and farther from Earth than 200 R_E in the quiet-time-averaged tail.

The histograms in Figure 3 indicate that the location of the x-line in the time-averaged tail might more usefully be thought of as the point of equal occurrence frequency of positive and negative B_z in a tail where the x-line moves over a fairly large distance. The trajectory of ISEE-3 evidently spanned the range where the x-line resides more than 50% of the time. The occurrence frequency histograms shown in Figure 4 imply the x-line is beyond the orbit of the moon more than 95% of the time and inside the apogee of ISEE-3 approximately 70% of the time. These remarks need to be qualified by the phrase "as measured along the first deep-tail trajectory of ISEE-3." It is possible that the figure-8 shape of the ISEE-3 trajectory has caused changes transverse to the tail axis to be

ISEE-3 HELIUM MAGNETOMETER

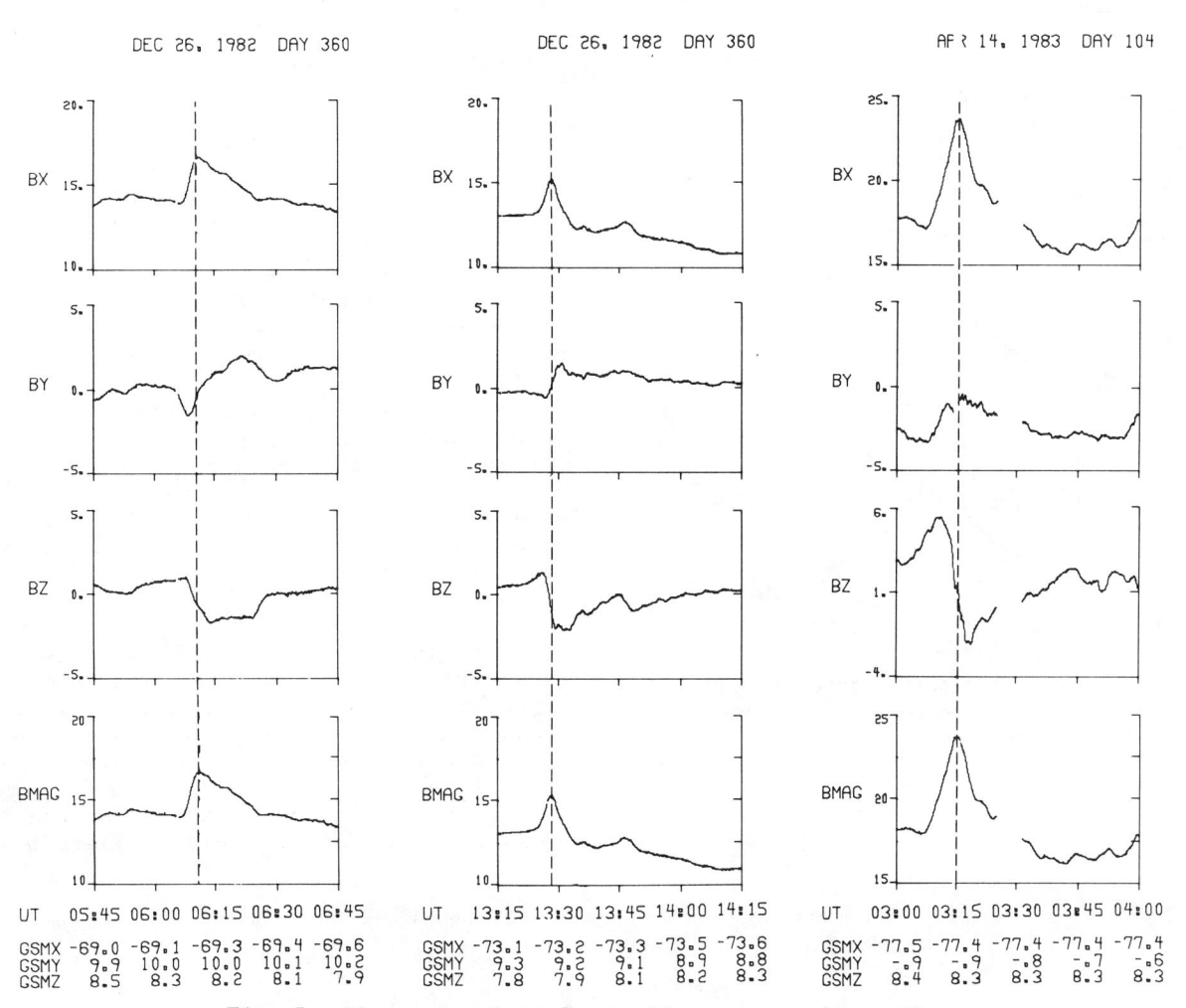

Fig. 5. Three examples of traveling compression regions.

misidentified as changes along the tail axis. The ambiguity may be removed when the data from the second deep-tail pass are analyzed.

Transient Events

ISEE-3 encountered three types of discrete, transient magnetic features. Each had a characteristic duration less than half an hour and each was associated with substorms. Plasmoids, which were well described by Hones [this meeting], are one of the three types. We will concentrate on the other two types. One of these appears to be the lobe signature of passing plasmoids. The nature of the third type has not been determined.

Figure 5 shows three examples of the "passing plasmoid" type. The duration of each segment is one hour. These events are observed in the tail lobes. Following Slavin et al., [1984] we will call them traveling compression regions (TCR's). They are characterized by a distinctive compression pulse which usually lasts five to fifteen minutes. Since the lobe field is predominantly oriented in the x-direction, the x-component profile duplicates the compression in field strength. B_y and B_z also change as the pulse passes. Whereas the y-component does not change in the same way from event to event, the change in the z-component is always the same. Prior to the event, B_z is positive. In the event sequence, B_z first increases then decreases through zero, then more slowly and often irregularly returns to a positive value less than its pre-event value. The compression in B_x peaks during the positive-to-negative excursion of B_z. The sequence is the same in both tail lobes. An occurrence rate of a few TCR's per day is not uncommon in the ISEE-3 data, but there were also days with no recorded occurrences.

Freely propagating compression and rare-

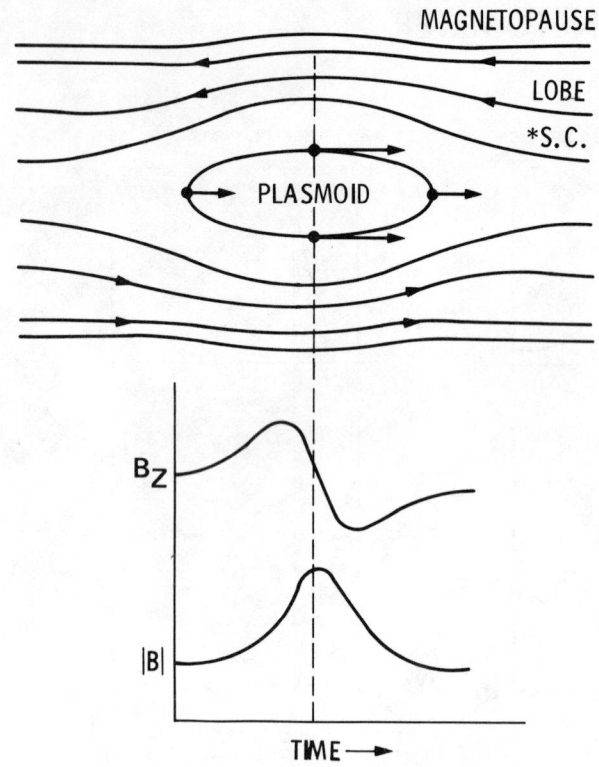

Fig. 6. A model to explain traveling compression regions as peripheral encounters with passing plasmoids.

faction waves seem unable to produce such features. A wave would cause the changes in B_x and B_z to be in-phase, and not ninety degrees out-of-phase as observed. Solitons too can be rejected as causes of TCR's. Solitons would leave the final state the same as the initial state, which is not observed.

Figure 6 presents an explanation which is consistent with the known features of TCR's. A plasmoid passing down the tail compresses and distorts the field in the tail lobes. The model predicts the observed B_z signature and the observed timing of the B_z signature relative to the compression pulse. The figure shows that the B_z signature is the same in both lobes, as observed. The decay of the B_z signal also reflects known properties of plasmoids. The train of a plasmoid is often irregular and the post-plasmoid field is stretched more tailward [Hones, this volume].

Traveling compression regions are associated with substorms. As an example, Figure 7 shows when the peak of an TCR passed ISEE-3 relative to a midlatitude Pi 2 event recorded in the AFGL midlatitude magnetometer chain [H. J. Singer, private communication, 1983]. The Pi 2 event, which marks a substorm, began six minutes before the peak in the TCR at ISEE-3. At this time the geocentric distance of ISEE-3 was 78 R_E. If the signal was launched in both directions from the distance usually cited, 20 R_E down the tail, then an upper limit on the velocity of propagation of the signal is 58 R_E in six minutes, or 1000 km s^{-1}. This is at the upper end of the speeds found for plasmoids. Therefore, the

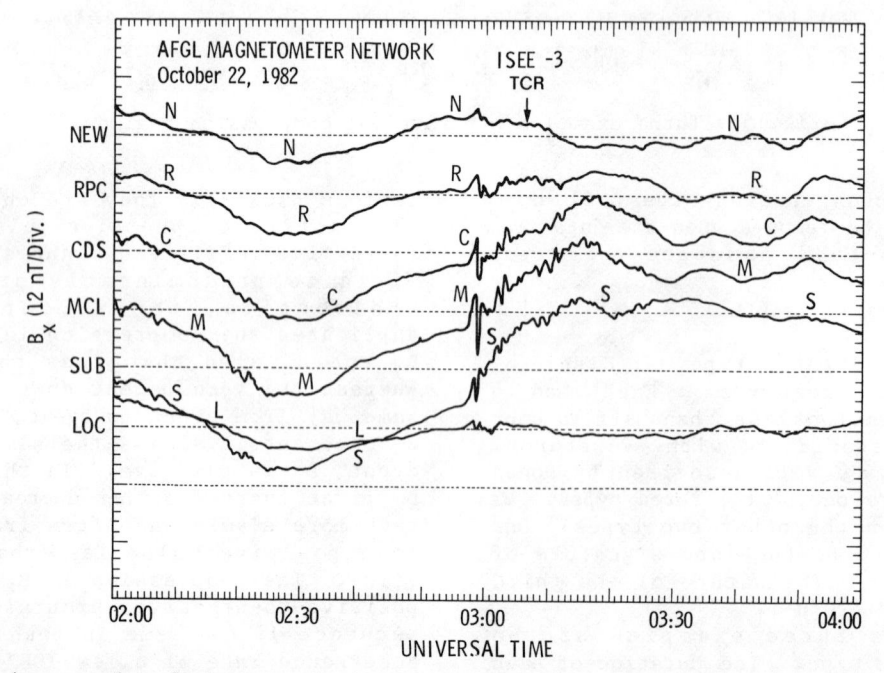

Fig. 7. An example of the association between traveling compression regions and substorms. The Pi-2 event marking a substorm is evident midlatitude magnetic data from the AFGL magnetometer network.

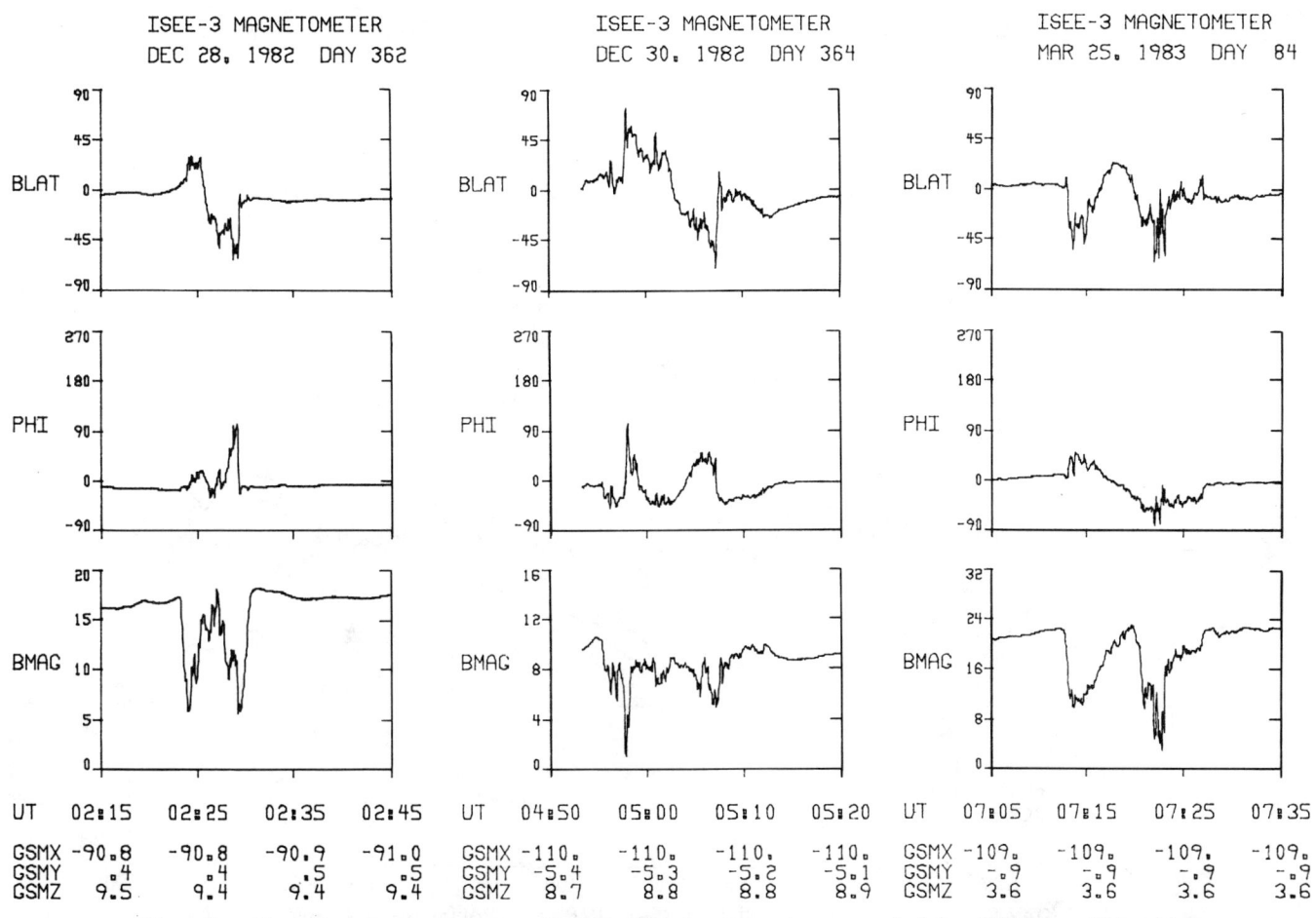

Fig. 8. The three strong-core-field, bipolar events recorded by ISEE-3. Note the y-component bipolarity in the third case.

substorm association in general and the timing in in this instance agree with the suggestion that passing plasmoids cause TCR's.

The third type of transient magnetic feature observed by ISEE-3 resembles plasmoids. The magnetic signature of a passing plasmoid is a "bipolar" oscillation in B_z (first an increase then a negative excursion - the same as for TCR's) and a "weak core field" (the field strength is minimum in the middle of the plasmoid). Events of the new type are bipolar in either B_z or B_y and they have strong core fields. The axis of the toroidal field which produces the bipolar field signatures in plasmoids runs in the y-direction. The axis of the new class of structures runs in the x-direction, parallel to the tail axis. Three of the new type of magnetic structures have been identified in the ISEE-3 data. This is less than ten percent of the bipolar events analyzed. The remaining 90+ percent of the events have the usual features of plasmoids (with two exceptions described below).

The three examples of the new magnetic structures are shown in Figure 8. Each data segment lasts one-half hour. All events occurred when ISEE-3 was between 90 R_E and 110 R_E geocentric

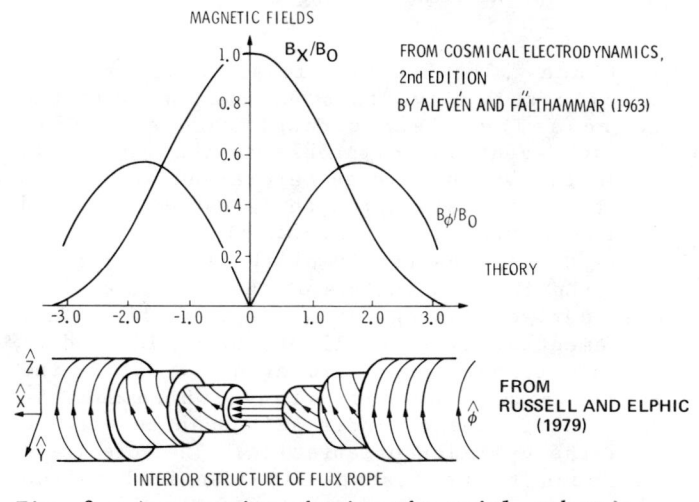

Fig. 9. A composite showing the axial and azimuthal field components given by a particular axially symmetric force-free-field model (Alfven and Falthammar, 1963) and a sketch of an idealized model for magnetic flux ropes (Elphic and Russell, 1983). The distances are normalized to the length unit $B_oc/4\pi i_o$, where B_o and i_o are the field strength and current amplitude at the center of the structure.

246 ISEE 3 MAGNETIC FIELD OBSERVATIONS

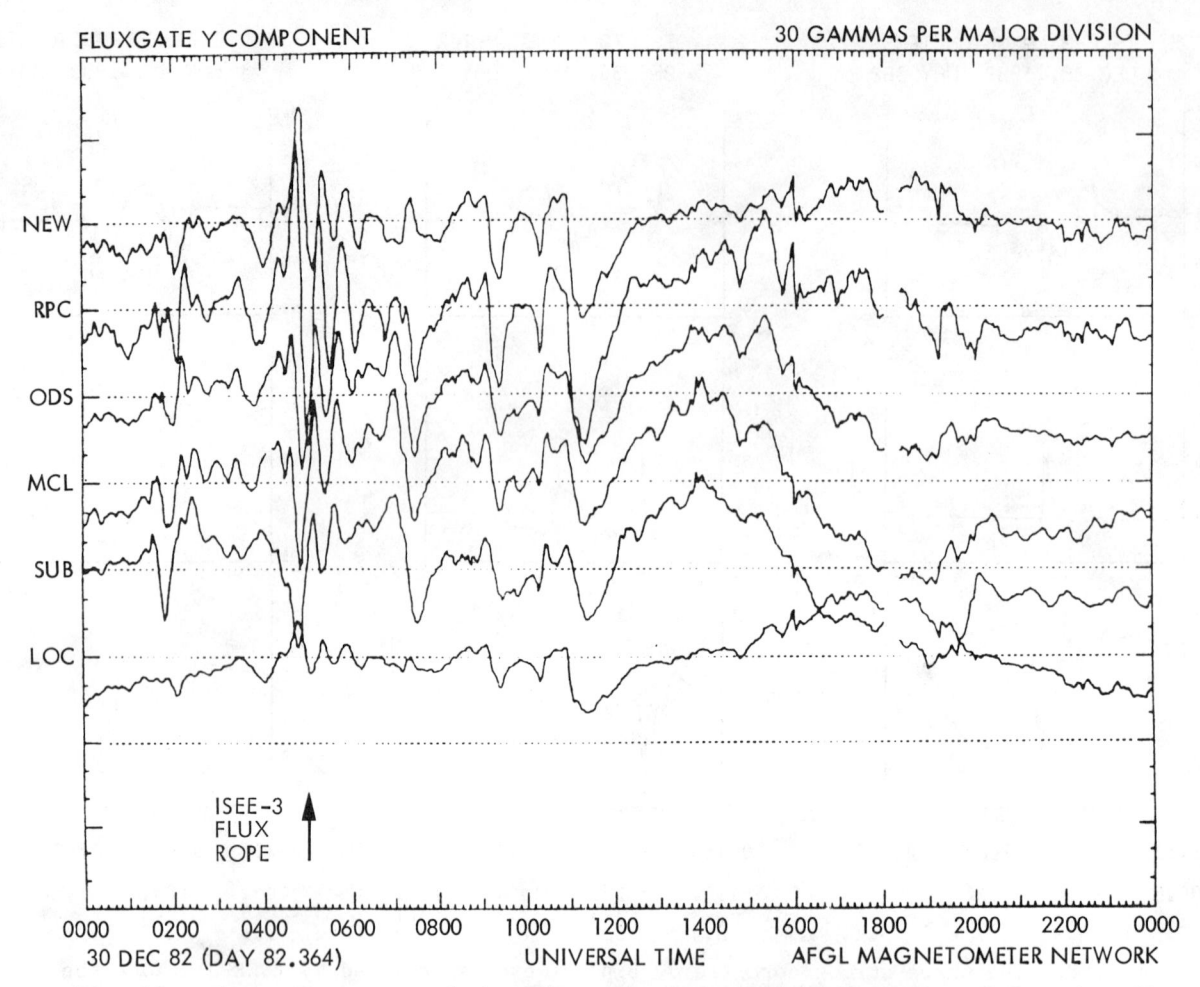

Fig. 10. The y-components of the AFGL magnetometer network station on December 30, 1982. Note the strong wave activity centered at 0500 uT. This is also the time of the ISEE-3 strong-core field event.

distance. The bipolar signature is in B_z in the first two events shown and in B_y in the third. The field strength at the middle of each event is comparable to the lobe field strength, which is seen before and after each event. The field strength is markedly reduced at the extremes of the bipolar excursions.

The bipolar character of the field suggests that the basic structure of these features is a cylinder enclosing a toroidal field. This statement applies equally to plasmoids and to the new strong-core-field events. But in the latter case, the core field, which must be parallel to the cylinder axis, can be used to find the cylinder orientation. The core field is found to be essentially identical to the adjacent lobe field. Therefore if these features are indeed cylindrical, they are x-aligned.

The magnetic signature of the strong core features is qualitatively the same as that of the magnetic flux ropes in the ionosphere of Venus [Russell and Elphic, 1979]. The field near the center of a flux rope is strong and directed parallel to its axis. The field near the periphery of a flux rope is weak and directed azimuthally to its axis. We expect the analysis by Elphic and Russell [1983] of the structure of flux ropes at Venus applies qualitatively to the strong core field structures in the geotail. A notable feature of the flux rope models is the existence of a component of current parallel to the magnetic field. The parallel current exceeds the transverse current in peak value and dominates the center of the flux rope. The flux rope is therefore to a large extent "force-free".

An idealized sketch of a flux rope is given in Figure 9 [Russell and Elphic, 1979]. The figure also shows the parallel and azimuthal components of the magnetic field predicted by a particular axially symmetric force free field model [Alfven and Falthammer, 1963]. The correspondence between the two models is evident. The correspondence between the model force-free field and the field in the interior of the strong-core-field events in the geotail is also evident.

The flux rope model in Figure 9 can be used to show how a bipolar signal can appear in

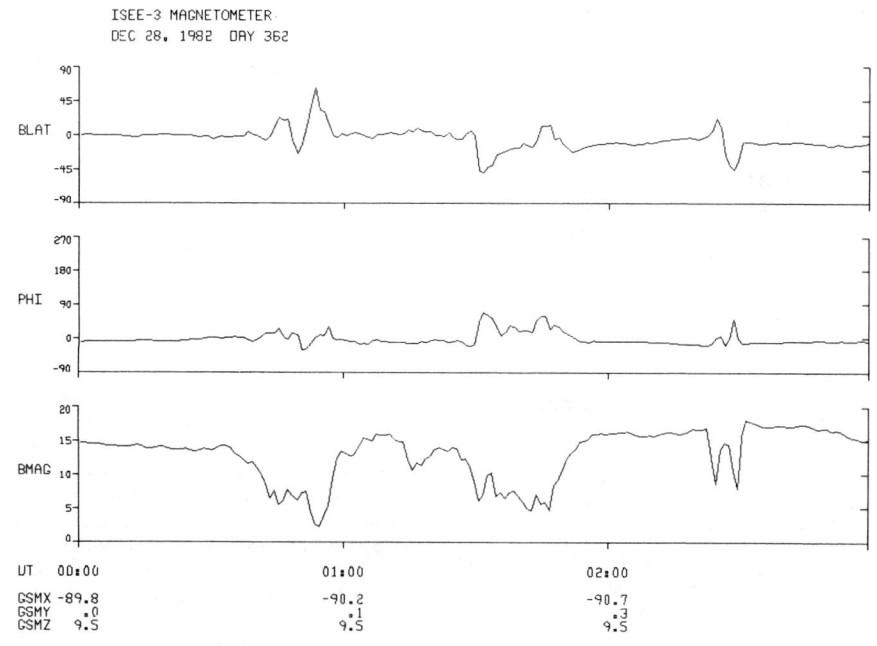

Fig. 11. An interval containing two reversed-bipolar plasmoid events (indicating earthward moving plasmoids) and one strong-core-field, bipolar event.

either B_z or B_y if the axis of the rope is x-aligned. If the tube crosses a spacecraft by moving in the y-direction, a B_z bipolar signature will result. Similarly, a north-south motion of the tube across a spacecraft will make a bipolar signal in B_y.

Mid-and high-latitude magnetograms were highly disturbed throughout the days on which the three strong core field events were observed by ISEE-3 [R. L. McPherron, personal communication, 1983]. The Greenland chain and Tromsø magnetograms for December 28, 1982 indicate that bay activity was in progress at 0200 UT, shortly before ISEE-3 observed the event at 0230 UT. Geomagnetic activity reached a peak at 0300 UT and recovered at 0340-0350 UT. Mid-latitude North American magnetograms indicate that substorm onsets occurred at 0200 and 0300 UT.

The ISEE-3 event observed on December 30, 1982 at 0500 UT was accompanied by dramatic wave activity. Figure 10 shows the y-component of the AFGL magnetometer chain. At 0500 UT, these stations were near the midnight meridian. Note that the large amplitude (~45 nT)) wave activity reaches a maximum at the time of the ISEE-3 event. Disturbances were also recorded at Tromsø and all the Greenland chain magnetometer stations. These records indicate that a substorm began at 0400 UT, intensified at 0130 and was followed by more activity at 0500 UT.

During the event at 0730 on March 25, 1983, wave activity was again recorded by the Mid-latitude and East-West chain stations. This activity began at 0300 UT and lasted until about 1100 UT. The maximum amplitudes (~50 nT) were observed from 0600-0830 UT, near the time of the ISEE-3 event.

As a final item, we note the occurrence of two plasmoid signatures which have a reversed sense of the bipolar signal. That is, B_z first decreases through zero, then increases. The events are shown in Figure 11. They occurred within one hour of each other. A reversed bipolarity in a plasmoid model implies the plasmoid moved earthward. The plasma data for the first event in fact reveal an earthward flow. The data for the second event are ambiguous. The entire three hour interval covered by the figure is highly interesting. The third bipolar feature is the first of the three strong-core-field events shown in Figure 8. The interpretation of this sequence of events does not follow naturally from the same plasmoid model which accounts for the other ninety percent of the cases. However, we note the existence of a model by Galeev [1982] which produces earthward moving plasmoids. Qualitatively it seems that if reconnection ceases prior to merging of all plasma sheet field lines, then a reverse slingshot effect pulls the plasmoid up the plasma sheet-lobe boundary.

Acknowledgments. This report represents one aspect of research carried out by the Jet Propulsion Laboratory for NASA under contract NAS7-100. The work carried out at UCLA was supported in part by a grant from NSF (ATM 81-10455).

One of the authors (JAS) acknowledges the support of the NRC Research Associateship Program.

The ISEE-3 plasma data used in this study were provided by the plasma science team at Los Alamos National Laboratory.

References

Alfven, H., and C.-G. Falthammar, Cosmical Electrodynamics, Oxford University Press, London, 1963.

Bame, S. J., R. C. Anderson, J. R. Asbridge, D. N. Baker, W. C. Feldman, J. T. Gosling, E. W. Hones, Jr., D. J. McComas, and R. D. Zwickl, Plasma regimes in the deep geomagnetic tail: ISEE-3, Geophys. Res. Lett., 10, 912, 1983.

Behannon, K. W., Mapping of the earth's bow shock and magnetic tail by Explorer 33, J. Geophys. Res., 73, 907-930, 1968.

Elphic, R. C., and C. T. Russell, Magnetic flux ropes in the Venus ionosphere: Observations and models, J. Geophys. Res., 88, 58-72, 1983.

Galeev, A. A., Magnetospheric tail dynamics, in Magnetospheric Plasma Physics, edited by A. Nishida, Center for Academic Publications, Japan, Tokyo, and D. Reidel Publishing Co., Dordrecht, 1982, pp. 143-196.

Howe, H. C., and J. H. Binsack, Explorer 33 and 35 plasma observations of magnetosheath flows, J. Geophys. Res., 77, 3334-3344, 1972.

Mihalov, J. D., D. S. Colburn, R. G. Currie, and C. P. Sonnet, Configuration and reconnection of the geomagnetic tail, J. Geophys. Res., 73, 943, 1968.

Russell, C. T., and R. C. Elphic, Observations of magnetic flux ropes in the Venus ionosphere, Nature, 279, 616, 1979.

Slavin, J. A., B. T. Tsurutani, E. J. Smith, D. E. Jones, and D. G. Sibeck, Average configuraion of the distant (< 200 R_e) magnetotail: Initial ISEE-3 magnetic field results, Geophys. Res., Lett., 10, 973, 1983.

Slavin, J. A., E. J. Smith, B. T. Tsurutani, D. G. Sibeck, H. J. Singer, D. N. Baker, J. T. Gosling, E. W. Hones, and F. L. Scarf, Substorm associated traveling compression regions in the distant tail: ISEE-3 geotail observations, submitted, Geophys. Res. Lett., 1984.

Tsurutani, B. T., J. A. Slavin, E. J. Smith, R. Okida, D. E. Jones, Magnetic structure of the distant Geotail from -60 to -220 R_E; ISEE-3, in press, Geophys. Res. Lett., 1983.

Questions and Answers

D. N. Baker: The "nonconforming" events (i.e., the atypical plasmoid signatures) that you showed all seemed to be at $\lesssim 100\ R_E$ geocentric distance. Is this a correct impression of the overall data set and, if so, is this "near-earth" character significant for your interpretation of the phenomenon?

Also, from your estimate of $\beta \sim 3$ in the plasma sheet what would you estimate the plasma sheet ion temperature to be?

Siscoe: You are correct. The three nonconforming events occurred around $100\ R_E$ distance. We do not know what these features are. We can not say whether they were observed in a narrow distance range because they are confined to that range, or what seems more likely, because when ISEE-3 was in that distance range, geomagnetic activity was especially high and these features are associated with high geomagnetic activity.

With regard to the next part of your question, pressure balance across the plasma sheet-lobe boundary requires an ion temperature of approximately 1500 eV based on a plasma sheet density of $0.1\ cm^{-3}$. In this estimate the contribution of the electrons is ignored because $T_e/T_i \ll 1$ in the plasma sheet where it has been measured. A more general expression of pressure balance that presumes nothing about number density n and the ion/electron temperature difference is $(k/e)n(T_e + T_i) \approx 150\ eV\ cm^{-3}$.

Reiff: You showed a plasmoid that was observed at ISEE-3 only 6 minutes after Pi-2's at the ground. That implies to me that the neutral line associated with the plasmoid must be fairly far downstream (if you observe it at $200\ R_E$ with $V_A = 1/6\ R_E/s$ and V of the plasmoid $= 1/12\ R_E/s$, then the X line must have been formed at $114\ R_E$ down the tail). Would this be a contracted oval substorm? It seems like this is evidence of motion of the X line even during relatively quiet conditions.

Also, if northward merging occurs, one might expect to see portions of the tail peeled off. It would look (I guess) like northward B_z + tailward flow. Any evidence for this?

Siscoe: At the time of the event to which you refer, ISEE-3 was only 78 R_E down the tail. A remote site of plasmoid formation is therefore not needed in this case, although such may occur.

With regard to your second question, I do not know if northward flows with positive B_z are observed. The Los Alamos group is responsible for the systematic survey of plasma behavior. Merging of the type you described would connect magnetosheath and lobe field lines. Therefore evidence of the associated flow would have to be sought in lobe plasma, and this might present a problem, because of low densities.

Russell: The force-free structures you see in the tail, similar to Venus flux ropes, could be the tailward ends of large flux transfer events, could they not? In some sense then they are tail fragments, connected to the polar cap as ordinary lobe magnetic flux but maintaining their individual identity because of their twist.

Siscoe: No obvious objection to your interpretation occurs to me. But other possibilities also seem plausible. For example, they might be analogous to wing-tip vortices trailing from the east and west extremities of passing plasmoids.

BOUNDARY LAYERS OF THE EARTH'S OUTER MAGNETOSPHERE

T. E. Eastman and L. A. Frank

Department of Physics and Astronomy, The University of Iowa,
Iowa City, Iowa 52242

Abstract. The magnetospheric boundary layer and the plasma-sheet boundary layer are the primary boundary layers of the earth's outer magnetosphere. The magnetospheric boundary layer occurs everywhere near the outer magnetospheric boundary or magnetopause. Any plasma, momentum or energy transport from the solar wind to the magnetosphere is carried out by way of this boundary layer. More than 98% of the high-β plasma of the shocked solar wind or magnetosheath is deflected around the magnetosphere. Although the plasma flow is predominantly tangential to the magnetopause surface, some 1-2% of the oncoming solar wind plasma gains entry into the magnetosphere and initially provides a source for the magnetospheric boundary layer. This boundary layer is generally intermediate in number density, mean energy and flow speed with respect to its high-β source region and the low-β hot plasma of the frontside outer magnetosphere.

On the tailward side, the plasma-sheet boundary layer is also intermediate in number density and mean energy with respect to the low-β lobe region and the medium to high-β central plasma sheet. This boundary layer is identified by high-speed ion beams which are dominantly field-aligned and flowing sunward and, occasionally, antisunward. Counter-streaming ion beams are also frequently observed within the plasma-sheet boundary layer. Low-energy ion beams, likely of ionospheric origin, are also commonly observed within this boundary layer.

Recent satellite observations of these two boundary layer regions indicate that they provide for more than 50% of the plasma and energy transport in the outer magnetosphere although they constitute less than 5% by volume. Relative to the energy density in the source regions, plasma in the magnetospheric boundary layer is predominantly de-energized whereas plasma in the plasma-sheet boundary layer has been accelerated.

The reconnection hypothesis continues to provide a useful framework for comparing data sampled in the highly dynamic magnetospheric environment. Reconnection provides for fairly tractable models in which dissipative effects are localized to a "diffusion" region so that the idealized MHD relations can be assumed valid throughout most of the system. However, observations of the boundary layers of the outer magnetosphere suggest that dissipative effects here are ubiquitous. For this and other reasons, steady-state reconnection is not generally applicable to the earth's magnetosphere. However, observations of "flux transfer events" and other detailed features near the boundaries have been recently interpreted in terms of non-steady-state reconnection. Alternative hypotheses are also being investigated. More work needs to be done, both in theory and observation, to determine whether reconnection actually occurs in the magnetosphere and, if so, whether it is important for overall magnetospheric dynamics.

Introduction

The magnetospheric boundary layer and the plasma-sheet boundary layer are the primary boundary layers of the earth's outer magnetosphere. As shown in Figure 1, the magnetospheric boundary layer occurs everywhere near the outer magnetospheric boundary or magnetopause. In various regions, other terms are used such as low-latitude boundary layer, plasma mantle or magnetotail boundary layer. Any plasma, momentum or energy transport from the solar wind to the magnetosphere is carried out by way of this boundary layer. More than 98% of the high-β plasma of the shocked solar wind or magnetosheath is deflected around the magnetosphere. Although the plasma flow is predominantly tangential to the magnetopause surface, some 1 to 2% of the oncoming solar wind plasma gains entry into the magnetosphere and initially provides a source for the magnetospheric boundary layer. This boundary layer is generally intermediate in number density, mean energy and flow speed with respect to its high-β source region and the low-β hot plasma of the frontside outer magnetosphere. Plasma in this boundary layer generally has a tailward flow component everywhere except for the

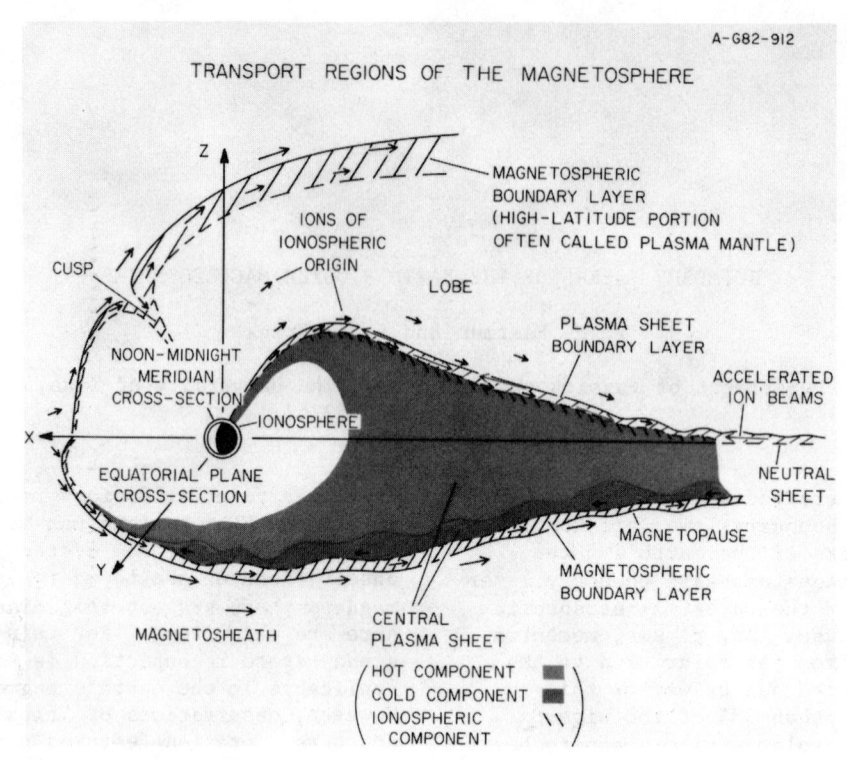

Fig. 1. A schematic of the primary transport regions of the earth's magnetosphere.

more turbulent flows in the exterior cusps.

The plasma-sheet boundary layer constitutes an interface region between the lobes and the plasma sheet. It is also intermediate in number density and mean energy with respect to the low-β lobe region and the medium to high-β central plasma sheet. Ion beams are commonly observed in this boundary layer. These beams are flowing sunward and, occasionally, antisunward. Counter-streaming ion beams are also frequently observed there. Low-energy ion beams, likely of ionospheric origin, are frequently observed flowing tailward within both the plasma-sheet boundary layer and the lobe regions.

Signatures of Magnetospheric Plasma Regimes

Ion and electron velocity distributions sampled in-situ provide a convenient method for classifying the various regions of the earth's magnetosphere. The University of Iowa's plasma analyzers on board the ISEE spacecraft provide full three-dimensional plasma measurements using seven pairs of channeltrons for ion and electron measurements from 1 ev to 45 keV. Two survey plots of these data are presented in Figure 2 in the form of energy-time spectrograms to illustrate the identification of various plasma regimes. The four panels of the spectrograms display, from top to bottom, averages of responses sampled by the ion detector centered on the equatorial plane. Four quadrants of view are used for these averaged responses and they include sunward (v), duskward (>), antisunward ($\wedge$) and dawnward (<) quadrants, respectively. These survey plots also display azimuthally averaged electron spectra in the bottom spectrogram. Values of $\log_{10}$ of the energy in units of eV are given along the ordinate with time in hours (UT) marked along the abscissa. Spacecraft coordinates are given in solar ecliptic coordinates with Θ_{SE} and ϕ_{SE} denoting solar-ecliptic latitude and longitude, respectively, and R_E denotes geocentric radial distance in units of earth radii.

Magnetospheric Boundary Layer

A crossing of the frontside boundary region is shown in Figure 2A with data obtained by ISEE 1 on day 229 of 1978. A distinct magnetopause transition occurs at 0142 to 0143 UT based on simultaneous magnetometer measurements (courtesy of Dr. C. T. Russell, UCLA). Prior to crossing the magnetopause on this inbound pass, magnetosheath plasmas are sampled characterized by ions with energies up to ~ 2 keV flowing antisunward at speeds of 150 to 180 km/s as shown especially in the first and fourth panels. After crossing the magnetopause, ion distributions similar to that of the magnetosheath but with reduced density and flow speeds (i.e. lower average kinetic energy) continue to be observed.

Three-dimensional response arrays or energy-phase angle (E-ϕ) plots as shown in Figures 3 and 4 provide a convenient representation of the distribution functions. Each plot shows detector responses for one

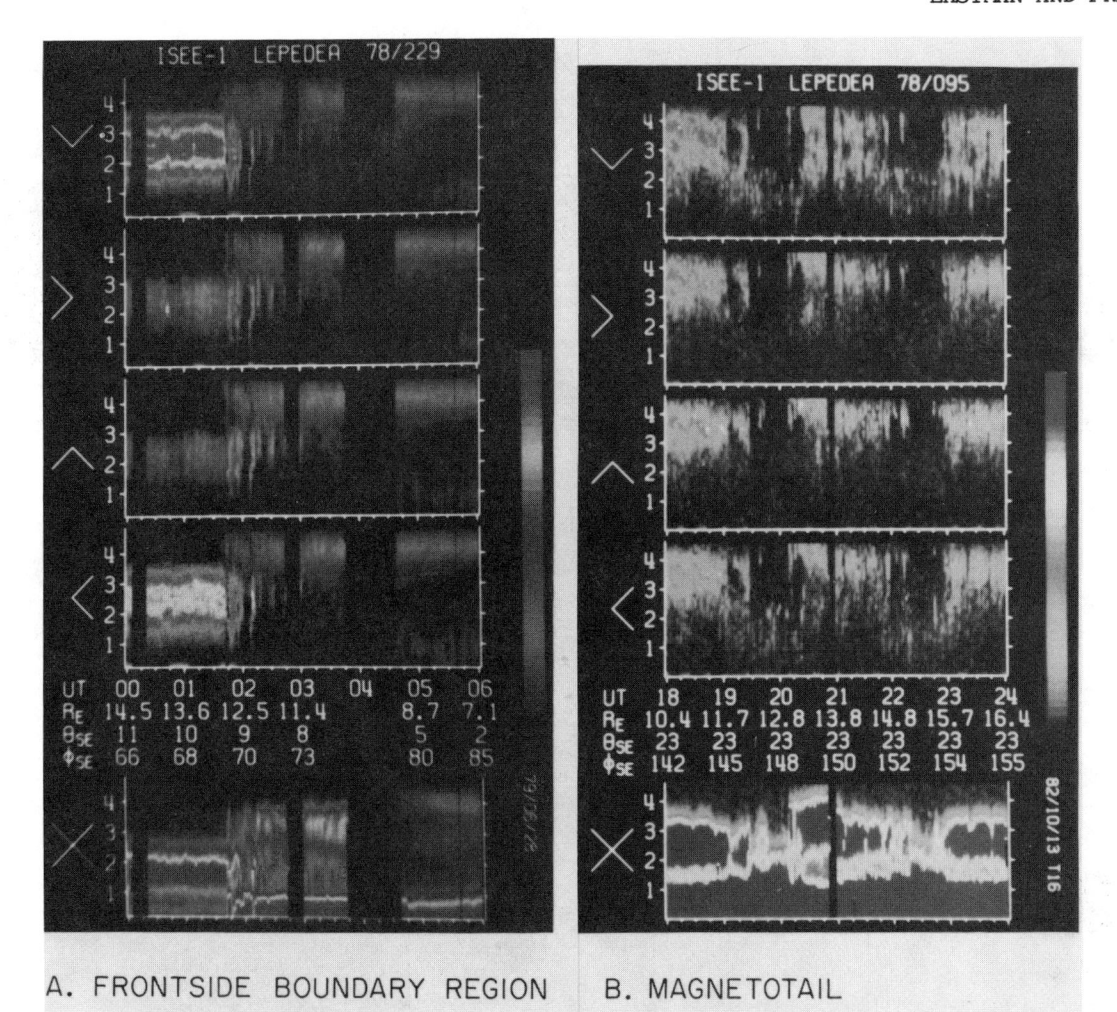

Fig. 2. A. Energy-time spectrogram of the plasma data sampled on 17 August 1978 (day 229) during an inbound crossing by ISEE-1 of the earth's frontside magnetospheric boundary.
B. Energy-time spectrogram of plasma data sampled during an outbound pass of the earth's magnetotail. During this six-hour period on 5 April 1978 (day 95), ISEE-1 moves from 10.4 to 16.4 R_e (earth radii) radial distance at local times of ~ 21 to 23 hours.

128-sec instrument cycle. Detailed use of this energy-spin angle format is illustrated in Eastman et al. (1984a). Values of $\log_{10}$ of the particle energy in units of eV from ~ 1 eV to 45 keV are given along the ordinate. Each frame displays observations from 16 azimuthal sectors (abscissa). Since the spin axis is nearly perpendicular to the ecliptic plane, the abscissa may be interpreted as solar ecliptic longitude of flow direction from 0° (sunward direction) to 360°. Tailward-directed ion flow appears near the center of each plot whereas sunward-directed flow appears along both the left and right-hand borders in this folded scale.

Spectrogram plots of energy versus spin angle are shown in Figures 3A and 3B for the day 229 magnetopause crossing that clearly show the similarity of ion distributions observed in the magnetosheath to those observed in the magnetospheric boundary layer (compare especially the detector responses from 10 eV to ~ 5 keV). Although the plots in Figure 3 are only shown for the detector centered on the ecliptic plane, the full three-dimensional measurements confirm this close similarity in ion distributions sampled on each side of the magnetopause (Eastman et al., 1984a).

The nearby outer magnetosphere as shown in Figure 3C has a higher mean energy of ~ 6 keV and a nearly isotropic angular distribution in contrast to the low mean energies of ~ 200 eV observed in the magnetosheath. The magnetosheath is shocked solar wind plasma which has even lower temperatures of a few tens of eV before crossing the bow shock. Plate 3B shows the "hot" magnetospheric plasma combined with a low-energy component that resembles the magnetosheath plasma in its ion velocity distribution. The superposition of low-energy and high-energy ion components is characteristic of the boundary layers of the magnetosphere, and examples are shown in both Figures 3 and 4.

When crossing the earth's outer magnetospheric boundary, the electrons display a similar change

FRONTSIDE BOUNDARY REGION, ISEE-1 DAY 229 OF 1978

A. MAGNETOSHEATH (0131 UT) — **B. MAGNETOSPHERIC BOUNDARY LAYER** (0151 UT) — **C. OUTER MAGNETOSPHERE** (0155 UT)

MAGNETOTAIL, ISEE-1 DAY 84 OF 1978

D. CENTRAL PLASMA SHEET (1629 UT) — **E. PLASMA-SHEET BOUNDARY LAYER** (1635 UT) — **F. LOBE** (1648 UT)

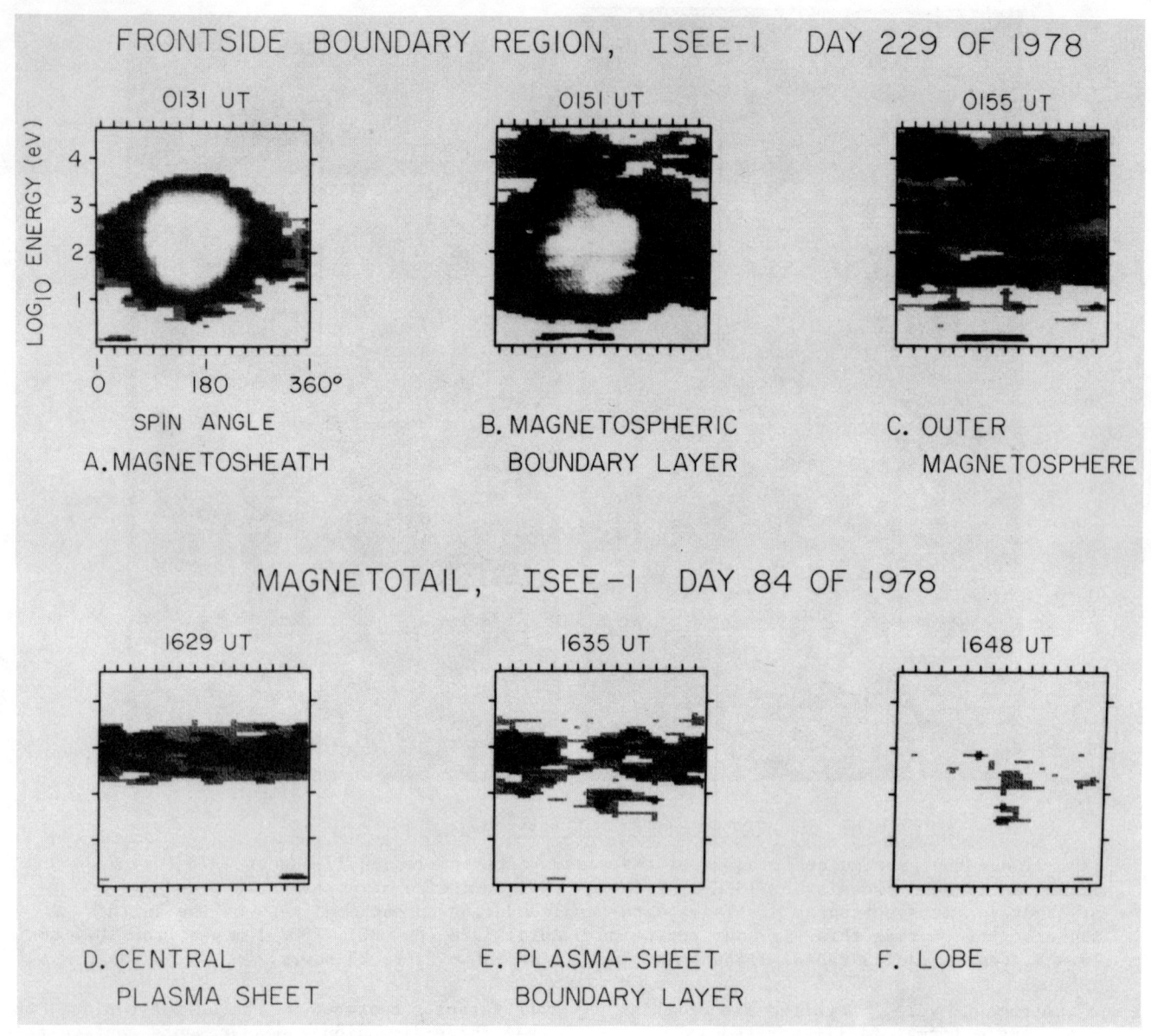

Fig. 3. Ion velocity distributions are shown here in energy-spin angle plots of the ecliptic plane detector response. Dark shading corresponds to higher response except for the center of A and B where the highest response is again white. The scales shown on plot A apply to all energy-spin angle plots in Figures 3, 4 and 8. Plots A-C apply to the frontside boundary crossing of Figure 2A and plots D-F show sample distributions from a crossing of the plasma-sheet boundary layer that occurred during an extended quiet period.

in spectra and mean energy. The top four panels of Figure 2A show ion responses whereas the bottom panel shows electron responses that are averaged over all azimuthal angles. Both the ion and electron responses show a low-energy and a high-energy component within the boundary layer near 0150 UT. Three-dimensional samples of the particle distribution functions show that these components are related closely to the magnetosheath and magnetospheric populations, respectively, in terms of their spectra and pitch-angle distributions. Unlike the ions, the electrons preserve a more isotropic distribution because of their relatively short bounce period (10-20 sec at 200 eV).

Within the boundary layer the electrons often exhibit a field-aligned anisotropy which corresponds to a field-aligned current. For the example shown in Figure 2A, electrons from 20 to 300 eV show a strong field-aligned anisotropy near 0150 UT. Direct numerical integration of the velocity distribution results in a current

MAGNETOTAIL ION DISTRIBUTIONS

ISEE-1 LEPEDEA SPECTROGRAMS, 5 APRIL 1978

ION DISTRIBUTION FUNCTIONS

CENTRAL PLASMA SHEET

PLASMA SHEET BOUNDARY LAYER

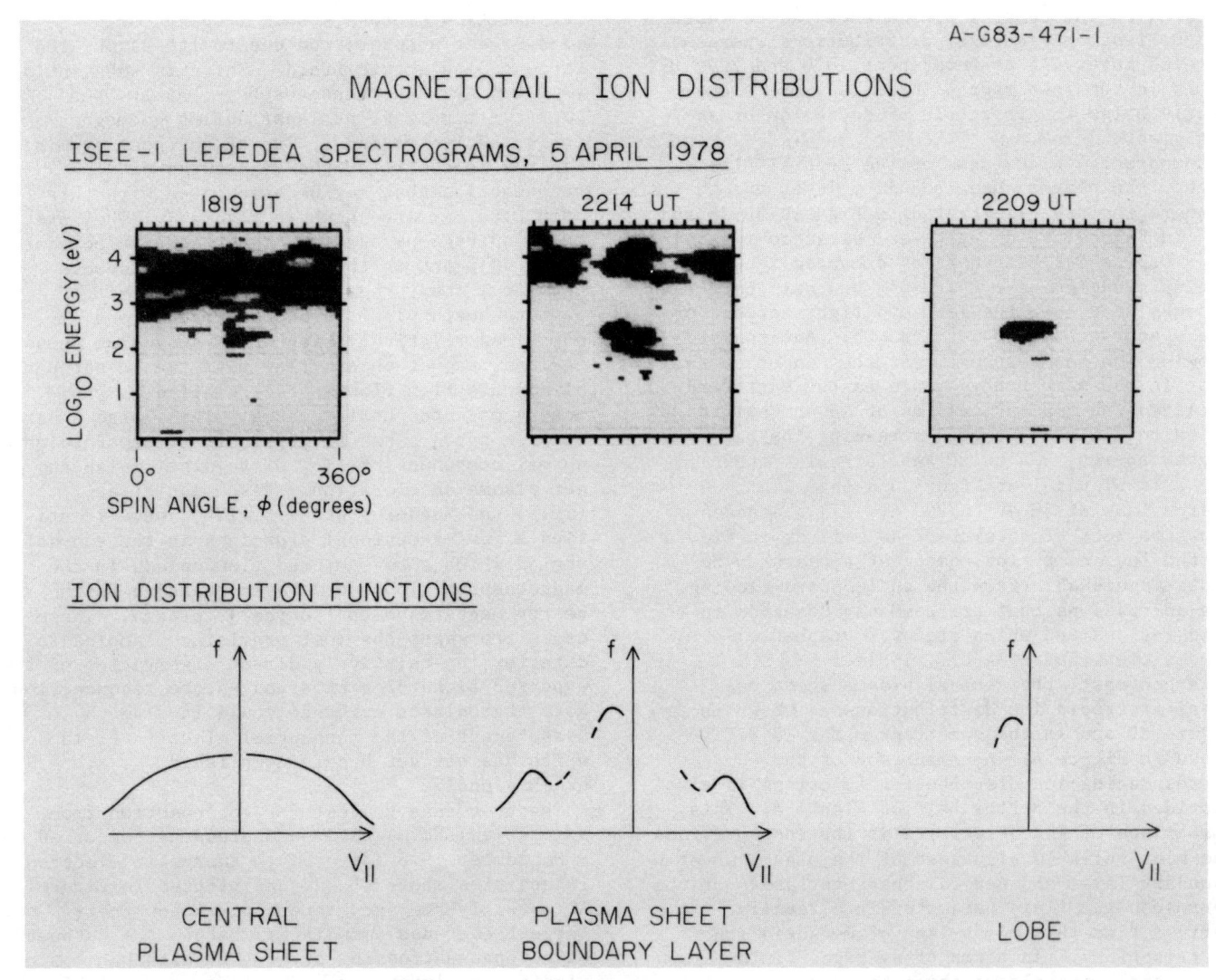

LOBE

Fig. 4. Samples of ion velocity distributions are shown here for a crossing of the plasma-sheet boundary during a magnetically active period ($K_p = 3^+$). For each region, the corresponding reduced distribution $f(v_\parallel)$ is shown.

density of $\sim 4 \pm 2 \times 10^{-7}$ A/m2 which is close to values predicted by the kinetic magnetopause and boundary layer model developed by Roth (1979). Within this same boundary layer interval near 0150 UT, the ion bulk flow is directed at $\sim 110°$ with respect to the ambient magnetic field direction. Polarization electric fields, $\vec{E} = -\vec{V} \times \vec{B}$, are established by this cross-field plasma flow in the magnetospheric boundary layer and depolarizing currents are expected to be produced along the highly conductive field lines that link the boundary layer region to the low-altitude cusp region and the ionosphere (Eastman et al., 1976; Sonnerup, 1980). These measurements thus provide direct evidence of the primary signatures for a MHD dynamo generator process operating within the magnetospheric boundary layer.

Plasma-sheet Boundary Layer

As shown in Figure 1, there is another boundary layer region within the earth's magnetotail called the plasma-sheet boundary layer. Signatures of this region include enhanced anisotropies of the particle distribution functions, high-energy ion beams and enhanced broadband electrostatic noise. The high-speed ion beams observed in the plasma-sheet boundary layer are dominantly field-aligned and flowing sunward and, occasionally, antisunward. Counter-streaming ion beams are also frequently observed. Low-energy ion beams, likely of ionospheric origin, are also commonly observed within this boundary layer. Figure 2B is a survey plot of ion and electron data sampled in the magnetotail. Before 1905 UT, ISEE-1 is located in the central plasma sheet as indicated by the isotropic angular distributions whereas, at time intervals centered near 1950 and 2240 UT, it is in the lobe region where ions are either nearly below the threshold of detection or at energies < 1 keV.

Comparative signatures of the central plasma sheet, the plasma-sheet boundary layer and the lobe region are illustrated in Figures 3D-3F and 4. In Figure 3E, an earthward-directed ion beam at ~ 0.3 to 2 keV appears as enhanced responses for spin angles near 0° (360°) and this beam thus appears near both the left and right borders of the spectrogram plot for 1635 UT. Antisunward-flowing ion beams appear near the center of the plot in the same energy range as the earthward-directed ion beam as well as at lower energies of 50 eV to 200 eV. Counter-streaming ion beams at higher energy, ~ 4 to 20 keV, are also shown in the 2214 UT panel of Figure 4 along with lower energy ions at 50 eV to 300 eV. In the cases shown in both Figures 3D-3F as well as in Figure 4, the low-energy ion component appears to be indistinguishable from the antisunward-flowing, low-energy ions that are commonly observed in the lobe regions and which are of ionospheric composition (Sharp et al., 1981).

In contrast, the central plasma sheet has quasi-isotropic ion distributions as shown in Figure 3D and in the spectrogram for 1819 UT shown in Figure 4. A comparison of the corresponding ion distribution functions is also provided in the bottom half of Figure 4. This comparison of ion velocity distributions provides the most reliable signature of the plasma-sheet boundary layer and several characteristics of the corresponding fluid parameters are readily derived from this comparison of measured ion distributions. In terms of averaged fluid parameters, the plasma-sheet boundary layer is intermediate in number density and mean energy with respect to the low-β lobe region and the medium to high-β central plasma sheet.

The plasma-sheet boundary layer occurs during both quiet and active periods and is a persistent and pervasive boundary region. The plasma-sheet boundary crossing shown in Figures 3D-3F occurred after 10 hours of $K_p = 0$ conditions. However, the plasma-sheet boundary layer during this quiet period has ion beams at much lower energies (0.1 - 2 keV) than the more active period ($K_p = 3^+$) used for the example in Figure 4. During the active period, the ion beams have energies of ~ 4 to 20 keV and are indicative of significantly more acceleration of the source plasma at this time.

Possible Reconnection Signatures near the Frontside Magnetospheric Boundary

As noted above, magnetosheath plasma that crosses the magnetopause boundary to contribute to the magnetospheric boundary layer is primarily de-energized in that process. In contrast, the reconnection process should provide for significant acceleration due to its associated tangential electric field. There is substantial evidence that the magnetosphere has an open magnetic topology, at least during geomagnetically active periods. Two of the most promising regions to check for the signatures of reconnection that may be associated with this open topology are shown in Figure 1. They are the cusp regions and the magnetospheric boundary layer adjacent to the sub-solar magnetopause boundary. Near these regions the oncoming magnetosheath plasma flow is subsonic and sometimes relatively stagnant. Away from these regions, magnetosheath flow near the magnetopause always has high flow speeds relative to the nearby magnetospheric plasma. This large shear in flow speed makes it difficult to separate any normal component of flow or to distinguish any net plasma acceleration. Paschmann et al. (1979) and Sonnerup et al. (1980) have identified a few exceptional crossings in the sub-solar region which show enhanced flow speeds in the magnetospheric boundary layer relative to the nearby magnetosheath (source?) plasma. These cases represent the most promising examples to date for the relatively direct observation of the expected signatures of steady-state reconnection. Note that direct evidence would involve the measurement of the tangential electric field which has not yet been accomplished unambiguously.

The best case to test for reconnection from available ISEE magnetopause crossings occurred on 8 September 1978 (day 251). Energetic electron intensities above 45 keV are plotted in Figure 5. Azimuthally-averaged intensities show their largest decrease immediately before the outbound magnetopause crossing at 0044 UT although significant intensity reductions occur within the last six minutes before crossing the boundary. Two intervals sampled prior to 0044 UT are marked as "boundary layer"; at these times, ion spectra characteristic of the magnetosheath are observed although they are shifted to higher energies corresponding to a net acceleration (Eastman et al., 1982). Within the reconnection model these accelerated ions should be on open field lines. Pitch-angle distributions of the energetic electrons, however, show relatively symmetric peaks near 90° and 270° (see plots 2-5 in the upper-right-hand corner of Figure 5).

If the high-speed ion flow observed in the boundary layer is on open field lines, then the energetic electrons should be very rapidly lost with a time scale on the order of their bounce time (~ 1.5 s). On the contrary, significant intensity levels of energetic electrons are present during the high-speed flow interval as shown in Figure 5. Furthermore, their intensity rapidly decreases near the magnetopause as expected for a transition from closed to open field lines. The intensity decrease during the last 30 s before the magnetopause encounter

ENERGETIC ELECTRONS (≥ 45 keV)
AVERAGE INTENSITIES AND ANGULAR DISTRIBUTIONS

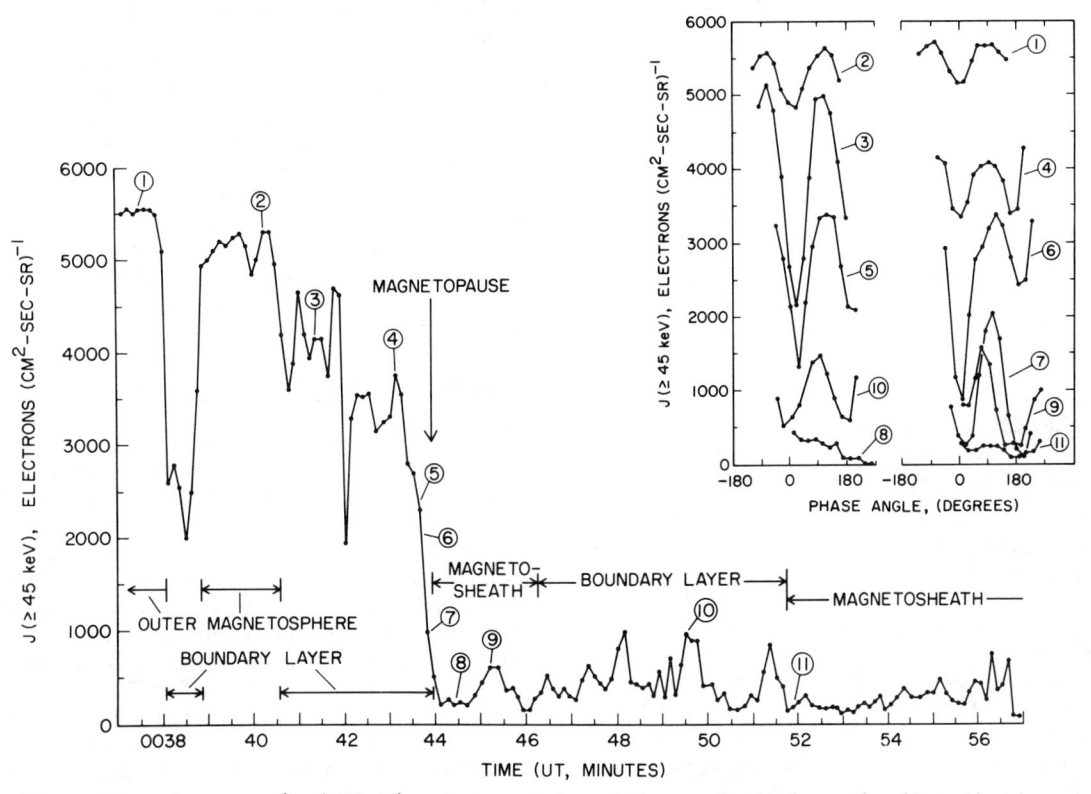

Fig. 5. Energetic electron (≥ 45 keV) average intensities and pitch angle distributions sampled with the ISEE-1 LEPEDEA's Geiger-Mueller tubes. These measurements span the period 0037 to 0057 UT on 8 September 1978. Spin-phase distributions sampled in the ecliptic plane are given at upper right for each numbered period (from Eastman et al., 1982).

occurs with only a 10% change in field magnitude which is inconsistent with simple mirror trapping of these electrons with a magnetic mirror placed in the magnetosheath. Rapid diffusion near the magnetopause is probably a major factor in producing these intensity variations. Without an adequate explanation for observing such energetic electron signatures on open field lines, we tentatively conclude that the boundary layer flow observed during this event is primarily on closed field lines. By itself, these observations are not inconsistent with the reconnection hypothesis as applied to this event although they do suggest a need to consider time-dependent processes.

With the possible exception of some recent observations of "flux-transfer events", the 8 September 1978 ISEE crossing of the magnetopause combined with a small sample of less suitable test cases constitutes our primary (nearly direct) evidence for reconnection in the earth's outer magnetosphere. Reconnection probably plays some role in magnetospheric dynamics and it may even play a major role. However, we need to carefully consider the various competing processes that are likely operating near the magnetospheric boundaries as we attempt to separate those signatures that are uniquely related to reconnection. Some examples include finite ion-gyroradius effects, diffusion, acceleration due to inductive and charge-separation electric fields, effects of curvature and gradient drift, effects associated with non-symmetric pressure tensors, and strong magnetic turbulence.

As noted above, reconnection is expected to result in high-speed plasma jetting away from localized reconnection sites. The velocity of the jet, which is expected to occur primarily in the boundary layer (Yang and Sonnerup, 1977), can be estimated by adding the Alfven velocity, $\vec{V}_a$, based on the difference magnetic field, to the observed magnetosheath flow velocity, $\vec{V}_{MS}$. Thus, the velocity in the boundary layer, $\vec{V}_{BL}$, is predicted to be

$$\vec{V}_{BL} = \vec{V}_{MS} + \vec{V}_a = \vec{V}_{MS} \pm \frac{\vec{B}_{BL} - \vec{B}_{MS}}{(4\pi n_i m_i)^{1/2}}$$

where n_i and m_i denote the number density and mass of source ions in the magnetosheath. The sign of $\vec{V}_a$ depends on the sign of the normal component of the magnetic field at the disconti-

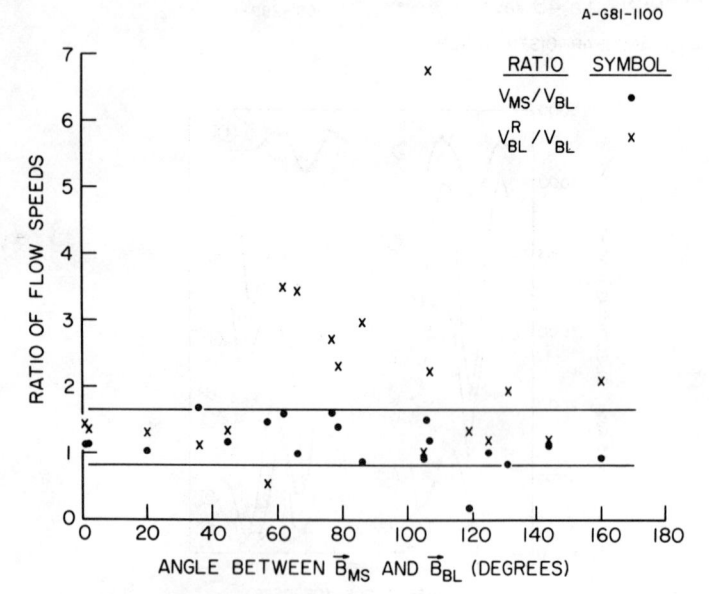

Fig. 6. The ratio of flow speeds as a function of the angle between the magnetosheath and boundary layer field vectors, $\vec{B}_{MS}$ and $\vec{B}_{BL}$, respectively. The reconnection hypothesis predicts that $V_{BL}^R/V_{BL} \simeq 1$ where V_{BL}^R denotes the predicted boundary layer speed and V_{BL} denotes the observed boundary layer speed. Our results show that observed flow speeds for most boundary layer crossings are closely correlated with magnetosheath flow speeds. Furthermore, only six cases have V_{BL}^R/V_{BL} values within 1.25 ± 0.42 (bounded by two horizontal lines) and have an angle less than 36° between $\vec{V}_{BL}$ and $\vec{V}_R$ (from Eastman et al., 1983a).

nuity. This result is limited by the spatial and temporal variations of actual observing conditions. However, comparisons of these predicted velocities for a large number of magnetopause crossings should reveal some significant correlation with observed boundary layer flow velocities provided the plasma and field observations are all treated in three dimensions. It is also assumed that there are no significant tangential flow gradients near the magnetopause. This comparison was done for 17 boundary layer cases selected on the basis of data quality and uniform spatial distribution over the frontside magnetospheric boundary (Eastman et al., 1984a). The results of this comparison are shown in Figure 6 in which flow speeds measured in the magnetospheric boundary layer are compared with flow speeds measured in the nearby magnetosheath plasma as well as predictions based on the steady-state reconnection description. The ratio of observed magnetosheath to boundary layer flow speed, V_{MS}/V_{BL}, is bounded to within 1.25 ± 0.42 whereas the ratio of predicted to observed boundary layer flow speed lies outside of this range in 10 out of 17 cases. Independent of the outcome of our continuing analyses of the 8 September 1978 crossing and related high-speed flow events, our results show that the boundary layer is generally losing plasma, momentum and energy relative to the adjacent magnetosheath source plasma. Thus, at least away from the subsolar region, the magnetospheric boundary layer acts primarily as a generator in which $\vec{E}\cdot\vec{J} < 0$. Scudder (1984) has recently shown that the sign of $\vec{E}\cdot\vec{J}$ for rotational shear layers can be of either sign depending on the tangential Mach number of the external plasma flow. Further study is needed to evaluate the role of reconnection in such shear layers and the extent to which the boundary layer occurs on open vs. closed field lines.

Spatial and Temporal Variations

In-situ observations of plasmas and fields within the earth's magnetotail provide an excellent means of testing theories such as reconnection which can provide predictions of large-scale plasma flows and changes in field topology. The main difficulty in interpreting the available observations is that satellites provide only single-point measurements within a very dynamic system so that spatial and temporal variations are easily confused.

Occasionally some measurements provide the necessary information to uniquely separate these spatial and temporal variations. An example of this was provided by Williams (1981) who reported on observations from the medium-energy-particles experiment, on-board the ISEE spacecraft, that provided three-dimensional measurements of energetic particles from 20 keV up to the Mev range in fixed energy channels. Williams (1981) analyzed ion beams sampled in the plasma-sheet boundary layer and found some cases with a distinct velocity-dispersion signature. In pitch-angle space, the beams evolved into a "smoke-ring" as particles at lower pitch angles first passed the spacecraft leaving a remnant at higher pitch angles. These cases of temporal ion beams were consistent with single-particle motion along field lines from a source ~ 100 R_e (earth radii) away from the earth.

The most commonly discussed spatial regions in the magnetotail are the central plasma sheet and the lobe regions. Dual spacecraft observations provided the first unique identification of the plasma-sheet boundary layer as a distinct spatial region (DeCoster and Frank, 1979). This identification of the plasma-sheet boundary layer as a distinct spatial region is supported by recent ISEE observations of ion velocity distributions sampled as the spacecraft passes from the central plasma sheet into the lobe region. On every crossing of this transition region, a sequence of ion distributions is observed with signatures that are exemplified by Figures 3D-3F and Figure 4. The ion beams often persist through several cycles of the plasma instrument and the "smoke-ring" effect that characterizes the higher-energy ion beams dominated by the velocity dispersion effect is not usually observed at energies below ~ 25 keV. Further analysis is needed to evaluate how common the velocity dispersion signature is for higher energies. Thus, the plasma-sheet boundary layer is a distinct spatial region within the earth's magnetotail within which significant spatial and temporal variations occur.

Evidence for Reconnection in the Earth's Magnetotail

Possible evidence for reconnection in the earth's magnetotail has been the subject of numerous observational studies over the past decade. However, reasonably unique and convincing evidence has continued to be just as elusive as it has been for observations near the frontside magnetospheric boundary as discussed above. Recently, a particularly clear event was selected for intensive study which, on preliminary analysis, showed all the basic signatures expected for reconnection within the neutral-line model. The event occurred on 22 March 1979 and it has been part of the Coordinated Data Analysis Workshop activity known as CDAW 6. Using data from 10 spacecraft both near earth and in the outer magnetosphere along with ground-based observations, it was possible to develop a very detailed time history of the substorm which developed on that day. Figure 7 illustrates the expected sequence of events for three time intervals based on a reconnection model involving only spatial motions of X- and O-lines. The onset of a magnetospheric substorm occurred at 1055 UT. Fortunately, the ISEE spacecraft did not pass directly into the lobe region as so often happens near onset so that useful measurements of plasma flows were then possible during this critical period.

Using the neutral-line model as described by Hones (1979) and references therein, the X-line moves tailward of both spacecraft with earthward-directed ion flow observed at both spacecraft. Next, a new X-line forms near earth resulting in tailward-directed ion flow at both spacecraft as the old x-line convects tailward. Finally, the new X-line moves tailward and is briefly located between the two spacecraft resulting in oppositely directed flow at ISEE-1 and ISEE-2. In particular, net earthward ion flow is observed at ISEE-1 based on measurements by the University of Iowa and Los Alamos/Max-Planck Institute plasma instruments. All of these observations are described reasonably well by the neutral-line model except for one critical problem. ISEE-1 is located tailward of ISEE-2. Thus the direction of flow is opposite to that predicted by this application of the reconnection model.

A relatively complex model like the neutral line model should not be discounted on the basis

Fig. 7. A schematic diagram of anticipated plasma flows and field topology associated with applying the neutral line model to the substorm event of 22 march 1979.

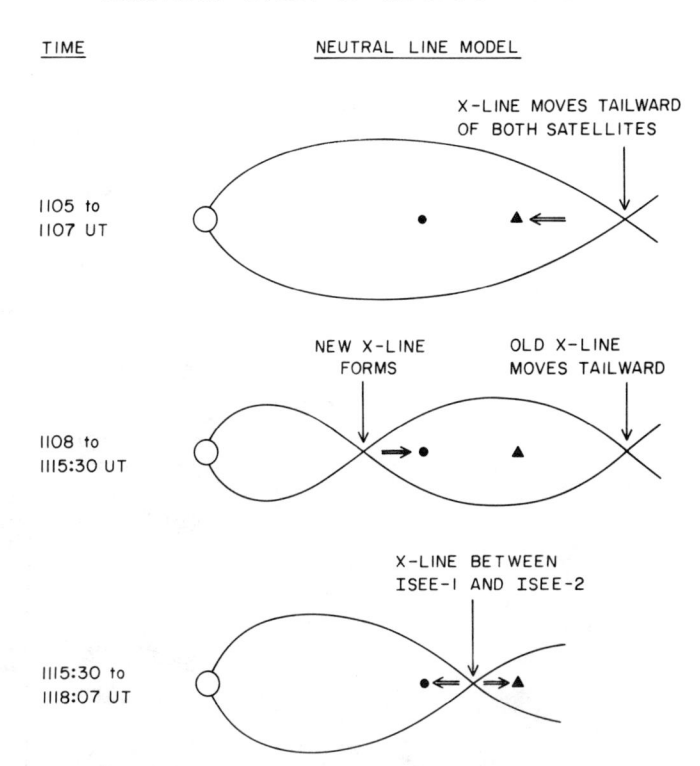

of only a few such puzzling inconsistencies. Reconnection represents a class of problems which are very complex in their application to the interpretation of observations in a region as dynamic as the earth's outer magnetosphere. Additional aspects of agreement and disagreement with the neutral line model as applied to this event are described more fully by Fritz et al. (this volume). More complex scenarios have been developed to interpret the 22 March 1979 event within a reconnection model with multiple X- and O-lines. For this event, however, such interpretations require time variations in topology in addition to spatial variations which can then be used to "explain" almost any conceivable set of observations. During this event, counter-streaming ions were observed by the plasma instruments near 1100 UT, a basic signature of the plasma-sheet boundary layer. Effects associated with that boundary layer complicate the observed flow signatures. Spatial movements of the plasma-sheet boundary past the ISEE spacecraft can explain many features of the 22 March 1979 event without introducing more complex scenarios, whether or not they involve reconnection (Dr. C. Huang, private communication, 1984).

258 BOUNDARY LAYERS

PLASMA SHEET BOUNDARY NEAR ONSET

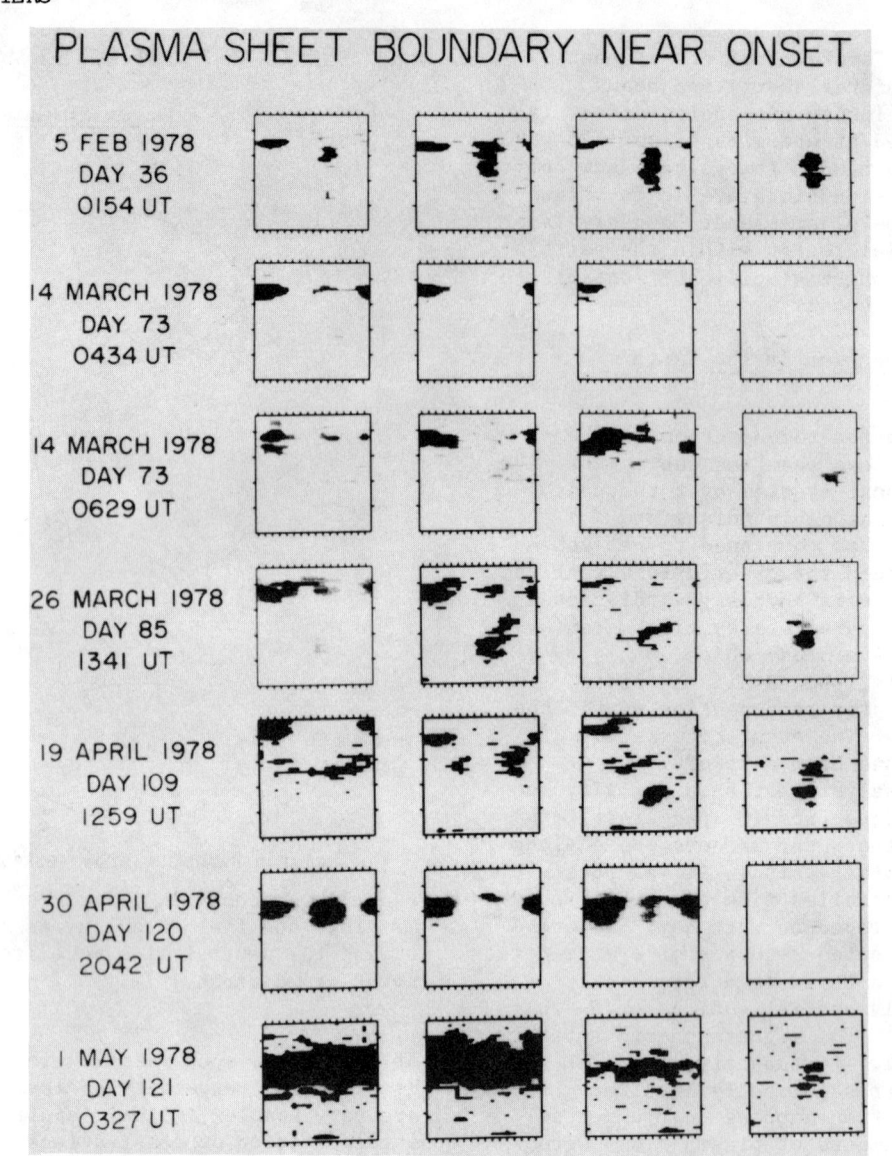

Fig. 8. Energy-phase (E-φ) plots of ion responses from the LEPEDEA equatorial plane detector for seven examples of plasma sheet "dropout". All cases include the last three instrument cycles in high-data-rate (128-sec instrument cycles) sampled within the plasma-sheet boundary layer. The first sample of lobe plasma is shown in the right-hand column and the time of this transition is listed for each event. The center of each plot corresponds to antisunward flow whereas sunward flow corresponds to φ = 0°. Ion flow with duskward or dawnward components occurs with phase angle ranges of 0 ≤ φ ≤ 180° and 180° ≤ φ ≤ 360°, respectively.

Importance of the Boundary Layer Regions

Some of the problems in interpreting data sampled in the earth's outer magnetosphere are related to the presence of boundary layers. For example, seven crossings of the plasma-sheet boundary layer near the onset time of magnetospheric substorms are shown in Figure 8 in the form of energy-spin angle spectrograms similar to those used in Figures 3 and 4. Several of these cases show counter-streaming ions sometimes in conjunction with antisunward-flowing ions at lower energy, e.g., see the second panel for 26 March 1978. Each successive panel for each crossing period represents data sampled during consecutive 128-sec instrument cycles. In each case, more isotropic ion distributions characteristic of the central plasma sheet were sampled at earlier times. The characteristic signatures of the plasma-sheet boundary layer are present during every one of these crossings from the plasma sheet into the lobe. Similarly, as the spacecraft re-enters the plasma sheet, as so often happens during substorm recovery phase, the plasma-sheet boundary layer is again observed on every encounter.

During such boundary crossings near onset and during recovery phase there is one important difference. The counter-streaming ion beams which are at low-energy (below ~ 5 keV) in the 30 April 1978 crossing, have an azimuthally-averaged spectrum which is very similar to the spectra sampled immediately before in the central plasma sheet. This close relationship of average spectra in the plasma-sheet boundary layer and the adjacent central plasma sheet applies to all cases shown in Figure 8 and is a common feature of that boundary region during both quiet and active periods (Eastman et al., 1984b). The plasma sheet boundary moves past the ISEE spacecraft near onset resulting in plasma sheet "dropout". Later, when the plasma sheet is encountered again, particle distributions sampled in the plasma-sheet boundary layer and central plasma sheet are at much higher energies. Particle distributions sampled prior to "dropout" near onset have enhanced energies only when previous activity has already contributed to higher energies in the plasma sheet. High-speed flows sometimes observed near onset are associated with ion beams in the plasma-sheet boundary layer (Eastman et al., 1984b).

The higher energy ion beams sampled in the plasma-sheet boundary layer during recovery times indicate that the acceleration process producing these ion beams provides for larger acceleration during active periods even though the same type of ion beam occurs during both quiet and active periods. This difference is also evident for the quiet period crossing shown in Figure 3D-3F in contrast to the more active period shown in Figure 4. The examples shown in Figure 8 also show the overlap of low-energy, tailward-streaming ion beams from the lobe into the plasma-sheet boundary layer.

MHD models of reconnection only provide predictions concerning the fluid parameters of the plasma including density, velocity and temperature. The multi-component ion distributions observed in the plasma-sheet boundary layer do have well-defined moments of their velocity distributions. However, the proper interpretation of those fluid parameters in terms of cause-effect relations is very much complicated since convection, gradient-drift, etc. can result in ion distributions locally sampled that comprise components with very different sources. For example, the low-energy, antisunward-flowing ions near 100 eV on 26 March 1978 are likely of ionospheric origin and are related to similar ion beams observed in the adjacent lobe. At 5 to 10 keV, high-energy ion beams are observed that are field-aligned and flowing sunward. These ions have been accelerated somewhere tailward of the spacecraft. An antisunward-flowing ion beam is also present that probably resulted from the near-earth mirroring of these same sunward-flowing ion beams. For such multi-component ion distributions in the plasma-sheet boundary layer, it is clearly necessary to evaluate cause-effect relationships for each component in addition to

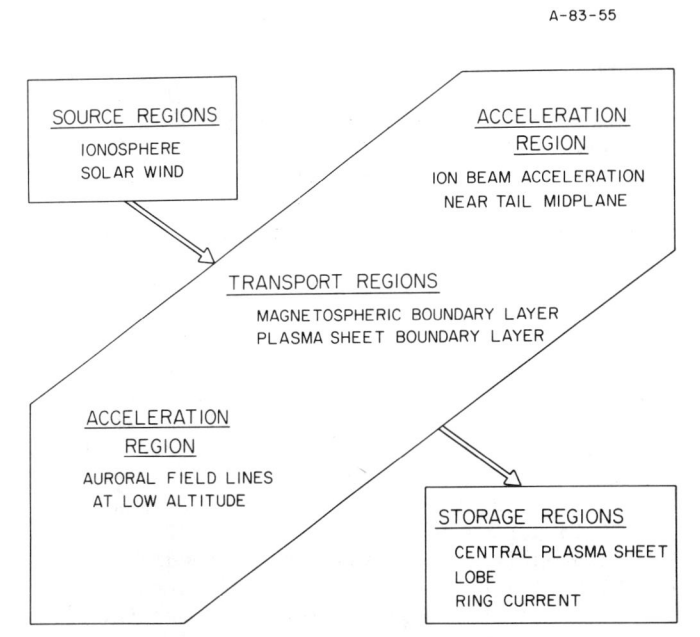

Fig. 9. The functional relationship of the various magnetospheric plasma regimes. There is a close relationship between the acceleration and transport regions via the magnetospheric boundary layers.

describing the fluid parameters of the system.

Transport Regions of the Earth's Magnetosphere

Taken together, the magnetospheric boundary layer and the plasma-sheet boundary layer constitute less than 5% by volume of the magnetosphere although they provide for more than 50% of the plasma and energy transport (Eastman et al., 1984b). The primary relationships of source, transport and storage regions are summarized in Figure 9 in terms of the spatial regions of the magnetosphere shown in Figure 1. Relative to the energy density of plasmas in the source regions, the magnetospheric boundary layer is predominantly de-energized whereas the plasma-sheet boundary layer has been accelerated.

High-speed ion flow also appears to be limited to the boundary regions. For example, available ISEE observations in the earth's magnetotail within a radial distance of ~ 22 Re indicate that high-speed ion flows are observed only in association with transitions from the plasma sheet into the lobe region, or at least in association with close encounters with that boundary (Eastman et al., 1984b). Ion beam acceleration may be closely associated with conditions of a thinned plasma sheet in which the boundary region extends in close to the "neutral" sheet where the sign of B_x reverses. Under such conditions, current-sheet acceleration is likely to play a significant role in particle acceleration (Lyons and Speiser, 1982). Such current-sheet acceleration may also involve reconnection as discussed by Cowley (1980) although these processes are not equivalent. Field-aligned potentials may also develop along the boundary resulting in direct acceleration (DeCoster and Frank, 1979).

The acceleration and transport regions are combined in Figure 9 to indicate their close relationship. In-situ satellite observations in the outer magnetosphere suggest that the boundary layers are critical regions for magnetospheric dynamics.

We have found several common characteristics of the boundary layers:
(1) Significant ion anisotropies with single beams or counter-streaming ion beams which often have positive slopes in the reduced velocity distribution.
(2) Significant spatial gradients in plasma and field parameters (e.g., ∇n, ∇B, ∇P), frequently with medium β (~ 1) conditions.
(3) Significant field-aligned currents carried primarily by 50 eV - 2 keV electrons (Frank et al., 1981).
(4) Enhanced broadband electrostatic noise (Gurnett et al., 1976; Gurnett and Frank, 1977), and
(5) The frozen-in-field approximation as well as the simplified form of Ohm's law are not reliable for scale lengths and plasma parameter ranges that are often observed in the boundary layer regions (Eastman, 1979).

The last two signatures suggest that some diffusive processes are likely important near the boundary layers. Plasma waves may result in significant scattering for both electrons (Tsurutani and Thorne, 1983) and ions (Gary and Eastman, 1979). Wave-particle interactions may also contribute to the general evolution of ion beams observed within the plasma-sheet boundary layer from highly collimated, monoenergetic beams into the broad, quasi-isotropic ion distributions characteristic of the central plasma sheet (Grabbe and Eastman, 1983). All of these effects need to be investigated further and separated from effects that may be uniquely associated with the reconnection process.

Haerendel et al. (1978) first suggested that reconnection may occur in the exterior cusp regions as a localized, transient process. Later, using ISEE magnetometer data, Russell and Elphic (1978) identified signatures of localized current filaments which they interpreted as "flux transfer events" (FTEs) in which flux tubes draped across the magnetopause result from non-steady, impulsive reconnection. Recent reviews by Scholer (1983) and Saunders (1983) provide an update on these recent developments. Assuming that FTEs are signatures of reconnection, then they appear to be the primary way in which reconnection occurs near the frontside magnetospheric boundary. One outstanding question is whether some FTE signatures are really associated with flux transfer at all as opposed to finite current filaments as discussed by Lemaire (1977).

In conclusion, our observations indicate that the boundary layers of the earth's magnetosphere are also its primary transport regions and that acceleration processes in the outer magnetosphere are closely linked to these boundary layers. There are several sources of free energy available in these boundary layers for driving plasma waves which can, in turn, affect the particle distributions. The magnetopause and magnetospheric boundary layer as well as the plasma-sheet boundary layer are typically marginally stable or unstable. If reconnection occurs in association with such conditions then it probably occurs often near these boundaries wherever the local field conditions are appropriate. Thus, more work is needed to evaluate the proper signatures for non-steady-state or sporadic reconnection. However, comparisons with observations will continue to be complicated by the probable signatures of competing processes.

Acknowledgments. We are grateful for some valuable input provided by Dr. C. Y. Huang of the University of Iowa. This research was supported in part by the National Aeronautics and Space Administration under contract NAS5-26257 and grants NGL 16-001-002, NAG-295 and by the Office of Naval Research under contract N00014-76-C-0016.

References

Cowley, S. W. H., Plasma populations in a simple open model magnetosphere, Space Sci. Rev., 26, 217, 1980.
DeCoster, R. J. and L. A. Frank, Observations pertaining to the dynamics of the plasma sheet, J. Geophys. Res., 84, 5099, 1979.
Eastman, T. E., E. W. Hones, Jr., S. J. Bame, and J. R. Asbridge, The magnetospheric boundary layer: Site of plasma, momentum and energy transfer from the magnetosheath into the magnetosphere, Geophys. Res. Lett., 3, 685, 1976.
Eastman, T. E., The plasma boundary layer and magnetopause layer of the earth's magnetosphere, Ph.D. thesis, Univ. of Alaska, Fairbanks, 1979.
Eastman, T. E. and L. A. Frank, Observations of high-speed plasma flow near the earth's magnetopause: evidence for reconnection?, J. Geophys. Res., 87, 2187, 1982.
Eastman, T. E., B. Popielawska, and L. A. Frank, Three-dimensional plasma observations near the outer magnetospheric boundary, submitted to J. Geophys. Res., 1984a.
Eastman, T. E., L. A. Frank, W. K. Peterson, and W. Lennartsson, The plasma-sheet boundary layer, accepted for publication in J. Geophys. Res., 1984b.
Frank, L. A., R. L. McPherron, R. J. DeCoster, B.

G. Burek, K. L. Ackerson, and C. T. Russell, Field-aligned currents in the magnetotail, J. Geophys. Res., 86, 687, 1981.

Fritz, T. A., D. N. Baker, R. L. McPherron, and W. Lennartson, Implications of the 1100 UT March 22, 1979 CDAW 6 substorm event for the role of magnetic reconnection in the geomagnetic tail, 1984 (this volume).

Gary, S. P. and T. E. Eastman, The lower hybrid drift instability at the magnetopause, J. Geophys. Res., 84, 7378, 1979.

Grabbe, C. L. and T. E. Eastman, Generation of broadband electrostatic noise by ion beam instabilities in the magnetotail, submitted to J. Geophys. Res., 1983.

Gurnett, D. A., L. A. Frank, and R. P. Lepping, Plasma waves in the distant magnetotail, J. Geophys. Res., 81, 6059, 1976.

Gurnett, D. A. and L. A. Frank, A region of intense plasma wave turbulence on auroral field lines, J. Geophys. Res., 82, 1031, 1977.

Haerendel, G., G. Paschmann, N. Sckopke, H. Rosebauer, and P. C. Hedgecock, The frontside boundary layer and the problem of reconnection, J. Geophys. Res., 83, 3195, 1978.

Heikkila, W. J., Exit of boundary layer plasma from the distant magnetotail, Geophys. Res. Lett., 10, 218, 1983.

Hones, E. W., Jr., Transient phenomena in the magnetotail and their relation to substorms, Space Sci. Rev., 23, 393, 1979.

Lemaire, J., Impulsive penetration of filamentary plasma elements into the magnetospheres of the Earth and Jupiter, Planet. Space Sci., 25, 887, 1977.

Lyons, L. R. and T. W. Speiser, Evidence for current sheet acceleration in the geomagnetic tail, J. Geophys. Res., 87, 2276, 1982.

Paschmann, G., B. U. O. Sonnerup, I. Papamastorakis, N. Sckopke, G. Haerendel, S. J. Bame, J. R. Asbridge, J. T. Gosling, C. T. Russell, and R. C. Elphic, Plasma acceleration at the earth's magnetopause: Evidence for reconnection, Nature, 282, 243, 1979.

Roth, M., La structure interne de la magnétopause, Aeronomica Acta, Belgian Institute of Aeronomy, A-No. 221, 1980.

Russell, C. T., and R. C. Elphic, Initial ISEE magnetometer results: magnetopause observations, Space Sci. Rev., 22, 681, 1978.

Saunders, M. A., Recent ISEE observations of the magnetopause and low latitude boundary layer: A review, J. Geophys., 52, 190, 1983.

Scholer, M., Energetic particle signatures near magnetospheric boundaries, J. Geophys., 52, 176, 1983.

Scudder, Jack D., Fluid signatures of rotational discontinuities at the earth's magnetopause, submitted to J. Geophys. Res., 1984.

Sharp, R. D., D. L. Carr, W. K. Peterson, and E. G. Shelley, Ion streams in the magnetotail, J. Geophys. Res., 86, 4639, 1981.

Sonnerup, B. U. O., Theory of the low-latitude boundary layer, J. Geophys. Res., 85, 2017, 1980.

Sonnerup, B. U. O., G. Paschmann, I. Papamastorakis, N. Schopke, G. Haerendel, S. J. Bame, J. R. Asbridge, J. T. Gosling, and C. T. Russell, Evidence for magnetic field reconnection at the earth's magnetopause, J. Geophys. Res., 86, 10049, 1981.

Tsurutani, B. T. and R. M. Thorne, Diffusion processes in the magnetopause boundary layer, Geophys. Res. Lett., 9, 1247, 1982.

Williams, D. J., Energetic ion beams at the edge of the plasma sheet: ISEE 1 observations plus a simple explanatory model, J. Geophys. Res., 86, 5507, 1981.

Yang, C.-K., and B. U. O. Sonnerup, Compressible magnetopause reconnection, J. Geophys. Res., 82, 699, 1977.

Questions and Answers

Birn (Comment): I would like to point out that the so-called current sheet acceleration model of Lyons and Speiser is not necessarily different from a model based on particle accleration at a distant neutral line and a subsequent deformation of the distribution function as the particles move earthward along the field lines. The Lyons and Speiser model consists of particles entering the plasma sheet from the lobes and this plasma flow across the separatrix between lobe and plasma sheet is exactly the definition of reconnection given by Vasyliunas.

Speiser (Comment): In my model it is one-dimensional, and so there really is no separatrix uniquely defined. Conceptually it is like a reconnection model with a neutral point or line some distance downstream, so it is consistent with a reconnection model. However as the x-line and separatrices are not specifically defined in this model, to be fair, one could also say the model is consistent with a closed model where the electric field is created by viscous processes.

Eastman: Independent of whether reconnection is associated with establishing the field topology required for current-sheet acceleration, that process can operate whether or not reconnection is actively occurring so, in that sense, the two processes are distinct. Consider a plasma encountering a current sheet with a finite normal field component B_z combined with a linear variation of the tangential field B_x through the layer (with B_x changing sign). As noted by Lyons and Speiser (1982), plasma acceleration will occur under these conditions in the presence of an electric field E_y This electric field thus accelerates the particles as they oscillate across the current sheet until they are finally ejected from it. The first adiabatic invariant is not conserved in the current-sheet acceleration process. An important feature of the Lyons and Speiser (1982) model is that measured source distributions can be input and numerical analysis using the model leads to predicted output ion distributions. We are currently applying measured three-dimensional ion distributions, both for input and output, in evaluating the comparative role of current-sheet acceleration versus other acceleration processes such as direct field-aligned acceleration through an electrostatic potential. Models for MHD reconnection, however, provide no prediction concerning the distribution functions since only predictions about bulk plasma parameters such as density, velocity and temperature result from such models.

Nishida: Can't you distinguish the tailward flows in the lobe and in the plasma sheet by using the flux level as a criterion? Presumably the flux is significantly lower in the lobe.

Eastman: I presume that you intend "plasma sheet" to include both the central plasma sheet and the plasma-sheet boundary layer. Our observations indicate that the tailward flow component at low energy within the plasma-sheet boundary layer cannot, in general, be distinguished from the ion streams observed in the lobe. However, due to the added high-energy ion component, the total flux level and plasma density in the "plasma sheet" is always higher than in the lobe region. There probably is no absolute flux level which could be used as a criterion for region identification. The best criteria known to us at the present time are in terms of characteristics of the ion velocity distributions as shown in Figures 3, 4 and 8.

Reiff: The velocity dispersion seen at the edges of the plasma sheet is essentially identical to that which is observed at the cusps and plasma mantle, with one crucial difference: on the dayside the injection is onto open field lines, and the most energetic particles (electrons and ions) escape to the magnetosheath, leaving only the less-energetic plasma behind. Thus it is not evidence against merging that the plasma mantle or even parts of the low-latitude boundary layer are slower than the adjacent magnetosheath. The energetic electron signatures must be used, and they are suggestive that the low-latitude boundary layer is often on closed field lines, but that can't be used to argue that merging isn't occurring elsewhere.

Eastman: I would not want to imply that not finding the expected signatures of reconnection in a given circumstance should be considered as evidence that reconnection does not occur. Similarly, any failure in attempting to falsify the reconnection hypothesis cannot be taken as evidence for it.

In so far as velocity dispersion process operates in the cusp and plasma mantle and in the plasma-sheet boundary layer, it proceeds in essentially the same way. Plasma enters the source region in a spatially localized manner and then propagates away along trajectories that are predominantly field-aligned. At the same time, an energy-dependent cross-field drift leads to spatial separation which is transverse to the propagation boundary. In both cases the original source region is on open field lines and the ultimate destination for the plasma is the central plasma sheet or inner magnetospere, except for any portion of the plasma mantle that is merely "lost" downstream. Heikkila (1983) has described how this latter possiblity may actually proceed.

Energetic particle observations provide an excellent tracer of field topology although they do not provide unique or conclusive results on the question of open versus closed field lines. Symmetric pitch-angle distributions with high intensities indicate closed topology whereas asymmetric streaming distributions with low intensity suggest an open topology. On this basis (for ions or electrons) the various regions of the earth's outer magnetosphre can be roughly ordered according to the extent to which the energetic particle signatures show these signatures of closed versus open topology.

Region	Topology
central plasma sheet and ring current	closed (always?)
frontside magnetospheric boundary layer	
tailward portion of the magnetospheric boundary layer	
plasma-sheet boundary layer	
lobe regions	
plasma mantle	open (always?)

Vasyliunas: (1) It is a little-known fact that Petschek's model does predict a greatly increased flow speed away from the X line at the edges of the post-reconnection flow region when the tail lobe magnetic field is not uniform but increases toward the earth (see Vasyliunas, Rev. Geophys. Space Phys., 13, 303, 1975, Fig. 7 and associated discussion).

The observation that the largest earthward flows occur near the boundaries of the plasma sheet thus not only does not contradict reconnection but in fact is one of its expected signatures.

(2) Concerning tangential momentum balance at the magnetopause, the prediction $\Delta V = \pm V_A$ applies only to the part of the boundary layer immediately adjacent to the magnetopause, where the plasma has entered locally. Near the flanks, where most of the boundary layer plasma presumably has entered in regions sunward of the observation point, there is no predicted local relation for the average velocity of the entire boundary layer.

Eastman: (1) Yang and Sonnerup (1977) have provided calculated plasma and field parameters for their revised version of the Petschek model. Measured plasma parameters for typical crossings of the frontside boundary layer are not consistent with predictions of that model (Eastman, 1979). The same calculations could be used to represent the plasma-sheet boundary layer as the slow expansion fan. However, the predicted temperature profile is then nearly opposite to what is measured since temperature values increase as one progresses towards the central plasma sheet.

The Petschek model as discussed in Vasyliunas (1975) is a purely symmetric model which is not applicable to the frontside boundary layer. In the magnetotail, most symmetric MHD reconnection models lead to high-speed ion flow near the tail midplane, contrary to our observations. As suggested, the Petschek model can provide for the accelerated flow along a slow shock surface (plasma-sheet boundary?) along the direction of increasing magnetic field. Along the tailward direction, towards decreasing magnetic field, the flow will again peak near the tail midplane. This configuration may apply in the vicinity of neutral lines forming in the deep tail; however, ion observations near the midplane of the distant magnetotail will be needed to test for this possibility.

(2) Dr. Vasyliunas correctly notes that portions of the boundary layer plasma sampled farther from the magnetopause are associated with source locations that are successively farther sunward from the spacecraft along the magnetopause boundary. This involves an inherent limitation of single-point observations made from spacecraft. However, there are at least two considerations that aid in overcoming this limitation for properly inferring the source ion distribution.

1. Checks of the momentum balance condition were made using the first full instrument cycles of data samples on each side of the magnetopause so that the time interval for comparison is ≤ 5 minutes. Comparisons of three-dimensional ion distributions also show a close relationship between the magnetosheath and boundary layer populations suggesting relatively local plasma entry. Temporal variations in magnetosheath flow would map to both temporal and spatial variations in the boundary layer. However, the boundary layer distributions change more in terms of density than in flow direction. Thus effects associated with reconnection should be separable whenever $\vec{V} + \Delta\vec{V}$ differs significantly from $\vec{V}$.

2. With a sufficiently large set of cases, the effect of temporal variations of the source population should average out with respect to the "ideal" source velocity that should be used for evaluating the predicted $\Delta\vec{V}$. Also, large temporal variations will be reflected in large temporal variations in the measured boundary layer distributions which can be evaluated by comparing successive instrument cycles. As noted above, the major effect involves density variations as opposed to variations in the flow directions that are expected from the reconnection hyopthesis.

>35 keV ION OBSERVATIONS FROM ISEE-3 IN THE DEEP TAIL

P. W. Daly, T. R. Sanderson, K.-P. Wenzel

Space Science Department of ESA
ESTEC, Noordwijk, The Netherlands

In October 1982, ISEE-3 made its first pass through the geomagnetotail at a distance of 60 - 90 R_e from the Earth. During the next pass from the end of December to March 1983 the spacecraft reached a down-tail distance of 220 R_e. We present here data from the DFH experiment on board ISEE-3 during these deep tail excursions. The instrument measures ion intensities at 8 energies between 35 and 1600 keV, and at 24 directions, allowing excellent determination of the particle distribution in three-dimensional velocity space.

The dominant feature of the deep tail ion distribution is that they are strongly peaked in the tailward direction. In many instances, the flux at fixed energy is proportional to $\exp(\alpha \cos\psi)$, where ψ is the angle between the ion velocity and some axis of symmetry. This axis is shifted from the GSE-X axis by a few degrees, in the sense expected from the solar wind aberration. The anisotropy factor α is often independent of energy, and varies between 2 and 3. Figure 1 shows data from such an event, from October 22, 1982, at 80 R_e down tail. The data points are plotted as numbers which indicate the energy channel (1 = 35.0 – 56.5 keV; 6 = 384 – 620 keV). The lines drawn through the data points are the result of a fit consisting of three ion species, protons, helium and oxygen, each one a Maxwellian distribution transformed along the symmetry axis by 660 km/s. The fitted temperatures are 9, 26 and 100 keV respectively. In this fitted distribution, channels 1 and 2 are dominated by protons, 3 and 4 by helium and 5 and 6 by oxygen. The resulting helium to proton ratio at 33 keV/Q is 5%, consistent with observations made by composition experiments (G. Gloeckler, private communication).

In the period from January 1 - March 1, 1983, corresponding to GSE X distance of -125 to -220 R_e, there are 10 cases of earthward streaming, two of which last an hour, and the rest of which are no more than a quarter hour.

Comparisons of our data with the magnetospheric plasma regions identified by the Los Alamos electron data for the first 12 hours each of Jan. 24 and 29, show that ion distributions identical to those found in the plasma sheet (high intensity, tailward streaming) are also seen at times in the boundary layer and lobe. This indicates that the energetic ion sheet is in fact wider than the plasma sheet. During the period of earthward streaming in the electrons on January 29, UT 0910 -1010, the energetic ions also exhibited a distribution corresponding to a bulk motion earthward of ~200 km/s. This can be taken as evidence that, at least occasionally, the reconnection X-line can be further down-tail than 200 R_e.

A full report of these results is given in Daly et al. [1982].

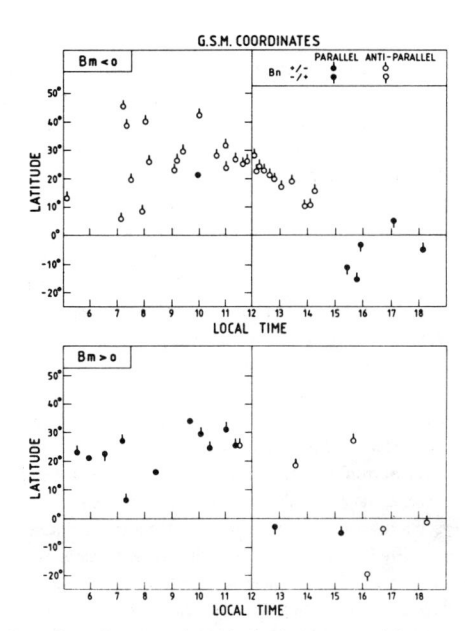

Figure 1. Data from 6 energy channels, 3 telescopes, and 8 sectors plotted as differential flux against the cosine of the angle to the symmetry axis. The lines drawn through the data points result from a fit to proton, helium and oxygen distributions, as described in the text.

References

Daly, P. W., T. R. Sanderson and K. P. Wenzel, Streaming ion distributions in the distant geomagnetic tail: the effect of ion composition mixing, *J. Geophys. Res.*, to be submitted, 1983.

Cowley, S. W. H., R. J. Hynds, I. G. Richardson, P. W. Daly, T. R. Sanderson, K. P. Wenzel, J. A. Slavin and B. T. Tsurutani, Energetic ion regimes in the deep geomagnetic tail: ISEE-3, *Geophys. Res. Lett.*, to be published, 1984.

THREE-DIMENSIONAL COMPUTER MODELING OF DYNAMIC RECONNECTION IN THE MAGNETOTAIL: PLASMOID SIGNATURES IN THE NEAR AND DISTANT TAIL

Joachim Birn

Los Alamos National Laboratory
Los Alamos, NM 87545

Abstract. Two- and three-dimensional computer models of the dynamics of the magnetosphere and in particular the magnetotail have shown, that the basic features of the idealized linear or steady state reconnection theory are still found in time dependent and spatially more complicated configurations such as the magnetotail, which basically resembles a plane sheet pinch but in addition has small magnetic field components perpendicular to the sheet, field line flaring and variations along both directions parallel to the current sheet. These basic features are the formation of a magnetic neutral X-line or separator, where two surfaces separating magnetic fluxes of different topology intersect, with the generation of an electric field along the separator and the production of strong plasma flows parallel to the current sheet away from the separator in opposite directions.

In addition, the computer models of magnetotail dynamics have produced many large scale features that are directly observed or deduced from observation in relation with magnetospheric substorms. Among those features are: the thinning of the plasma sheet, the formation of a plasmoid, a region of closed magnetic loops detached from Earth, which moves tailward at a speed of several hundreds of km/sec, and the generation of field-aligned currents.

In view of the recent discovery of plasmoid signatures in the distant magnetotail at about 200 R_E from ISEE-3 satellite measurements, we discuss the properties of the plasmoid in the computer simulations, in particular its topology, spatial extent and speed, the current system associated with it and its local appearance at a fixed location in space. Furthermore, we discuss the conversion of the energy flux around the separator, current deviations and the occurrence of field-aligned currents and their generation by shear flows.

Introduction

MHD computer models have been used with great success to model reconnection in the earth's magnetosphere and magnetotail. They have been able to show that indeed reconnection can be a very powerful converter of magnetic energy into kinetic energy by a topological change of the magnetic field configuration connected with the formation of magnetic neutral lines. The typical flow pattern around the X-type neutral line or separator predicted by steady state models [Sweet, 1958; Parker, 1963; Petschek, 1964] was found for a variety of different initial and boundary conditions and resistivity models [e.g. Ugai and Tsuda, 1977; Leboeuf et al., 1978; Sato and Hayashi, 1979; Birn, 1980; Lyon et al., 1981; Birn and Hones, 1981; Brecht et al., 1982; Forbes and Priest, 1983; Sato et al., 1983]. The major drawback of these large scale models so far is that the diffusion process that is necessary to enable reconnection is based on some more or less ad hoc model of anomalous resistivity. Whereas this probably has not much effect on the large scale spatial structures of the models, resistive MHD theory has its limitations when finer scales in space and time are considered. In particular, the time scale of the instabilities depends basically on the more or less arbitrary resistivity and the resistive MHD models cannot tell what actually happens within the so-called diffusion region, where deviations from the ideal Ohm's law $\underline{E} + \underline{v} \times \underline{B} = 0$ are important.

There are, however, still many features and structural details, not present in the simplified reconnection theories, that can be explained by a time-dependent MHD model using more realistic two- and three-dimensional geometries. In this paper we will discuss those features on the basis of three-dimensional MHD simulations of reconnection in the magnetotail. In view of the discovery of plasmoid signatures in the distant tail [Hones et al., 1983] the emphasis will be on the properties of the plasmoid found in the computer simulations, its topology, spatial extent and speed, the current system associated with it and its local appearance at different fixed locations. Furthermore, we will discuss current deviations in the reconnection region, the generation of field-aligned currents, and the energy flux in the reconnection region.

The Model, Initial and Boundary Conditions

The computer model uses an explicit leapfrog scheme to solve the non-linear time-dependent MHD equations including constant resistivity. [For more details, see Birn and Hones, 1981; Birn, 1976]. All quantities in the actual numerical calculations and the corresponding dimensionless numbers used in most of the figures are normalized by B_N, the lobe field strength, ρ_N, the density difference between center plasma sheet and lobe, and L_N, the plasma sheet half width (as defined e.g. by $B_x \propto \tanh z/L_N$), all taken at the near-earth boundary $x = 0$, or combinations thereof, e.g. $v_N = B_N/(\mu_0\rho_N)^{1/2}$, $t_N = L_N/v_N$, and $p_N = B_N^2/\mu_0$.

The initial configurations are self-consistent three-dimensional equilibrium models of the magnetotail [Birn, 1979] as shown in Figure 1. The figure shows a quarter cross-section of the tail with the earth somewhere to the left. Full lines

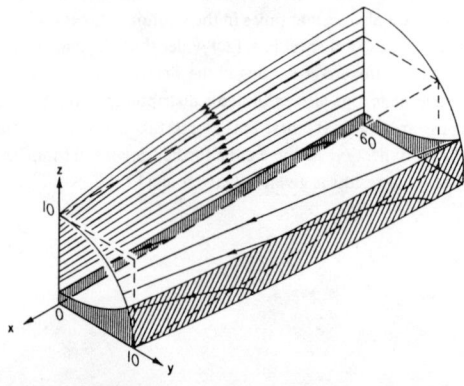

Fig. 1. Schematic representation of a quarter tail section used as initial configuration. Solid lines with arrows represent magnetic field lines in the midnight meridian plane z=0 and on some surface close to the magnetopause. The hatched region indicates the plasma sheet.

Table 1. Parameters of the different computer runs

Run	Plasma sheet broadening λ	Lobe density n_L	Magnetic Reynolds number S
A	~2	0.25	200
B	1	0.1	200
C	~2	0.05	200
D	~2	0.25	1000

with arrows indicate magnetic field lines and the hatched region indicates the plasma sheet. The width of the plasma sheet is defined here not by the transition from closed plasma sheet field lines to open lobe field lines but by the scaling distance L_N on which, for instance, the plasma pressure drops by some factor 2 or 3. The model still basically resembles a plane one-dimensional sheet pinch. There are, however, small, but possibly important, deviations. The first one is the presence of a small normal magnetic field component in the z direction, which is positive in the neutral sheet and becomes negative in the lobes. This is connected with a flaring of the lobe field lines and a decrease of field strength with distance from the earth. The second one is the presence of a y component of the magnetic field which leads to a flaring also in the y direction adding to the decrease in field strength. The third one is a variation (here increase) of plasma sheet thickness and correspondingly the normal field B_z across the tail in the y direction toward dawn and dusk. This is optional and we will compare its effect by comparison with simulations starting from a configuration without such variation (but still presence of B_z and B_y and the corresponding flaring). The actual computation system was a rectangular box (indicated by dashed lines). The size of the box was 10 units in the y and z directions and 64 units in the x direction, all expressed in terms of the typical scale length (plasma sheet half width) in the z direction. The number of grid cells was $16 \times 10 \times 16$ in x, y, and z. A non-uniform cell size in the z direction was used to get a better resolution of the plasma sheet and the singular tearing layer at the neutral sheet within. Assuming a scale length of 2 to 3 R_E the system is close to the actual size of the magnetotail with a length comparable to the distance recently explored by the ISEE 3 satellite. Symmetry was imposed at the boundary planes y=0 and z=0. At the other boundaries we kept the normal magnetic field fixed and set the velocity equal to 0. This is of course not realistic, it corresponds, however, to the boundary condition usually used for stability analysis and is probably the most stable one.

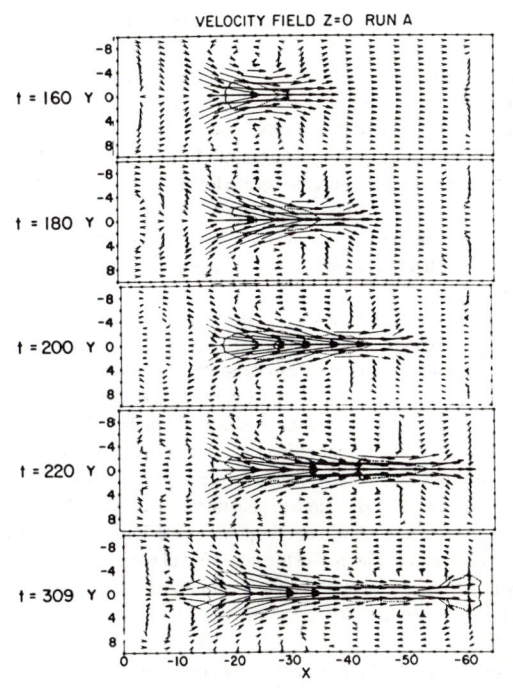

Fig. 3. Flow vectors and magnetic neutral lines (dotted lines) in the x,y plane for run A. After Birn and Hones (1981).

The Time-Dependent Evolution

The time-dependent evolution of the system is initiated by the occurrence of (anomalous) resistivity which causes the system to diffuse slowly. Out of the perturbations grows the tearing-like instability on a faster time scale. We will discuss the evolution and the influence of several parameters in the model on the basis of four different runs (see Table 1). These parameters are:

1) The increase of the plasma sheet thickness and the magnitude of B_z with $|y|$ expressed by the factor λ.
2) The lobe density n_L; in the numerical code, a finite lobe density, somewhat higher than usual at locations within about 60 R_E, was imposed in order to limit the wave propagation speed which determines the maximum possible time step; the value of n_L given in Table 1 is normalized by the difference between the maximum plasma sheet density and the lobe density. Since constant temperature is assumed, n_L also represents the lobe pressure normalized in the same way as the density.
3) The resistivity, which influences the time scale of the unstable growth and possibly also the wave length of the most unstable mode. The parameter S in Table 1 is the magnetic Reynolds number or Lundquist number which is equal to the inverse resistivity in normalized units.

To demonstrate the spatial variation of the four different runs, let us use the scaling parameters $B_N = 50$ nT and $v_N = 1000$ km/s for cases A, C, and D and $B_N = 25$ nT, $v_N = 500$ km/s for case B. These parameters yield a lobe magnetic field strength of about 15 nT at the center location in x and 7 nT at the far end for cases A, C, and D, and of 16 nT and 13 nT, respectively, for case B. The corresponding ion density values (assuming that the ions are predominantly protons) at the near earth boundary, at the center, and at the distant end are 1.49 cm^{-3}, 0.40 cm^{-3}, and 0.32 cm^{-3}, for cases A and D, 1.31 cm^{-3}, 0.62 cm^{-3}, and 0.42 cm^{-3} for case B, and 1.25 cm^{-3}, 0.17 cm^{-3}, and 0.08 cm^{-3} for case C, with

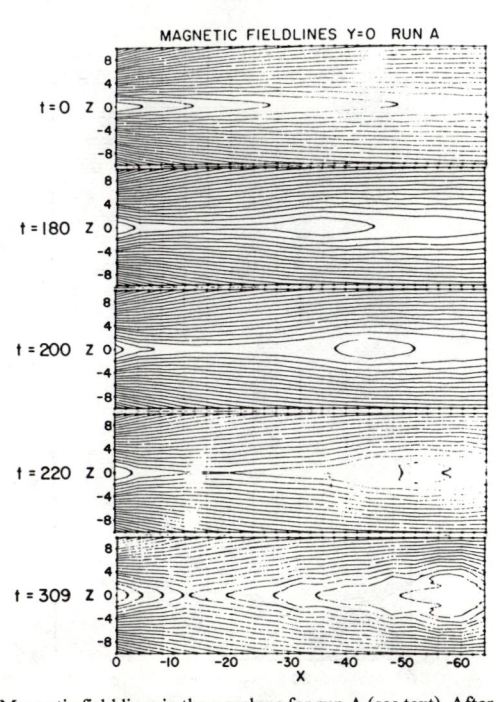

Fig. 2. Magnetic field lines in the x,z plane for run A (see text). After Birn and Hones (1981).

266 THREE-DIMENSIONAL COMPUTER MODELING

lobe densities of 0.30 cm^{-3} for case A and D, 0.12 cm^{-3} for case B, 0.06 cm^{-3} for case C.

As a reference case we use the evolution of the fields as published by Birn and Hones [1981], case A of Table 1, shown in Figures 2 and 3. Figure 2 shows magnetic field lines in the midnight meridian plane (y=0) for different times (in units of a typical Alfvén travel time of about 10-15 sec). We can see plasma sheet thinning, and the formation of closed magnetic loops, the plasmoid that subsequently moves tailward. The structures within the plasmoid may or may not be related to reflections at the far boundary which was assumed to be closed in this run. We will see later that there are characteristic features probably not caused by boundary effects.

The evolution of the velocity field is shown in Figure 3 in the equatorial plane z = 0 by arrows representing velocity vectors. The earth is again to the left. The maximum length of the vectors corresponds roughly to the typical Alfvén speed of the order of 1000 km/sec. The dotted line indicates the neutral line, which consists of X-points in its earthward part and 0-points in the tailward part of a closed line. We can see that reconnection and the occurrence of fast flow are restricted in the y direction to about half the width of the tail or less. To demonstrate the cause for this result we compare this figure with the corresponding figure resulting from run B without the plasma sheet thickening toward the flanks of the tail, shown in Figure 4 in the same kind of representation as in Figure 3. We see that now reconnection and fast flow extends across the whole tail even though we used the very restrictive boundary conditions $\underline{v} = 0$ at the boundary in y.

A characteristic feature of the velocity pattern is the fact that strong earthward flow earthward from the main X-line becomes pronounced much later than the tailward flow. This can be explained mostly by the fact, that the near-earth boundary was closed in the present simulations [see Birn, 1980]. The stronger earthward flows which appear in Figure 4 could suggest, however, that the initial plasma sheet shape may also play some role.

The magnetic field evolution of case B is shown in Figure 5 again in the x,z plane. It looks very similar to the evolution of the reference case (Figure 2). Remarkable, however, is the more pronounced dent at the near earth end of the

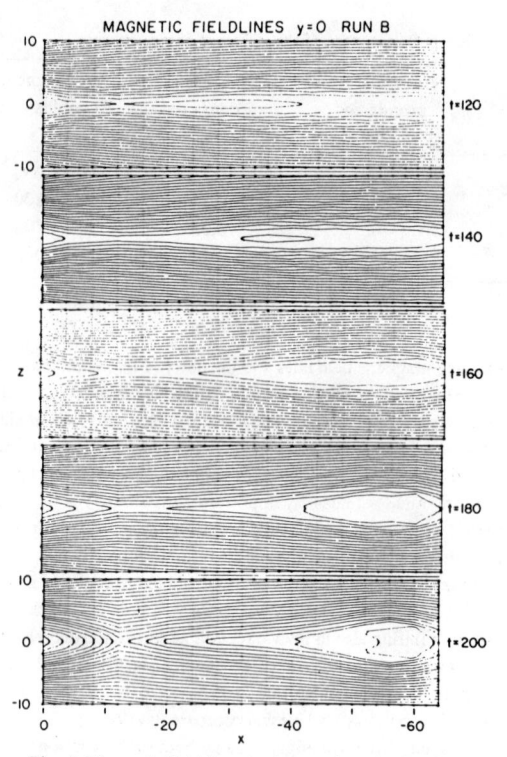

Fig. 5. Magnetic field lines in the x,z plane for run B.

plasmoid. This is most likely not a result of reflections at the tailward boundary. A very similar result was obtained by Sato and Hayashi [1979] in a simulation of driven reconnection with open boundary conditions. The cause of this dent is a reversal of the current direction in this region which will be discussed later. Notice that a satellite crossing this region would see an apparent neutral sheet crossing.

Now let us study the effect of the enhanced lobe density. Figure 6 shows the

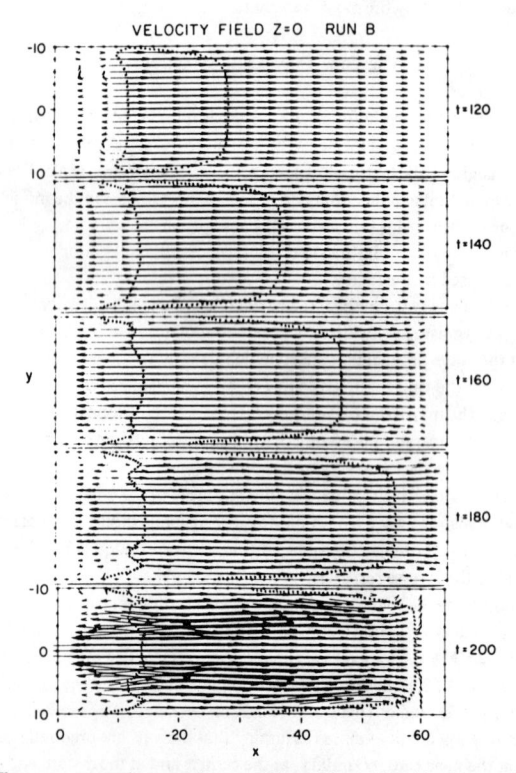

Fig. 4. Flow vectors and magnetic neutral lines (dotted lines) in the x,y plane for run B.

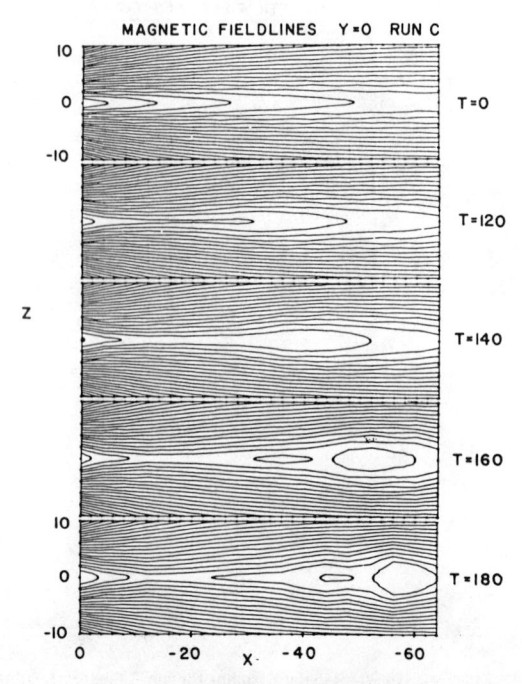

Fig. 6. Magnetic field lines in the x,z plane for run C.

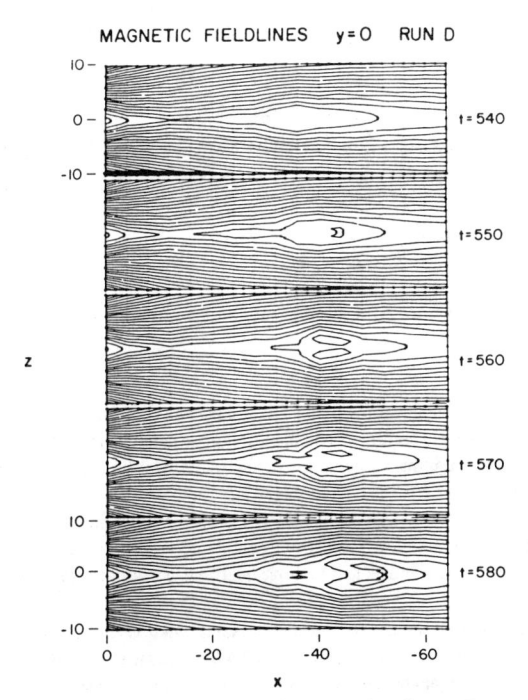

Fig. 7. Magnetic field lines in the x,z plane for run D.

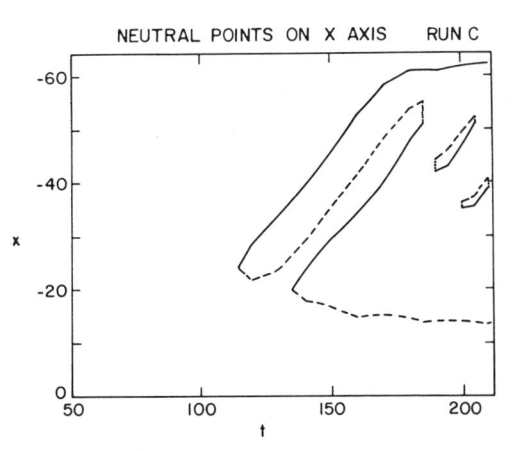

Fig. 8. Location of neutral points on the x axis as a function of time for case C. Solid lines depict X-point locations and dashed lines depict 0-point locations (normalized units).

magnetic field evolution for case C with reduced lobe density. It is again very similar to the previous figures. Figure 6, however, shows more clearly the characteristic wavelength in the x direction leading to a multiple island structure. This waviness was also present in the first example, case A, but could not be seen in the magnetic field figure whereas it was absent in case B. The waves are moving tailward as a part of the plasmoid structure as illustrated by Fig. 7, which shows the location of neutral points on the x-axis as a function of time, for case C. The speed of the plasmoid and the phase speed of the wave stay fairly constant at about 600 km/sec (using $v_N = 1000$ km/sec) despite the changing flow speed. This value is consistent with the typical plasmoid speed of about 700 km/sec concluded by Hones et al. [1983]. The wavelength is very close to the wavelength of the mode that is expected to grow fastest in the linear tearing theory of a one-dimensional current sheet for a magnetic Reynolds number S=200 as used for the simulations discussed so far. This wavelength should increase with increasing magnetic Reynolds number proportional to $S^{1/4}$. We have therefore performed another run with S=1000, run D of Table 1, also to see the effect on the temporal evolution. A part of the evolution is shown in Figure 8 in the same format as Figures 2, 5, and 6. The first thing to point out is the time scale of the figures shown at the right side. Because of the slower diffusion it takes much longer for the instability to grow from the "diffusion noise."

The next thing to look for is the wavelength of the structures in the x direction. It is similar to that in Figure 6, actually even a little bit shorter. This indicates that the wavelength of this characteristic wave is not simply the typical resistivity dependent wavelength of the fastest tearing mode as in usual linear tearing theory. The absence of the waviness in case B indicates that the wavelength in x might be related to the finite equilibrium scale length in the y direction imposed by the thickening of the plasma sheet toward the dawn and dusk flanks. The phase speed of the wave, equivalent to the plasmoid speed, in case D was only 270 km/sec as compared to 450 km/sec in case A and 600 km/sec in case C (using $v_N = 1000$ km/sec in all cases). This seems to indicate a dependency of the phase speed on the lobe density but possibly also on the resistivity. Without further investigation it is not clear whether the waves are part of purely resistive modes or represent superposed ideal MHD modes which produce multiple neutral lines only as a side effect when resistivity allows it. The fact that stable waves of similar periods of 5-10 minutes are a common feature in magnetotail observations [Hones et al., 1978] could support the latter view. As the waves originate in the near earth reconnection region and travel tailward, they are certainly not due to reflections at the distant boundary.

The third feature present in run D is the filamentation in the z direction which starts as a dent in the near earth end of the plasmoid and leads to the formation of multiple islands also in the z direction. It is difficult from single satellite observations in a configuration like this to identify "neutral sheet" crossings. It is very unlikely that these features are caused by boundary reflections because the fast flow velocities do not reach the right boundary before the time of the last frame.

Now, let us see what the magnetic field evolution would look like at some fixed locations within the system. Figure 9 shows the evolution of the magnetic field strength (bottom) and the inclination of the field with time at a location y=0 in the midnight meridian plane at $z = 1.1 R_E$ above the neutral sheet (z = 0) for case C. We have assumed that the initial scale length (plasma sheet half width) is $L_N = 2 R_E$ which accounts for some gradual thinning before onset of the unstable evolution. This location is thus well within the original plasma sheet. The

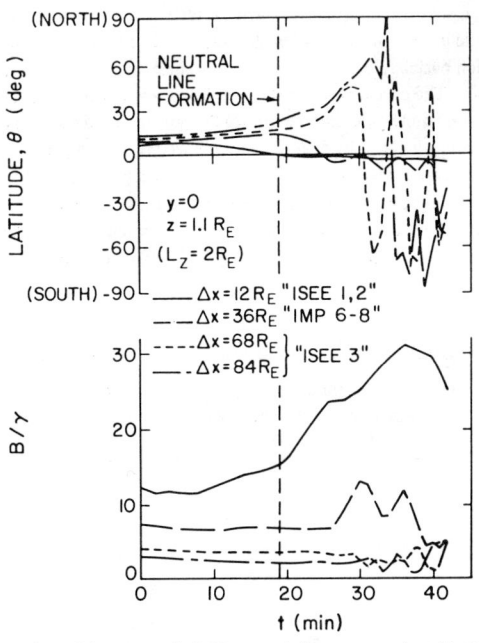

Fig. 9. Evolution of the magnetic field strength (bottom part) and latitude for run C at different locations indicated in the figure.

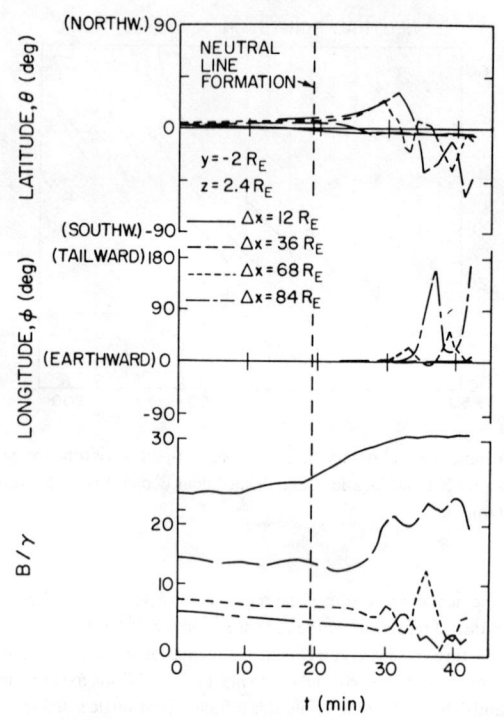

Fig. 10. Evolution of the magnetic field latitude λ, longitude φ, and magnitude B for run C at different locations indicated in the figure.

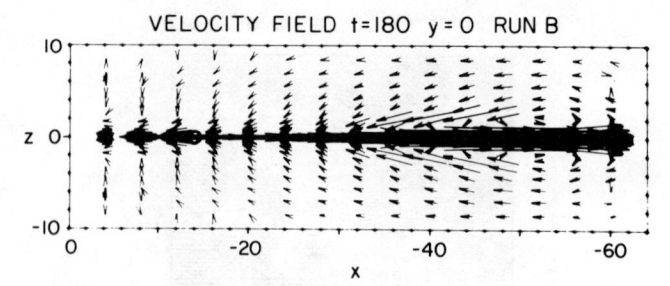

Fig. 11. Velocity vectors in the x,z plane at t = 200 (normalized units) for case B.

magnetic field has been scaled by some unit, which can be chosen arbitrarily, to resemble the actually observed values. The different curves give the evolution at different locations in x tailward from the main X-line. The solid lines correspond to a distance of 12 R_E representing maybe what ISEE 1 or 2 would observe within this system, the dashed lines correspond to a distance of 36 R_E closer to typical IMP 6, 7, and 8 locations, and the dotted and dash-dotted lines correspond to distances of 68 R_E and 84 R_E, respectively, not quite as far as the most distant observations by ISEE 3, but qualitatively in that direction.

Let us start with the solid and dashed curves. The inclination shows that there is only a slight southward dipping of the field starting at about the formation of the neutral line in the nearest location. At the same time the field strength starts to increase which basically shows that the plasma sheet thins and a satellite enters the lobe region. The observations in the far tail as shown by the dotted and dash-dotted curves are quite different: we can see first an enhanced northward field which can become inclined by a large angle followed by strongly southward field marking the arrival of the plasmoid. We can also see a periodicity of roughly six minutes which can be related to multiple neutral lines. At the same time the field strength within the plasmoid is reduced and highly variable. This result is very similar to the actual observations. Lui et al. [1977] pointed out that magnetic signatures on IMP at substorm expansion and plasma sheet thinning mostly consist only of a slight southward dipping. Indeed, a satellite has to be very close to the neutral sheet in this model to see strongly southward field. On the other hand, the plasmoid signatures in the far tail as seen by ISEE 3 [Hones et al., 1983] are remarkably similar to those in the computer model: the strong northward inclination followed by strong southward inclination. Even the quasi-periodicity of 5 to 10 minutes and multiple neutral line passages are a common feature.

Figure 10 shows a similar evolution at the same locations in x but at some higher distance above the neutral sheet, z = 2.4 R_E, and away from the center plane at y = 2 R_E, again for case C. For these locations we also show the azimuthal direction φ of the field, with φ = 0 corresponding to earthward field and φ = 180° corresponding to tailward field. Note that there are two apparent neutral sheet crossings at the most tailward location. Since the location of the "observations" and the position of the neutral sheet at z=0 are fixed in the model, this apparent crossing is caused by the dents in the field structures mentioned earlier. Otherwise the signatures are similar to those of Figure 9. Only the field does not get as strongly northward and southward.

Let us summarize the results of this part:
1) We find a cross-tail extent of the reconnection region of about 8 to 15 R_E for the most realistic configurations, strongly dependent on the equilibrium configuration.
2) The field signatures in the near and in the distant tail are very similar in many structural details to those actually observed.

Energy Flux and Current Patterns

An interesting question is that of the way the energy flows and about its conversion. Table 2 shows some typical contributions to energy inflow and outflow around the separator for case C. There are three contributions considered here, the Poynting flux $\underline{S} = 1/\mu_o \underline{E} \times \underline{B}$, the kinetic energy flux $1/2 \rho v^2 \underline{v}$, and the enthalpy flux $(u + p) \underline{v} = 5/2 \, nkT \, \underline{v}$, which represents the convection of internal, or thermal energy u including deformation effects. This definition makes sense also in an isothermal model, if the isothermal equation of state can be considered the consequence of large thermal conductivity along field lines rather than that of a very large number of internal degrees of freedom of a gas. The heat flux cannot be calculated directly in the present model because of the isothermal assumption. Typical values for each contribution are shown in Table 2 for the three spatial directions. The inflow from the z direction consists mainly of Poynting flux as is expected and the outflow in the x direction, which is much more intense is mainly convection of thermal and kinetic energy. So far, we have the typical conversion picture we are used to from two-dimensional theory. By looking at the inflow from the y direction we see that there is a significant enthalpy flux, which is more intense than the inflow from the z direction, although of course less intense than the outflow.

Another interesting feature is demonstrated by Figure 11 which shows velocity vectors in the x,z plane for case B. The center part of the figure, which is unfortunately not well to be seen because of too much overlap of the velocity arrows, represents the normal flow pattern away from the X-line or separator, marked by the circle. At the boundary of this region tailward from the separator,

Table 2. Energy inflow and outflow around the separator (X-line) for case C

	Inflow from ± z	Inflow from ± y	Outflow in x earthward	Outflow in x tailward
	(erg/cm² sec)	(erg/cm² sec)	(erg/cm² sec)	(erg/cm² sec)
Poynting flux	.021	.018	.003	.001
Bulk kinetic energy flux	.002	.008	.147	.431
Enthalpy flux	.006	.090	.451	.480

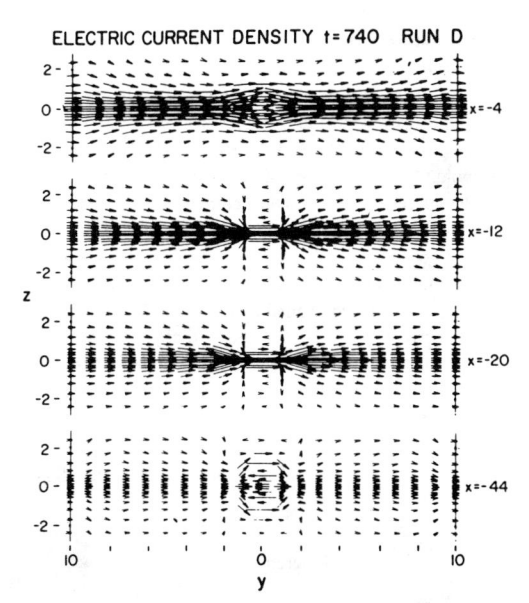

Fig. 12. Projections of electric current density vectors for case D in different cross-sections of the tail as indicated by the locations in x tailward from the near-earth boundary (normalized units).

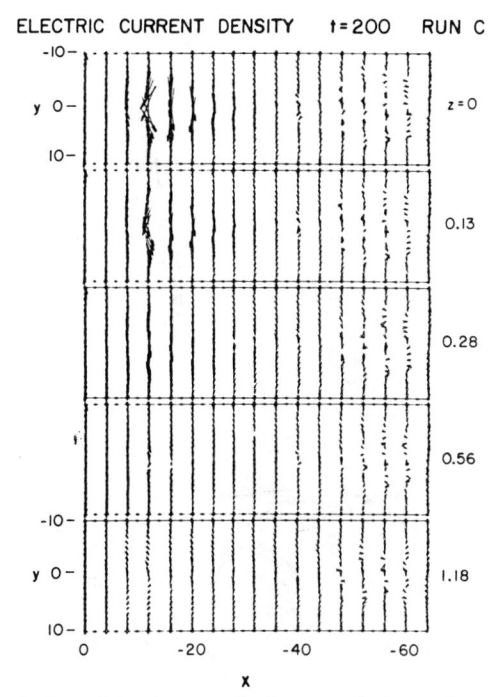

Fig. 14. Projections of electric current density vectors in the x,y plane and several other planes as indicated on the right side for case C.

however, strong earthward flow occurs. This represents another inflowing energy flux mainly along the magnetic field from the tail.

The next figures shall demonstrate the typical current pattern connected with the plasmoid and the reconnection region around the separator. Figure 12 shows projections of the electric current density vectors in cross-sections of the tail at different values of x shown at the right hand side for case D. Only the inner part with the plasma sheet is shown in each case. The look direction is tailward. The top panel is earthward from the separator, the second one close to the separator, and the two bottom ones are tailward from the separator. One can see a strong current concentration at the separator and even at some distance tailward. Earthward from the separator and even more pronounced in the distant tail the current is deviated around the center region. In the distant tail the current inside the plasmoid even changes direction causing the earlier mentioned magnetic field "dents." The same signatures can be found even in case B, where reconnection occurred across almost the whole tail, which shows that the current inversion in the center of the bottom part of Figure 12 is not just a numerical artifact. The splitting of the cross tail current into two layers is closely related to the slow shocks in Petschek's [1964] steady reconnection model. These layers lie indeed inside the separatrix as indicated in Figure 13 which shows magnetic field lines in the x, z plane with the separatrices as dashed lines and the current density maxima for constant x connected by dotted lines, for case B. The current layers coincide approximately with the flow vorticity layers where the plasma flow direction changes from tailward to earthward, shown in Figure 11.

Typical current deviations in the equatorial x,y plane and several other parallel planes are demonstrated by Figure 14 for case C. These deviations produce earthward currents on the dawn side and tailward currents on the dusk side earthward from the neutral line (dashed line). The same current deviation is also found in case B with reconnection across the whole tail. These current deviations do not directly lead to field-aligned currents in the same direction, because there is a rotation of the magnetic field direction in the same direction as the current deviation [Birn and Hones, 1981]. A field-aligned current system that is actually found, which is very similar in all cases, is shown in Figure 15 for case B by

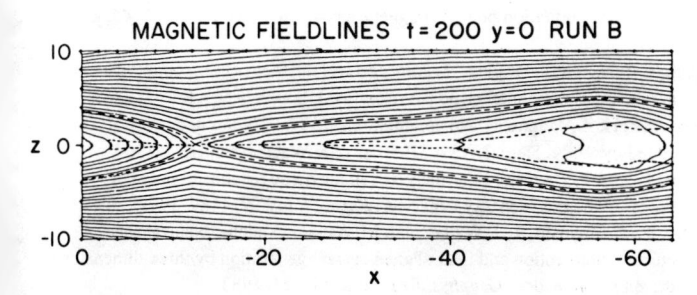

Fig. 13. Magnetic fieldlines in the x,z plane at t = 200 for case B. The dashed line represents the separatrix and the dotted line represents the current density maxima for constant x.

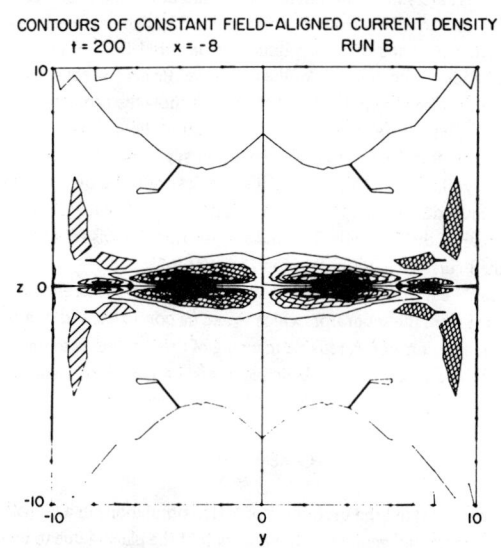

Fig. 15. Contours of constant field-aligned current density in the cross-section x = -8 at t = 200 (normalized units) for case B. Single hatching indicates earthward field-aligned currents and cross-hatching indicates tailward field-aligned currents.

THEORETICAL 3-D MAGNETIC CONFIGURATION
(T = 180) RUN A

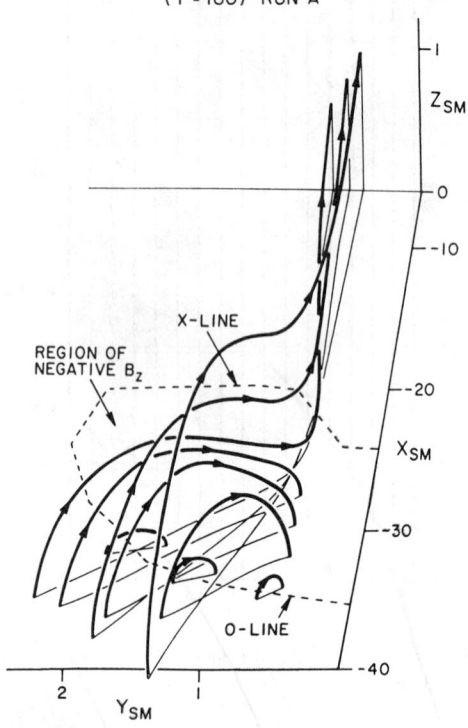

Fig. 16. Three-dimensional representation of magnetic fieldlines for case A after Hones et al. (1982). Projections of the magnetic fieldlines into the x,y plane are shown by light lines.

contour lines of constant parallel current density in a cross-section of the tail at x=−8 earthward from the separator. Earthward currents are indicated by single hatching tailward currents by cross-hatching. The main field-aligned current system in this cross-section has the signatures of so-called region 2 currents [Iijima and Potemra, 1976] as found near the earth, earthward on the duskside and tailward on the dawnside. They are surrounded by some oppositely directed currents corresponding to the region 1 currents observed close to the earth. These currents, however, are smaller in magnitude and are not found in all our simulations. Figure 16, taken from Hones et al. [1982], demonstrates how the main field-aligned current system is produced in our model. The figure shows magnetic field lines on the duskside of the tail above the neutral sheet as seen from the tail in the direction of the earth. Thin lines show the projections of the field lines into the equatorial plane z=0. We see that at the X-line the lowest field line closest to the neutral sheet is convected the most toward the midnight meridian plane y=0, whereas higher field lines are less affected by this convection. The inward convection along the X-line from the flanks of the tail therefore produces a shear of the magnetic field and it is this shear that is responsible for the field-aligned currents with the "region 2" signatures.

The "region 2" currents in the present simulations occur inside the plasma sheet earthward from the separator, which would be consistent with the common view of their generation site. A reliable mapping of those currents to the ionosphere, however, requires a realistic selfconsistent global model including ionospheric boundary conditions and a realistic resistivity model.

Conclusions

We have demonstrated the usefulness of MHD simulations in a realistic three-dimensional magnetotail geometry. Energization of the plasma due to conversion of magnetic into kinetic energy is found without any external driving force. The computer simulations have produced many additional features which can be found in satellite observations in the magnetotail, most recently by using the ISEE 3 satellite at distances up to 220 R_E. Among those features are the spatial limitation of reconnection in the cross-tail y direction, a finite scale of the plasmoid structure in the x direction along the tail usually connected with the appearance of multiple neutral lines, and filamentary structures of the cross-tail current in the z direction caused by characteristic current deviations through the edges of the plasmoid. A characteristic difference between magnetotail field signatures in the distant tail and those in the nearer tail of the model showed a remarkable resemblance to actual observations.

The current system of the plasmoid and at the reconnection site were discussed and it was demonstrated that velocity shear arising in the three-dimensional reconnection model produced field-aligned currents with the signatures of the observed "region 2" currents. The full observed field-aligned current system, however, was not obtained, most likely because the boundary conditions of the model did not include a realistic interaction of the magnetotail with the solar wind and with the ionosphere.

Acknowledgment. This work was done under the auspices of the US Department of Energy.

References

Birn, J., The resistive tearing mode by a two-dimensional finite difference method, in *"Computing in Plasma Physics and Astrophysics,"* edited by D. Biskamp, P. 4, Garching, W. Germany, 1976.

Birn, J., Self consistent magnetotail theory: General solution for the quiet tail with vanishing field-aligned currents, *J. Geophys. Res., 84,* 5143, 1979.

Birn, J., Computer studies of the dynamic evolution of the geomagnetic tail, *J. Geophys. Res., 85,* 1214, 1980.

Birn, J., and E. W. Hones, Jr., Three-dimensional computer modeling of dynamic reconnection in the geomagnetic tail, *J. Geophys. Res., 86,* 6802, 1981.

Brecht, S. H., J. G. Lyon, J. A. Fedder, and K. Hain, A time dependent three-dimensional simulation of the earth's magnetosphere: reconnection events, *J. Geophys. Res., 87,* 6098, 1982.

Forbes, T. G., and E. R. Priest, On reconnection and plasmoids in the geomagnetic tail, *J. Geophys. Res., 88,* 863, 1983.

Hones, E. W., Jr., G. Paschmann, S. J. Bame, J. R. Asbridge, N. Sckopke, and K. Schindler, Vortices in magnetospheric plasma flow, *Geophys. Res. Lett., 5,* 2069, 1978.

Hones, E. W., Jr., J. Birn, S. J. Bame, G. Paschmann, and C. T. Russell, On the three-dimensional magnetic structure of the plasmoid created in the magnetotail at substorm onset, *Geophys. Res. Lett., 9,* 203, 1982.

Hones, E. W., Jr., D. N. Baker, S. J. Bame, W. C. Feldman, J. T. Gosling, D. J. McComas, R. D. Zwickl, J. Slavin, E. J. Smith, and B. T. Tsurutani, Structure of the magnetotail at 220 R_E and its response to geomagnetic activity, submitted to *Geophys. Res. Lett.,* 1983.

Leboeuf, J. N., T. Tajima, C. F. Kennel, and J. M. Dawson, Global simulation of the time-dependent magnetosphere, *Geophys. Res. Lett., 5,* 609, 1978.

Lui, A.T.Y., C. -I. Meng, and S. -I. Akasofu, Search for the magnetic neutral line in the near-earth plasma sheet, 2, systematic study of IMP 6 magnetic field observations, *J. Geophys. Res., 82,* 1547, 1977.

Lyon, J. G., S. H. Brecht, J. D. Huba, J. A. Fedder, and P. J. Palmadesso, Computer simulation of a geomagnetic substorm, *Phys. Rev. Lett., 46,* 1038, 1981.

Parker, E. N., The solar flare phenomenon and the theory of reconnection and annihilation of magnetic fields, Astrophys. J. Suppl. Series, 8, 177, 1963.

Petschek, H. E., Magnetic field annihilation, in *"The Physics of Solar Flares,"* edited by W. N. Hess, NASA SP-50, p. 425, Washington D.C., 1964.

Sato, T., and T. Hayashi, Externally driven magnetic reconnection and a powerful magnetic energy converter, *Phys. Fluids, 22,* 1189, 1979.

Sato, T., T. Hayashi, R. J. Walker, and M. Ashour-Abdalla, Neutral sheet current interruption and field-aligned current generation by three-dimensional driven reconnection, *Geophys. Res. Lett., 10,* 221, 1983.

Sweet, P. A., The neutral point theory of solar flares, in *"Electromagnetic Phenomena in Cosmic Physics,"* edited by B. Lehnert, p. 123, Cambridge Univ. Press, 1958.

Questions and Answers

Hones: Is the earthward flow you showed above and below the plasma sheet related to the boundary at the down-stream end of your model?

Birn: The flows you mentioned are, at least partly, caused by vorticity at the edges of the fast tailward flow within the plasma sheet. These return flows are stronger in case B, where the plasma sheet did not thicken toward dawn and dusk and reconnection took place across the whole tail, such that plasma could not be supplied from the dawn and dusk flanks. Reflections at the down-tail boundary of my model, however, may also play some role, in particular at later stages of the evolution.

Lui: In your simulation of tail reconnection, you have shown that plasma at the location immediately earthward of the neutral line flows tailward. Do you know what force drives the plasma at that location tailward?

Birn: The main reason is the reduction of earthward directed jxB forces due to a reduction of the normal magnetic field component B_z. This reduction of B_z takes place even prior to the formation of the neutral line, when B_z actually changes sign, and it overcompensates the simultaneous increase of the cross-tail current density.

Moore: What is the 3-D structure of the plasmoid around the legs of the "O" line?

Birn: What you call "legs" of the O-line are apparently the parts close to where the O-line joins the x-line and where the neutral line is roughly in the x direction. The closed loops of magnetic field lines which surround the neutral line in this region lie in planes oriented also at small angles to the x-direction. The angle increases somewhat for smaller loops approaching the O-line.

Speiser: 1) In the vicinity of the X-line (upstream and downstream) how does the electric field vary as a function of position parallel to the plasma sheet (x,y) as well as a function of z within the plasma sheet? (As a function of time?) Is any z variation a function of the resistivity model? 2) For the reversed current filaments in the plasmoid, you showed that three current sheets, for example, might be formed. Can you comment on the expected observational signatures and how these might differ from that of a single current sheet (that might, for example, oscillate past the satellite position)?

Birn: 1) The electric field, which consists mainly of its cross-tail component E_y is strongly peaked at the x-line. It remains roughly constant in some region around the x-line. The typical scale lengths on which it falls off from the peak value are about 8-10 in the x-direction, 1 in the y direction, and 0.3-0.5 in the z-direction, all in units of the original plasma sheet half-width. The scale length in the y-direction, however, strongly depends on the initial configuration. If the plasma sheet initially does not thicken toward the dawn and dusk flanks, the scale length for the electric field in the y-direction approaches the tail radius, which is 10 in the present model. The electric field strength increases strongly with time. The variation seems to be closely related to the thinning of the plasma sheet and thus only indirectly dependent on resistivity. 2) Satellite crossings through the reversed current filaments are characterized by weak fields between the first and last field reversal. They can therefore be distinguished from neutral sheet crossings from one lobe to the other, but not from neutral sheet flappings across the satellite which do not lead to encounters of stronger lobe-like fields of the opposite side.

EXTERNALLY DRIVEN MAGNETIC RECONNECTION

Raymond J. Walker and Tetsuya Sato[1]

Institute of Geophysics and Planetary Physics University of California, Los Angeles, CA 90024

Abstract. We have simulated externally driven magnetic reconnection by solving the magnetohydrodynamic equations in an initially plane current sheet. Both two dimensional (2D) and 3D versions of the model have been developed. In this model, we postulate that reconnection in the tail is triggered by a local compression of the plasma sheet which results from an invasion of the solar wind into the magnetotail. Thus we start the simulation by introducing flow from the lobes normal to the plasma sheet. When resistivity is generated in a local region of the neutral sheet reconnection develops and magnetic energy is converted into plasma bulk flow. As the reconnection proceeds, the cross tail current is concentrated in two thin slow shock layers. On the downstream side of the slow shocks strong plasma flows away from the reconnection region are generated. The flows near the equator are normal to **B** while those in the slow shocks are along **B**. Near the equator the flows exceed the local magnetosonic velocity. In addition, plasma from regions adjacent to the reconnection region is drawn into the reconnection region thereby creating appreciable flows ($\sim .2\ V_A$) in the Y (cross tail) direction. Our simulation results, also, demonstrate that the night side substorm current system is a natural consequence of driven magnetic reconnection. The dawn to dusk cross tail current is interrupted locally and field aligned currents are generated. The field aligned current flows towards the ionosphere on the morning side and away from the ionosphere in the evening. The field aligned currents flow in a narrow band at the outer edge of the plasma sheet. The extent of this field aligned current system is limited in the Y direction with the largest currents near the edges of the reconnection region. In the equatorial plane, current vortices form connecting the reconnection region and the region of reconnected field lines. In the region of reconnected field lines the current is now from dusk to dawn rather than dawn to dusk. The $\mathbf{J} \times \mathbf{B}$ force in this region opposes the flow from the reconnection region. The area of dusk to dawn current also is the region where the flow becomes super magnetosonic and is characterized by a rapid decrease in pressure. The sharp decrease in pressure is a fast shock. The super magnetosonic flow is maintained by this sharp pressure gradient.

I. Introduction

Magnetic reconnection in the tail is widely believed to be the direct cause of magnetospheric substorms. Based on observational evidence, a phenomenological model of substorms in the tail has been developed. In this near-earth neutral line model, reconnection on the dayside magnetopause is followed by the formation of a new neutral line in a localized region in the near-earth ($\sim 10\text{-}15\ R_E$) plasma sheet (McPherron et al., 1973; Hones, 1973). On the earthward side of the neutral line, the plasma sheet thickens and the field becomes more dipolar. The earthward plasma sheet is characterized by earthward flow and a northward magnetic field.

[1] Also at Institute for Fusion Theory, Hiroshima University, Hiroshima 730, Japan.

On the tailward side of the neutral line, the plasma sheet thins as a magnetic bubble flows tailward away from the Earth (Pytte et al., 1976; Hones et al., 1976). During the tailward flow the normal magnetic field component in the plasma sheet should be southward.

Several observational studies have inferred a localized transient field aligned current system near midnight during substorms (Akasofu and Meng, 1969; McPherron et al., 1973; Bostrom, 1974) (Figure 1). This current flows into the ionosphere on the morning side and away on the evening side and is believed to connect to the ionospheric westward electrojet (see the recent review by Kamide (1982) and Baumjohan (1982) for a complete list of references). Atkinson (1966) suggested that such a current system would result from localized reconnection in the magnetotail and Sato (1982) has developed a model for the resulting field aligned currents based on his 2D simulation of forced reconnection. It should be noted that Akasofu and Kan (Kan and Akasofu, 1978; Akasofu, 1980) on the other hand, have argued that reconnection is not necessary to generate the substorm current systems. They have proposed a current interruption model in which the cross tail current is diverted into the ionosphere without reconnection.

Several numerical magnetohydrodynamic (MHD) models of reconnection in the tail have been developed. These include studies of tearing mode reconnection (Birn and Hones, 1981; Sato and Walker, 1982; and Forbes and Priest, 1983) and studies of driven magnetic reconnection (Sato, 1979; and Sato et al., 1983a,b). Birn and Hones (1981) have presented a three-dimensional (3D) simulation of tail dynamics. Starting with Birn's (1979)

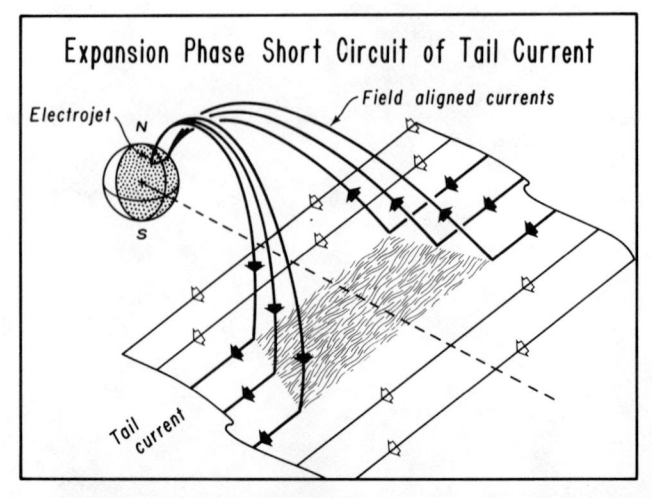

Fig. 1. Schematic drawing of tail current interruption and the generation of field aligned currents during substorms (courtesy of R.L. McPherron).

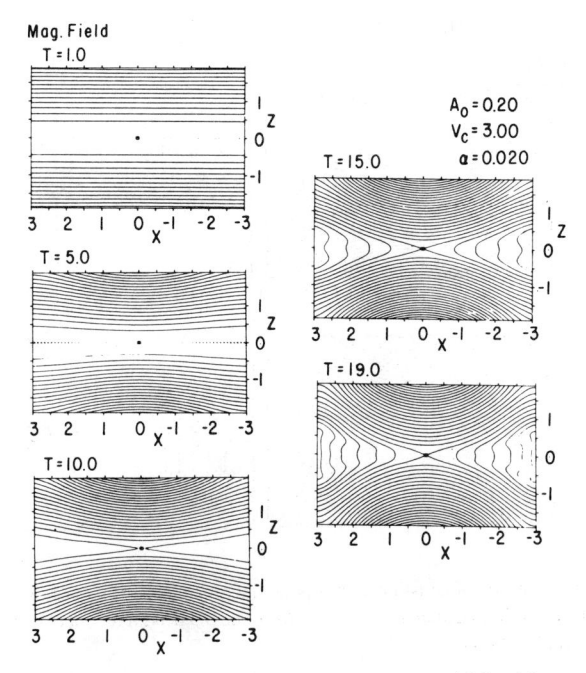

Fig. 2. Time series of contours of constant vector potential for driven magnetic reconnection in two dimensions (Sato, 1979).

self-consistent 3D tail models, they solved the MHD equations throughout the tail. In this calculation, the tearing mode instability was driven by a sudden increase in resistivity. The calculation reproduced many of the features of the near-earth neutral line model. After the resistivity was turned on, the plasma sheet began to thin and an X-type neutral line formed. A bubble formed tailward of the neutral line and began moving down the tail. Birn and Hones also observed field aligned currents in their model substorm. The field aligned currents were directed tailwards on the dawnside and toward the Earth on the duskside. The polarity of these currents is opposite to that of Figure 1. They suggested that their current system was responsible for the substorm associated changes in the region 2 currents of Iijima and Potemra (1976).

There is controversy over the importance of the resistive tearing mode in tail reconnection. As discussed above, the simulations of Birn and Hones suggest that tearing is important and can lead to appreciable flows in the tail. Sato and Hasegawa (1982) have argued that the resistive tearing mode instability saturates at a flow speed which is too low to be important during substorms. Ugai (1982), also, has argued that tearing does not lead to the explosive reconnection which is needed to account for substorm observations.

In the second type of tail model, reconnection was assumed to be triggered by a local compression of the plasma sheet as a result of an invasion of solar wind into the magnetotail (Sato, 1979; Sato et al., 1983a,b). In this driven reconnection model, reconnection results from flow from the lobes normal to the neutral sheet. In this paper, we report on the results of the driven reconnection calculations. In particular, we will discuss the changes in the magnetospheric currents which result from driven reconnection with emphasis on the generation of field aligned currents. We also will discuss the plasma flows which result from driven reconnection. These results will be compared with the results of the tearing mode model and observations.

II. Simulation Model

We have simulated externally driven reconnection by solving the single fluid magnetohydrodynamic equations in a tail-like plasma and field configuration. Both two-dimensional (2D) (Sato, 1979; Sato and Hayashi, 1979) and 3D (Sato et al., 1983a,b,; Sato and Hayashi, 1982) versions of the model have been developed. In the following discussion, we will describe the 3D model noting the places where it differs from the 2D model. In order to maximize the spatial resolution of the model, we started with a very simple neutral sheet configuration. We adopted a Harris (1962) type magnetic field configuration. The magnetic field and plasma configurations we used are

$$\mathbf{B}(x,y,z) = (B_o \tanh(z/L), 0, 0)$$

$$\rho(x,y,z) = \rho_o$$

$$T(x,y,z) = T_o \operatorname{sech}^2(z/L)$$

where $\mathbf{B}$ is the magnetic field, ρ is the mass density, T is the temperature, B_o, ρ_o, T_o are the constants and L is the half width of the neutral sheet in the z (north-south) direction.

As noted above, we assumed that reconnection is triggered by a non-uniform compression of the plasma sheet somewhere in the tail accompanied by inflation of the magnetic flux in the lobe region. In order to compress the plasma sheet, we injected a magnetized plasma (Poynting flux, kinetic energy flux and thermal energy flux) through the north and south boundary planes ($z = \pm L_z$) of the simulation box. Initially, the incoming mass flow pattern was such that

$$\rho v_z(z = \pm L_z) = \mp \frac{A_o}{4}\left(1 + \cos\frac{\pi x}{L_x}\right)\left(1 + \cos\frac{\pi y}{L_y}\right) \quad (1)$$

where v_z is the z-component of the velocity, L_x, L_y, L_z are dimensions of the simulation box and A_o is a constant. ρv_x and ρv_y were initially set to zero. Later they were adjusted so that the mass flow vector was always perpendicular to $\mathbf{B}$ at the boundary. The inflow pattern in (1) enabled us to reduce the physical domain of the simulation box. We did this by assuming symmetry (or anti-symmetry when appropriate) about the equator (z=0), the noon-midnight meridian (y=0) and the neutral line (x=0). Thus we

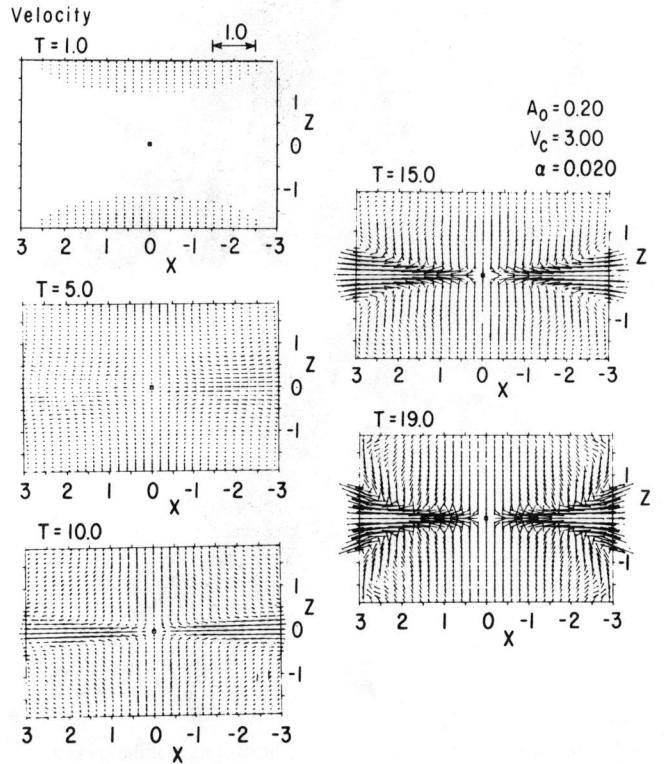

Fig. 3. Time series of plasma flow vectors corresponding to the magnetic field configuration in Figure 2 (Sato, 1979).

274   EXTERNALLY DRIVEN MAGNETIC RECONNECTION

were able to reduce the simulation domain by a factor of eight and thereby to increase the resolution of the model. The outflow and side boundaries (x = ± L_x and y = ± L_y) were assumed to be free boundaries through which plasmas can freely enter or exit. In the 2D model closed boundaries with B_z = 0 and v_x = 0 also were used.

Recent particle simulations have demonstrated that anamalous resistivity can be generated by the lower hybrid drift instability (Winske and Liewer, 1978; Tanaka and Sato, 1981a) and that it is more strongly generated and lasts much longer in the presence of an external force that compresses the plasma sheet (Tanaka and Sato, 1981b). The resistivity is roughly proportional to the square of the diamagnetic drift velocity (Huba et al., 1978). Thus we have adopted the resistivity (η) model

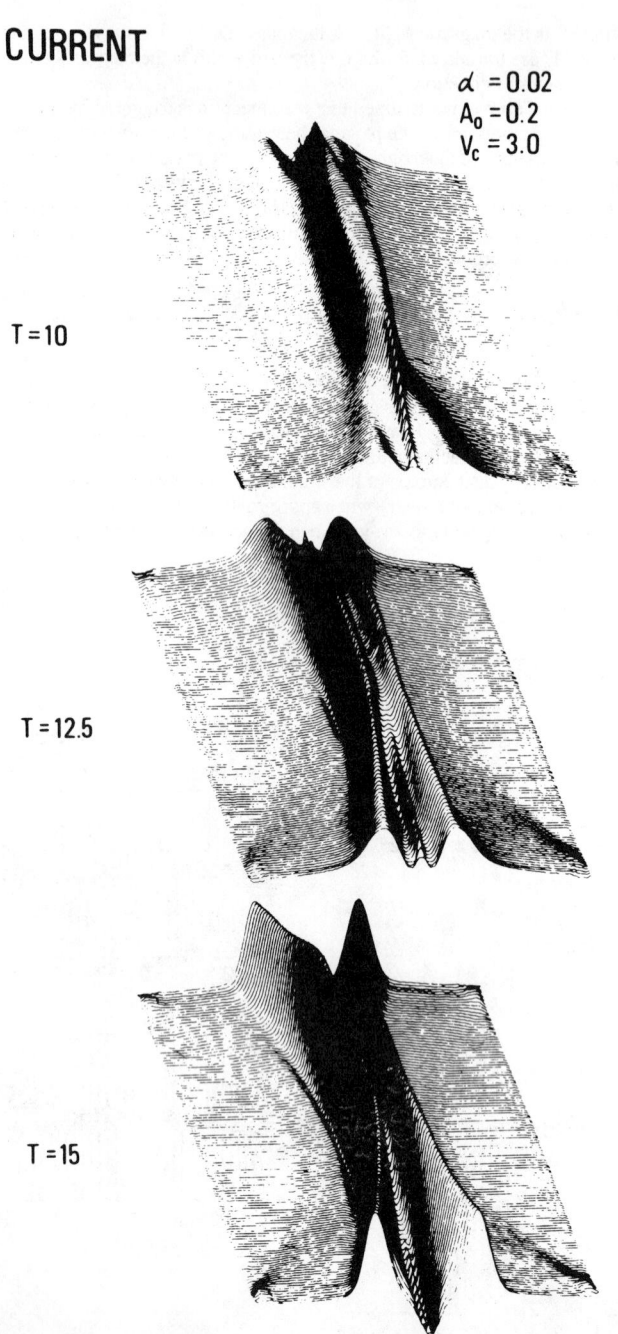

Fig. 4. Three-dimensional display showing the evolution of the cross tail current during the reconnection event in Figures 2 and 3. The x-direction is from top to bottom and the z-direction is from left to right (Sato, 1979).

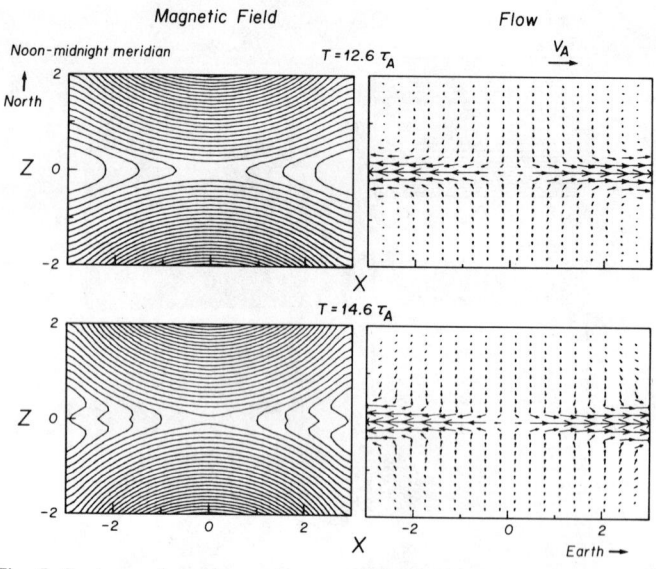

Fig. 5. Contours of constant vector potential and plasma flow vectors in the noon-midnight meridian plane (y=0) from the 3D driven reconnection run (Sato et al., 1983a).

$$\eta = \alpha(V_D - V_C)^2 \text{ for } V_D > V_C$$
$$\eta = 0 \qquad \text{otherwise} \qquad (2)$$

where V_D is the diamagnetic drift velocity and α and V_C are constants. V_D = J/ne where J is the neutral sheet current density, n is the plasma density

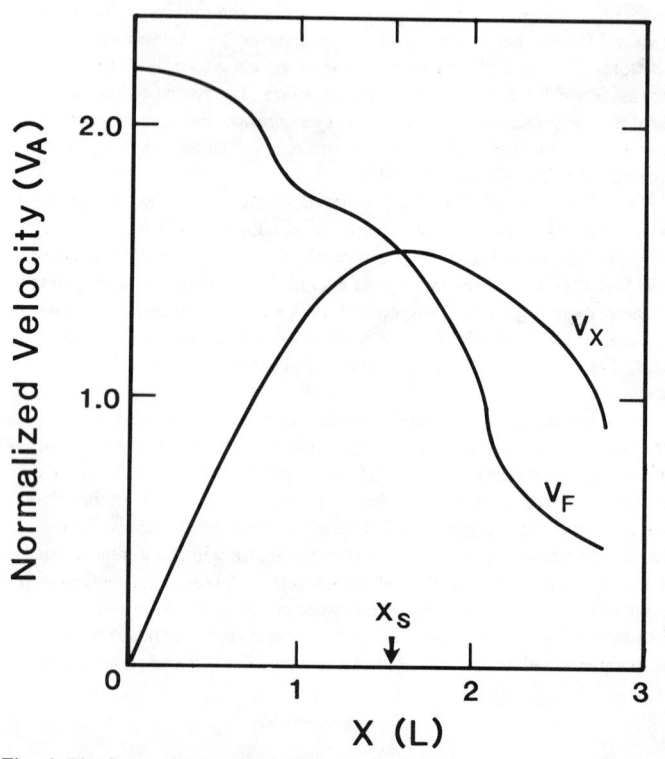

Fig. 6. The flow velocity (v_x) as a function if distance along the x axis. The neutral line is at x=0. The local magnetosonic velocity is labeled V_F (Sato et al., 1983b).

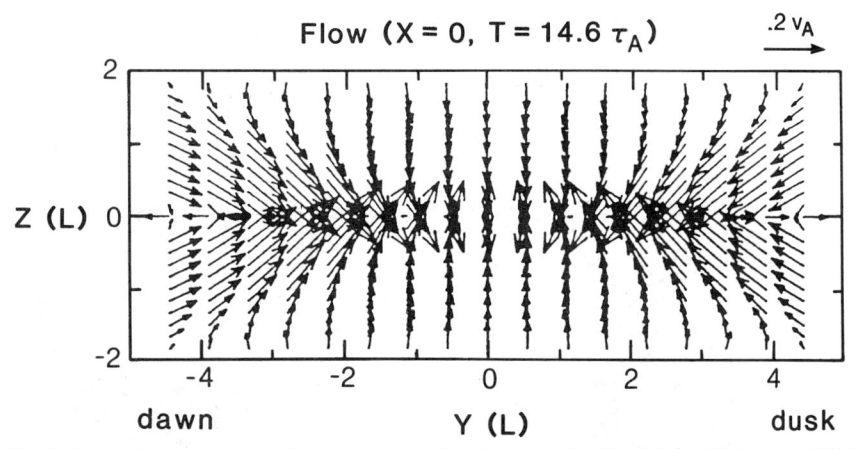

Fig. 7. Flow velocity vectors in the y-z (dawn-dusk) plane at x=0 at T = 14.6 τ_A (Sato et al., 1983b).

and e is the electron charge. It should be noted that at least under the assumption that the resistivity is dependent on the neutral sheet current the reconnection process does not depend sensitively on the functional form of the resistivity (Sato and Hayashi, 1979).

The simulation system was a rectangular box with dimensions $L_x = 3L$, $L_y = 5L$ and $L_z = 2L$ ($L_x = 3L$, $L_z = 2L$ in 2D). This was implemented on a $41 \times 40 \times 51$ point grid (74×83 in 2D). The two-step Lax-Wendroff method was employed.

In the actual calculations, all variables were normalized to the following parameters: L, length; $V_A \equiv B_0/(\mu_0\rho_0)^{1/2}$, velocity; B_0, magnetic field; ρ_0, mass density; $B_0/\mu_0 L$, current; $B_0^2/2\mu_0$, pressure; B_0^2/μ_0, energy density; $\mu_0 L V_A$, resistivity; and $B_0 V_A$, electric field. In the computer runs presented in this paper we set $A_0 = 0.02$, $d = 0.02$ and $V_c = 3$ while B_0, T_0, and ρ_0 were normalized to 1. V_c and V_D are expressed in units of $V_{D_0} = B_0/(\mu_0 L n_0 e)$ the initial central neutral sheet drift speed.

III. Simulation Results

Plasma Flow and Slow Shock Formation

Figures 2, 3 and 4 from the 2D calculation demonstrate many of the salient features of driven reconnection (Sato, 1979). Figure 2 shows a time series of magnetic field lines and Figure 3 shows the corresponding plasma flow vectors. The T=1 panel gives an indication of the plasma flow pattern through the top and bottom of the simulation box while the middle panel (T=5) shows the convection pattern which is established prior to the onset of reconnection. By T=10 reconnection has started and the outgoing plasma has begun to be accelerated. As reconnection proceeds, the plasma flow is greatly enhanced in the downstream region ($V \approx V_A$) and recon-

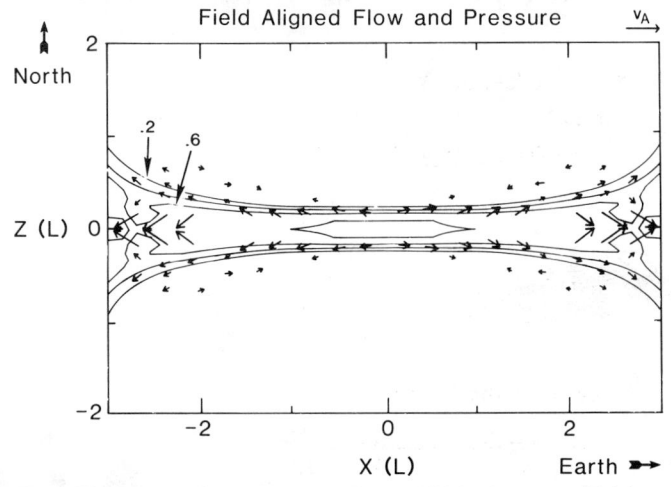

Fig. 8. Field aligned flow vectors larger than .1 V_A in the noon-midnight meridian at T = 14.6 τ_A. Contours of constant plasma pressure have been superimposed on the flow vectors.

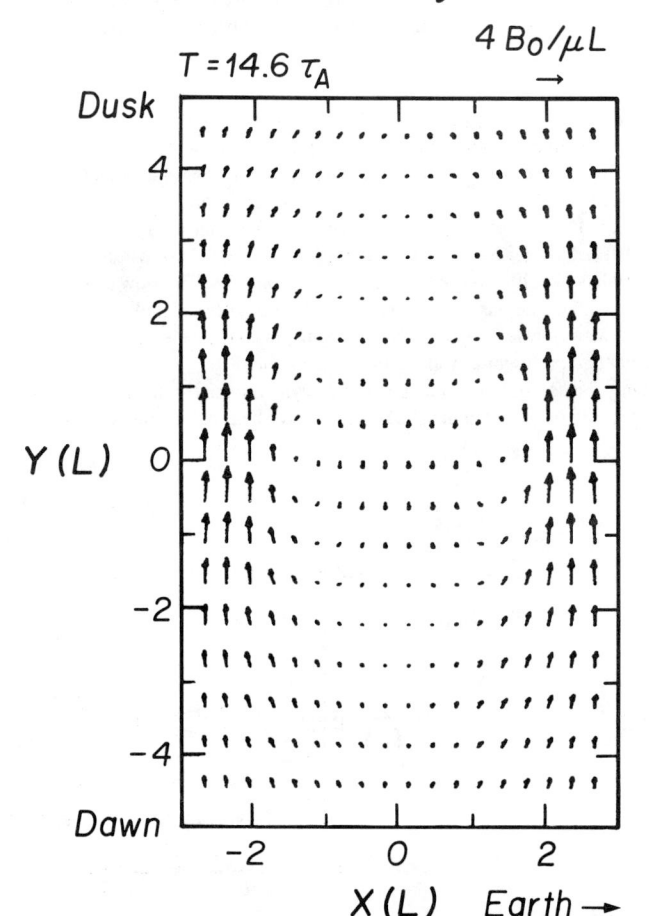

Fig. 9. Current density vectors in a plane parallel to the equator (z=.36 L) at T = 14.6 τ_A (Sato et al., 1983a).

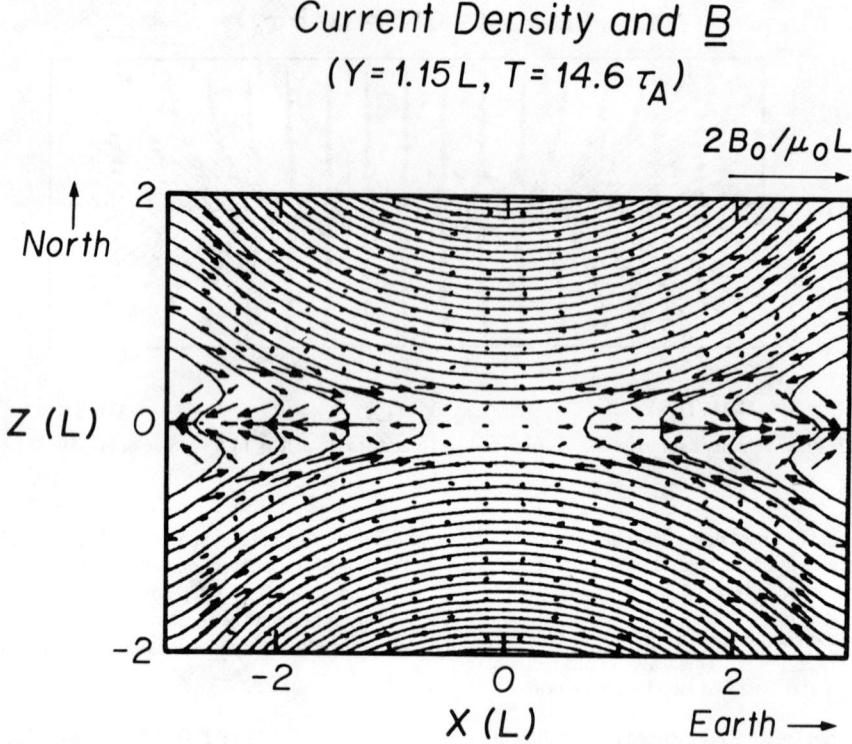

Current Density and $\underline{B}$
$(Y = 1.15 L, T = 14.6 \tau_A)$

Fig. 10. Current density vectors in a plane parallel to the noon-midnight meridian on the dusk side of the tail (y=1.15 L) at T = $14.6\tau_A$. Magnetic field lines have been superimposed on the current density vectors (Sato et al., 1983a).

nected field lines are carried away from the neutral point.

Figure 4 gives a three-dimensional graphic display of the current density. The top and bottom of each panel correspond to the side boundaries of the simulation box and the left and right ends correspond to the input boundaries. The current is initially peaked in the center of the figure (T=10) but at later times the current sheet is separated into two sheets. Sato (1979) has identified these two sheets as slow shocks. The slow shocks coincide with the demarcation zone between the accelerated and non-accelerated flows in Figure 3. Note that the plasma in the slow shocks is accelerated preferentially along the magnetic field.

Only the compressional magnetohydrodynamic modes, the fast and slow modes, can be included in the 2D model. The shear Alfvén mode which carries field aligned currents cannot be included. Thus in order to model field aligned currents, we had to expand the model to three dimensions (Sato et al., 1983a,b).

In Figure 5, magnetic field lines and flow vectors from the 3D calculation have been plotted in the central meridian of the tail (y=0) for two times. Prior to T = 14.6 τ_A the acceleration rate increased with time while after T = 14.6 τ_A it saturated and remained nearly constant. In the noon-midnight meridian the results are qualitatively very similar to those from the 2D simulation. In particular, a slow shock structure appears and plasma is accelerated away from the neutral line. Figure 6 shows the flow speed distribution along the x axis. The curve marked V_F gives the distribution of the magnetosonic speed. The flow becomes super Alfvénic beyond about

FIELD ALIGNED CURRENT DENSITY
$(X = 3.0 L, T = 14.6 \tau_A)$

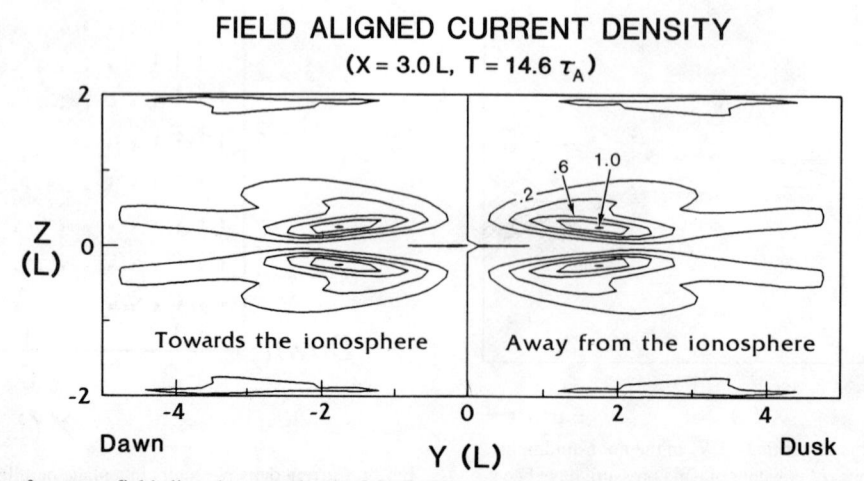

Fig. 11. Contours of constant field aligned current density $[(\mathbf{J} \cdot \mathbf{B})/B]$ in the y-z plane at x=3.0 L at T = 14.6 τ_A (Sato et al., 1983b).

x=1 and exceeds the local magnetosonic speed beyond x_s. The local magnetosonic Mach number eventually reaches approximately 2. It is interesting to note that the absolute flow velocity reaches a maximum at x_s and then decreases. This indicates that a deceleration mechanism is operating for $x > x_s$.

The flow vectors in the y-z plane at x=0 are plotted in Figure 7. Reconnection is occurring in the central part of the figure. On the sides of the reconnection region the flow has an appreciable y-component as plasma is drawn into the x-line. v_y is as large as $0.2V_A$. This is a three-dimensional effect since such flows are not allowed in the 2D model.

In discussing Figure 3 we noted that much of the flow in the slow shock region was along **B**. The field aligned component of the flow from the 3D calculation is plotted in the noon-midnight meridian in Figure 8. Only those flow vectors for which the field aligned component is larger than 0.1 V_A are included. Contours of constant plasma pressure have been superimposed on the flow vectors. The flow is primarily field aligned in the region of the slow shocks which are characterized by the sharp pressure gradient. $V_\parallel/V$ is between 0.50 and 0.90. Large flows normal to **B** are observed only at the equator. The flow vectors on reconnected field lines near the edge of the simulation box point toward the equator since the field lines are bent in this region (see Figure 5).

Tail Current Interruption and Field Aligned Currents

In Figure 9, the cross tail current density has been plotted in a plane parallel to the equator and 0.36L above it. The current is almost completely interrupted in the central region. A substantial part of the current is diverted around the center. The x-component of the current earthward of the reconnection region is towards the Earth on the dawn side and away from the Earth on the dusk side. Since **B** is primarily in the x-direction this indicates the presence of field aligned currents. This can be seen more clearly in Figures 10 and 11.

Magnetic field lines have been superimposed upon current density vectors in a plane on the dusk side which is parallel to the noon-midnight meridian and 1.15L from it (Figure 10). There are two sets of current vectors of interest. First near the outer edge of the reconnected field lines, current density vectors with large field aligned components are observed. Contours of field aligned current density are plotted in the y-z plane at the outer edge (x=3.0L) of the simulation box in Figure 11. The high latitude field

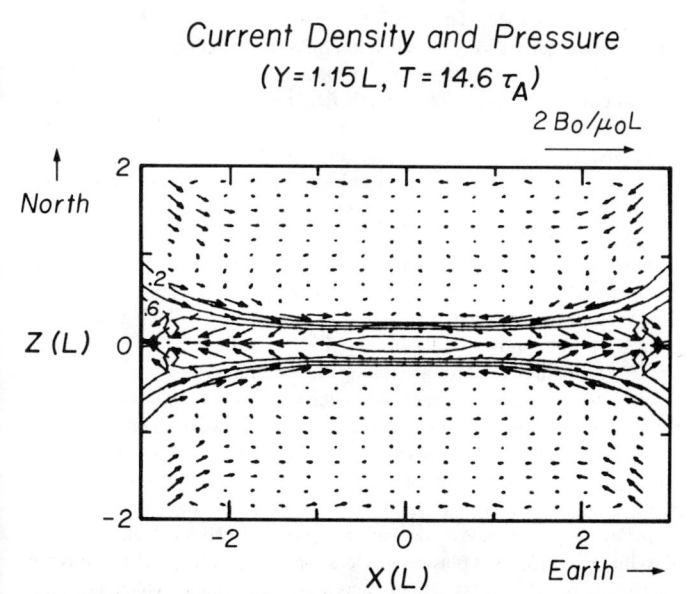

Fig. 12. Same as Figure 10 except plasma pressure contours have been superimposed on the current density vectors (Sato et al., 1983a).

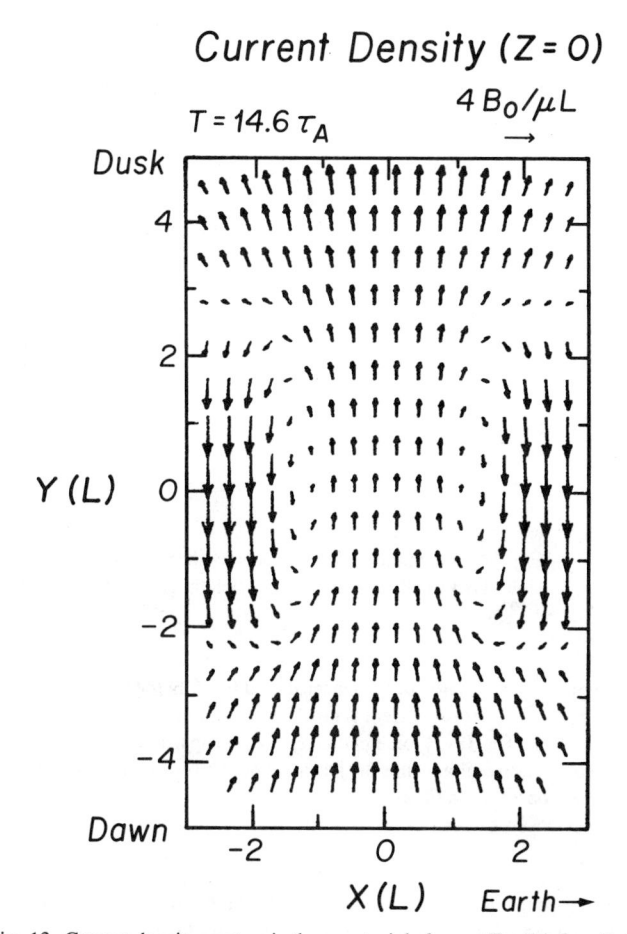

Fig. 13. Current density vectors in the equatorial plane at T = 14.6 τ_A (Sato et al., 1983a).

aligned currents in Figure 10 appear as two pairs of contours. One pair in the northern hemisphere and one pair in the southern hemisphere. On the dawn side the field aligned current flows from the magnetotail toward the ionosphere while on the dusk side it flows from the ionosphere toward the magnetotail. There are two additional features of note in Figure 11. First in a small region the field aligned current density is as large as the initial neutral sheet current density. This indicates that locally the neutral sheet current is almost completely interrupted. Second, as we saw in Figure 10, the field aligned currents flow above the equator and on the sides of the reconnection region.

That the field aligned currents reside on the outer edge of the model plasma sheet is emphasized in Figure 12 where pressure contours have been superimposed on the current density vectors at y=1.15L. To the accuracy of the model, the field aligned currents occur in the sharp pressure gradient which characterizes the outer edge. Again, this is the region of the slow shock.

The second set of interesting current vectors in Figure 10 lies just equatorward of the field aligned currents. Here the current flows earthward and toward the equator. However, this large earthward current density does not indicate an earthward field aligned current since the field has changed direction in this region. In contrast to the field farther from z=0, the field in this region is mainly north-south. Most of this current is normal to the field. Field aligned currents here are small and away from the Earth (Figure 11). Thus the earthward pointing currents represent a diversion of part of the cross tail current toward the Earth. In Figure 13 the current density vectors have been plotted in the equatorial plane (z=0). The earthward flow can be seen in current vortices which have formed between the reconnec-

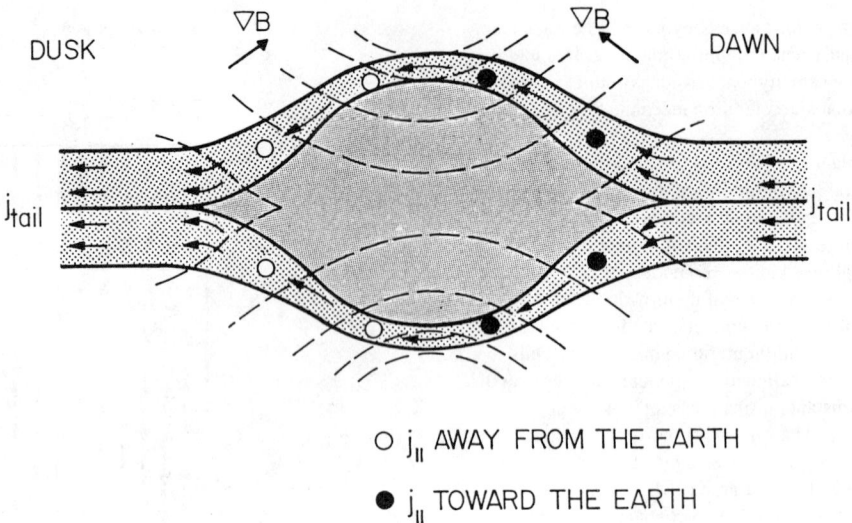

Fig. 14. Schematic drawing of plasma sheet cross section during reconnection showing the diversion of the cross tail current into the slow shocks and the generation of field aligned currents (Sato, 1982).

tion region and the region of reconnected field lines. In the region of reconnected field lines the current is now from dusk to dawn rather than dawn to dusk. The $\mathbf{J} \times \mathbf{B}$ force in this region thus opposes the flow from the reconnection region. Thus the plasma should be decelerated in this region as was observed in Figure 6.

IV. Discussion

Although the driven reconnection model is highly simplifed, it can aid us in understanding many of the features of substorms in the tail. In particular, our simulation results show that rapid flows both earthward and tailward of the neutral line and the nightside substorm current system are natural consequences of driven magnetic reconnection.

In our model, the tail current is locally interrupted by three-dimensional externally driven reconnection and a field aligned current system is generated with currents toward the ionosphere on the morning side and away from it on the evening side. This current presumably closes in the westward electrojet. Note, we have not included the ionosphere in this calculation since we used open boundary conditions. This current system is consistent with the current wedge model (Figure 1) most frequently used to interpret both magnetic field and STARE radar observations obtained near midnight during substorms (see the reviews by Kamide (1982) and Baumjohan (1982)).

The field aligned currents in the magnetosphere can be calculated by solving

$$B\frac{\partial}{\partial s}\left(\frac{J_\parallel}{B}\right) = c\rho \frac{d}{dt}\left(\frac{\Omega}{B}\right) + \frac{2}{B}\mathbf{J}_\perp \cdot \nabla B - \mathbf{J}_{in} \cdot \frac{\nabla N}{N} \quad (3)$$

with

$$\Omega = (\mathbf{B} \cdot \nabla \times \mathbf{v})/B$$

$$\mathbf{J}_\perp = \mathbf{J}_D + \mathbf{J}_{in}$$

and

$$\mathbf{J}_D = c\mathbf{B} \times \nabla P/B^2$$

$$\mathbf{J}_{in} = c\mathbf{B} \times \left(\frac{\rho}{B^2}\frac{d\mathbf{v}}{dt}\right)$$

where $\mathbf{J}_\parallel$ and $\mathbf{J}_\perp$ are the parallel and perpendicular current densities, B is the magnetic field, ρ is the mass density, Ω is the field aligned component of vorticity, $\mathbf{J}_{in}$ is the inertia current, N is the density, P is the pressure and v is the velocity (Hasegawa and Sato, 1979; Sato and Iijima, 1979). The partial derivative on the left hand side of (3) is along $\mathbf{B}$. The first term in (3) represents a discharge current due to convective motion and may be important for region 1 currents (Sato, 1982). The second term represents a diversion of magnetospheric current in a region of magnetic gradients. The term has been used to explain the region 2 field aligned currents by diversion of the ring current (Sato and Iijima, 1979). The third current source originates where there is a density gradient in the direction of the inertia current and is usually negligibly small in the magnetotail (Sato, 1982).

Sato (1982) has argued that the second term in (3) is responsible for the substorm field aligned current system. His prediction is sketched in Figure 14. He argued that as reconnection develops, the cross tail current tends to concentrate in the two slow shock layers. Furthermore, the field magnitude at the shocks increases because of the pile up of field which is driven towards the neutral sheet. Thus $\mathbf{J}_\perp \cdot \nabla B > 0$ in the dawn sector and $\mathbf{J}_\perp \cdot \nabla B < 0$ in the dusk sector giving the substorm field aligned current system.

Current density vectors in the y-z plane at x = 3L are plotted in Figure 15 superimposed on contours of magnetic field magnitude. As expected the cross tail current is diverted through the slow shocks. Grad B is toward the center and top (north) and bottom (south) so that $\mathbf{J}_\perp \cdot \nabla B > 0$ at dawn and $\mathbf{J}_\perp \cdot \nabla B < 0$ on the dusk side as predicted. If we overlay Figure 11 on Figure 15, we see that the field aligned currents occur on the edges of the slow shocks where $\mathbf{J}_\perp \cdot \nabla B \neq 0$.

In order to evaluate the importance of the first term in (3) for generating the observed field aligned currents we have examined the flow vectors. The flow vectors are consistent with a field aligned vorticity which gives a field aligned current opposite to that observed (i.e., see Figure 7). Therefore term (1) is not important for generating these currents.

We noted in discussing Figure 6 that the velocity peaked at $x=x_s$ where the flow became super magnetosonic. In Figure 16, we have plotted the corresponding distributions of the pressure (p) and the reconnected field (B_z). There are three features of interest. The first is the diffusion region (x $< x_D$) where the reconnected field increases linearly with distance as predicted by the steady state reconnection theory (Petscheck, 1964; Vasyliunas, 1975). The plasma acceleration is almost linear in x in this region (Figure 6). The theory also predicts a second region in which the pressure should be constant such as that observed for $x_D < x < x_s$. This is

CURRENT DENSITY AND |B|
($X = 3.0\,L$, $T = 14.6\,\tau_A$)

Fig. 15. Current density vectors at $x=3.0\,L$ at $T = 14.6\,\tau_A$. Contours of constant B have been superimposed on the current density vectors (Sato et al., 1983b).

called the "wave" region. The third feature occurs at x_s where there is a sharp decrease in the plasma pressure. B_z also decreases slightly at x_s. Sato et al. (1983b) have interpreted this as evidence for a fast shock. The supermagnetosonic flow is maintained by this pressure gradient.

The resistivity model used in (2) is proportional to the square of the neutral sheet current and was chosen to model the anomalous resistivity from the lower hybrid drift instability. Recent theories indicate that lower hybrid drift anomalous resistivity does not penetrate to the center of the plasma sheet but may cause a steepening of the current profile which may ultimately excite other instabilities (Drake, 1984). We do not believe our results would change significantly if the resistivity model were changed. Sato and Hayashi (1979) have shown that the major features of driven reconnection do not change qualitatively for other resistivity models which are proportional to various powers of the current. In particular the slow shocks which are necessary to generate the field aligned currents were found in all cases. It is interesting to note that the current system in the model quickly evolves to one in which most of the current and hence the resistivity is near the edge of the plasma sheet (see Figure 4).

The term driven when applied to reconnection does not have the same meaning as when it is used to describe phenomenological models of substorms. People describing substorms use the term driven to refer to a model in which the energy input rate from the solar wind to the magnetosphere equals the energy output rate during the substorm. Externally driven reconnection is not a model of this type of passive system. Driven reconnection is an active energy converter in which magnetic energy is converted into plasma thermal energy and flow (Sato, 1979; Sato et al., 1983b). The rate at which energy is released is much faster than the rate at which energy is input into the model. The driven reconnection is thus similar to the unloading model of substorms in which magnetic energy stored in the tail lobes is converted into plasma energy. It is interesting to note that the energy conversion is greater in our 3D models than in the 2D models. This occurs because magnetic and thermal energy is drawn into the reconnection region from the sides as well as the top and bottom (Figure 7).

Both driven reconnection and the tearing mode reproduce some of the observed features of substorms (Sato, 1979; Birn and Hones, 1981; Sato and Walker, 1982; Sato et al., 1983a,b) and it is possible that processes analogous to both operate in the tail during substorms. The substorm observations suggest an explosive onset of reconnection which the tearing mode doesn't provide (Sato and Hasegawa, 1982; Ugai, 1982). Therefore a process analogous to driven reconnection may be necessary to start the substorm process. Sato and Walker (1982) have shown that in the presence of flow parallel to the neutral sheet the tearing mode can generate larger perturbation flow velocities than it can without parallel flow. Thus once the reconnection has started the tearing mode may become important. Tearing mode reconnection may help explain the complex flow and field signatures reported during substorms (Birn and Hones, 1981; Sato and Walker, 1982).

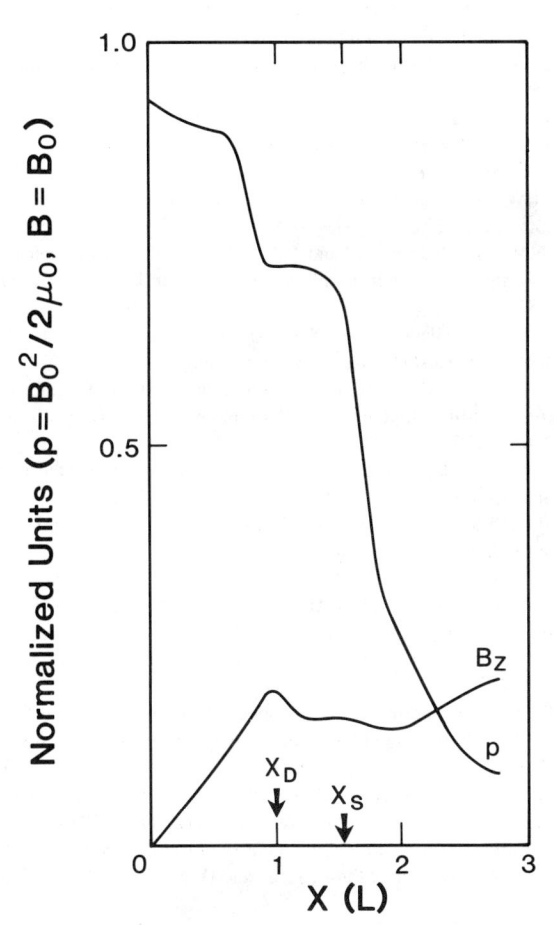

Fig. 16. Pressure and B_z along the x axis ($y=0$, $z=0$) versus x (Sato et al., 1983b).

Acknowledgements. We would like to thank R. L. McPherron for providing Figure 1. This work was supported by NASA Solar Terrestrial Theory Program Grant NAGW-78 and Air Force Contract F196-28-82-K0019. The work in Japan was supported by grants-in-aid from the Ministry of Education, Science and Culture.

References

Akasofu, S.-I., What is a magnetospheric substorm?, in *Dynamics of the Magnetosphere*, S.-I. Akasofu, ed., D. Reidel Publ. Co., Dordrecht, Holland, 447, 1980.

Akasofu, S.-I. and C.-I. Meng, A study of polar magnetic substorms, *J. Geophys. Res., 74,* 293, 1969.

Atkinson, G., A theory of polar magnetic substorms, *J. Geophys. Res., 71,* 5157, 1966.

Baumjohan, W., Ionospheric and field aligned current systems in the auroral zone: A concise review, *Adv. Space Res.*, 1982, in press.

Birn, J., Self-consistent magnetotail theory: General solution for the quiet tail with vanishing field aligned currents, *J. Geophys. Res., 84,* A9, 5143, 1979.

Birn, J. and E.W. Hones, Jr., Three-dimensional computer modeling of dynamic reconnection in the geomagnetic tail, *J. Geophys. Res., 86,* A8, 6802, 1981.

Bostrom, R., Ionosphere-magnetosphere coupling, *Magnetospheric Physics*, B.M. McCormac, ed., D. Reidel Publ. Co., Dordrecht, Holland, 45, 1974.

Drake, J.F., Tearing mode and anomalous transport processes *Geophys. Mono. Series*, this volume, 1984.

Forbes, T.G. and E.R. Priest, On reconnection and plasmoids in the geomagnetic tail, *J. Geophys. Res., 88,* A2, 863, 1983.

Harris, E.G., On a plasma sheet separating regions of oppositely directed magnetic fields, *Nuovo Cimento, 23,* 115, 1962.

Hasegawa, A. and T. Sato, Generation of field aligned currents during substorms, in *Dynamics of the Magnetosphere*, S.-I. Akasofu, ed., D. Reidel Publ. Co., Boston, 529p., 1979.

Hones, E.W., Jr., Plasma flow in the plasma sheet and its relation to substorms, *Radio Sci., 8,* 979, 1973.

Hones, E.W., Jr., S.J. Bame, and J.R. Asbridge, Proton flow measurements in the magnetotail plasma sheet made with Imp-6, *J. Geophys. Res., 81,* 227, 1976.

Huba, J.D., N.T. Gladd and K. Papadopoulos, Lower-hybrid drift wave turbulence in the distant magnetotail, *J. Geophys. Res., 83,* 5127, 1978.

Iijima, T. and T.A. Potemra, The amplitude distribution of field aligned currents at northern high latitudes observed by Triad, *J. Geophys. Res., 81,* 2165, 1976.

Kamide, Y., The relationship between field aligned currents and the auroral electrojets: A review, *Space Sci. Rev., 31,* 127, 1982.

Kan, J.R. and S.-I. Akasofu, A mechanism for current interruption in a collisionless plasma, *J. Geophys. Res., 83,* 735, 1978.

McPherron, R.L., C.T. Russell, and M.P. Aubry, Satellite studies of magnetospheric substorm on August 15, 1968, 9. Phenomenological model of substorms, *J. Geophy. Res., 78,* 3131, 1973.

Petscheck, H.E., Magnetic field annihilation, *NASA Spec. Publ., SP-50,* 425, 1964.

Pytte, T., R.L. McPherron, M.G. Kivelson, H.I. West, Jr., and E.W. Hones, Jr., Multiple-satellite studies of magnetospheric substorms: Radial dynamics of the plasma sheet, *J. Geophys. Res., 81,* 5921, 1976.

Sato, T., Strong plasma acceleration by slow shocks resulting from magnetic reconnection, *J. Geophys. Res., 84,* A12, 7177, 1979.

Sato, T., Auroral physics, in *Magnetospheric Plasma Physics*, A. Nishida, ed., Center for Academic Publ., Japan and D. Reidel Publ. Co., Tokyo, 1982.

Sato, T. and T. Hayashi, Externally driven magnetic reconnection and a powerful magnetic energy converter, *Phys. Fluids, 22,* 1189, 1979.

Sato, T. and T. Iijima, Primary sources of large-scale Brickeland currents, *Space Sci. Rev., 24,* 347, 1979.

Sato, T. and A. Hasegawa, Externally driven magnetic reconnection versus tearing mode instability, *Geophys. Res. Lett., 9,* 1, 52, 1982.

Sato, T. and T. Hayashi, Three-dimensional simulation of sheromak creation and distribution, Inst. for Fusion Theory, Hiroshima University preprint HIFT-66, 1982.

Sato, T. an R.J. Walker, Magnetotail dynamics excited by the streaming tearing mode, *J. Geophys. Res., 87,* A9, 7453, 1982.

Sato, T., T. Hayashi, R.J. Walker, and M. Ashour-Abdalla, Neutral sheet current interruption and field aligned current generation by three-dimensional driven reconnection, *Geophys. Res. Lett., 10,* 3, 221, 1983.

Sato, T., R.J. Walker, and M. Ashour-Abdalla, Driven magnetic reconnection in three-dimensions: Energy conversion and field aligned current generation, submitted to *J. Geophys. Res.*, 1983b.

Tanaka, M. and T. Sato, Simulations on lower hybrid drift instability and anomalous resistivity in the magnetic neutral sheet, *J. Geophys. Res., 86,* 5541, 1981a.

Tanaka, M. and T. Sato, Multiple excitation of lower hybrid drift waves in the neutral sheet, *Phys. Rev. Lett., 47,* 714, 1981b.

Ugai, M., Spontaneously developing magnetic reconnection in a current sheet system under different sets of boundary conditions, *Phys. Fluids, 25,* 6, 1027, 1982.

Vasyliunas, V.M., Theoretical models of magnetic field line merging, 1, *Rev. Geophys. Space Phys., 13,* 303, 1975.

Winske, D. and P.C. Liewer, Particle simulation studies of the lower hybrid drift instability, *Phys. Fluids, 21,* 1017, 1978.

Questions and Answers

Dungey: When you want to model the field-aligned currents driven from the tail and not just reconnection, I believe you need more physics. This could be the Hall effect, but preferably something more elaborate like Stan Cowley did more than ten years ago. I think it is time to put in the Cowley modifications. What do you think?

Walker: I agree we need to include more physics as we build more sophisticated models of the tail. Cowley's theory is a good starting point for these studies. However, for the Hall effect I believe the changes will be on a different spatial scale than we have considered here.

Hones: Does the dynamic pressure (or time-variations of it) on the rapidly (spatially) expanding magnetopause near the dawn-dusk meridian constitute something like the driving force in your model?

Walker: A key feature of the driven reconnection model is the establishment of a convection system in which the plasma sheet thins (see Figure 3). We also require that this system be spatially nonuniform. The change in the dynamic pressure and corresponding changes in the cross-section shape of the tail will contribute to this nonuniformity.

Priest: The distinction between spontaneous and driven reconnection may be smaller than you have been suggesting. If the boundary conditions are free enough, the linear tearing mode may develop in its nonlinear phase into a state of Petschek-Sonnerup reconnection that is similar to your driven reconnection. However, one difference is that in the driven case you will just release the energy at the small rate that it is being fed in, whereas spontaneous reconnection can release much more energy that has been previously stored over a long time, and it can release it at a much faster rate.

Walker: Our simulations have demonstrated that the driven reconnection process is not a simple passive pump but rather is an active energy converter in which magnetic energy in converted into plasma thermal energy and flow (see Sato et al. 1983b). The rate at which energy is released is much faster than the rate at which energy is input into the model. We have called the system "driven" because a driving force is required to bring the system to the energy release stage. We do not feel that it has been demonstrated that spontaneous reconnection (i.e., reconnection without a growth phase or independent of the IMF) can rapidly release energy. The studies of Sato and Hasegawa (1982) and Ugai (1982) indicate that tearing is not an effective energy converter. The important question is what happens when the tail configuration changes from a stable configuration to one more tail-like under the influence of the IMF.

Rostoker: In your contention that $\bar{J}_\perp \cdot \nabla B$ is the important term in the expression of Hasegawa and Sato (1979), you showed a "cartoon" where you indicated the presence of a ∇B parallel (or anti-parallel) to $J_\perp$. Could you indicate to me the origin of this ∇B?

Walker: Recall that in our model flux is being driven towards the neutral sheet. This results in a pile up of the field at the slow shocks. This is clearly seen in the simulation data where the gradient is toward the center and northward in the northern hemisphere (see Figure 15 in the text).

Vasyliunas: I would like to point out that the expression for the field-aligned current density given by Hasegawa and Sato and by Sato and Tijima is incomplete: there are two additional terms containing the vorticity perpendicular to the magnetic field (the complete equation is given in Vasyliunas, in Magnetospheric Currents, ed. T. Potemera, AGU Geophysical Monograph, in press). Hasegawa and Sato assumed that the vorticity is parallel to the magnetic field, an assumption that is manifestly incorrect (a simple counterexample is a corotating plasma in a dipole field, where the vorticity is parallel to the rotation axis everywhere).

Walker: Thank you for pointing out the two additional terms in the expression for field aligned currents which was derived by Hasegawa, Sato and Iijima. Sato has argued that the perpendicular component of the vorticity gives a smaller contribution to the field aligned current than the parallel component. In your example of the simple corotating plasma there are both parallel and perpendicular components to the vorticity. I should emphasize that the $J_\perp \cdot B$ term is the most important one in our case. The predictions of the theory and the simulation observations agree very well.

COMPUTER MODELING OF FAST COLLISIONLESS RECONNECTION

J. N. Leboeuf, F. Brunel and T. Tajima

Institute for Fusion Studies, University of Texas
Austin, Texas 78712

J. Sakai

Department of Applied Mathematics and Physics, Faculty of Engineering, Toyama University
Toyama 933, Japan

C. C. Wu and J. M. Dawson

Department of Physics, University of California
Los Angeles, California 90024

Abstract. Particle simulations of collisionless tearing, reconnection and coalescence of magnetic fields for a sheet-pinch configuration show that reconnection is Sweet-Parker like in the tearing and island formation phase. It is much faster, or even explosive, in the island coalescence stage. Island coalescence is the most energetic process and leads to large ion temperature increase and oscillations in the merged state. Similar phenomena have been observed in equivalent MHD simulations. Coalescence and its effects, as observed in our simulations, may explain many of the features of solar flares and coronal X-ray brightening.

I. Introduction

Computer modeling of magnetic field reconnection, including island coalescence, has been tackled mainly with collisional MHD codes. Few kinetic studies of reconnection have been reported so far [Dickman and Morse, 1969; Amano and Tsuda, 1977; Katanuma and Kamimura, 1980; Terasawa, 1981; Hamilton and Eastwood, 1982; Leboeuf, Tajima and Dawson, 1981 and 1982; Tajima, 1982] even though in space plasmas collisionless tearing modes are believed to be one of the most effective mechanisms for magnetic field line reconnection. This is particularly true for the Earth's magnetospheric tail where the particles' mean free path is very large. Dickman, Morse and Nielson [1969] used a magnetostatic code to study tearing modes in the Astron fusion device. They found that the Astron plasma layer first develops tearing modes, but at later times the wavelength of this mode increases by coalescence until the plasma is completely reassambled. Amano and Tsuda [1977] were the first to study forced reconnection with an electromagnetic code. They forced a flow towards the initial magnetic neutral sheet and observed the formation of an x-point. They also remarked that Joule heating in the diffusion region was not sufficient for the explosive energy release such as that observed in astrophysical and geophysical phenomena. The study of Katanuma and Kamimura [1980] involves using a magnetostatic code to study the nonlinear evolution of collisionless tearing modes. They verified the Drake-Lee [1977] theory of tearing. They did observe island coalescence in the case of multi-mode tearing. However, no discussion of the energetics of the interaction was given. Again using a magnetostatic code, Hamilton and Eastwood [1982] realistically modeled the geomagnetic tail and confirmed the stabilizing influence of a small magnetic field normal to the sheet of the tearing mode. Finally, Terasawa [1981] used a reduced Darwin model, with electrostatics neglected, and followed the ions only. He verified the explosive tearing mode theory of Galeev, Coroniti and Ashour-Abdalla [1978]. The question is whether adding the electrons and the electrostatics will modify his conclusions.

Our program of kinetic simulations of collisionless reconnection was primarily motivated by the laboratory experiments of W. Gekelman and R. L. Stenzel [Stenzel and Gekelman, 1979; Gekelman and Stenzel, 1981]. The tools used consist of magnetostatic and electromagnetic finite size particle simulation models with two spatial dimensions only. This paper will mainly be a review of our own work, and some comparisons with similar MHD simulations. Applications to the physics of solar flares will also be discussed.

By having a current flow in two strips perpendicular to the plane of the simulation, with the current ramped in time much as it is in the experiments of Gekelman and Stenzel, we are able to pass through the successive stages of current sheet formation in between the two strips, tearing of the current sheet to form magnetic islands and finally magnetic island coalescence. The onset of coalescence occurs in an explosive fashion. Recent MHD simulations of reconnection and coalescence [Brunel, Tajima, and Dawson, 1982] exhibit comparable behaviour. Addition of a magnetic field parallel to the strips (a toroidal field in tokamaks) introduces incompressibility and prevents the fast reconnection that otherwise occurs [Tajima, 1982]. The tearing phase leads to almost no energization of the particles' kinetic energy. Magnetic island coalescence, on the other hand, leads to a large increase in ion temperature. The oscillations exhibited by the ion temperature in the merged phase have also been observed in an MHD simulation of island coalescence [Wu, Leboeuf, Tajima, Dawson, 1980]. These oscillations resemble what is reported of the solar gamma ray amplitude oscillations [Forrest et al., 1982] associated with loop coalescence in solar flares [Tajima, Brunel, Sakai, 1982].

II. Computer Model

Our collisionless particle simulations follow the evolution of a plasma configuration which is unstable against the tearing and subsequent coalescence instabilities. The electromagnetic code [Leboeuf, Tajima, Dawson, 1982] is two-and-one-half (two space, x and y, and three velocity and field, x, y, and z dimensions) dimensional and periodic in both the particles and field quantities. The model includes electrons and ions, with ion-to-electron mass and temperature ratios of $M_i/m_e = 10$ and $T_i/T_e = 1/2$ respectively in a benchmark case. Both species of particles are given Maxwellian velocity distributions in all three directions at time t = 0. Four particles per unit cell are typically used. The particles are loaded uniformly in space so that the density is uniform at t=0. We focus on rectangular system sizes $L_x \times L_y = 128\Delta \times 32\Delta$ and $256\Delta \times 16\Delta$, where Δ is the unit grid spacing. For the first, $\Delta = \lambda_e$, the electronic Debye length and the speed of light $c = v_{te}$, the electron thermal velocity, so that the collisionless skin depth $\delta = c/\omega_{pe} = 5\lambda_e$. For the second, $\Delta = 2\lambda_e$, with collisionless skin depth $c/\omega_{pe} = 6\lambda_e$ and the ratio of thermal velocity to speed of light $v_{te}/c = 1/6$.

The magnetic configuration is established by external current strips placed at y = 0 and y = L_y which extend along x. The current flows in the z-direction. To avoid infinite magnetic energy, the return path of the current is chosen to be through the plasma. At t = 0, the current is zero in the strips and rises sinusoidally from zero at t = 0 to a maximum at the quarter period after which it is kept constant (crowbar phase), with a rise time greater than or equal to the magnetosonic transit time from center to plates. By varying the strength of the currents in the strips, different plasmas are obtained. For the 128 × 32 case, the plasma β = 0.2, for maximum magnetic field and average density. This entails electron and ion Larmor radii of $1.3\lambda_e$ and $2.6\lambda_e$ respectively, and an Alfvén velocity $v_A = 1.22 v_{te}$. For the 256 × 16 case, the maximum field is such that β = 0.06, $\rho_e = .85\lambda_e$, $\rho_i = 1.90\lambda_e$. The Alfvén speed is $v_A = 2.2v_{te}$. When a constant toroidal field is imposed in the z-direction it is such that $0.2 \leq B_t/B_p \leq 4$ where B_p is the maximum poloidal field. Straightforward calculations of the magnetic Reynolds number from the finite size particles collision frequency yield S = 800. However since the current flow is in the ignorable z direction, the actual collisionality may be regarded as much smaller than the in-plane value and S greater than the above value.

It should be noted that our model is limited by the periodic boundary conditions imposed on both the particles and fields. Moreover, the system sizes used only describe a few collisonless skin depths, ion Larmor radii and Debye lengths. We are actually looking at rather microscopic x-points and o-points.

III. Magnetic Reconnection

We discuss the compressible cases without toroidal field first and those with toroidal field next. The "collisionless" reconnection rate in both cases is measured and compared with theory and similar MHD simulations.

The evolution of the field lines in the 128 × 32 case ($B_t=0$) is shown in Fig. 1. The corresponding plasma current density is displayed in Fig. 2. The various stages of evolution consist of a current sheet formation at the center of the system, break-up of the current sheet to form a chain of x-points and o-points, i.e. establish a sheet pinch configuration, swelling of the islands by reconnection and finally coalescence of the islands. All of these phenomena happen within 2 to 4 Alfvén times. As the current increases in the strips, an equivalent amount of current is returned through the plasma. (Note the external current is maintained throughout the simulations.) Pinching occurs and induces a flow ($V_y = \pm J_p^z \times B_x^e$) through the x-points and into the o-points where the plasma remains trapped.

The attractive force between the so-formed plasma filaments induces coalescence. The external circuit is coupled to the plasma and is the source of free energy. The events described are forced on the plasma by the external circuit and in that sense we are looking at forced reconnection.

In the 256 × 16 case ($B_t=0$), the change in magnetic topology is best illustrated by the plasma current density. Tearing of the long and narrow current sheet induces 16 islets. They eventually coalesce pairwise down to one island, as shown in Fig. 3.

Tajima [1982] found that when a constant toroidal field B_z of strength B_t is added to the above configuration things happen in a qualitatively similar fashion for $B_t \lesssim B_p$, the maximum poloidal field, as when $B_t = 0$ in terms of topological changes. In particular the violent coalescence instability is still seen.

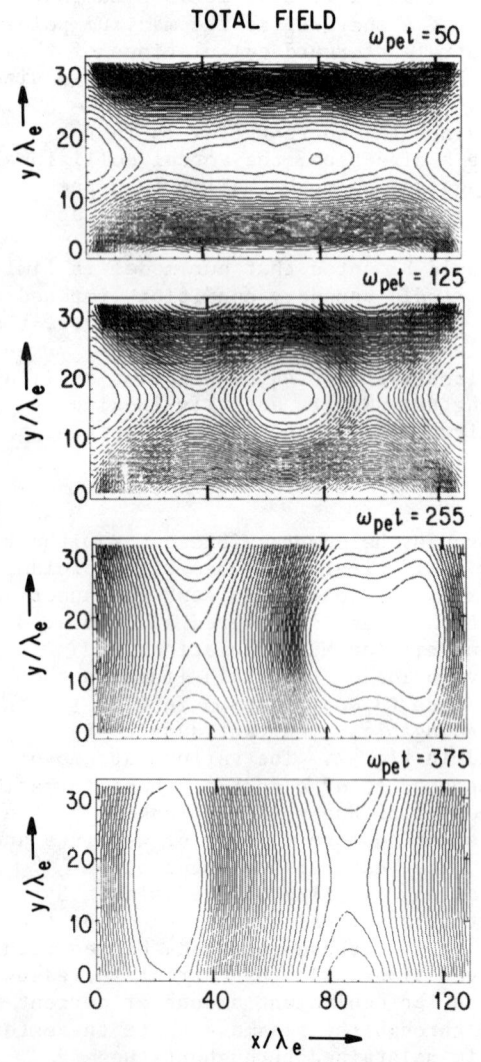

Fig. 1. Time evolution of the magnetic field lines in the 128 × 32 case with $B_t/B_p = 0$. (Leboeuf, Tajima, Dawson, 1982)

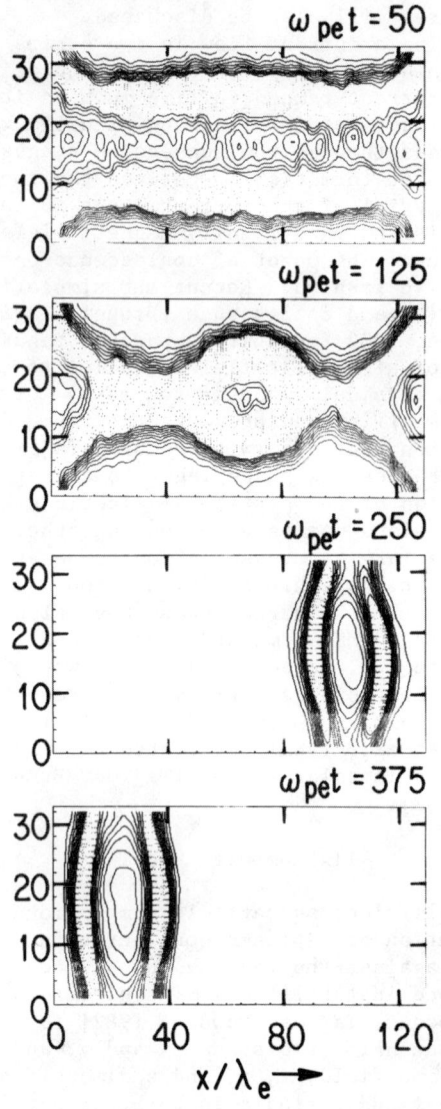

Fig. 2. Time evolution of the current density in the 128 × 32 case with $B_t/B_p = 0$. The dotted contours indicate regions of maximum plasma current. (Leboeuf, Tajima, Dawson, 1982)

When $B_t \gtrsim B_p$, however, the current sheet still tears into many islets but the coalescence instability does not occur (within the simulated time scales). For $B_t > B_p$ i.e. strong field perpendicular to the plane of the simulations, the plasma is strongly magnetized ($\rho_e \lesssim \Delta$, $\rho_i \lesssim 3\Delta$) and cross-field motion of the particles is strongly impaired except through E × B spatial diffusion, i.e. the plasma is nearly incompressible since $\nabla \cdot V = \nabla \cdot (E \times B) = 0$. In this case many islets lead to turbulence of tearing modes. Each mode has a lifetime roughly the inverse of the linear growth rate. A renormalized turbulence theory was constructed

PLASMA CURRENT DENSITY
$\beta = 0.06$

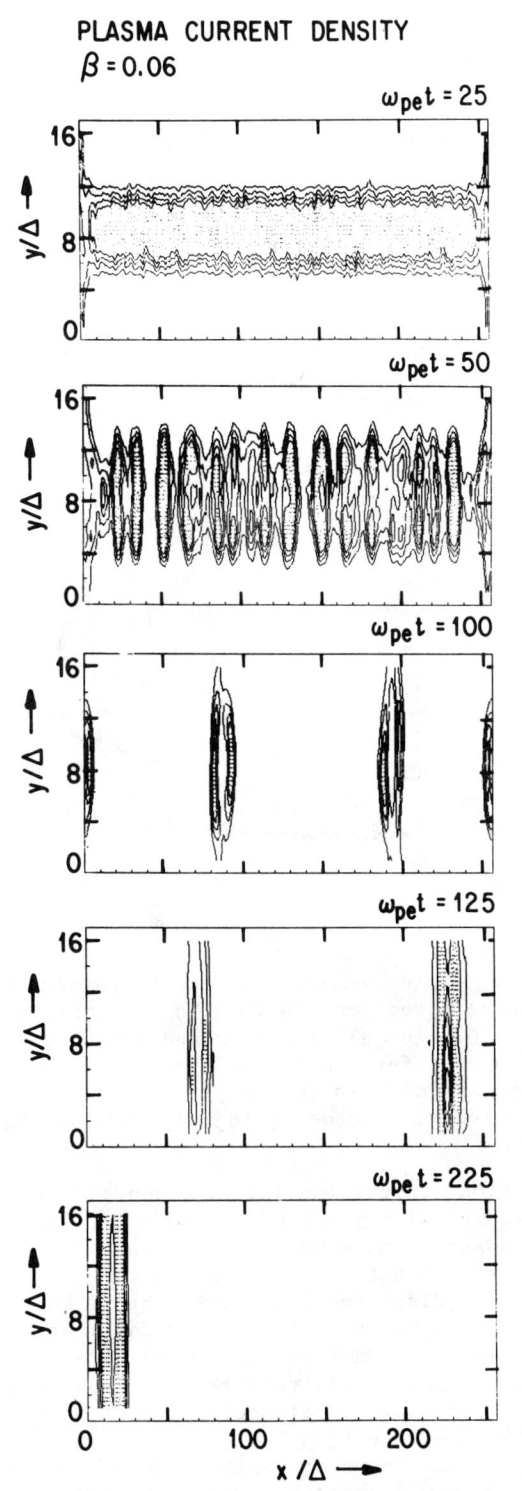

Fig. 3. Time evolution of the current density in the 256 × 16 case with $B_t/B_p = 0$. (Leboeuf, Tajima, Dawson, 1982)

tearing turbulence the magnetic fluttering-induced electron response adds the essentially new physics.

Measurements of the poloidal flux trapped in one island or islet (private flux) as a function of time for $B_t/B_p = 0.2$ (128 × 32 case) and $B_t/B_p = 2$ (256 × 16 case) are shown in Figs. 4a and 4b respectively. In the weak toroidal field case, a linear phase, which encompasses the island swelling stage, precedes the explosive coalescence one. For the strong toroidal field case, the rise is linear in time over ten times longer time scales and the flux increase is 10^{-2} times smaller.

MHD simulations of magnetic reconnection driven either by external currents pinching the plasma by Brunel, Tajima and Dawson [1982] or by the coalescence instability of an equilibrium chain of magnetic islands by Bhattacharjee, Brunel and Tajima [1983] show similar behavior. It is found that fast magnetic reconnection may consist of more than one stage. After the Sweet-Parker phase [Parker, 1979] is established for an Alfvén time, a faster second phase of reconnection takes over if the plasma is compressible. The Sweet-Parker flux is

$$\psi_{sp}(t) = \eta^{1/2} B_p(y=a) \left(\frac{n_i}{n_e}\right)^{1/2} \left(\frac{v_A}{L}\right)^{1/2} t , \quad (1)$$

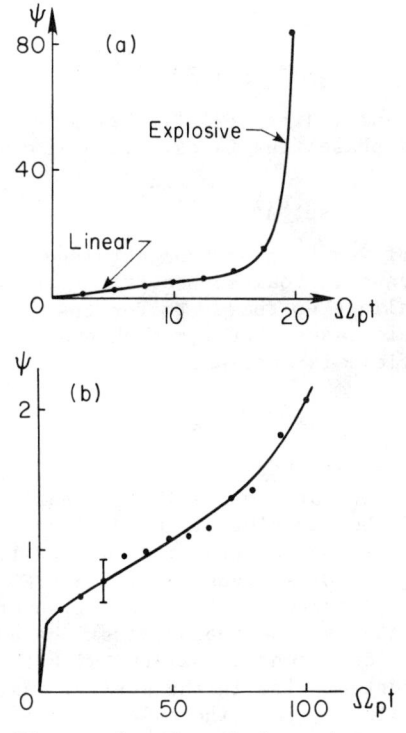

Fig. 4. Time evolution of the private poloidal flux in a) the 128 × 32 case with $B_t/B_p = 0.2$ and b) in the 256 × 16 case with $B_t/B_p = 2$. Ω_p represents the electron gyrofrequency measured with respect to the maximum poloidal field. (Tajima, 1982)

by Tajima [1982] based on this observation. The flux increases with a time exponent smaller than predicted by Rutherford [1973] or Drake and Lee [1977] and may be explained by this turbulent tearing mode theory. In the collisionless

where n_e and n_i are the densities outside and inside the current channel, a the current channel width, η the resistivity, L the length of the reconnecting region, B_p, the poloidal field and v_A the Alfvén velocity. Our basic equations to describe the system have been reduced in order to gain reasonable analytical expressions and straightforward understanding of the underlying physics:
(outside)

$$\frac{\partial \psi}{\partial t} = \underset{\sim}{v}_\perp \times \underset{\sim}{B}_p, \quad (2)$$
$$p_e^2 + B_e^2/8\pi = p_i^2 + B_i^2/8\pi \sim p_i, \quad (3)$$

(inside)

$$\frac{\partial \psi}{\partial t} = \eta \nabla_\perp^2 \psi \approx \eta B_p/a, \quad (4)$$
$$n_e L v_\perp = n_i a u. \quad (5)$$

The left-hand side of Eq. (5) is the particle flux outside the separatrix, while the right-hand side is that within the separatrix. The first equations [Eq. (2) and Eq. (4)] are the magnetic flux equations, i.e. Faraday's law, while the second equations [Eq. (3) and Eq. (5)] are related to the equations of motion of the plasma. As is expected, the outside solution should be the MHD solution according to Eq. (2). The velocity u in the internal layer was determined to be

$$u \sim v_{Ai} = B_e/(4\pi M n_i)^{1/2} \quad (6)$$

by Brunel and Tajima [1983]. Using Eqs. (2)-(6) the second phase flux is calculated to go as

$$\psi = \psi_{sp}(t_A)(t/t_A)^{n_i/n_e}. \quad (7)$$

Scalings of $\psi \propto t^4$ have been obtained in the pinching cases. Coalescence driven reconnection yields scalings up to $\psi \propto t^2$ for the compressible cases, but $\psi \propto t$ when a large toroidal field is applied.

IV. Energetics

We concentrate here on cases with $0 \lesssim B_t/B_p \lesssim 0.2$ or compressible situations. A summary of the behavior of the various components of the temperatures for the 128 × 32 case with $B_t = 0$ is given in Fig. 5. No sizable increase is detected in the current sheet formation and tearing phases, just the adiabatic compression component associated with the external current rise in the strips. As the islands coalesce, fast increase in the temperatures is apparent. Most of the increase is concentrated in the x-direction, where the ion temperature in the merged phase is on average 30 times its initial level. The ions achieve a higher temperature than the electrons in all three directions. Note the large

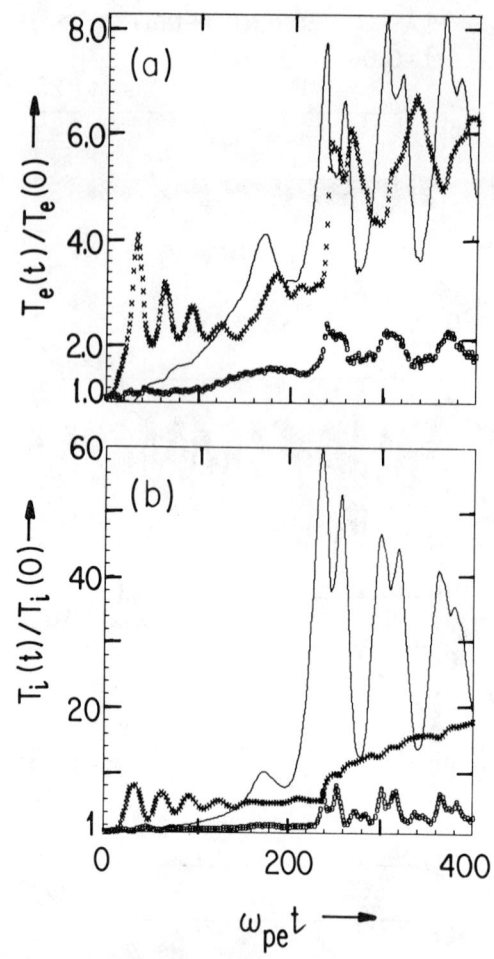

Fig. 5. Time evolution of the temperatures in all three directions in the 128 × 32 case with $B_t/B_p = 0$. for a) electrons and b) ions. The full curve refers to the x-direction, the crosses and circles to the y and z ones respectively. (Leboeuf, Tajima, Dawson, 1982)

oscillations in the ion and electron temperatures with a period $\tau = 60/\omega_{pe}$ in the merged phase. The momentum distribution functions of electrons and ions in the x-direction displayed in Fig. 6 exhibit bulk heating and symmetric high energy tails at late times. As shown in Fig. 7, an even more pronounced increase in temperature is obtained from the successive coalescence events of the 256 × 16 case with $B_t = 0$.

Results for the compressible 128 × 32 case with $B_t/B_p = 0.2$ are displayed in Fig. 8. Fig. 8a is a plot of ion temperature versus time. Again sharp increase upon coalescence followed by oscillations is apparent. Fig. 8b, which represents the ion distribution function in the x-direction after coalescence, shows bulk heating and symmetric tails. Fig. 8c is a plot of the ion distribution function along the toroidal field. It shows three regimes. First the bulk, then the exponential section of two islands takes place within 1 to 2 Alfvén times. The magnetic energy contained in the islands is explosively released into kinetic energy as seen in Figs. 6, 7 and 8. The amount of available potential energy W_c by attracting two toroidal current rods I of radius a with separation L is

$$W_c \simeq -2I^2/c^2 \ln(L/a) . \quad (8)$$

Our simulations show that about 1/6 of the energy W_c was transferred to kinetic energy upon coalescence in the 128 × 32 case with $B_t = 0$. This amount of energy conversion is about two orders of magnitude above that during the tearing process. The oscillations in temperature observed in the 128 × 32 cases are found to be magnetosonic ones with a frequency $\omega = kv_A$, where $k = 2\pi/a$, a the current channel width and v_A calculated according to the magnetic field measured at the island. These temperature oscillations can be attributed to the overshooting of the two coalescing and colliding current filaments. Once the two filaments merge, they are bound by the common flux and the resulting island shape oscillates from prolate to oblate on the time scale of the temperature oscillations. The colliding plasmas cause turbulent flows within the final island and the originally directed energy is eventually dissipated into heat. The turbulent mixing in the island also causes the bulk heating observed on the distribution functions.

Similar features have also been observed in 2-D Eulerian MHD simulations of the coalescence instability of magnetic islands formed by nonlinear tearing modes [Wu, Leboeuf, Tajima and Dawson, 1980]. A sheet-pinch configuration is modeled. Initially small perturbations consisting of two linearized eigenfunctions are imposed: one with a wavelength $\lambda = L/2$ and a smaller perturbation with $\lambda = L$, L being the system size. The perturbation with L/2 gives rise to the formation of two magnetic islands at the stage of the nonlinear tearing mode. These islands then interact with each other due to the presence of the smaller perturbation with wavelength $\lambda = L$; they coalesce and merge into larger units. The time evolution of the fluid kinetic energy for an MHD case with S = 2000 is displayed in Fig. 9b, alongside a temperature plot for the 128 × 32 case with $B_t = 0$, as shown in Fig. 9a. The initial increase in energy up to $t = 120\tau_A$ corresponds to the exponential growth of the L/2 perturbation with linear growth rate $\gamma_{L/2} = 9.6 \times 10^{-3}$. By $t = 720\tau_A$, the perturbation with wavelength $\lambda = L$ ($\gamma_L = 1.2 \times 10^{-2}$) has reached a large enough amplitude to trigger coalescence. During coalescence, the fluid energy rises almost exponentially with growth rate $\gamma_c = 3.8 \times 10^{-2}$ ($\gamma_c > \gamma_L, \gamma_{L/2}$). While the kinetic energy only accounted for 0.04% of the magnetic energy up to $t = 720\tau_A$, it accounts for 10% of it in the

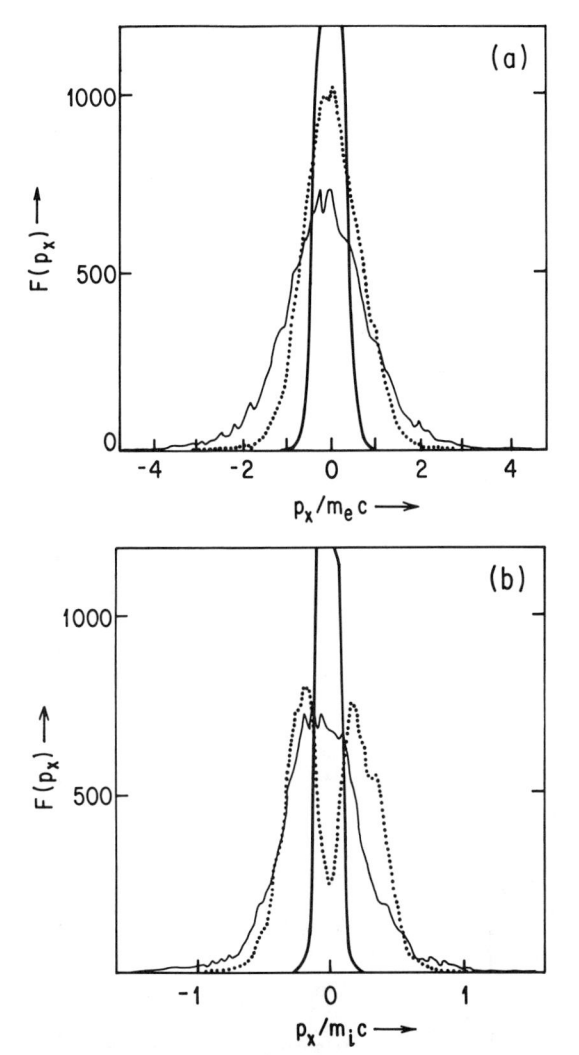

Fig. 6. Momentum distribution functions in the x-direction for the 128 × 32 case with $B_t/B_p = 0$. and for a) electrons and b) ions at $\omega_{pe} t = 50$ (full bold curve: island formation stage), $\omega_{pe} t = 225$ (dotted curve: coalescence phase), $\omega_{pe} t = 325$ (full thin curve: merged state). The momenta are normalized to $m_e c$ for the electrons and $M_i c$ for the ions. (Leboeuf, Tajima, Dawson, 1982)

$f_2(p_z) = \exp(-p_z/p_o)$ and third the flat distribution extending up to the relativistic factor $\gamma \sim 2$ in the relativistic region, where $p_o^2/2M_i = 10x$ (bulk temperature). The bulk ion heating in the x-direction is attributed to the "adiabatic heating"; the exponential heating in the x and z directions are inductive in nature. The hot flat long tails in the x and z directions are due to acceleration by the magnetosonic shock [Tajima et al., 1983]. The maximum energy may be estimated based on the Alfvén Mach number. Most of the energy of the particles still belongs to the bulk component, however.

We have seen that the total flux reconnection

merged phase, 220 times the saturation value for the island formation stage. In the merged phase, the kinetic energy presents oscillations whose period $\tau = 80\tau_A$, with τ_A defined with respect to the asymptotic field, roughly equivalent to oscillations at the Alfvén frequency determined with the field at the island. These oscillations and the larger energy gain upon coalescence are similar to what we observe in the particle simulations.

V. Application to Solar Flares

Recent direct observations in soft X-rays [Howard and Svetska, 1977] of interconnecting coronal loops spur the theorist to consider loop coalescence as an important process for solar flares and coronal X-ray brightening phenomena.

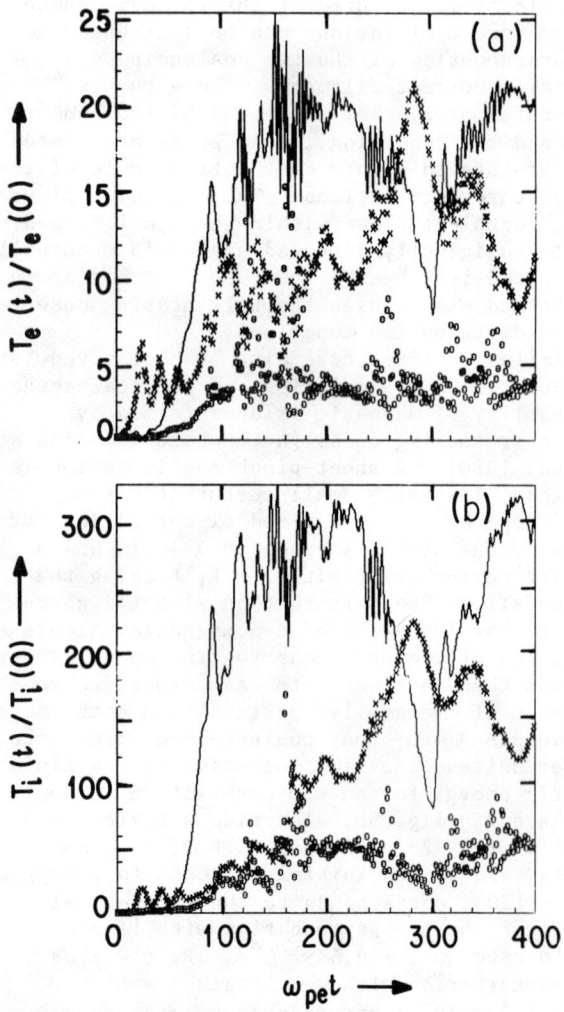

Fig. 7. Time evolution of the temperatures in all three directions in the 256 × 16 case with $B_t/B_p = 0$. for a) electrons and b) ions. The full curve refers to the x-direction, the crosses and circles to the y and z ones respectively. (Leboeuf, Tajima, Dawson, 1982)

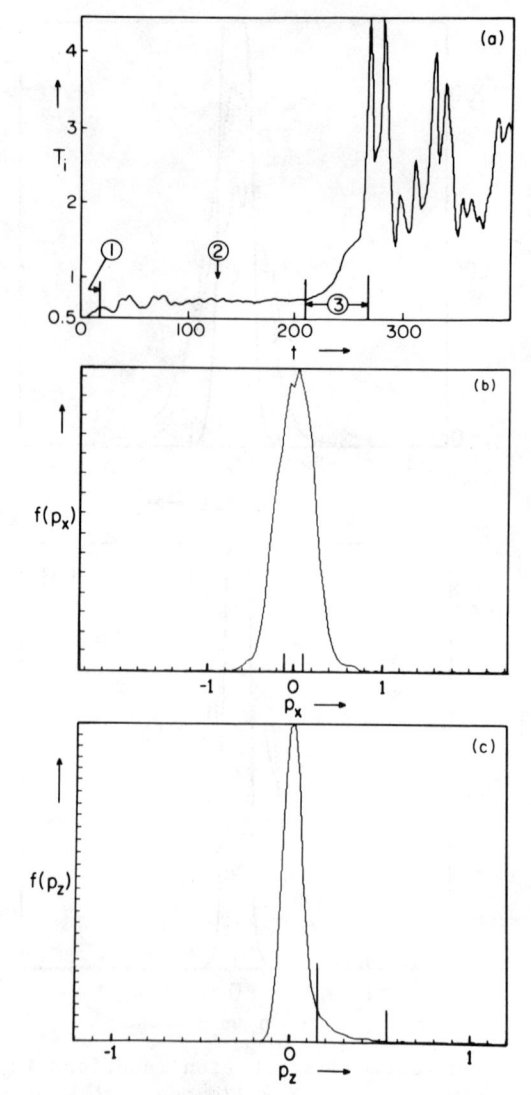

Fig. 8. Various data from 128 × 32 case with $B_t/B_p = 0.2$. a) Time evolution of the ion temperature. Phases 1 and 2 are the tearing growth and saturation phases, phase 3 is the explosive coalescence phase. The period of temperature oscillations in the merged phase is $\tau \sim \tau_A$. b) ion distribution function in the x-direction and in the merged state. c) Ion distribution function in the z-direction and in the merged state. The thermal momenta are indicated by tickmarks near $p = 0$. Momenta are normalized with respect to $M_i c$. (Tajima, Brunel, Sakai, 1982)

Another recent observation [Forrest et al., 1982] of amplitude oscillations in gamma-ray emission from the impulsive phase of a solar flare adds curiosity and an important clue to the underlying physical process. The nonlinear development of the coalescence instability of the current loops might provide a coherent explanation of the above observations [Tajima, Brunel and Sakai, 1982]. Some of the results

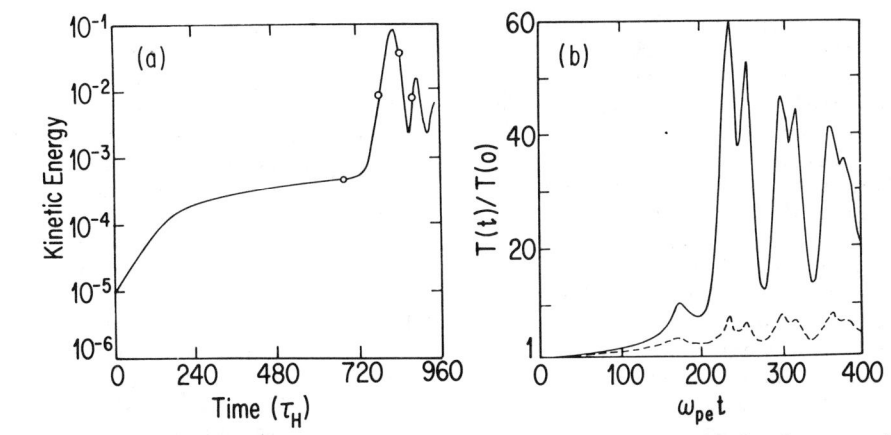

Fig. 9. Comparative data from particle simulations and MHD simulations. a) Ion (full curve) and electron (dotted curve) temperatures in the x-direction for the 128 × 32 case with $B_t/B_p = 0$. plotted as a function of time. b) Time evolution of the kinetic energy for the MHD simulation of island coalescence induced by nonlinear tearing modes with S = 2000. Note the oscillations on both at $\tau \sim \tau_A$, with the poloidal field measured at the island. (Wu, Leboeuf, Tajima, Dawson, 1980)

presented here offer a quantitative and natural explanation of such known characteristics as the impulsive nature of flares, the time scale of the impulsive phase, intense heating by flares, and formation of the high energy tails on the particle distributions.

The following scenario has been proposed by Tajima, Brunel and Sakai [1982]. The flare loop slowly expands after it emerges from the photosphere as the toroidal field curvature of the loop makes the centrifugal motion. In time, the toroidal current J_t builds up, increasing the poloidal magnetic field B_p. As the poloidal field B_p reaches the critical value that is of the order of magnitude B_t, the adjacent flare current loops can now coalesce rapidly facilitated by the fast reconnection process governed by Eq. (7), the faster second phase. Such a fast coalescence of flare loops proceeds explosively once in its nonlinear regime in a matter of one or two Alfvén times, releasing more than one-tenth of the magnetic energy into (ion) kinetic energy. For the flare loop magnetic field (100 Gauss) with current rod size ($a = 10^8$ cm), the energy density is $W_c \sim 0.5 \times 10^{20} \ln(L/a) \sim 1.5 \times 10^{20}$ erg/cm and the energy available in length $d = L(\sim 10^9 \text{cm})$ is $E = 1.5 \times 10^{29}$ erg for $a = 10^8$, $d = L = 10^9$ and $E = 1.5 \times 10^{31}$ erg for $a = 10^9$, $d = L = 10^{10}$. The released ion energy is $E_{ion} \sim E/6$ and is in between 2×10^{28} and 2×10^{30} erg due to coalescence. This amount of energy is in the neighborhood of the solar flare energy [Sturrock, 1980]. These energies can be released during the impulsive phase as well as during the main phase.

With this magnetic field, the Alfvén time is of the order of 1-3s, which is approximately the time scale for fast coalescence. The time scale for the impulsive phase is observed to be of the order of a few seconds. The sudden nature of the impulsive flare phase [Sturrock, 1980] is thus explained by increasing the field aligned current and by the faster second phase reconnection in the course of coalescence. The field aligned particle distribution $f(p_z)$ of Fig. 8 should represent approximately the energy observed in gamma rays from the flare loop interface with the photosphere where the energetic particles react with dense photospheric nuclei. The X-ray spectra represent the electron energy distribution, which also shows the oscillatory characteristics in parallel with the ions characteristics. Observation of these radiation spectra by Chupp, Forrest and Suri [1975] shows that the soft X-ray energy domain (up to 400 keV) and hard X-ray domain (up to 7 MeV) have different distribution characteristics: in the hard X-ray domain (700 keV-7 MeV) the energy spectrum is exponential. This type of characteristic seems to match the simulation results of Fig. 8, where the particle distribution breaks into the bulk, the $\exp(-p_z/p_0)$ domain (energy up to a typical temperature 10-50 times of the bulk temperature), and the flat low-population relativistic domain. The amplitude of the oscillation (~ 1 Alfvén time) and its more minute characteristics, resemble what is reported of the solar gamma ray amplitude oscillations [Forrest et al., 1982].

VI. Discussion

We have examined through collisionless particle simulations some of the phenomena associated with current sheet formation, tearing of the sheet to form a chain of x-points and o-points and finally island coalescence. The analysis of these phenomena is far from complete. Nevertheless, the measured reconnection rates in the island formation stage can be matched with theoretical results from an analysis which is a modification of the Sweet-Parker reconnection rate by plasma compressibility. The effective resistivity is supplied in the collisionless case by wave-particle interactions and turbulent electron orbit modifications. The coalescence instability leads to an explosive increase of the reconnection rate. The consequences of tearing on the plasma are minimal in terms of particle energy gains. The consequences of the coalescence instability are large ion temperature increases and large temperature oscillations in the merged phase. The energy increase is accounted for by the loss of potential energy of the attracting current filaments, i.e. loss of poloidal magnetic energy. The oscillations in temperature are explained simply by the oscillations of the merged island at its magnetosonic frequency. It has been interesting to note that similar phenomena are observed in MHD simulations of reconnection and island coalescence even though both types of simulations cover vastly differing spatial scales. Finally, the nonlinear development of the coalescence instability seems to account for the impulsive nature of some types of solar flares, their time scale, intense plasma heating by flares and formation of high energy tails on the particle distributions.

Acknowledgments. This work was supported by the National Science Foundation grant ATM 82-14730 and Department of Energy, Office of Fusion Energy grant DE-FG05-80-ET-53088.

References

Amano, K., and T. Tsuda, Reconnection of magnetic field lines by clouds-in-cells plasma model, J. Geomag. Geoelectr., 29, 9, 1977.

Bhattacharjee, A., F. Brunel, and T. Tajima, Magnetic reconnection driven by the coalescence instability, Phys. Fluids, 26, 3322, 1983.

Brunel, F., T. Tajima, and J. M. Dawson, Fast magnetic reconnection processes, Phys. Rev. Lett., 49, 323, 1982.

Brunel, F., and T. Tajima, Confinement of a high-beta plasma column, Phys. Fluids, 26, 535, 1983.

Chupp, E. L., D. J. Forrest., A. N. Suri, High energy gamma-ray radiation above 300 keV associated with solar activity, in Solar Gamma-, X-, and EUV Radiation, Ed. by S. R. Kane, p. 341, (Reidel, Dordrecht, Holland), 1975.

Dickman, D. O., R. L. Morse, and C. W. Nielson, Numerical simulation of axisymmetric, collisionless, finite-β plasma, Phys. Fluids, 12, 1708, 1969.

Drake, J. F., and Y. C. Lee, Nonlinear evolution of collisionless and semicollisional tearing modes, Phys. Rev. Lett., 39, 453, 1977.

Forrest, D. J., E. L. Chupp, J. M. Ryan, C. Reppin, E. Rieger, G. Kanbach, K. Pinkau, G. Share and G. Kinzer, Evidence for impulsive ion acceleration during the 0312 UT flare of 1980 June 7, in Proceedings of the 17th International Cosmic Ray Conference, Paris, France, 1981, to be published.

Galeev, A. A., F. V. Coroniti, M. Ashour-Abdalla, Explosive tearing mode reconnection in the magnetospheric tail, Geophys. Res. Lett., 5, 707, 1978.

Gekelman, W., and R. L. Stenzel, Magnetic field line reconnection experiments, 2. Plasma parameters, J. Geophys. Res., 86, A2, 659, 1981.

Hamilton, J. E. M., J. W. Eastwood, The effect of a normal magnetic field component on current sheet stability, Planet. Space Sci., 30, 293, 1982.

Howard, R., and Z. Svetska, Development of a complex of activity in the solar corona, Solar Phys., 54, 65, 1977.

Katanuma, I., and T. Kamimura, Collisionless tearing instabilities, Phys. Fluids, 23, 2500, 1980.

Leboeuf, J. N., T. Tajima and J. M. Dawson, Magnetic x-points, islands coalescence and intense plasma heating, in Physics of Auroral Arc Formation, Ed. by S. I. Akasofu and J. R. Kan, p. 337, AGU, Washington, D. C., 1981.

Leboeuf, J. N., T. Tajima and J. M. Dawson, Dynamic magnetic x-points, Phys. Fluids, 25, 784, 1982.

Parker, E. N., Cosmical Magnetic Fields, Chapter 15, Clarendon Press, Oxford, 1979.

Rutherford, P., Nonlinear growth of the tearing mode, Phys. Fluids, 16, 1903, 1973.

Stenzel, R. L., and W. Gekelman, Experiments on magnetic field line reconnection, Phys. Rev. Lett., 42, 1055, 1979.

Sturrock, P. A., Flare models, in Solar Flares: A Monograph from Skylab Solar Workshop II, Ed. P. A. Sturrock, p. 411, Colorado Associated University Press, Boulder, 1980.

Tajima, T., Tearing and Reconnection, in Fusion Energy - 1981 (International Centre for Theoretical Physics, Trieste, 1982), p. 403. International Atomic Energy Agency, Vienna, Austria, 1982.

Tajima, T., F. Brunel and J. Sakai, Loop coalescence in flares and coronal x-ray brightening, Astrophys. J., 258, L45, 1982.

Tajima, T., F. Brunel, J. Sakai, L. Vlahos and M. Kundu, The coalescence instability in solar flares, Proceedings of the IAU Conference on Unstable Current Systems in Astrophysical Plasmas, Ed. by M. Kundu, IAU, in press, 1983.

Terasawa, T., Numerical study of explosive tearing mode instability in one-component plasmas, J. Geophys. Res., 86, 9007, 1981.

Wu, C. C., J. N. Leboeuf, T. Tajima and J. M. Dawson, Magnetic islands coalescence and intense plasma heating, University of California at Los Angeles, Plasma Physics Group Report No. PPG-511, 1980.

Questions and Answers

Priest: Your numerical simulations are of great interest, but you need to satisfy many observational criteria before applying them seriously to the solar flare. For example, if you want to interpret a flare as tearing and coalescence when the toroidal field is too great, you must demonstrate that there is an instability threshold with the field being stable until it is surpassed. Indeed, Mok and Van Haven have suggested that line tying of field lines in the dense photosphere is so effective that tearing cannot occur at all for most fields. Also, comparing your bursty ion temperature with a graph of bursty x rays is not sufficient. Have you demonstrated that the electric fields and particle numbers are adequate when scaled up to solar parameters?

Lebouef: We have given circumstantial evidence that the coalescence instability could give rise to phenomena similar to those observed during impulsive flares. I agree more work needs to be done.

Schindler: As shown by recent analytical work, adiabatic electrons have a stabilizing effect for intermediate wavelengths in the presence of a magnetic field component normal to the current sheet. Has this effect been seen in the simulations you discussed?

Lebouef: No, we have not considered the effect of a normal component of the magnetic field. It is not clear that a normal component will stabilize the coalescence instability, however.

Vasyliunas: The ratio of length and time scales to the characteristic parameters (gyroradii, plasma periods, etc.) is very much smaller in your simulations than in space and astrophysical plasmas, and non-MHD effects are therefore relatively much more important. To what extent, then, can your simulation results be scaled to apply to the magnetosphere or the sun?

Lebouef: There is not much difference between the phenomena observed in MHD simulations under similar conditions and those observed in equivalent kinetic simulations.

THE NONLINEAR TEARING MODE

G. Van Hoven and R. S. Steinolfson

Department of Physics, University of California, Irvine, California 92717

Abstract. The nonlinear behavior of the tearing instability is investigated with numerical solutions of the resistive, incompressible, MHD equations. The initial state for the nonlinear computations is provided by the linear instability, with the amplitude selected such that the nonlinear terms just equal the dominant linear term in one of the equations at some location in the spatial grid. Typical simulations are described for a magnetic Lundquist number S of 10^4 and wavelength parameters $\alpha(= 2\pi a/\lambda$, where a is the shear scale and λ the instability wavelength) from 0.05 to 0.5. In all cases, the nonlinear mode initially evolves at the linear growth rate, followed by a period of reduced growth. Another common feature is the formation of secondary flow vortices, near the tearing surface, which are opposite in direction to the initial linear vortices. At high S and low α these vortices result in the creation of a new island centered at the initial X-point. The one constant-Ψ solution investigated had markedly different behavior from the remaining nonconstant-Ψ solutions. Not only was its growth reduced (approximately an order of magnitude less over the same time period) but, whereas the nonconstant-Ψ computations showed a reduction by about 20% of the initial magnetic energy in the shear layer, the constant-Ψ simulation indicated a reduction of magnetic energy two orders of magnitude smaller. The island width of the nonconstant-Ψ solutions became larger than twice the width of the shear layer.

Introduction

The primary astrophysical application of magnetic reconnection involves its use as a mechanism for the release of stored magnetic energy, as exemplified by solar flares and geomagnetic substorms. A prime candidate for the realization of dynamic reconnection is the resistive magnetic tearing mode, a spontaneous instability of a stressed magnetic field. In view of the perceived importance of this process, it is unfortunate that there is a dearth of relevant computations of its large-scale and large-amplitude temporal behavior. Previous nonlinear studies (Van Hoven and Cross, 1973; Schnack and Killeen, 1978) have treated multiple tearing layers, or relatively short wavelengths which are known to saturate at low levels. The former authors showed that compressibility does not have a large effect on this relatively slow instability.

In this paper, we describe an effort to move beyond these limitations. We have performed a series of nonlinear computations of tearing-mode development which achieve higher values of the magnetic Reynolds number and larger wavelengths than previously considered. As we show in what follows, the behavior of the instability in this "astrophysical" limit exhibits some differences from previous results.

Formulation

The nonlinear phase of the tearing mode (Furth et al., 1963) is studied in slab geometry using incompressible, constant-resistivity, MHD theory. The initially stationary plasma, with uniform thermodynamic properties, is embedded in a force-free, constant-amplitude magnetic field. A linear mode, at its maximum linear growth rate γ, provides the initial state for the nonlinear computation. We present results for a Lundquist number S (ratio of the resistive time to the hydromagnetic time) of 10^4 and values of the wavelength parameter $\alpha(= 2\pi a/\lambda$, where a is the shear scale and λ the disturbance wavelength) of 0.05, 0.13, and 0.50. [A larger parameter range and additional computational results are considered by Steinolfson and Van Hoven (1984)]. The mode with $\alpha = 0.5$ is a constant-Ψ [weak reconnecting-field (B_y) variation with y out to the peak of the inflow velocity (Furth et al., 1963), a condition thought to be synonymous with weak nonlinear growth (Rutherford, 1973)] solution in the linear regime, while the other two are nonconstant-Ψ, and the $\alpha = 0.13$ mode corresponds to maximum linear growth (Steinolfson and Van Hoven, 1983).

The linear theory predicts a chain of X-points and islands in the magnetic field lying along the tearing surface (x-axis) in our geometry)

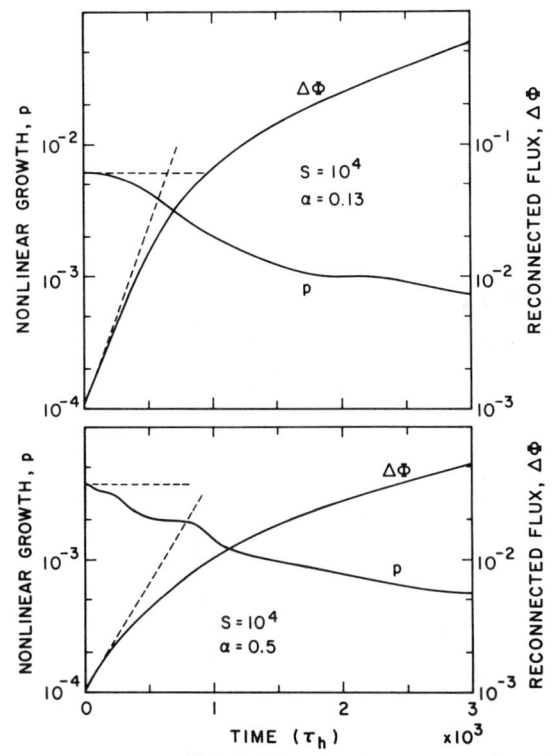

Fig. 1. Nonlinear Evolution of the growth rate $p=\gamma\tau_h$ and reconnected magnetic flux vs time in units of the hydromagnetic period.

near y = 0. We isolate one wavelength of this initial disturbance and do not allow it to interact with adjacent wavelengths. Because of the symmetries involves, our computation only extends from the center of one island (x = 0) to the adjacent X-point (x_{max}) and from the tearing surface (y = 0) to a relatively large distance (y_{max}). The distance y_{max} is large enough that the perturbation is essentially negligible and decaying exponentially with y. Symmetry boundary conditions are applied at the remaining three boundaries. An expanding grid is used in the y-direction, with minimum spacing near y = 0, in order to resolve the tearing layer. The nonlinear equations are solved numerically using a fully-implicit, alternating-direction procedure.

Numerical Results

The evolution of two of the modes, as measured by reconnected flux and nonlinear growth rate $[p \equiv \gamma\tau_h \equiv \partial\Delta\Phi/\partial t]$, is shown in Fig. 1, where time is measured in units of the hydromagnetic time $\tau_h = a/v_A$. The dashed curves represent continued linear growth. The long wavelength mode ($\alpha = 0.05$) evolves almost identically to the $\alpha = 0.13$ solution, in terms of these quantities, with somewhat (a few percent) more flux reconnection at the final time. Although the two nonconstant-Ψ modes display comparable nonlinear behavior, they are in sharp contrast to the considerably smaller reconnected flux for the constant-Ψ mode (Rutherford, 1973).

The total magnetic energy, per unit distance perpendicular to the tearing plane and averaged over one wavelength, removed from the magnetic fields is tabulated in the first row of part A of Table I. By contrast, the energy initially in the shear layer is 5.5×10^8 ergs/cm^2 [$aB^2/8\pi$]. As a percentage of the initial shear-layer energy, the total magnetic energy removed from the fields is 21, 12, and 0.07 for the modes with $\alpha = 0.5$, 0.13, and 0.5, respectively. The second and third rows in Part B show the percent of the total energy that was removed from the x- and z-components. Longer wavelength modes remove more energy from the z-component, while none is removed from the z-component for

TABLE 1. Energy balance. Magnetic energy release (ergs/cm^2) computed for reference B = 7.3 G, a = 10^7cm; the remaining energies are given in percents.

wavenumber (α)	0.05	0.13	0.5
A. magnetic energy release:	1.2×10^8	7.0×10^7	3.8×10^5
B. energy source:			
magnetic (x)	89.7	97.0	100.0
magnetic (z)	10.3	3.0	0.0
C. energy budget:			
magnetic (y)	8.0	16.2	49.2
magnetic (z)	0.0	0.0	2.6
kinetic (x)	4×10^{-2}	3×10^{-2}	8×10^{-2}
kinetic (y)	6×10^{-4}	5×10^{-4}	3×10^{-4}
kinetic (z)	3×10^{-3}	3×10^{-4}	5×10^{-5}
thermal	92.0	83.8	48.2

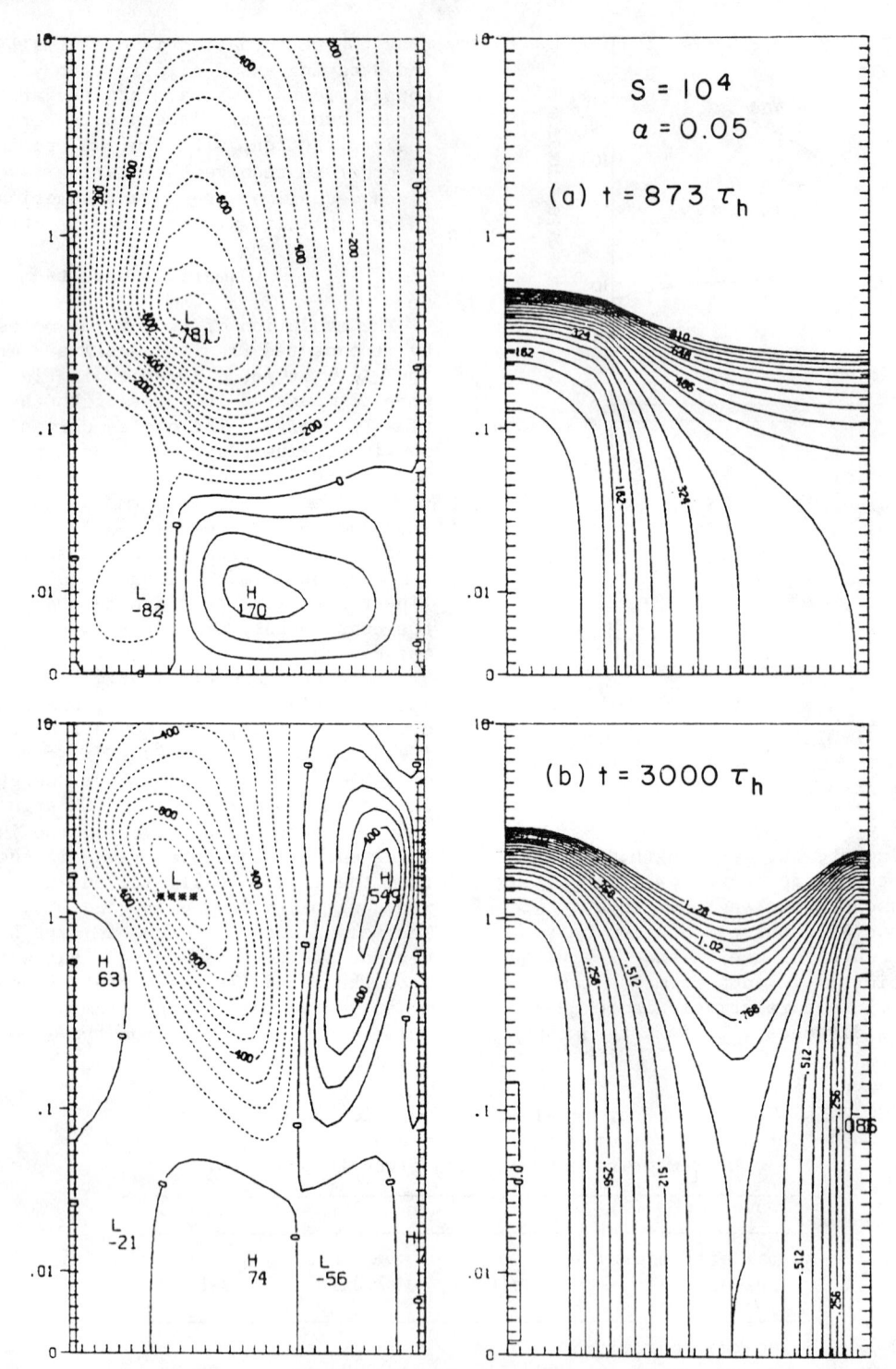

Fig. 2. Formation of a new magnetic island. The stream function (left) is given in units of 10^{-5} av_A and the magnetic flux (right in units of aB_o. The abscissa is the x/λ axis, ranging from the original 0 to X points, and the ordinate is the (expanded for small values) y/a axis.

the constant-Ψ solution (energy actually transfers into this component). The available energy in part A is distributed among the various components as shown (on a percentage basis) in part C. Since the energy equation is not included in this incompressible computation, the thermal energy (from Joule dissipation) is assumed to make up the difference between the amount released and that which appears in magnetic and kinetic forms. Note that more energy goes into heating as the disturbance wavelength increases, with less into the y-component of the magnetic field.

For all three of the modes, a secondary flow vortex, oriented oppositely in direction to the initial linear vortex, forms near the X-point, as illustrated in Fig. 2(a) which has a non-linear y-scale. The velocity is parallel to the flux-function contours on the left while some of the magnetic field lines near the tearing surface are shown on the right. The dashed flux-function contours indicate a clockwise vortex (the distorted linear vortex), and solid curves represent a counter-clockwise vortex. The two modes not shown in this figure continue evolving for the duration of the calculation with the qualitative spatial behavior in Fig. 2(a). However, for the long wavelength mode, the secondary flows become strong enough to alter the basic magnetic topology and cause the formation of an additional magnetic island centered at the linear X-point [Fig. 2(b)].

Conclusion

A primary result in these computations is that the nonlinear evolution generally differs from one region of parameter space to another and, hence, a typical characterization of the evolution in the nonlinear regime is not possible. Some general statements that do apply to all solutions are: (1) The nonlinear spatial distributions of the physical variables differ substantially from the linear behavior; (2) Once nonlinear effects become important, the growth slows considerably from the linear rate; (3) Neither the linear growth rate nor the nonlinear growth is a good predictor of the nonlinear performance of a particular mode in terms of magnetic energy conversion; and (4) More of the stored magnetic energy is converted to thermal energy (by resistive dissipation) as disturbance wavelength increases (>90% at α = 0.05). Comparisons of these results with those of other nonlinear computations are described in Steinolfson and Van Hoven (1984).

Acknowledgments. This work was supported by the Solar Terrestrial Theory Program of NASA and the Atmospheric Sciences Section of NSF. Acknowledgment is also made to the Institute of Geophysics and Planetary Physics at the Los Alamos National Laboratory, which provided funding under their grant program, and to the National Center for Atmospheric Research, which is sponsored by NSF, for the use of their computer facilities.

References

Furth, H. P., J. Killeen, and M. N. Rosenbluth, Finite-Resistivity Instabilities of a Sheet Pinch, Phys. Fluids 6, 459, 1963.

Rutherford, P. H., Nonlinear Growth of the Tearing Mode, Phys. Fluids 16, 1903, 1973.

Schnack, D. D., and J. Killeen, Linear and Nonlinear Calculations of the Tearing Mode, Theoretical and Computational Plasma Physics (Vienna: International Atomic Energy Agency), p. 337, 1978.

Steinolfson, R. S., and G. Van Hoven, The Growth of the Tearing Mode: Boundary and Scaling Effects, Phys. Fluids 26, 117, 1983.

Steinolfson, R. S., and G. Van Hoven, Nonlinear Evolution of the Resistive Tearing Mode, Phys. Fluids 27, , 1984.

Van Hoven, G., and M. A. Cross, Energy Release by Magnetic Tearing: The Nonlinear Limit, Phys. Rev. A7, 1347, 1973.

ON THE CAUSE OF X-LINE FORMATION IN THE NEAR-EARTH PLASMA SHEET:
RESULTS OF ADIABATIC CONVECTION OF PLASMA-SHEET PLASMA

G. M. Erickson

Department of Space Physics and Astronomy, Rice University, Houston, TX 77251

Abstract. Self-consistent, static-equilibrium solutions are presented for two-dimensional magnetospheric-magnetic-field configurations with isotropic thermal pressure. These solutions include a dipole field and are not restricted to the asymptotic theory. Adiabatic convection of plasma sheet flux tubes is modeled as a series of static-equilibrium solutions in which flux tubes conserve their PV^γ as they convect, which resulted in time dependent magnetospheric configurations. Specifically, it is found that a deep minimum in the equatorial B_z develops in the inner plasma sheet, thereby causing the magnetic-field configuration to become more stretched and tail-like in time. These results suggest X-line formation in the inner plasma sheet as a consequence of lossless, adiabatic convection of plasma sheet flux tubes.

Introduction

X-line formation is now regarded as one of the primary candidates for the basic physical process to account for the observed energy dissipation associated with magnetospheric substorms. It is clear that changes in the solar wind can trigger magnetospheric substorms [see, e.g., Akasofu, 1980]. However, it is not clear that magnetospheric substorms are always associated with specific changes in the solar wind. For this reason a substorm mechanism which does not necessarily require an external trigger for X-line formation is attractive. Schindler [1974] offered the following scenario for the substorm mechanism. During the substorm growth phase, free energy is accumulated in the tail, and the tail becomes more and more unstable to perturbations that try to create neutral lines. The presence of a normal magnetic field component in the current sheet inhibits the instability. But when the plasma-sheet becomes sufficiently thin and/or the normal magnetic field component becomes sufficiently small, some breakup mechanism, such as the ion-tearing mode, becomes operative leading to X-line formation and substorm onset [see also Nishida and Nagayama, 1973; Russell and McPherron, 1973; Hones, 1977]. Erickson and Wolf [1980] have presented the argument that approximately lossless, adiabatic, earthward convection of plasma-sheet plasma on closed field lines is necessarily time dependent. They hypothesize that if earthward convection occurs in the tail, even if solar wind conditions were steady, magnetotail field lines would become more and more stretched and tail-like in time resembling the growth phase scenario above. Schindler and Birn [1982] and Birn and Schindler [1983] have self-consistently modeled the quasi-static evolution of tail-like configurations within the framework of the asymptotic theory. They found that in the absence of unrealistic plasma loss, earthward convection is time dependent and drives the tail toward instability.

As discussed by Erickson and Wolf [1980] it is the presence of the earth's dipole field which prevents approximately lossless, adiabatic convection of plasma-sheet flux tubes from proceeding in a time-independent manner. Specifically, it is the fact that plasma cannot expand very far along a line of force in the presence of a dipole field which causes the tail configuration to become more stretched and tail-like as convection proceeds. The asymptotic theory excludes some $(\vec{B}\cdot\nabla)\vec{B}$ magnetic tension terms in the force balance of the system and, therefore, cannot accurately represent the earth's dipole field and its critical role as the endpoint of plasma-sheet flux tubes. In this paper, preliminary results are presented of self-consistent modeling of quasi-static convection of plasma-sheet flux tubes in two dimensions. These models include a dipole field and are not restricted to the asymptotic theory. The results of this modeling confirm the results of the previous efforts mentioned above. Also, the results lend support to the Schindler [1974] growth phase scenario, although they do not exclude the possibility of other substorm scenarios [e.g., Coroniti and Kennel, 1972; Akasofu, 1980; Atkinson, 1980].

Quasi-Static Convection

To model convection of plasma-sheet flux tubes we make the following assumptions: (1) plasma-sheet ions are in bounce equilibrium, (2) thermal

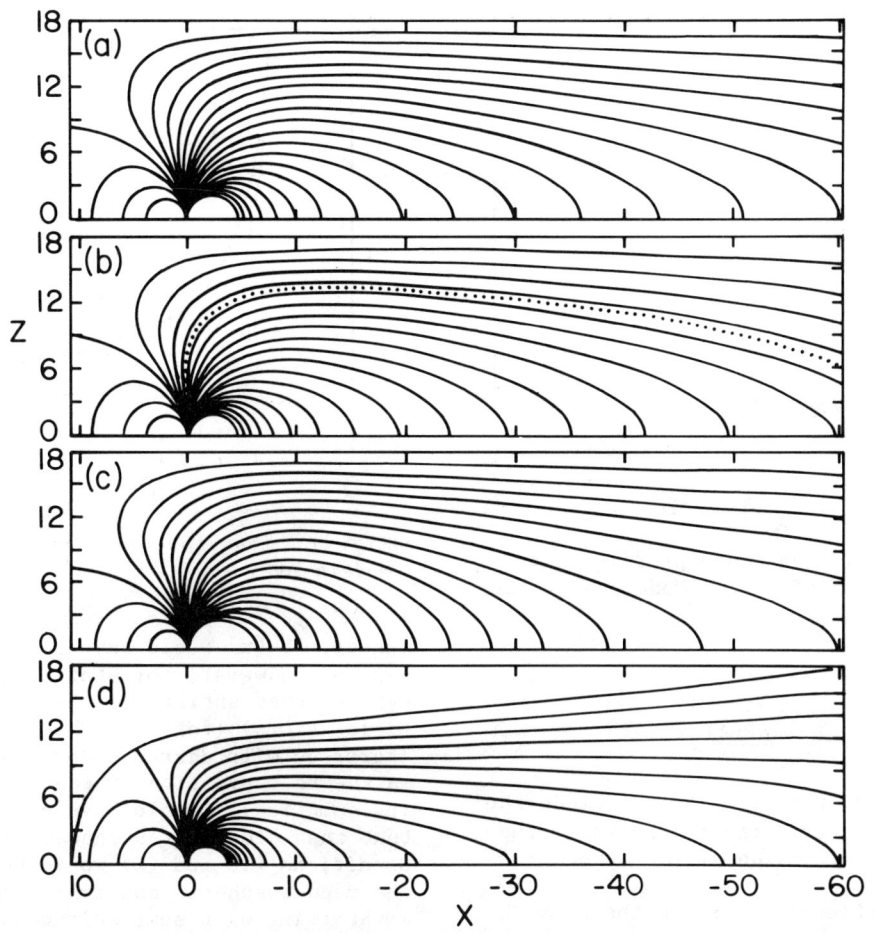

Fig. 1. Some examples of self-consistent, two-dimensional, magnetospheric-magnetic-field configurations. Tail field lines are shown containing equal amounts of magnetic flux. The amount of magntic flux passing through the right boundary is the same in each example. (a) Model A. Using the appropriate boundary conditions (P and A at the magnetopause, P in the equatorial plane, $B_x (z = 0) = 0$, and one-dimensional force balance at the right boundary) a modified (P = constant for $0 \geq x \geq -4.5 \ R_E$) Fuchs-Voigt [1979] model with k = 1.54 is obtained. The Fuchs-Voigt anlaytic models feature an exponential decline of the physical quantities down the tail. (b) Model B. This model is the same as Model A except that the height of the plasma-sheet (dotted line) is chosen as 6 R_E at x = -60.5 R_E. (c) Model C. In this model the equatorial pressure declines as $|x|^{-1 \cdot 2}$, based on the observations of Behannon [1968]. (d) Model D. This model has the same equatorial pressure as Model A, but the tail magnetopause is flared (5.7°), and the dayside magnetopause is rounded.

plasma pressure is isotropic, (3) inertial forces are small compared to pressure gradients, and (4) convection is lossless. Under these assumptions we seek static-equilibrium solutions of the momentum equation and Maxwell's equations,

$$\vec{J} \times \vec{B} = \nabla P \quad (1)$$

$$\vec{J} = \frac{1}{\mu_0} \nabla \times \vec{B}, \quad (2)$$

and

$$\nabla \cdot \vec{B} = 0. \quad (3)$$

In two dimensions, the magnetic vector potential is $\vec{A} = A\hat{e}_y$ (GSM coordinates in units of earth radii are used throughout) with $B_x = -\partial A/\partial z$, $B_z = \partial A/\partial x$, and (1)-(3) are rewritten as

$$\nabla^2 A = -\mu_0 \frac{dP(A)}{dA} - m \frac{\partial}{\partial x} \delta(x)\delta(z), \quad (4)$$

with

P,A = constant along field line.

In (4), the delta function represents is the dipole source. Numerical solutions of (4) are obtained on a grid displaced from the origin, and A is decomposed into its dipole, A_d, and plasma current source, A_j, parts. Hence, we actually

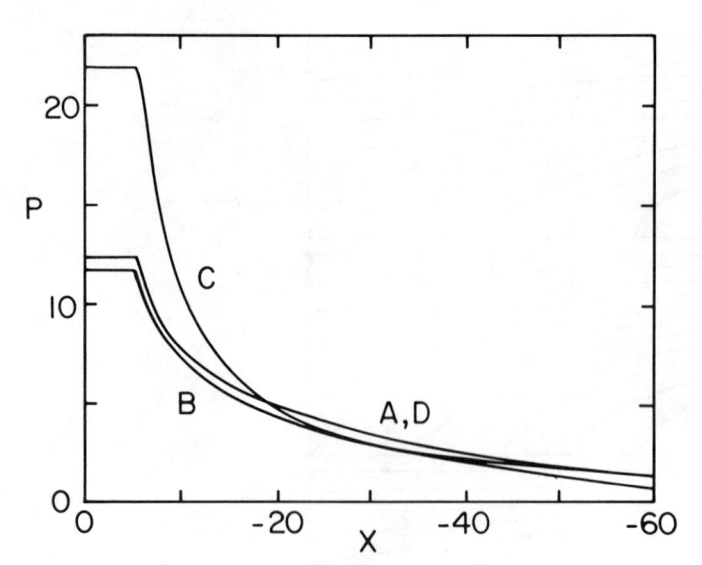

Fig. 2. The equatorial pressure profiles for Models A-D. Note that dP/dx for Models A, B, and D is the same.

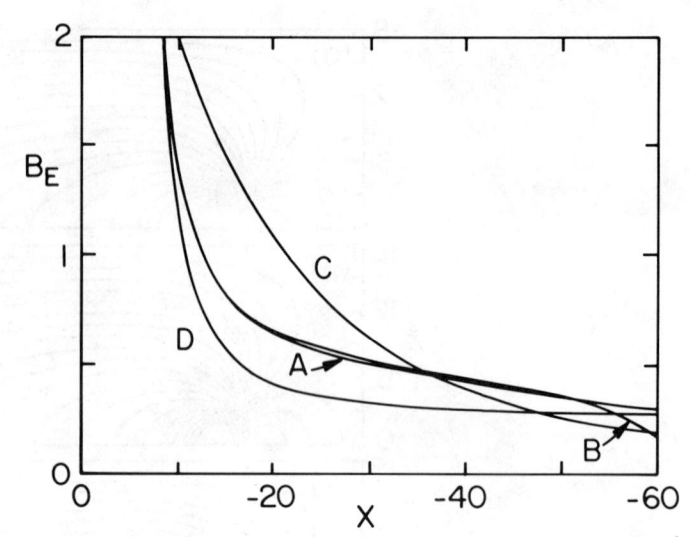

Fig. 3. The equatorial magnetic field, B_e, for Models A-D.

find solutions of

$$\nabla^2 A_j = -\mu_0 \frac{dP(A)}{dA}. \qquad (5)$$

Adjustable parameters of these models include the equatorial plasma pressure, the thickness of the plasma-sheet, and the location of the magnetopause.

By the frozen-in-flux theorem and the fact that in two dimensions both P and A are constant along a magnetic field line, we can use A to tag the plasma as it convects. Also, under the assumptions, the plasma convects so as to keep its PV^γ constant, where γ is the adiabatic index (which we chose to be 2), and

$$V = \int_{f\ell} ds/B \qquad (6)$$

is the volume of a flux tube of unit magnetic flux. Thus, convection is modeled as a time-sequence of static-equilibrium solutions of (5) such that each solution shares the same $PV^\gamma(A)$. That is, we require that

$$PV^\gamma(A) = \text{independent of time}. \qquad (7)$$

Results

Some examples of self-consistent, two-dimensional magnetospheric-magnetic-field configurations are shown in Figure 1 for different pressure distributions, plasma-sheet heights, and magnetopause locations. Figures 2-5 show the equatorial pressure distribution, the equatorial magnetic field, $PV^2(x)$, and $PV^2(A)$, respectively, for these models. Note that in order for approximately lossless, adiabatic convection to occur in some region of the tail in a time-independent manner, $PV^\gamma(x)$ would have to be constant in that region. However, for static-equilibrium configurations that satisfy (1) and include a dipole field, $PV^\gamma(x)$ (for $\gamma \geq 5/3$) is not constant but increases with distance down tail. Thus, as earthward convection proceeds, flux tubes entering some region would have a larger particle content than did their predecessors in the region. Conditions (1) and (7) would then require that the magnetospheric configuration vary in time.

Starting with some self-consistent solution of (1) or (5) at $t = 0$, an electric field $E_{mp}(t)$ is imposed at the magnetopause allowing magnetic flux to enter the magnetopause and forcing earthward convection of plasma-sheet flux tubes. In practice, the value of A at the magnetopause, A_{mp}, is changed, the equatorial plasma pressure is adjusted, and new solutions of (5) are

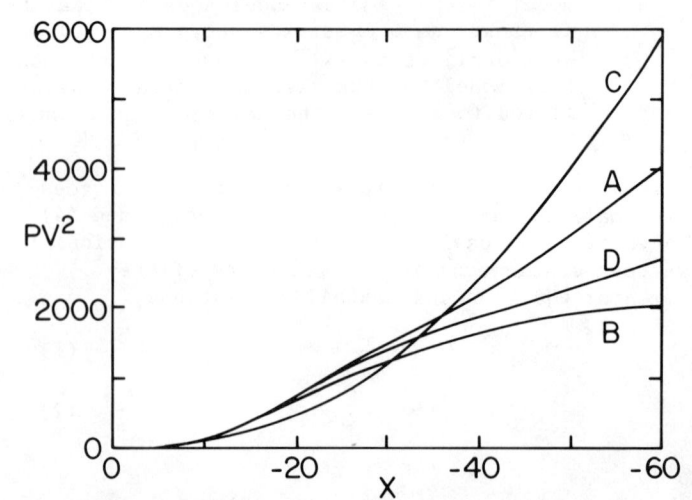

Fig. 4. PV^2 vs. x for Models A-D. A constant $PV^\gamma(x)$ is required for steady, lossless, adiabatic convection.

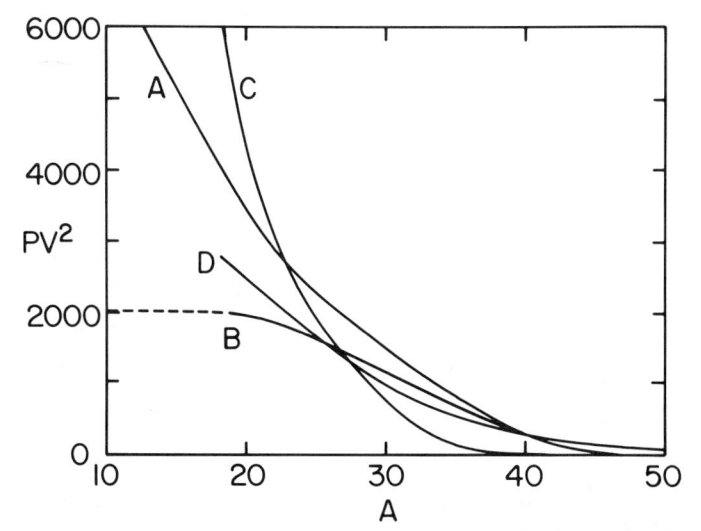

Fig. 5. PV^2 vs. A for Models A-D. The approach of the top of the plasma-sheet to the equatorial plane implies a neutral line tailward of the right boundary in Model B. When Model B is convected, it is assumed that flux tubes of constant PV^2 enter the right boundary represented by the dashed-line continuation of the $PV^2(A)$ curve.

obtained until (5) and (7) are satisfied simultaneously. Thus a time sequence of static-equilibrium solutions of (1)-(3) is constructed with time parameterized by the value of A_{mp}. Forcing Models A and B to convect in this manner resulted in Models A' and B' shown in Figures 6 and 9. In Model B' modeling was performed only out to $x = -60.5\ R_E$ where the height of the plasma-sheet was chosen so that $PV^2(x)$ had zero slope there at $t = 0$. Beyond $x = -60.5\ R_E$, PV^2 was assumed constant. For $t > 0$, the height of the plasma-sheet was chosen such that the amount of magnetic flux contained between the top of the plasma-sheet and the magnetopause was constant in time. For quiet times and a constant electric field of 10^{-4} V/m at the magnetopause, the three "snapshots" of Figures 6 and 9 for $A_{mp} = 0, -12$, and -22 correspond to $t = 0, 1.5$, and 2.7 hrs, respectively. For an electric field of 5×10^{-4} V/m the respective times would be $t = 0, 0.3$, and 0.54 hrs. The conversion of the other physical quantities from the units used to SI units is given in the Appendix.

As expected, the adiabatic convection of plasma-sheet flux tubes resulted in time-dependent configurations, consistent with the earlier

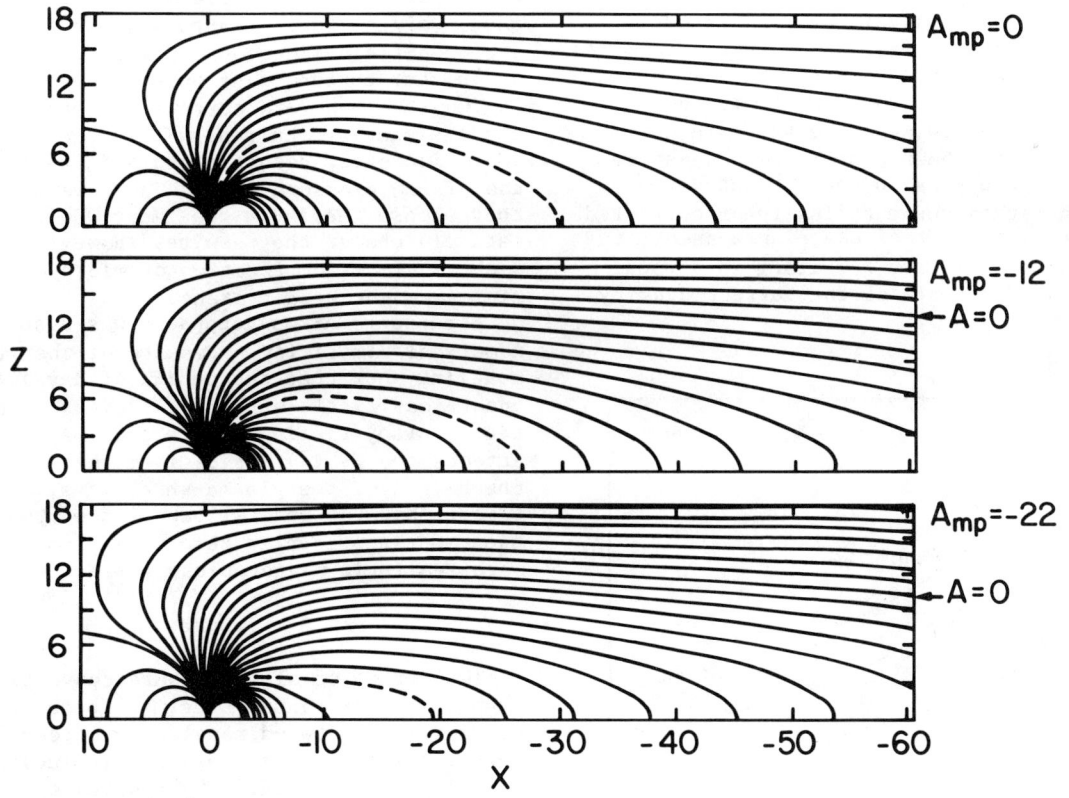

Fig. 6. Model A' showing the results of convecting Model A under the constraint that $PV^2(A)$ remains unchanged (as explained in the text). The two-dimensional magnetic field configuration if shown for $A_{mp} = 0$ (top panel), $A_{mp} = -12$ (middle panel), and $A_{mp} = -22$ (bottom panel). Tail field lines are shown containing equal amounts of magnetic flux. The $A = 0$ field line is labeled, and the dashed field line is $A = 30$ for reference.

conclusions of Erickson and Wolf [1980] and Schindler and Birn [1982] that steady-state convection is theoretically unlikely in the magnetotail. Being unable to convect in a steady state, the field strength in the lobe increases as flux is piled up, and the configuration becomes more tail-like in time. Figures 7 and 10 show the evolution of the equatorial pressure distribution for Models A' and B'. The drift of plasma around the earth was accounted for by defining the equatorial pressure at the origin to be the máximum pressure of the system for all times. Thus the pressure was allowed to reach this value but not exceed it. Figures 8 and 11 show the evolution of the equatorial magnetic field B_e for Models A' and B'. Note the minimum in B_e that develops (and gets deeper with time) near $-10\ R_E$, corresponding to the stretching of the field in the near-earth part of the plasma-sheet as higher and higher content flux tubes are convected into the region. Also, the plasma current density has approximately doubled in this region from $A_{mp} = 0$ to $A_{mp} = -22$. It is in this near-earth plasma-sheet region that the tail appears least stable.

Admittedly, the manner in which the drift of plasma around the earth is treated does not seem very realistic. This feature was chosen merely for numerical simplicity and convenience. Condition (7) was strictly enforced tailward of the P = constant region, while it was not enforced inside the P = constant region. This resulted in a sharp inner edge of the current sheet with current density increasing from zero to its peak value within only a few R_E. The westward currents in the plasma-sheet contribute a negative B_e earthward of the peak in the current density. As higher and higher content flux tubes convected into the near-earth part of the plasma-sheet, the westward current increased resulting in a local B_e minimum just earthward of the current density

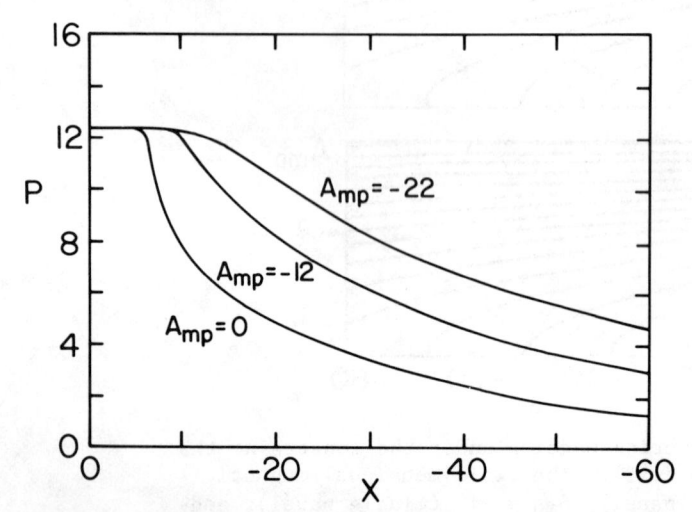

Fig. 7. The equatorial plasma pressure for Model A' for A_{mp} = 0, -12, and -22. The end of the P = constant region is at x = -8.5 R_E for $A_{mp} \leq -12$.

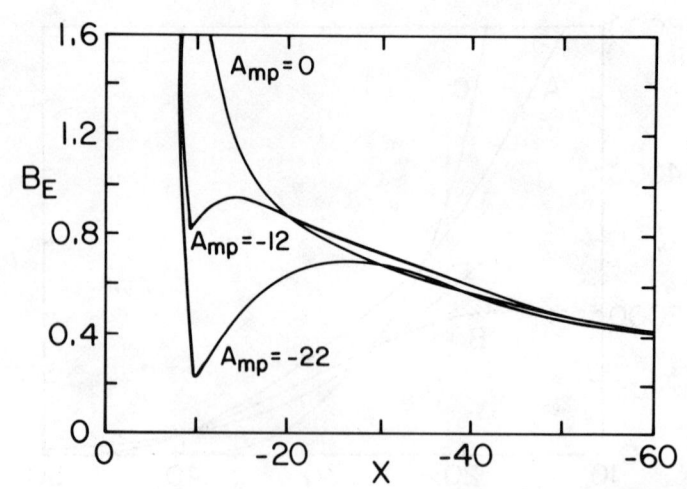

Fig. 8. The equatorial magnetic field for Model A' for A_{mp} = 0, -12, and -22.

maximum. In a more realistic treatment the drift of plasma out of the noon-midnight meridian plane and around the earth would still occur within only a few R_E, but the artificial constraint that the pressure not exceed a certain constant value would be removed. In that case a peak in the plasma pressure might develop just tailward of the inner edge of the plasma-sheet. The current, $-dP/dA$, would change sign at this peak in a closed-field-line configuration. As higher and higher content flux tubes convect into the region, the westward current tailward of the peak and the eastward current within the inner edge would increase, resulting in a B_e minimum near the plasma pressure peak. Thus, we would expect that a more realistic treatment would not qualitatively change the results. However, details such as the exact location or width of the B_e minimum might be different.

A rectangular magnetopause was also chosen for numerical simplicity. Results of the static modeling show that the choice of dayside magnetopause has little effect on the tail configuration. Also, flaring of the tail magnetopause affects the PV^γ (A) curves in much the same way as the height of the plasma-sheet does. Thus we expect that the general behavior of the convection models presented here will also occur for more realistic magnetopause shapes.

Summary

The effect of slow, sunward convection in a magnetospheric plasma-sheet has been investigated using computed two-dimensional, force-balanced magnetic field configurations, including the earth's dipole field. The results confirm the earlier conclusions of Erickson and Wolf [1980] and Schindler and Birn [1982] that approximately lossless, adiabatic convection of plasma-sheet flux tubes is a time-dependent process. In this process the magnetotail becomes more stretched and tail-like as convection proceeds, resulting

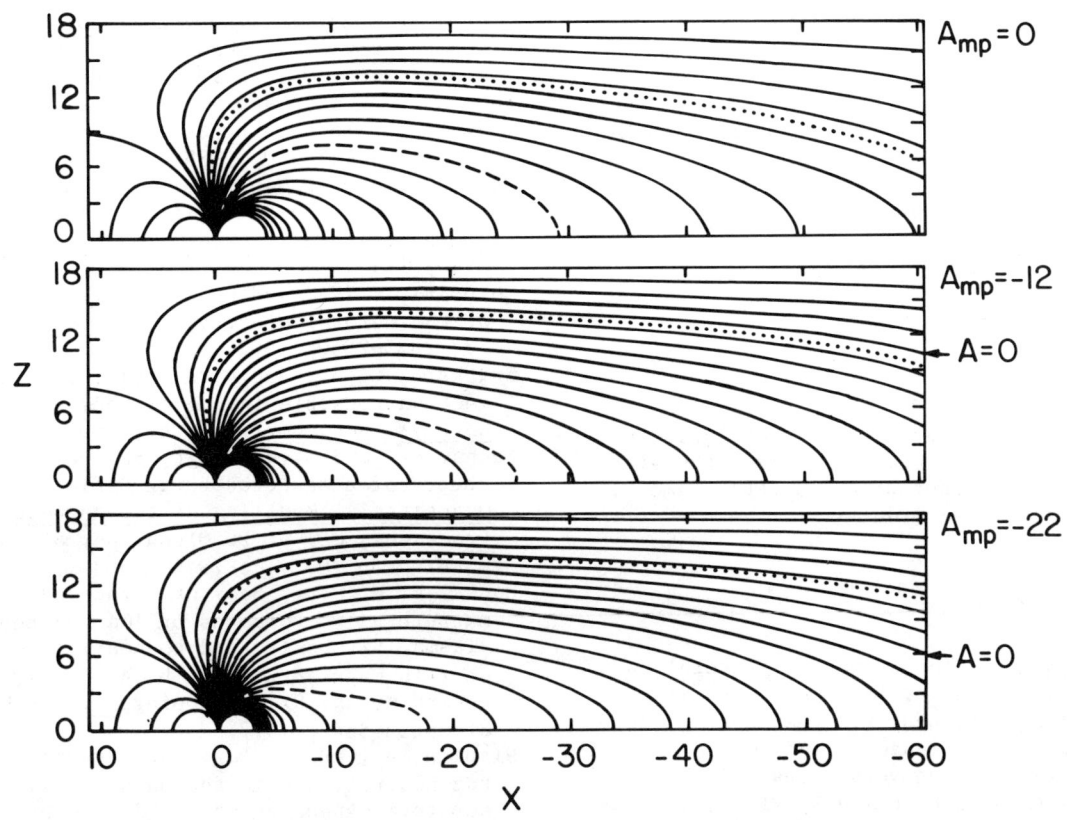

Fig. 9. Model B' shows the results of convecting Model B in the same manner as Model A with the additional condition that the total flux contained between the top of the plasma-sheet (dotted line) and the magnetopause is constant. The snapshots shown are displayed in the same manner as Figure 6.

in configurations more unstable to perturbations which try to create neutral lines. The results support the Schindler [1974] growth-phase scenario, although they do not exclude other possibilities. Preliminary results suggest that forcing plasma-sheet plasma to adiabatically convect sunward leads to a buildup of magnetic energy in the magnetotail. A minimum in the equatorial field strength develops in the near-earth part of the plasma-sheet, and that minimum rapidly deepens as convection proceeds, which obviously suggests formation of a near-earth X-line.

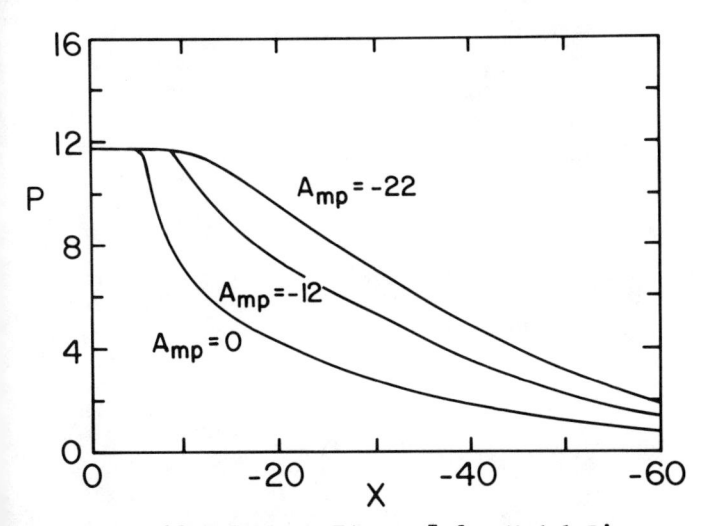

Fig. 10. Same as Figure 7 for Model B'.

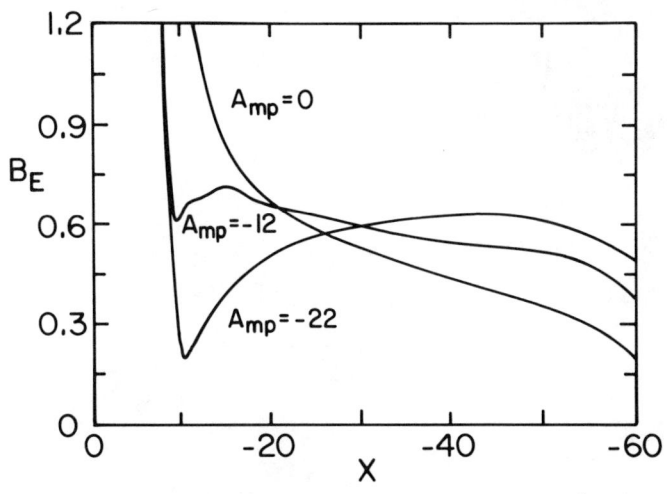

Fig. 11. Same as Figure 8 for Model B'.

Appendix

The unit of distance used here is R_E (6.38 × 10^6 m). The other physical quantities are given in arbitrary units which can be converted to MKS units as follows:

$$P \rightarrow \delta P/(\mu_0 R_E^4),$$

$$A \rightarrow \delta^{1/2} A/R_E,$$

$$B \rightarrow \delta^{1/2} B/R_E^2,$$

$$V \rightarrow V/(\delta^{1/2} R_E^3).$$

Time is parameterized as the value of A_{mp}, the value of A on the magnetopause. Given the electric field at the magnetopause, $E_{mp}(t)$, time (in seconds) is determined from

$$\int_0^t E_{mp}(t')dt' = -A_{mp}.$$

Acknowledgments. The author is grateful to Richard Wolf and Hannes Voigt for their helpful suggestions and continuous interest throughout this project, and to Michael Heineman and Ken Yates for stimulating conversations. This work was supported in part by the U.S. Air Force Geophysical Laboratory under contract F19628-83-K-0016 and by the National Science Foundation under grants ATM81-20391 and ATM82-06026.

References

Akasofu, S.-I., The solar wind-magnetosphere energy coupling and magnetospheric disturbances, Planet. Space Sci., 28, 495-509, 1980.

Atkinson, G., The expansive phase of the magnetospheric substorm, in Dynamics of the Magnetosphere, ed. S.-I. Akasofu, pp. 461-481, D. Reidel Publ. Co., Dordrecht, Holland, 1980.

Behannon, K. W., Mapping of the earth's bow shock and magnetic tail by Explorer 33, J. Geophys. Res., 73, 907-930, 1968.

Birn, J., and K. Schindler, Self-consistent theory of three-dimensional convection in the geomagnetic tail, J. Geophys. Res., 88, 6969-6980, 1983.

Coroniti, F. V., and C. F. Kennel, Polarization of the auroral electrojet, J. Geophys. Res., 77, 2835-2850, 1972.

Erickson, G. M., and R. A. Wolf, Is steady convection possible in the earth's magnetotail?, Geophys. Res. Lett., 7, 897-900, 1980.

Fuchs, F., and G.-H. Voigt, Self-consistent theory of a magnetospheric B-field model, in Quantitative Modeling of the Magnetospheric Processes, ed. W. P. Olson, pp. 86-95, AGU, Washington, D.C., 1979.

Hones, E. W., Jr., Substorm processes in the magnetotail: Commends on 'On hot tenuous plasma, fireballs, and boundary layers in the earth's magnetotail' by L. A. Frank, K. L. Ackerson, and R. P. Lepping, J. Geophys. Res., 82, 5633-5640, 1977.

Nishida, A., and N. Nagayama, Synoptic survey for the neutral line in the magnetotail during the substorm expansion phase, J. Geophys. Res., 78, 3782-3798, 1973.

Russell, C. T., and R. L. McPherron, The magnetotail and substorms, Space Sci. Rev., 15, 205-266, 1973.

Schindler, K., A theory of the substorm mechanism, J. Geophys. Res., 79, 2803-2810, 1974.

Schindler, K., and J. Birn, Self-consistent theory of time-dependent convection in the earth's magnetotail, J. Geophys. Res., 87, 2263-2275, 1982.

NUMERICAL SIMULATION OF THE DAYSIDE RECONNECTION

M. Hoshino and A. Nishida

Institute of Space and Astronautical Science
4-6-1 Komaba, Meguro Tokyo 153 Japan

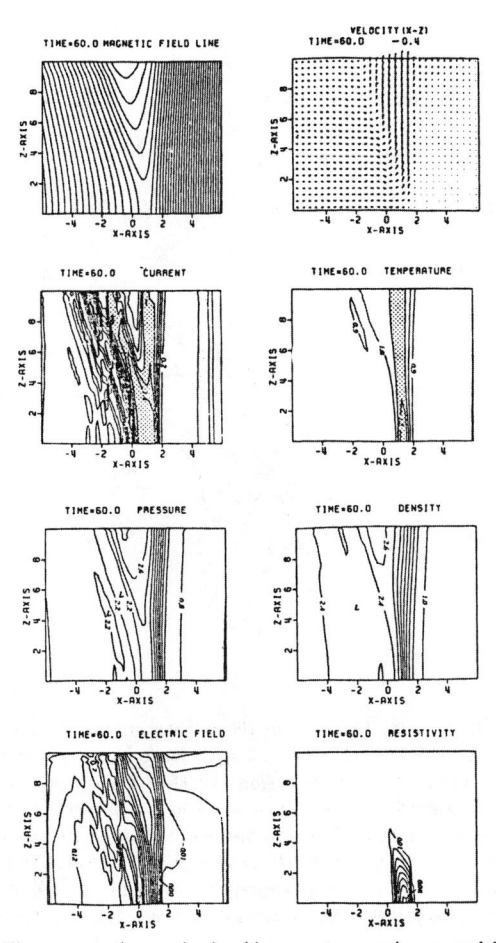

Figure 1. The reconnection on the dayside magnetopause in our model calculation. The magnetosheath is on the left side, and the magnetosphere is on the right hand side. From top to bottom the figure shows magnetic field lines and the plasma flow vectors, contour maps of current density and temperatures, contour maps of plasma pressure and density, and contour maps of electric field and resistivity.

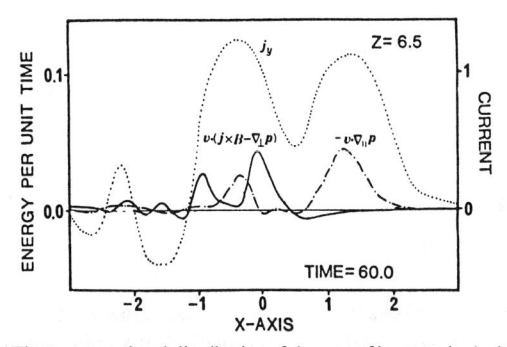

Figure 2. The cross-sectional distribution of the rate of increase in the kinetic energy perpendicular and parallel to the magnetic field, respectively. The current density is also given as a reference. The left hand current peak corresponds to the slow shock and the right one is the slow expansion.

To examine the basic characteristics of reconnection on the dayside magnetopause, we have numerically studied the reconnection process at an interface where the total pressure is in balance but the thermal pressure is higher on one side than on the other. In our scheme, boundary condition is assumed to be free boundary, and reflection of the perturbations originating from inside the simulation region is suppressed by placing an absorbing region.

We present the outline of our simulation result. At the initial stage the resistivity is zero everywhere and the plasma is at rest, i.e., electric field does not exist. Then we assume that the anomalous resistivity arises suddenly in the localized region centered at the interface. Once the anomalous resistivity is locally generated, the magnetic field begins to diffuse, and at the same time the plasma starts to move near the neutral point. The plasma flow toward the neutral point is generated by the fast mode rarefaction wave. As time elapses further, the pressure difference from the magnetosheath to the magnetosphere drives a strong flow across the interface. Plasma is heated by the Joule heating as it traverses the diffusion region where the resistivity is anomalously high. As time proceeds further the slow shock and the slow expansion fan develop from the diffusion region. Figure 1 shows the simulation result at this stage. At the slow shock the magnetic field and the flow velocity change their direction sharply. The slow expansion fan is the magnetosphere-side termination of the magnetosheath plasma that has flowed inward along the reconnection field lines, and there the plasma pressure and density drop to the magnetospheric level.

Further we compare the energy exchanges by forces perpendicular and parallel to the magnetic field, respectively. It is seen in Figure 2 that, (1) the acceleration parallel to B is prominent at the slow shock and at the slow expansion fan, particularly at the latter, and (2) the acceleration perpendicular to B has two separate peaks at the front and end of the slow shock which has a finite width in the present simulation.

We have carried out simulations also for the case where the interplanetary and the magnetospheric field line are not exactly anti-parallel. Since our purpose is mainly to see how the efficiency of the plasma entry and acceleration depends on the relative orientation of two fields, we have assumed simply that the reconnection line is directed at half angle between the two field directions. It is found that the acceleration rate is small when the two fields have the same polarity, as compared to when they have opposite polarities.

This material is reported in full in Hoshino and Nishida [1983].

References

Hoshino, M. and A. Nishida, Numerical Simulation of the dayside reconnection, *J. Geophys. Res. 88*, 6926, 1983.

Questions and Answers

Sonnerup: The Levy-Petschek-Siscoe model has a rotational discontinuity, not a slow shock, upstream of the slow expansion fan. Question 1: Does the direction of the tangential magnetic field reverse across your shock? Question 2: How is it possible for the slow expansion waves to remain behind the shock, since they propagate faster than the flow speed behind the shock?

Hoshino: Answer 1: *The direction of the tangential magnetic field does not reverse across our shock. We verified that our shock is a slow mode shock with the following reasons: (1) the entropy increases across our shock, (2) the inflow plasma velocity is faster than the slow mode velocity, but less than the Alfvén velocity. Answer 2: The Levy-Petschek-Siscoe model assumes that no plasma is present on the magnetosphere side so that MHD wave cannot propagate on the magnetosphere side. Because of this, the magnetic field on the magnetosphere side cannot bend. The plasma at the front of the slow expansion fan flows along the magnetic field in the rest frame, but the erosion (namely, the opening of the magnetosphere field lines at the diffusion region) proceeds toward negative X (magnetosphere-side) so that the slow expansion fan is inclined toward positive X (magnetosheath-side). On the other hand, in our model the plasma is present and MHD wave can propagate on the magnetosphere side so that fast mode rarefaction wave makes the magnetic field lines on the magnetosphere side tilted. Because of this, the slow mode wave can propagate toward negative X, and the location of the magnetosphere-side front of the expansion fan is determined by the combination both of this slow mode wave and of the plasma velocity directed to negative X and positive Z (northward direction). In this way the magnetosphere-side front of the slow expansion fan is formed in the negative X region. On the other hand, another magnetosheath-side front of the slow expansion wave propagates to the direction of the positive X, which is the same direction as for slow shock wave. If the diffusion occurs on a line of zero width, this latter front of the fan would coincide with the slow shock. However, in our model the diffusion region has a finite size, so that these two features do not coincide.*

COMMENTS ON SIMULATION OF ANOMALOUS RESISTIVITY

J. W. Dungey

Blackett Laboratory, Imperial College
London SW7 2BZ, England

Quantitative. Assuming $E_y \sim 1$ mV/m, electron acceleration $\approx 3.5 \cdot 10^5$ km/s². If electron mean speed $\bar{v}_e (= j/ne) >$ sound speed, acoustic noise is likely, so assume $\bar{v}_e \sim 10^3$ km/s. Then electrons must be accelerated for only a few milliseconds. If $B_z = 1$ nT, $\Omega_e = 176$ radians/s so, near the neutral line, noise is needed to oppose acceleration. For noise power of 10^{-6} (V/m)²/Hz, the velocity diffusion coefficient can be expressed as $3.5 \cdot 10^5$ eV/s. Plausibly this is enough, but much better theory/simulation is needed to be sure.

Neutral Sheet Simulation. Tanaka and Sato [1981] obtained interesting results, but disadvantage of *periodic boundary conditions*: there is no escape for plasma from the central sheet.

Time Step. Time step is limited by the plasma period. The spectrum is unknown, but the most important frequency range could be well below ω_p. Brown and Dungey [1983] discuss neglect of displacement current to allow bigger time steps. $\underline{E}$ is then determined by curl curl $\underline{E} = -\mu_0\ \partial \underline{j}/\partial t$ and $\partial \underline{j}/\partial t$ is dominated by a term independent of $\underline{E}$ and a term $\epsilon_0 \omega_p^2 \underline{E}$, which facilitates solution.

Development of Linear Theory In the same two-dimensional geometry as Tanaka and Sato, with the unperturbed state stratified, uses the method of Robertson et al. [1981] in terms of particle energy W and canonical momentum P_y. A simple condition determines the direction of wave-particle energy exchange for resonant particles in an element $dWdP_y$. This is a generalisation of the energisation of a wave by an overtaking beam.

The same approach with Larmor radius expansion is used to evaluate fluid modelling for "background" particles. Anisotropy of the pressure can be estimated and its effect on waves has previously been studied [Dungey, 1982].

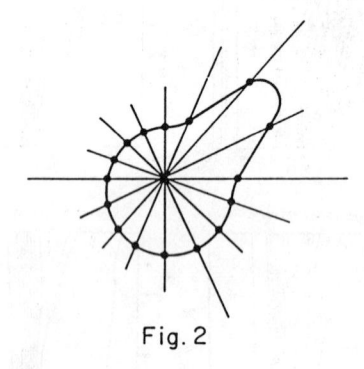

Fig. 1

Fig. 2

Velocity Distributions Linear theory shows that sharp structure may result from resonance or from short wavelength in relation to Larmor radius, and the same can be argued for the nonlinear Vlasov equation. Particle modelling is well established, but alternative methods should be sought, remembering that four quantities per particle are needed. For water bag methods the most daunting trouble is illustrated in Fig. 1 (both figures are in velocity space). A brutal cure is illustrated in Fig. 2. The water bag is represented by points, each constrained to stay on a line through the origin, and perhaps constrained not to pass through the origin.

No computing has been done.

References

Brown, M. G. and J. W. Dungey, Economising plasma simulation by total neglect of the displacement current, *J. Comp. Phys. 52*, 205, 1983.

Dungey, J. W., A formulation for computation of a class of collision-free plasmas in two dimensions, *J. Plasma Phys. 28*, 141, 1982.

Robertson, C., S. W. H. Cowley, and J. W. Dungey, Wave-particle interactions in a magnetic neutral sheet, *Planet. Space Sci. 29*, 399, 1981.

Tanaka, M. and T. Sato, Simulations on lower hybrid drift instability and anomalous resistivity in the magnetic neutral sheet, *J. Geophys. Res. 86*, 5541, 1981.

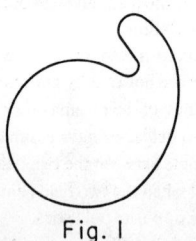

RECONNECTION DURING THE FORMATION OF FIELD REVERSED CONFIGURATIONS

Richard D. Milroy

Mathematical Sciences Northwest, Inc.
Bellevue, Washington 98004

Abstract. Theoretical and experimental studies of magnetic field line reconnection during the formation of a field reversed configuration is reviewed. Good agreement is found between the predictions of an MHD computer model and experimental observations of field line reconnection at the θ-pinch ends during formation. Observations of an additional small-scale reconnection along the magnetic field null during the radial implosion is explained with more detailed calculations based on a hybrid (zero inertia electron fluid and PIC ions) computer model.

I. Introduction

A field reversed configuration (FRC) is an attractive potential fusion concept due to its compact size, high power density, and simple machine geometry. The plasma is confined in a closed magnetic field configuration, which has poloidal field components only ($B_\theta=0$), as illustrated in Figure 1. Here, R is the major radius, r_S is the separatrix radius, and r_C is the coil radius. FRCs, which appear stable to all gross MHD instabilities, have been produced. Such configurations decay over a diffusive timescale due to particle diffusion across the closed field lines and the resistive decay of the reverse bias magnetic flux.

Field reversed configurations are formed in a field reversed θ-pinch by the following technique (illustrated in Figure 2): (i) an initial reverse bias field is frozen into a cold pre-ionized plasma; (ii) the current in the θ-pinch coil is quickly reversed, producing a large forward bias field which causes the plasma to implode radially; (iii) the oppositely directed magnetic field lines reconnect near the θ-pinch ends, forming a closed field configuration; and (iv) the magnetic forces at the ends of the configuration cause it to contract axially until a 2-D equilibrium is reached. Clearly, reconnection of oppositely directed magnetic field lines near the θ-pinch ends is the essence of this FRC formation technique.

In this paper the reconnection of magnetic field lines near the θ-pinch ends and possible small-scale reconnection in the central core during the formation of FRCs is discussed. In Section II, 2-D MHD computer simulations [Milroy and Brackbill, 1982] of reconnection near the θ-pinch ends are compared with results from the FRX-B [Armstrong et al., 1981] TRX-1 [Armstrong et al., 1982] and HBQM [Sevillano et al., 1982] experiments. In Section III, results from a 2-D hybrid (particle ion and fluid electron) computer model [Hewett and Seyler, 1981], which predicts the rapid small-scale reconnection along the reversal layer, are reviewed. A brief summary and conclusions are presented in Section IV.

II. Reconnection At The θ-Pinch Ends

Milroy and Brackbill [1982] have studied the formation of an FRC in the FRX-B experiment with a two-dimensional MHD code. These results will be summarized and, in addition, more recent results on FRC formation in TRX-1 and HBQM will be discussed.

The MHD code is based on a two-dimensional (r-z) time-dependent dynamic model, as described in Milroy and Brackbill [1982]. Separate electron and ion temperatures are calculated, and the effect of unequal parallel and perpendicular thermal conduction is included. A vacuum field solver is used to calculate the magnetic field away from the plasma. The vacuum field region includes the external coils, allowing for a proper representation of coil geometry.

Resistive particle-field diffusion is calculated using classical resistivity plus an additional resistivity to account for microinstabilities. To establish the appropriate initial conditions for the post-implosion phase (including the deposition of magnetic field energy in the plasma), the empirical anomalous resistivity developed by Chodura [1975] has been used during the radial implosion.

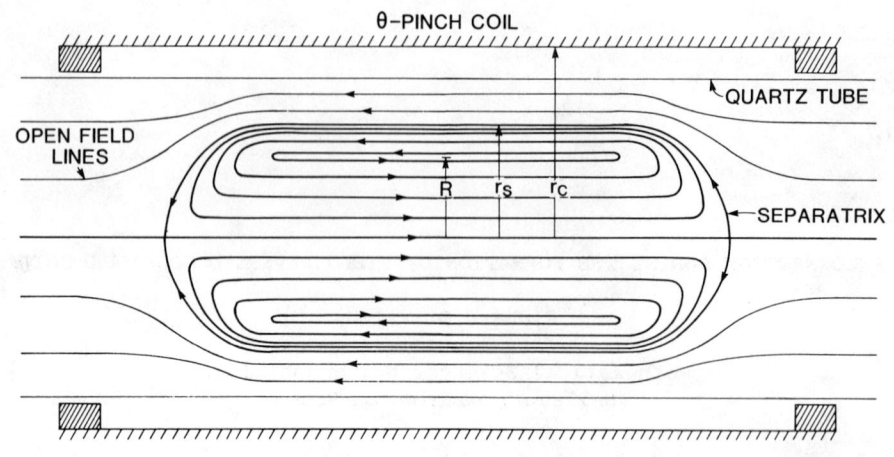

Fig. 1. Schematic of a field-reversed configuration in field-reversed θ-pinch.

$$\eta_{CH} = \frac{m_e \nu_{CH}}{ne^2} \quad \text{sec} \quad [1]$$

$$\nu_{CH} = C_c \omega_{pi} \left[1 - \exp\left[-\left| \frac{v_E}{fv_s} \right| \right] \right]$$

where v_E is the electron drift velocity and v_s is the sound speed. The adjustable parameters C_C and f have values of 1 and 3, respectively, for the present calculations. After the radial implosion phase (but during the magnetic reconnection and subsequent times), a smaller additional resistivity based on the saturation of the lower-hybrid-drift instability [Davidson and Krall, 1977] is assumed.

$$\eta_{LHD} = 7.2 \times 10^{-4} \, C_{LH} \left[\frac{|B|}{n} \right] \left[\frac{v_E}{v_i} \right]^2 \quad \text{sec} \quad [2]$$

where B(G) is the magnetic field strength, $n(cm^{-3})$ is the plasma density, and v_i is the ion thermal velocity. C_{LH} is an adjustable parameter which is set to one unless explicitly stated otherwise.

Figure 2 shows magnetic field lines and marker particles at several times for a typical FRX-B case. The assumed set of simulation parameters is summarized in Table 1. A passive mirror at the ends of the θ-pinch coil, created by decreasing the inner diameter of a small portion at the coil ends, was used to encourage the reconnection and form a more symmetric FRC. At about 1 μsec, the radial implosion is complete and the plasma parameters along the reversal layer are: plasma density, $n_e = 6 \times 10^{15}$ cm^{-3}; electron temperature, $T_e = 200$ eV; and ion temperature, $T_i = 150$ eV. The magnetic Reynolds number, $R_m = 4\pi/(\eta c^2)LV_A$, is approximately 1000, where L is the length of

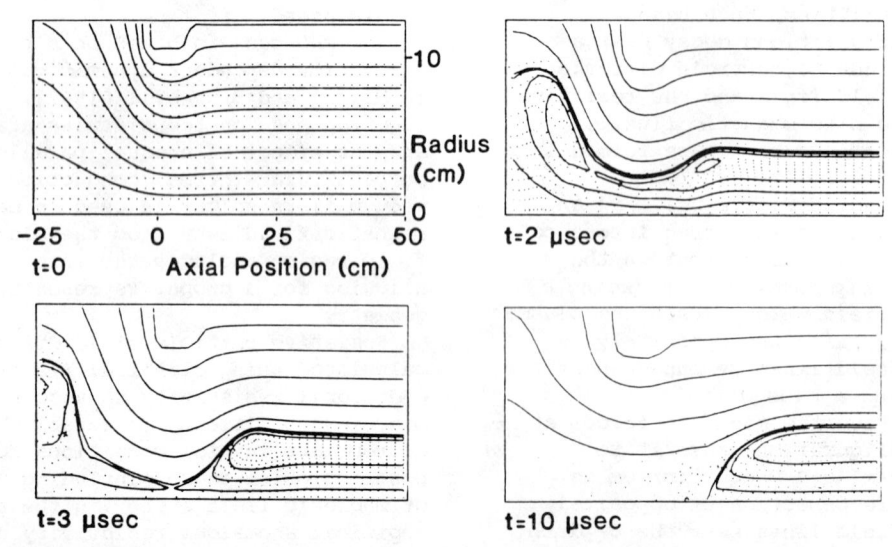

Fig. 2. Magnetic field lines and marker particles from an MHD simulation of FRC formation in FRX-B.

the reconnection region (≈25 cm) and v_A is the Alfvén velocity. Reconnection begins near the θ-pinch ends at about this time and proceeds in a manner similar to that described by Sato and Hayashi [1979]. Some important differences, however, include the cylindrical geometry, a limited amount of reverse bias flux to reconnect, and more complex external boundary conditions. Reconnected field lines near the magnetic reversal layer have very sharp curvature and exert force on the plasma sufficient to reverse the normal outflow. This is illustrated in Figure 3a, which shows velocity vectors at a time of 2 μsec. Endloss (outward motion) combined with axial contraction (inward motion) on closed field lines causes a continuous flow of plasma away from the magnetic x-point. Thus, the field gradients are made steeper and the current density is locally intensified. This is illustrated in Figure 3b, which is a vector representation of the magnetic field profile at a time of 2 μsec. Sharper magnetic field gradients cause faster diffusion both in the usual manner and through increased resistivity. Reconnection is complete (the separatrix reaches the axis of symmetry) at about 3.3 μsec and the FRC contracts axially to form a 2-D equilibrium. The rate of reconnection, expressed in terms of w/v_A, is approximately 0.1 where w is the radial inflow velocity near the x-point. This is in agreement with the Petschek [1963] model of $w/v_A \approx \pi/(4\ln[R_m])$.

The calculated sensitivity of reconnection to several assumed parameters including the

TABLE 1. FRX-B Computer Simulation Parameters

Geometry

  Coil Length, 100.0 cm
  Coil Radius, 12.5 cm
  Wall Radius, 10.0 cm
  Radius of Coil in End Mirror, 11.0 cm
  Length of End Mirror, 7.0 cm

Magnetic Field History

  Initial Bias Field, -0.11 T
  Quarter Cycle Time, 2.3 μsec
  Peak Magnetic Field, 0.9 T
  Time at Crowbar, 3.36 μsec
  Magnetic Field at Crowbar, 0.65 T

Initial Plasma Conditions

  Temperature, 1 eV
  Ion Density, 7.7×10^{14} cm^{-3}

Anomalous Resistivity Model

  t <1.0 μsec: Chodura with $C_C = 1$, f = 3
  t ≥1.0 μsec: Lower Hybrid with $C_{lh} = 1$

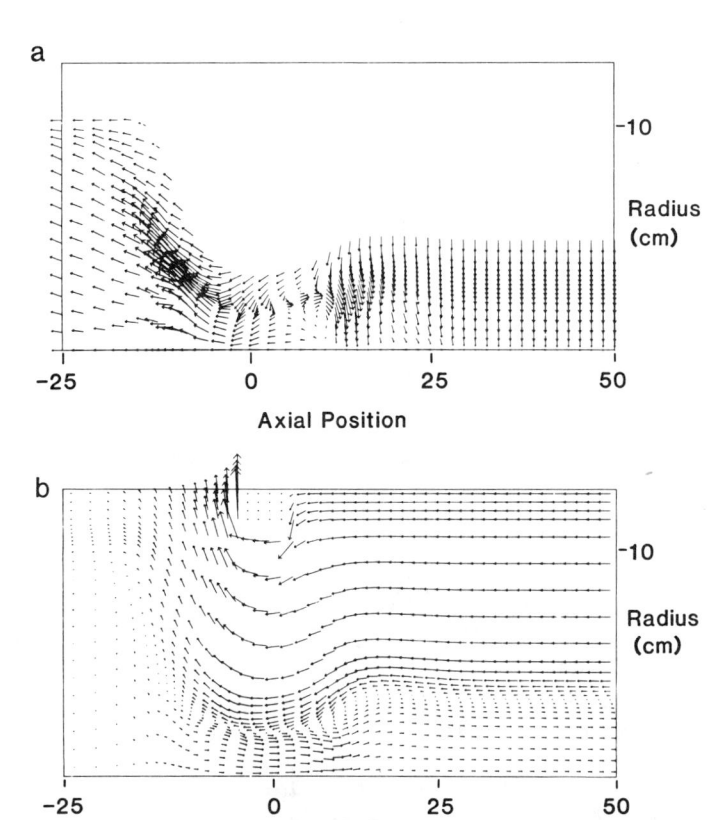

Fig. 3. Velocity vectors (a) and vector representation of magnetic field profile (b) from the simulation shown in Fig. 2.

magnitude of the resistivity factor, C_{LH} in Equation [2], the magnitude of the reverse bias field, the initial plasma density, and the coil shape are discussed in Milroy and Brackbill [1982]. The rate of reconnection is found to be insensitive to the magnitude of the resistivity in agreement with Sato and Hayashi [1979]. Due to the dynamic nature of the reconnection process, when the constant C_{LH} is decreased the field gradients become sharper and the current density intensifies to compensate. Setting C_{LH} to 0 leads to very sharp gradients near the x-points, making it difficult to resolve with a finite grid. Full numerical resolution requires a much finer grid and significantly more computer time and has not been attempted. However, it appears likely that such a calculation would lead to a solution with a reconnection rate comparable to those with $C_{LH} \cong 1$. Increasing the initial reverse bias field, which increases the amount of reverse bias flux to reconnect, is found to increase the reconnection rate. This is due to increased magnetic forces leading to a more rapid evacuation of the plasma from the vicinity of the x-points. Increasing the initial plasma density reduces the Alfvén velocity and, correspondingly, slows the reconnection rate.

BIAS = −560 G   P₀ = 7 mTorr

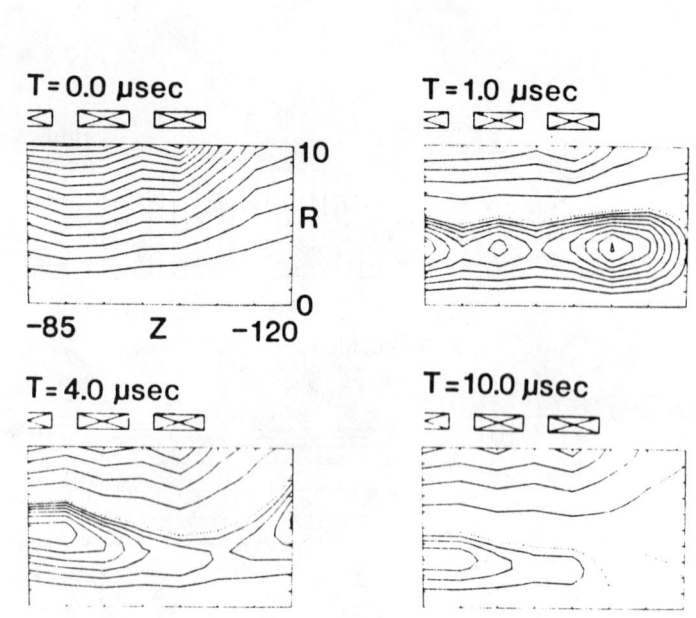

Fig. 4. Magnetic field lines reconstructed from magnetic probe measurements near the θ-pinch ends in the HBQM device. The filling pressure is 7 mTorr. The magnetic field has a bias of −0.56 kG, a peak forward field of about 4.7 kG, and a quarter cycle time of 500 ns.

The passive mirrors at the coil ends can have a significant effect on the reconnection. For a calculation identical to that in Figure 2 except with a straight coil, the field lines do not reconnect until most of the plasma is lost through the open ends. This is in agreement with FRX-B experiments in which the passive mirrors were removed and no reconnection was observed [Armstrong et al., 1980]. For other calculations with an increased amount of trapped bias flux, however, the passive mirrors do not appear to significantly alter the reconnection rate.

Unfortunately, probe measurements of the magnetic field structure in the reconnection region are difficult and have not been made for the FRX-B experiment. Consequently, a detailed comparison between the code and experiment cannot be made. However, Cochrane et al. [1981] have used an axial array of B_z probes to measure the magnetic field on axis near the ends of the θ-pinch. Numerical simulations are in good agreement with experimental observations on several important points including: (a) the axial position that the x-point initially appears on axis; (b) the time at which the x-point initially appears on axis; and (c) the velocity of the x-point subsequent to reconnection. This, plus the agreement on the effects of coil shape, is convincing evidence that numerical calculations are accurate.

Reconnection can be expected to take place only when the magnetic field and plasma conditions are such that the ends of the resulting FRC are sufficient to cause an axial contraction, or at least to prevent an axial expansion. It has been shown that for an elongated FRC in equilibrium, axial force balance can be expressed as

$$\beta = 1 - \frac{1}{2}\left[\frac{r_s}{r_c}\right]^2 \qquad [3]$$

where

$$\beta \equiv \frac{1}{\pi r_s^2} \int_0^{r_s} \beta \, dA$$

is the cross-sectional average of the plasma $\beta \equiv P/P_m$ near the axial midplane. Here P is the plasma pressure and P_m is the maximum pressure defined at the neutral layer (B = 0). If, after the radial implosion, β is less than the value given by Equation 3, the FRC will contract until an equilibrium is reached and the equality is satisfied. On the other hand, if β is greater than the value given by Equation 3, complete reconnection cannot take place until the plasma pressure decreases due to particle or energy losses and the equality is satisfied. Prior to reaching a 2D equilibrium, β is sensitive to the amount of trapped reverse bias flux. For large amounts of trapped flux, β is small, leading to strong magnetic forces and a rapid reconnection followed by a violent axial implosion. For smaller amounts of bias flux, the reconnection is slower and the axial implosion may be almost nonexistent.

Detailed probe measurements of the field configuration have been made for the HBQM experiment, and field line plots for the first 10 μsec are shown in Figure 4. This experiment has the same radial dimensions as FRX-B; however, it differs in a number of important parameters. Most notably, a much weaker bias field is used and the field reversal is very rapid (quarter cycle time of 0.5 μsec). Also, a segmented coil with no passive mirror is used. In this experiment, field lines are observed to reconnect very rapidly beyond the coil ends, but the configuration then expands axially and reconnects again near 8 μsec. The initial reconnection beyond the coil ends takes place in the neutral gas beyond the θ-pinch ends. However, axial forces apparently are not sufficient to confine the plasma, so it expands axially. When the internal pressure drops so that Equation 3 can be satisfied, a second reconnection takes place and an FRC is formed.

In the TRX-1 experiment, two additional coils at each end of the θ-pinch are used to control the timing of field line reconnection. A d.c. "cusp coil" (current in the opposite direction to the forward θ-pinch coil current) inhibits the reconnection, and a "trigger coil" between the cusp and θ-pinch coils forces a rapid reconnection when it is fired. This method is illustrated in Figure 5, which shows the calculated magnetic field lines from an MHD simulation. In this case, the trigger coils were fired at 1 μsec, leading to complete reconnection at about 2.5 μsec. By varying the time at which the trigger coils are fired, the time at which the field lines reconnect can be controlled over a limited period. In both experiments and simulations, it is found that reconnection will take place spontaneously even when the trigger coils are not fired. The time of spontaneous reconnection depends on several parameters, the most important of which are the quantity of trapped reverse bias flux, the field magnitude under the cusp coil, and the plasma density. For typical TRX-1 operating conditions, this time is in the 5-6 μsec range. If neither the cusp coil or trigger coil is activated (θ-pinch only), simulations predict spontaneous reconnection about 1-2 μsec earlier, showing that the cusp coils do delay the onset of reconnection. Experimentally, the reconnection time can be inferred from excluded flux measurements, as illustrated in Figure 6 for triggered and spontaneous reconnection. The excluded flux is proportional to r_s^2 and becomes small near the ends when reconnection is complete and the FRC contracts axially out of the end region. Good agreement has been found

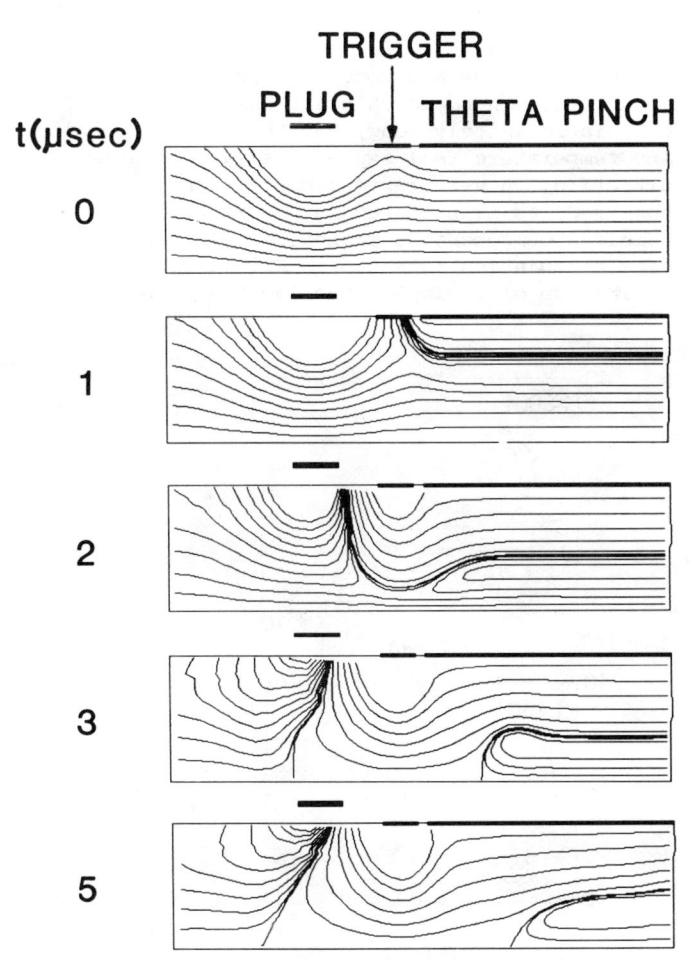

Fig. 5. Magnetic field lines from an MHD simulation of FRC formation in TRX-1. The initial density is 8.7×10^{14} cm^{-3}. The magnetic field has a bias of -2.5 kG, a peak forward field of 10 kG, and a quarter cycle time of 3.33 μsec.

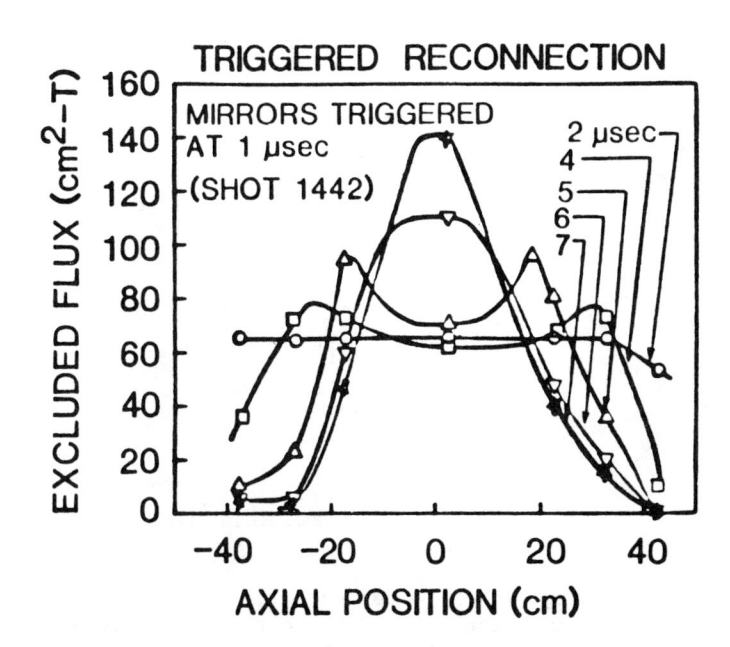

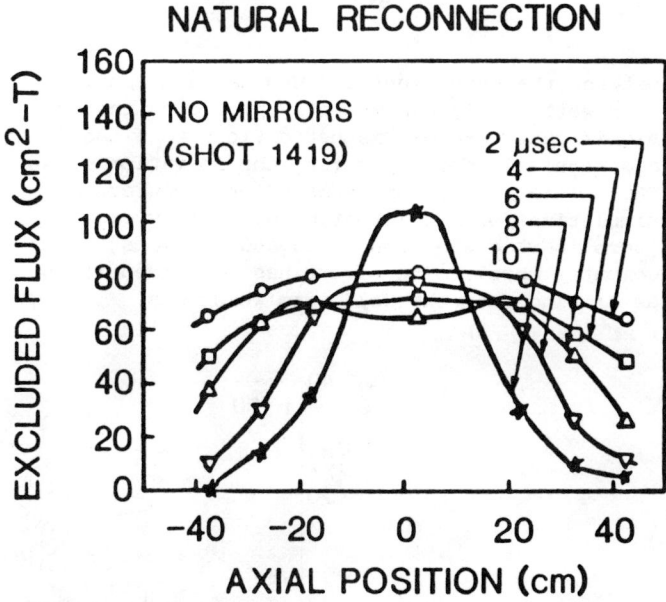

Fig. 6. Experimental excluded flux as a function of axial position and time in TRX-1. In (a) the trigger coils are fired at 1 μsec; in (b) they are not fired at all.

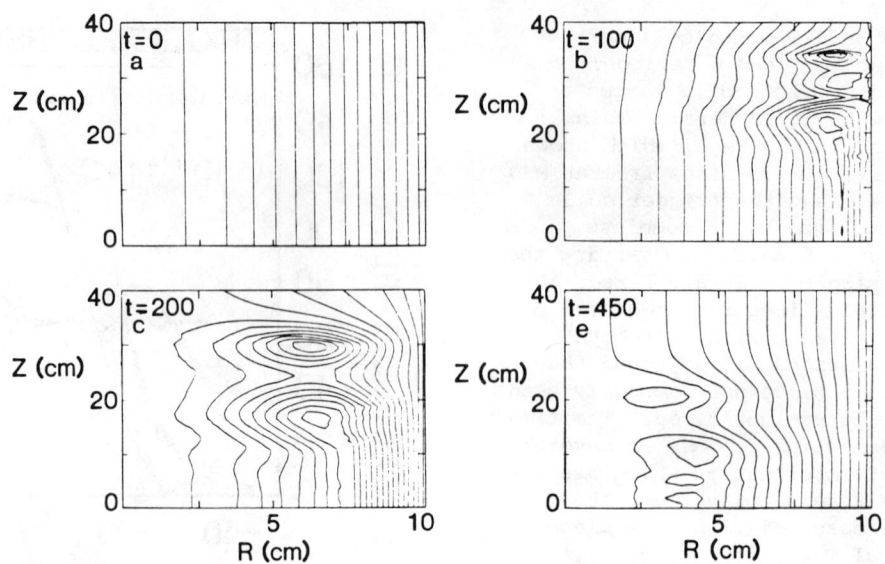

Fig. 7. Poloidal magnetic field lines (rA_θ) from a hybrid computer simulation of FRC formation. The flux contours are equally spaced between the minimum and maximum values at that particular time.

between the experiment and MHD calculations.

Hewett [1983] has recently predicted that a significant toroidal magnetic field (B_θ) should spontaneously develop during the formation of an FRC. This prediction is based on computer simulations of FRC formation using a hybrid (zero inertia electron fluid and PIC ions) computer model. The effect has been identified as arising from the Hall term which was neglected in the MHD model previously described. In these calculations, the θ-pinch coil is infinitely long, but the initial density and temperature fall exponentially in the axial direction. A very rapid reconnection is observed, as illustrated in Figure 7. The toroidal component of magnetic field for the same calculation is shown in Figure 8.

The use of a simple Ohms law, which neglects

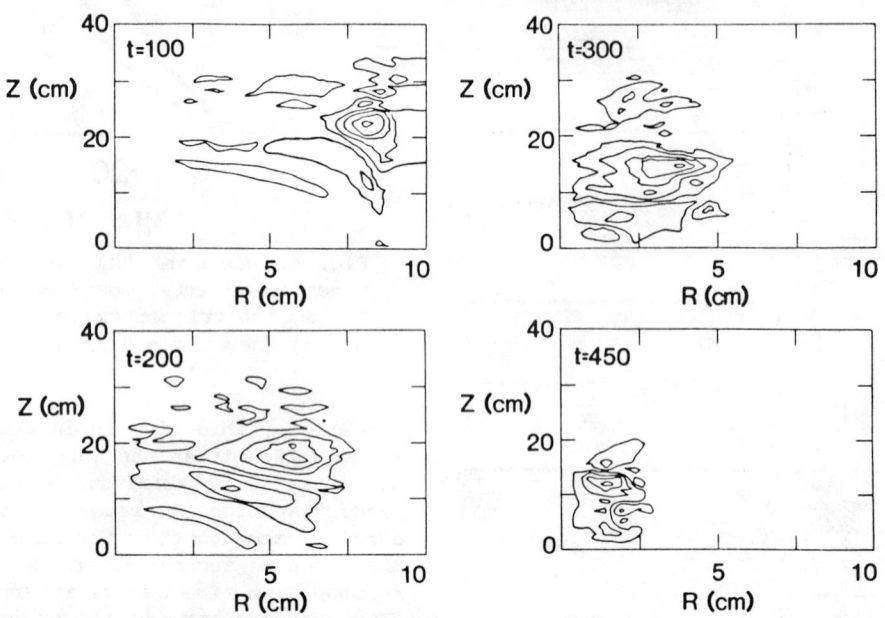

Fig. 8. Contours of B_θ from the same calculation as in Fig. 7. At t = 100 ns, the maximum B_θ is 241 G and the minimum is -1.13 kG located at the center of the closed loops at r = 8, z = 12, and at r = 8, z = 22, respectively. The contours are equally spaced between these values. At t = 200 ns, the maximum B_θ is 910 G and the minimum is -1.62 kG located at r = 3.5, z = 11, and at r = 6, z = 17,

the Hall term as well as other tensor conductivity effects, is a serious limitation of the MHD model. The inclusion of these effects would no doubt significantly influence the details of the reconnection process. The possible spontaneous generation of significant torroidal magnetic fields during the formation of an FRC could have a strong influence on its subsequent evolution. In order to investigate this effect as a function of coil geometry and initial conditions, the MHD model is being modified to include a generalized Ohm's law [Braginskii, 1965]. The effects of a tensor conductivity, the Hall term, the $\vec{\nabla}P_e$ term, and the thermal force will be included, as outlined by Brackbill and Goldman [1983].

III. Rapid Small-Scale Reconnection Along The Reversal Layer

There is experimental evidence that, in addition to the magnetic field line reconnection at the ends, there also is some rapid small-scale reconnection during the radial implosion. By using an internal probe array, Irby et al. [1979] have directly measured the internal magnetic field structure and observed the formation and coalescence of magnetic islands on timescales of the order of the Alfvén time across the reversal layer. Seveillano et al. [1982] have observed similar island formations, as illustrated in Figure 4. This rapid small-scale reconnection along the reversal layer also is found in the numerical computer simulations of Hewett and Seyler [1981].

During FRC formation, significant phenomena can occur in the vicinity of field nulls, which is not correctly described by the MHD equations. To include the effects, Hewett and Seyler [1981] have employed a two-dimensional (r-z) time-dependent hybrid simulation model [Hewett, 1980] which includes the proper ion dynamics at the field null. The ion component of the plasma is represented by an ensemble of "particles" that are advanced in time by the PIC technique, while the electrons are represented as a zero inertia fluid. The model is applied to a short periodic axial section of a long θ-pinch with initial conditions of $n = 3 \times 10^{14}$ cm^{-3}, of D_2, $T_e = T_i = 1$ eV, and a bias magnetic field of -1.5 kG. The external magnetic field rises according to an assumed 45 kV implosion voltage, as in FRX-B.

The calculated poloidal flux contours are shown at several simulation times in Figure 9. By 65 ns, the poloidal flux has significant axial perturbations which rapidly develop into small magnetic islands. These islands, in turn, coalesce into relatively large magnetic islands. Hewett and Seyler [1981] have identified the rapid development of axial perturbations (Figure 9a) as being due to the kinetic-ion, fluid-electron version of the

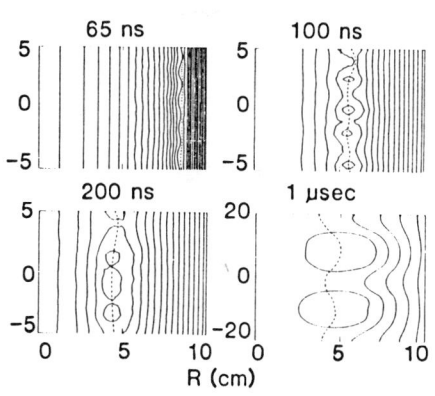

Fig. 9. Contours of poloidal flux (rA_θ) from a hybrid computer simulation of the field reversal phase of FRC formation at (a) t = 65 ns, (b) 100 ns, (c) 200 ns, and (d) 1 μsec. The model is applied to a periodic section of a long θ-pinch.

Kruskal-Schwarzschild (K-S) instability. This is a magnetized plasma instability analogous to the hydrodynamic Rayleigh-Taylor instability and is driven by ion acceleration in a region where the density gradient is favorable for instability. The rapid small-scale island formation (Figure 9b) is due to an nonlinear kinetic effect. Distortions in the reverse bias field, where the K-S mode is unstable, tend to focus the reflected ion beam into localized regions on the positive-bias side of the magnetic null. Secondary reflections then distort the positive bias field lines such that the ions are more sharply focused into clumps near the magnetic null. This process is illustrated in Figure 10. After several bounce times, clumps of enhanced density form with reconnected field lines surrounding them. A toroidal magnetic field of about 1.2 kG magnitude and with a similar axial scale length is observed to be spontaneously generated at the same time. The toroidal field enhances the axial focusing and is absolutely essential for the early reconnection. The small-scale islands rapidly grow in amplitude and then coalesce into relatively large magnetic islands. The rapid spatial variation of the field leads to an enhanced flux loss rate at the magnetic null.

IV. Summary and Conclusions

The reconnection of magnetic field lines during the formation of FRCs has been reviewed. Detailed experimental data on the actual reconnection process (magnetic probe measurements of the field configuration, for example) is scarce; however, existence of the formed FRC is proof that magnetic field lines do reconnect at the ends. End reconnection has been simulated for a number of different experiments with an MHD code and, where possible, the results have been compared with

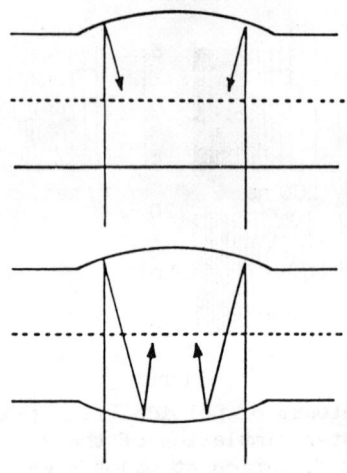

Fig. 10. Schematic illustrating the focusing of reflected ions into regions of enhanced density.

the experiments. For all comparisons, the agreement is very good, indicating that the MHD solution captures most of the important physics of the end reconnection process. Hybrid computer simulations that account for ion kinetic effects, as well as the Hall term in the generalized Ohm's Law, show that significant B_θ should be generated during the reconnection process, an effect not observed in the MHD calculations. Hybrid simulations show that kinetic effects, combined with the spontaneous generation of a small B_θ variation, can lead to rapid small-scale reconnection along the magnetic field null during the radial implosion phase of FRC formation. Small scale magnetic islands rapidly form along the field null and then coalesce into larger islands.

References

Armstrong, W.T., J.C. Cochrane, J. Lipson, R.K. Linford, K.F. McKenna, A.G. Sgro, E.G. Sherwood, R.E. Siemon, and M. Tuszewski, FRC studies on FRX-B, in "Proceedings of the Third Symposium on Physics and Technology of Compact Toroids in the Magnetic Fusion Energy Program," Los Alamos National Laboratory Report No. LA-8700-C, 1980.

Armstrong, W.T., D.G. Harding, E.A. Crawford, and A.L. Hoffman, Flux-trapping during the formation of field-reversed configurations, Phys. Fluids 25, 2121, 1982.

Armstrong, W.T., R.K. Linford, J. Lipson, D.A. Platts, and E.G. Sherwood, Field-reversed experiments (FRX) on compact toroids, Phys. Fluids 24, 2068, 1981.

Brackbill, J.U., and S.R. Goldman, Magnetohydrodynamics in laser fusion: Fluid modeling of energy transport in laser targets, Communications on Pure & Applied Mathematics, Vol. XXXVI, p. 415, 1983.

Braginskii, S.I., Transport processes in a plasma, Reviews of Plasma Physics, Vol. I, M.A. Leontovich, Editor, Consultants Bureau Enterprise, p. 205, 1965.

Chodura, R., A hybrid fluid-particle model of ion heating in high Mach number shock waves, Nucl. Fusion 15, 65, 1975.

Cochrane, J.C., W.T. Armstrong, J. Lipson, and M. Tuszewski, Observations of separatrix motion during the formation of a field-reversed configuration, Los Alamos National Laboratory Report No. LA-8716-MS, 1981.

Davidson, R.C., and N.T. Krall, Review of anomalous transport in high temperature plasmas, Nucl. Fusion 17, 1313, 1977.

Hewett, D.W., A global method of solving the electron-field equations in a zero-inertia-electron-hybrid plasma simulation code, J. Comp. Phys. 38, 378, 1980.

Hewett, D.W., Spontaneous development of toroidal magnetic field during formation of the field-reversed theta-pinch, Accepted for publication in Nuclear Fusion, 1983.

Hewett, D.W., and C.E. Seyler, Reconnection phenomena during the formation phase of field-reversed experiments, Phys. Rev. Lett. 46, 1519, 1981.

Irby, J.H., J.F. Drake, and H.R. Griem, Observation and interpretation of magnetic field line reconnection and tearing in a theta pinch, Phys. Rev. Lett. 42, 228, 1979.

Milroy, R.D., and J.U. Brackbill, Numerical studies of a field-reversed theta-pinch plasma, Phys. Fluids 25, 775, 1982.

Petschek, H.E., in "The Physics of Solar Flares," AAS-NASA Symposium, NASA SP-50 (ed. W.N. Hess), p. 425, Greenbelt, MD, 1963.

Sato, T., and T. Hayashi, Externally driven magnetic reconnection and a powerful magnetic energy converter, Phys. Fluids 22, 1189, 1979.

Sevillano, E., H. Meuth, and F.L. Ribe, Driven-mirror formation of a two-cell field-reversed configuration, in "Proceedings of the Fifth Symposium on the Physics and Technology of Compact Toroids," pp. 12-15, Bellevue, WA, 1982.

Questions and Answers

Priest: What are the plasma beta and magnetic Reynolds number for your simulation and how do they compare with the laboratory values? Also, how many grid points and what type of code do you use?

Milroy: In equilibrium, axial force balance forces the average plasma beta at the axial midplane to $<\beta> = 1-1/2(r_s^2/r_c^2)$ where r_s is the plasma separatrix radius; r_c is the coil radius and β is defined as plasma pressure over external field pressure. Typically r_s/r_c is .4 to .6 so $<\beta> = .8 - .9$. Typical magnetic Reynolds numbers in both simulations and the laboratory are about 1000. The code is a dynamic, time-dependent MHD model that originated with J.U. Brackbill. It employs a continuously rezoned Lagrangian mesh with an adaptive algorithm developed by Brackbill and Saltzman. This mesh automatically concentrates in regions of high current density to resolve regions of interest like the x-point. Typically a 45×25 point grid is employed.

DRIVEN MAGNETIC RECONNECTION DURING THE FORMATION OF A TWO-CELL FIELD-REVERSED CONFIGURATION

E. Sevillano and F. L. Ribe

University of Washington, Seattle, WA 98195

Abstract. The formation of a two-cell Field-Reversed Configuration has been accomplished with the addition of three independently-driven coils at the ends and near the central plane around the axis of a theta pinch. The reconnection process in the region between the cells has been studied with internal magnetic field probes. It is found that the reconnection in this region (x-circle) consists of two distinct phases. During the initial phase flux is dissipated slowly. This phase is followed by a rapid reconnection process which eventually leads to the formation of two independent cells. The resistivity at the x-circle is a few times classical in the initial phase and an order of magnitude greater during the rapid reconnection process. Recent theory and magnetohydrodynamic particle simulations by Brunel, et al. predict a sudden increase in magnetic reconnection similar to that observed here as the reconnected field lines near the diffusive layer assume a transverse, radial component and are convected axially with the fluid. In an alternative qualitative model radial convection of low density plasma to the diffusive layer near the x circle causes depletion of current carriers and consequent onset of anomalously high resistivity, leading to an increased reconnection rate.

1. Introduction

In this paper, we report measurements of magnetic reconnection at an x-point (circle) on the cylindrical field null in a field-reversed configuration (FRC) formed in a low-density ($\sim 10^{15}$cm^{-3}) theta pinch.

Extensive laboratory studies of reconnection at a neutral current sheet were carried out by Baum, et al. [1973], and by Baum and Bratenahl [1976, 1977]. Using a double inverse pinch wherein two parallel current carrying rods accelerated preionized argon plasma (n $\approx 10^{14}$-10^{15}cm^{-3}, $T_e \approx 2$ eV) away from their centers to impinge on a neutral layer between rods, reconnection was observed, converting "private" rod magnetic flux to common reconnected flux around both rods. Anomalous resistivity in the current layer two orders of magnitude greater than classical was observed, as were X-rays generated by runaway electrons in impulsive flux transfer events [Bratenahl and Baum, 1983]. They were able to observe the x-point magnetic structure, as well as an x-shaped current density attributed to a Petschek slow shock [Petschek, 1964]. More recently, Stenzel and Gekelman [1979], using a large, low-density (n $\approx 10^{12}$cm^{-3}, $T_e \approx 5$ eV) device with parallel current-carrying plates with axial current along with x-line, employed magnetic and current probes to demonstrate the angular divergence of the reconnected field lines, as well as gradients of magnetic intensity and peaked transverse current density to characterize a Petschek shock structure. The resistivity observed was only moderately enhanced over the classical value.

Sato and Hayashi [1979] and Birn and Hones [1981], among others, used MHD computational simulation in reversed-field geometry similar to that of the Earth's magnetic tail to observe reconnection driven by a localized plasma influx perpendicular to the neutral current sheet (Sato and Hayashi), as well as by anomalous resistivity driven by the plasma current. The reconnection led to later stages to narrow parallel plasma jets (Sato and Hayashi) or plasmoids [Hones, 1980] ejected from the x-point along the neutral axis. More recently Brunel, Tajima and Dawson [1982] have carried out MHD particle simulations in initially planar geometry with reconnection of the antiparallel magnetic field being driven by external magnetic fields from a rod perpendicular to the field lines. Their model had constant resistivity corresponding to magnetic Reynolds number S ≈ 2000 for the current-width dimension. They found a slow phase of annihilation of the antiparallel fields, followed by faster second phase, characterized by a Petschek-like angular magnetic-line region with reconnected perpendicular field lines moving on-axis away from the central diffusive region.

2. Apparatus and Measurements

In experiments of the type reported here (see for example, Armstrong, et al. [1981]) field-reversed configurations (FRC's) with closed

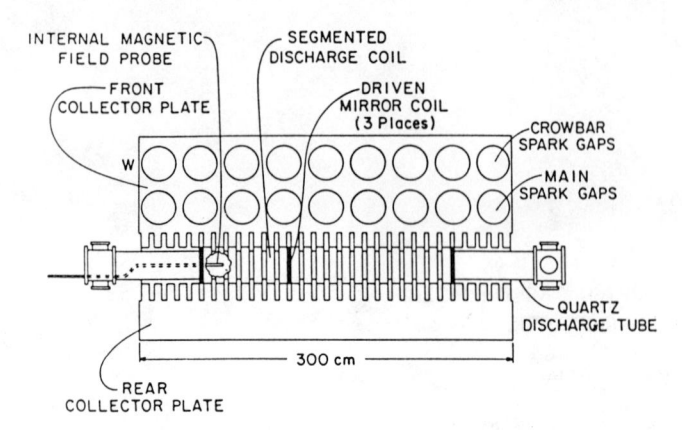

Fig. 1. Top view of the HBQM showing the driven magnetic mirror coils and probe. The positions of the mirror coils are marked black.

field lines are produced by applying a θ-pinch compression field to preionized plasma, magnetized by an initial field (reversed bias) in the opposite direction. The result in our case is normally a 2-m long FRC where reconnection of the antiparallel field lines at the ends is aided by two driven end mirrors. A central pulsed mirror produces two FRC's by inducing reconnection near the middle of the 2-m FRC. In the 2-m FRC the last closed field lines are bounded by a separatrix outside of which open field lines leave the compression coil axially.

The High Beta Q Machine (HBQM) theta pinch [Knox, et al, 1982; Sevillano, 1983] used in the experiments (Fig. 1) has a 220-cm long, 22-cm ID compression coil, segmented into rings every 10 cm with a void fraction of 33 percent. The compression magnetic field rises to its peak value of 4.7 kG (without axial bias) in 400 nsec and is passively crowbarred with an L/R decay time of 40 μsec. The reversed bias field is applied before the compression field with a rise time of 25 μsec.

Deuterium filling pressures of 5 and 7 mTorr and a bias field of -560 G were used. Under these conditions excluded-flux measurements indicate that internal magnetic probes do not significantly alter the plasma behavior. The separately pulsed magnetic mirrors (Fig. 1) had a rise time of 1.7 μsec, providing a magnetic field on-axis in vacuum of 900 G, and were located at z = 100 cm, -30 cm, and -110 cm, where z = 0 is the axial central plane of the device.

Four internal B_z probes axially separated by 10 cm were placed inside a 0.6-cm OD quartz sheath. The sheath was inserted from one end of the theta pinch (Fig. 1) and could be positioned at any radial position and any axial position with z<0, giving measurements at eight axial positions at 5 cm intervals and at 10 radial positions at 0.5-1.5 cm intervals. The radial magnetic field profiles were fitted using a higher order polynomial and the value of the flux function $\psi(r,z) = \int_0^r B_z(r',z) 2\pi r' dr'$ was obtained by a discharge-to-discharge accumulation of data using a digital data acquisition system.

3. Experimental Results

Constant-flux plots are presented in Figs. 2 and 3 for initial filling pressures p_0 = 7 mTorr and 5 mTorr and a bias field B_b = -560 G at several times during the discharge. The magnetic mirrors were triggered at t = -1.7 μsec so that their peak would coincide with the initiation of the main discharge. The position of the FRC separatrix is denoted by a dashed line; the contours are drawn every 15 kMx inside the separatrix and every 75 kMx outside. The central reconnection separatix, or x-circle, is indicated by an x in Fig. 2 and 3. The positions of the theta-pinch coil segments are shown by the crossed

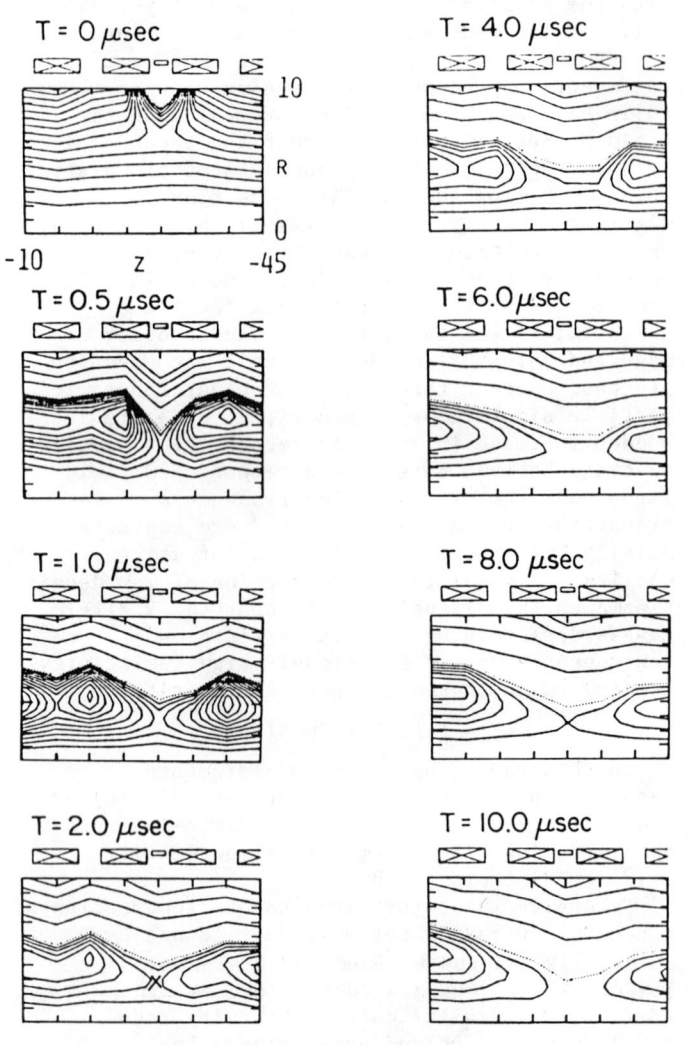

Fig. 2. Flux contours in the region near the center of the coil showing the formation of a two-cell configuration. The mirror coils were triggered at t = -1.7 μsec. The filling pressure was 7 mTorr and the bias field was -560 g.

rectangles; the position of the central mirror coil is denoted by the small rectangle at z = -30 cm. The axial chain of magnetic islands produced by the coil segmentation at early times coalesces after about 2 μsec. The reconnection events to be analyzed occur at later times.

The rate of reconnection was quantified by measuring the trapped reversed flux at z = -30 cm which corresponds to the position of the central mirror coil and at z = -10 cm in the straight-field-line region of the 2-m FRC. The trapped flux is plotted in Fig. 4 at these two axial locations for 7 mTorr. Figure 5 shows the same data for 5 mTorr. The effect of the mirror coil can be clearly seen starting at t = -1.7 μsec by a rapid decrease of the trapped flux at the

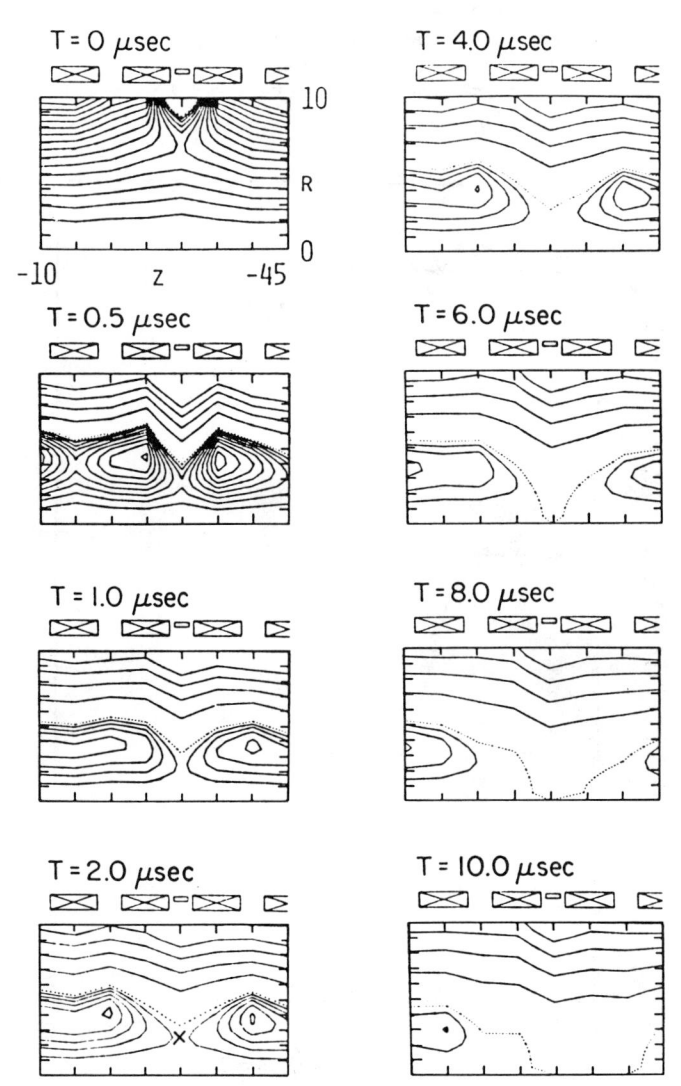

Fig. 3. Flux contours in the region near the center of the coil showing the formation of a two-cell configuration. The mirror coils were triggered at t = -1.7 μsec. The filling pressure was 5 mTorr and the bias field was -560 g.

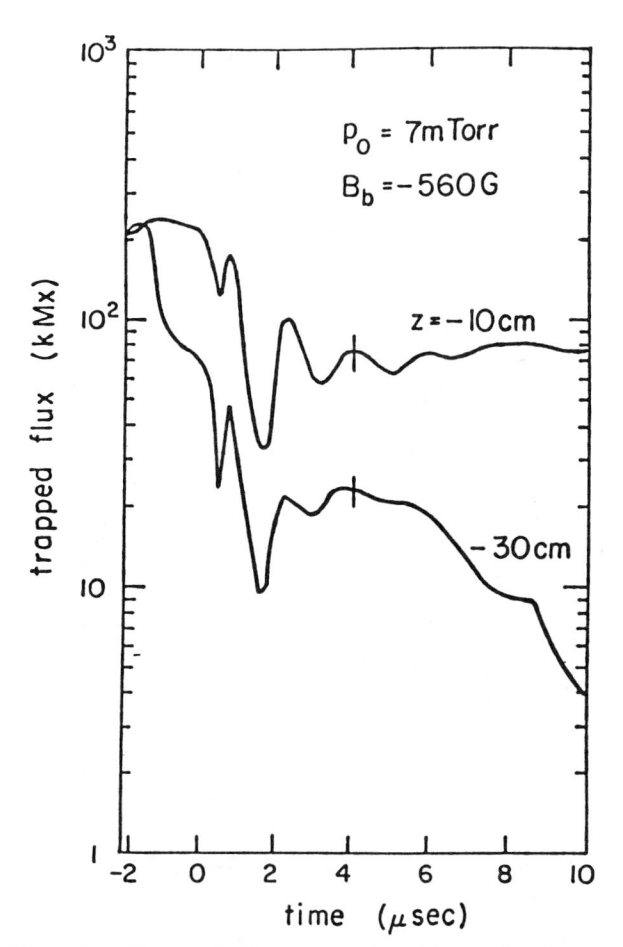

Fig. 4. Trapped flux as a function of time at two axial positions. The mirror coil was located at the z = -30 cm position. The filling pressure was 7 mtorr and the bias field was -560 G. The estimated errors are represented with bars.

z = -30 cm position. The graph for z = -10 cm in Fig. 4 indicates the undisturbed quasisteady equilibrium of the 2-m FRC that exists for approximately 20 μsec. In the 5-mTorr case of Fig. 5 the quasisteady equilibrium duration is shorter. After the initial rapid flux decay during the implosion there are some oscillations and finally the flux reaches an approximate plateau value. In the region of the mirror a further, more rapid, decrease in the flux is then observed to occur at 5.5 μsec as the formation of a two-cell configuration takes place.

Log-log plots of the data at the position of the mirror are shown in Fig. 6 for the two filling pressures considered. The abscissa is $t-t_o$ where t_o is defined as the time when the plateau phase is reached. The value for 7mtorr is t_o = 2.2 μsec and for 5 mTorr t_o = 1.6 μsec. A sharp change in slope is observed in both cases. Figure 7 is a log-log plot of reconnected flux for p_o = 7 mTorr, obtained by subtracting the flux after $t-t_o$ = 3.2 μsec from the plateau value.

316   TWO-CELL FIELD-REVERSED CONFIGURATION

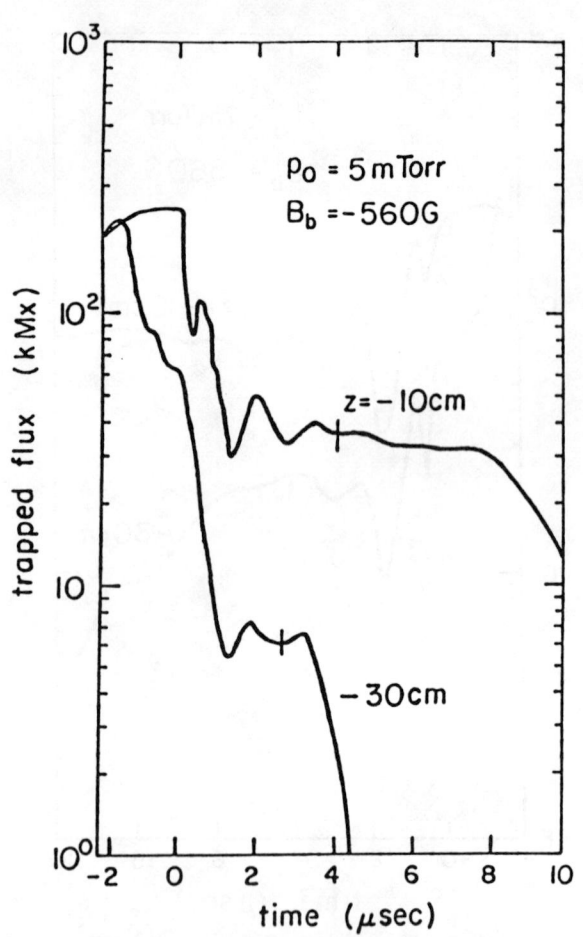

Fig. 5. Trapped flux as a function of time at two axial positions. The mirror coil was located at the z = -30 cm position. The filling pressure was 5 mTorr and the bias was -560 G. The estimated errors are represented with bars.

4. Interpretation of Results

4.1 Radial Diffusion Balanced by Axial Flow on Perpendicular Field Lines

The reconnection process in the region near the x-circle can be described by two distinct processes. First we assume a straight resistive current layer of radius R, width 2a <<R and length 2L where antiparallel magnetic lines diffuse and annihilate. The layer width stabilizes when the radial inward flow of plasma at velocity w balances the axial plasma flow along field lines [Sweet, 1956; Parker, 1963, 1979]. In the region exterior to the diffusive layer the resistivity is not important and the flux rate of change $\dot{\psi}$ is determined by the radial fluid velocity:

$$\dot{\psi} = wB2\pi R \quad , \qquad (1)$$

where we assume cylindrical geometry. Balance of the radial inflow with the longitudinal outflow at velocity v gives

$$\rho_e wL = \rho_i v\ell \quad , \qquad (2)$$

where ρ is the mass density and the subscripts e and i refer to the regions outside and inside the current layer. Combining Eqs. (1) and (2) we obtain [Brunel, et al., 1982]

$$\dot{\psi} = 2\pi RBv\rho_i\ell/\rho_e L \quad . \qquad (3)$$

For a compressible plasma the flux reconnected in the diffusive layer may pile up in the region of the x-circle. The flux lines in this region become tapered with angle α to the axis. This tapering can be seen in the experimental flux contours of Figs. 2 and 3 at times near t = 2 μsec. This tapering effectively shortens the length of the diffusive layer to L = αa, leading to increased $\dot{\psi}$. Plasma and flux are convected away axially at velocity v by reconnected flux lines which have a perpendicular radial component near the axis.

As a new flux tube reconnects near the x-circle

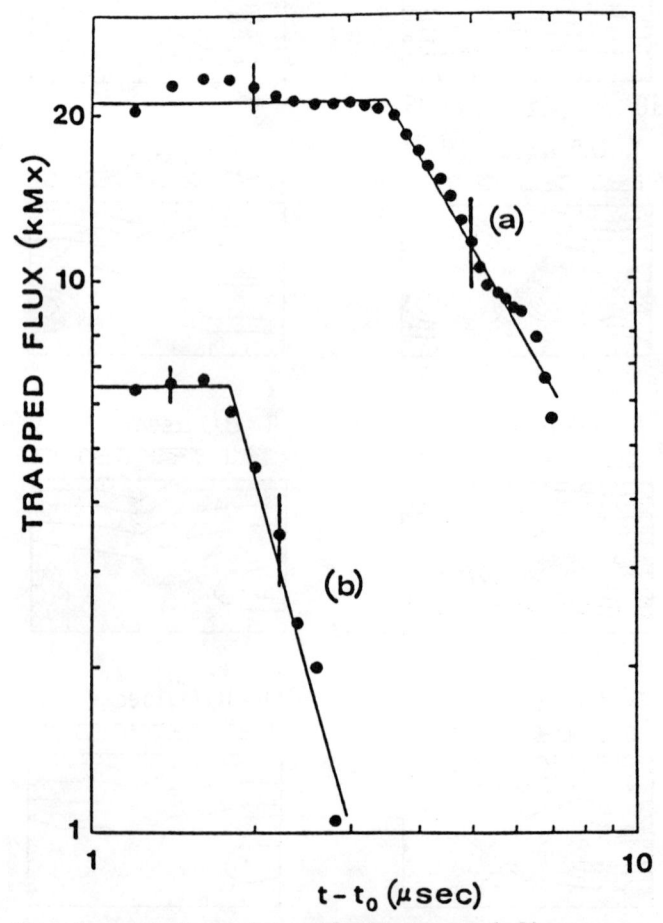

Fig. 6. Log-Log plots of the trapped flux as a function of time at z = -30 cm (the mirror position) for the two filling pressures studied. Curve (a) is for 7 mTorr; (b) for 5 mTorr. Note the sudden change in slope indicating the onset of the second phase of rapid reconnection.

its plasma density goes up to ρ_i, the characteristic density of the internal region. At some distance along the z axis greater than L this flux tube is located outside the diffuse internal region. There, the magnetic field is B and the density is ρ_e. The angle α is $\alpha = B_r/B$, where B_r is the average radial magnetic field in the diffusion layer. This field is given by $B_r = \psi/2\pi RL_t$ where ψ is the total reconnected flux and L_t is the length along z traveled by the ψ flux lines. Setting $L_t = L + vt$ we obtain (Brunel, et al. 1982)

$$L = vt\psi_c/(\psi-\psi_c) \quad , \qquad (4)$$

where ψ_c is $2\pi RaB$. Brunel et al. assume that the fast phase of reconnection sets in when the diffusive length is shortened by the angularity of the lines such that axial convection on perpendicular lines matches radial convention into the diffusive layer. Equating (4) and (3) gives for the reconnected flux:

$$\psi = \psi_0 (t/t_0)^{\rho_i/\rho_e} + \psi_c \quad , \qquad (5)$$

where $\psi(t_0) = \psi_0$.

Qualitative agreement with this model can be seen in Fig. 6 which shows an abrupt change in slope of reversed trapped flux for each of the two cases. The reconnected flux $\psi_1-\psi_{tr}$ is plotted versus $t-t_0$ for 7 mTorr in Fig. 7. Here $\psi_1 = 22.5 \pm 1.5$ kMx corresponds to the average value of the trapped flux during the plateau phase and ψ_{tr} is the trapped flux (from the axis to the x-circle). Density measurements were not possible in the region between FRC cells because of interference by the mirror coils. Thus the ratio ρ_i/ρ_e in Eq. (5) could not be determined experimentally. However the value of the exponent of the fast phase in Fig. 7 is 4 ± 1, indicating $\rho_i > \rho_e$. This would be expected since ρ_e represents plasma on open field lines and ρ_i that in the originally confined region. The most important feature of the model is that it predicts the onset of a second, fast phase of reconnection in the region of the x-circle, a result which is also obtained experimentally.

The merging velocity w of the field lines was obtained experimentally from data similar to those of Figs. 2 and 3 which give the motion of the separatrix radius r_s during the fast phase. The velocity directly obtained from the motion of the flux surface at t = 6 μsec is $\dot{r}_s = 2.7 \times 10^5$ cm/sec. This value can be compared for consistency with that obtained for w using Eq. (1). At t = 6 μsec the value $\dot{\psi} = 4.5$ kMx/μsec is obtained from Figs. 4 and 6. From magnetic field profiles in the mirror region we find B = 1.3 kG and R = 2.5 cm, giving w = 2.2×10^5 cm/sec in satisfactory agreement with $\dot{r}_s$. (The Alfven velocity calculated using n = 0.7×10^{15} cm^{-3} is $v_A = 7 \times 10^6$ cm/sec). We note that the magnetic Reynolds numbers ($8\pi a v_A/c^2\eta$) for the p_0 = 7mTorr

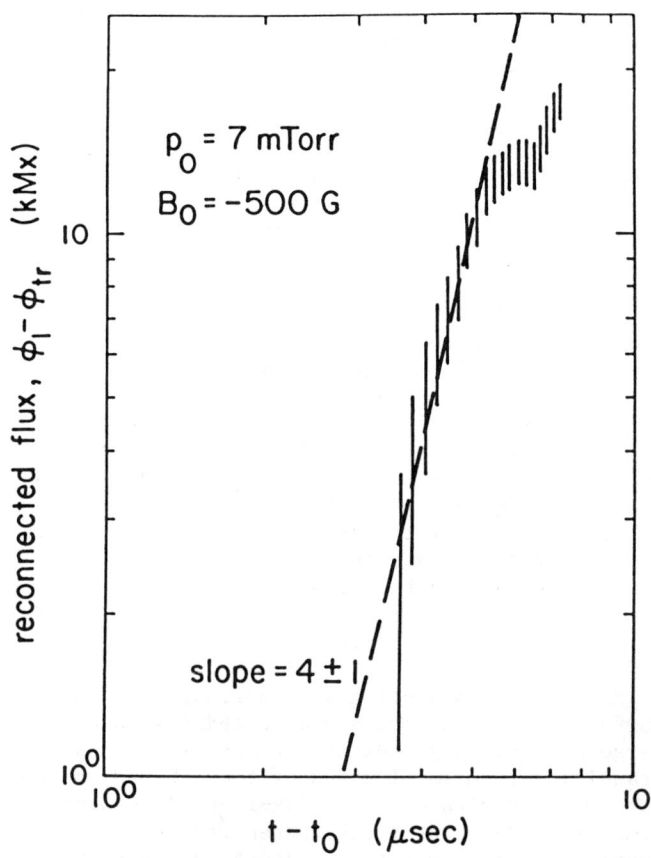

Fig. 7. Log-log plot of the reconnected flux as a function of time at z = -30 cm (the mirror position) for 7 mTorr filling pressure for the fast reconnection phase.

data are 150 for the slow phase and 50 for the fast phase, using a = 1cm and the η values from Section 5.

4.2 Change of Resistivity by Radial Convection of Current Carriers

Both in Fig. 2 and Fig. 3 the breaks in $\dot{\psi}$ (at 5.5 μsec and 3.2 μsec) occur approximately when the separatrix approaches the x-circle where the onset of the sudden increase in $\dot{\psi}$ is measured. As a qualitative second interpretation it is reasonable to assume that a corresponding lower density of current carriers will be convected to the x-circle, leading to increased, anamolous resistivity in the diffusive layer and therefore an increased $\dot{\psi}$.

5. Plasma Resistivity

The values for the resistivity η in the region of the x-circle before and during the second phase of the reconnection were calculated using the equation

$$\dot{\psi} = 2\pi r \eta c J_\theta \quad , \qquad (6)$$

where R is the radius of the field null. For the case of 7 mTorr the resistivity during the fast reconnection phase is obtained at t = 7 μsec, i.e., 1.5 μsec into the fast phase. The value for ψ at this time, obtained from Figs. 4, and 6, is ψ = 7.5 kMx/μsec. At R = 2.7 cm, the current density obtained from the magnetic field profile at this time from $J_\theta = \partial B_z/\partial r$ is J_θ = 1.5 kA/cm^2. We neglect a small contribution (about 10%) from $\partial B_r/\partial z$ and note that the convection term in Ohm's law arising from v×B is absent on the x-circle. We thus determine η = 3.7x10^{-15}sec. This is larger than the resistivity prior to the second phase when similar measurements give η = 6.5x10^{-16} sec. The classical resistivity is approximately 2x10^{-16} sec corresponding to $T_e \approx$ 80 eV [Sevillano, 1980]. A similar analysis for 5 mTorr at R = 1.7 cm where ψ = 5 kMx/μsec at t = 4 μsec, gives J_θ = 0.8 kA/cm^2 and η = 8x10^{-15}sec for the fast phase. Again in this case a large increase in the value of the resistivity during the second phase is seen.

6. Conclusions

Magnetic reconnection at a cylindrical neutral surface between oppositely directed fields at large magnetic Reynolds number exhibits a sudden transition from small rates of flux change to much larger rates with angular flux-line structures similar to those of the Petschek reconnection model. During the fast reconnection phase the resistivity at the field null in the diffusive layer is greater than classical by an order of magnitude. Before the fast reconnection phase the resistivity is a few times the classical value.

References

Armstrong, W.T., R.K. Linford, J. Lipson, D.A. Platts and E.G. Sherwood, Field-reversed experiments (FRX) on compact toroids, Phys. Fluids 24, 2068, 1981.
Baum, P.J., A. Bratenahl, M. Kao and R.S. White, Plasma instability at an x-type magnetic neutral point, Phys. Fluids 16, 1501, 1973.
Baum P.J. and A. Bratenahl, Laboratory solar flare experiments, Solar Phys. 47, 331, 1976.
Baum, P.J. and A. Bratenahl, On reconnection experiments and their interpretation, J. Plasma Phys. 18, 257, 1977.
Birn, J. and E.W. Hones, Jr., Three-dimensional computer modeling of dynamic reconnection in the geomagnetic tail, J. Geophys. Res. 86, 6802, 1981.
Bratenahl, A. and P.J. Baum, A plasmoid release mechanism that could explain the substorm's impulsive earthward diversion of cross-tail current, this volume, 1984.

Brunel, F., T. Tajima and J.M. Dawson, Fast magnetic reconnection processes, Phys. Rev. Lett. 49, 323, 1982.
Hones, E.W., Jr., Plasma flow in the magnetotail and its implications for substorm theories, in Dynamics of the Magnetosphere, S. I. Akasofu, Ed., P.545, D. Reidel Publ. Co., Dordrecht, Holland, 1980.
Knox, S.O., H. Meuth, F.L. Ribe and E. Sevillano, Reversed-field flux-trapping in a low-compression theta pinch, Phys. Fluids, 25, 1982.
Parker, E.N., Sweet's mechanism for merging magnetic field in conducting fluids. J. Geophys. Res. 62, 509, 1957.
Parker, E.N., Solar-flare phenomena and the theory of reconnection and annihilation of magnetic fields, Ap. J. Suppl. Ser. 27, 177, 1963.
Parker, E.N., Cosmical Magnetic Fields, Their Origin and Their Activity, Clarendon Press, Oxford, 1979, Ch. 15.
Petschek, H.E., Magnetic field annihilation, in The Physics of Solar Flares, AAS-NASA Symposium on the Physics of Solar Flares, NASA Spec. Pub. Sp-50, 425, 1964.
Sato, T. and T. Hayashi, Externally driven magnetic reconnection and a powerful magnetic energy converter, Phys. Fluids 22, 1189 1979.
Sevillano, E., Magnetic reconnection in field-reversed configurations, Ph.D. Thesis, University of Washington, Seattle, 1983. University Microfilms, 300 North Zeeb Road, Ann Arbor, MI 48106.
Stenzel, R.L. and W Gekelman, Experiments on magnetic field-line reconnection, Phys. Rev. Lett. 42 1055, 1979.
Sweet, P.A., The neutral point theory of solar flares, in Electromagnetic Phenomena in Cosmical Physics, p. 123, Proc. IAU Symposium No. 6, Stockholm, 1956.

Questions and Answers

Rothwell: Have you scaled the laboratory ions to the magnetosphere, i.e. the ion gyroradius divided by the scale size of the experiment compared with the same ratio for either a 10 keV proton or oxygen ion in a 10 nT lobe field and a scale dimension of 10^3 km?

Sevillano: I have not made the comparison but I can give you typical physical parameters of our experiment. The ion gyroradius is about 1 cm and the radial dimension of the discharge tube is 10 cm. Our working gas is deuterium.

Hones: What is your interest in dividing the plasmoid into two plasmoids? Is it relevant to trying to move plasmoids?

Sevillano: A multiple-cell field reversed configuration has the advantage of multiple-mirror confinement on the open field lines. Using many of these cells it should be possible to obtain better confinement thus reducing the sharp density gradients and transport. We did not attempt to translate plasmoids although in principle using the center mirror we could have had one plasmoid out of each end.

RECONNECTION IN SPHEROMAK FORMATION AND SUSTAINMENT

James H. Hammer

Lawrence Livermore National Laboratory
University of California
Livermore, CA 94550

Abstract. Reconnection phenomena observed in Spheromak magnetic confinement experiments are reviewed. Many features of the observations are consistent with a model based on three-dimensional magnetic relaxation while preserving global magnetic helicity.

Introduction

The Spheromak is a magnetic confinement device that is being explored in both the U.S. and Japanese fusion programs. It is a member of the "Compact Torus" family of magnetic structures characterized by a set of closed, nested toroidal flux surfaces but without any coils, transformer cores, etc., protruding through the hole in the torus. The Spheromak is closely related to the Reversed Field Pinch (RFP) in that most of the magnetic field is produced by plasma currents flowing along the magnetic field lines (a near force free field) rather than by external coils. The Spheromak has magnetic field components of comparable strength in both the toroidal (azimuthal) and poloidal (in the plane perpendicular to the azimuthal unit vector) directions. The large internal magnetic energy in the Spheromak makes it rich in magnetohydrodynamic phenomena and reconnection, in particular, plays an important role in the formation, resistive decay and instability processes.

Formation

Spheromaks have been formed successfully in a surprisingly large number of ways. Several completely distinct methods have worked: inductive formation using a flux core [Yamada et al., 1981]; combination Z and theta pinch formation [Goldenbaum et al., 1980]; and magnetized plasma gun formation [Alfvén et al., 1960, Jarboe et al., 1980, Jarboe et al., 1983, Turner et al., 1981, and Turner et al., 1983]. Each of these methods, illustrated in Figs. 1, 2, 3, involves reconnection in some fashion. In the preformation state poloidal field lines are wrapped around or penetrating solid bodies. During formation, plasma is created and the toroidal component of the field is introduced. Magnetic forces cause the flux surfaces to distort and eventually tear off to form the desired set of closed, nested surfaces. Highly dynamical Spheromak-like structures have also been produced using a conical theta pinch [Wells, 1962 and Nolting et al., 1973].

Magnetic probe data has confirmed the existence of closed surfaces and hence demonstrates that some form of reconnection has occurred. Figure 4 shows a poloidal flux surface map from a Z-theta pinch experiment following formation [Goldenbaum et al., 1980] and Fig. 5 shows the evolution of contours of toroidal magnetic field (B_T x radius is approximately a function of poloidal flux alone for this slow formation technique) for a flux core experiment [Yamada et al., 1981]. In Fig. 6, the axial magnetic field, B_z, is shown as a function of axial position, z, along the axis of symmetry at various times for a plasma gun experiment [Jarboe et al., 1980]. The field in the entrance region between the gun muzzle and the flux conserving container decays rapidly (Fig. 6a), corresponding to reconnection of the field as depicted in Fig. 3. The field in the Spheromak following reconnection (Fig. 6b) decays on a much slower time scale. In one gun experiment [Turner et al., 1983] the reconnection of field lines in the entrance region was forced to occur on an even more rapid time scale by a fast, single-turn pinching coil wrapped around the entrance region. Figure 7a shows the time behavior of the magnetic field due to the pinching coil at a point on the symmetry axis in the middle of the entrance region, with the plasma gun turned off. Figure 7b shows the field due to the plasma gun alone and Fig. 7c shows the combined field with both the plasma gun and pinching coil. The reconnection with the pinching coil on occurs at a rate 6 times faster than the unforced rate, i.e., in about 10 μsec, which is still many times the radial Alfvén transit time.

(a)

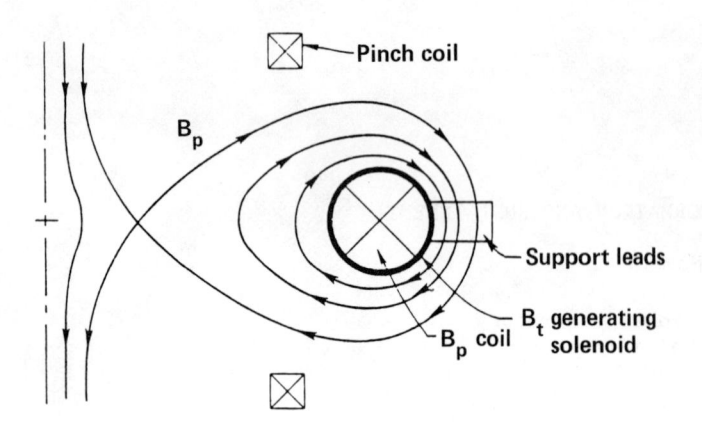

(b)

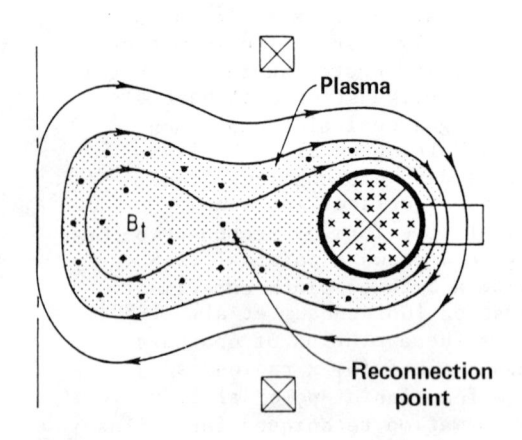

(c)

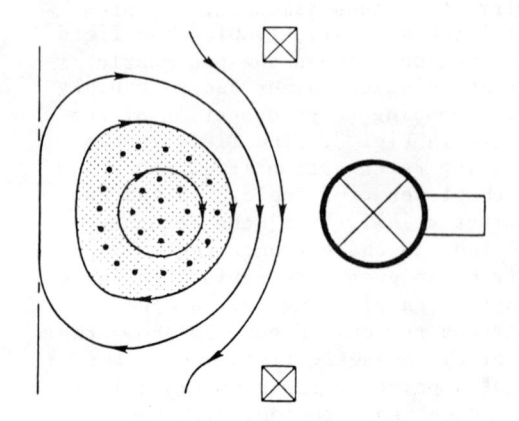

Fig. 1. Inductive Spheromak formation using a flux core. Poloidal magnetic field surrounding the toroidal core is pre-established. Toroidal field is induced into a plasma by the currents flowing in the core and the plasma pinches radially inward, reconnecting and forming a Spheromak.

In the "classical" view of these formation methods ("classical" refers to the dominant view held several years ago) the reconnection is axisymmetric and two-dimensional in nature. A consequence of this view is that the poloidal magnetic flux contained in the resulting Spheromak should be less than or equal to the poloidal flux linking the formation device. In many instances, however, the poloidal flux measured in the Spheromak does not conform to the classical prediction and may be as much as ten times the initial poloidal flux [Alfvén et al., 1960, Jarboe et al., 1980, Jarboe et al., 1983, Turner et al., 1981, and Turner et al., 1983]. This phenomenon, known as "flux amplification" implies that three-dimensional effects are playing a big role: toroidal flux is being twisted into poloidal flux, breaking symmetry during formation.

Flux amplification can be understood quantitatively by applying Taylor's relaxation theory [Taylor, 1974], originally developed for the RFP. The theory maintains that during the MHD turbulent formation process, many of the flux invariants of ideal MHD are broken by reconnection, leaving only the global magnetic helicity, $K = \int \underline{A} \cdot \underline{B} \, dV$, invariant. The magnetic energy of the system, $W = \int B^2/2 \, dV$, is then minimized subject to the constraint that K alone is conserved. The helicity is unambiguously

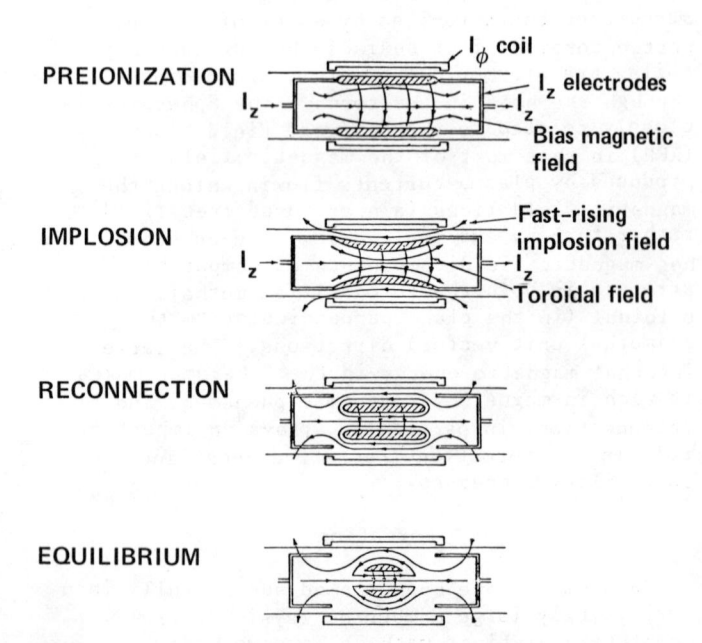

Fig. 2. Z-theta pinch Spheromak formation. The symmetry axis is at right angles to the axis in Fig. (1). After pre-establishing a bias field in the z direction, a current is drawn in the z direction by electrodes, producing a toroidal field. The azimuthal, single turn coil is then energized, producing a field that opposes the bias field. Reconnection and relaxation to a Spheromak equilibrium follows.

defined if the gauge is restricted to cases where $\oint \underline{A} \cdot \underline{dl}$ = flux enclosed by the curve within the plasma volume, and if no flux penetrates bounding surfaces. To understand the special significance of the helicity, it is worthwhile to briefly review its relationship to

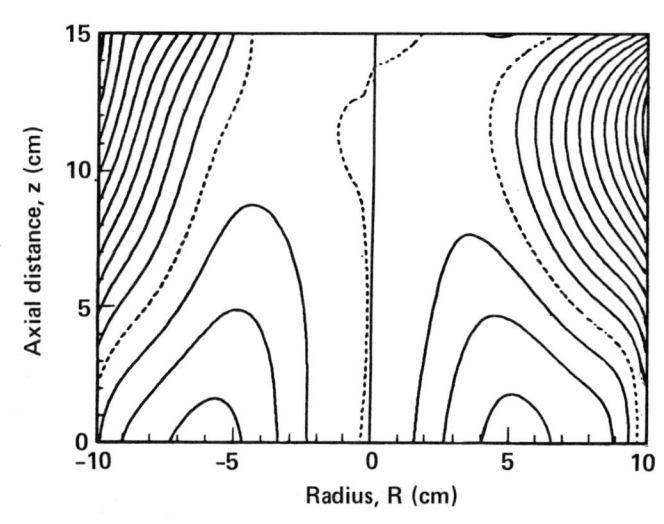

Fig. 4 Poloidal flux map from a z-θ pinch formation experiment.

the topological linkage of the magnetic field.

Consider an infinitesimal flux tube with a half twist in it as shown in Fig. 8a. It's assumed, for simplicity, that there is no internal twist to the tube, i.e., the lines do not spiral about the center of the tube. We can evaluate K for the tube,

$$K = \int \underline{A} \cdot \underline{B} \, dV = \oint \underline{A} \cdot \underline{dl} \, BdA = d\psi \oint \underline{A} \cdot \underline{dl}$$

where $d\psi$ is the flux within the tube and the integration path is shown in Fig. 8b. The integral can be evaluated by adding and substracting pieces to the curve as shown in Fig. 8c. The first of the curves in Fig. 8c gives no contribution to the integral since it encloses no magnetic flux. This can be seen by breaking the curve up into a number of closed loops between the reference plane and the curve. The second curve in Fig. 8c encloses the flux tube and hence gives $\oint \underline{A} \cdot \underline{dl} = d\psi$ and so the helicity is

$$K = d\psi^2.$$

Similar arguments for two tubes of flux $d\psi_1$, $d\psi_2$ allow evaluation of the helicity. If the tubes are linked $K = 2 \, d\psi_1 \, d\psi_2$, whereas if they are unlinked $K = 0$. More complex structures such as knots, etc., can be evaluated by simply projecting the three dimensional curves onto a plane and examining the crossings. The helicity thus corresponds to a net twist, linkage or "knotted-ness" of magnetic flux.

If infinitely conducting plasma fills the tubes, then the dynamical equations predict that the helicity of each tube separately will remain constant in time. A small amount of resistivity, however, will allow reconnection of tubes with each other, without destroying global properties.

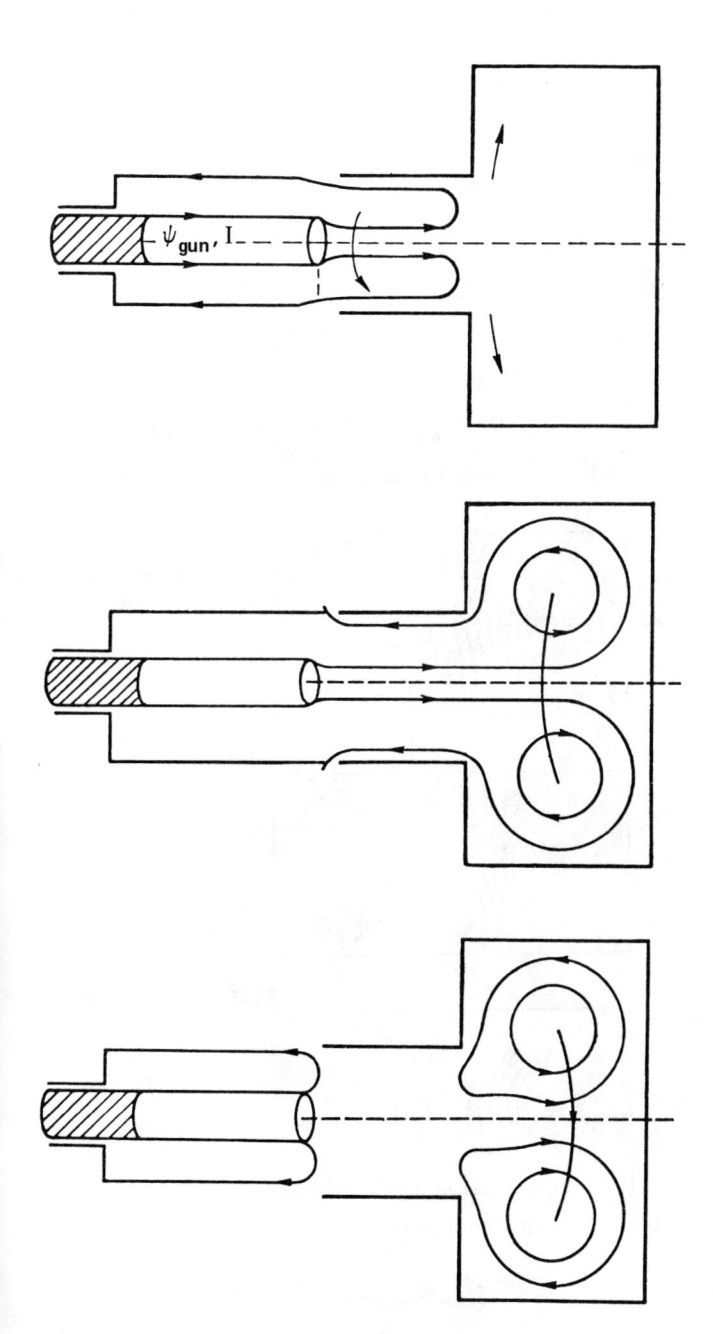

Fig. 3. Coaxial plasma gun formation of a Spheromak. Poloidal flux threading the coaxial gun is pre-established. A current is drawn across the gun electrodes, producing a toroidal field that stretches the poloidal lines through an entrance region and into a flux conserving chamber. The field in the entrance region reconnects, leaving an isolated Spheromak.

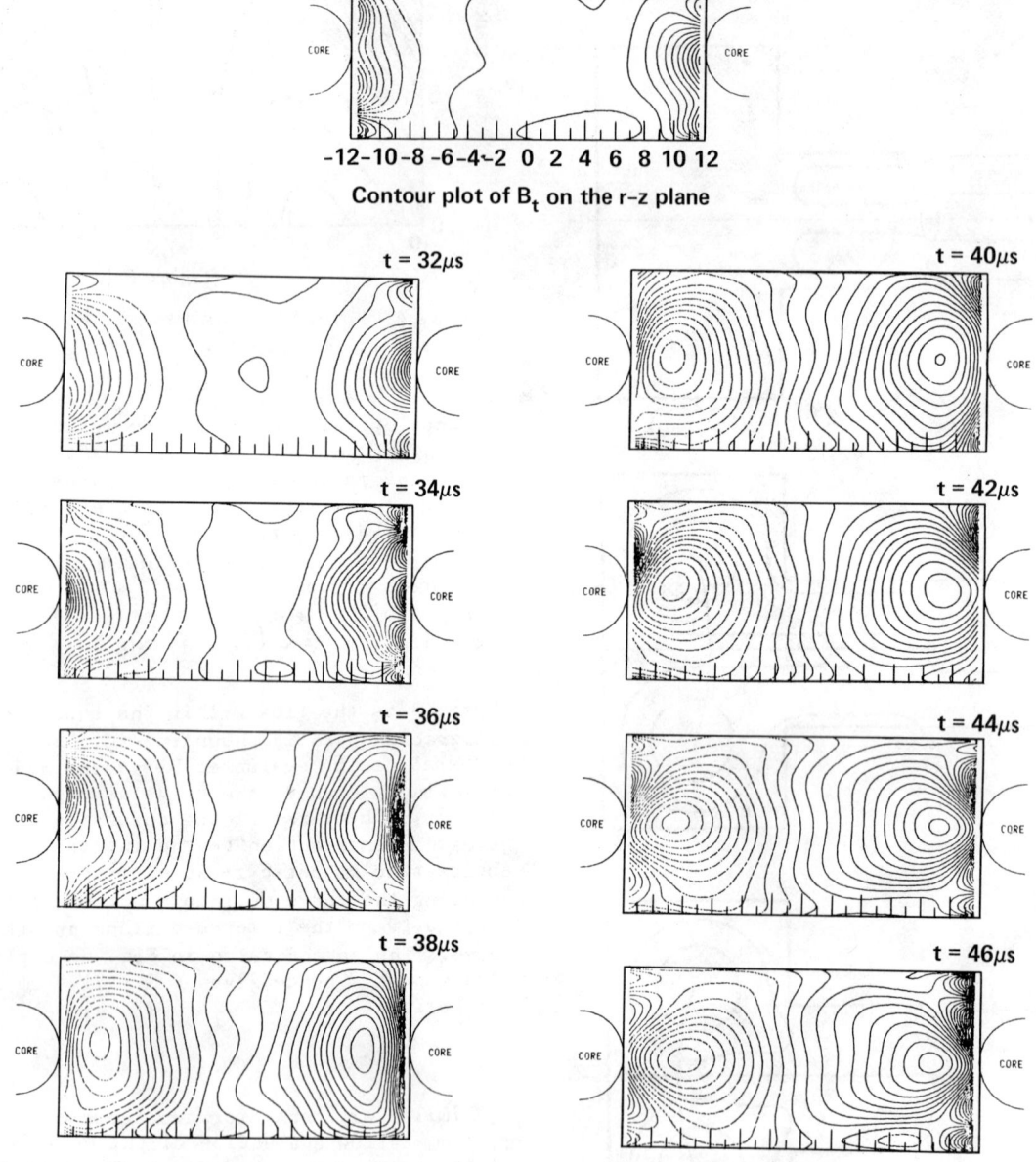

Fig. 5. Contours of radius x B_T for different times during Spheromak formation in a flux core experiment.

Fig. 9 shows how reconnection can transform one type of helicity, in this case a simple knot, into another, a series of loops. Note that while the topology alters, the total value of the helicity is preserved. If many tubes are present, knots and twists, etc., can move from one tube to another by reconnection and local conservation of the helicity on each tube will be destroyed. In the case of complete intermixing and reconnection, only the global helicity, the helicity summed over all the tubes is preserved. The generalization of K to the case of arbitrary gauge and to flux penetrating bounding surfaces is discussed in the appendix.

Fig. 10 illustrates schematically how these ideas apply to flux amplification in the plasma

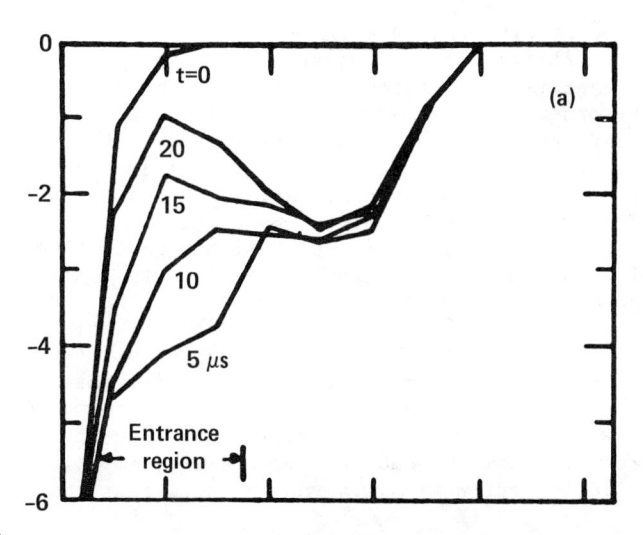

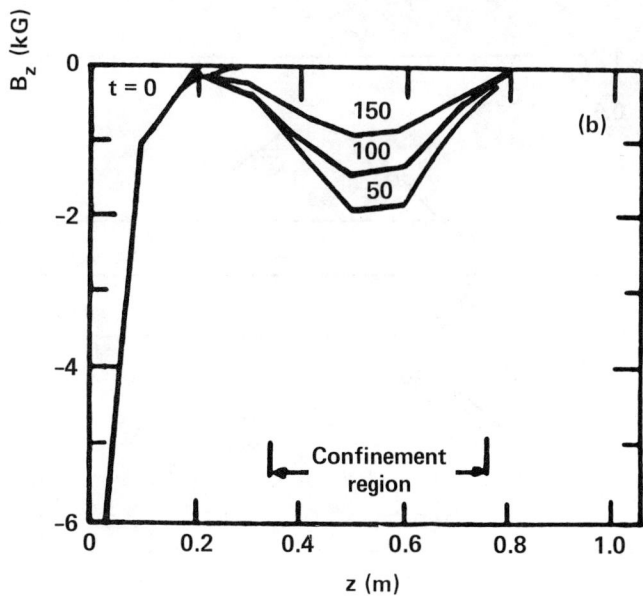

Fig. 6. The upper figure shows decay of the field in the entrance region of a plasma gun experiment. The lower figure shows the slower decay of the Spheromak field. All times are in microseconds.

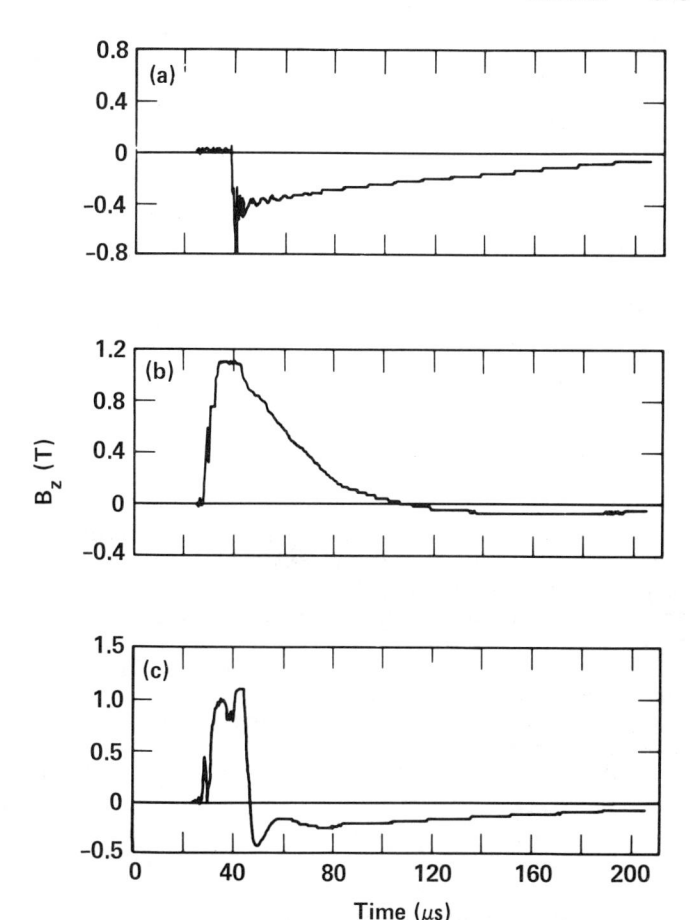

Fig. 7. Axial field in the entrance region of a plasma gun experiment: (a) shows the field due to a pinching coil alone; (b) shows the field due to the plasma gun alone; (c) shows the combined fields.

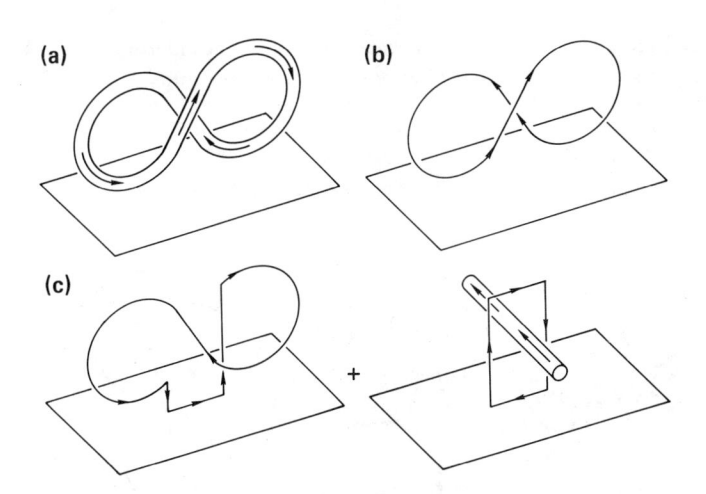

Fig. 8. (a) A flux tube with a half twist above a reference plane; (b) the integration path for evaluating $\int \underline{A} \cdot \underline{dl}$; (c) the integration path split into two pieces; the first encloses no flux and the second encloses the flux tube.

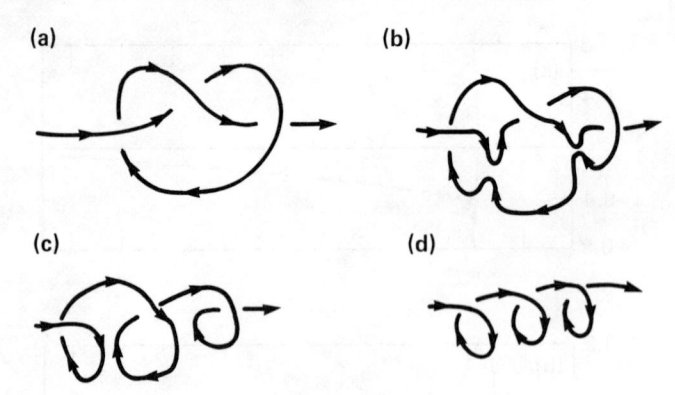

Fig. 9. (a-d) Transformation of a simple knot into three twists by reconnection.

gun formation of a Spheromak. The plasma column can deform in a kink-like motion and twist toroidal flux into the poloidal plane. Reconnection then allows for relaxation to an axially symmetric state.

The relaxed state in the Taylor theory is described as having minimum magnetic energy for the given total helicity. An immediate consequence is that

$$\nabla \times \underline{B} = \mu \underline{B}$$

where μ is a constant in space. In Fig. 11, the solutions to this equation (solid lines) are compared to the observed field profiles [Turner et al., 1983] showing good agreement. Similar agreement is found for each of the formation methods including the conical theta pinch [Nolting et al., 1973]. Figure 12 shows the predicted values of poloidal flux in the Spheromak for a given input of helicity

$$K_{in} = 2\left(\int V \, dt\right) \psi_g$$

where V is the voltage across the electrodes and ψ_g is the poloidal flux threading the gun. Quantities linear in the field such as the poloidal flux scale roughly as the square root of K, so ψ_P is plotted versus $(\int V \, dt)^{1/2}$ with $\psi_g^{1/2}$ held fixed in Fig. 12a and versus $\psi_g^{1/2}$ with $(\int V \, dt)^{1/2}$ held fixed in Fig. 12b. The theoretical curves are fit to the data by allowing for ohmic loss during injection ($K = \beta K_{in}$ with $\beta \leq 1$) and a low energy cutoff when the capacitor bank energy is insufficient to stretch the gun flux into the flux conserver.

Resistive Decay and Sustainment

During the resistive decay of the Spheromak, the profiles can drift away from the preferred (minimum energy) state. Figure 13 shows the toroidal and poloidal field versus time during the decay phase of the gun produced Spheromak at LASL. At about 130 μsec, a fairly abrupt change occurs, bringing the ratio of toroidal to poloidal fields back to the ratio in the minimum energy state (the dashed line). The circles show results from a 1 1/2-dimensional transport calculation. The readjustment of fields and fluxes imply that three-dimensional motions and reconnection are returning the Spheromak to the preferred state [Yamada et al., 1982]. The ability to adjust itself in this fashion during the decay has led to suggestions that a Spheromak could be maintained in steady state, a very advantageous property from the controlled fusion viewpoint. A source of helicity could steadily replace the helicity lost due to resistive decay, and the relaxation mechanism would distribute the fluxes appropriately (in a strictly two-dimensional, axisymmetric state, the Spheromak is doomed to a finite lifetime by Cowling's theorem [Cowling, 1934]). Recent results from the LASL plasma gun experiment, illustrated in Figs. 14 and 15, show that this

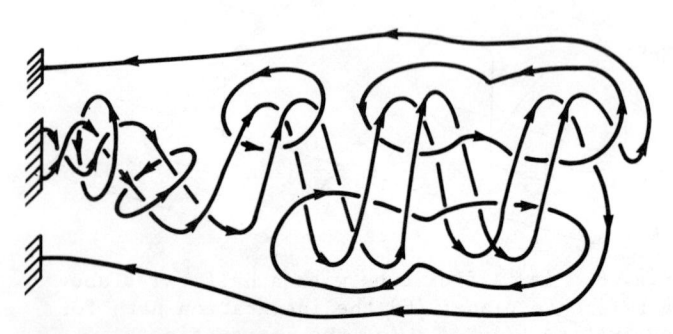

Fig. 10. Flux amplification in plasma gun formation of a Spheromak.

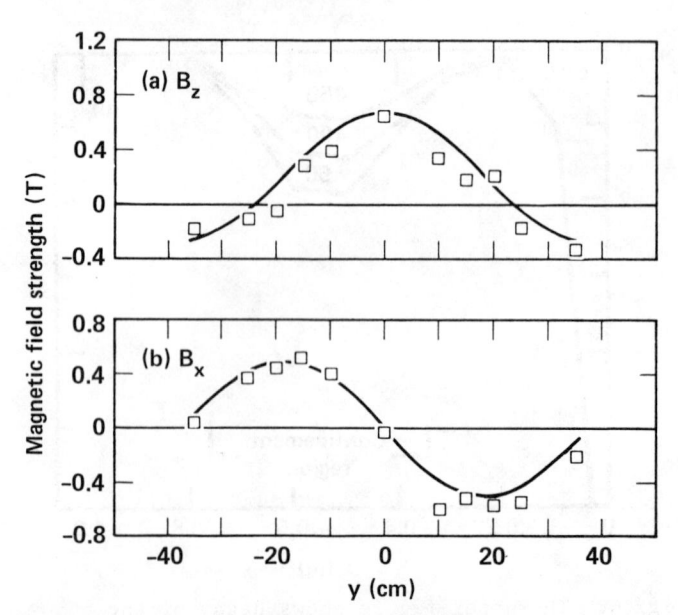

Fig. 11. Comparison of experimental field profiles (boxes) to theoretical, force free profiles (solid curves) in a plasma gun experiment.

is indeed possible. The plasma gun is used as a long-pulse helicity source, in either of the two modes shown in Fig. 14a,b. The measured poloidal field in these two cases is shown in Fig. 15. The field is sustained for several

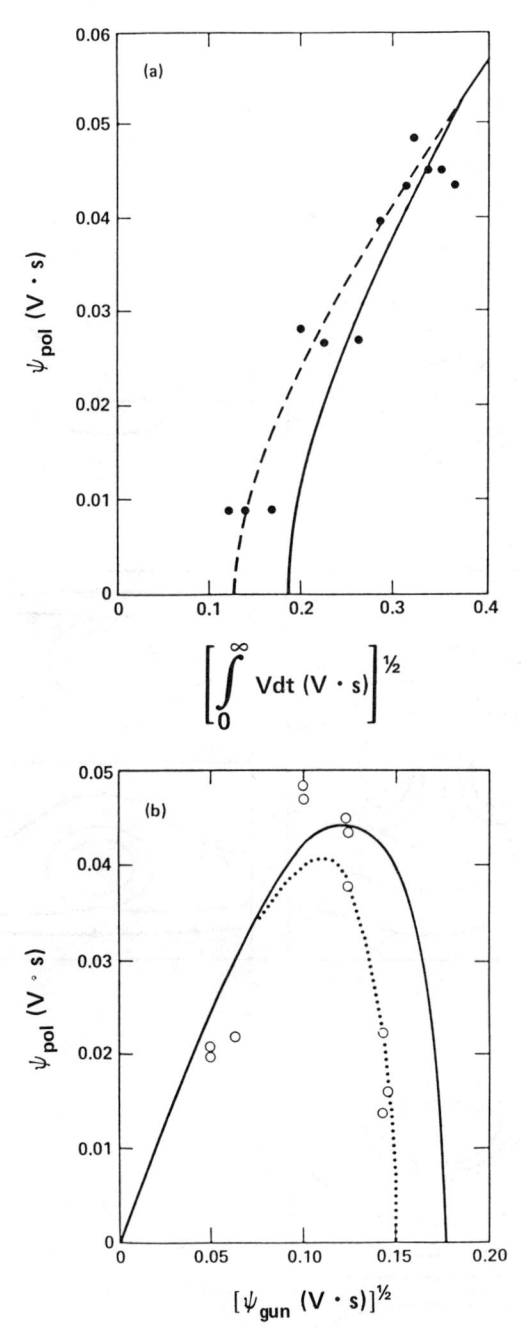

Fig. 12. Observed poloidal flux in the Spheromak produced by a plasma gun in volt seconds as a function of (a) volt seconds introduced through the gun insulator with ψ_{gun} = .011 V sec and (b) poloidal flux in the plasma gun with $\int Vdt$ = 0.1 V sec. Solid curves are theoretical predictions using different fitting parameters (see Turner et al., 1983 for details).

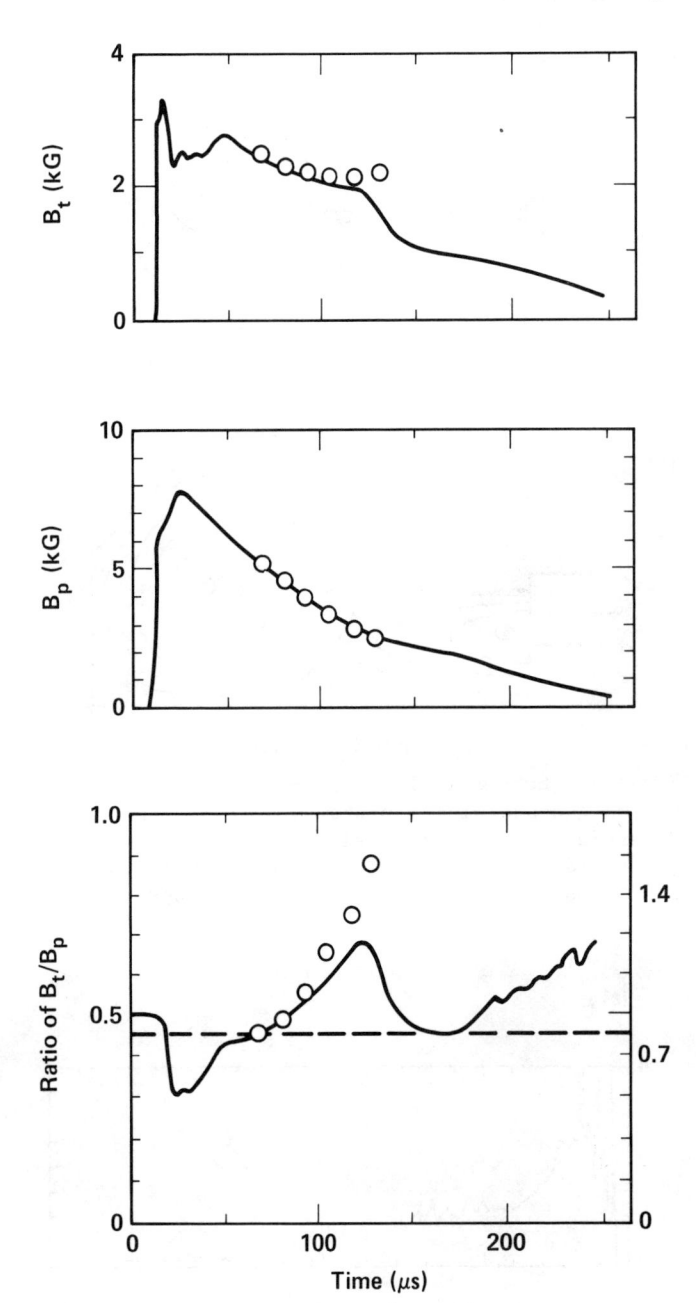

Fig. 13. Decay of fields in a plasma-gun-produced Spheromak.

times the resistive decay time and is limited only by the time scale of the power supply.

Stability

The Spheromak, while in some sense a preferred MHD state, is not unconditionally stable. A particularly dangerous instability is the "tilt" mode [Rosenbluth et al., 1979], where the Spheromak tumbles about an axis at right angles to the symmetry axis if the plasma is not sufficiently oblate (shorter in the axial direction than in the radial direction) or if

326 SPHEROMAK FORMATION AND SUSTAINMENT

(a)

Normal sustained

(b)

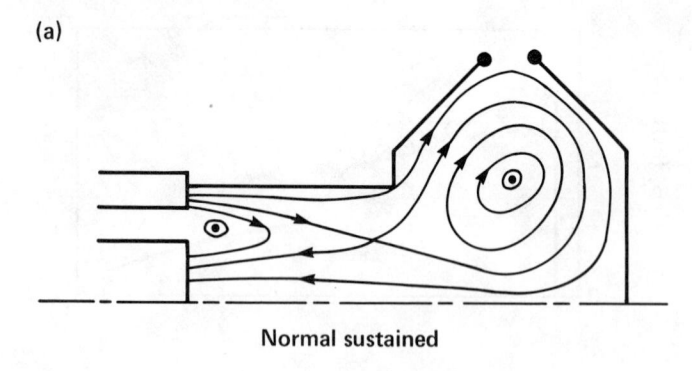

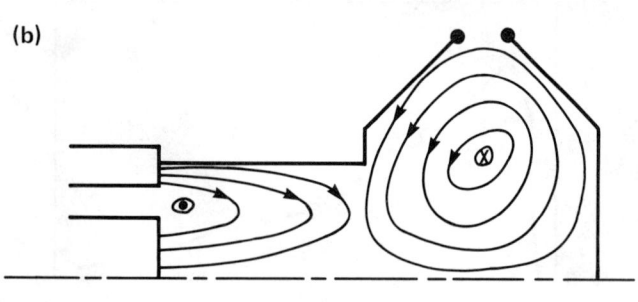

180° flipped sustained

Fig. 14. Schematic field line patterns in a plasma-gun-sustained Spheromak. In (a) the Spheromak has the same orientation as the plasma gun fields. In (b) the Spheromak is rotated 180° about an axis coming out of the page.

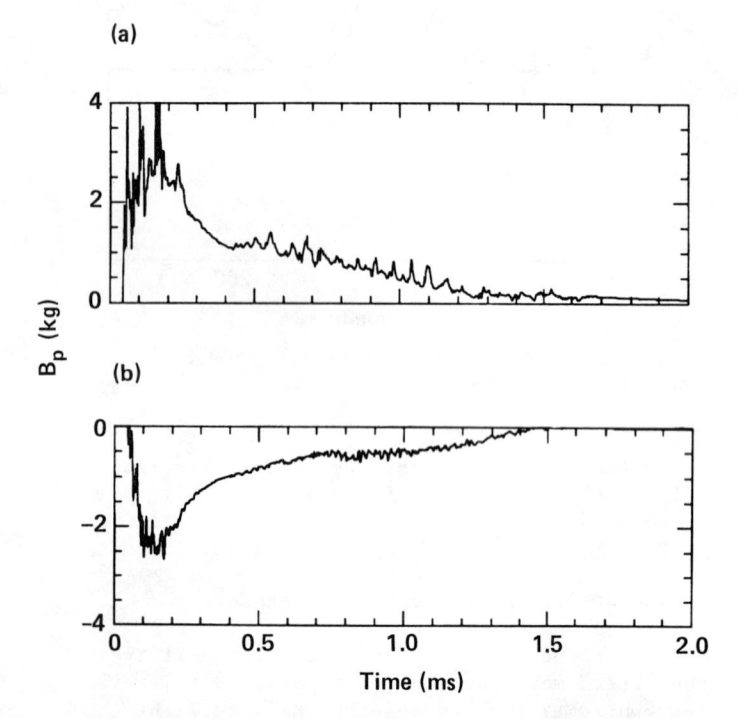

Fig. 15. Observed poloidal field vs. time for the (a) normal, and (b) flip sustained modes illustrated in Fig. 14.

conducting walls are too distant. In an externally applied field, the Spheromak field can reconnect with applied field as it tilts, as shown schematically in Fig. 16. Measured field profiles [Jarboe et al., 1983] consistent with

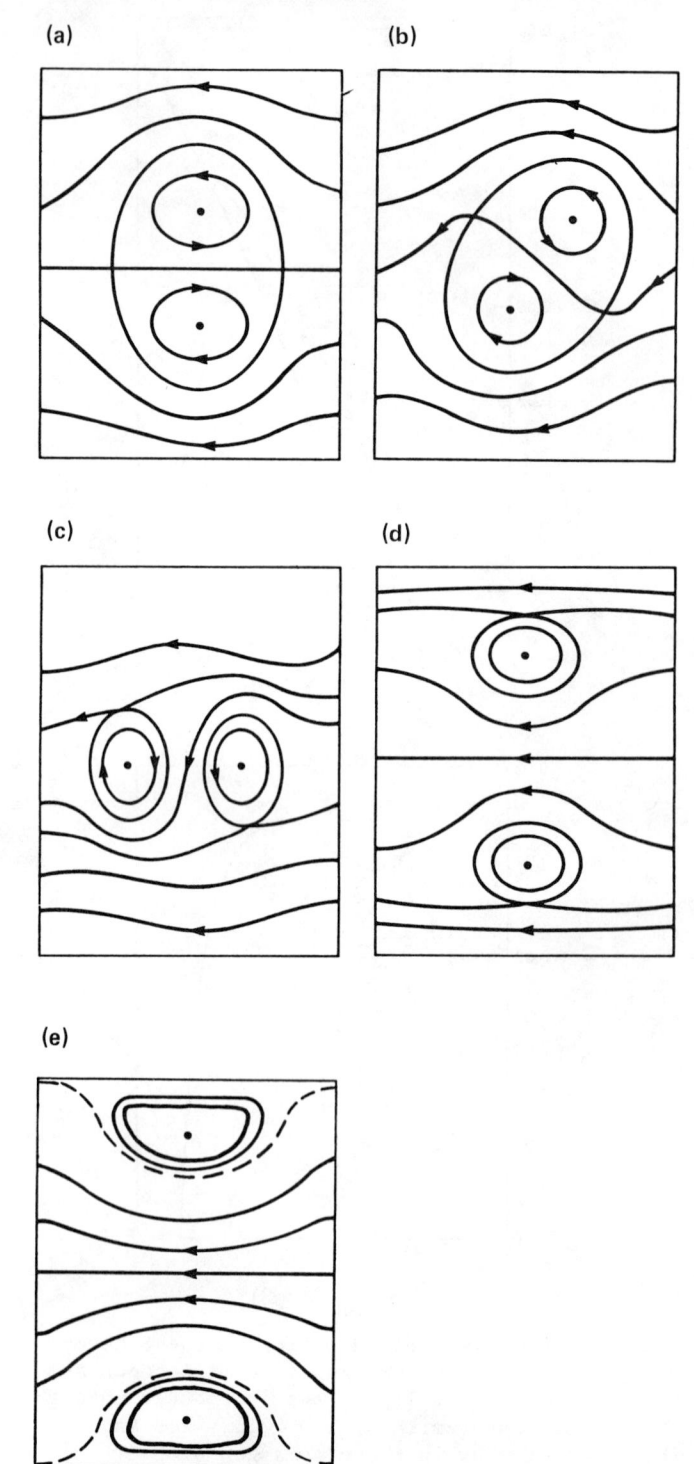

Fig. 16. (a-e) Field line patterns during tilting and reconnection of a Spheromak.

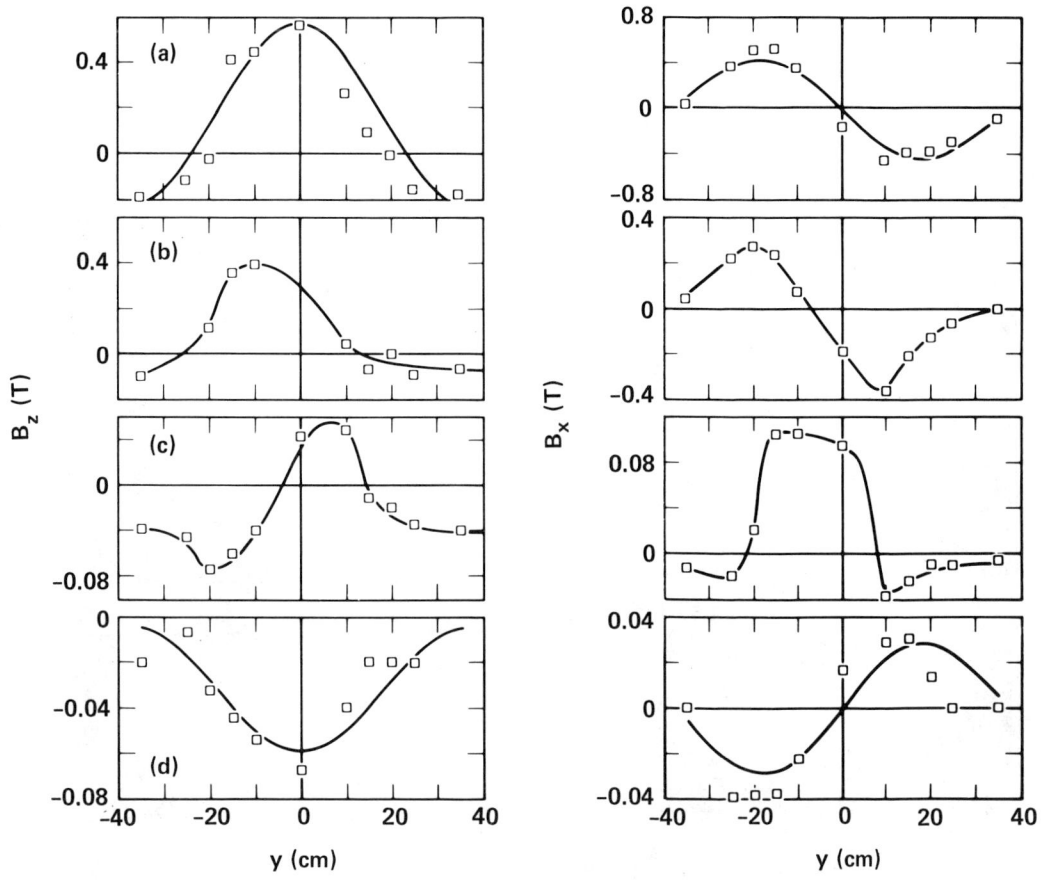

Fig. 17. (a-d) Observed field profiles in a plasma gun experiment consistent with Fig. 16 (a-d).

the illustrations in Fig. 16a-d are shown in Fig. 17 where y is a coordinate along a diameter through the center of the device, z is along the symmetry axis and x is normal to y and z. Figure 18 shows the measured fields in the final state consistent with Fig. 16e, i.e., a Spheromak flipped 180° with the external flux completely reconnected and passing through the center of the torus.

Figure 19 shows contours of radius x B_T for a flux-core-produced Spheromak undergoing a tilt instability. In this case, the Spheromak tilts about 90° before reconnection completely destroys the configuration [Sato et al., 1983].

Conclusion

Reconnection plays an essential role in all phases of existence for the Spheromak. It is present during the formation and slow decay, as well as the rapid destruction of the Spheromak when an instability occurs. Reconnection in an intrinsically three-dimensional fashion raises the intriguing possibility that a Spheromak may be sustained indefinitely against resistive losses by the injection of magnetic helicity.

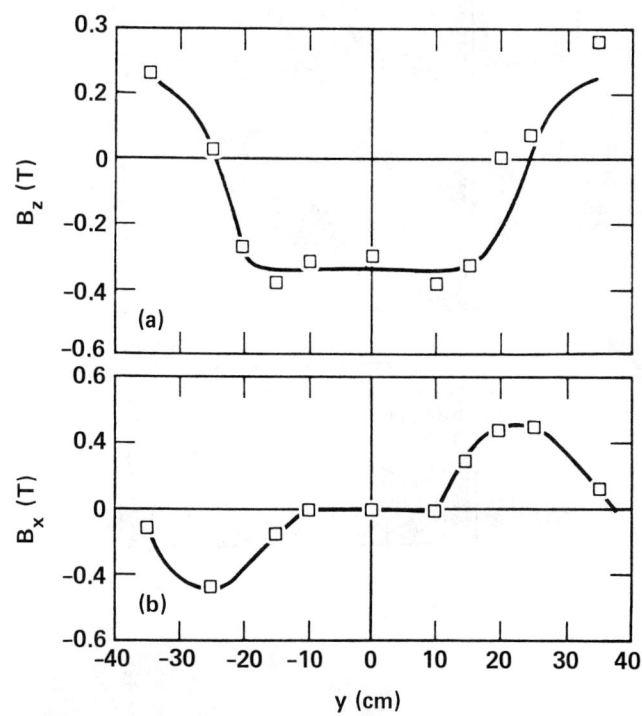

Fig. 18. Observed field profiles in a plasma gun experiment consistent with Fig. 16 (e).

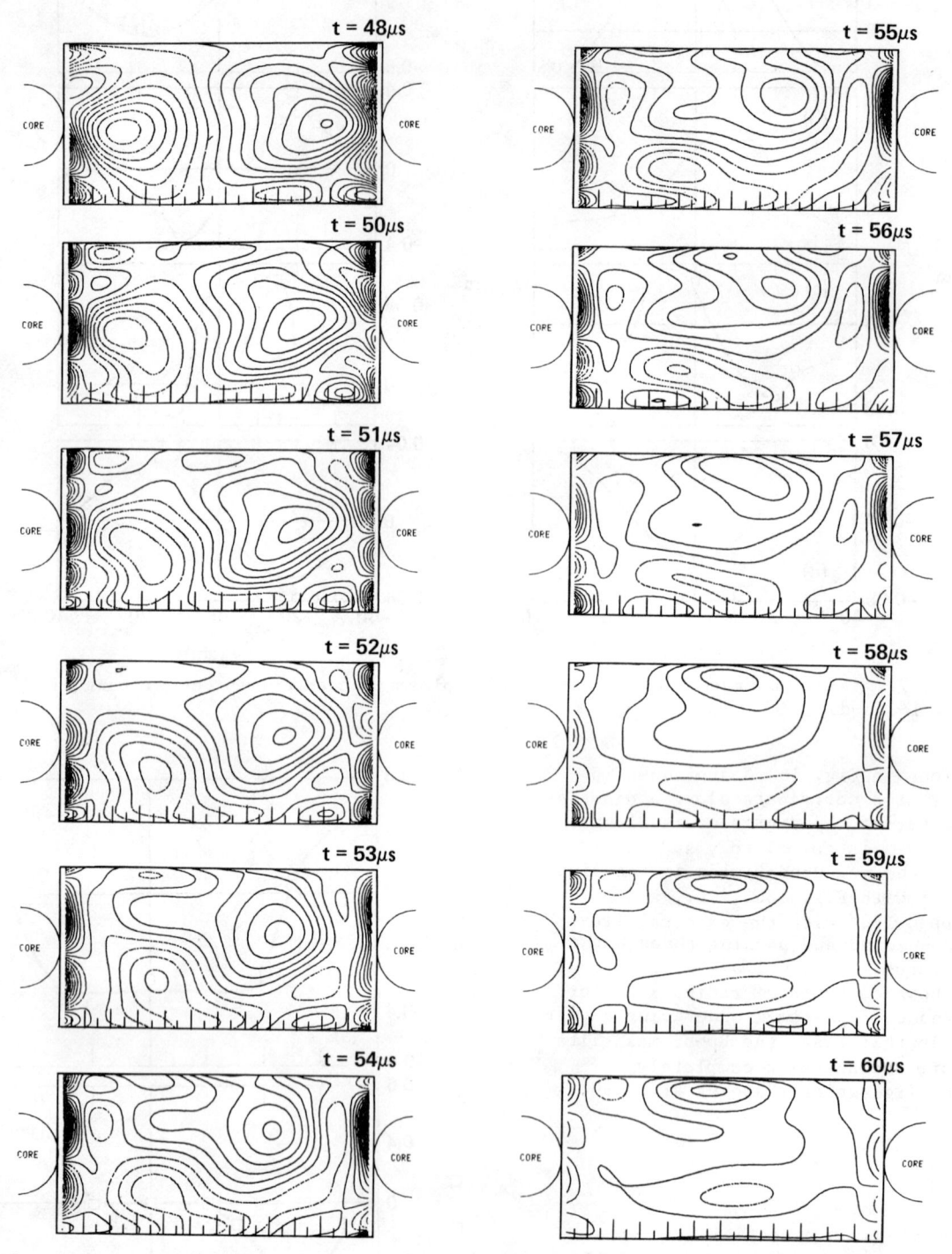

Fig. 19. Contours of radius x B_T for a Spheromak undergoing a tilt instability in a flux core experiment.

APPENDIX

Some confusion exists regarding gauge invariance properties of the helicity as well as the appropriate generalization when magnetic field lines intercept the conducting boundaries. To clarify these issues consider the time derivative of the usual definition of the helicity

$$\dot{K} = \left(\int \underline{A} \cdot \underline{B} \, dV\right)^{\cdot} = \int \underline{\dot{A}} \cdot \underline{B} \, dV + \int \underline{A} \cdot \underline{\dot{B}} \, dV . \quad (A1)$$

The volume under consideration is assumed to be enclosed by conducting surfaces, possibly partitioned by infinitesimally thin insulators to allow for injection of magnetic flux. Field lines are allowed to intercept the walls. For the time being we restrict ourselves to gauges satisfying $\oint \underline{A} \cdot \underline{dl}$ = enclosed magnetic flux within the plasma volume. This restriction implies

$$\underline{E} = -\underline{\dot{A}} - \nabla\phi , \quad (A2)$$

where $\oint \nabla\phi \cdot \underline{dl} = \dot{\psi}_{ex}$ = time rate of change of flux outside the plasma volume that is enclosed by the curve. Using Faraday's Law, $\underline{\dot{B}} = -\nabla \times \underline{E}$, Gauss's Law, $\nabla \cdot \underline{B} = 0$, and integrating by part gives

$$\dot{K} = -2 \int \underline{E} \cdot \underline{B} \, dV - \int_{surface} \phi \, \underline{B} \cdot \underline{ds}$$
$$+ \int_{surface} \underline{A} \times \underline{E} \cdot \underline{ds} . \quad (A3)$$

The surface integral over ϕ includes surfaces where ϕ may be multivalued when $\dot{\psi}_{ex} \neq 0$. If the volume is filled with conducting plasma, then $\underline{E} \cdot \underline{B} = 0$ (except in the neighborhood of reconnection regions, which presumably constitute a small portion of the volume at large magnetic Reynold's numbers) and the first term can be neglected. If no field lines penetrate surfaces, $\underline{B} \cdot \underline{ds} = 0$, and if $\dot{\psi}_{ex} = 0$, then the second term vanishes. If no insulators are present, the condition $\underline{E} \times \hat{n} = 0$ at conducting surfaces causes the third term to vanish and therefore $\dot{K} = 0$, the usual form of the helicity is preserved. In the more general case, including insulators and allowing for $\underline{B} \cdot \underline{ds} \neq 0$, $\dot{\psi}_{ex} \neq 0$, consider the condition $\underline{E} \times \hat{n} = 0$ on conductors:

$$\underline{E} \times \hat{n} = -(\underline{\dot{A}} + \nabla\phi) \times \hat{n} = 0 . \quad (A4)$$

Equation A4 implies that on the conducting surface

$$\phi(t,\underline{x}) = \Phi(t,\underline{x}_0) - \int_{\underline{x}_0}^{\underline{x}} \underline{\dot{A}} \cdot \underline{dl} \quad (A5)$$

where $\underline{x}_0$ is an arbitrary point on the conducting surface. The function Φ can be different on each conducting surface separated by an insulator. Using Eq. (A5) in Eq. (A3) (neglecting the $\underline{E} \cdot \underline{B}$ term again by assuming the volume is filled with conducting plasma) we find

$$\dot{K} = \iint_{conductor} \int_{\underline{x}_0}^{\underline{x}} \underline{\dot{A}} \cdot \underline{dl} \, \underline{B} \cdot \underline{ds} - \sum_i \Phi_i(t,\underline{x}_0) \int_{conductor} \underline{B} \cdot \underline{ds}$$

$$- \int_{\substack{multivalue \\ surfaces}} \phi \, \underline{B} \cdot \underline{ds} + \int_{surface} \underline{A} \times \underline{E} \cdot \underline{ds} \quad (A6)$$

where the sum on i corresponds to summing over the number of distinct conducting surfaces separated by insulating gaps.

Since the elements of flux intercepting the conductors, $\underline{B} \cdot \underline{ds}$, are constant in time, the first term in Eq. (A6) can be brought over to the left-hand side:

$$\dot{K}' = -\sum_i \Phi_i \int_{conductor} \underline{B} \cdot \underline{ds} - \int_{\substack{multivalue \\ surface}} \phi \, \underline{B} \cdot \underline{ds} + \int_{surface} \underline{A} \times \underline{E} \cdot \underline{ds} \quad (A7)$$

where

$$K' = \int \underline{A} \cdot \underline{B} \, dV - \int_{conductor} \left(\int_{\underline{x}_0}^{\underline{x}} \underline{A} \cdot \underline{dl}\right) \underline{B} \cdot \underline{ds} \quad (A8)$$

Note that K' is independent of the choice of $\underline{x}_0$:

$$K'(\underline{x}_1) = K'(\underline{x}_0) + \int_{\underline{x}_0}^{\underline{x}_1} \underline{A} \cdot \underline{dl} \int_{conductor} \underline{B} \cdot \underline{ds} , \quad (A9)$$

However, since the volume is totally enclosed, $\int \underline{B} \cdot \underline{ds} = 0$ for the entire surface and $K'(\underline{x}_1) = K'(\underline{x}_0)$. K' is also gauge invariant:

$$K'(\underline{A} + \nabla\phi) = \int \underline{A} \cdot \underline{B} \, dV + \int \nabla\phi \cdot \underline{B} \, dV \quad (A10)$$

$$- \iint \int^{\underline{x}} \underline{A} \cdot \underline{dl} \, \underline{B} \cdot \underline{ds} - \int \phi \, \underline{B} \cdot \underline{ds}$$

$$= K'(\underline{A}) + \int_{surface} \phi \, \underline{B} \cdot \underline{ds} - \int_{surface} \phi \, \underline{B} \cdot \underline{ds}$$

$$= K'(\underline{A})$$

where the surface integrals now include any surfaces where ϕ may be multivalued.

Next consider the right-hand side of Eq. (A7). The term $\underline{A} \times \underline{E}$ gives a contribution only in the insulating gaps. For simplicity, consider for the moment only a single gap:

$$\int \underline{A} \times \underline{E} \cdot d\underline{s} = \oint \underline{A} \cdot \underline{dl} \left(-\int \underline{E} \cdot d\underline{x}\right) \quad (A11)$$
$$\equiv V\psi$$

where V is the voltage across the gap, and ψ is the flux inside the plasma volume that is linked by the gap. The first term in Eq. (A7) gives a contribution when the gap divides the boundary into distinct conducting regions,

$$\sum \Phi_i \int \underline{B} \cdot d\underline{s} = \Phi_1 \psi_1 + \Phi_2 \psi_2 \quad (A12)$$

where $\psi_{1,2}$ are the fluxes through the two distinct conducting regions. Flux conservation gives $\psi_2 = -\psi_1 \equiv \psi$, so the term becomes

$$\sum \Phi_i \int \underline{B} \cdot d\underline{s} = -\psi(\Phi_1 - \Phi_2) \quad (A13)$$

From Eq. (A5) $\Phi_1 - \Phi_2 = \phi_1 - \phi_2 = -\int_{gap} \underline{E} \cdot d\underline{x} = V$ so

$$-\sum \Phi_i \int \underline{B} \cdot d\underline{s} = \psi V . \quad (A14)$$

When the gap does not divide the boundary into two distinct conducting regions (e.g., a poloidal gap in a torus) then the first term is zero. The second term $-\int \phi \underline{B} \cdot d\underline{s} = -\psi_{ex}\psi = V\psi$ in this case so the result is the same for either type of gap. Adding the terms on the right-hand side gives

$$K' = 2\psi V . \quad (A15)$$

For many gaps, the right-hand side of (A15) becomes simply $2 \sum_i \psi_i V_i$ = sum over all gaps of the flux linking the gap times the voltage drop across the gap. In the case ψ_i = constant we then have

$$K'(t) - K'(0) = 2 \sum \psi_i \int V_i \, dt . \quad (A16)$$

The integral $\int V \, dt$ is the amount of flux injected through the gap which links the flux ψ, analogous to the simple case of two linked flux tubes where the helicity is twice the product of the linked fluxes. The generalized helicity K' given in Eq. (A8) is therefore seen to be the correct extension to the case when flux penetrates conducting walls, and to arbitrary gauges, as has been suggested by J. B. Taylor.

The extra term, $\int (\int^x \underline{A} \cdot \underline{dl}) \underline{B} \cdot \underline{ds}$, appearing in Eq. (A8) has a simple interpretation in the cases where $\oint \underline{A} \cdot \underline{dl}$ = flux enclosed within the plasma volume. The integral could be rewritten as a sum over flux bundles penetrating the wall, $d\psi = \underline{B} \cdot d\underline{s}$

$$\int \left(\int^x \underline{A} \cdot \underline{dl}\right) \underline{B} \cdot d\underline{s} = \int d\psi \int_\alpha^\beta \underline{A} \cdot \underline{dl} \quad (A17)$$

where $\underline{\alpha}$, $\underline{\beta}$ are the locations where the flux bundle $d\psi$ enters and exits the surface. Now consider a generalized system that includes the space exterior to the plasma-containing volume. The flux bundle, $d\psi$, follows a path through the exterior space from $\underline{\beta}$ to $\underline{\alpha}$, and we can write

$$\underbrace{\int_\alpha^\beta \underline{A} \cdot \underline{dl}}_{\text{surface}} + \underbrace{\int_\beta^\alpha \underline{A} \cdot \underline{dl}}_{\text{exterior}} = \oint_{\text{exterior}} \underline{A} \cdot \underline{dl} . \quad (A18)$$

The right-hand side of Eq. (A18) vanishes by the gauge constraint so we can write

$$K' = \int_{plasma} \underline{A} \cdot \underline{B} \, dV - \int d\psi \int_\alpha^\beta \underline{A} \cdot \underline{dl} \quad (A19)$$

$$= \int_{plasma} \underline{A} \cdot \underline{B} \, dV + \int d\psi \underbrace{\int_\beta^\alpha \underline{A} \cdot \underline{dl}}_{\text{exterior}}$$

$$= \int \underline{A} \cdot \underline{B} \, dV \bigg|_{\text{extended volume}} .$$

Calculating the generalized helicity, K', is therefore equivalent to extending the volume of integration for the usual definition of K to include the closure of all flux loops penetrating the conducting boundaries (note, however, that the gauge constraint refers to flux within the plasma volume only, so that extraneous flux linkages in the extended volume are not included in K'). This is often a useful way to calculate K' in complex geometries. Since K' is a topological property of the fields, the exterior region can be idealized as long as the linkages with the fields in the plasma volume are preserved.

This work was performed under the auspices of the U. S. Department of Energy by the Lawrence Livermore National Laboratory under contract number W-7405-ENG-48.

Acknowledgments. I would like to acknowledge helpful discussions with C. W. Hartman, T. R. Jarboe, J. B. Taylor, and W. C. Turner. The data presented in this paper were provided by G. C. Goldenbaum, M. Yamada, A. Janos, W. C. Turner, A. R. Sherwood, T. R. Jarboe, and C. W. Barnes. This work was performed under the auspices of the U.S. Department of Energy by the Lawrence Livermore National Laboratory under contract number W-7405-ENG-48.

References

Alfvén, H., L. Lindberg, and P. Mitlid, J. Nucl. Energy Part C, 116 (1960).
Cowling, T. G., Mon. Notic. Roy. Astron. Soc. 94, 39 (1934).
Goldenbaum, G. C., J. H. Irby, Y. P. Chong, and

G. W. Hart, Phys. Rev. Lett. 44, 393 (1980).
Jarboe, T. R., et al., Phys. Rev. Lett. 45, 1264 (1980).
Jarboe, T. R., I. Henins, A. R. Sherwood, Chris W. Barnes, and H. W. Hoida, Phys. Rev. Lett. 51, 39 (1983).
Nolting, Jindra, and Wells, J. Plasma Physics 9, 1 (1973).
Rosenbluth, M. N. and M. N. Bussac, Nucl. Fus. 19, 489 (1979).
Sato, T. and C. Hayashi, Phys. Rev. Lett. 50, 38 (1983).
Taylor, J. B., Phys. Rev. Lett. 33, 1139 (1974).
Turner, W. C., et al., J. Appl. Phys. 52, 175 (1981).
Turner, W. C., et al., Phys. Fluids 26, 1965 (1983).

Wells, D. R., Phys. Fluids 5, 1016 (1962).
Yamada, M., et al., Phys. Rev. Lett. 46, 188 (1981).
Yamada, M., et al., Plasma Physics and Controlled Nuclear Fusion Research, Vol. II, 265, Baltimore, Maryland (1982).

Questions and Answers

Vasyliunas: From the definition of the helicity K it is evident that K is independent of the choice of gauge for $\underline{A}$ if and only if the magnetic field is entirely contained within the volume over which K is defined, with zero normal component of $\underline{B}$ at the bounding surface. Does the concept have any applicability to an open magnetic configuration, where field lines may cross the boundary?

Hammer: The helicity is a meaningful concept in the case with field penetrating surfaces if care is used in choosing the gauge. For instance the Coulomb gauge, $\bar{A} = -\int_{-\infty}^{t} \bar{E} dt$, works if it can be evaluated. Any definition that preserves the topological property of the helicity can be used.

THE ROLE OF MAGNETIC RECONNECTION PHENOMENA IN THE REVERSED-FIELD PINCH

D. A. Baker

Los Alamos National Laboratory
Los Alamos, New Mexico 87545

Abstract. The reversed-field pinch (RFP), an axisymmetric toroidal magnetic confinement experiment, has physics rich in the area commonly called field line reconnection or merging. This paper reviews the topics where reconnection plays a vital role: (a) RFP formation and the phenomenon of self-reversal, (b) RFP sustainment in which the RFP configuration has been shown to be capable of maintaining itself for times much longer than earlier predictions from classical resistive MHD theory, (c) steady state current drive in which "dynamo action" and associated reconnection processes give rise to the possibility of sustaining the configuration indefinitely by means of low frequency ac modulation of the toroidal and poloidal magnetic fields, (d) the effects of reconnection on the formation and evolution of the magnetic surfaces which are related to the plasma containment properties. It appears that all phases of the RFP operation are intimately related to the reconnection and field regeneration processes similar to those encountered in space and astrophysics.

1. Introduction

The reversed field pinch (RFP) stems from the earliest Z-pinches [Cousins and Ware, 1951] following the studies of Bennett, [1934]. In the 50's much research was done on the Z-pinch, particularly at laboratories in Los Alamos and Livermore in the U.S., Harwell and Aldermaston in the U.K., and Fonteray-aux-Roses in France. A good reference list for this early work is available in [Glasstone and Loveberg, 1960]. Later history and physics discussions of the toroidal pinch concept are outlined in [Baker and DiMarco, 1975], [Bodin and Keen, 1977], [Bodin and Newton, 1980] and [Baker and Quinn, 1981]. The basic pinch idea is to use the self-magnetic fields of a current plasma column to contain the plasma. The simple Z-pinch concept was plagued with problems of MHD instability and the modern RFP is the result of experimental and theoretical studies which have led to grossly stable pinch configurations.

The objectives of the paper will be to discuss specific topics in which the space physics related concepts of field line reconnection or merging, flux annihilation or generation, and dynamo action are relevant in the RFP physics. These topics include RFP formation and self-reversal, RFP sustainment, steady state current drive, and effects on field line topology which affect plasma containment.

2. The Reversed-Field Pinch Configuration

The modern day reversed-field pinch used for fusion-oriented plasma containment studies is typified by the Los Alamos ZT-40M experiment [Baker, et al., 1983] shown in Figure 1. Currently there are four additional RFP experiments in which extensive results are available: OHTE in the U.S. [Tamano, et al., 1983], HBTX-1A in the U.K. [Bodin, et al., 1983], ETA BETA II in Italy [Antoni, et al., 1983], and TPE-1R(M) in Japan [Ogawa, et al., 1983]. There are several other RFP experiments which are recently operating or are scheduled to begin operation in 1984: the REVERSATRON at the University of Colorado; ZT-P in Los Alamos; REPUTE-I, Tokyo; STP-3, Nagoya; HIT-1, Hiroshima; STE-RFP, Kyoto; and a small experiment at the Tokyo Institute of Technology.

The modern RFP utilizes a thin, toroidal metal vacuum liner nested inside a thick highly conducting shell. The liner is resistive and the shell has gaps in the poloidal (short way around the torus) and toroidal (long way around the torus) directions to allow the pulsed fields to enter the plasma region inside the liner. In experiments using very slowly applied toroidal fields, the gap in the shell which extends in the toroidal direction can be omitted. Outside the shell are poloidal and toroidal windings which produce the necessary magnetic fields to (1) induce the toroidal electrical field to drive the toroidal current that produces the poloidal confining field B_θ, (2) provide an initial toroidal stabilizing field B_ϕ, and (3) provide equilibrium controlling fields as needed during startup and sustainment. In the present experiments the windings are energized from capacitor banks. The reversed field configuration is a combination of toroidal and poloidal fields producing a set of nested surfaces whose field lines possess a high degree of shear (radial variation of magnetic field line pitch) as the fields from different magnetic flux surfaces are compared (Figure 2.). The RFP differs from the tokamak by having the highly sheared magnetic field whose toroidal field reverses in the outer region of the discharge and by having a thick conducting shell whose purpose is to provide global MHD stability, whereas the tokamak relies on a very strong unidirectional toroidal magnetic field. A general schematic comparison of the RFP and tokamak field components along the midplane of the torus is given in Figure 3. Magnetic cores are often used to increase the transformer coupling from the pulsed poloidal field coils (multi-turn transformer primary) and the plasma (single turn secondary).

3. Reconnection During RFP Formation

3.1 Formation by Reconnection of Large Amplitude Helical Disturbances

The earliest experiments on fast-current-rise toroidal pinches were formed inside a toroidal flux conserving shell and demonstrated the ability of a toroidal current-carrying plasma, having an initially unidirectional toroidal stabilizing field, to re-arrange itself into a pinched discharge whose toroidal field is increased internally and reversed in the outer regions of the pinch. See, for example, [Colgate, et al., 1958]. This self-reversal process, observed in the early fast pinches, (microsecond time scales) was explained qualitatively as follows. An axisymmetric toroidal pinch first forms as shown in Figure 4a. The highly pinched plasma column is MHD unstable to a helical perturbation and kinks into a large amplitude helix as shown in Figure 4b. This helical, multi-turn, solenoid-like current generates an increased component of toroidal field, thus increasing the toroidal flux in the interior region. Since the conducting boundary conserves the total toroidal flux on the time scales of these experiments, eddy currents are induced in the wall which produce a negative toroidal flux in the outer portion of the discharge thus maintaining a constant total toroidal flux inside the conducting shell. At this point the pinches tend to return toward axisymmetry. A closer look at the process is shown in a sector of the torus in Figure 5. The field of the kinked plasma in Figure 5b has helical flux surfaces and a helical x-point locus (separator). Thus the plasma can return towards axisymmetry by field line

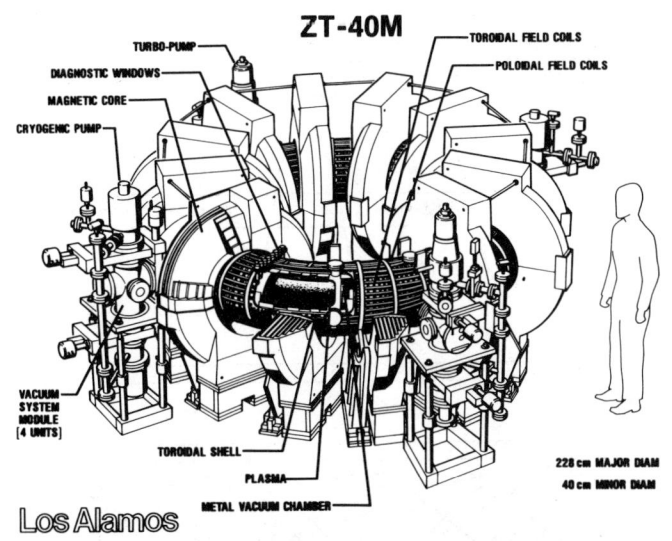

Fig. 1. Drawing of the Los Alamos ZT-40M reversed field pinch experiment.

reconnection at the x-point and by a diffusion and smoothing of the plasma column. The ideal smoothed out configuration of Figure 5c now has a reversed field where the toroidal field was initially all in the same (forward) direction (Figure 5a). It is clear that field-line reconnection is a necessary ingredient of forming an axisymmetric pinch by self-reversal since rapid changes in field line topology and changes in the flux inside and outside separatrix surfaces are involved.

The simplified qualitative picture just described obviously involves plasma dynamics, reconnection and diffusion processes. The actual quantitative theoretical description of self-reversal is currently under intense study using powerful two and three dimensional (3-D) MHD computer codes. The earliest RFP results from 3-D numerical computations were reported by Sykes and Wesson, [1977]. They verified the growth of a helical unstable mode which produced the field reversal. As the calculation proceeded, a second helical mode appeared with half the wavelength of the first one. This was followed by resistive tearing, reconnection and a return to near axial symmetry. These first calculations used a straight linear pinch simulated in a rectangular cross-section conducting box. The calculation did verify the early concepts that self-reversal can be produced by the growth of a helical disturbance and a return to symmetry by reconnection processes. This work has been recently extended with more powerful computer codes [Schnack, et al., 1983; Aydemir and Barnes, 1983a, 1983b; Holmes, et al., 1983]. Experimental evidence for field reversal by helical modes is summarized by Bodin and Newton [1980].

3.2 The Taylor Relaxation Model

A model which predicts that a current-carrying plasma column located inside a cylindrical (coordinates, r,θ,z) flux conserving shell, with a suitably high value

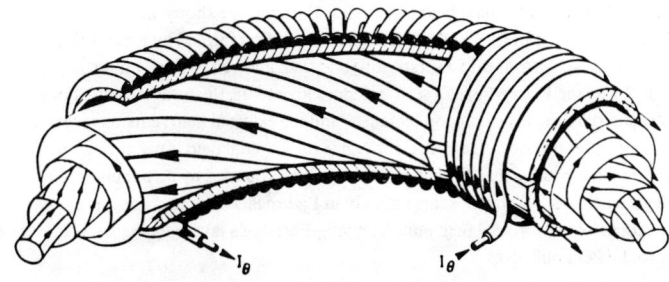

Fig. 2. Schematic showing the toroidal field winding, the slotted conducting shell, and the highly-sheared field lines lying on nested flux surfaces of a reversed field pinch.

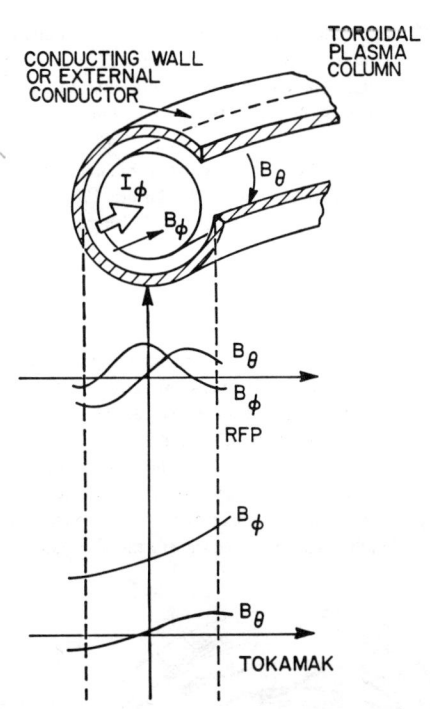

Fig. 3. Comparison of tokamak and the reversed field pinch fields. The poloidal B_θ and the toroidal B_ϕ field components along the midplane of the torus are shown.

of the ratio of toroidal current to toroidal flux, will relax to a reversed-field configuration has been proposed [Taylor, 1974, 1975, 1976]. The basic premise in this description is that a given configuration of magnetic field having an initial value of magnetic helicity $K = \int_{vol} \mathbf{A} \cdot \mathbf{B} dV$ (where $\nabla \times \mathbf{A} = \mathbf{B}$) and longitudinal magnetic flux $\phi = \int_A B_z dA$ will evolve by magnetic reconnection processes to a state of lowest magnetic field energy $W_m \propto \int_{vol} B^2 dV$ keeping the value of K and ϕ constant. The use of the magnetic helicity as an invariant was used earlier in the astrophysics literature in related arguments leading to the prediction of force-free field configurations [Woltjer, 1958, 1959, 1960]. The magnetic helicity is an exact invariant for any closed flux tube in a perfectly conducting fluid. The Taylor hypothesis is that the localized reconnection processes due to a small amount of dissipation will change the field line topology but will leave the total global magnetic helicity in the plasma volume conserved on the time scale of the relaxation, while the magnetic energy is not. This hypothesis leads to a force-free configuration satisfying $\nabla \times \mathbf{B} = \mu \mathbf{B}$ where μ is a constant. In terms of easily measured dimensionless parameters $\Theta = B_{\theta wall}/B_{zave}$ and $F = B_{zwall}/B_{zave}$ the model predicts cylindrically symmetric Bessel function solutions $B_z = B_0 J_0(2\Theta r/a)$ and $B_\theta = B_0 J_1(2\Theta r/a)$ for $\Theta < 1.55$ and a helically symmetric state for $\mu a \gtrsim 3.1$ (a is the shell radius). For $\Theta \gtrsim 1.2$ the field is reversed. The lowest energy states lie on a locus in F–Θ space as shown in Figure 6. In practice, the Taylor model is a qualitative guide to the RFP self-reversal behavior. Experimentally the F–Θ curve normally lies above that predicted as shown in Figure 7. The actual experiments deviate from the idealized Taylor model in several aspects as shown in Table 1. In spite of these differences the model has been a very useful guide for the states resulting from the reconnection relaxation processes.

The Taylor analysis has generated much interest and many authors have studied variations and modifications of the original analysis. A correction to the original work has been published [Reiman, 1980, 1981]. Extensions of the model from cylindrical to toroidal geometry have been made [Miller and Turner 1981; Faber, White, and Wing, 1982a, 1982b; Edenstrasser, 1983a]. Arguments relating to why the magnetic helicity should decay slowly compared to the magnetic energy have been advanced [Montgomery, Turner, and Vahala, (1978)]. An extensive statistical mechanical study using incompressable MHD and the K and ϕ invariants predict a state having fluctuations about the Taylor state [Turner, 1983a, 1983b]. The Taylor work has spawned many other papers which discuss minimum energy states obtained using differing constraints or geometries, some of which allow a non-zero plasma pressure and/or fluid flow: [Rosenbluth and Bussac, 1979; Sudan, 1979; Finn, Manheimer, and Ott, 1981; Marklin and Bondeson, 1980; Bondeson, et al., 1981; Erlebacher, 1981; Bhattacharjee, et al., 1980, 1982; Brandenburg, 1982; Edenstrasser and Schuurman, 1983b; Finn and Antonson, 1983; Turner, 1983c]. A related analyses which maximizes entropy instead of minimizing energy has also appeared [Hameiri and Hammer, 1982].

3.3 Toroidal Flux Generation

The generation of positive and negative toroidal flux is involved in the explanation of self-reversal phenomenon in short current risetime pinches (Sec. 3.1). A corresponding generation of flux has also been demonstrated clearly for a very slowly rising current (~14 ms risetime) in ZT-40M where, unlike the earlier experiments, the high temperature (~0.3 keV) and long duration current pulse (~15 ms) have precluded internal magnetic probe measurements. The measurement is made by means of a toroidal flux pickup loop surrounding the vacuum liner, as shown in Figure 8. The measured waveforms of toroidal current, the

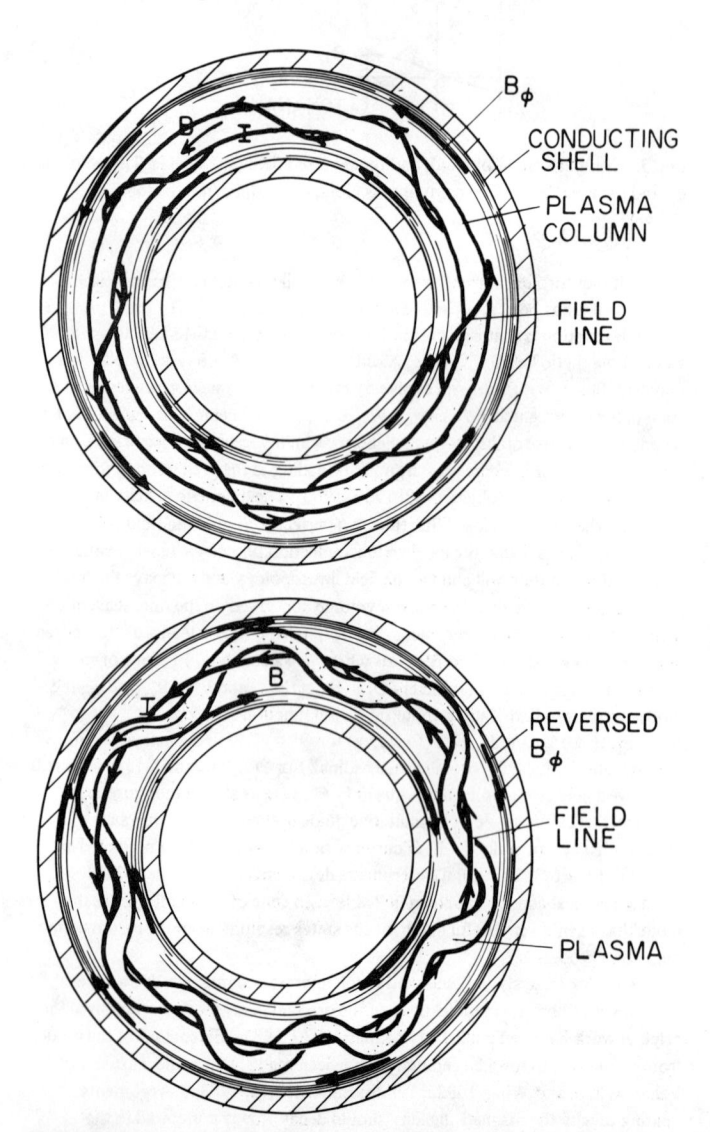

Fig. 4. Top view of a toroidal pinch inside a flux conserving shell. (a) pinched plasma and a sample field line; (b) plasma column kinked into a helix, strengthening B near the minor axis and reversing B outside.

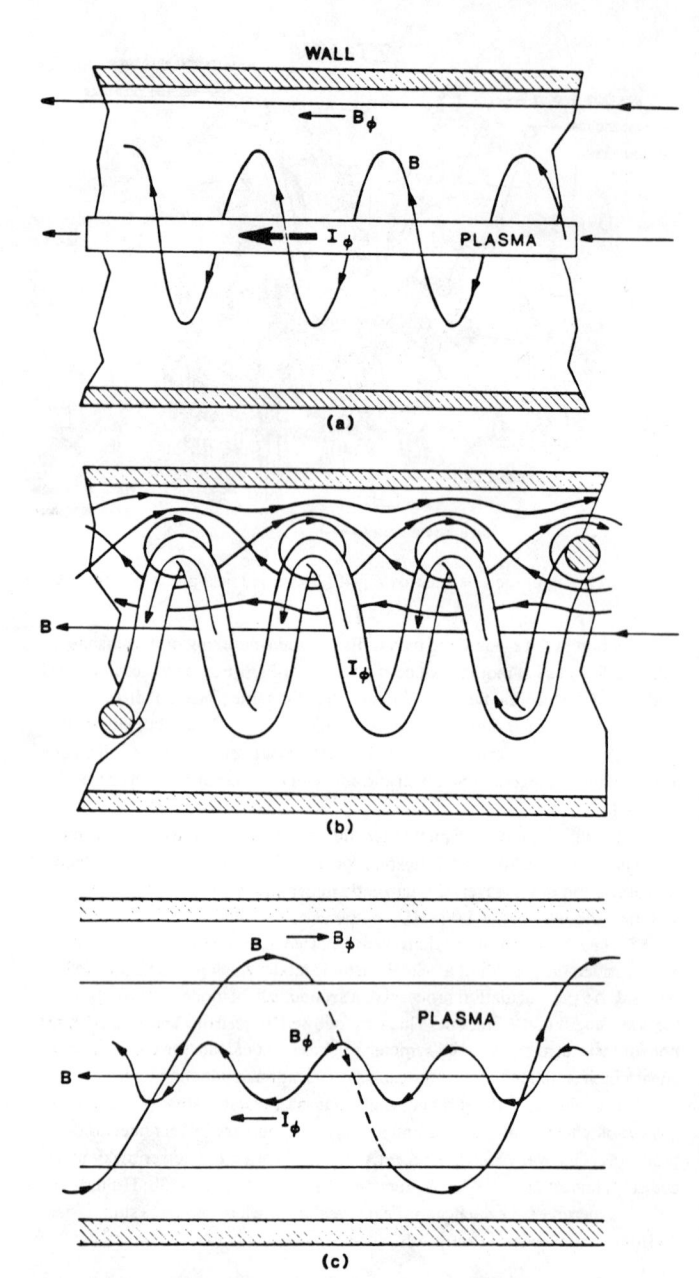

Fig. 5. Demonstration of self-reversal of the magnetic field by plasma kinking and field merging. (a) pinched plasma; (b) pinch after kinking into a helix; separatrices and x-points have formed; (c) plasma column with reversed field configuration after field line reconnection and return to symmetry.

toroidal flux, and toroidal field just outside the liner, are shown in Figure 9 [Phillips, et al., 1983]. The RFP configuration is first formed by a rapid rise in the current to ~70 kA in 0.75 ms (startup) and then allowed to slowly increase to ~170 kA (see Figure 9a). As seen in Figure 9b and 9c, the total toroidal flux as measured by the external loop increases and the external field remains reversed during the entire slow rise of current. Since the toroidal field is negative on the outside, the postive flux ϕ^+ in the discharge is surrounded by an annular region of negative flux ϕ^- as shown schematically in Figure 8. The dotted line represents the locus of the toroidal field null. Applying Faraday's law to the region inside the toroidal field null gives

$$\dot{\phi}^+ = -\oint_{null} \mathbf{E} \cdot d\boldsymbol{\ell} \, . \qquad (1)$$

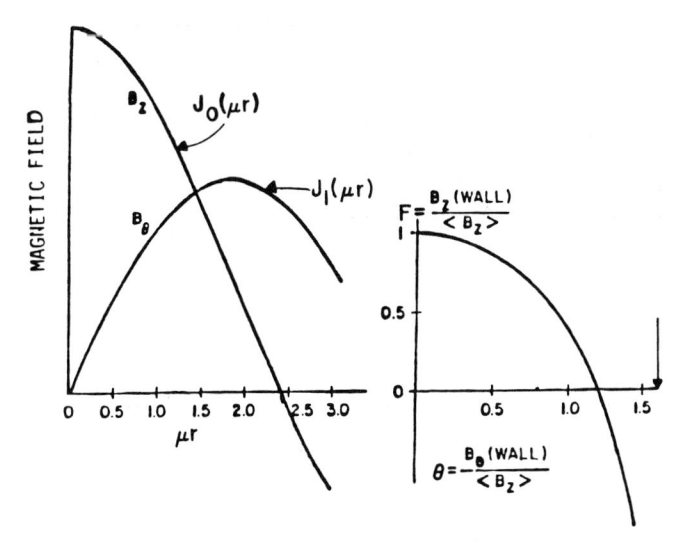

Fig. 6. Left: Bessel function field profiles for the Taylor lowest energy state. Right: F–Θ diagram showing field reversal for Θ ≳ 1.2 and the threshold (arrow) for the formation of helical lowest energy states at Θ ≃ 1.55.

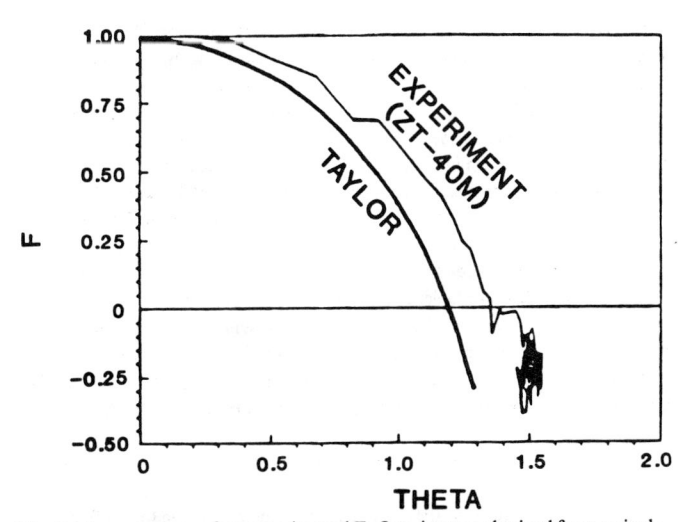

Fig. 7. A comparison of an experimental F–Θ trajectory obtained from a single discharge in the ZT-40M experiment with the Taylor prediction. The trajectory starts at F = 1 and moves to higher Θ and lower F values as the discharge current increases and the reversed field state is formed.

Similarly, for the negative flux annulus

$$\dot{\phi}^- = \oint_{null} \mathbf{E}\cdot d\boldsymbol{\ell} - \oint_{liner} \mathbf{E}\cdot d\boldsymbol{\ell} . \quad (2)$$

Adding the two equations we obtain the total $\dot{\phi}$ sensed by the flux loop

$$\text{loop voltage} = \dot{\phi}^+ + \dot{\phi}^- = -\oint_{liner} \mathbf{E}\cdot d\boldsymbol{\ell} . \quad (3)$$

The first terms in the right hand sides of Eqs. (1) and (2) represent the equal annihilation or generation rates of positive and negative toroidal flux at the toroidal field null and cancel when the two equations are added. Annihilation and generation correspond to the mean poloidal E at the reversal point being positive or negative, respectively. The remaining term [rhs of Eq. (3)] represents the rate of entering or leaving of negative flux at the liner boundary. Since the field is always reversed at the boundary for the slow rise of current in Figure 9, the net flux can only increase by the loss of negative flux at the boundary, volume, i.e., $\oint_{liner} \mathbf{E}\cdot d\boldsymbol{\ell} < 0$. The fact that the flux loop measures an increase in the total toroidal flux does not guarantee that positive flux is generated i.e. that $\Delta \phi^+ > 0$, since negative flux can be removed from the system through the liner. (For example the toroidal field null can move toward the liner while the values of reversed field and ϕ^+ remain the same). To demonstrate the generation of positive flux (and the equal amount of negative flux) during the slow rise of current, in the absence of internal field measurements, one must show that more negative flux is removed at the boundary than was initially present just after the RFP was formed during the startup i.e. the total measured increase of toroidal flux must exceed $|\phi^-_{initial}|$. This demonstration follows from a theorem [Caramana and Moses, 1983a] which puts an upper limit on the ratio of negative to total flux required for equilibrium.

$$\frac{|\phi^-|}{\phi_{total}} \leq \frac{1}{2}\left[(F^2 + \Theta)^{1/2} - 1\right] , \quad (4)$$

where F and Θ are as defined in the last Section. When this upper limit is calculated for the discharge conditions just after startup F = –0.2 and Θ = 1.6 for Figure 9 it is found that the observed 150% increase of net flux is five times the limit on $|\phi^-_{initial}|$ imposed by Eq. (4) [Caramana and Baker, 1983b]. Therefore a large amount of positive and negative flux is generated during the slowrise of current. This demonstrates that flux is generated not only during the rapid startup of an RFP (Sec 3.1) but also during a very slow rise (multi-milliseconds) of current. This interesting flux generation effect, which is the opposite of the well known field annihilation due to resistive dissipation, has been called "the dynamo effect" in analogy to the field generation by solar and terrestrial dynamos [Moffat, 1978].

4. Sustainment of the RFP Configuration

4.1 Predictions of a Simple Symmetric Ohm's Law Model

In the early history of the reversed field pinch, it was generally believed that the RFP configuration once produced would, of necessity, decay by resistive diffusion (see for example [Robinson, et al., 1972]). Indeed if a cylindrically symmetric pinch were formed and a simple Ohm's law were valid

$$\mathbf{E} = \overleftrightarrow{\eta}\cdot\mathbf{J} - \mathbf{v}\times\mathbf{B} , \quad (5)$$

($\overleftrightarrow{\eta}$ is the resistivity tensor, **J** the current density, and **v** the plasma velocity) then this conclusion would be correct. This follows from Eq. (1) when one evaluates the right hand side. For $d\boldsymbol{\ell}$ along **B**, which is totally poloidal at the toroidal field null, one may write

$$\oint_{null} \mathbf{E}\cdot d\boldsymbol{\ell} = \oint [\overleftrightarrow{\eta}\cdot\mathbf{J} - \mathbf{v}\times\mathbf{B}]_{\parallel}\cdot d\boldsymbol{\ell} = \oint \eta_{\parallel} J_\theta d\ell > 0 , \quad (6)$$

where ∥ denotes a component parallel to **B**. The last inequality follows from the fact that $\eta_{\parallel} > 0$ for a collisional resistivity, $\mu_0 J_\theta = -dB_z/dr$ (from the Maxwell equation $\nabla\times\mathbf{B} = \mu_0\mathbf{J}$) and $dB_z/dr < 0$ where $B_z = 0$ (the field reversal radius). Thus $\eta_{\parallel}J_\theta$ is positive and from Eq. (1) $\dot{\phi}^+ < 0$ and the positive flux must decay.

TABLE 1. Deviations From the Taylor Model in Real Experiments

Taylor	Experiment
Plasma surrounded by perfect conductor	Double wall, closest one resistive, gaps in outer shell, field windings.
Passive relaxation.	Driven system (toroidal E field)
Zero plasma pressure.	$\beta_p = 2\mu_0\bar{P}/B^2_{\theta wall} \sim 10\%$
High conductivity plasma throughout the volume	Cold plasma next to wall, gas released from the wall.

The interesting feature of recent experiments is that constant current RFP discharges have been obtained for lifetimes much exceeding the predictions of classical theory and Ohm's law. It now appears, from this strong experimental evidence, that the RFP configuration can be sustained as long as the current and plasma density are maintained. The flat-topped current of the present ZT-40M experiment is maintained in present experiments by a toroidal electric field produced by transformer action. The plasma density can be replenished when needed by gas injection. A sample flat-topped current discharge in ZT-40M sustained 20 ms is shown in Figure 10. Classical local ohmic calculations for cylindrical symmetry predict that, for the conditions of this discharge, the positive toroidal flux would decay in a few milliseconds [Caramana and Baker, 1983b]. One is thus led to the conclusion that the discharge cannot be described by a symmetric local Ohm's law model. In general, this implies, for the sustained and slowly rising current RFP, one or both of the following: (1) the local Ohm's law description is not valid, (2) the pinch is not symmetric.

4.2 The Mean Field Theory for a Turbulent Dynamo

A popular explanation of field generation and sustainment in the RFP was first advanced by Gimlett and Watkins, [1975] borrowing from the mean-field MHD used to describe dynamo theories for the earth, sun and other conducting, rotating bodies in astrophysics. (See, for example, [Krause and Rädler, 1980].) Just as a local Ohm's law, symmetric RFP cannot exist in steady state, neither can the fields of a rotating conducting object retain steady axisymmetric field in the presence of an Ohm's law description. This conclusion is implied by a theorem due to Cowling [1934]. The essence of mean-field approach is to assume that there are mean $<\ >$ and fluctuating δ components for $\mathbf{B} = <\mathbf{B}> + \delta\mathbf{B}$ and $\mathbf{v} = <\mathbf{v}> + \delta\mathbf{v}$, where $<\delta\mathbf{B}> = <\delta\mathbf{v}> = 0$. When these are substituted into Ohm's law Eq. (5) and the resulting equation averaged, one obtains a modified Ohm's law for the mean electric field

$$<\mathbf{E}> = <\overset{\leftrightarrow}{\eta}\cdot\mathbf{J}> - <\mathbf{v}>\times<\mathbf{B}> - <\delta\mathbf{v}\times\delta\mathbf{B}> \quad . \quad (7)$$

Taking components parallel to $<\mathbf{B}>$ along the mean toroidal field null and, for simplicity, neglecting the fluctuations in resistivity gives

$$<\mathbf{E}>_\parallel = \eta_\parallel<\mathbf{J}_\parallel> - <\delta\mathbf{v}\times\delta\mathbf{B}>_\parallel . \quad (8)$$

The new term arising from the fluctuations in Eq. (8) need not vanish even though the mean values of the fluctuating components of plasma velocity and magnetic field $\delta\mathbf{v}$, $\delta\mathbf{B}$ are zero. For the simplest case of isotropic turbulence, the $<\delta\mathbf{v}\times\delta\mathbf{B}>$ term gives a contribution $\alpha<\mathbf{B}>$ to the mean electric field, and is called the "alpha

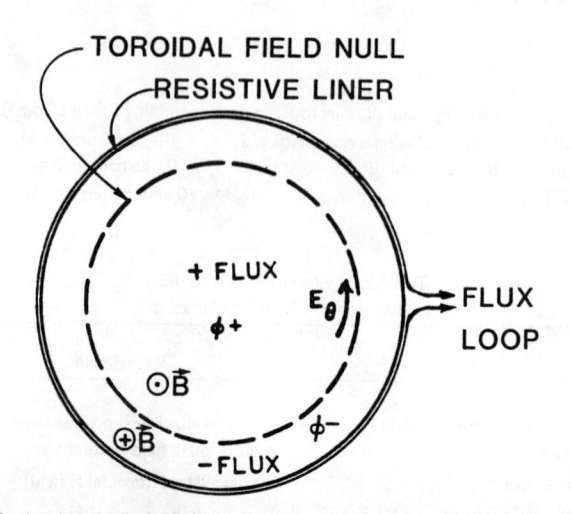

Fig. 8. A schematic cross section of the ZT-40M discharge showing the positive and negative flux regions and the location of the flux loop that measures the net toroidal flux inside the resistive vacuum liner.

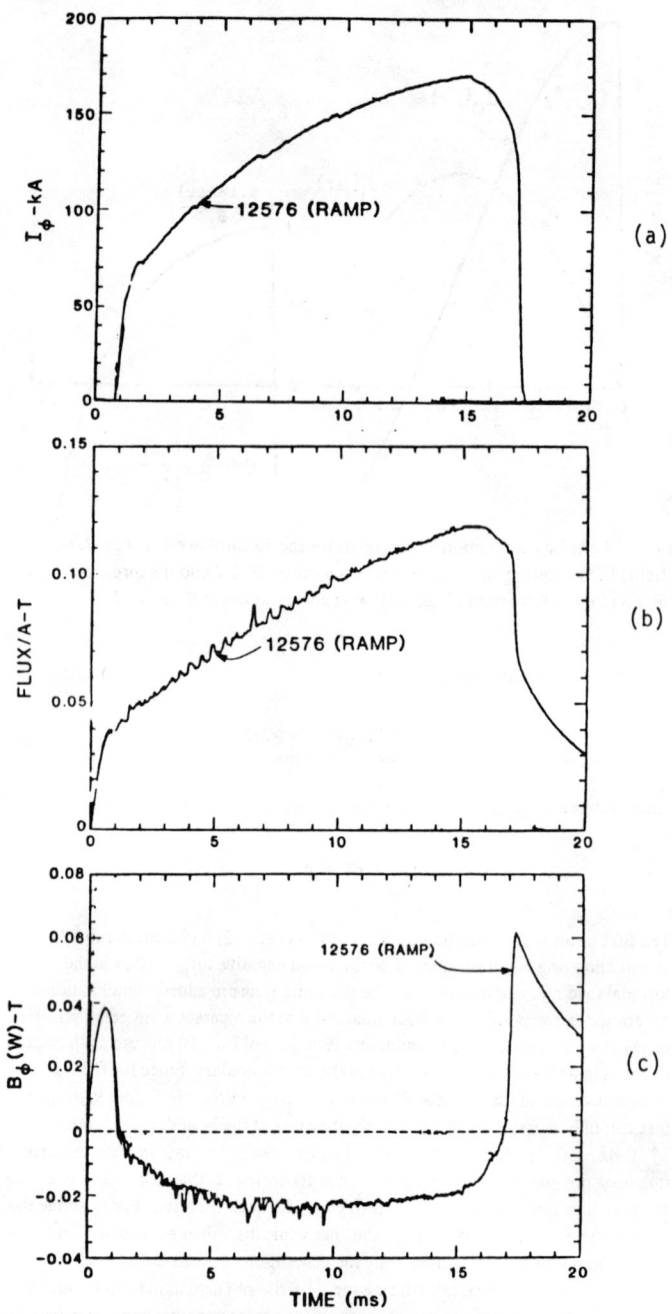

Fig. 9. Waveforms for a slowly-rising current in ZT-40M. (a) toroidal current; (b) toroidal flux; (c) toroidal field just outside the vacuum liner.

effect". Studies and extensions of the alpha effect concept in the RFP context have been made [Schaffer, 1982; Gerwin and Keinigs, 1982; Keinigs, 1983]. If the correlations and amplitudes of $\delta\mathbf{v}$ and $\delta\mathbf{B}$ are suitable, the two terms on the right hand side of Eq. (8) may cancel or even have a sum with the opposite sign from that of a conventional collisional Ohm's law. Since Faraday's law is linear, the mean values of flux are related to the mean electric field by Faraday's equation with no new terms. One then has a possible way around the decay implied by Eqs. (1) and (6) since the mean poloidal E can now be zero or negative and the mean flux can remain constant or grow even though there is dissipation present. There does, of course, have to be energy supplied to the system to retain a steady state against losses. This model is not by itself complete in that it *postulates* the existence of suitable fluctuations in the magnetic field and plasma velocity. It can be called the kinematic dynamo since a self-consistent plasma

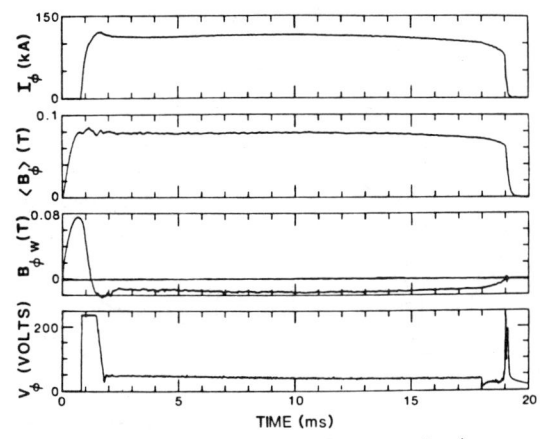

Fig. 10. Toroidal current I_ϕ, average toroidal field $\langle B_\phi \rangle$ (flux/cross-section area), external toroidal magnetic field $B_{\phi w}$ and toroidal voltage V for a long lived ZT-40M discharge.

dynamics following Newton and Maxwell yet needs demonstration.

Recent calculations with three-dimensional codes have been employed aiming to delineate self-consistent dynamo action involving reconnection and turbulent processes. A sample of such calculations of the reconnecting surfaces of an RFP configuration are shown in Figure 11 [Caramana, Nebel, and Schnack, 1983]. The corresponding plasma flow is shown in Figure 12.

The sequence of events displayed in Figure 11 is as follows: (1) a resistively unstable RFP configuration is produced by the ohmic heating which overpeaks the current on the interior of the plasma column; (2) a first reconnection occurs which leads to a helically deformed state; (3) a second reconnection occurs which causes the flux to grow around a new magnetic axis and increases the magnetic shear leading to a stable configuration; (4) ohmic heating can then repeat the current profile and the whole process repeats. The first reconnection is of the rapid Sweet-Parker type scaling as $\eta^{1/2}$ and described by the Kadomtsev [1975] model. The second reconnection scales as η and proceeds on a slower time scale. This is in contrast to the single type of rapid reconnection returning the plasma to axisymmetry used in tokamak descriptions [Kadomtsev, 1975]. The above calculations suggest a periodic "sawtooth" behavior not unlike that which has been observed on ZT-40M when operated at high Θ values [Watt and Nebel, 1983; Nebel, 1983]. These sawtooth events produce positive increments in the toroidal flux and are identified with individual dynamo events. Analogous events are apparent on the flux trace for a slow current rise (Figure 9). It is noted that the above calculations are in contrast to the model of Hutchinson [1982] that uses reconnection arguments for the RFP based on the Kadomtsev model for the entire process.

4.3 Helical Ohmic States

The possibility of steady state RFP having a helically symmetric plasma column with a steady plasma flow pattern and satisfying Ohm's law was suggested by the computer calculations of Sykes and Wesson, [1975]. The conditions for such a state were explored by Gimblett [1980].

Preliminary studies of the problem of whether it is possible to set up a stationary ohmic helical state by the reconnection associated with the resistive tearing mode have been made [Dagazian 1980a, 1980b]. Numerical and analytic work on this problem have been reported [Schnack, 1980; Dobrott and Schnack, 1983; Aydemir and Barnes, 1983a, 1983b]. The last authors report that 2-D and 3-D computer calculations have demonstrated steady states which are maintained against resistive diffusion by the dynamo action of large helical flows.

4.4 Models With Islands and/or Stochasticity of the Magnetic Field Lines

An important topic is the possible breaking up of the nested flux surfaces of a toroidal equilibrium by magnetic field perturbations. It is known that rather small non-axisymmetric field errors can resonate with the helical field lines leading to a change in the field topology [Kerst, 1962]. These errors can produce small "islands" of nested surfaces, each with its own magnetic axis. As the field perturbations are made larger, portions of the error fields with different harmonic content can interact with the main fields and with each other to produce a region where there are no well-behaved flux surfaces and the field lines wander chaotically [Rosenbluth, et al., 1966; Walker and Ford, 1969; and Spencer, 1980]. Such behavior is of much interest in the fusion field because of its effects on plasma containment. Internal plasma perturbations and reconnection processes can lead to such "ergodic" behavior as has been observed in 3-D computer modeling [Schnack, et al., 1983; Aydemir 1983a]. The effect of such changes in field topology on transport and plasma containment is an active area of research.

A tangled discharge model to explain the self-reversal and sustainment of an RFP has been advanced [Rusbridge, 1977; Miller, 1983]. This model retains Ohm's law but assumes that the field lines behave stochastically over the entire plasma volume. EMF's on the interior cause charge separation and electrostatic fields which drive the current along the **B** line against the applied electric field in the reversed magnetic field region.

The possible generation of magnetic islands by reconnection in an RFP has led

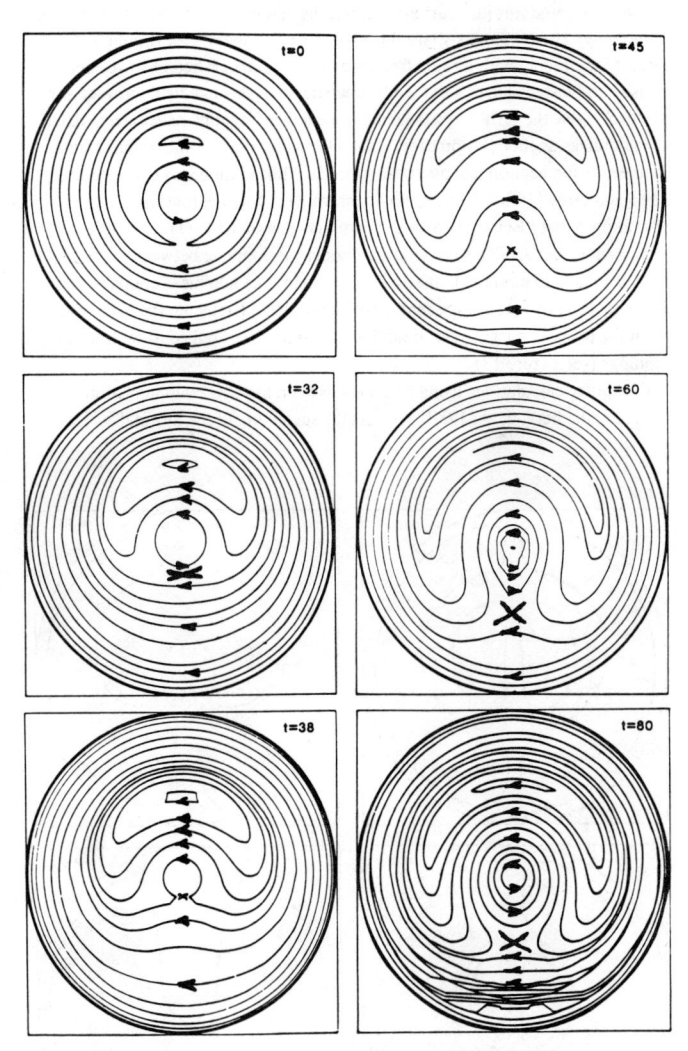

Fig. 11. Results of a single helicity calculation. A cross-section of the reconnection of the helical flux surfaces. The first reconnection proceeds from t=0 to =45 and the second reconnection thereafter. Time is in units of the Alfvén time. The arrows correspond to the direction of the auxiliary field $\mathbf{B}* = \mathbf{B} - (r/r_s)B_\theta(r_s)\theta - B_z(r_s)\hat{z}$ where r_s is the radial position of the reconnection x-point [Kadomtsev, 1975].

to a model for sustaining the configuration [Jacobson, 1984a]. The model makes use of the change in the magnetic surface topology produced by a periodic radial perturbing field, and the space charge electric field resulting from ion transport to drive the required currents. A second mechanism [Jacobson, 1983] uses a rectification process produced by modulating the electrical resistivity in the presence of recurring magnetic islands.

Very recently a model has been advanced [Jacobson and Moses, 1984b] which proposes RFP sustainment by replacing Ohm's law with a kinetic theory making use of a Fokker-Planck equation. The collision term is modified with a term used to describe transport in an assumed stochastic field [Rosenbluth and Rechester, 1978]. Reversed-field solutions in slab geometry have been obtained numerically under the assumption that the fields are force-free.

4.5 RFP Sustainment With Oscillating Fields

Even though the sustainment of the RFP configuration against dissipative field diffusion appears no longer to be a problem, there is the fact that present toroidal RFPs use an inductive toroidal electric field to maintain them. Since this field is produced by the transformer action associated with continuously increasing the flux in the central hole of the torus, this method of necessity limits the duration of the unidirectional toroidal current because the magnetic field cannot be increased indefinitely. The tokamaks share this property and schemes for driving a direct current indefinitely with a radio frequency field or particle beams have been devised and tested. Thanks to the plasma relaxation through field line reconnection processes, the RFP has a potential scheme for a steady state current drive which utilizes ac modulation of the currents on the toroidal and poloidal field windings using economical audio frequencies. This method was suggested by Bevir and Grey [1980]. The possibility arises from the fact that the rapid relaxation of the discharge keeps the configuration on an F-Θ trajectory as discussed in Sec. 3.2. This constitutes a nonlinear coupling between the poloidal and toroidal field circuits. Unlike a linear system, this nonlinear coupling allows the generation of a dc component of toroidal current to sustain the discharge when the proper phasing of ac modulating currents are applied to each set of field windings (see Figure 13).

Computer calculations using a model based on the F-Θ coupling [Johnston, 1981; Schoenberg, et al., 1982] predicts the sustaining of the toroidal current by

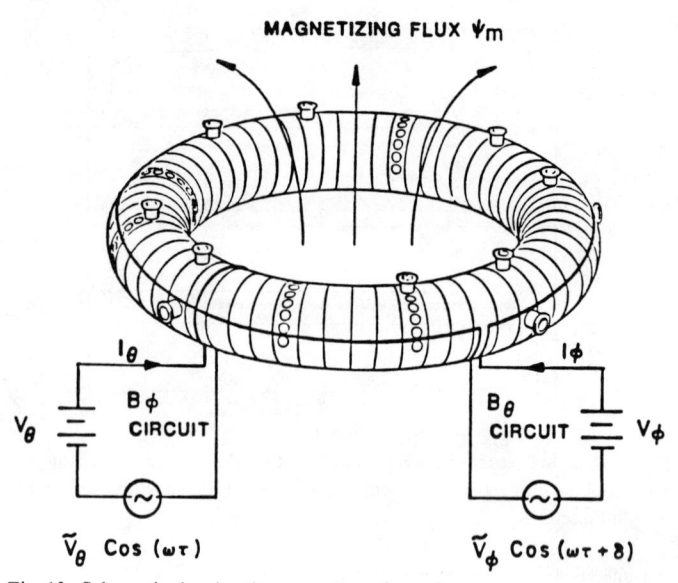

Fig. 13. Schematic showing the connection of ac voltage sources to produce a steady state dc component current in the RFP.

this technique [Schoenberg, et al., 1983a]. The results of such a calculation showing a steady dc component of current and constant mean-field flux are shown in Figure 14. Preliminary experiments in which each winding of the ZT-40M experiment was individually ac modulated have confirmed the main premises on which this technique is based [Schoenberg, 1983b]. The full test to produce a net unidirectional current by this technique on ZT-40M is planned.

5. Conclusion

It is impressive how many of the present-day RFP concepts and experiments are intimately related to the rapid field line reconnection processes. The startup, sustainment and containment are vitally controlled and modified by the relaxa-

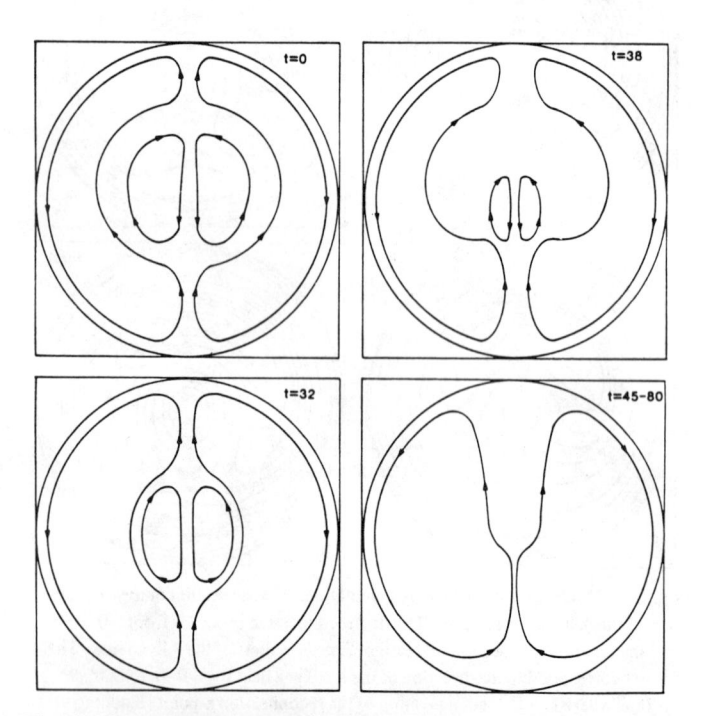

Fig. 12. Plasma flow patterns corresponding to Fig. 11.

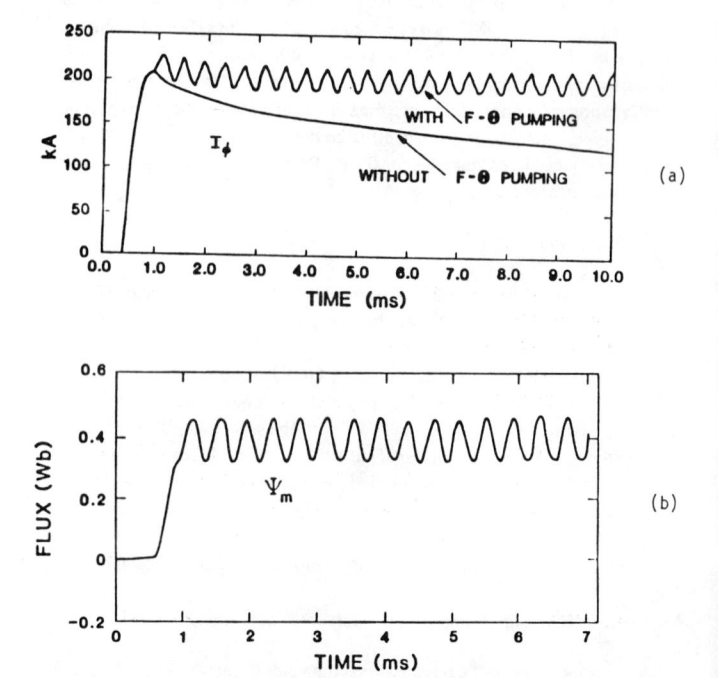

Fig. 14. Computer demonstration of dc current drive with ac voltages on the field windings. (a) current; (b) poloidal flux through the hole in the torus.

tion of the plasma configuration through the field line reconnection phenomenon. Two areas are particularly outstanding examples where concepts having origins in the astrophysics area have greatly enhanced the laboratory study of the RFP; namely the mean-field dynamo and the plasma relaxation to a lowest energy state.

The phenomenon of reconnection in a conducting plasma leads to an area rich in mechanisms for flux generation and RFP sustainment in the presence of collisional dissipation. As further work clarifies which of the many possible mechanisms can be made self-consistent and in agreement with experimental observations, the results will have considerable impact to both the space and laboratory phyics communities. It is clear that further and closer collaboration between researchers of space and astrophysics plasmas and those associated with laboratory-produced plasmas would greatly benefit both disciplines. It is hoped that this review with its extensive bibliography will aid in motivating this cooperation.

Acknowledgements. The author is indebted to the space physics and reversed-field pinch communities whose research and ideas are reviewed in this paper. I particularly acknowledge my colleagues in the Los Alamos RFP experimental and theoretical groups for their contributions.

References

Aydemir, A. and Barnes, D. C., Three Dimensional Numerical Studies of Reversed-Field Pinch, *Bull. Amer. Phys.* Soc. 28, 1230, 1983a.

Aydemir, A. and Barnes, D. C., Sustained Self-Reversal in the Reversed-Field Pinch, *University of Texas at Austin Report,* IFSR#102, 1983b.

Antoni, V., et al.., Studies on High-Density RFP Plasmas in the Eta Beta II Experiment, *Plasma Physics and Controlled Nuclear Fusion Research 1982*, International Atomic Energy Agency, Vienna, Vol. I, 619-640, 1983.

Baker, D. A. and DiMarco, J. N., The LASL Reversed-Field Pinch Program Plan, *Los Alamos Scientific Laboratory report* LA-6177-MS, Los Alamos, New Mexico, 1975.

Baker, D. A. and Quinn, W. E., The Reversed-Field Pinch, Chapter 7 of *Fusion*, Vol. I, Part A, E. Teller, Editor, Academic Press, Inc., New York, NY, 1981.

Baker, D. A., et al., Performance of the ZT-40M Reversed-Field Pinch with an Inconel Liner, *Plasma Physics and Controlled Nuclear Fusion Research 1982*, International Atomic Energy Agency, Vienna, Vol. I, 587-595, 1983.

Bennett, W. H., Magnetically Self-Focussing Streams, *Phys. Rev. 45*, 890-897, 1934.

Bevir, M. K. and Gray, J. W., Relaxation, Flux Consumption, and Quasi-Steady State Pinches, *Proc. Reversed Field Pinch Theory Workshop,* Los Alamos National Laboratory report LA-8944-C, 176, 1980.

Bhattacharjee, A., Dewar, R., and Monticello, D. A., Energy Principle With Global Invariants for Toroidal Plasmas, *Phys. Rev. Lett. 45*, 347-350, 1980 (errata in *Phys. Rev. Lett. 45*, 1217, 1980).

Bhattacharjee, A. and Dewar, R., Energy Principle With Global Invariants, *Phys. Fluids 25*, 887-897, 1982.

Bodin, H. A. B. and Keen, B. E., Experimental Studies of Plasma Confinement in Toroidal Systems, *Rep. Prog. Phys. 40*, 1415-1565, 1977.

Bodin, H. A. B. and Newton, A. A., Reversed-Field Pinch Research, *Nucl. Fusion 20*, 1255-1324, 1980.

Bodin, H. A. B., et al., Results from the HBTX-1A Reversed-Field Pinch Experiment, *Plasma Physics and Controlled Nuclear Fusion Research 1982*, International Atomic Energy Agency, Vienna, Vol. I, 641-657, 1983.

Bondeson, A., Marklin, G., An, Z. G., Chen, H. H., Lee, Y. C., and Liu, C.S., Tilting Instability of a Cylindrical Spheromak, *Phys. Fluids 24*, 1682-1688, 1981.

Brandenburg, J. E., A Theory of the Relaxation of Finite Beta Toroidal Plasma, *Lawrence Livermore National Laboratory Report* No. UCRL-87096, 1983. (Submitted to Phys. Fluids)

Caramana, E. J. and Moses, R. W., General Characteristics of the Reversed Field Pinch Equilibria With Specified Global Parameters, submitted for publication *Nucl. Fusion*, 1983a.

Caramana, E. J. and Baker, D. A., Dynamo Effect in Sustained Reversed Field Pinch Discharges, submitted for publication *Nucl. Fusion*, 1983b.

Caramana, E. J., Nebel, R. A., and Schnack, D. D., Nonlinear, Single Helicity Magnetic Reconnection in the Reversed-Field Pinch, *Phys. Fluids 26*, 1305-1319, 1983c.

Colgate, S. A., Ferguson, J. P., Furth, H. P., The Toroidal Stabilized Pinch, *Proc. U.N. Conf. on Peaceful Uses of Atomic Energy 2nd, 32*, 129-139, 1958.

Cousins, S. W., Ware, A. A., Pinch Effect Oscillations in a High Current Toroidal Ring Discharge, *Proc. Phys. Soc. B, 64*, 159-166, 1951.

Cowling, T. C., The Magnetic Field of Sunspots, *Monthly Notices of the Royal Astro. Soc. 94*, 39-48, 1934.

Dagazian, R. Y., Helical Ohmic States for RFPs, *Bull. Amer. Phys. Soc. 25*, 865, 1980a.

Dagazian, R. Y., On A Helical Ohmic State for Reversed Field Pinches, Proc. of the Reversed-Field Pinch Theory Workshop, *Los Alamos National Laboratory report* LA-8944-C, 123-128, 1980b.

Dobrott, D., Schnack, D. D., Steady-Flow in Resistive MHD With Helical Symmetry, *Bull. Amer. Phys. Soc. 28*, 1191, 1983.

Edenstrasser, J. W., Nalesso, G. F., and Schuurman, W., Finite-Beta Minimum Energy Equilibria of RFPs, Screw Pinches and Tokamaks, *Nuclear Inst. and Methods 207*, 75-85, 1983a.

Edenstrasser, J. and Schuurman, W., Finite Beta Minimum Energy Equilibria of Weakly Toroidal Discharges, *Phys. Fluids 26*, 500-507, 1983b.

Erlebacher, G., A Variational Method for the Evolution of Toroidal Plasmas, *Bull. Amer. Phys. Soc. 26*, 1055, 1981.

Faber, V., White, A. B., and Wing, G. M., An Analysis of Taylor's Theory of Toroidal Plasma Relaxation, *J. Math. Phys. 23*, 1524-1537, 1982a.

Faber, V., White, A. B., and Wing, G. M., Flux-Free States in Taylor Relaxation of a Toroidal Plasma, *Los Alamos National Laboratory report* LA-UR-82-3362, 1982b.

Glasstone, S. and Lovberg, R. H., *Controlled Thermonuclear Reactions*, Chap. 7, Van Nostrand-Rienhold, Princeton, New Jersey, 1960.

Finn, J., Manheimer, W., and Ott, E., Spheromak Tilting Instability in Cylindrical Geometry, *Phys. Fluids 24*, 1336-1341, 1981.

Finn, J. M. and Antonson, Jr., T. M., Turbulent Relaxation of Plasmas With Flow, *Phys. Fluids 26*, 3540-3552, 1983.

Gerwin, R. and Keinigs, R., Dynamo Theory: Can Amplification of Magnetic Field Profiles Arise From a Cross-Field Alpha Effect?, *Los Alamos National Laboratory report* LA-9290-MS, 1982.

Gimblett, C. G., Watkins, M. L., MHD Turbulence Theory and Its Implications for the Reversed-Field Pinch, *Proc. 7th European Conf. on Controlled Fusion and Plasma Phys.*, Lausanne, 1, 103, 1975.

Gimblett, C. G., Some Necessary Conditions for a Steady State Reversed Field Pinch, Proc. of the Reversed Field Pinch Theory Workshop, *Los Alamos National Laboratory report* LA-8944-C, 254, 1980.

Hameiri, E. and Hammer, J., Turbulent Relaxation of Compressible Plasmas, *Phys. Fluids 25*, 1855-1862, 1982.

Holmes, J. A., Carreras, T. C., Hender, T. C., Hicks, H. R., and Lynch, V. E., Nonlinear Evolution of Resistive Modes in Reversed Field Pinches, *Bull. Amer. Phys. Soc. 28*, 1230, 1983.

Hutchinson, I. H., Helical Reconnections of the Reversed Field Pinch, *Bull. Amer. Phys. Soc. 27*, 1033, 1982.

Jacobson, A. R., private communication, 1983.

Jacobson, A. R., A Possible Plasma-Dynamo Mechanism Driven by Particle Transport, *Phys. Fluids 27*, 7-9, 1984a.

Jacobson, A. R. and Moses, R.W., Nonlocal DC Electrical Conductivity of a Lorentz Plasma in a Stochastic Magnetic Field, submitted to *Phys. Rev.*, 1984b.

Johnston, J. W., A Plasma Model for Reversed Field Pinch Circuit Design, *Plasma Phys. 23*, 187-201, 1981.

Kadomtsev, B. B., Disruptive Instability in Tokamaks, *Sov. J. Plasma Phys. 1*, 389-391, 1975. (Translated from *Fiz. Plasmy 1*, 710, 1975)

Keinigs, R. K., A New Interpretation of the Alpha Effect, *Phys. Fluids 26*, 2558-2560, 1983.

Kerst, D. W., The Influence of Errors on Plasma-Confining Magnetic Fields, *J. Plasma Phys.* (J. Nucl. Energy, Part C) 4, 253-262, 1962.

Krause, F. K. and Rädler, K. H., *Mean-Field Magnetohydrodynamics and Dynamo Theory,* Pergamon Press, New York, New York, 1980.

Marklin, G. and Bondeson, A., Minimum Energy States of a Resistive Discharge with Confined Plasma Pressure, *Bull. Amer. Phys. Soc. 25*, 1027, 1980.

Miller, G. and Turner, L., Force-Free Equilibria in Toroidal Geometry, *Phys. Fluids 24*, 363-365, 1981.

Miller, G., Steady State Discharges, submitted for publication to *Phys. Fluids*, 1983.

Moffat, H. K., *Magnetic Field Generation in Electrically Conducting Fluids,* Cambridge Univ. Press, 1978.

Montgomery, D., Turner, L., and Vahala, G., Three-Dimensional Magnetohydrodynamic Turbulence in Cylindrical Geometry, *Phys. Fluids 21*, 757-764, 1978.

Nebel, R. A., Nonlinear Tearing, Toroidal Flux Regeneration, and Sawtooth Oscillations in the Reversed-Field Pinch (RFP), *Bull. Amer. Phys. Soc. 28*, 1188, 1983.

Ogawa, K., et al., Experimental and Computational Studies of Reversed-Field Pinch on TPE-1R(M), *Plasma Physics and Controlled Nuclear Fusion Research 1982*, International Atomic Energy Agency, Vienna, Vol. I, 575-585, 1983.

Phillips, J. A., et al., Toroidal Current Ramping in ZT-40M, *Bull. Amer. Phys. Soc. 28*, 1098, 1983.

Reiman, A., Minimum Energy State of a Toroidal Discharge, *Phys. Fluids 23*, 230-231, 1980.

Reiman, A., Taylor Relaxation in a Torus of Arbitrary Aspect Ratio and Cross-Section, *Phys. Fluids 24*, 956-963, 1981.

Robinson, D. C., Crowe, J. E., Gowers, C. W., Nalesso, G. F., Newton, A. A., Verhage, A. J. L., and Bodin, H. A. B. Controlled Fusion and Plasma Physics (Proc. 5th European Conf. Grenoble, 1972) Euratom-CEA, Grenoble, 2, 47-58, 1972.

Rosenbluth, M. N., Sagdeev, R. Z., Taylor, J. B., and Zaslavski, G. M., Destruction of Magnetic Surfaces by Magnetic Field Irregularities, *Nucl. Fusion 6*, 297-300, 1966.

Rosenbluth, M. N. and Rechester, A. B., Electron Heat Transport in a Tokamak With Destroyed Magnetic Surfaces, *Phys. Rev. Lett. 40*, 38-41, 1978.

Rosenbluth, M. N. and Bussac, M. N., MHD Stability of a Spheromak, *Nucl. Fusion 19*, 489-498, 1979.

Rusbridge, M. G., A Model of Field Reversal in the Diffuse Pinch, *Plasma Phys. 19*, 499-516, 1977.

Schaffer, M. J., A Plasma Model with Dynamo for Sustained Reversed Field Pinches, *General Atomic Report # GA-A16759*, 1982.

Schoenberg, K. F., Gribble, R. F., and Phillips, J. A., Zero Dimensional Simulations of Reversed-Field Pinch Experiments, *Nucl. Fusion 22*, 1433-1441, 1982.

Schoenberg, K. F., Gribble, R. F., and Baker, D. A., Oscillating Field Current Drive for Reversed-Field Pinch Discharges, submitted to *J. Appl. Phys.*, 1983a.

Schoenberg, K. F., Buchenauer, C. J., Massey, R. S., Melton, J. G., Moses, R. W., Nebel, R. A., and Phillips, J. A., F-Theta Pumping and Field Modulation Experiments on a Reversed Field Pinch, to be published *Phys. Fluids*, 1983b.

Schnack, D. D., Dynamical Determination of Ohmic States of a Cylindrical Pinch, *Proc. of the Reversed-Field Pinch Theory Workshop*, Los Alamos National Laboratory report LA-8944-C, 118, 1980.

Schnack, D. D., Caramana, E. J., and Nebel, R. A., Three-Dimensional MHD Simulation of Large Scale RFP Dynamics, *Bull. Amer. Phys. Soc. 28*, 1229, 1983.

Spencer, R. L., Magnetic Islands and Stochastic Field Lines in the RFP, *Proc. of the Reversed-Field Pinch Theory Workshop*, Los Alamos National Laboratory reporrt LA-8944-C, 129-134, 1980.

Sudan, R. N. Stability of Field Reversed, Force-Free Plasma Equilibria With Mass Flow, *Phys. Rev. Lett. 42*, 1277-1281, 1979.

Sykes, A. and Wesson, J. A., *Eighth European Conference on Controlled Fusion and Plasma Physics*, Prague, Vol. I, 80, 1977.

Taylor, J. B., Relaxation of Toroidal Plasma and Generation of Reverse Magnetic Fields, *Phys. Rev. Lett. 33*, 1139-1141, 1974.

Taylor, J. B., *Plasma Physics and Controlled Thermonuclear Reserch*, (Proc. 5th Int. Conf. Tokyo 1974) 2, 161-167, 1975.

Taylor, J. B., Relaxation of Toroidal Discharges, *Pulsed High Beta Plasmas*, Pergamon Oxford, 59-67, 1976.

Tamano, T., et al., Pinch Experiments in OHTE, *Plasma Physics and Controlled Nuclear Fusion Research 1982*, International Atomic Energy Agency, Vienna, Vol. I, 609-618, 1983.

Turner, L., Statistical Magnetohydrodynamics and Reversed-Field Pinch Quiescence, *Nucl. Instrum. Methods 207*, 23-33, 1983a.

Turner, L., Statistical Mechanics of a Bounded, Ideal Magnetofluid, *Ann. Phys. 149*, 58-161, 1983b.

Turner, L., Analytic Solutions of Curl B = Lambda B Having Separatrices for Geometries With One Ignorable Coordinate, to be published in *Phys. Fluids*, 1983c.

Walker, C. H. and Ford, J., Amplitude Instability and Ergodic Behavior for Conservative Nonlinear Oscillator Systems, *Phys. Rev. 188*, 416-432, 1969.

Watt, R. G., Nebel, R. A., Sawteeth, Magnetic Disturbances, and Magnetic Flux Regeneration in the Reversed-Field Pinch, *Phys. Fluids 26*, 1168-1170, 1983.

Woltjer, L., A Theorem on Force-Free Magnetic Fields, *Proc. Nat. Acad. Sci., 44*, 489-491, 1958.

Woltjer, L., Hydromagnetic Equilibrium II: Stability in the Variational Formulation, *Proc. Nat. Acad. Sci., 45*, 769-771, 1959.

Woltjer, L., On The Theory of Hydromagnetic Equilibrium, *Rev. Mod. Phys. 32*, 914-915, 1960.

Questions and Answers

Vasyliunas: Dynamo action implies that there is energy being added to the magnetic field to balance the loss by ohmic heating. How important is the amount of energy involved, in comparison with the total energy budget of the device, and are there any constraints on dynamo models from energy availability considerations?

D. A. Baker: In the present state of our knowledge of dynamos there are no energy constraints since one can postulate that any amount of energy needed can, in principle, be supplied by the toroidal current circuit. Any added energy shows up as a higher toroidal loop voltage.

R. Moses: The energy associated with ohmic heating is about classical. The energy associated with dynamo drive is about 40% of the ohmic amount. These estimates are based on a plasma resistance obtained using the magnetic helicity from a modified Bessel Function model of the fields which matches the experimental F-Theta diagram.

RECONNECTION IN TOKAMAKS

V. K. Paré

Oak Ridge National Laboratory, Oak Ridge, Tennessee 37831

Abstract. In a tokamak, the superposition of an externally applied toroidal magnetic field and the field of a toroidal current induced in the plasma results in field lines that go helically around the torus. The equilibrium structure consists of toroidal surfaces nested around a toroidal magnetic axis. Often a plasma instability will cause an "island" (a separate set of nested surfaces with a helical magnetic axis) to form by reconnection. These structures usually rotate toroidally and can be observed as oscillations in local x-ray emission. Resistive MHD theory shows that one type of island can grow and – in a second reconnection – "take over" from the original magnetic axis. Sequential two-dimensional images generated from arrays of x-ray detectors show both the rotation and the growth of these islands and also the "takeover."

I. Introduction

The tokamak concept is currently the leading contender for development into a fusion power reactor. For this reason many research tokamaks are being operated with elaborate sets of diagnostic devices, and much theoretical effort is being devoted to understanding the observed phenomena. The occurrence of magnetic reconnection during dynamic instability processes in tokamaks has been well established by the agreement with experiment of theories in which reconnection is implicit. Recently it has also become possible to visualize the process directly in time sequences of x-ray images of tokamak plasmas.

We describe briefly the magnetic configuration of a tokamak and the design of diagnostic systems used for observing phenomena involving reconnection. We then discuss the behavior of a particular instability and show how its evolution in space and time has been studied both theoretically and experimentally.

II. Magnetic Configuration of a Tokamak

The basic elements of a tokamak are shown in Fig. 1. End losses are precluded by the toroidal magnetic field. Since such a field must vary inversely with major radius, the ∇B drift would destroy confinement in a simple toroidal system. This drift can be cancelled by making the field lines go helically around the torus, so that a particle spends on the average equal amounts of time drifting toward and away from the midplane. The distinguishing feature of a tokamak is that the helicity is introduced by the poloidal magnetic field of a toroidal current induced in the plasma itself.

The structure of the magnetic fields can be discussed with the aid of the definitions in Fig. 2, which represents a tokamak cut in half along its major axis, exposing two poloidal planes. At a location (R_1, Z_1) the poloidal flux function Ψ is the integral of the poloidal field crossing a surface whose periphery is a circle of radius R_1 and height Z_1, centered on the major axis. The contours of constant Ψ form tubular surfaces when swept around in the toroidal direction; these surfaces by definition contain the helical field lines and are called flux surfaces. The line of maximum Ψ is known as the magnetic axis.

Because of the high mobility of particles along field lines and because the plasma pressure must be constant on a flux surface, plasma parameters such as temperature and density are also normally uniform over a flux surface. If the distribution of toroidal current over a poloidal plane has simple topology (decreasing monotonically in all directions from a maximum), then Ψ also has a simple topology of nested closed contours. The global helicity is specified by q, the ratio of toroidal to poloidal transits of a field line. It is a flux surface quantity. Since the toroidal current density is normally peaked in minor radius, q has its minimum value at the magnetic axis and increases toward the outside. If q is an irrational number, a single field line covers a magnetic surface ergodically. However, if q is rational, $q = m/n$, the field line closes on itself without covering the surface. Such magnetic surfaces are called singular surfaces.

Singular surfaces are important because they are "weak spots" where instabilities can develop. A radial magnetic perturbation having poloidal mode number m and toroidal mode number n resonates with the helicity at the singular surface $q = m/n$ and can cause a change of magnetic topology, generating an "island" in the "contour map" of the poloidal flux function. The island extends around the torus with the helicity of the flux surface at which it is resonant.

III. Diagnostics for Magnetic Structures in Tokamaks

The high energy densities in most tokamak plasmas preclude the use of physical probes. It has not thus far been possible to develop another diagnostic that will, by direct measurement, map the current distribution or poloidal field within the plasma.

A very useful substitute exists, however: the plasma emits x rays whose emission power density can be observed as a function of space and time by relatively inexpensive semiconductor detectors. The emission is a complex function of plasma density, temperature, and impurity content, but all these parameters are expected to be constant over a flux surface. Thus the x-ray emission can give information on the flux surface shape and topology, even though absolute values cannot be inferred.

The geometry of the x-ray detector system now operating on the ISX-B tokamak is shown in Fig. 3. There are 3 arrays containing a total of 80 detectors. Each detector in an array is masked so as to have a uniform sensitive area, and all the detectors in an array view the plasma through a common collimating slit. A detector with its

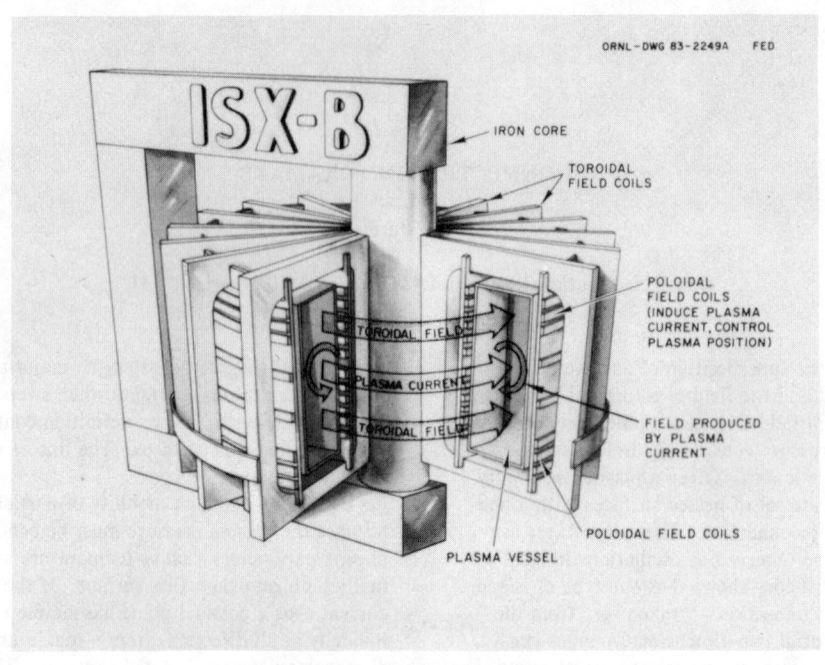

Fig. 1. Basic elements of a tokamak.

associated current amplifier produces a signal proportional to the x-ray power incident on the detector, within the photon energy range of roughly 0.5 to 3 keV. The bandwidth is at least 150 kHz, so that rapid variations can be followed. Each detector is connected to its own digitizer. The digitizers can convert at rates up to 512 kHz, and there is sufficient local digital memory to store 8192 data values from each detector.

The signal in a slit-collimated detector is proportional to the line integral of emission power density along the detector's line of sight in the plasma. Various efforts have been made to develop algorithms that will, as in computer-assisted medical tomography, convert these signals into a two-dimensional map of emission density in a poloidal plane. For a plasma, whose parameters may vary rapidly with minor radius r but should be relatively smooth functions of poloidal angle θ, it is appropriate to represent the x-ray emission power density by an expansion of the form

$$E(r,\theta) = \sum_{m=0}^{M} [f_m(r)\cos(m\theta) + g_m(r)\sin(m\theta)] \ .$$

With three arrays in the same poloidal plane, as in Fig. 3, a maximum harmonic number $M = 3$ can be employed, yielding realistic representations of noncircular equilibria and low-order instability modes [Navarro et al., 1981].

Fig. 2. Definition of ψ, the poloidal flux function.

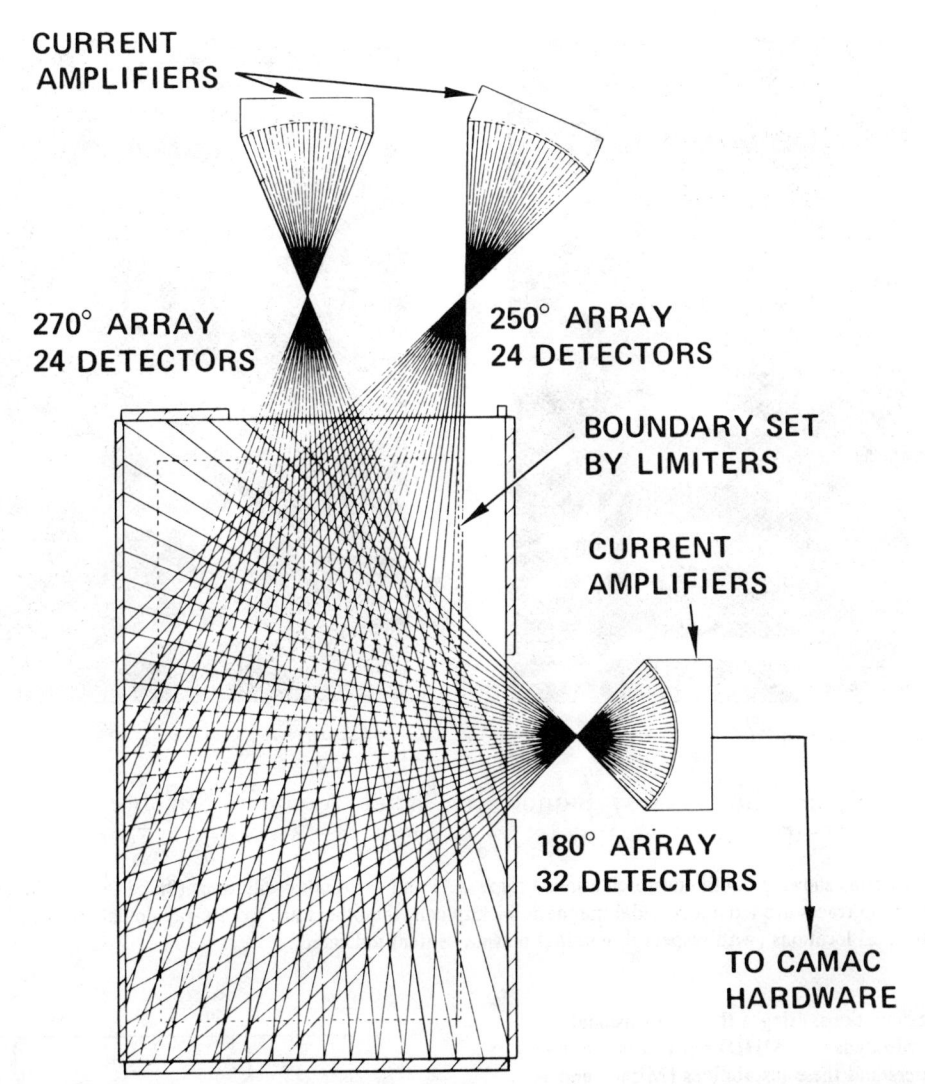

Fig. 3. Lines of sight of slit-collimated x-ray detectors installed on the ISX-B tokamak. The rectangular hatched outline is a cross section of the plasma vessel as shown in Fig. 1. The major axis is to the left in this figure.

IV. Instabilities and Reconnection in Tokamaks

Early x-ray detection systems contained modest numbers of detectors. They revealed two types of periodic behavior in the interior of tokamak plasmas – continuous, nearly sinusoidal oscillations and "sawtooth" oscillations [von Goeler et al., 1974]. One or the other might be observed, depending on subtle differences in discharge conditions. Examples of signals from both types of discharges are shown in Fig. 4.

The continuous oscillations, with frequencies in the kilohertz range, had previously been detected at the plasma edge with poloidal magnetic pickup coils. It was recognized that these oscillations represented island structures that were rotating, either poloidally or toroidally. The rotation is caused by kinetic phenomena not clearly related to the physics of the islands themselves, but it is fortuitous in that it allows a small number of detectors to sample the entire poloidal structure of the instability. Thus the poloidal and toroidal mode numbers m and n could be determined by comparing the electrical phases of the signals from different detectors having appropriately positioned lines of sight.

The sawtooth oscillation is the more common of these phenomena in tokamaks. As seen by a detector looking near the center of the plasma, each cycle of the oscillation consists of a slow (1- to 10-ms) rise followed by a drop that occurs in less than 0.2 ms. The rise is associated with an increase in central plasma temperature. The drop is preceded by a growing oscillation (5 to 20 kHz) representing a rotating instability with $m = 1, n = 1$ mode numbers. For detectors whose lines of sight pass outside a certain minor radius, known as the inversion radius, the phase of the sawtooth is reversed – an approximately linear decrease in signal is followed by a sharp rise. At successively larger radii the inverted sawtooth is observed in progressively more rounded and delayed form. An example of sawtooth waveforms is shown in Fig. 5. Clearly the drop in the central x-ray signal represents a sudden transfer of heat energy from the region within the inversion radius to the region immediately outside it; the deposited energy then diffuses as a heat pulse to the edge of the plasma.

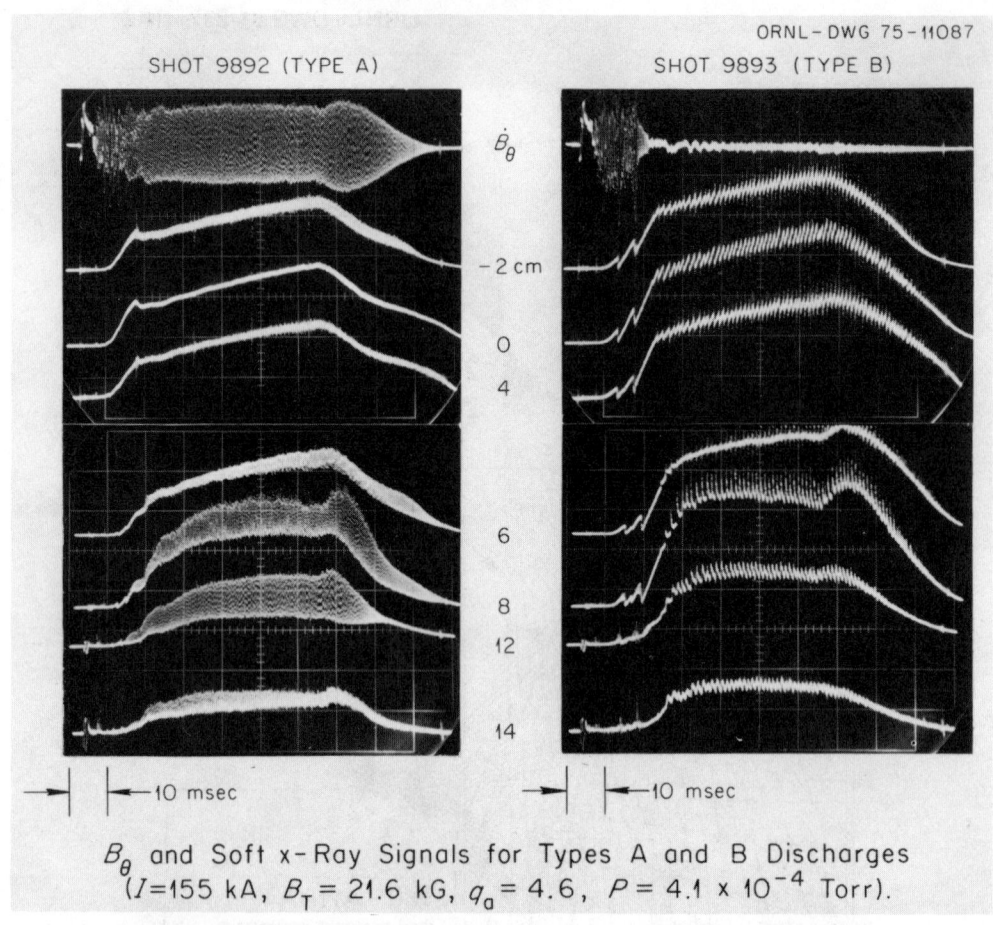

Fig. 4. Waveforms showing continuous oscillations ("type A") and sawtooth oscillations ("type B") in the ORMAK tokamak. The top traces are from a poloidal magnetic pickup coil; the others are from x-ray detectors looking vertically past the horizontal locations (with respect to nominal plasma center) indicated.

Extensive numerical calculations, using a three-dimensional (3-D), resistive magnetohydrodynamic (MHD) representation, have been done in order to understand these instabilities [Mirnov and Semenov, 1971; Kadomtsev, 1975; Sykes and Wesson, 1976; Waddell et al., 1976; Dnestrovskii et al., 1977; White et al., 1977; Jahns et al., 1978]. It is expected that over a large part of the range of tokamak plasma parameters heating of the central part of the plasma will, by reducing resistivity, sharpen the peak in the current distribution. This heating is responsible for the rising phase of the sawtooth signal. The concentration of current drives q below 1 in the center, creating a $q = 1$ magnetic surface. The calculations show that the ($m = 1$, $n = 1$) tearing mode that is thus resonant has typically a high growth rate. This mode causes a change in magnetic topology, forming an island. The island, within which the current and pressure distributions are relatively flat, grows inward in radius from the $q = 1$ surface, "stealing" flux surfaces from the original magnetic structure that existed within $q = 1$. During this process the field lines must be transferred, by reconnection, from the old flux surfaces to the new ones. It is the rotation of this island structure that produces the growing oscillation in the x-ray signals.

Finally, in a second change of magnetic topology, the original magnetic axis, with its associated heat energy, is pushed out of the plasma, and the energy is redistributed around the region just outside $q = 1$. The plasma then contains nested flux surfaces as it did

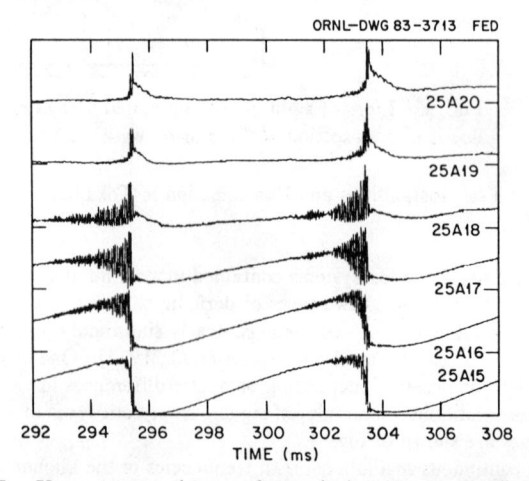

Fig. 5. X-ray "sawtooth" waveforms during 16 ms of a discharge in the ISX-B tokamak. The line of sight of detector 25A15 passes just outside the magnetic axis, while that of detector 25A18 is approximately at the "inversion radius." The onset and growth of the ($m = 1$, $n = 1$) tearing mode are revealed, through toroidal rotation, by the 15-kHz oscillation. At about 303.6 ms the original magnetic axis, with its associated heat energy, is expelled; the energy appears outside the inversion radius.

before formation of the island; however, the central current and temperature have been reduced so that there is no longer a $q = 1$ surface. Thereafter, heating of the central region increases the current there, reintroducing a $q = 1$ surface and thus allowing the cycle to repeat.

Development and improvement of resistive MHD codes have continued, with the objectives of including additional physical effects and of computing plasma quantities that are more easily observable than the flux function. Two codes in use at Oak Ridge National Laboratory are RST [Lynch et al., 1981], which implements toroidal geometry, and KITE [Hicks et al., 1984], which employs cylindrical geometry but more accurately models the variations of resistivity and density with temperature.

Figure 6 shows six stages in the growth of the ($m = 1, n = 1$) tearing mode as calculated by the KITE code. Expulsion of the original magnetic axis takes place just prior to $t = 3.72$, the time of the fifth plot. The RST code contains provisions for simulating the

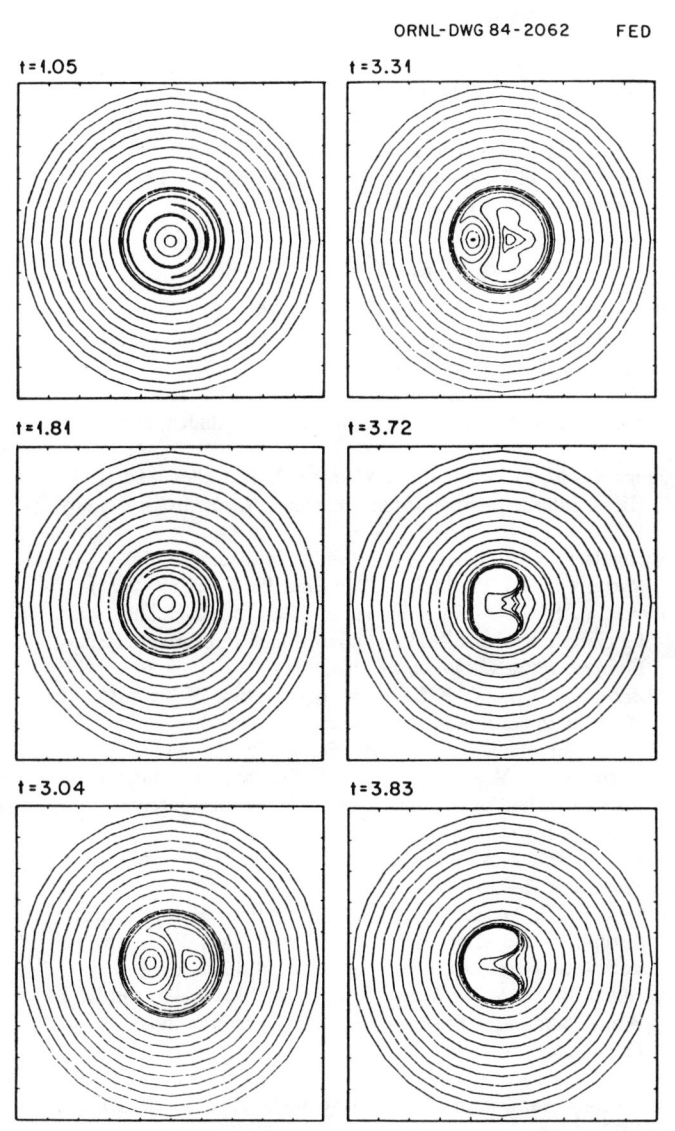

Fig. 6. Sequence of stages in the growth of the ($m = 1, n = 1$) tearing mode in a tokamak plasma calculated by the resistive MHD code KITE. Contours of the poloidal flux function are shown. The time scale is in arbitrary units.

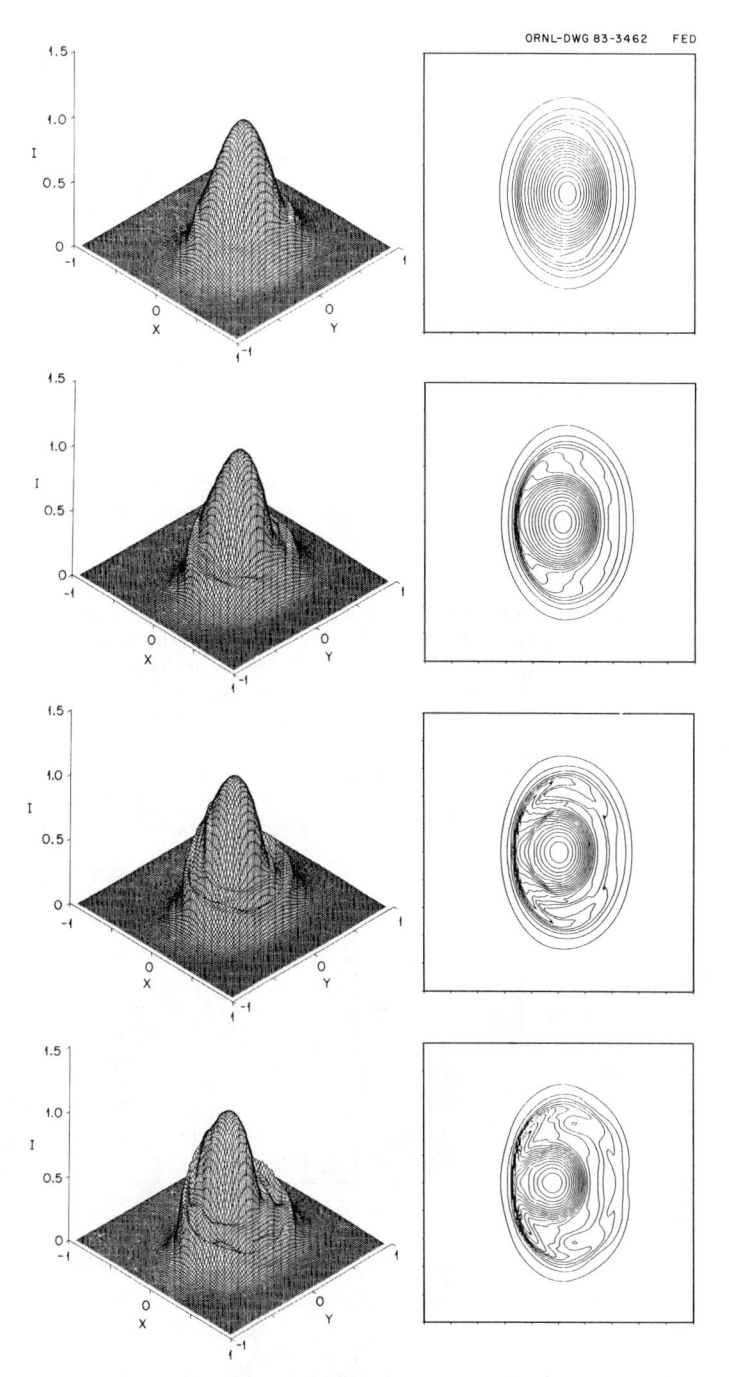

Fig. 7. Sequence of stages in the growth of the ($m = 1, n = 1$) tearing mode in an ISX-B-like plasma, calculated by the resistive MHD code RST. The 3-D and corresponding contour plots represent simulated x-ray emission power density over a poloidal plane, with no toroidal rotation assumed.

x-ray emission power density, and some results are shown as both 3-D and contour plots in Fig. 7. Since the pressure and temperature distributions in the region of the island are relatively flat, the simulated x-ray emission shows no island, but simply a displacement of the emission maximum as the island grows.

To compare the experimentally observed x-ray emission with these results, it is helpful to suppress the toroidal rotation. This has

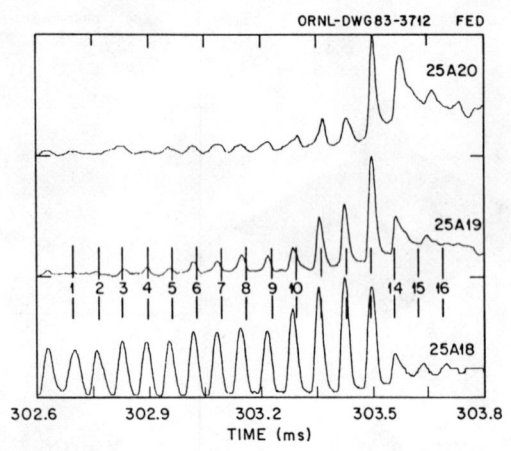

Fig. 8. Expansion of three of the waveforms shown in Fig. 5. The numbered time points are those for which the tomographic images in Fig. 9 were produced.

been done for the ISX-B discharge pictured in Fig. 5 by generating a series of tomographic images at the times marked on the expanded waveforms shown in Fig. 8. The selected times are at the maxima of signals from detectors viewing vertically some distance out from the original magnetic axis and correspond to "freezing" the toroidal rotation at a phase 180° from that depicted in Figs. 6 and 7.

Figure 9 is a montage of the 16 resulting images. It can be seen that the displacement of the emission maximum (representing the original magnetic axis) accelerates rapidly toward the end of the process. This behavior is predicted by the MHD codes.

The sequence in Fig. 9 represents every 34th data point over only a fraction of the time sequence recorded by the data acquisition system. To take fuller advantage of the data, a motion picture has been made that shows the mode growth superimposed on the toroidal rotation.

V. Summary

Calculations with several different computer codes based on the resistive MHD equations have shown that ($m = 1$, $n = 1$) tearing modes in tokamak plasmas grow by magnetic reconnection. The observable behavior predicted by the codes has been confirmed in detail from the waveforms of signals from x-ray detectors and recently by x-ray tomographic imaging.

Acknowledgments. B. A. Carreras and J. L. Dunlap have made important contributions to the clarity and accuracy of this paper. V. E. Lynch and L. Garcia have been generous in providing results of the MHD code calculations. This research was sponsored by the Office of Fusion Energy, U.S. Department of Energy, under Contract No. W-7405-eng-26.

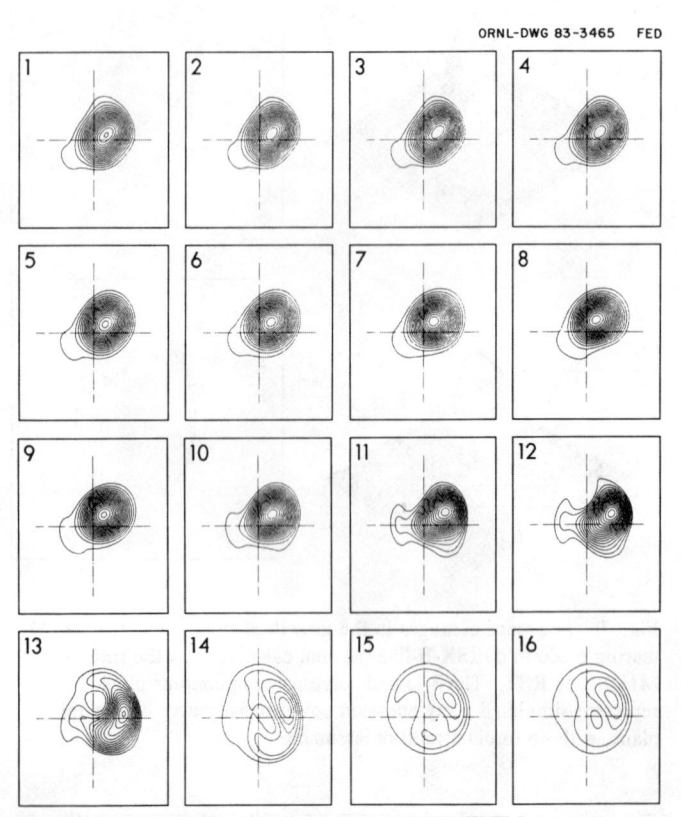

Fig. 9. Sequence of tomographic images of x-ray emission power density in a poloidal plane of the ISX-B tokamak. The numbers on the images correspond to the time points indicated in Fig. 8.

References

Dnestrovskii, Yu. N., S. E. Lysenko, and R. Smith, Numerical simulation of internal-mode relaxation oscillations in a tokamak, *Sov. J. Plasma Phys.* 3, 9, 1977.

Hicks, H. R., L. Garcia, and B. A. Carreras, 3-D nonlinear MHD calculations with drift and thermal effects included, to be published.

Jahns, G. L., M. Soler, B. V. Waddell, J. D. Callen, and H. R. Hicks, Internal disruptions in tokamaks, *Nucl. Fusion* 18, 609, 1978.

Kadomtsev, B. B., Disruptive instability in tokamaks, *Fiz. Plazmy* 1, 710, 1975 (*Sov. J. Plasma Phys.* 1, 389, 1975).

Lynch, V. E., B. A. Carreras, H. R. Hicks, J. A. Holmes, and L. Garcia, Resistive MHD studies of high β tokamak plasmas, *Comput. Phys. Commun.* 24, 465, 1981.

Mirnov, S. V., and I. B. Semenov, Investigation of the instabilities of the plasma string in the Tokamak-3 system by means of a correlation method, *Sov. J. At. Energy* 30, 22, 1971.

Navarro, A. P., V. K. Paré, and J. L. Dunlap, Two-dimensional spatial distribution of volume emission from line integral data, *Rev. Sci. Instrum.* 52, 1634, 1981.

Sykes, A., and J. A. Wesson, Relaxation instability in tokamaks, *Phys. Rev. Lett.* 37, 140, 1976.

von Goeler, S., W. Stodiek, and N. Sauthoff, Studies of internal disruptions and $m = 1$ oscillations in tokamak discharges with soft-x-ray techniques, *Phys. Rev. Lett.* 33, 1201, 1974.

Waddell, B. V., M. N. Rosenbluth, D. A. Monticello, and R. B. White, Non-linear growth of the $m = 1$ tearing mode, *Nucl. Fusion* 16, 528, 1976.

White, R. B., D. A. Monticello, M. N. Rosenbluth, and B. V. Waddell, Saturation of the tearing mode, *Phys. Fluids* 20, 800, 1977.

A PLASMOID RELEASE MECHANISM THAT COULD EXPLAIN THE SUBSTORM'S IMPULSIVE EARTHWARD DIVERSION OF CROSS-TAIL CURRENT

A. Bratenahl and P. J. Baum

Institute of Geophysics and Planetary Physics
University of California, Riverside, CA 92521

Abstract. The sudden interruption and earthward diversion of cross-tail current in association with substorms is a concept of long standing, but the detailed mechanism which causes it has remained unclear. This uncertainty, which helps prolong the debate over just what it is that causes substorms, might be resolved along with the debate itself, if greater theoretical attention were given the subject of time-dependent reconnection. Be that as it may, an elementary and perhaps obvious deduction is that whenever the reconnection rate increases, current is diverted into two types of propagating wave systems, rarefaction upstream, compressive downstream. Laboratory reconnection experiments have demonstrated a sudden and large amplitude process of this kind, and it is now reasonably certain that it is the result of an abrupt decrease in the density of inflowing plasma. In the Hones model of magnetotail dynamics, plasmoid detachment occurs when the reconnection process progresses through the last closed field line of the plasma sheet boundary, i.e., from the high density plasma sheet into the low density tail lobes. To the extent the laboratory evidence may be relevant we would have to conclude that the substorm is a result of the plasmoid detachment process. In this case, the impulsive current diversion would drive a blast wave earthward and forcibly eject the plasmoid tailward. Propagation of the large amplitude rarefaction waves into the tail lobes would be caused by the extraction of the magnetic energy needed to drive the earthward and tailward processes. But the essential question remains: What has theory to say about the effect of sudden density decreases on the rate of reconnection?

Introduction

Closely correlated in time with substorms at the earth, and highly suggestive of a cause-effect relationship, a complex, transient dynamic process takes place in the magnetotail. The plasmoid model of Hones [1976; 1979; 1980, see also Hones, this volume, Fig. 4] [Hones et al., 1982], is quite remarkable in the way it organizes a wide variety of observational data into a single conceptual image and recently, strong additional support came from ISEE-3 observations of tailward convecting plasmoid structures in the magnetotail at ~230 Re (see Hones, this volume).

Reconnection, the subject of this conference, is, of course, the process involved in the production and detachment of plasmoids [Birn, 1980; Birn and Hones, 1981]. Reconnection is also the cause of substorms according to the widely believed reconnection-substorm model [Atkinson, 1966, 1967; McPherron, 1979; McPherron et al., 1973; Russell and McPherron, 1973; Russell, 1974]. The purpose of this paper is two-fold: first, to present evidence suggesting that the particular form of time-dependent reconnection involved in plasmoid detachment can produce dynamical effects that fit remarkably well into an updated version of the reconnection-substorm model; second, to point out an important new direction for theoretical work, having the potential to strengthen substantially the concept suggested by the evidence.

Any dynamical model of the magnetosphere must be built upon a physical foundation consistent with the dynamo nature of its interaction with the magnetized solar wind [Reiff et al., 1981] and the resultant, potentially unstable stress balance equilibria maintained in the magnetotail [Schindler, 1974, 1975, 1976, 1980]. The magnetotail is observed to be a permanent, albeit dynamically changing feature. Obviously then, if the continuous but generally nonsteady buildup of tensional stress and free energy storage is not matched 'pari pasu' by release mechanisms, one must expect episodic releases. Indeed, theoretical analysis strongly suggests that even under a steadily applied solar wind stress, stable equilibrium solutions may be rendered virtually nonexistent by the internal magnetospheric convection which is necessarily induced [Siscoe and Cummings, 1969; Hill and

Reiff, 1980a,b; Erickson and Wolf, 1980; Schindler and Birn, 1982; Birn and Schindler, 1983]. In this case, periodic releases (relaxation oscillations) would be expected (Bratenahl and Baum, 1976). But in any case the plasmoid detachment process involves time-dependent reconnection in a form that, in all likelihood, includes as a special feature, the sudden decrease in density of inflowing plasma. Time-dependent reconnection to date has received far too little theoretical attention, and this special feature, virtually none at all. Under these circumstances, we shall have to rely heavily on evidence and insights derived from our own experience with laboratory reconnection experiments [Bratenahl and Yeates, 1970; Baum and Bratenahl, 1980, and references therein], especially the more recent developments [Beeler, 1980] and our interpretation of them [Bratenahl and Baum, 1983].

An Updated Version of the
Reconnection Substorm Model

Essentially all the near-earth magnetic signatures of substorms seem to be explainable one way or another in terms of the rapid development of a current circuit known as the substorm current loop or current wedge. The magnetic signatures we refer to are those which are due to: the enhancement and morphological changes in the westward polar electrojet (PEJ); the impulsive dipolarization of the inner nightside magnetosphere; and the injection of hot plasma sheet plasma into and enhancing thereby, the ring current system.

The substorm current loop extends from the roughly central ~7Re portion of the near-earth plasma sheet all the way to the auroral ionosphere where it enhances the westward polar electrojet simultaneously in both hemispheres. The current flows earthward in the dawn side of the wedge and tailward in the dusk side. The current circuit closes in the plasma sheet but its flow direction there is opposite to the normal dawn-dusk plasma sheet current. Therefore, its development constitutes a major disruption and diversion of the pre-existing current system.

Atkinson [1966; 1967] seems to have been the first to suggest that the formation and rapid development of the current loop might be the result of impulsive reconnection ("impulsive recombination") at the plasma sheet end of what he called the "recombination slot" (the wedge). His reconnection substorm model developed quickly [McPherron et al., 1973; McPherron, 1979], and gained wide acceptance, now forming an integral part of the Hones plasmoid model of magnetotail dynamics. Nevertheless, some strong skepticism persists [Akasofu, 1980, 1981; Kan et al., 1980]. One good reason at least for this is that surprisingly little attention has been given to a rather critical issue: what is the physics of impulsive reconnection; and in the context of magnetospheric physics, can it produce the effects that have been attributed to it? We believe we can give an affirmative answer albeit based on qualitative arguments and evidence from an analogue experiment in the laboratory. To prepare the way for this, some refinements of the reconnection substorm model need first to be considered.

It has been known for some time [Heppner et al., 1971] that the westward PEJ is associated with an equatorward electric field, and it is, therefore, predominantly a Hall current, $\vec{J}_H$. But Hall currents are dissipationless, $\vec{E}\cdot\vec{J}_H=0$, and are produced by whatever it is that drives the Pedersen current, $\vec{J}_p$, at right angles to it through a medium with a tensor conductivity such as the ionosphere. The Pedersen current, flowing across the narrow PEJ strip, and parallel to the equatorward electric field ($\vec{E}\cdot\vec{J}_p>0$), is alone responsible for ohmic heating of the strip and resultant increase in tensor conductivity. To understand the substorm enhancement of the westward Hall current PEJ, one must understand the generator mechanism that drives the Pedersen current. In the consideration of energy transfer from its storage as magnetic free energy in the tail lobes, and its subsequent conversion to other forms in the substorm process we note that $\vec{E}\cdot\vec{J}>0$ holds in the reconnection process itself. Therefore the Pedersen current generator ($\vec{E}\cdot\vec{J}<0$) must be located somewhere else within the current wedge, between the X-line and the ionosphere. Note that Kan and Akasofu [1976] have already proposed a generator mechanism to drive field aligned sheet currents and V-shaped potential structures associated with auroral arcs. Our motivation here, however, is different. We see the primary need for the generator is to account for the Pederson current, and in a way that fits the plasmoid detachment process. The production of V-potential structures associated with auroral arcs is then a very natural but secondary consequence.

In most prior work, the wedge current system has been thought to consist only of field-aligned line-currents connecting the eastern and western ends of the PEJ with the corresponding ends of the x-line. This arrangement, which fits neatly into the cross-tail current disruption-diversion picture actually performs a passive function: it simply provides for the unimpeded flow of the Hall current. On the other hand, the Pedersen current circuit, flowing across the PEJ strip must be continuous with field-aligned sheet-current structures extending completely across the wedge in the manner discussed by Rostoker and Boström [1976]: current flowing earthward on the poleward sheet and tailward on the equatorward sheet. This current must then close by passage from the latter to the former through a generator region where an oppositely directed electric field is somehow maintained at the expense of kinetic energy. It is, perhaps not so obvious how this arrangement fits into the cross-tail current disruption-diversion picture but we suggest that the following scenario (Figure 1) may help make it clear nevertheless. Following the example of Rostoker and Boström [1976] we look for an EMF due to a charge separation electric field. In this case it is due to the acceleration of plasma by an earthward propagating compressive wave. An additional contribution comes later from enhancement of the normal magnetosphere convection, particularly the part that eventually flows sunward around the dawn side of the earth (Figure 1 near E, the inner edge of the plasma sheet). We must emphasize that this process is highly time dependent.

Hill and Reiff [1980b] point out that, as the rate increases in the temporal development of reconnection, compressive waves are propagated earthward and tailward, and rarefaction waves, northward and southward into the tail lobes.

It should be noted that all four of these wave systems carry current in circuits that close across the plasma sheet in a direction opposite to the normal dawn-dusk cross-tail current, and hence represent diversions of that current system. Moreover, in time-dependent reconnection they are driven by an EMF which is the line integral of E along the x-line or separator and its value is equal to the rate of increase of magnetic flux whose field lines close within either of the two outflow sectors defined by the separatrix.

The rarefaction waves are generated by the increasing rate of convective inflow of magnetic flux and plasma and accompany the withdrawal of energy from progressively more remote regions of the tail lobes. The compressive waves are similarly generated by the increasing outflow rate. The tailward waves feed the growing plasmoid and develop the force that pushes it down the tail, ultimately expelling it at altogether when the separatrix of last closed field lines is crossed. Our interest here, however, concerns the earthward-propagating compressive waves.

As they move earthward the waves continuously accelerate the plasma they encounter ahead, and in the process they slow down and tend to be overtaken by waves coming from behind. In the extreme case of impulsive reconnection which we shall shortly come to, a double shock structure or blast wave is expected to develop: a forward-facing fast shock, separated from a backward-facing slow shock by an expanding region of compressed hot plasma. The blast wave in Figure 1 is the region between C and D, bounded by the shocks (the heavy dashed lines).

Looking at the blast wave from front to back the magnetic profile would show an upward jump in magnetic field at both shocks. However, as

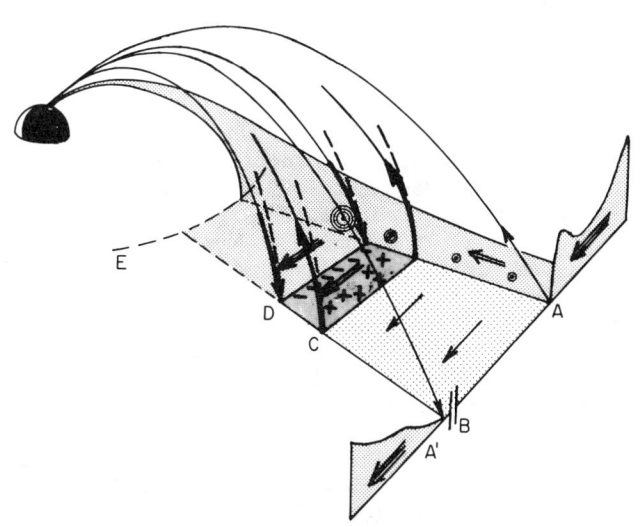

Fig. 1. Schematic view of the developing current wedge shortly after plasmoid detachment from the x-line (AA') adjacent to the modified plasma sheet profile. The region between the dashed lines is the blast wave: slow mode shock at C, fast mode shock at D, E marks the location of the plasma sheet inner edge before substorm onset. The arrows represent currents.

the blast wave approaches the inner edge of the plasma sheet (Figure 1 at E), there is a point, generally well outside synchronous orbit where the increase in ambient field ahead is sufficient to wipe out the field jump in the forward shock, and it loses its fast mode property. The remaining field jump was beautifully documented for the first time by the ATS-6 and OGO-5 pair of spacecraft [Russell, 1974; his Figure 17]. On the basis of abundant evidence gathered since [Moore et al., 1981; Baker et al., this volume and references therein] which includes detailed polarization studies of Pi-2 bursts [Lester et al., 1983], there can be little doubt as to what causes the impulsive dipolarization of the magnetic field within the substorm wedge, and the injection of hot plasma into the trapped-particle ring current system [Young, 1983]. The timing of all these events with midlatitude ground based magnetic signatures provides nearly conclusive evidence that all these low latitude substorm effects are parts of a single process, the blast wave effect.

Now if one is willing to accept this part of the overall substorm scenario, then the more speculative part dealing with its high latitude aspects, namely the Pedersen current generator and its relation to the enhancement of the westward PEJ, should not require a large mental leap. First it should be noted that the earthward wave system in the wedge is driven to a large extent by the $J_\perp$ across the wedge. Second, the strong acceleration of ambient plasma by the forward-facing wave (with the accompanying development of an adverse pressure gradient) will produce a polarization electric field due to charge separation: the heavy ions will lag tailward behind the electrons so that the resulting electric field is directed toward the earth. The field-aligned sheet currents (Figure 1, heavy arrows), continuous with the ionospheric Pedersen current, are the resulting depolarization currents which are load-limited by two effects: the Ohmic Pedersen conductivity across the PEJ strip; and v-shaped electrostatic structures [Carlquist and Bostrom, 1970; Gurnett, 1972; Swift et al., 1976] that are likely to develop at moderate altitudes on upward current-carrying sheets due to over-current instabilities. These structures account for auroral arcs through energization of downward-flowing electrons. Moreover, since the blast wave is likely to be turbulent [Swift, 1977], the polarization EMF is likely to be irregular with corresponding irregularities in the field-aligned sheet currents. This in turn will imprint similar irregularities in the Ohmic heating and hence the Pedersen conductivity within the PEJ strip in both auroral zones. Thus arcs should develop first on the equatorward edge (downward directed electrons), spreading poleward later on as observed. Let us turn now to the process of development and detachment of plasmoids.

Observations [Hones et al., 1982] and numerical modeling [Birn, 1980; Birn and Hones, 1981] and theory [Pellinen, 1979; Heikkila, 1983] support the concept of development and detachment of magnetic islands or plasmoids through a time-dependent reconnection process somewhere in the 10-25 Re distance range down the tail. On a purely theoretical basis, moreover, this distance range has been found to be particularly sensitive to the development of a reconnection x-line [Zwingman, 1983; see also Erickson, this volume].

The process of plasma sheet thinning, which provides the necessary precondition for the ion tearing mode [Schindler, 1980] which initiates the reconnection process, has been provided a plausible theoretical model basis by Hill and Reiff [1980b]. Progressing through nonlinear stages within the closed plasma sheet field lines the process becomes dominated by a single x-line o-line pair, forming thus a growing magnetic island or plasmoid.

It is estimated that it takes about 5-10 minutes for the reconnection process to reach the last closed field line boundary of the plasma sheet and begin to invade the very low density plasma on open field lines in the tail lobes. At this point, the plasmoid, no longer linked to the earth by closed field lines, becomes detached, and is forcibly ejected down the tail.

On the basis of laboratory experiments to be discussed in the next section, we are very confident in suggesting that at the moment the plasmoid is detached, the rapid decrease in density of the inflowing plasma will trigger what we call an impulsive flux transfer event (IFTE) or impulsive reconnection in which the voltage drop along the x-line could increase by an order of magnitude or more. This would almost certainly initiate the earthward propagation of the blast wave responsible for the substorm phenomena we have just finished discussing. However, before this catastrophic event, the preceding 5-10 minutes of the developing plasmoid through time-dependent reconnection contains most of the same earthward ingredients in a much milder, shock-free form. We suggest that this phase is responsible for the well-known precursor effects leading to the catastrophic onset of the expansion phase, in particular, the sudden brightening of a single auroral arc, characteristically at the equatorward border of what should become the enhanced westward PEJ strip. Even the momentary dimming of that arc [Pellinen and Heikkila, 1978] just before expansive phase onset might be explained by the same magnetic field effect that alters the character of the forward-facing shock in the blast wave.

One additional concept involved with the reconnection current-diversion process is worth considering and it involves both pairs of diverted current circuits, rarefaction and compression. Along the edges of the wedge, there is a transition on conduction mode. In the undisturbed plasma sheet, the current is carried by both ions and electrons. The diverted current, however, is driven by the inductive electric field, and is carried mostly by electrons. In the transition between these regions, there must be compensating divergences in both the ion and electron currents, negative at the anode side and positive at the cathode side [Atkinson, 1967]. This should result in a substantial buildup of plasma at the dawn edge of the wedge and a corresponding depletion at the dusk edge (Figure 1, see plasma sheet profile near points A, A'). There are at least three resulting effects that are likely to be important: (i) double layer development (Figure 1 at point B; see also next section) would be favored on the dusk side of the x-line; (ii) the x-line would tend to elongate in the duskward direction; (iii) the excess plasma buildup on the dawn side might tend to modify the fast mode shock at a greater distance from the earth on that side and at a lesser distance on the dusk (depletion) side. Items (i) and (ii) could account for the westward expansion of the PEJ, and item (iii), the "fault line effect" [Lezniak and Winkler, 1970] enhancing the asymmetry of plasma injection and the ring current.

Time-Dependent Reconnection in the Laboratory

The Double Inverse Pinch device (DIPD), (Figure 2a), consists basically of a pair of insulation-covered conducting rods which pass through the lower of two large disc-shaped electrodes and connect to the upper one. External capacitor banks are applied between the rods and the lower electrode. During the 13 μsec first quarter cycle rise of parallel currents in the two rods, the following events are observed.

(i) A pair of outward-expanding cylindrical inverse pinches are initiated by current discharge over the insulator surfaces in preionized argon plasma. High density plasma is maintained in the pinch current sheets by their snowplow action into the stationary plasma ahead. But it is essential to note also that behind the expanding pinches, an outward flow of low density plasma from the rods is maintained by ionization of occluded gas that is continually stripped off the insulator surfaces. The fact is that after initial breakaway of the pinch current systems, each subsequent increment of pinch current must also first develop over the insulator surface before it propagates away as an elementary (compressive) hydromagnetic wave. The current increment then adds itself to its pinch current system only after the elementary wave overtakes it.

(ii) The inverse pinches collide at the center of the device after ~2μsec. Their currents then merge together in an expanding oval of compressed plasma, and leave behind at the collision site a compressed ridge of high density plasma and pinch current, sandwiched between the incoming streams of low density plasma. The ridge plasma is ejected downstream at high velocity in a process that starts at its extremities but then quickly propagates upstream toward the collision site which is an x-line in the 2-D magnetic field and a double stagnation line in the plasma flow. At sufficiently low initial filling density, the combination of inflow and outflow quickly evolves into the Petschek (1964) structure of quasi-stationary slow mode shocks, a quiescent reconnection mode that persists for the next 4μsec, (Fig. 2b). During this interval, the system performs like a transformer (see Bratenahl and Baum, 1983 for more detailed analysis). The primary consists of the rod currents, I_p with an equal return on the outer oval cylinder of compressed plasma (Figure 2c). The field of this current is $\vec{B}_p$. The secondary is the current system I_s associated with the Petschek system. Its return is in the opposite direction on the outer oval (Figure 2d). The field of this current system, $\vec{B}_s$ adds to the two flux cells where field lines link just one or the other of the two rods, but subtracts from the third flux cell where field lines link both rods. The total field is, $\vec{B}_{tot} = \vec{B}_p + \vec{B}_s$ (Figure 2b).

(iii) After 4μsec of steadily increasing I_s, the quiescent mode is abruptly terminated by an IFTE. This is an "explosive" process resembling the failure of a reservoir dam, (Figure 3). Large amplitude compressive blast waves propagate downstream and large amplitude expansion fans propagate upstream (Figure 3, Third panel, upper

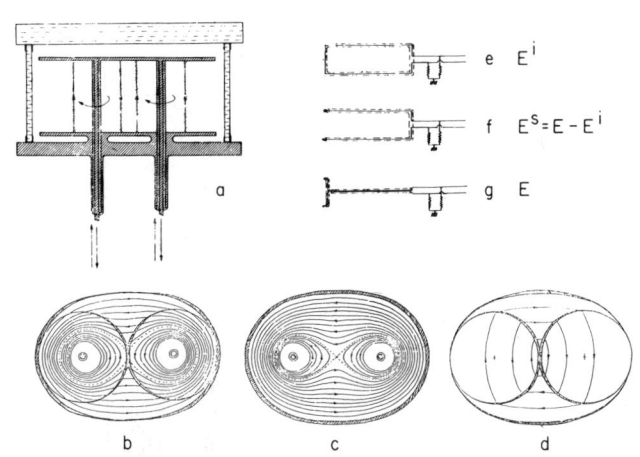

Fig. 2. a) The DIPD; b) Schematic of the field and currents before IFTE; c) The same after IFTE and the large amplitude waves have dispersed; d) The Petschek wave system of currents and field. In the transformer analogy, c) is the primary, d) is the induced secondary. e)-g) show the probes that, in the strictly 2-D magnetic field symmetry, permit direct measurement of $\vec{E}^i$, $\vec{E}^s$, and $\vec{E}$.

and lower trace, respectively). If a line integral of $\vec{B}_{tot}$ is taken along a path enclosing but lying beyond the region yet reached by the wave disturbances, the total current enclosed gives no indication whatever that IFTE has taken place. On the other hand the current system I_s, locally diminishes very rapidly (Figure 3, bottom panel). The implication of this is quite obvious: at IFTE, the original I_s system is disrupted and diverted into the large amplitude waves.

The basic cause of this disruption is equally obvious. During the quiescent period of reconnection, the I_s system was supported by the initial high density ridge of plasma. This condition, however, could not be maintained indefinitely by inflow of low density plasma replacing the steady pumpout of high density plasma. The inevitable sudden drop of density on the x-line (Figure 3, bottom panel), has the effect of opening a switch in an inductive circuit, the current is rerouted on whatever nearby path is available. But with stored flux and available free energy upstream, the diverted current is propelled away, accompanied by a surge of inductive electric field $\vec{E}^i = -\partial\vec{A}/\partial t$ (Figure 3, top panel, short dashed lines) where A is the vector potential (Coulomb gauge).

There is an obvious macroscopic analogy here with what might be expected to occur in the case of the relatively high density plasma sheet when reconnection suddenly starts to invade the very low density plasma in the tail lobes. It seems possible that this analogy might even carry over somewhat into the microscale. When IFTE was first discovered in the DIPD, all the available evidence suggested that the over-current

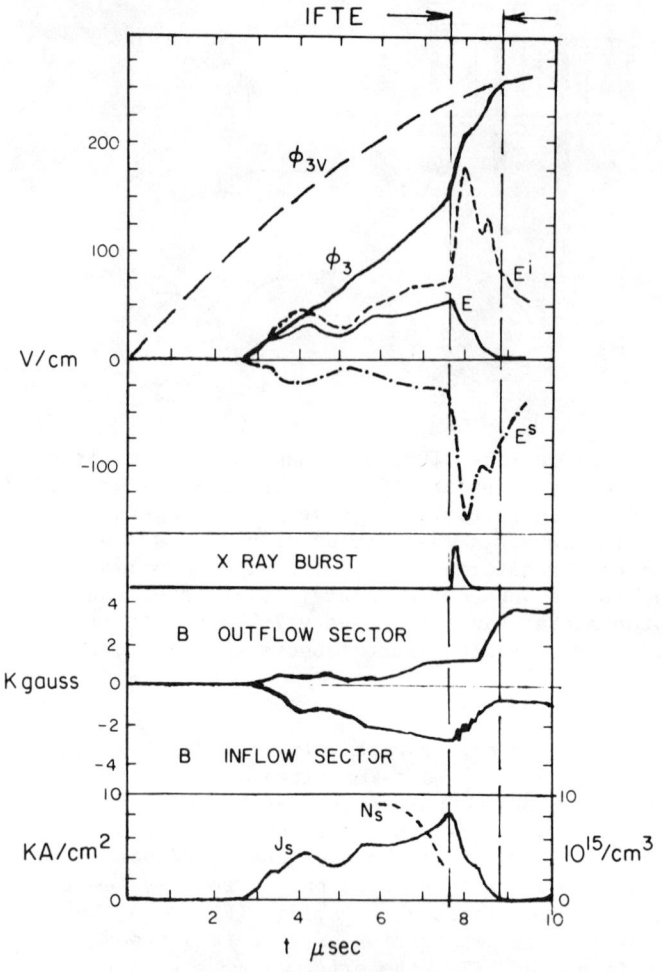

Fig. 3. Schematic of typical IFTE in the DIPD. Top panel: $\vec{E}^i$, $\vec{E}^s$, $\vec{E}$ (also shown: the magnetic flux $\phi 3$ outside the dotted line separatrix of Figure 2b; ϕ_{3v} shows hypothetically what that flux would be if there were no induced secondary current system). Second panel: the x-ray burst. Third panel: the behavior of B at points 1 cm from the x-line: upper trace, the blast wave; lower trace, the upstream rarefaction wave. Bottom panel: x-line current and plasma densities.

instability (shortage of sufficient current carriers) took the form of a sudden manifestation of anomalous resistivity. Ion acoustic noise was actually observed [Baum and Bratenahl, 1974] also x-ray production at the anode [Baum, et al., 1973] suggesting an initial electron runaway (Figure 3, second panel). Alfven [1969] thought otherwise, urging that we look for an explosive double layer. Ten years were to pass before suitable probes could be devised that could separately measure $\vec{E}^i$, $\vec{E}^s = -\vec{\nabla}\phi_s$ (where ϕ is the scalar potential and $\vec{E} = \vec{E}^i + \vec{E}^s$; Figure 2e,f,g, [see also Stenzel and Gekelman, 1979]. By observing the behavior of these quantities at points distributed over 80% of the x-line (the double probes could not be placed closer than 1 cm to the electrodes), Beeler [1979] found, that the ratio E/J along the x-line remained essentially constant (in Figure 3, compare E trace, first panel with behavior of J_s, bottom panel): there was no indication whatever of anomalous resistivity. Quite the contrary, it was found that prior to IFTE, the 30% excess of $\vec{E}^i$ over $\vec{E}$ is offset by an opposing $\vec{E}^s$. During IFTE, $\vec{E}^i$ jumps to a value 2-3 times $\vec{E}$ and yet again the excess of $\vec{E}^i$ over $\vec{E}$ is offset by an opposing $\vec{E}^s$. Now since the total electrostatic voltage drop between the electrodes is determined by the observed, but almost negligible, IR drop over the 10 cm path length of outer return path plasma oval, we must conclude that somewhere in the unexplored regions close to the electrodes, $\vec{E}^s$ must reverse sign i.e, be in the same direction as $\vec{E}^i$ and be very large. If not an actual double layer, at least we must have an abnormally large electrode sheath.

This is the basis for the suggestion in the previous section that the explosive formation of a double layer (Figure 1, at point B) may be the crucial element in impulsive reconnection following an abrupt decrease in inflowing plasma. The situation within the plasma sheet current wedge may not be so very different. On the rapid time scale of the presumed IFTE, the conditions are indeed favorable for over-current instability to take the form of double layer formation, and this possibility should be seriously considered. In the DIPD the pulse of $\vec{E}^i$ is typically characterized by at least two peaks (Figure 3, top panel). Explosive double layers are notoriously short lived (intermittent). This could easily explain the multiple poleward leaps in the PEJ that are frequently observed [Wiens and Rostoker, 1975; Nagai et al., 1983].

Conclusions

The Hones plasmoid model seems to be fully capable of producing, directly or indirectly, nearly all the presently observed phenomena associated with substorms. This conclusion, however is based to a considerable extent on evidence and insights gained from laboratory experiments exhibiting impulsive reconnection. The DIPD experiment, of course, does not replicate magnetotail dynamics, but merely suggests ideas that appear to be applicable and worth pursuing further. In addition, this conclusion is based on a modified version of the more conventional conception of the current wedge. This is made necessary in order to account for the Pedersen current generator, the impulsive nature of the dipolarization process and injection of hot plasma into the ring current system, all in a manner consistent with impulsive reconnection.

With these provisos in mind, it would seem that a concerted theoretical effort should be directed toward a better understanding of: (i) time-dependent reconnection, especially the effect of a sudden density decrease in the inflowing plasma with he possibility of explosive double layer formation; (ii) the nature of hydromagnetic blast waves, their ability in the plasma sheet to serve as the Pedersen circuit current generator, and how such a blast wave might be modified as it advances into the steadily increasing magnetic field of the earth.

Acknowledgments. The authors would like to express their deep appreciation to one of the referees for a very thoughtful critique. We gratefully acknowledge the support of the Institute of Geophysics and Planetary Physics, Los Alamos National Laboratory, grant number 82-101.

References

Akasofu, S.-I., What is a Magnetospheric Substorm?, in Dynamics of the Magnetosphere, Ed. S.-I. Akasofu, D. Reidel, Dordrecht, Holland, 447, 1980.

Akasofu, S.-I., Magnetospheric Substorms: Newly Emerging Model, Planet. Space Sci., 29, 1069, 1981.

Alfven, H., Private Communication, 1969.

Atkinson, G., A Theory of Polar Substorms, J. Geophys. Res., 71, 5157, 1966.

Atkinson, G., An Approximate Flow Equation for Geomagnetic Flux Tubes and its Application to Polar Substorms, J. Geophys. Res., 72, 5373, 1967.

Baum, P. J., A. Bratenahl, Spectrum of Turbulence at a Magnetic Neutral Point, Phys. Fluids, 17, 1232, 1974.

Baum, P. J., A. Bratenahl, Magnetic Reconnection Experiments, Advances in Electronics and Electron Physics, Ed. L. Marton and C. Marton, Academic Press, N. Y., 54, 1, 1980.

Baum, P. J., A. Bratenahl, R. S. White, X-ray and Electron Spectra from the Double Inverse Pinch Device, Phys. Fluids, 16, 226, 1973.

Beeler, R. G., The Impulsive Flux Transfer Event in the Double Inverse Pinch device, Ph. D. Dissertation, Physics Department, Univ. of Calif. at Riverside, 1979.

Birn, J., Computer Studies of the Dynamic Evolution of the Geomagnetic Tail, J. Geophys. Res., 85, 1214, 1980.

Birn, J., E. W. Hones, Three-Dimensional Computer Modeling of Dyamic Reconnection in the Geomagnetic Tail, J. Geophys. Res., 86, 6802, 1981.

Birn, J., K. Schindler, Self-consistent Theory of Three-dimensional Convection in the Geomagnetic Tail, J. Geophys. Res., 88, 6969, 1983.

Bratenahl, A., P. J. Baum, On Flares, Substorms, and the Theory of Impulsive Flux Transfer Events, Solar Physics, 47, 345, 1976.

Bratenahl, A., P. J. Baum, Laboratory Experiments on Reconnection in Current Sheets, in I. A. U. Symposium No. 107, Ed. M. R. Kundu, College Park, Md., Aug. 8-11, 1983, in press, D. Reidel, Dordrecht, Holland, 1983.

Bratenahl, A., C. M. Yeates, Experimental Study of Magnetic Flux Transfer at the Hyperbolic Neutral Point, Phys. Fluids, 13, 2696, 1970.

Carlquist, C.-G. and R. Boström, Space Charge Regions Above the Aurora, J. Geophys. Res. 75, 7140, 1970.

Erickson, G. M., R. A. Wolf, Is Steady Convection Possible in the Earth's Magnetotail?, Geophys. Res. Lett., 7, 897, 1980.

Gurnett, D. W., Electric Fields and Plasma Observations in the Magnetosphere, Critical Problems of Magnetospheric Physics, Ed. E. R. Dyer, Proc. COSPAR/IAGA/URSI Symp., Madrid, Spain, ICUSTP Secretariate, Nat. Acad. Sci., Washington, DC, 123, 1972.

Heikkila, W. J., The Reason for Magnetospheric Substorms and Solar Flares, Solar Physics, 88, 324, 1983.

Heppner, J. P., J. D. Stolarik, E. M. Wescott, Electric-Field Measurements and the Identification of Currents Causing Magnetic Disturbances in the Polar Cap, J. Geophys. Res., 76, 6028, 1971.

Hill, T. W., P. H. Reiff, On the Cause of Plasma-Sheet Thinning During Magnetospheric Substorms, Geophys. Res. Lett., 7, 177, 1980a.

Hill, T. W., P. H. Reiff, Plasma-Sheet Dynamics and Magnetospheric Substorms, Planet. Space Sci., 28, 363, 1980b.

Hones, E. W., The Magnetotail: Its Generation and Dissipation, in Physics of Solar Planetary Environments, Ed. D. J. Williams, AGU, II, 558, 1976.

Hones, E. W., Transient Phenomena in the Magnetotail and Their Relation to Substorms, Space Sci. Review, 23, 393, 1979.

Hones, E. W., Plasma Flow in the Magnetotail and its Implications for Substorm Theories, in Dynamics of the Magnetosphere, Ed. S.-I. Akasofu, D. Reidel, Dordrecht, Holland, 545, 1980.

Hones, E. W., J. Birn, S. J. Bame, G. Paschmann and C. T. Russell, On the Three-Dimensional Magnetic Structure of the Plasmoid Created in the Magnetotail at Substorm Onset, Geophys. Res. Lett., 9, 203, 1982.

Hughes, W. J., Pulsation Research During the IMS, Rev. Geophys. Space Phys., 20, 641, 1982.

Kamide, Y., S.-I. Akasofu, Global Distribution of Pedersen and Hall Currents on the Electric Potential Pattern During Moderately Disturbed Period, J. Geophys. Res., 86, 3665, 1981.

Kan, J. R. and S.-I. Akasofu, Energy Source and Mechanism for Accelerating the Electrons and Driving the Field-Aligned Currents of the Discrete Auroral Arc, J. Geophys. Res., 81, 5123, 1976.

Kan, J. R., S.-I. Akasofu, L. C. Lee, Physical

Processes for the Onset of Magnetospheric Substorms, in Dynamics of the Magnetosphere, Ed. S.-I. Akasofu, D. Reidel, Dordrecht, Holland, 357, 1980.

Lester, M., W. J. Hughes, H. W. Singer, Polarization Patterns of Pi-2 Magnetic Pulsations and the Substorm Current Wedge, J. Geophys. Res., 88, 7958, 1983.

Lezniak, T. W. and J. R. Winkler, Experimental Study of Magnetospheric Motions and the Acceleration of Energetic Electrons During Substorms, J. Geophys. Res., 75, 7075, 1970.

McPherron, R. L., Magnetospheric Substorms, Rev. Geophys. Space Phys., 17, 657, 1979.

McPherron, R. L., C. T. Russell, M. P. Aubry, Satellite Studies of Magnetospheric Substorms on August 15, 1968. 9. Phenomenological Model for Substorms, J. Geophys. Res., 78, 3131, 1973.

Moore, T. E., R. L. Arnoldy, J. Feynman, D. A. Hardy, Propagating Substorm Injection Fronts, J. Geophys. Res., 86, 6713, 1981.

Nagai, T., D. N. Baker, P. R. Higbie, Development of Substorm Activity in Multiple-onset Substorms at Synchronous Orbit, J. Geophys. Res., 88, 6994, 1983.

Pellinen, R. J., Model for the Onset of a Magnetospheric Substorm, Planet. Space Sci., 27, 19, 1979.

Pellinen, R. J., W. J. Heikkila, Observations of Auroral Fading Before Breakup, J. Geophys. Res., 83, 4207, 1978.

Petschek, H. E., Magnetic Field Annihilation, NASA SP-50, U. S. Government Printing Office, p.425, 1964.

Reiff, P. H., R. W. Spiro, T. W. Hill, Dependence of Polar Cap Potential Drop on Interplanetary Parameters, J. Geophys. Res., 86, 7639, 1981.

Rostoker, G., R. Bostrom, A Mechanism for Driving the Gross Birkeland Current Configuration in the Auroral Oval, J. Geophys. Res., 81, 235, 1976.

Russell, C. T., The Solar Wind and Magnetospheric Dynamics, in Correlated Interplanetary and Magnetospheric Observations, Ed. D. E. Page, D. Reidel, Dordrecht, Holland, 1, 1974.

Russell, C. T., R. L. McPherron, The Magnetotail and Substorms, Space Science Reviews, 11, 111, 1973.

Schindler, K., A Theory for the Substorm Mechanism, J. Geophys. Res., 79, 2803, 1974.

Schindler, K., Plasma and Fields in the Magnetospheric Tail, Space Science Reviews, 17, 589, 1975.

Schindler, K., Similarities and Differences Between Magnetospheric Substorms and Solar Flares, Solar Physics, 47, 91, 1976.

Schindler, K., Macroinstabilities in the Magnetotail, in Dynamics of the Magnetosphere, Ed. S.-I. Akasofu, D. Reidel, Dordrecht, Holland, 311, 1980.

Schindler, K., J. Birn, Self-Consistent Theory of Time-Dependent Convection in the Earth's Magnetotail, J. Geophys. Res., 87, 2263, 1982.

Siscoe, G. L., W. D. Cummings, On the Cause of Geomagnetic Bays, Planet. Space Sci., 17, 1295, 1969.

Stenzel, R. L. and W. Gekelman, Experiments on Magnetic Field Line Reconnection, Phys. Rev. Lett., 42, 1055, 1979.

Swift, D. W., Turbulent Generation of Electrostatic Fields in the Magnetosphere, J. Geophys. Res., 82, 5143, 1977.

Wiens, R. G., G. Rostoker, Characteristics of the Development of the Westward Electrojet During the Expansive Phase of Magnetospheric Substorms, J. Geophys. Res., 80, 2109, 1975.

Young, D. T., Near-Equatorial Magnetospheric Particles from 1 eV to 1 MeV, Rev. Geophys. Spc. Phys., 21, 402, 1983.

Zwingman, W., Self-consistent Magnetotail Theory: Equilibrium Structures Including Arbitrary Variation Along the Tail Axis, J. Geophys. Res., 88, 9109, 1983.

Questions and Answers

Vasyliunas: The principle of gauge invariance implies that the "inductive electric field" $-\partial A/\partial t$ cannot be measured as a physical quantity but is merely a mathematical construct. What you call the measured inductive field $\bar{E}^i$ seems to be merely the average value of the physical electric field $\bar{E}$ around a closed loop = $\oint \bar{E}\cdot d\ell/L$ (L = length of loop), and nothing is gained by giving the electric field in the integral any special name.

Bratenahl: I am aware that this is a controversial subject but it is important to get it settled. That is one of my reasons for coming here, it is especially important in time-dependent reconnection. It is simply not correct (Konopinski, 1981) to say that the rotational and irrotational contributions to $\bar{E}$ are not separately measurable. In the Coulomb gauge, the vector and scalar potentials from which $\bar{E}^i$ and $\bar{E}^s$ may be individually derived, can be uniquely determined point by point through global Poisson-type integrals for any real system (sources bounded within a finite space). Gauge invariance, on the other hand is a completely separate issue. It is simply the mathematical device (Cragin and Heikkila, 1981) which provides for the simplification of the otherwise sometimes awkward, unsymmetrical set of electrodynamic equations when they are expressed in terms of the potentials $\bar{A}$ and ϕ. Although you are indeed correct in saying that $\oint \bar{E}\cdot d\ell/L = \oint \bar{E}^i \cdot d\ell/L$, the point by point differences in the integrands can be very different. In fact, they can be profoundly affected by the behavior of $\bar{E}^s$.

LABORATORY EXPERIMENTS ON CURRENT SHEET DISRUPTIONS, DOUBLE LAYERS, TURBULENCE AND RECONNECTION

W. Gekelman and R. Stenzel

Department of Physics
University of California
Los Angeles, CA 90024

Laboratory plasmas offer a unique opportunity to study the interaction of magnetic fields with plasmas. In particular, the diffusion region of the reconnection problem can be properly modeled since its scale length is much smaller than that of the entire convection region. Whereas in space, it is difficult to locate and analyze the field reversal region, in the laboratory one can perform repeated in-situ measurements of fields, particles and waves with high resolution in time and space. Such experiments are presently being performed at UCLA. Since the major results have been presented in recent publications this extended abstract serves to summarize the findings and to point out the relevant references.

By imposing an X-type neutral magnetic field of increasing strength in time on a large collisionless plasma column we observed the self-consistent formation of a classical neutral sheet [Stenzel and Gekelman 1981]. The associated plasma flow leads to jetting velocities approaching the Alfvén speed [Gekelman et al., 1982]. The plasma is heated and compressed but the energy density nkT maximizes near the outflow regions of the neutral sheet rather than in the central stagnation region [Gekelman and Stenzel, 1981]. Comprehensive measurements of fields, flows, density and temperature were combined in the generalized Ohm's law in order to determine the resistivity which was found to be anomalously large and spatially inhomogeneous [Stenzel et al., 1982]. The complex spatial resistivity profile led some observers to comment on measurement errors [Baum and Bratenahl, 1983] but our subsequent observations of runaway electrons in the current sheet [Stenzel et al., 1983] showed, more fundamentally, that the fluid description of the plasma in the diffusion region is an over simplification and should be replaced by kinetic theory.

In spite of the macroscopic stability of the neutral sheet the diffusion region contains a rich spectrum of microinstabilities. Most extensive investigations have been performed on magnetic fluctuations [Gekelman et al., 1982b; Gekelman and Stenzel, 1984]. From two-probe cross-correlation functions Fourier analyzed in time and space the fluctuation spectrum $S_{(\omega, \vec{K})}$ has been identified to follow the dispersion surfaces of oblique whistler waves. Furthermore, the polarization of individual modes has been confirmed to be right-hand circular. Although the wave intensity decreases from the lower hybrid to the electron gyrofrequency recent higher frequency scans indicated magnetic fluctuations near cyclotron harmonics. These are thought to be associated with electromagnetic cyclotron harmonic waves in high beta plasmas [Khaladze et al., 1972]. Electrostatic wave analysis has revealed ion acoustic turbulence [Stenzel and Gekelman, 1981b] and enhanced microwave emission due to mode conversion from unstable electron plasma waves [Whelan and Stenzel, 1981].

In order to understand the origin of the various microinstabilities the particle distribution functions have been investigated. Using a novel directional velocity analyzer [Stenzel et al., 1983b] the electrons are observed to exhibit tails of runaway electrons accelerated by the electric field along the separator [Stenzel and Gekelman, 1984]. These velocity-space anisotropies are confined to the current sheet and are not observed in the adjacent convection region. They do not only provide a source of free energy for exciting instabilities but are also important for the transport properties. For example, the energetic electron tail makes a major contribution to the neutral sheet current and energy flow along the separator. The plasma resistivity is not determined by the bulk electron temperature but by inertial effects and wave-particle interactions of the spatially varying particle tails.

Particle accelerations, heating and energy transport undoubtedly involve wave-particle interactions in a collisionless plasma. Observations of heat transport have been initiated by studying electron temperature fluctuations with a new array probe [Wild et al., 1983a]. Spatial cross-correlation measurements revealed that electron heating occurs in bursts arising from current bursts. The heat pulses spread out from the current sheet across the magnetic field at an enhanced thermal conductivity.

Using the directional velocity analyzer in conjunction with a high-speed computer for mass data processing it is now possible to map diffusion processes in phase space $(\vec{v}, \vec{r})$. Observations indicate that a perturbation of the distribution function in the current sheet by an obstacle (e.g. satellites, moons) relaxes on a spatial scale short compared with interparticle collision lengths in a turbulent plasma [Wild et al., 1983b]. These microscopic kinetic processes have to be further investigated in order to explain the "anomalous" fluid properties.

Macroscopic disruptions of a current sheet are of primary concern for rapid reconnection events such as substorms and flares. By increasing the current density spontaneous impulsive disruptions of the current sheet have been observed [Stenzel et al., 1983c]. These involve a redistribution of the current profile from a sheet to two channels. Magnetic energy is released in this process which manifests itself by an inductive voltage drop inside the plasma at the location of the current disruption. A nonstationary potential double layer is formed during the disruption. Particle beams are generated and the plasma is thinned in the perturbed current layer. Onset and recovery of these spontaneous disruptions are explained by the nonlinear interaction between the global circuit properties and the local plasma behavior.

The location of the magnetic energy storage, the energy transport and its conversion into kinetic energy are crucial aspects of dynamic reconnection processes. New disruption experiments have been initiated where the bulk of the magnetic energy is stored locally inside the plasma and suddenly released by a controlled current disruption [Stenzel and Gekelman, 1983]. Circulating currents corresponding to large amplitude magnetic waves are set up which tear the current sheet, convect and dissipate the excess magnetic energy.

References

Baum, P. J. and A. Bratenahl, Comments, *J. Geophys. Res.* 88, 506, 1983.

Gekelman, W. and R. L. Stenzel, Magnetic field line reconnection experiments 2. Plasma parameters, *J. Geophys. Res.* 86, 659, 1981.

Gekelman, W., R. L. Stenzel and N. Wild, Magnetic field line reconnection experiments 3. Ion acceleration, flows, and anomalous scattering, *J. Geophys. Res.* 87, 101, 1982a.

Gekelman, W., R. L. Stenzel and N. Wild, Magnetic field line reconnection experiments, *Physica Scripta T2/2*, 277, 1982b.

Gekelman, W. and R. L. Stenzel, Magnetic field line reconnection experiments

6. Magnetic turbulence, *J. Geophys. Res.* (in press), 1984.

Khaladze, T. D., D. G. Lominadze and K. N. Stepanov, Dispersion of cyclotron waves in a plasma, *Sov. Phys.-Tech. Phys.* 17, 196, 1972.

Stenzel, R. L. and W. Gekelman, Magnetic field line reconnection experiments 1. Field topologies, *J. Geophys. Res.* 86, 649, 1981a.

Stenzel, R. L. and W. Gekelman, Magnetic field line reconnection experiments, *Proceedings of the XVth International Conference on Phenomena in Ionized Gases, Vol. III*, p. 46, Minsk, USSR, July 14-18, 1981b.

Stenzel, R. L., W. Gekelman and N. Wild, Magnetic field line reconnection experiments 4. Resistivity, heating, and energy flow, *J. Geophys. Res.* 87, 111, 1982.

Stenzel, R. L., W. Gekelman, and N. Wild, Electron distribution functions in a current sheet, *Phys. Fluids 26*, 1949, 1983a.

Stenzel, R. L., W. Gekelman, N. Wild, J. M. Urrutia, and D. A. Whelan, Directional velocity analyzer for measuring electron distribution functions in plasmas, *Rev. Sci. Instrum. 54*, 1302, 1983b.

Stenzel, R. L., W. Gekelman and N. Wild, Magnetic field line reconnection experiments 5. Current disruptions and double layers, *J. Geophys. Res. 88*, 4793, 1983c.

Stenzel, R. L. and W. Gekelman, Perturbations of current sheets during magnetic reconnection, *Bull. Am. Phys. Soc. 28*, 1093, 1983.

Stenzel, R. L. and W. Gekelman, Laboratory experiments on current sheet disruptions, double layers, turbulence and reconnection, in *Proceedings of the IAU Symposium No. 107, College Park, MD, August 8-11, 1983*, D. Reidel Publ. Co., 1984.

Whelan, D. A. and R. L. Stenzel, Electromagnetic wave excitation in a large laboratory beam-plasma system, *Phys. Rev. Lett. 47*, 95, 1981.

Wild, N., R. L. Stenzel and W. Gekelman, Electron temperature measurements using a 12-channel array probe, *Rev. Sci. Instrum. 54*, 935, 1983a.

Wild, N., R. L. Stenzel and W. Gekelman, Experimental modelling of satellite wakes in auroral arcs, *Geophys. Res. Lett. 10*, 682, 1983b.

Questions and Answers

Vasyliunas: From the electron parameters shown in one of your figures, it appears that the relative ion-electron drift speed, J/ne, is nearly comparable to the electron thermal speed. In the magnetotail, on the other hand, J/ne is typically much smaller than the electron thermal speed. Thus it is not clear to what extent features like double layers, which may depend on exceeding critical current densities, can be scaled from the laboratory to the magnetotail.

Stenzel: Indeed, the scaling from the laboratory to the magnetotail is a difficult task and the point you raised is important to consider. The threshold condition $vd/ve \geq 1$ holds for the classical one-dimensional, stationary double layer in unmagnetized plasmas. In the laboratory we observe three-dimensional, impulsive double layers in a high beta plasma with thresholds as low as $vd/ve \simeq 0.1$. It is theoretically not clear what the threshold of a possible double layer in the magnetotail might be. This, unfortunately, compounds the difficulties for making predictions, but it does not rule out a qualitative comparison between the laboratory and space.

MAGNETIC RECONNECTION IN DOUBLETS

Torkil H. Jensen

P. O. Box 85608, San Diego, CA 92138

Doublet [Ohkawa, 1968] is a magnetic configuration with the prospect of being able to confine plasmas for thermonuclear fusion. Because of this prospect Doublet research was carried out at GA for a number of years.

The Doublet configuration is axisymmetric and it employs a strong toroidal magnetic field. The poloidal magnetic field forms one hyperbolic magnetic axis located in the midplane and two elliptic axes above and below the midplane. Thus a separatrix with a cross sectional shape of a figure "8" is formed. Experimentally, Doublets were found subject to two sudden, mostly axisymettric deformations. During one of these the plasma split into two separate plasmas, each with one elliptic axis much like two separate tokamak plasmas. During the other the three magnetic axes tended first to merge at the midplane and secondly the plasma became unstable toward motion either up or down. These deformations involve magnetic reconnections. In the first, field lines which initially link all three magnetic axes reconnect to link only one of the elliptic magnetic axes while the reverse is true for the second type of deformation.

A reasonable, conceptually simple theoretical description of these observations was found [Jensen and Thompson, 1978; Jensen and McClain, 1978, 1982; McClain and Jensen, 1981]. A resistive MHD theory for linear, axisymmetric perturbations showed that the plasma was near marginal stability before one of the deformations took place; furthermore the mode in question has the property of causing either of the reconnections observed dependent on the sign of the initial perturbation which in the context of linear theory grows exponentially. This linear perturbation theory fails however to explain two features of the observed deformations. One is that the linear growth rates of the instability are so small that it may not explain the observed rapid deformations. Another is that when an experiment is repeated under one set of circumstances, only one of the deformations was observed; in the context of linear theory one might expect that random fluctuation might lead randomly to one of the two deformations. A nonlinear perturbation theory [Jensen and McClain, 1982] taking into account quadratic perturbation terms may explain both features. The nonlinear terms always lead to an increased growth rate of the instability making it "explosive" in the sense that the amplitude becomes unbounded in a finite time. Furthermore this destabilization takes place only for one sign of the instability. For the other sign the nonlinear terms are stabilizing. The sign for which destabilization takes place depends on the unperturbed equilibrium only. Thus, the linear and the nonlinear theories provide a reasonable explanation of the observed features. According to these theories the instability may be stabilized with proper arrangements of the experimental circumstances; this was also supported by experimental observations.

Under circumstances where the plasma is stable, partial reconnection of the type mentioned may be driven in an oscillatory fashion by the external circuits. Since reconnections are accompanied by resistive dissipation, such forced reconnections may be used for heating the plasma. A theory [Jensen et al., 1981] for such heating of the plasma suggests practicality for fusion applications even in the limit of vanishing plasma resistivity.

Acknowledgment. Work supported by Department of Energy, Contract DE-AT03-76-ET51011.

References

Jensen, T. H. and W. B. Thompson, Low frequency response of a resistive plasma to axially independent or axisymmetric perturbations, *J. Plasma Phys. 19*, 227, 1978.

Jensen, T. H. and F. W. McClain, Numerical parameter study of stability against resistive axisymmetric modes for doublets, *J. Plasma Phys. 20*, 61, 1978.

Jensen, T. H., F. W. McClain and H. Grad, Low-frequency heating of doublets, *J. Plasma Phys. 25*, 133 1981.

Jensen, T. H. and F. W. McClain, Nonlinear stability of doublets against axisymmetric resistive MHD modes, *J. Plasma Phys. 28*, 495, 1982.

McClain, F. W. and T. H. Jensen, Stability of Doublet III plasmas against axisymmetric resistive MHD modes, *J. Plasma Phys. 26*, 431, 1981.

Ohkawa, T., Multipole configuration with plasma current, *Kakuyugo-Kenkyo 20, 6*, 557, 1968.

SOME COMMENTS ON SOLAR RECONNECTION PROBLEMS

Ronald G. Giovanelli

CSIRO Division of Applied Physics
Lindfield, NSW 2070, Australia

Reconnection of magnetic flux tubes is a wide-spread phenomenon both on and inside the Sun. Today I have time to refer to only two unsolved examples, one for the theoreticians and one for the observers.

The solar convection zone, some 200 000 km deep, contains vast numbers of flux tubes, all in pressure equilibrium with the non-magnetic surroundings. It is unlikely that reconnection between such tubes occurs by Petschek's mechanism in the convection zone. However, there is another reconnection process which *will* operate. From time to time, gas motions drag flux tubes into contact, their axes intersecting mostly at significant angles (Figure 1a). The conditions for reconnection are established and maintained provided the gas motions continue to drag the tubes into contact. Normally the reconnection rate is not less than that in the usual Petschek process, i.e. 0.01 to 0.1 V_A. But occasionally the angle between the axes may be very small whilst the tubes are being dragged into contact, or twisted around one another (Figure 1b). Piddington [1976] has pointed out that the components of the forces of magnetic tension which tend to draw reconnected field lines apart are then very weak and reconnection is slow. Twisted tubes may well be reconnected in a manner resembling spot welding. They may spiral in either sense, and are believed responsible for building flux ropes by a sequence of these processes. Eventually the flux ropes float to the surface to produce further sunspots. This virtually non-reconnection process is a vital part of the solar cycle, and can be observed in part from its surface consequences. I would like to ask theoreticians to help by studying it in as great detail as possible.

Probably everyone is satisfied that reconnection produces flares though the detailed field structure responsible has been elusive. Hoping for progress towards observing the magnetic configuration in flares, I organized a session on the geometry of magnetic fields in active phenomena at the Patras IAU meeting last year. It turned out that virtually nothing was known of the details of fields in flares. For that matter, the situation was not much better in respect of fields in prominences. The two groups studying these (the H. A. O. team and Leroy's group in France) had quite differing interpretations. This was not surprising. The two methods were different (HAO, Stokes polarimeter: Leroy, Hanle effect), Leroy's angular resolution being about 5″ and the HAO's about 6″ to 8″. As we shall see, these are inadequate to yield the required information about field structure. But why is this relevant to this Conference? I, for one, am convinced that reconnection occurs in prominences, and that it will be far easier to study the phenomenon there than in flares. I recommend a massive programme to do this. Flares can come later. The new HAO Stokes polarimeter and the French THEMIS project should provide excellent tools with which to undertake such work, but they must be used properly.

Let us see what is involved by considering the properties of quiescent prominences. Many long years ago, Lucien and Marguerite d'Azambuja [1948] found that over one-third of all prominences are born in spot groups. They can be seen on the Sun's disk as absorption markings on monocromatic Hα images, where they are known as filaments. Over some months they drift polewards, being tilted gradually towards parallels of latitude by differential rotation. Figure 2 shows typical filaments at a time of high solar activity. These structures are usually called *quiescent* prominences. They have intricate geometric structures. The d'Azambujas' low-resolution spectroheliograms showed them looking like arches of a bridge (Figure 3, right). At the ∼5″ to 8″ resolution used to study prominence fields, the prominences would not look very different from this.

I was intrigued 20-25 years ago by a huge montage on the wall of K. O. Kiepenheuer's office, which showed a prominence on the disk. It appeared like an array of shorter, spaced structures all tilted at about the same angle (perhaps up to 45°) from the overall filament axis. Some excellent pictures of this phenomenon are shown in Figures 4 and 5. The prominence axis separates regions of opposite magnetic polarities, and the tranverse structures seen in Hα lie presumably along lines of force. Sturrock [1972] had stressed earlier that shear parallel to the axis would produce a systematic tilting of field lines, and this would be at least in qualitative agreement with what is found. So far, so good. It would be nice to trace the field lines back to the photosphere. Do you know that no trace of magnetic footpoints has ever been found, no matter how the Hα structures have been extrapolated? Perhaps we do not know yet what we should be looking for.

The most astounding features of quiescent prominences is that they disappear occasionally by blowing off the Sun. Figure 6 shows an interesting example studied in detail by Marie McCabe of the University of Hawaii. The prominence,

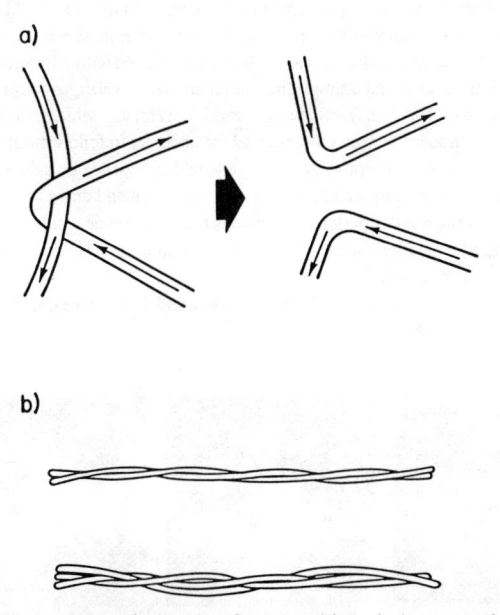

Fig. 1. Reconnection of subsurface flux tubes. (a) Typical reconnection of flux tubes dragged together at a substantial angle. (b) Lack of reconnection when almost parallel flux tubes are wound around one another.

358 SOLAR RECONNECTION PROBLEMS

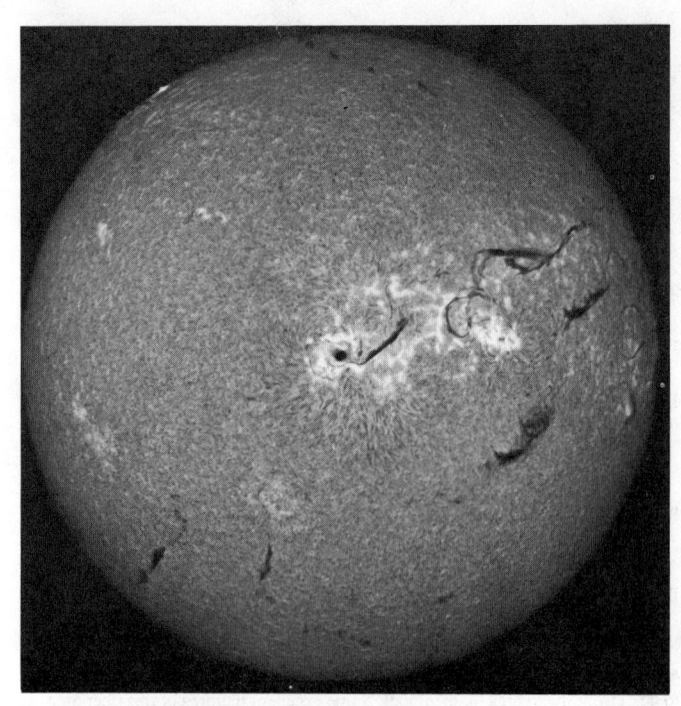

Fig. 2. Monochromatic image of the sun in Hα. The filaments appear as long dark lines. (Sacramento Peak Observatory photograph.)

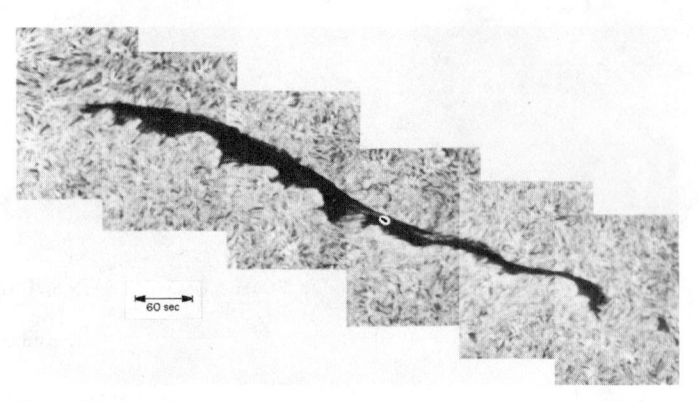

Fig. 4. Filament on the disk. An Hα image showing projections on the side further from the limb and a smooth edge on the side nearer the limb. The fine structures are inclined at substantial angles to the filament axis. (Sacramento Peak Observatory photograph.)

just visible near the limb at upper left on 11, 12 and 13 March, suddenly disappeared between 1919 and 1938 UT on March 13. The coronagraph at 1959 UT shows the filament material rising above the limb to appear as a splendid eruptive prominence with much fine structure. There is also another filament nearby (top right in Figure 6a) which disappeared before 1927 UT, March 12. In two such cases out of three, the filament reappears 2 or 3 days later, almost as if nothing had happened. But in about one-third of the events, the filament disappears permanently. So there are alternative final results for what appears to be a single phenomenon. Why?

My own observations of eruptuve prominences began in the pre-World War II days at the then Commonwealth Solar Observatory. The Hale spectrohelioscope then in use had limited angular resolution, but was equipped with a fine grating monochromator. It also had a simple line-shifter so that wavelength could be changed by up to ±10Å in a fraction of a second. This facility which, tragically, has been lost in modern instruments, enabled the observer to follow motions with great delicacy. A typical eruptive prominence would commence by a darkening of the Hα structures and by simultaneously developing up and down motions of increasing amplitude. After perhaps 20 minutes the prominence would start to blow up. Large two-ribbon flares sometimes accompanied the disappearance of the filament, but not always.

Of course Sturrock's theory of flares comes to mind. But before jumping to conclusions we should look at the geometry of these prominences more closely. Figure 7 is a photo of a quiescent prominence at the limb. It indicates our disinterest in prominences that we do not know whether this is the typical appearance of all quiescent prominences or only some. What would it look like on the disk? A great deal could have been discovered about the geometry of these prominences from a systematic study of their structures from limb to limb. Would they be the same in prominences about to erupt? We solar astronomers must bear a joint blame for not pressing for such studies over the past dozen years. There is a low, curved arch which joins the chromosphere at either end of these presumably transverse structures. Above this is a close array of fine vertical streamers whose upper ends are at somewhat irregular heights. The low ends seem to join the curved arch. Gas falls slowly down these streamers. It is not clear to me just what the motions are in the curved arch, but convention would say that they are downwards at both ends. A word of warning—an hour of good observations is far better than relying on even long-established convention.

While Hα structures usually lie along field lines, the absence of an Hα structure does not imply the absence of a field line. Fields probably extend widely outside many such structures. Even closer than 1500 km to the base of the photosphere, the gas pressure is probably inadequate to confine prominence fields, and these must extend sideways and lower. Surface currents may well preserve the identity of individual tubes of force, and stronger gas concentrations and appropriate

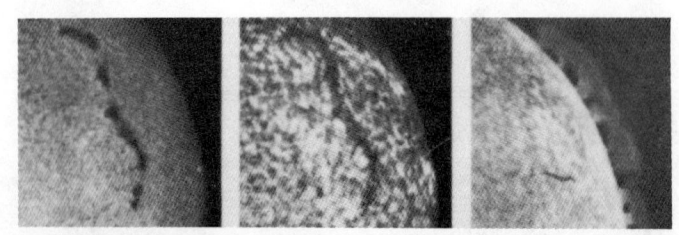

Fig. 3. A quiescent prominence on the disk and at the limb. Low resolution Hα and K₃ images obtained by M. and L. d'Azambuja. (Meudon Observatory photographs.)

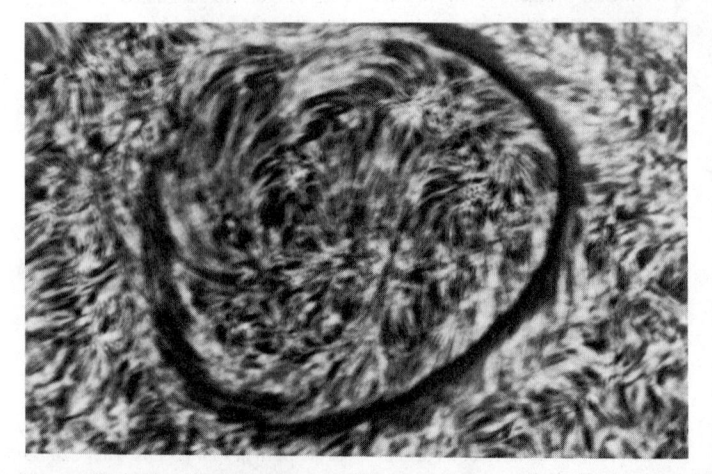

Fig. 5. Filament on the disk. An unusual filament bent into almost circular shape by large-scale photospheric motions. The individual fine structures are at considerable angles to the axis of the filament. (Sacramento Peak Observatory photograph.)

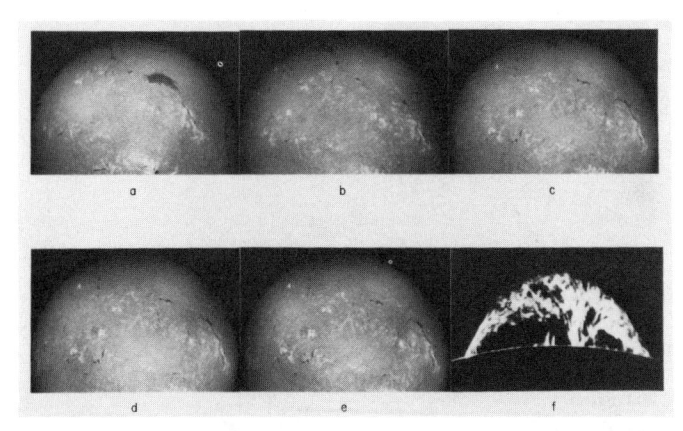

Fig. 6. A sequence of five Hα images of a disk filament which erupted and disappeared on 13 March, 1970 and a coronograph picture showing the filament material as it rose into the corona above the solar limb. Universal times are: (a) 1903, 11/3; (b) 1927, 12/3; (c) 1903, 13/3; (d) 1919, 13/3; (e) 1938, 13/3; (f) 1959, 13/3. (Photographs by courtesy of Marie McCabe, Institute for Astronomy, University of Hawaii.)

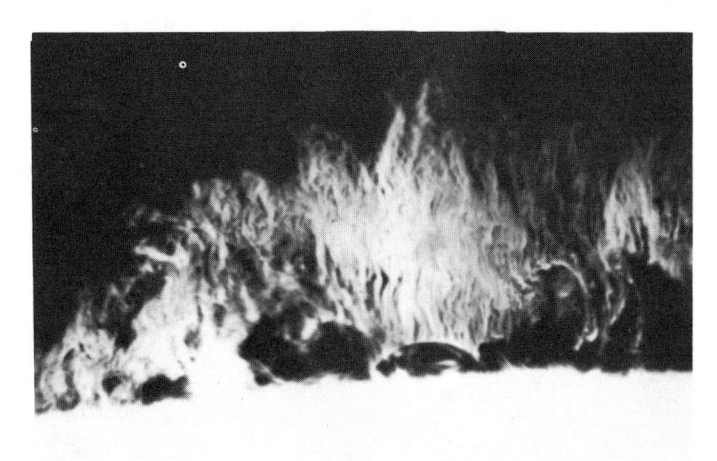

Fig. 7. Quiesent prominence on the limb. A high-resolution Hα image showing arrays of almost-vertical streamers ending apparently in arches. (Sacramento Peak Observatory photograph.)

temperatures can make them visible. The field probably extends below the arch to levels of the order of 1000 km, forming a magnetic canopy over a non-magnetic region. Direct measurements of photospheric fields below typical arches would help in studying this phenomenon. I cannot think of a magnetic structure which could give rise to both the arch and the vertical array of fine structures above. Do these really represent vertical field lines connecting with the arch? Or is the gas there, denser than its surroundings, just falling down under its own weight, pulling field lines down with it? And how do they terminate? We require daily observations of quiescent prominences, from limb to limb, so as to establish their typical three-dimensional structures. Only after the field geometry is established is any further work on the theory of prominences and related reconnection phenomena justified.

The task is difficult, even for the quiescent case. Since the structures are seen best in Hα, this may be the best line for magnetic observations. Unfortunately Hα is broad and has weak splitting, $g \sim 1$. The line profile varies from feature to feature, so in the case of the HAO group the full Stokes profiles must be recorded. It will be essential to use an area detector array. Two things reduce the difficulty of the task: in some cases averages will be permissible over large numbers of detector elements; in others, exposure times can be fairly long.

To study reconnection, eruptive filaments and prominences must be observed systematically with all the power and resolution that can be devised. Early warning should be available, and it may be possible to develop some simple monitor based on the darkening or motion of the filament. Finally, the precise location of the region observed must be known from simultaneous Hα images.

A query arises as to the value, in such studies, of space observations such as are planned with the Shuttle and SOT. If the mission is limited to about a week, as seems probable, it is highly unlikely that an adequate set of observations could be obtained differentiating between prominences which recur or disperse after eruption. The main programme must be carried out from ground-based observatories, though SOT observations would provide valuable information on the structure of prominences and their fields at the highest resolution.

Despite the difficulties involved, it is a fantastically good project.

It has been a unique experience for me to speak to a conference in this way (by video tape). I would rather have been present, but that has been impossible. I appreciate the opportunity greatly, and wish to express my gratitude to Ed Hones for providing me with it. Thank you all for your attention.

References

d'Azambuja, M. and Mme. L., Annales de l'Observatoire de Paris, Section d'Astrophysique a Meudon, Tome VI, Fascicule VII, 1948.
Piddington, J. H., Solar magnetic fields and convection VI: Basic properties of magnetic flux tubes, *Astrophys. Space Sci. 45*, p. 47, 1976.
Sturrock, P. A., Magnetic models of solar flares, in *Solar Activity Observations and Predictions,* (McIntosh and Dryer, eds.) p. 163, MIT Press, Cambridge, 1972.

DRIVEN AND NON-DRIVEN RECONNECTION; BOUNDARY CONDITIONS

W. I. Axford

Victoria University of Wellington
Private Bag, Wellington, New Zealand

I have enjoyed this conference very much and feel that considerable progress has been made in our understanding of reconnection and particularly in the degree of acceptance of the general principles of reconnection by both space and laboratory physicists. In fact, I felt somewhat diffident about the conference beforehand, as the Gordon conference held a few years ago was much less successful in this respect and I have had the feeling that during the last 10 years the subject had even gone backwards to some extent.

The main topics I want to emphasize in this discussion are contained also in the paper, "Magnetic Field Reconnection," that I have submitted for publication in the proceedings. They concern the importance (or otherwise) of resistivity to the reconnection process. I have argued that provided the electrical conductivity is finite, reconnection is not inhibited provided the system wants to evolve in such a direction. In an initially static situation, the conductivity may determine the growth time of an instability which leads to large-scale reconnection (the "tearing" mode) but the subsequent reconfiguration will proceed in general more rapidly (the Alfvén wave travel time across the system being a rough estimate of the time scale if there are no other strong stabilizing influences). If, however, the system is not static, but is driven in such a way that reconnection is promoted by having oppositely-directed field lines pushed toward each other, then one might not have to wait for finite conductivity to permit field lines to slowly diffuse toward each other and reconnect in a current layer of a given thickness. Instead, the thickness may decrease until the diffusion time becomes suitably short and reconnection proceeds at the rate the system demands.

Dayside reconnection on the magnetopause would seem to be an example of driven reconnection (it is somewhat similar to the Sonnerup-Priest stagnation point flow problem), however, it is complicated by the difficulty of removing the reconnecting plasma and field with the overall configuration controlled to such an extent by the pressures exerted by the solar wind along the whole magnetopause. Reconnection evidently proceeds relatively easily in this case but seems to be intermittent for reasons which are not necessarily connected with the effective resistivity of the medium.

In the tail of the magnetosphere the situation is quite different, but also complicated. Substorms may be considered as a kind of tearing mode instability problem in which the tail evolves through a series of configurations, one of which is eventually tearing-mode unstable with a growth time determined by whatever resistivity is available. However, the tail configuration is driven continuously by internal stresses and accordingly the thickness of the region of field reversal in the middle of the plasma sheet can become as small as necessary to select a tearing rate which is as fast as the system can accept. In this sense there is a difference between driven and non-driven reconnection in that the changes induced in the latter case may take place faster than the growth time of instabilities corresponding to any instantaneous configuration.

The onset of reconnection in the magnetotail is probably connected with the growth phase of substorms and the thinning of the plasma sheet and is not necessarily associated with the occurrence of the more dramatic events which one usually considers to be the onset of a substorm. The latter appear to be associated not with the onset of reconnection but the transition to the occurrence of reconnection in the tail lobes, where the Alfvén speed is perhaps 10 times higher than in the plasma sheet. Whether or not this should be described as an instability is a matter of taste but it is fair to say that Akasofu might be right, in a certain sense, in claiming that the magnetosphere is basically a driven system. However, I cannot help remarking that 20 years ago he convinced me that, in fact, substorms are a more spontaneous phenomenon in general and not simply a direct response to external influences from the interplanetary medium as Hines and I believed at the time.

Rostoker: Akasofu is saying that both processes are operative and he believes one is dominant. He passed this (Figure 1) on to me, which came from the last Los Alamos meeting, and he tries to emphasize that, in fact, the substorm involves both processes. And, in fact, the picture he draws in there is not too far away from the kind of thing we believe right now, I think.

Axford: I wish he were here to clarify the matter. Now let me touch on the question of "current interruption" as a cause of flares and substorms. There has been a trend, in recent years, to attempt to do plasma physics in terms of currents and electric fields, usually ignoring the fact that the plasma has mass, momentum, and energy. This approach works perfectly well in some situations, for example the ionosphere, where simplifying assumptions can reasonably be made concerning the magnetic field configuration and the dynamics of the plasma itself is not an important issue. In general, however, one does not get very far with this approach and those who use it tend to have difficulties in describing and analyzing quite simple problems. The magnetotail is a place where one must begin with a hydromagnetic analysis and consider aspects such as plasma flow and pressure, stress balance and so on. There are, of course, currents and electric fields, but, as nicely described in Dungey's book, *Cosmic Electrodynamics* they are in a sense secondary quantities which can be determined once the others are known. Quite a few people have been tempted to say that a substorm is simply a current diversion and evidently feel that they have understood something. In fact this is no more than a statement that the magnetic field changes and does not shed any light on the matter. It is more important to know why the current had its original configuration, which depends on the distribution of plasma, its motion and pressure, and also on the state of the interplanetary medium and the ionosphere; that is, the currents are basically determined by the required $\underline{j} \times \underline{B}$ forces. These quantities, $(p, \rho, \underline{u}, \underline{B})$ in turn, presumably contain the elements which determine whether a substorm is going to occur or not and it is unrewarding to simply claim, for example, that a large scale change in the current distribution must be due to a sudden appearance of anomalous resistivity. Ohm's law in a plasma is a useful way of determining the electric field if the magnetic field and plasma velocity are known but it does not control the current.

Dungey: What causes electric fields?

Axford: As I said, you have written a good book on this. In general in a plasma, $\underline{E} + \underline{v} \times \underline{B} \approx 0$ and one can therefore regard $\underline{v}$ and $\underline{B}$ (and in consequence the plasma pressure and density) as "causing" electric fields - if the question has any point. It might be argued that perhaps a large $E_\parallel$ could be produced by anomalous resistivity, but this is hardly likely to cause current in the

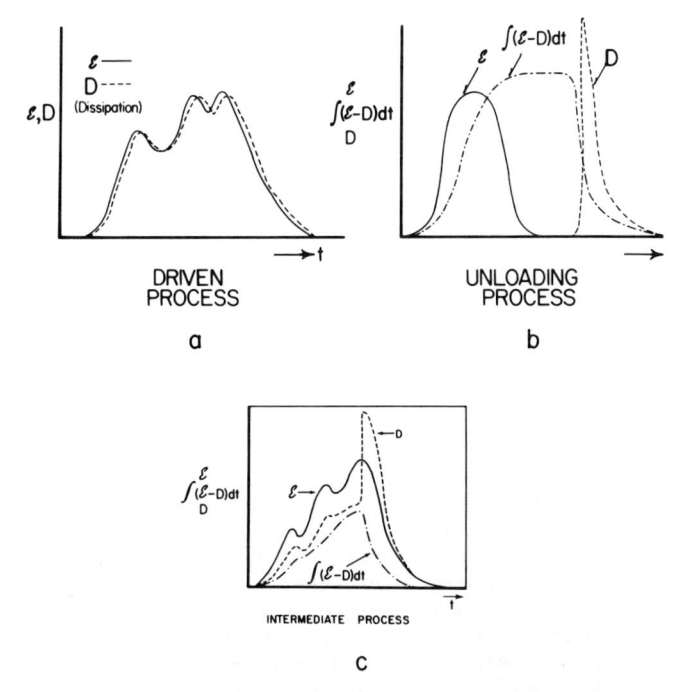

Fig. 1. Differences between a driven system (a) and an unloading process (b) are schematically shown by using an energy input function $\epsilon(t)$ and an output (dissipation) function $D(t)$. In (c), an intermediate process between (a) and (b) is shown. (From Akasofu, 1980).

magnetotail to be "diverted" into the ionosphere to cause a substorm (it is however a reasonable basis for a solar flare theory in the manner of Alfvén, Carlquist and Colgate).

Dungey: I agree that (those quantities) tell you what the current is. Then, knowing the current, resistivity tells you the electric field.

Axford: In general the current and resistivity make a negligible contribution to the electric field except near a neutral point where Ohm's law does not have a simple form.

Baum: May I point out that the first equation (plasma moments) is the sum of the ion and electron equations of motion and the second (Ohm's law) is the difference of these equations. I really do not see that one is more fundamental than the other.

Axford: I think it is useful to read the appropriate chapter of Dungey's book and consider the implications in the circumstances of the magnetotail.

For me, the most interesting new results reported at the meeting were those concerning the composition of the plasma sheet and field-line tracing by energetic particles, which were presented by the Los Alamos group. There is plenty of evidence that auroral particles are mainly solar (from helium-3 and doubly-ionized helium-4) but one was not able to be sure about the whole plasma sheet since there is a copious source of ionospheric ions (singly-ionized oxygen, for example). It is very interesting to learn that the plasma sheet oxygen is in fact solar, which suggests that the plasma sheet originates mainly at relatively large ($\gtrsim 50\ R_E$) distance from the earth.

Energetic particles provide perhaps the only means we have of tracing the topology of field lines in the magnetotail and of determining whether or not reconnection has occurred. The effectiveness of the procedure has been well confirmed at this conference, in particular in showing that plasmoids with closed field lines are found in the magnetotail as part of a substorm and (possibly) in demonstrating neutral point acceleration. The latter is important because it is difficult to probe, or even find, a neutral point directly, and neutral point acceleration itself may be of significance in other connections (the first phase acceleration of solar flares, for example).

Hones: I'd like to address your comment that you're leaning toward the view that the substorm may, indeed, be driven. I think that there's certainly a change in character of energy dissipation that is clearly identifiable, the breakup of the auroras. When you have suitable instruments on a satellite in the magnetotail, you can see the flow start tailward on a reproducible time sequence, clearly related to a sharp increase in dissipation at the earth. But there is, before that, a so-called growth phase that was invented by McPherron 13 years ago, and that is the driven feature, I think. But what really identifies a substorm is the sudden arc brightening.

Axford: Sure, but the question is just whether you consider that an instability or not.

Reference

Akasofu, S.-I., What is a magnetospheric substorm?, in *Dynamics of the Magnetosphere* (S.-I. Akasofu, ed.) p. 447, D. Reidel Publ. Co., Dordrecht, Holland, 1980.

COMMENTS ON NUMERICAL SIMULATIONS

T. Sato

Hiroshima University, Institute for Fusion Theory
Hiroshima 730, Japan

I'd like to comment on a couple of things about numerical simulation. One is just about the philosophical discussion we have had, that is, spontaneous or driven. This is a very interesting discussion to me because philosophy is sort of a motivating force for us theorists. The other thing is the numerical or technical one. Frankly, I didn't want to touch on the technical matter because this should be a common sense one for those who are working at numerical simulation. But since many people take numerical simulation results at their face value, I would like to remind you of the reality hidden behind them.

First, I wish to point out that the meaning of "driven" in driven reconnection is different from that defined by Schindler or Akasofu. My definition is closer to Axford's definition. In Figure 1 the horizontal axis is time and the vertical axis is free energy. In the spontaneous case, for some unpredicted reason an excess energy of the system is suddenly released at a certain point. However, one does not answer how such an unstable state far beyond a stable limit is realized in the magnetotail. In the driven case, there is a definite energy buildup phase starting from a stable state; namely, energy in the black box increases from a stable level subject to an external source. When the state has reached a certain position, the energy is released suddenly. The difference between driven and spontaneous is whether the cause (plasma flow) to trigger reconnection is specified or reconnection is triggered unpredictably. Another difference is that in driven reconnection the reconnection rate is dependent on the speed of the external plasma flow, but in spontaneous reconnection the rate is dependent on the internal condition such as the resistivity. From the purely theoretical point of view, this is a very interesting and challenging problem. Thus, a more elaborate study is required in the future.

Let's go on to the numerical problem. A couple of key points related to reconnection simulation are given in Figures 2-4. The first one is the resistivity, and the second one is the mesh size. As you know in the tearing mode instability and the resistivity driven reconnection, the reconnection rate is given by E_R. When we normalize this reconnection rate by $V_A B_0$ (V_A is the Alfvén velocity and B_0 is the magnetic field), then it becomes inversely proportional to the magnetic Reynolds number. Thus, the reconnection rate is a function of magnetic Reynolds Number. In the numerical study of the tearing mode instability, therefore, at least we have to repeat simulations by changing the magnitude of resistivity and compare the results to clarify how sensitive to the resistivity the nonlinear evolution is. This is the minimum requirement for the study of the tearing mode, because usually an unrealistically small Reynolds number like 100 must be used in the simulation. No matter how elegantly the simulation was done, therefore, the result would not be convincing if only one example was given. The result may be correct. But a parametric study is an unavoidable task when one wishes to do simulation on the tearing mode and resistivity driven reconnection.

In the driven case, when we normalize the reconnection rate by $V_A B_0$, then it becomes V_{EX}/V_A. Therefore, we have to check whether driven reconnection is really a function of V_{EX}/V_A or not. Here I am going to show some results which we did some years ago. We have changed the magnitude of the resistivity, and also changed the driving force, V_{EX}. The two left hand panels of Figure 3, [Sato and Hayashi, 1979] show the case when we changed the resistivity (defined by

$\eta = \alpha(j_{NS} - j_c)^2$ by an order of magnitude for a fixed driven force. The horizontal axis shows time and the reconnection rate is given vertically. The onset time of reconnection is different when we change the resistivity. But note that the final reconnection rate is almost constant. The right top panel of Figure 3 represents the case when the driving force ($A_0 = E_R^D$) is changed by fixing the resistivity. The right bottom panel shows the input flow flux versus the reconnection rate. It would be obvious that in the driven case, the result is certainly dependent upon the driving force. Having done such a parametric study, we can reach a convincing conclusion.

What I want to do here is not so much to display our result as to say that at least such a parametric study must be done for any case. Otherwise, the result is not physically convincing, no matter how it looks similar to a naturally occurring phenomenon.

Now we go on to the last part, shown in Figure 4; that is the mesh size. This is also a key factor in simulations. For example, in the tearing mode case, the mesh size must be taken to be much smaller than the resistive layer width which is given by λ. This factor L is the neutral sheet width, or the current layer width. The resistive layer width (λ) is usually much smaller than the neutral sheet width. Tearing instability occurs when the eigenfunction has a complicated structure shown in the middle of Figure 4. Because of this minute structure in the eigenfunction in the resistive layer, the system becomes unstable. Let us define the whole system size by a small letter, ℓ. For example, take $\ell = 20 R_E$, L = 2 R_E, and the Reynolds number is, very conservatively, 100, though actually much bigger than 100. Suppose this lambda to be 0.2 R_E, then the mesh size delta should be 0.1. Actually, it must be less than 0.1. In the present case, therefore, mesh number must be larger than 200. So, if you want to make a simulation for this configuration, at least, you have to have 200 mesh points. Strictly, this must be 2000 or so. This is the point I wish to emphasize. Namely, we have to define very fine-grained grids if we want to simulate the tearing mode instability with accuracy. If mesh size is larger than lambda, judging from our simulation done for such cases, the results end up with very large numerically enhanced acceleration. In order to have a convincing result, therefore, the delta value must be smaller than lambda. I would say that even if we take delta smaller than lambda, numerical artifact acts to enhance the plasma flow speed. Therefore the choice of mesh size is essential in numerical simulations, particularly, of tearing mode instability, although most simulations violate this condition, namely, $\Delta > \lambda$.

In conclusion, since we are interested in energy conversion process, we have to construct a numerical model with the greatest care about the key parameters which govern the energy conversion process rather than simply obtaining a result. We can get anything when we let the computer make the calculation. A simulation study is not merely a calculation. It must be based on a deep physical insight.

Editor's note: As is stated in the Preface, the authors of all contributions and comments in this section were given the opportunity to modify their transcribed texts in a minimal fashion, being asked to change the basic information content

SPONTANEOUS VS DRIVEN

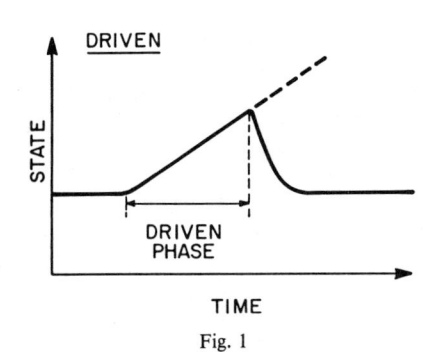

Fig. 1

as little as possible. The author of the present commentary chose to revise, in particular, his Figure 1 and the corresponding definitions of driven and spontaneous processes. Through this alteration his definition of "driven" became much closer to the definition of "spontaneous" as given in the next commentary (by K. Schindler), except for the difference in what determines the time scale of the fast evolution and, for example, the reconnection rate after the sudden transition from a slow "driven" state to a dynamic state. Another consequence of the revision was that the initial part of the discussion after this commentary lost

KEYS TO NUMERICAL SIMULATION

- **RESISTIVITY**

    TEARING MODE INSTABILITY
    RESISTIVITY-DRIVEN RECONNECTION
    RECONNECTION RATE E_R

    $$E_R = \eta j_{NS} \leq \eta j_{NSO}$$

    $$E_R^T = \frac{E_R}{V_A B_0} \leq \frac{\eta}{V_A B_0} \frac{B_0}{\mu_0 L} = \frac{\eta}{\mu_0 L V_A} = \frac{1}{R_M}$$

    DRIVEN RECONNECTION

    $$E_R = E_{ex} = B_0 V_{ex}$$

    $$E_R^D = \frac{E_R}{V_A B_0} = \frac{B_0 V_{ex}}{V_A B_0} = \frac{V_{ex}}{V_A}$$

    WHERE j_{NSO} IS THE INITIAL NEUTRAL SHEET CURRENT AND V_{ex} IS THE DRIVING FLOW SPEED

Fig. 2

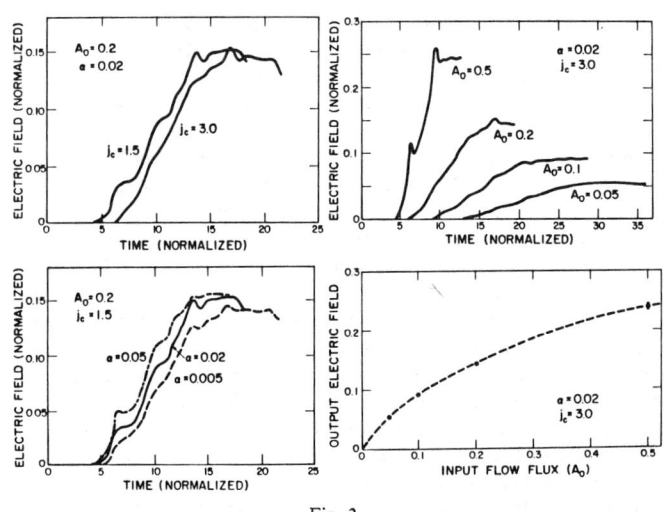

Fig. 3

- **MESH SIZE (Δ)**

    TEARING MODE INSTABILITY

    $\Delta \ll \lambda$ (RESISTIVE LAYER)

    $\sim \frac{1}{\sqrt{R_M}} L$ (L, NEUTRAL SHEET WIDTH)

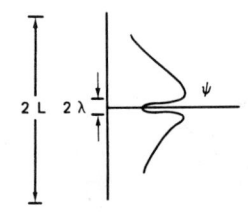

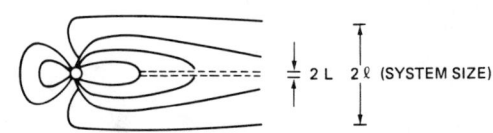

    $2L$ 2λ ψ

    $2L$ 2ℓ (SYSTEM SIZE)

    E_x

    $\begin{cases} \ell = 20 R_E \\ L = 2 R_E \\ R_M = 10^2 \rightarrow \lambda = 0.2 R_E \end{cases}$

    THEN $\Delta = 0.1 R_E$

    $N = \frac{\ell}{\Delta} = 200$ (MESH NUMBER)

    IF Δ IS LARGER THAN λ,

    THEN

    UNUSUALLY LARGE NUMERICAL ACCELERATION OF PLASMA OCCURS!!

    $\Delta > \lambda$

    EVEN IF $\Delta \lesssim \lambda$,

    NUMERICALLY ENHANCED ACCELERATION IS UNAVOIDABLE!!

Fig. 4

its basis. That part of the discussion therefore does not appear here.

Rothwell: I have one comment about the simulation codes. I've been watching here the data that we got from space which is excellent, and I've been watching the data that we've gotten from ground and laboratory experiments was also excellent. I think the simulation codes can be used as a bridge between those two by putting in the different parameters for the two cases and seeing how they scale from one region to the other. In fact, you could even use the ground-based experiments, calibrate the code, see what effects are real in the computer simulation, and then apply it to space. We should not restrain ourselves to a single code because I think different laboratory experiments scale differently in the MHD and single particle regimes.

Sato: What we wish to do in the future is certainly that we can solve the whole process self-consistently. But what we can do now is to do the best we can do under the actual constraints such as the limit of computer memory and speed. The role of numerical simulation is to elucidate the basic physical process rather than to simply simulate a realistic phenomenon, at least at the present stage.

Rothwell: Well, what I see is that you've got a certain set of parameters for space. Now, what I'm suggesting is to put some of your efforts into inserting parameters for the lab experiments to see if you get the same phenomena coming out. For example, for single ion motion, you can scale the gyroradius to the dimensions of the lab and to those of space. Also, one wants to know how the ion gyro frequency compares to the collision frequency. That would give you an indication whether you could scale the single particle motion. If you have the same phenomena coming out, then you know you can make comparisons between the laboratory experiment and the plasma sheet.

Priest: One point about numerical simulation is that we've only just started to realize how much structure there should be in the magnetic field. Often, we have very strong numerical dissipation in our codes which completely scrubs out this very small scale structure and thereby removes a lot of the important physics of what's going on. I would draw your attention to very important recent papers by Matthews and Montgomery in the *Journal of Plasma Physics* and by Frisch and their collaborators using spectral methods where they have no numerical dissipation. They show that at high magnetic Reynolds number, the current forms extremely fine filamentary structures. And I believe we should really try and investigate these in more detail.

Reference

Sato, T. and T. Hayashi, Externally driven magnetic reconnection and a powerful magnetic energy converter, *Phys. Fluids 22* 1189, 1979.

DEFINITION OF SPONTANEOUS RECONNECTION

K. Schindler

Ruhr-Universität Bochum
Institut für Theoretische Physik IV
4630 Bochum 1, F.R.G.

Well, I found in this conference that everybody has his own view on driven versus spontaneous, so let me give you my view. Yesterday afternoon, I went to the library at the expense of missing some important talks here. I thought it was about time to see what other people think about spontaneous versus driven, and I think the most natural region to go to is fluid dynamics. It doesn't make sense to go to general dictionaries, because, for instance, even the term "work" is different in physics than in ordinary life. Let me say at the beginning that I very much agree with Sato's definition he just gave of "spontaneous."

First of all, I found that there is a very close link between "spontaneous" and "instability," just as I pointed out on Monday, where I used the word "instability" to define spontaneous reconnection. That was confirmed during my search here. One of the prominent examples for instability is the thermal convection instability. And I looked at this example a bit more closely, and I will give you the reason below why this is the closest analogy. Just to remind you, if you heat a fluid layer from below, it takes a certain Rayleigh number to make it unstable. Beyond the onset point you find qualitatively new features. That is called "spontaneous," and this is a bit more than semantics. It's a new qualitative property that appears and it is spontaneous although we have an energy flux through the system. It's a misconception, I think, to call this "driven" pointing at the energy flux through it. Of course, the convection would not exist without this energy flux. But what makes it "spontaneous" is that without any particular external signal, a new qualitative feature appears. And this is what is called an "instability" and "spontaneous." And the occurrence of the instability is, of course, predictable. I mean, there's nothing unpredictable about it because you can compute the critical Rayleigh number, at which it occurs. There is one complication in reality, but it still does not make it unpredictable. The onset may be not always right at the critical point, but you have a broad band; by some external fluctuations, you can make the instability go off earlier. Also, you can make it metastable if you have some additional stabilizing effects that suddenly go away. And that is very important in the magnetosphere.

So, what is "driven"? I couldn't find in this whole context the term "driven," but maybe I haven't looked well enough. But there was a term, "forced convection," but even that is an instability. There is one instability forcing another. I suspect what one would call "driven" is just making convection by pushing the fluid up at the appropriate location. That would be the driven process. But as I say, I might not have looked well enough.

Let me now tell you why I'm fond of this analogy. This is not my idea, this comes from a recent paper by Maschke and Samarito [1982]. They point out that the equations of resistive tearing are very similar to the Boussinesque approximation in fluid dynamics used to describe the convection instability. In this analogy, the magnetic diffusion equation is, of course, the heat conduction equation, and the equation of motion is the Navier-Stokes equation. And the bifurcation aspects of these two problems are very similar as Maschke and Samarito have shown.

From these considerations I got a little reassured of what distinction should be made in the field of the magnetosphere. If we have a smooth energy transport into the magnetosphere and suddenly we have this qualitatively new feature (change of B-topology) coming up; then, using this terminology we don't have a choice other than calling this spontaneous or unstable, if you like. If we "tell" the system where it should make its neutral line and where it should make its plasmoids, then, it is driven. And this provides a very clear-cut observational distinction. I should emphasize the difference I see is a qualitative difference, not only a quantitative one.

Finally I think that my earlier discussion of metastable states comes very close to Ian Axford's view about quasi- or pseudo-instability. I call it metastable because we have seen on Monday that there may be stabilization by the presence of adiabatic electrons. This stabilizing effect, however, can be reduced by other processes, ionospheric or fluctuations, and that can make this system unstable. That is then a situation where the tearing mode is extremely fast and has "explosive" character. The onset without such speed-up is very dull. That may be an additional feature on top of what Ian Axford said about the tearing mode.

References

Maschke, E. K. and B. Samarito, *Physica Scripta T2/2*, 410, 1982.

ASTROPHYSICAL IMPLICATIONS OF RECONNECTION

Jonathan Arons

Astronomy Department, University of California
Berkeley, California 94720

In some sense, I'm a hired gun, come both to learn things and to somehow, in 15 minutes, tell you how this is all relevant to the rest of the universe. As I think you all realize, this is a large place, with a lot of different things going on. So first, let me try to make some general remarks about what I think I've learned. Also, I'll give you some examples of things from my own interests in which you'll see the ways in which astrophysicists talk about reconnection, but without the appearance of reconnection as a definite process with its own signature in observations of astronomical systems. Instead, it shows up as a piece of models within the general problem of how magnetized systems become energized. Normally, I'm not happy about making general homilies because I think such activities should be left to theologians, not physicists. For once, however, I'll try.

It's clear that the basic physics of reconnection is something we really have got to learn from the earth and the lab and to some degree from the sun. In such a complex array of physical problems, there's no possibility of going off to the remote sensing of objects from stars to quasars and expect to understand the dynamics of reconnection in a quantitative manner, if we did not already have a clear knowledge of the physics from more direct experiment and associated theory. I have been especially impressed by the level of experimental detail needed to understand reconnection. Both the space and laboratory experimenters assemble space and time resolved measurements of electric fields, magnetic fields, mass densities, flow velocities, pressures, and detailed velocity space distributions of the fast particles present. That's on the order of a 20 dimensional parameter space that you have to worry about. However, the basic lesson is, as has been known for at least 30 years and is implicit in Giovanelli's original papers, that the free energy stored in a stressed but hydromagnetically stable (in ideal MHD sense) magnetic configuration can be released by some dissipative process. I mean dissipation rather generally, inductance or impedance in very general usages of the terms, the way, for example, electron beam people mean impedance when the current flow is limited by the electrons' inertia. Such impedance can have a role in dissipating stored magnetic energy just as interesting as the direct creation of entropy due to resistivity. Any of these processes can lead to making accessible the energy tied up in magnetic stress, which is otherwise not accessible in ideal MHD (if the system was unstable in ideal MHD, its configuration would be changing long before the "dissipative" processes could work, so that one becomes uninterested in the "slow" reconnection process). So before one gets into the study of the effects of reconnection on a system, or any particular realization of the reconnection process itself, either you or nature already have set up an "equilibrium" configuration, be it static equilibrium or a flowing stationary state—this latter includes statistically stationary, where small scale time dependent fluctuations of small or large amplitude form a substructure to a larger scale steady structure in which energy is stored.

All this is conventional wisdom. The main thing I got out of this meeting as compared to previous meetings on reconnection is that there is now rather general agreement that the individual reconnections, through which energy is released, are usually quite bursty. That is, rather rarely does one see the theorists' idealization of steady reconnection flow on a scale large compared to the local dissipation region (although still small compared to the overall system size—in principle, locally steady reconnection flows could fit into globally unsteady systems). On the other hand, the detailed physics of locally unsteady reconnection flows is conceptually close to that of the steady state models—only the calculations of the x-point structure, the Hall current flows, etc. become inordinately harder. So in terms of principles, I haven't seen anything vastly new this week, but in terms of what actually happens at reconnection sites, it's clear that enormous advances have occurred in the last five years.

For an astrophysicist to make use of the experimental and theoretical knowledge of reconnection as a set of processes in modeling observations of remote sources, a knowledge of the free energy sources is needed, as well as a usable accounting of the various modes into which reconnection flows channel that energy. One sink for the free energy in the magnetic field is the bulk motion of the magnetized fluid. The results presented at this meeting suggest to me that this is the dominant resting place of the energy released by reconnection in the terrestrial magnetosphere. If this were all, one would be very depressed at ever finding unambigous evidence for reconnection in the rest of the universe, since the simple existence of flows, usually observed by Doppler shifts of spectral lines, can be counted on to give rise to competing models for the energy source of the flow, all explaining the same data. Fortunately, reconnection flows also give rise to thermal heating and to fast particle acceleration, whose time resolved structure might be a more precise indicator of the (usually inductive?) electric fields present in reconnection events. We've heard rather little at this meeting about particle acceleration in the magnetosphere and an awful lot about the bulk flow characteristics of reconnection, largely because the research tends to focus on the dominant form of the energy; of course, the fast particle properties are heavily used in diagnosing the medium, but there was far less mention of the acceleration of these particles than I expected before I came. For the astrophysicist, this issue cannot be dropped even though the magnetospheric evidence suggests that reconnection is not a big league particle accelerator, simply for the diagnostic reasons described above. In remote sources, one could hope to investigate the acceleration of such particles by time and spatially resolved studies of synchrotron emission, as has been done and is being done in studies of supernova remnants in the interstellar medium, where time variable, localized regions of particle acceleration are seen.

The solar problem illustrates how the relative partition of energy sinks can change as one goes to systems beyond the earth. Here, we seem to get a lot more thermal heating and/or electron beam acceleration, with bulk motion in flares and prominences (perhaps) an indirect consequence of the deposition of energy in the other two forms. Perhaps this difference comes about because gravitational confinement of (cool) plasma controls the inertial properties of solar loops, while gravity doesn't matter a hill of beans in the terrestrial environment. There is a big difference of detail in whether the energy goes into thermal plasma heating and thermal X-ray emission, or whether it is mostly in directed electron beams which create X-rays as the beams stop, but either way, the effect of reconnection seems quite different in the sun, and leads to dissipation of the magnetic energy into photons at a level that causes the astronomer to have more optimism about seeing the consequences of reconnection in the rest of the universe.

In both the terrestrial and the solar examples, the approach to the study of reconnection is clear. We know about the magnetic structure and therefore the configuration in which free energy is stored in magnetic form from "direct" measurement. For the solar case, of course, the measurement is not direct, it's in photons, with all the complications of interpretation alluded to by Giovanelli in his videotape: you must first understand what Hα is doing if you are going to try to infer a magnetic field through solar imagery in Hα. Nevertheless, the space and time resolution available is huge compared to what we're used to in the rest of astrophysics. Because of the details available in both the terrestrial and solar problems, people are much more willing to try global MHD simulations, both ideal and resistive, to tell us about what the structure is like at places and times where direct observation isn't available and to assess the free energy available for energy release, with some confidence that only one set of input parameters and boundary conditions is relevant. This excludes the plethora of modelling calculations you might have to do to assess the possible relevance, if you didn't know the configuration in advance from experiment. The biggest loophole in such modeling is our ignorance of the microscopic transport processes and how they couple to the macroscopic configurations, as was apparent in the work presented by Drake and by Leboeuf. Here, advances will come as the ability to do long time step plasma simulation, coupled to the hydromagnetic models, progresses in the next few years. This sort of work is needed, whether or not you are committed to the religion of reconnection rates always being able to adjust to whatever the hydromagnetic problem wants to impose.

Theoretical work in the terrestrial and solar problems seems to be focussed mostly on models of the individual reconnection events, rather than on the "global" issue of how these events control the general average power dissipation in the magnetosphere and in the solar corona. This is as it should be, given the high space and time resolution of the data and our ignorance of how the detailed physics really works. For example, we don't yet have an answer to the question, raised by Drake's work, on whether the reconnection flow automatically controls the dissipation layers by changing their spatial scale, or whether the layers can react back and change the flow, by propagating slow shocks out to infinity again. Furthermore, the popularity of lower hybrid drift as a dissipation mechanism is clearly limited by the $k.B = 0$ assumption; in general geometry, the growth of instabilities driven by current flow must include the coupling to whistlers, which drastically alters the character of these phenomena.

An understanding of these microscopic questions is essential to seeing how the underlying physics connects to the overall global power dissipation as mediated by reconnection. This subject was hardly mentioned here, since the gross energetics of the terrestrial magnetosphere and of the solar corona have been known for a long time—the essential issues now are in the nature of the microscopic processes, the individual bursty events. But from the remote sensing point of view, feeding back one's understanding of the individual events into how the system would look if you viewed it from a great distance is the essential problem for the astrophysicist. As an illustration, consider the differing consequences of a steady state versus an inductive model for the reconnection electric field for the acceleration of high energy particles, whose presence is known by the synchrotron radiation from the relativistic electrons and/or from the π^0 gamma rays emitted as protons pass through matter. If one applies a steady state reconnection model for such particle acceleration, a substantially larger system, and/or greater magnetic field strength, and/or larger flow velocity will be inferred to give the requisite particle acceleration rate than would be required if bursty, inductive fields were known to give rise to the acceleration. Sometimes the difference can be tested, such as in the x-ray observations with millisecond time resolution of x-rays from Cygnus x-1, and sometimes one can't tell the difference, as in the models proposed many years ago for neutral sheet acceleration in the Crab Nebula. In such circumstances, one really wants to know from local experiment and its associated theory what to expect from a plasma with parameters and boundary conditions known either directly from the photon observations, or from ideal MHD models. That's why we need to know the nature of the individual reconnection events in the terrestrial magnetosphere and in the sun in detail—if these are well understood theoretically, then we will be in a position to extend the associated theory to more exotic environments with some expectation of predictive success.

Let me illustrate with some remarks on astrophysics. We often know flow velocities, densities and temperatures from spectroscopy, sometimes in a surprising amount of detail. For example, the millimeter interferometry people are now down to looking at molecules that are as close as 50 to 100 AU from a protostar in Orion, where they find supersonic outflows and even hints at the magnetic structure from the polarization of the masers in the flow. Now, this is huge compared to microscopic things like gyro-length and Debye length in the partially ionized plasma, but is quite sufficient to strongly constrain ideal MHD models, and thus tell us something about the free energy sources needed to drive reconnection flows. Bruce Draine, in an amusing model recently published in the Astrophysical Journal, suggests this outflow is due to the outward magnetic pressure of a toroidal magnetic field, created as a rotating, collapsing protostar winds up the initially poloidal flux linking the collapsing core of a dense cloud to the envelope. Despite the winding of the field through many thousands of turns, he assumes no dissipation, in particular, no tearing of the field whatsoever. This type of modelling provides a ripe ground for the application of reconnection concepts, in particular, for study of tearing modes in a partially ionized plasma at the places in the sheared field where $k.B = 0$. So far, most astrophysicists are not even aware that questions of this sort exist.

The example just given is a good one for how reconnection enters astrophysics. First, the observers pose a theoretical problem in sufficient detail that the theorists respond with an ideal fluid model in which mass, energy and momentum are conserved in the large; usually, more than one is created which is consistent with the facts. Later (usually much later) someone realizes that further dissipative processes must be added to the theory, which either change the models drastically (as I implied above would be the case for Drain'e scheme), or make an interesting addition that helps in understanding, and occasionally predicting, an improved data set. Unfortunately, our ability to non-linearly model reconnection events is too rudimentary to allow incorporation into models of this sort, so it just isn't done. I expect the advent of multi-dimensional fluid and MHD simulation into astrophysics, now in its infancy (in contrast to its maturity in the fusion program) will change this a lot; the usage of plasma simulation in astrophysics is still something for the future. Once reconnection events can be modeled, and the effects of individual events can be incorporated in the overall models of the observed flows, we will be in much better shape to answer questions on the role of reconnection in remote sources as a creator of energetic plasma flows, of plasma heating and beam acceleration, and as a means of accelerating fast particles, in a way that allows searches for the specific signatures of such events.

Priest: Certainly, in solar flares, it's clear that most of the energy released is extremely bursty. It's not smooth or steady state at all. That's why I'm particularly interested in the discovery of this new impulsive bursty regime of reconnection. I think this is closer to reality where you get pairs of x and o neutral points created by tearing and then annihilated by coalescence. This basic process goes on sporadically and repetitively.

Rostoker: This whole question of bursty, though, worries me because no one has assigned any time scale. One man's burst is another man's extremely slow and long-lived feature. And we have to be a little cautious as to how we interpret.

Priest: Well, I think the time scale for coalescence is a few Alfvén times. That's extremely fast compared to tearing times.

Rostoker: It depends what system you have and what the Alfvén velocity really is as to the kind of time scales you end up with.

Arons: Well, I think that last statement is the crux of it. In the language I was using, what you mean by a burst would be a rapid release of energy compared to the rate at which energy was used to set the system up. One example of this sort of thing shows up in the study of compact X-ray sources. Many of these objects emit X-rays because of gravitational accretion of plasma onto magnetized neutron stars. The relevant time scale for formation of the emission region is the time for plasma to fall freely from the magnetopause to the surface, typically on the order of one second. In one object, we see totally bursty emission on this time scale, suggesting some sort of dynamical instability, perhaps of a magnetosphere, is at work. But the great majority of the X-ray burst sources show steady emission for times orders of magnitude larger than the free-fall time, interrupted every few hours by large outbursts. This is a steady state on the hydromagnetic

368 ASTROPHYSICAL IMPLICATIONS

time scale, with some sort of instability whose threshold is only occasionally crossed. To explain this, one needs a longer "dissipation" time scale. So far, no one has shown how reconnection offers a plausible mechanism for such dissipation, either to get the overall time scale or the form of the bursts (rapid rise and slow fall) - instead, most people believe that the common garden X-ray burst sources are due to unstable fusion of the hydrogen and helium as it sits on the surface of the neutron star. Still, this model has its own problems, and perhaps a magnetic dissipation model will be found that does better.

Saunders: I think this point also forces us to remember that there is evidence for quasi-steady reconnection at the earth's dayside magnetopause as reported initially by Bengt Sonnerup and also further by Götz Paschmann at this meeting.

Arons: Yes, I'd like to get cleared up a little on that. My impression of the discussion of that day was they point to a substantial amount of more or less steady reconnection, but then it got to be like 20-25% of the data might be organized that way. And most of the rest of it seemed to be left aside which might or might not be organizable as a lot of flux transfer events. Is that a widely-believed view of how the data'll end up getting organized?

Bratenahl: I'd like to make a comment about the bursty X-ray emissions. It's not just the Alfvén time that counts, it's also the nature of the electric field involved. When things are changing rapidly, we must consider the inductive electric field and how plasma responds to it. Plasma reacts completely differently to inductive, as opposed to electrostatic (space charge) electric fields. In fact, plasma attempts to screen out inductive electric fields with oppositely directed electrostatic fields through production of space charges. This is a futile effort to the extent that the line integral of the inductive field cannot be quenched in this way.

Arons: Well, if I had time, I would show you some of that.

Bratenahl: It should influence, I think, the theoretical people to take a close look at time-dependent reconnection. I don't think it makes much sense to go on and on with the study of steady state reconnection. There is much more interesting dynamics in time-dependent reconnection.

Arons: Let me finish off with an example from my own interests in which many issues of interest to magnetospheric physicists recur in a transformed manner in an astrophysical problem. We have been pretty sure for some time that the X-ray emitting pulsars in binary star systems are magnetized neutron stars, drawing the power for their X-ray emission from gravitational accretion onto the star, with the magnetic field channeling the plasma onto the magnetic poles. The fact of pulsation is due to the rotation of the star. The channeling is loosely analogous to auroral precipitation in the limit of extremely strong scattering, so that no trapping and loss cone occurs and all the plasma flows along field lines in a fully collisional, fluid dynamic manner - the densities in the models are much higher than in planetary magnetospheres. In one class of objects, the plasma approaches the magnetosphere in a thin disk, with the gas orbiting the star in almost circular Keplerian orbits. At the inner edge of the disk, the plasma moves past the magnetic field and the magnetospheric plasma, since the Keplerian and co-rotation velocities don't agree at the magnetopause. Figure 1 shows an idealized form of the configuration, based on an investigation in progress by Chris McKee, Ralph Pudritz and myself. We have idealized the problem by looking at an aligned rotator, although we know the system should be oblique if there are to be pulses (of course, if Alex Dessler were here, he would immediately remind me that even an aligned rotator could pulse if it had a sufficiently strong surface anomaly, as is entirely possible). In our work, we assume the disk to have no large scale magnetic field of its own. However, if mass is to be transferred from disk to magnetosphere, so that accretion onto the star can occur, some dissipative process(es) must occur. In the absence of a large scale magnetic field in the disk, this dissipation is dominated by the Kelvin-Helmholtz instability, leading to a "viscous" interaction and an actual "diffusive" transfer of mass as

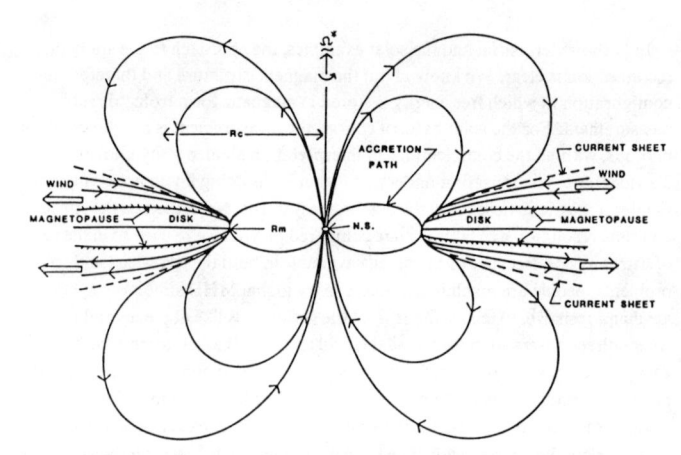

Fig. 1. A possible hydromagnetic model for accretion onto magnetized neutron stars from a disk. The magnetic and rotation axes are assumed to be aligned, for simplicity. Plasma flows along the accretion paths from the inner edge of the disk to the magnetic poles of the star, where its gravitational energy is released as X-rays. Field lines which would have closed outside the disk's inner edge are compressed in by the toroidal magnetopause current system. Mass transfer across the magnetopause occurs through diffusive mixing in excess of the Bohm rate, driven by the vortex shedding generated by the Kelvin-Helmholtz instability. At radii exceeding the corotation radius, where the angular velocities of the open field lines and of the disk are equal, plasma loaded onto the open field lines is centrifugally slung off, carrying away the angular momentum of the accreting plasma at the inner edge of the disk before it is brought into corotation with the neutron star. The viscous interaction at the disk's inner edge can also give direct extraction of angular momentum from the star by the wind, and both effects lead to significant changes in the sign and magnitude of the torque on the neutron star.

the instability causes the indefinite winding up of vortices in the field-plasma interface. This problem is directly analogous to the Axford-Hines picture of magnetospheric coupling to the solar wind, with a specific theory of the mass entry added which may have some relevance to the entry of plasma into the boundary layer on the flanks of the Earth's magnetosphere. The theory has some direct observational implications: in particular, plasma added to the magnetosphere exterior to the corotation radius, where the Keplerian angular velocity equals the stellar angular velocity, feels a net gravity outwards and pushes the field into an open configuration. The resulting wind (loosely akin to the winds proposed for the Jovian magnetosphere 10 years ago) carries away angular momentum in amounts sufficient to affect the spin rate of the underlying star, an observable parameter in many X-ray sources.

Reconnection enters our theory only in that near the corotation radius, the Kelvin-Helmholtz instability is weak since the velocity difference between disk and magnetosphere is small. Here, the next most important effect is reconnection of small scale loops with the large scale magnetospheric field, which allows sufficient transfer of mass to get the wind going in the first place. We estimate this simply by using an order of magnitude of the rate of flux transfer as would be expected on the Petscheck model; no further details are invoked. However, if accretion disks have large scale magnetic fields, as a few people have proposed, reconnection might take over from diffusive mixing driven by shear flow, and an even closer analogy to the geophysical environment might become of interest. So far, this altered scenario, and especially its observational distinctions, has not been investigated.

VALIDITY OF THE PETSCHEK MODEL

D. Biskamp

Max-Planck-Institut für Plasmaphysik
EURATOM Association, D-8046 Garching

I think there's really no need to stress the inherent importance of the Petschek model, especially at this meeting. Since its appearance in the early '60s, it has been widely accepted and there's virtually no explosive magnetic event that has not been explained in terms of this model. Nevertheless, in recent years, there has been increasing evidence that casts some doubt on the generality of this model. I would like to remind you of Dr. Priest's talk at this meeting. Therefore, I think it would be valuable to make some comments on these recent trends.

Let me begin by giving Petschek's original equations so that we know what we are talking about. These are the two dimensional incompressible MHD equations. Assuming homogeneous density $\rho_0 = 1$ we have only two scalar quantities, the flux function ψ describing the magnetic field and the stream-function ϕ describing the velocity field. The current density $j = \nabla^2\psi$ is only in the z-direction and so is the vorticity $\omega = \nabla^2\phi$. These quantities satisfy the equations derived from Ampere's law and the equation of motion

$$\frac{\partial \psi}{\partial t} + \vec{v} \cdot \nabla \psi = \eta j$$

$$\frac{\partial \omega}{\partial t} + \vec{v} \cdot \nabla \omega = \vec{B} \cdot \nabla j + \mu \nabla^2 \omega$$

The viscosity is usually neglected. In conventional units η is just the inverse magnetic Reynolds number. When talking about fast reconnection we always have in mind the limit of small resistivity or large Reynolds number.

These equations are well suited to numerical treatment on present-day computers. And it is, in fact, due to a number of numerical simulations that the objections to Petschek's model have arisen. These are simulations of spontaneous reconnection processes such as the island coalescence instability or the m = 1 resistive kink instability as well as computations of driven reconnection (for a review see Ref. 1). The common feature of these computations is that, contrary to expection, no Petschek-type configuration is generated for sufficiently small resistivity or large magnetic Reynolds number. Instead, there is a quasi-one-dimensional current layer of finite length to which two pairs of relatively weak slow shocks are attached. Decreasing the resistivity or increasing the enforced reconnection rate in a driven reconnection process leads to a further lengthening of the current layer until it reaches the global system size. So, what's wrong with Petschek's model and the related and refined models based on it? Let me say a few words on the assumptions inherent in Petschek's concept. The basic assumption is that there exists a hierarchy of spatial scales: a) the global system size, which depends on the particular process in question, e.g., a particular MHD unstable configuration; b) a quasi-self-similar ideal subsystem thereof including the neutral point, which is called the outer region in the reconnection models; and c) a small region just around the x-point, where the resistivity is important and which is called the diffusion region. Now, a solution of the problem can only be achieved by solving in the individual regions and then matching the solutions from inside to outside. That is, if you have a solution of the diffusion layer, it must be matched to the outer region. Now, the matching from the outer region to the global outside configuration is rather trivial because the latter just gives you the boundary conditions of the former. On the other hand, the matching of the diffusion region to the outer region is complicated.

What is the general idea of the Petschek diffusion region? Petschek didn't have very much to say about this, but later papers made substantial analysis of it (see, for example, Ref. 2) The Petschek diffusion region is in fact a small Sweet-Parker current layer, in which the plasma entering at a relatively low speed $u \cong Mv_A$, where M is the reconnection rate, is accelerated up to the Alfvén speed $v \cong v_A$, at which it continues to flow along the outer field reversal region, the region between the slow shocks. The reconnection rate is thus determined purely geometrically by the angle between the shocks, and you may have any value of M up to the order of unity. The main point is that Petschek's model assumes a smooth transition from the diffusion layer to the outer field reversal region.

If we now look more closely at the numerical computations, these consistently show that this is not the case. Let us consider the current profile in the diffusion region from a typical simulation of driven reconnection, see Fig. 1. There's a pronounced negative current region at both ends of the current layer which discontinuously connects to the current density in the outer region. Such behaviour has in principle to be expected if the velocity is close to the Alfvén speed. How does the flow in the diffusion region, especially at the edge, react to this negative current region here? When it hits this small region of pronounced negative current density, it is strongly decelerated and leaves the diffusion region with a velocity which is again much below the Alfvén speed, see Figs. 2 and 3. So the overall reconnection rate, even if the shocks form a large angle, has to be small. We may say that the negative current region acts like a plug on the diffusion region.

Let me give a rough qualitative interpretation using Ampere's law. I merely ask what Ampere's law prescribes for an x-point configuration or field reversal configuration to be maintained. So where is the current density high? It's high in the shock regions, but it is still much higher in the diffusion layer. If we have a diffusion layer which is just more or less round, then, of course, we would not have an x-point, but an o-point. So the plasma makes a compromise. It stretches the diffusion layer out. But if this just smoothly decays to zero, there would be nothing but an elongated o-point. So one needs some end-point corrections, small areas of negative current density, so that you may finally have the field-reversed configuration.

And now for a few words on what I did analytically. One can construct the self-similar solution by a power expansion in ηy in terms of the zeroth order current distribution $j_0(x)$. One may in principle choose any reasonable $j_0(x)$ profile. But, rather surprisingly, the simulations consistently show that $j_0(x)$ very nicely fits the famous $1/\cosh^2 x$ law, and it might be worthwhile to give some thought to why nature tends to prefer this profile. It is quite easy to calculate the first low-order terms in the expansion which already reveal important features of the longitudinal structure of the diffusion layer. But it seems improbable that the expansion may describe the complex behaviour at the transition into the outer region. And, in addition, there is the problem of matching. So before we can obtain a valid analytical solution, we have to go on discussing qualitatively, using the numerical results as guidelines. What happens to the diffusion region if η is

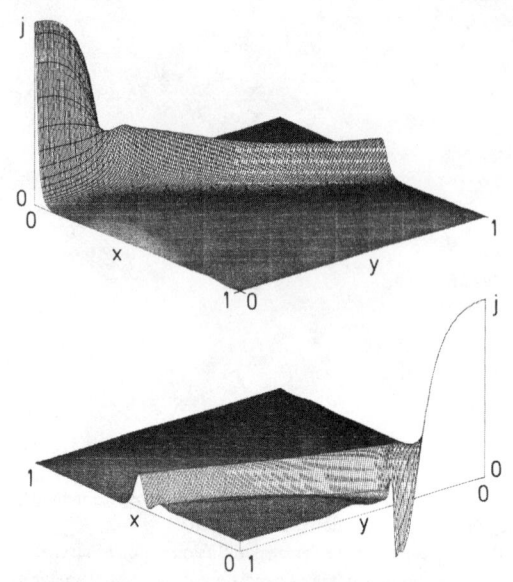

Fig. 1. Stereographic plots of the current density j from a numerical simulation of driven reconnection (only one quadrant of the full system is computed and displayed)

decreased? At all events reconnection will be (somewhat or much) more difficult. Hence magnetic flux will pile up in front of the layer, increasing the upstream Alfvén speed. The plasma therefore also assumes a higher speed along the layer, pushing the plug, discussed above, further out. This leads to a lengthening of the diffusion layer. In the simulations of steady-state-driven reconnection we find that the diffusion layer length scales roughly as $\eta^{-1/2}$. In spontaneously

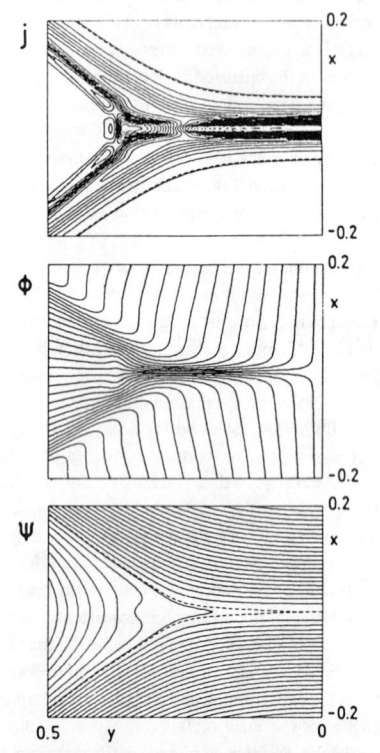

Fig. 2. Contour plots of current density j, streamfunction ϕ and flux function ψ displaying the behaviour in the diffusion region, taken from the same computer run as Fig. 1 (only half of the system is shown).

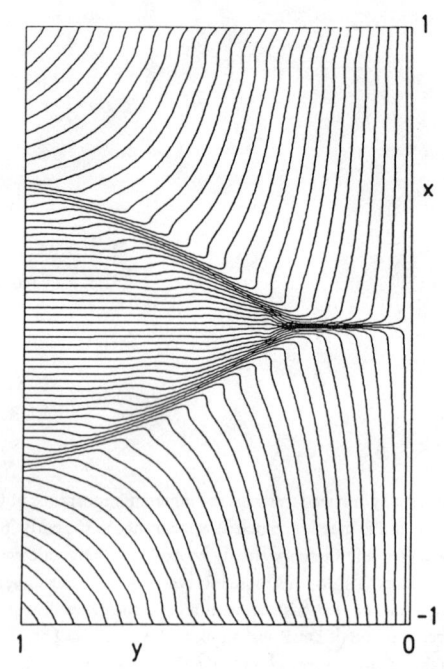

Fig. 3. Contour plot of ϕ as in Fig. 2, now displaying the global behaviour of the plasma flow.

reconnecting systems with a finite amount of free energy available, the reconnection rates are generally small and scale roughly as $\eta^{1/2}$.

Let me briefly comment on the stability of the diffusion region. Of course, such a current layer must finally become tearing-unstable, though it is stabilized over a wide parameter range owing to the strong inhomogenous flow. What happens in the case of instability depends on how symmetric the situation is. If it is rather unsymmetric, then you get little blobs that are convected out of the current layer, and this is the case Dr. Priest was referring to. If one has a rather symmetric configuration, one ends up with one big island; instead of the x-point, there is now an o-point and x-points at the end of the original current layer, and the overall reconnection rate drastically decreases. So, regardless of which of these processes really occurs, I think it is safe to say that the tearing mode does not substantially increase the reconnection rate.

In conclusion it thus appears that the classical Petschek model is not valid. Only if the large current density in the diffusion layer is avoided, by having a sufficiently high resistivity in the diffusion region, is a Petschek-type configuration set up. And that, in fact, is not too surprising, because the problem of the diffusion layer is thus avoided, and the outer solution can be written down analytically. One of the many questions you may ask is whether solar flares require anomalous resistivity? I don't know. In any case, nature is three-dimensional. And there's a good example in which explosive reconnection may well occur in the absence of a well-defined x-point. This is the standard model of the major disruption in tokamak plasmas, where fast reconnection is due to 3-dimensional MHD turbulence.

Vasyliunas: Just a comment. Petschek's model really does not assume a priori how the diffusion region varies as distinct from the slow shocks. The model treats a boundary layer which includes both the diffusion region and the region between the slow shocks, considered as a unit, and the model predicts average values for this boundary layer. And the current density is not higher in the diffusion region; the current density there is, in fact, less than the current density in the slow shocks.

Biskamp: This is a result where Petschek's model is in error.

Colgate: I completely agree with you and I'd love to use your last slide at the beginning of my talk. I want to discuss how you break the topology of 2 dimensions and bring it into 3 dimensions, just the point you bring up about the Petschek mechanism.

Biskamp: May I just come back to Vasyliunas' remark about the current density at the center. One can, of course, reduce the current density there by assuming an anomalous resistivity. This is what Sato did in his simulations, where, in fact, there is no enhancement of the current density. But this is not Petschek's original paper, and this is not what everybody quotes when saying that the fast reconnection does not depend on the resistivity. It does! Only if you assume that this current density is suppressed by local enhancement of the resistivity, does the whole concept work. Otherwise, it doesn't. Admittedly, an analytical proof is very difficult, but we are dealing with a fact that can no longer be disputed.

References

Biskamp, D., Resistive MHD processes, *Physica Scripta T2/2*; 405, 1982.

Vasyliunas, V. M., Theoretical models of magnetic field line merging, 1, *Rev. Geophys. Space Phys. 13,* 303, 1975.

SOLAR FLARES: AN EXTREMUM OF RECONNECTION

Stirling A. Colgate

Los Alamos National Laboratory
Los Alamos, New Mexico 87545

I will attempt to emphasize three points.
1. I believe that the solar flare is that particular astrophysical phenomenon that is the extremum of reconnection. No other phenomenon of which I am aware demands as rapid magnetic flux annihilation as is seen in the solar flare.
2. Plasma physics experiments can and should be performed in the laboratory that model reconnection as we observe it in astrophysics.
3. I believe that stochastic field lines derived from something similar to Alfvén wave turbulence are a necessary part of reconnection.

We performed experiments some 20 years ago at Lawrence Livermore Laboratory that gave a hint of what I believe is happening in the rapid reconnection in solar flares.

In this session Biskamp has just made the point that we need to "break the topology in the third dimension to explain reconnection." The experiments that we performed in the laboratory years ago were ones of mapping the flux surfaces and showed that we had broken just this symmetry; namely, that we no longer maintained a simple cylindrical geometry but had produced a more complicated one. This is similar to what Ed Hones has seen with high-energy particles in the magnetotail.

Solar Flare Topology

Many photographs in optical, EUV, and x-ray wavelengths support the general picture of a twisted loop of flux much like the original picture that Tommy Gold presented for the mechanism of the solar flare. From the magnetic field strength and the density and temperature derived from the emissivity observed in these loops, it is reasonable to infer that the magnetic field configuration is very nearly force free. It is likely that the twisted flux tube or force free field is the basic topology of unstable fields on the sun's surface. However, most modeling of reconnection is done in the limit similar to the Petschek mechanism where the fields are exactly opposite at the reconnection layer, i.e., a neutral point or neutral plane as in the magnetotail. The force free fields of the sun, on the other hand, imply a different topology, namely, one where there is a strong magnetic field in the direction of what would have been called "the ignorable coordinate" or the null surface of the models of opposed field reconnection. The addition of the uniform field throughout the opposed field configuration adds a further constraint to the fluid motions allowable in reconnection as we have just heard reviewed by Biskamp.

Shear and Field Topology

Once a magnetic field is added in the direction of the ignorable coordinate, i.e., the null line, then this becomes a sheared field. The field vector rotates as a function of the distance perpendicular to a flux surface. Most frequently in plasma astrophysics people have called this rotation of the magnetic vector "shear." In cylindrically symmetric geometry, shear leads to a gradient of the pitch. Pitch is defined for an axisymmetric configuration as the reciprocal of the number of turns of a field line per unit length. When the number of turns per unit length is different on each flux surface, then interchanges between two such surfaces is topologically constrained, similar to attempting to "thread" a nut and bolt of different pitch. This topological constraint was originally sought for confinement in thermonuclear fusion plasmas in axisymmetric systems. There are axisymmetric configurations of a helical field of constant pitch where, nevertheless, the field vector rotates as a function of radius. These configurations, "the screw symmetric pinch" are most unstable.

In axisymmetric fusion plasmas one might produce shear, either by internal plasma currents or external winding currents. In the case of coronal loops the option of external rigid current carriers is not available and so one speaks of a force free field with a plasma current that is parallel to the local field vector. This parallel current, $J_\parallel$, or field aligned current, as referred to in magnetospheric physics is, I believe, the origin of the instabilities that lead to rapid reconnection. This is because in laboratory plasma experiments the magnitude of this current was correlated with the ease with which the plasma escaped a given configuration. Before discussing the experiment that leads to this belief, I would like to outline some of the reasons for searching for a new and rapid mechanism of reconnection. This motivation is the behavior of the largest and most rapid solar flares.

Conditions of the August 4, 1972 Solar Flare

This flare released some 10^{31} ergs in several 10's of minutes, appeared as a region on the sun of $\sim 2 \times 10^9$ cm in major extent and was consistent with a flux tube of this length and radius of $\sim 2 \times 10^8$ cm. One assumes that an original magnetic field in this volume of space contained an energy several times greater than that which was released. Some magnetic field remained after the flare. One can then calculate the magnetic field strength such that $B^2/8\pi = 4 \times 10^4$ ergs cm^{-3} and find B of the order of a kilogauss. One can also calculate the inductance of the loop viewed as circuit element 2×10^9 cm long and return current at 2.5 times the original radius or $\sim 5 \times 10^8$ cm. This is 10 henrys only weakly dependent on geometry. The required parallel current of a force free field and this inductance necessary to store the released energy becomes $\sim 4 \times 10^{11}$ amps. This current creates the stress in the magnetic field that is needed to store the released energy. If one interrupts or changes a significant fraction of this current in the characteristic time of the energy release of the flare, namely 1000s, then the voltage developed along the coronal loop becomes $(d/dt)(LI) = 5 \times 10^9$ volts. If there were a reconnection mechanism that led to a dissipative resistivity that could explain the energy release and hence current interruption of such a flare, then very obviously a fraction of this voltage is available for accelerating the particles that are associated with such a flare. The acceleration of these particles is then a contribution to the resistance.

In particular the hard gamma rays (and neutrons) have been interpreted by Reuven Ramaty and Richard Lingenfelter as requiring the acceleration of 4×10^{33} protons greater than 30 to 50 MeV. If you assume that these protons are accelerated during 10^3 s, then this requires a proton current of 7×10^{11} amps about equal to the current necessary to maintain the stored energy of the acceleration with the electric fields of reconnection or, that is, current interruption. One is further led to the interpretation that almost all the parallel current has somehow been transferred to runaway protons by a mechanism not yet understood. The impedance presented to the electron flow must become high enough, presumably due to reconnection, to give rise to the necessary resistance or dissipation of the current of the flare.

Comparison of Laboratory Experiments and Solar Magnetic Fields

I wish to compare the mechanism of formation of the magnetic field configuration of a solar flare with the formation of a so-called stabilized pinch in the laboratory. The twisted or helical magnetic field configuration of the stabilized pinch or twisted coronal loop is predominantly an axial field on the inside and an azimuthal field on the outside. It is the same magnetic configuration presumed for solar flares; it is also measured in the laboratory in the case of a stabilized pinch. The difference is only one of time-scales because of the resistive diffusivity due to Coulomb collision processes. In the case of the solar flare, one starts with the emergence of a flux loop and active region. This flux loop may or may not be twisted, and may or may not have a constant number of turns per unit length during the period of emergence of roughly a week. The further twisting of a flux loop can take place because of the Coriolis force induced motions of the large scale eddies that are associated with the "feet" of a flux loop.

During the quiescent presumably stable period of formation, the flux loop emerges from a medium whose pressure is higher than that of the field. After it emerges into the relative vacuum of the corona, the particle pressure is confined by the field stress; that is, the pressure gradient forces are reversed. The diffusion taking place due to resistivity will be relatively modest because the diffusion coefficient corresponding to a resistivity of a plasma at a 10 to 100-eV temperature is relatively small on the space and time scales of a solar flare. For this temperature range the resistive diffusivity is 10^5 to 5×10^3 cm^2 sec^{-1} corresponding to a resistive age for the loop of 10^{11} to 10^{14} s. Consequently, the fluid motions on the time-scale of weeks can be expected to be represented by the perfect conductivity limit. Therefore, "line tied" motions are expected to accurately represent the topological variations observed. The loop of flux can therefore either be twisted further by the hydrodynamic motions of the surface layers of the sun, or the degree of twist may be held constant because no further work is performed by the solar fluid motions.

The Stabilized Pinch or Laboratory Simulation

The laboratory simulation of such a twisted flux tube is performed in an apparatus called a pinch tube. Here, the dimensions are the order of 10 cm diameter so that in order for a plasma to simulate the perfect conductivity limit, the time of formation of the configuration must be less than the time for the fields to diffuse some small fraction of the radius. Typically, one would hope to see the field configuration to be represented by the perfect conductivity approximation for dimensions greater than a centimeter or 20% of the radius. Hence for a diffusion coefficient of 10^5 cm^2 s^{-1}, derived from a plasma temperature of 10 eV typical of these experiments, the time-scale of pinch formation must be less than 10^{-5} s. To form the pinch configuration in this time therefore requires that the voltages used, $(d/dt)(LI)$ must be large, because the field must be large and time of formation small. The magnetic field must be large enough and plasma density low enough such that the Alfvén speed is high enough such that many traversals of Alfvén waves can take place during the pinch formation time. This is equivalent to the statement that the formation of the field topology must occur quasi-statically. Typical field strengths are then 5000 gauss with particle densities of 10^{15} cm^{-3}. The Alfvén speed becomes 3×10^7 cm s^{-1} and hence the traversal time of sound across the diameter is $1/3$ μs. Consequently in this approximation, if the pinch were formed in 5 μs, the process would be quasi-static in that there would be some 15 traversals of (magnetic) sound or of Alfvén waves across the system during the formation process.

The voltage required to "drive" this quasi-static formation process can be derived from the inductance, current, and time. A typical length is 40 cm so that the inductance becomes 2×10^{-7} henrys and the current = 5RB ~10^5 amps.

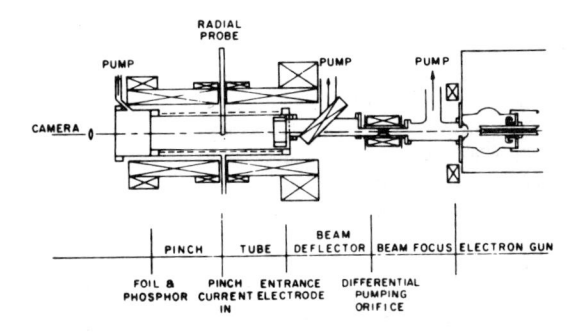

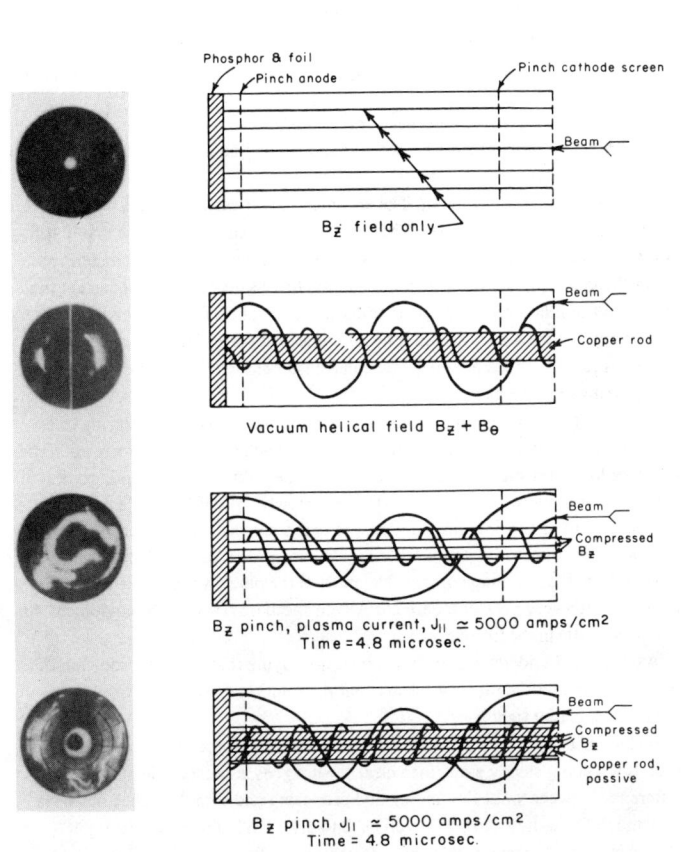

Fig. 1. The linear B_z pinch is probed with a pulsed 0.05 μs, high-energy, 500 keV, electron gun. The aluminum-coated phosphor beyond the pinch anode is photographed, and the beam pattern is displayed in the insert on the left. The vacuum B_z field spot size is small, ~1/30 of the diameter (10 cm). When a vacuum B_θ is added by current in the axial copper rod, the shear of the combined fields transforms the spot into an arc corresponding to the different pitch of the adjacent flux surfaces. Two pictures are spliced side by side, one taken with double B_θ showing twice the arc length. When a B_z pinch is made with $B_z \approx 5000$ gauss and $J_{||} \approx 2500$ A cm^{-2}, the nested-flux surfaces are destroyed and the beam is randomly distributed over the whole pinch. In the bottom picture, current flow in the plasma equals that in the central copper rod. The resulting large shear partially stabilizes the pinch. The beam patterns are the transforms of the tangled magnetic flux. [Birdsall et al., 1962; Colgate, 1978.]

force free configuration. Present measurements of gamma rays, as well as x-rays and microwave signals from flares imply that the acceleration of these particles must occur within less than several seconds of time. Such rapid acceleration makes stochastic, Fermi type acceleration mechanisms very difficult under these circumstances, and as a consequence, one is strongly induced to associate the

Consequently the voltage required for the formation of such a configuration in 5 μs is 5 kV. It is because of this voltage and current (500 MW) that capacitors are used to form the pinch. The capacitor does not "drive" plasma instabilities any more or less than the one-week time scale for the quasi-static emergence and evolution of a solar flare loop. In this case the energy of the flare is largely developed and stored below the solar surface before the flare loop emerges.

Plasma Conditions

The initial laboratory plasma conditions are assumed quiescent in the sense that the resistivity of the laboratory experiment corresponds to a plasma with a temperature of roughly 10 eV. The plasma number density in the case of the laboratory experiment is such that the runaway condition is marginally close to being met. In the case of the solar flare if the topological change due to its emergence occurs in a week, the resistive diffusion layer becomes 1.3×10^4 cm at $kT_e = 100$ eV. This current layer then also meets the runaway condition for an electron density $n \leq 10^9$ cm^{-3}. Hence some local regions of the solar flare should be similar to the laboratory runaway plasma conditions. Therefore, even though a laboratory plasma is "driven" with a capacitor bank it still forms slowly enough to model the quasi-static evolution of a solar flux tube. The microscale phenomena like plasma oscillations and electron cyclotron periods are so small compared to the formation time that the laboratory and astrophysical circumstances are similar.

Mapping of Magnetic Flux Tubes in a Laboratory Experiment

In Livermore in the early 60's we (Harold Furth and I) were particularly puzzled that the nested sheared flux surfaces predicted by hydromagnetic theory for the stabilized pinch configuration gave such exceedingly poor confinement of a plasma. This poor confinement was inferred from the low electron temperatures even in toroidal systems. We therefore felt it was necessary to map the flux surfaces to see whether the topological constraint inferred by shear was somehow or other broken by another phenomenon. Consequently we made a relatively small linear pinch tube, 10 cm in diameter and 40 cm long, and driven by two condenser banks, (1) to form the initial axial field and (2) a second to form the pinch. In addition, an electron beam probe was used to map the flux surfaces by pulsing a short, 10^8 s, intense several ampere and high energy 500 kV electron beam parallel to the magnetic field at various radii. The beam was mapped by the fields onto a phosphor and photographed from the opposite end. This is shown in Fig. 1. The mapping of the electron beam at the end plate is essentially an instantaneous mapping of the flux surfaces within the tube. When a rigid copper rod carried the current and the electron beam passed in vacuum, then the flux surface mapped was indeed the sheared magnetic field expected. The original beam circular spot mapped into a short coherent arc. On the other hand when a pinch was formed whose radius was roughly half the original, such that the axial field was compressed roughly fourfold and the external θ-field was correspondingly high, then the simple flux surfaces became stochastic as shown in the later pictures, Fig. 1. These flux surfaces became exceedingly distorted very early in time, namely several microseconds. After several more microseconds in time, half way through the rise of the field, the flux surfaces became totally tangled so that the high energy electron beam was scattered to the wall in less than the full length of the tube. We inferred from these measurements that for reasons then not understood that the magnetic surfaces were being completely destroyed by an unknown instability. We correlated the strength of this instability with the approach to runaway condition of the current carrying electrons. At that time and for many years since we have not understood a mechanism whereby the existence of the tangled field alone would cause energy to be fed into the distortion of the field associated with its stochasticity. It is my strong conviction that such a feedback mechanism most likely exists. If this is so, then the near runaway condition should establish the stochasticity that leads to the loss of the more energetic electrons carrying current of the plasma and hence connect the inside magnetic surfaces with the outside. The loss of these energetic electrons (or energetic ions necessary to maintain charge neutrality) corresponds to a net dissipation of magnetic field energy and hence to reconnection. Such a phenomenon would have a threshold for initiation associated with runaway that in turn may be triggered by a topological change induced by magnetohydrodynamic instabilities. Reconnection should be explored in the laboratory to help to explain it astrophysically.

I am indebted to Albert Petschek for discussions.

Axford: First of all, the model of Gold and Hoyle had two twisted flux tubes.
Colgate: I agree.
Axford: That's a minor point. The argument about external circuits, I think, is a slight misunderstanding here. What I was trying to say yesterday is that if you convert this to the sun and imagine that there are inductances and capacitances down there, storing a large amount of energy, there's no way the sun can get that energy out of it at a very fast rate in order to, say, make a flare. It can dribble the energy out just the way you said. The difference between a laboratory circuit and the sun is just the speed at which you can transfer energy. That's the point I was trying to make yesterday.
Colgate: Then let me point out. I think it's important to continue trying to do laboratory experiments. Stenzel's work has been such a major step forward in the first topology of just completely opposed field lines. To make the configuration of the pinch, you put the energy in slowly. Very slowly. And if you scale it by resistance and dimension, the same sort of time scale that you're using in the laboratory is equal to the one week, roughly, emergence, or month emergence, of a solar flare. It's not quite comparable; most of the pinches you make a little bit faster. But it's very slow compared to Alfvén speed traversal of the equipment. So it's quasi-static in the formation.
Axford: As I understand Biskamp, he is not saying that reconnection cannot occur; rather that the rate is arbitrary, but the solution doesn't look like Petschek's. Is that right?
Biskamp: I say, if you drive the system, you may enforce any reconnection rate. If it's going steady state, then clearly what goes in must come out. The difference between what I'm finding and proposing and what Petschek means is how much force- how strong do you have to drive? What is the power input? And that makes a real big difference. If you have a Petschek or similar type of situation then you don't pile up magnetic field.

References

Birdsall, D.H., S.A. Colgate, H.P. Furth, C.W. Hartman and R.L. Spoerlein, Particle motion on magnetic flux surfaces in stabilized and hardcore pinches, *Nucl. Fusion Suppl.*, Pt 3, p. 955, 1962.

Colgate, S.A., A phenomenological model of solar flares, *Ap. J. 221*, 1068, 1978.

EVIDENCE FOR THE OCCURRENCE AND IMPORTANCE OF RECONNECTION BETWEEN THE EARTH'S MAGNETIC FIELD AND THE INTERPLANETARY MAGNETIC FIELD

S. W. H. Cowley

The Blackett Laboratory
Imperial College of Science and Technology
London SW7 2BZ
United Kingdom

In line with the purpose of this session I want to begin with some appraisal. In particular I want to appraise the evidence which exists for the occurrence and importance of reconnection between the Earth's magnetospheric field and the interplanetary magnetic field (IMF). I remember talking to Jim Dungey about this subject a couple of years back, discussing particularly the controversy which had arisen concerning the interpretation of IMP 6, HEOS 2 and ISEE 1 and 2 observations at the dayside magnetopause. What Jim more or less said was that all this work was of course very interesting, but that he had really stopped worrying about the validity of the reconnection model of the magnetosphere after he had seen the results of Fairfield and Cahill [1966] in 1965. These results were the first to show that magnetospheric activity responds to southward turnings of the interplanetary field. However, most other magnetospheric physicists, being rather more of a skeptical bunch, took somewhat longer to be convinced, and have beavered away at the problem for almost another 20 years. All this activity has now resulted in a long list of items which have been cited as evidence for reconnection between the Earth's field and the IMF, much of which has been discussed at this meeting.

The evidence can be divided into three categories, the first of which I call "indirect evidence." This is evidence relating to the observed dependence of magnetospheric processes on the vector direction of the interplanetary field, and can itself be broken into two topics concerned with the IMF B_z and the IMF B_y components. The Fairfield and Cahill results which I've already referred to represent the start of investigations of the dependence of substorm occurrence and magnetic activity on the B_z component. Many subsequent studies have then shown how the magnetic activity indices such as Kp, AE, D_{ST} etc. also depend upon this component of the interplanetary field. The results of these studies have been sufficiently consistent and convincing that the basic B_z modulation is now almost taken for granted, while workers instead spend their time arguing about the 'best' parameter that can be pulled out of solar wind-IMF observations in order to predict magnetospheric activity. While being able to predict such activity, as measured for example by AE, is clearly important for some purposes, a more easily physically interpretable measure of magnetospheric conditions is the magnitude of magnetospheric convective flows. These can be conveniently parameterized by the total transpolar potential which can be measured across the polar cap by a low altitude spacecraft. The first paper on this subject by Pat Reiff and co-workers [Reiff et al., 1981] again showed that the z component of the field was a major factor in determining the flow. Similar related results had been published earlier by McDiarmid et al. [1978] from studies of the field-aligned current system which flows between the ionosphere and the magnetosphere.

Another consistent set of effects associated with IMF B_z was discovered in the late 1960s and early 1970s which were interpreted as relating to imbalances between the rate of dayside reconnection and subsequent substorm-associated reconnection in the tail [Cowley, 1982]. It was observed that when the interplanetary field turns southward, the dayside magnetopause is displaced inwards, the dayside cusp moves equatorward in response (together with a general equatorward expansion of the auroral zone), the magnetic flux in the tail lobes increases, and this is then followed by the occurrence of substorms. This sequence then has an obvious and consistent interpretation in terms of the reconnection picture.

There has also been a large number of IMF B_y-dependent effects observed in the Earth's magnetosphere, which to my mind are equally as suggestive as the B_z effects just discussed. We have recently identified phenomena at the dayside magnetopause which we take to be the direct initial causative mechanism leading to these effects [Cowley et al., 1983]. Specifically, we have found cases, identified by others as being examples where reconnection is occurring, where the east-west flows in the boundary layer just inside the magnetopause are actually reversed in sense compared with that in the adjacent magnetosheath, due to the magnetic stresses acting associated with the B_y component. This type of behaviour would of course be very difficult to understand in a diffusion model. The stresses resulting from IMF B_y lead to preferential eastward or westward flows in the dayside boundary layer depending on the sign of B_y, which reverse in sense between northern and southern hemispheres. These flows map down into the dayside cusp ionosphere where they have been observed both directly and via their associated current systems, the latter leading to the Svalgaard-Mansurov effect in high-latitude dayside magnetograms. There are also related flow asymmetries in the polar cap, the dusk and dawn auroral zones, and in the nightside Harang discontinuity region. Another immediate consequence of these effects is that the plasma mantle population observed in the tail lobes will also show an asymmetry [Cowley, 1981]. This effect was first found at the lunar distance by David Hardy and co-workers [Hardy et al., 1979], and Jack Gosling at this meeting has talked about the same effect now observed by ISEE-3 in the distant tail. A final directly related effect is what amounts to a partial penetration of the interplanetary field into the magnetosphere. It has been found that the field inside the magnetosphere tilts a little bit toward the direction of the B_y field outside, just as one might naively expect for a magnetically open magnetosphere. This effect was first discovered in the tail by Don Fairfield [1979], and we have subsequently found that it is also present at geosynchronous orbit [Cowley and Hughes, 1983].

Anyone not convinced by the above "indirect evidence" and still wishing to argue with the reconnection model of the magnetosphere then has to deal also with the second category of evidence which I call the "direct and *in situ* evidence." By "direct evidence" I mean a demonstration that if one follows a field line form the polar cap ionosphere that it eventually ends up in the solar wind, such as has been inferred from studies of solar energetic particle access to the polar caps [Paulikas, 1974]. Such a demonstration immediately means that reconnection must have occurred in the past, but doesn't actually prove that reconnection is going on when the observation is made. One could therefore still argue that the observed open flux is a fossil remnant of reconnection that

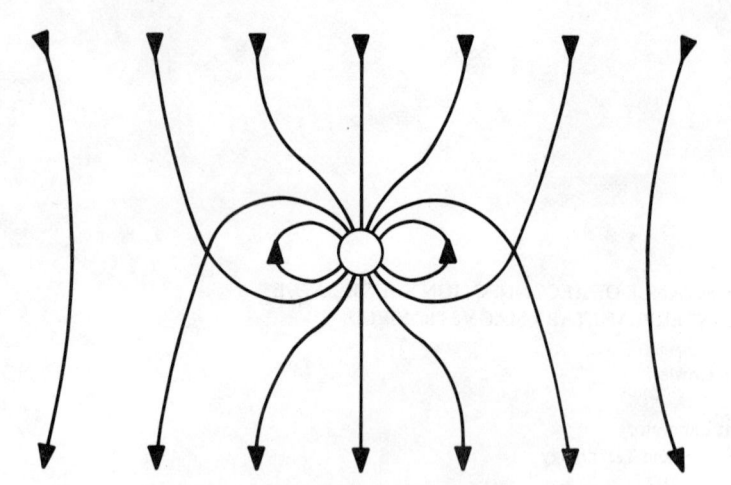

Fig. 1. Sketch of the magnetic topology of the open magnetosphere. A magnetic separator encircles the dipole at the intersection of the surfaces which bound the regions of "open," "closed," and "interplanetary" field lines.

occurred maybe in the Jurassic period which you still see. The important point is that there is no "interconnection" without "reconnection," but a demonstration that "interconnection" exists does not tell you that reconnection is occurring right now.

Turning now to the "*in situ*" observations of magnetospheric reconnection, there has been much discussion at this meeting of the recent observations which have been made at the dayside magnetopause and in the tail. Well-studied examples of quasi-steady dayside magnetopause reconnection exist which display the expected features [Sonnerup et al., 1981]. We have seen magnetosheath and ionospheric plasma acceleration at the magnetopause, together with the loss of energetic ring current particles along the open field lines into the magnetosheath, and the appearance of a measurable field component normal to the magnetopause current sheet. Observations of the accelerated magnetosheath ions at low altitudes in the dayside cusp were also discussed at this meeting by Pat Reiff. On the other hand dayside magnetopause reconnection also seems to occur, and possibly more frequently, as spatially and temporally localized "flux transfer events" (FTEs) which I will further discuss a little later.

With regard to reconnection in the geomagnetic tail, several papers at this meeting have amply demonstrated the abundance of evidence which exists for the association between substorms and near-Earth reconnection. However this is not wholly germane to this talk on reconnection and the open magnetosphere because in principle the observed 'pinching off' of the plasma sheet and plasmoid formation could also occur in a closed magnetosphere. In the ISEE-3 deep tail data, however, we seem to be seeing the effects of continuous as well as the sporadic substorm-associated tail reconnection. At $\sim 200\ R_E$ downtail, the plasma and energetic ion streaming in the plasma sheet is (with rare exceptions) continuously tailward as discussed here by Manfred Scholer and Pat Daly, while George Siscoe has described the ISEE-3 magnetometer observations which show that the average B_z threading the current sheet is negative at these distances as is then expected. The fact that reconnection may occur continuously, if not steadily in the tail may also relate to some observations that Tim Eastman told us about. He reported seeing ion beams on the outer surface of the plasma sheet in the near-Earth tail under essentially any magnetic conditions, and it seems likely that these beams are reconnection-associated [Cowley, 1980].

Having now looked at both the "indirect" as well as the "direct and *in situ*" evidence for reconnection between the Earth's field and the IMF, what now about the ultimate skeptic who demands to see "incontrovertible evidence"? This then firstly forces us to become legalistic and to say exactly what reconnection is. The definition I am going to adopt, and what I would recommend to you, was I believe first stated in a paper by Jim Dungey in 1978 where he says that reconnection is nothing other than an electric field along a magnetic separator [Dungey, 1978]. If such an electric field exists then reconnection by definition is occurring. As Götz Paschmann has already observed once at this meeting, however, this is clearly nearly impossible to observe. Firstly the separator has to be identified, which has never yet been convincingly done, and then a valid measurement of the electric field has to be made along it, which is also exceedingly difficult. Another "incontrovertible" way of showing that reconnection is occurring, perhaps somewhat less impossible, would be by showing that the amount of 'open' flux in the magnetosphere varies with time. The magnetic topology of the open magnetosphere is sketched in Figure 1. A magnetic separator encircles the dipole at the intersection of the surfaces which bound the regions of open, closed and interplanetary field lines. From Faraday's law the rate of change of open flux in the system is just the emf around the separator, i.e.

$$\frac{d\Phi}{dt} = -\oint_S \underline{E} \cdot d\underline{\ell}$$

where Φ is the amount of open flux, $\underline{E}$ is the electric field and $d\underline{\ell}$ is a line element along the separator S. So if the rate of change of flux is non zero, then the emf is non zero and reconnection must by definition be occurring. This argument contains no plasma physics, just Faraday's law. It should be noted, however, that if the amount of open flux is constant in time this does not necessarily show that reconnection is absent. In principle we could have equal and opposite voltages along different parts of the magnetic separator which add to zero, giving steady reconnection with a steady amount of open flux. Of course, there is plenty of evidence that the amount of open flux in the magnetosphere does change with time via the expansion and contraction of the auroral zone but in view of the difficulty of determining the exact location of the boundary of open and closed field lines maybe this wouldn't be counted as being incontrovertible.

I now want to turn briefly to consider future directions concerning magnetospheric reconnection, and clearly one important area over the next year or two is going to be the study of the detailed deep-tail data from ISEE-3. We have seen the initial fruits of those studies at this meeting, and it is important to see that this continues, to maximize the science return. It might also be worthwhile to dust off the old Pioneer 7 and 8 data and take a fresh look at that, since in my view much of the original analysis of that data was, with a few notable exceptions, rather inadequate.

Another area in which significant work remains to be done is, of course, reconnection at the dayside magnetopause. The first point I want to make echoes a comment made earlier by Mark Saunders, and that is that I don't think we should ignore the "quasi-steady" type of reconnection signature at the magnetopause, typified by the 8 September 1978 event, as a situation which is so rare that one can essentially forget about it as far as generating significant magnetospheric flows are concerned. Although the number of examples are relatively few (which may partly be due to overselection), I think Götz Paschmann had 17 on his latest slide, this is not necessarily inconsistent with the process being important. On the other hand FTEs are ubiquitously present at the dayside magnetopause whenever the IMF is southward pointing, and estimates of the flux transfer rate indicate that they are almost certainly an important voltage source. It is therefore appropriate that they should have been emphasized at this meeting.

There are also quite a lot of interesting directions for future research as far as FTEs are concerned. We know that their spatial structure is complicated, involving twisting and bending of the field, but the full 3-D structure of the field needs to be clarified. We also would like to know in some detail what causes this structure and how it evolves as the open tubes move over the dayside magnetopause and into the tail. We should therefore start thinking about some theory and modelling, and also perhaps look at some of the older spacecraft data sets which could extend the ISEE data out to higher latitudes. This is what Jim Dungey called a CLOSED program, standing for Close Look Over Sets of Existing Data.

We would also like to know the spatial and temporal spectrum of FTEs. Most of the events which have been published so far represent the large, well defined, $\sim 1\ R_E$ end of the spectrum which can be seen well away from the magnetopause, but often near to the magnetopause there is a proliferation of smaller scale magnetic perturbations as well which look just like small flux transfer events. It may therefore be that the magnetopause at times looks rather like spaghetti with some big FTEs and lots of little ones as well, covering most of the surface. A third future topic is the exact relationship between FTEs and the quasi-steady reconnection events. Richard Rijnbeek has already told us that all the latter events so far published have FTEs associated with them, sometimes very many of them. Perhaps the quasi-steady events simply represent times when magnetopause reconnection hasn't quite switched off between FTEs, so that accelerated flows are still seen at the magnetopause.

A final topic of future interest with regard to FTEs will be the attempt to identify their ionospheric signatures. In Figure 2 I have drawn a very rough sketch indicating what one might see in dayside high latitude ionospheric flows. Figure 2(a) is appropriate to steady dayside reconnection, and shows a fixed boundary between open and closed field lines mapping to the dayside magnetic separator, and a steady flow across that boundary from the dayside auroral zone into the polar cap. When flux transfer events are occurring, however, the open-closed field line boundary will not be steady, but will jump periodically equatorward as each FTE is formed. If reconnection ceases entirely between each FTE the boundary during these intervals will move poleward with the flow, before jumping equatorward again with the next FTE. On average, of course, the boundary can stay in the same place. The spatial scale of a FTE in the ionosphere can easily be estimated from magnetopause observations via flux conservation, and turns out to be ~200 km. This is quite a large dimension, comparable for example with the STARE field of view, but nevertheless represents only some fraction of the distance along the dayside boundary. Each FTE therefore affects only part of the boundary, which should thus have a corrugated appearance as I've shown in Figure 2(b). Within each corrugation the flow on the newly-opened flux tubes should abruptly change direction, e.g. to the north, in response to the stresses exerted by the magnetosheath flow and field. The twisting of the field we see in space may also relate to a local swirl in the flow within the corrugation and associated field aligned currents.

Before concluding this discussion there is one other topic which I will briefly mention, which has only been discussed once at this meeting, again by Pat Reiff. This is the possibility that a new mode of interaction occurs between the solar wind and the magnetosphere when the IMF is strongly northward, relating possibly to reconnection occurring poleward of the dayside cusps. I think we now have a fairly good idea as to how the Earth's magnetosphere works under normal conditions where B_z is either small or negative, at least in broad outline. I don't think we have got even that far yet for the 'quiet' magnetosphere when the IMF is strongly positive. I think, therefore, we should pay more attention to that subject in the future as well.

Luhmann: Would it be better to say that tail merging is spontaneous when you have a northward interplanetary field and driven in southward?

Cowley: I don't think I have anything useful to say about that, would anybody else like to comment?

Dungey: See if Janet has an answer.

Commentator: It is one of the questions we are really trying to solve, I guess.

McPherron: One curious question is why geomagnetic activity doesn't seem to reveal 'northward' reconnection in the same way as it does the southward. Is it just that we have the wrong measures of activity?

Cowley: When the field is strongly northward there does seem to be a considerable amount of magnetic activity but it's generally confined to very high latitudes. The field perturbations associated with the modified high latitude field-aligned current systems observed by low altitude spacecraft for strong positive B_z are not small by any means, and seem to increase in strength with positive B_z in the same way as the 'normal' field perturbations generally increase in strength with negative B_z. There does appear to be a relationship therefore between the magnitude of the new sort of flow cells that appear at very high latitudes for positive B_z, and the strength of the B_z component.

McPherron: So polar magnetometer chains ought to reveal this?

Cowley: You have to go poleward of about 80° invariant.

Eastman: Several observations that you have taken to be direct evidence for reconnection are actually indirect. From the definition of reconnection there are two things which certainly would be direct evidence. That is, you mentioned the tangential electrical field associated with reconnection, or the change in field topology, in the amount of open field lines. If we could measure those in some way, that would certainly be direct evidence. Now, how can you call an observation 'direct' evidence without being able to measure those?

Cowley: If you show that any flux connects between the Earth and the interplanetary medium, then you have shown that reconnection has occurred at least in the past. That is what I called 'direct' evidence. In order to show that reconnection is occurring now, you need to show that the amount of flux is changing with time. You would need to identify the boundary between open and closed field lines all around the polar cap and show that it moved such that the amount of flux changed with time. However, to tie that down completely is obviously very difficult.

Eastman: So in some sense all the rest of it is indirect evidence, whereas what you just described would be the direct evidence.

Cowley: It's what I would call incontrovertible evidence in terms of the legalistic definition of reconnection. However the weight of the other evidence, the "indirect", the "direct" and the "*in situ*" evidence that I have cited leaves in my mind no doubt that reconnection is occurring essentially at all times, that it's not fossil reconnection that we see when we infer the existence of open field lines from solar particle access.

Vasyliunas: A definition of reconnection actually refers to the plasma flow through the separatrix; then, given the fact that there's open flux now, all you need to prove that reconnection is occurring now is the fact that the solar wind is flowing. It's much simpler than the Jurassic period or whatever.

Cowley: I understand your definition in terms of the flow across the separatrix, but the main reason that I like the definition in terms of the electric field along the separator is that it immediately and directly relates "reconnection" to "interconnection" via Faraday's law. There is no interpretation that comes into the relationship except the Maxwell equation.

Colgate: Do you feel it would be pertinent for Stenzel to measure the electric field?

Cowley: There seems little doubt that there must exist an electric field along the separator in his experiment.

Luhmann: Is it known how the occurrence of plasmoids relate to the interplanetary field direction, or is the data too sparse?

Cowley: Well plasmoids relate to substorms, and substorms generally to intervals of negative B_z, I don't know I can say any more than that.

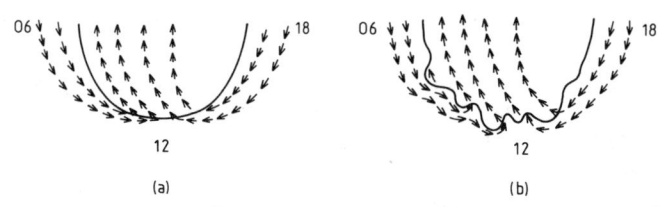

Fig. 2. Sketch of dayside northern hemisphere high latitude flows for (a) steady reconnection, (b) unsteady, localized FTE reconnection. The solid line shows the boundary between open and closed field lines, while the arrows show the plasma flow. The preferred sense of east-west flow shown for open flux tubes corresponds to the situation for positive IMF B_y.

References

Cowley, S. W. H., Plasma populations in a simple open model magnetosphere, *Space Sci. Rev., 26*, 217, 1980.

Cowley, S. W. H., Magnetospheric asymmetries associated with the Y-component of the IMF, *Planet. Space Sci., 29*, 79, 1981.

Cowley, S. W. H., The causes of convection in the earth's magnetosphere: a review of developments during the IMS, *Rev. Geophys. Space Phys., 20*, 531, 1982, and references therein.

Cowley, S. W. H., and W. J. Hughes, Observation of an IMF sector effect in the Y magnetic field component at geostationary orbit, *Planet. Space Sci., 31*, 73, 1983.

Cowley, S. W. H., D. J. Southwood and M. A. Saunders, Interpretation of magnetic field perturbations in the Earth's magnetopause boundary layers, *Planet. Space Sci., 31*, 1237, 1983.

Dungey, J. W., The history of the magnetopause regions, *J. Atmos. Terr. Phys., 40*, 231, 1978.

Fairfield, D. H., and L. J. Cahill, Jr., Transition region magnetic field and polar magnetic disturbances, *J. Geophys. Res., 71*, 155, 1966.

Fairfield, D. H., On the average configuration of the geomagnetic tail, *J. Geophys. Res., 84*, 1950, 1979.

Hardy, D. A., H. K. Hills, and J. W. Freeman, Occurrence of the lobe plasma at lunar distance, *J. Geophys. Res., 84*, 72, 1979.

McDiarmid, I. B., J. R. Burrows, and M. D. Wilson, Comparison of magnetic field perturbations at high latitudes with charged particle and IMF measurements, *J. Geophys. Res., 83*, 681, 1978.

Paulikas, G. A., Tracing of high-latitude magnetic field lines by solar particles, *Rev. Geophys. Space Phys., 12*, 117, 1974, and references therein.

Reiff, P. H., R. W. Spiro, and T. W. Hill, Dependence of polar cap potential drop on interplanetary parameters, *J. Geophys. Res., 86*, 7639, 1981.

Sonnerup, B. U. Ö, G. Paschmann, I. Papamastorakis, N. Sckopke, G. Haerendel, S. J. Bame, J. R. Asbridge, J. T. Gosling, and C. T. Russell, Evidence for magnetic field reconnection at the earth's magnetopause, *J. Geophys. Res., 86*, 10049, 1981.

NOW CONSIDER DIFFUSION

J. W. Dungey

Imperial College, London SW7 2BZ, England

I need only talk for a few minutes. I have been looking forward to this meeting for a long time. I enjoyed the anticipation and the actuality has come up to my expectations. I hope there's going to be another one before too long.

I want to talk about future work, but first I will reply to Stan Cowley's comment on my naivety in believing in the whole story to 99% confidence in '65, when I knew about Fairfield's results. I'm not as old as Stirling, but maybe I should put out a little of the wisdom of age. Does it matter whether you make the right judgement about theories? Yes, it does, particularly for experimentalists perhaps, but also for theorists. The work you do later depends on the judgement you've made on previous work. People have wasted a lot of time developing on insecure or even wrong foundations.

Now for future work. One mild surprise I have had is that we haven't heard more about diffusion, in two contexts. Gordon Rostoker is yet to come and he may talk about particles getting into the magnetosphere by diffusion. Lots of noise is observed and so diffusion must happen.

If time had not been short, I was planning to discuss in a handwaving way what sort of diffusion mechanisms one might consider. The other aspect of diffusion I was going to talk about is at the other end of things and is velocity diffusion, which is involved in anomalous resistivity. If you want to know what I think on that subject, you can read my extended abstract - again (Joke).

DEFINITION OF A SUBSTORM, PHYSICAL PROCESSES IN A SUBSTORM AND SOURCES OF DISCOMFORT

G. Rostoker

Department of Physics, The University of Alberta
Edmonton, Alberta, Canada T6G 2J1

I will start out by drawing your attention to a famous substorm that everyone is talking about. Five of us got together some time ago and tried to reach consensus. They were Bob McPherron, Syun Akasofu, Wolfgang Baumjohann, Yohsuke Kamide, and myself. We fought it out for two days in Münster and we did reach consensus; not only there, but a year later we even managed to write it all down without everyone objecting too much. It tends to happen that whenever individuals get up to express their views, they polarize. But they really aren't that polarized, and so let me give you Figure 1 on which we all agreed.

I want to talk about what a substorm is and then tell you some things that concern me a little bit. Let's suppose you have an interval of some continuous northward IMF for a rather long time (or, if you wish, $\epsilon = 0$) and then suddenly turn the IMF southward and then have an arbitrary length of time where it's just steadily southward. One then asks the question, what are you going to see? We all agree that, basically, the thing we're going to see first of all is a growth in the tail field energy storage. That is, the one thing the magnetosphere wants to do in response to the change in outside boundary conditions is to build a tail. But the energy may be pouring in too fast for that, and so the magnetosphere decides to dissipate some energy through I^2R heating and so geomagnetic disturbance and joule heating start to build up. Curiously enough, it builds up rather jerkily, and any of you who looked at growth phase phenomena, and even originally in Bob McPherron's work, will see what appear to be small signatures in the ground magnetograms with pi pulsations and all the things you characteristically associate with substorms. I might also add that should you be lucky enough to be right underneath one of the filaments, you'll see something which you would say is a big substorm, 200-300 nanoteslas not being uncommon. But heaven help you if you're a couple of hundred kilometers away because the current localization will produce a very small magnetic perturbation.

Returning to the response to a southward turning of the IMF (Fig. 1), if a case arises that the magnetosphere still can't cope with the energy through tail building and depositions in the ionosphere, then the symmetric ring current is going to grow rapidly. So there's kind of a hierarchy in my mind as one looks at this. Eventually you get enough I^2R heating because the current system builds up nice and slowly as a function of its large scale size and hence its self-inductance, and eventually you get to a state where if you were lucky enough to have this long period of southward IMF, everything would become stable at new levels. You'd have energy stored in the tail relative to the state from which you started. As well, the asymmetric ring current would be enhanced because it, of course, is part of the three-dimensional current system which involves the electrojet systems.

Ultimately, what happens after an hour or an hour and a half of reasonable southward field, is that a northward turning of the IMF will trigger what most people have been attracted to on magnetograms - namely, a sudden and rather spectacular change in auroral zone magnetic field, the so-called negative H bay. People generally call *that* the substorm. They point to *that* and say that's the substorm. In our view we have an overall process of which this is one component. Ultimately what happens here is the tail energy decreases because it's providing the requirements for the expansive phase and gradually the whole thing decays

away. To sort of give you a feeling for what I mean by this kind of behavior, let me mention first of all, it's very rare to be lucky enough to have an hour or so of good solid southward field without something else happening.

But - here's an example (Figure 2) of a group of events that I've looked at not too long ago. Notice B_z is northward until near ~0700 UT. It then turns southward and stays steadily southward until 0816 UT; then it pops northward and remains basically in the ecliptic plane or marginally northward. Figure 3 shows what happens on the ground with the magnetograms. These are an array of magnetograms covering the Alberta sector where the event was maximal. The dashed line is when the IMF turns southward, and the dotted line is where it turns northward. In the latter instance, the magnetogram shows a sharp response and everybody is attracted to that and calls it the substorm. But please - the substorm contains more than that. There are effects between ~0700 and 0816 UT - they don't look very big here, but in other events they are more attractive to the eye. These "smaller" effects are the signature of the driven system.

So having said that, let me now alert you to what the auroral electrojets associated with the driven system look like (Figure 4). We have an eastward electrojet and a westward electrojet which cuts up to the north of the eastward electrojet in the pre-midnight quadrant. This is what Syun Akasofu would call the driven system. When you have a substorm expansive phase of the type that people like to identify, a current wedge involving an intense ionospheric westward electrojet develops rather explosively in the pre-midnight region in the interface region between the eastward and westward electrojets. *But the driven system electrojets do not go away!* Maybe the borders move a little bit - the equatorward border may shift southward; perhaps the Harang discontinuity will shift southward and flutter around. But the driven system electrojets maintain their identity despite the development of the substorm current wedge near midnight. And my first dissident comment is that it's no fair to study the substorm phenomenon by looking at only one aspect of it - that is to study the substorm current wedge (which we associate with the explosive activity) without taking the driven system into account. Please remember that the substorm phenomena involve the driven and loading-unloading current systems which are relatively colocated near midnight. So my first point is - don't study one of these systems in isolation without asking what it might be doing to the other.

All of us, I think, in our business work by analogy and this process isn't restricted to our field alone. Virtually every bit of human understanding is built on something that happened before to which we try to draw an analogy and then bootstrap our way up into a better understanding. My analogy is the hydrodynamic analog. Sure, I know currents are important, and I know they modify things, but I start out with the basic thought in mind. Figure 5 shows a picture of a viscous fluid flowing past a cylinder (seen on end). As you make the flow faster and faster, a region develops in here which has circulation zones in it. Lovely convection patterns, very reminiscent of the kind of thing we talk about in our magnetospheric business. Well, it's no fair to show just the circulation pattern because that's a closeup of one part of a larger entity. Figure 6 shows the entire system. Close to the obstacle we have the region of closed circulation which is

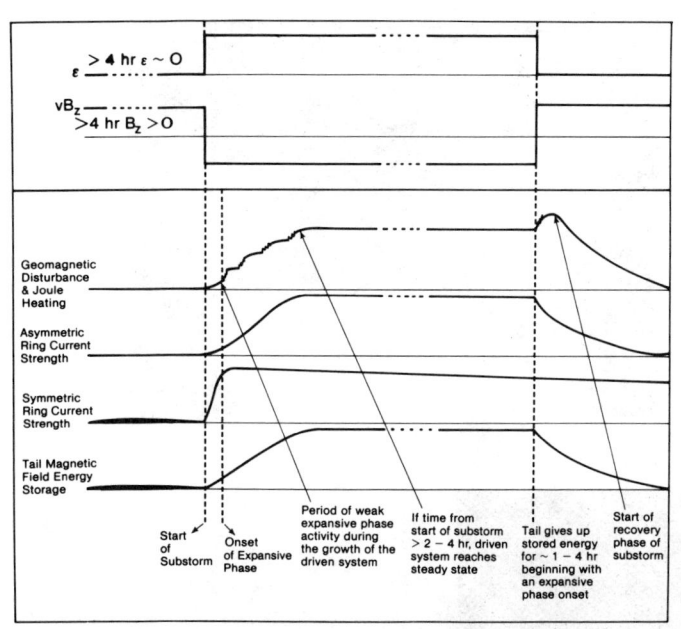

Fig. 1. Responses of geomagnetic parameters during the course of a magnetospheric substorm. The driven system currents and geomagnetic tail current may grow simultaneously. A reduction in energy input from the interplanetary medium may trigger unloading of tail energy (after Rostoker, Akasofu, Baumjohann, Kamide and McPherron).

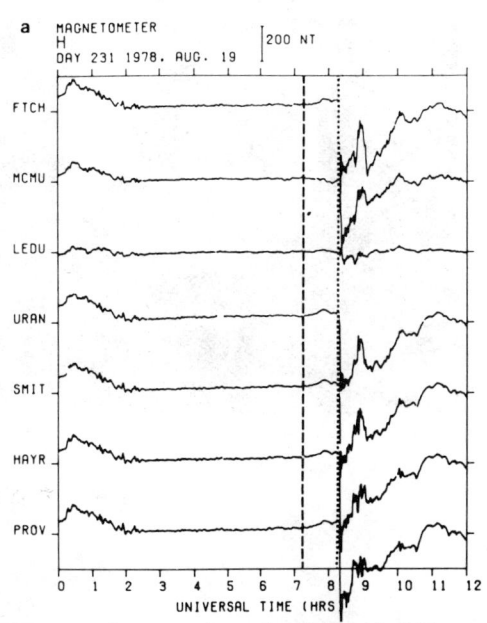

Fig. 3. Magnetograms from auroral zone showing growth of driven system eastward electrojet starting shortly after 0700 UT and expansive phase onset near 0816 UT (see Figure 2). The substorm expansive phase is considered to be triggered by the northward turning of the IMF [after Rostoker, 1983].

called by the hydrodynamicists *the separation bubble*. And behind it there's a laminar wake. I started out playing around with this many years ago, and if you sort of scale things, you find the separation bubble is of the order of two to three obstacle scale sizes in length.

Now for our magnetosphere the obstacle is about 40 earth radii in diameter. So you're looking possibly at the end of the separation bubble at 80 to 120 earth radii. Then comes the lengthy wake (which, in fact, if the flow velocity is large enough, produces a piece of modern art that's worth selling). For those of you who looked at the comet tail (Ed Hones presented it, as did Niedner) do you remember seeing little globs and blobs in what was an extended region behind? And the question is, What do you identify them as? Well, I think you can see what I'm getting at. Because when we look at the ISEE-3 data we ask ourselves what's being seen there. If we want to push ourselves a little bit with this hydrodynamic analog and say somewhere between say 80 and 120 earth radii or thereabouts is the end of the separation bubble, this, to me, would represent the end of our conventional magnetosphere containing the convection cells. Then, for the first pass, ISEE-3 (some 90 earth radii back) might very well see what we normally call the magnetotail. But I ask you to think a little bit how they interpret those passes at 220 earth radii. *What is the phenomenological character of a wake in an MHD situation?* And so, the second comment that I have which is, in the hydrodynamic analog, that you have both the separation bubble and a wake behind the obstacle. Don't discard the possibility that ISEE-3 might be seeing, in fact, a wake.

Now, the third point I want to come to. Obviously, my analog has directed me to a model of the magnetosphere which might be thought to be different than others. Figure 7 shows my picture of the magnetosphere in a projection in the equatorial plane. There is a shear zone between the low latitude boundary layer

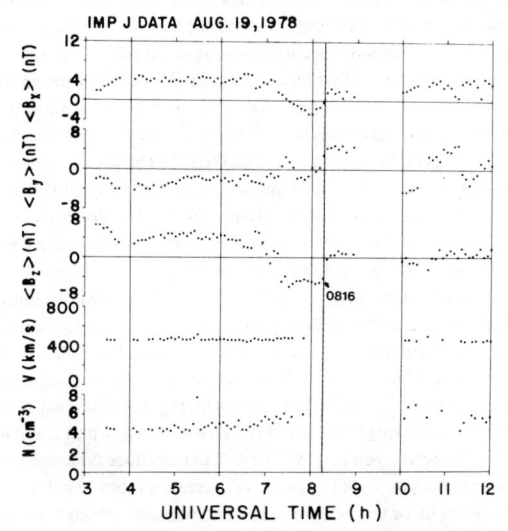

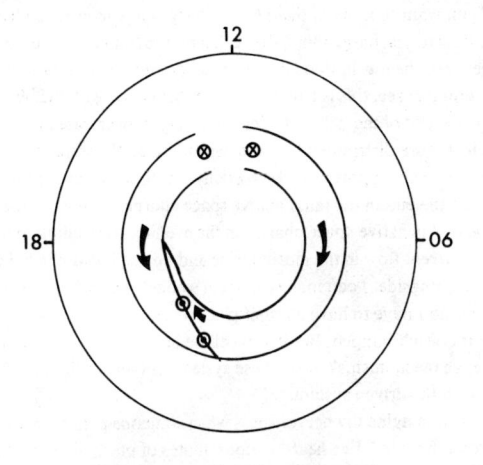

Fig. 2. Southward turning of IMF B_z shortly after 0700 UT followed by a sudden northward turning at 0816 UT [after Rostoker, 1983].

Fig. 4. Average configuration of the driven system auroral electrojets. These electrojets persist during substorm expansive phase activity.

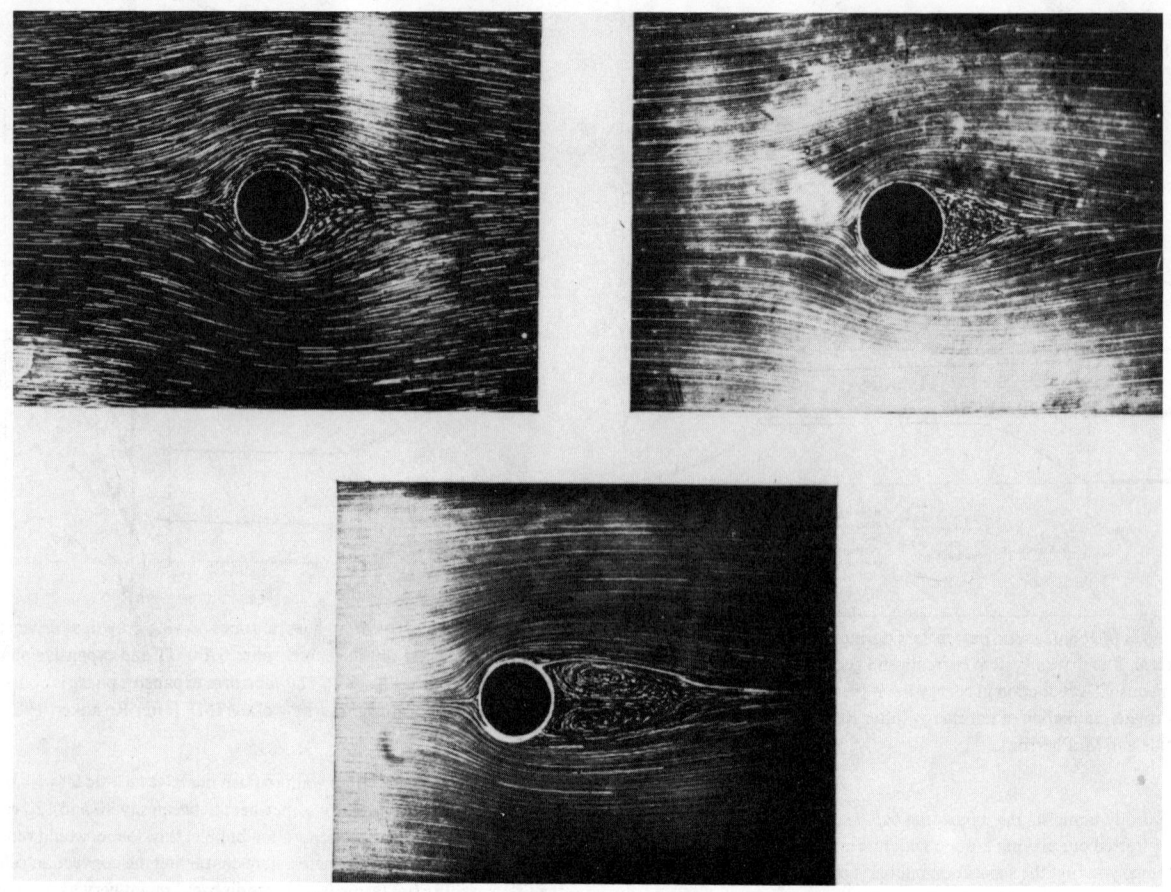

Fig. 5. Flow of viscous fluid past an obstacle at various flow velocities. Closed convection cells are clearly evident in the separation bubble behind the obstacle for the fastest flow (bottom panel). [From Lighthill, 1963].

and the central plasma sheet. The plasma sheet features earthward flow and the low latitude boundary layer anti-earthward flow. Ultimately plasma gets across the shear zone near the end of the tail, possibly by diffusion as suggested by Walter Heikkila. One may get anomalous mass transport because you can get a nice Kelvin-Helmholtz instability along this shear interface leading to lots of wave energy which can bounce particles around very nicely. These shear zones I like to think of in terms of space charge because it allows me to easily account for the basically dawn-to-dusk electric field across the central plasma sheet. Not only that, but if you want to think of these space charges as encouraging field-aligned currents to flow to discharge them, the region one currents map out to the interface between the low latitude boundary layer and the plasma sheet.

Now, as you can see, this is a nice, closed system here, with the plasma turning around near the end of the tail, sometimes in a very concentrated fashion and sometimes in a more distributed fashion, depending on the state of activity. You'll notice I have a lovely opportunity for vorticity due to curvature of the plasma flow lines near the end of the tail. Positive space charge is found on the morningside and negative space charge on the eveningside leading to downward field-aligned current flow in the morningside and upward field-aligned current flow on the eveningside. I can make a current wedge! And if I can make a wedge one way, why do I have to have another place where I can make a wedge? That doesn't say it doesn't happen. But if you believe in Occam's razor, do I need two different wedge mechanisms? Also, these systems shown in Figure 7 fit beautifully with the driven system.

Now you can imagine my nervousness when someone puts a plasmoid inside my "separation bubble." I've heard various quotes of plasmoid scale size - some involve half the extent of the tail. And then the plasmoid runs out through the back of the "separation bubble." What's it going to do to the shear zone responsible for the driven system? Well, I don't know. Maybe it's no problem at all, but I'm uncomfortable. I think you might be able to see why. Because as I told you, the driven system does not appear to be violently distorted through the generation of a substorm current wedge. The eastward jet doesn't disappear nor does the driven system westward electrojet. Everything looks pretty normal, except for the explosive wedge development in the region of the Harang discontinuity. And so I scratch my head a little bit and worry about that.

Perhaps that gives you a little bit of an idea of what my concern is regarding where a plasmoid - where a near-earth neutral line fits into the picture. I think I can understand the formation of a current wedge at the end of my "separation bubble" based on the flows in that region of space. But certainly things are flowing in, turning around, flowing the other way, and, I'm sure, going out in the opposite direction downtail on this side. This is hand-waving, you say. Perhaps. But too many things fit. You go and you calculate the vorticity in a shear zone like the one between the low latitude boundary layer and the central plasma sheet, given the boundary layer flows and the plasma sheet flows. You get a value (~ 0.03 s^{-1}), you map it down to the ionosphere, and you end up with what you see in auroral arcs (~ 20 s$^-$) which is sort of encouraging.

Let me conclude by telling you my interpretation of substorm signatures in the tail at the moment. One of the schools of thought - the one that I adhere to - suggests that all the action tends to occur in the boundary layer plasma sheet as observed in the magnetotail. (In fact, field lines threading the boundary layer plasma sheet map to the low latitude boundary layer.) And if you say that all the substorm is doing is making that boundary move up and down so that it may wash over the spacecraft, you may well be able to reproduce typical magnetotail signatures. I will take you back to one of the interesting moments of space physics, back in 1973, or thereabouts, when two schools of thought arose as to the origin of the interplanetary magnetic field. One was championed by John Wilcox, namely, the sector structure; that is, there were globs of field coming out

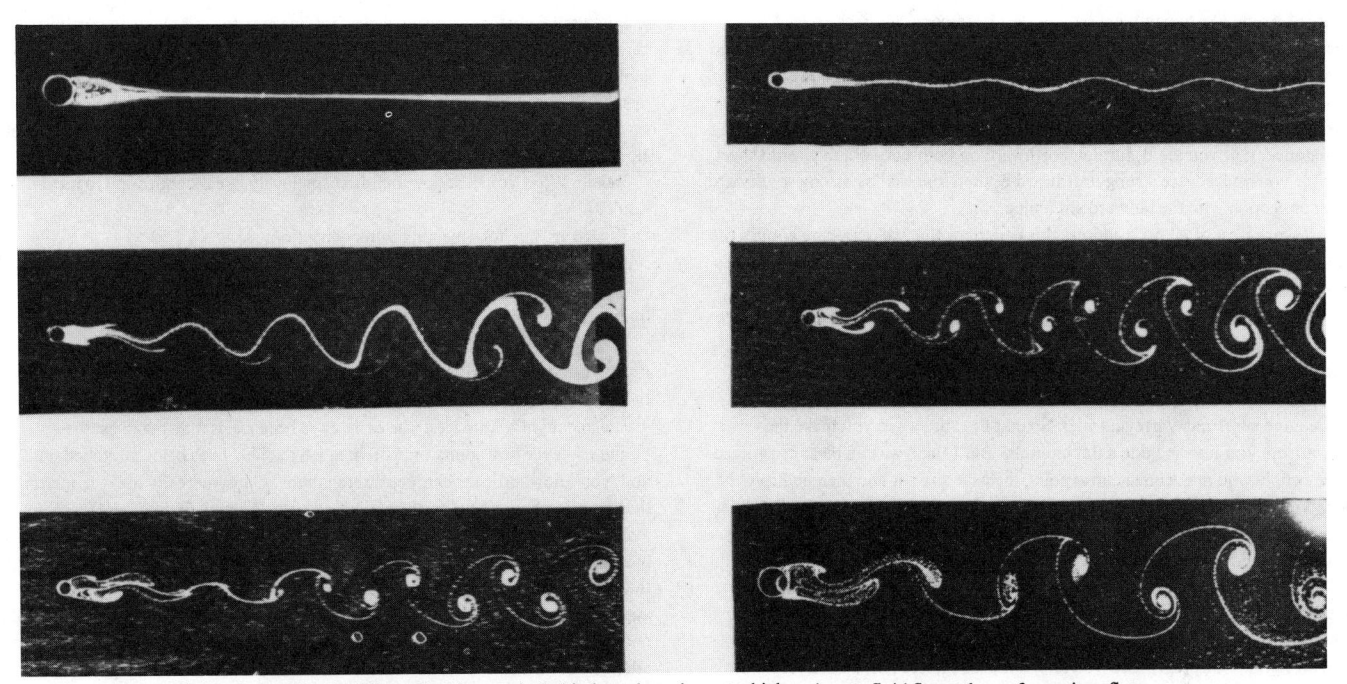

Fig. 6. Separation bubble and laminar wake behind an obstacle past which a viscous fluid flows shown for various flow velocities. The higher the velocity, the more unstable the wake becomes and the more distorted becomes the separation bubble. Our magnetotail should be considered as being composed of two parts - a separation bubble inside of 115 R_E and a laminar wake beyond 115 R_E [From Lighthill, 1963].

from the sun. As the sun rotates they pass by you, and you get the sector structure that way. In contrast, there was a second opinion for which Mike Schulz was a champion, namely, there was a current sheet roughly in the ecliptic plane and it was flapping up and down causing the IMF direction to flip back and forth. We know that Schulz's picture just got submerged until one definitive experiment was performed. Pioneer went up 16° above the ecliptic plane and destroyed the sector structure concept based on corotating structures once and for all in one fell swoop.

Well, I suggest to you that some magnetotail plasma and field signatures could be interpreted in two ways. You may, if you see a sudden change in time followed by another sudden change in time back to normality, either be seeing something passing by you as you sit in a moving earthward or anti-earthward homogeneous medium (the plasmoid interpretation). Or alternatively, you could have a boundary waving up and down. First you're on one side of it, then you're on the other side of it, or within it. And here you have nothing passing you by in the sense of earthward or anti-earthward motion, but nonetheless you get a particle or field signature as a function of time, which is of the same ilk. *I think that possibility has to be investigated.* And so there is a kind of request I would like to make. Namely, in the event that data may be explained in the context of more than one framework, both frameworks must be seriously studied until one or both of them are in violation of the data. And on that note I end my presentation.

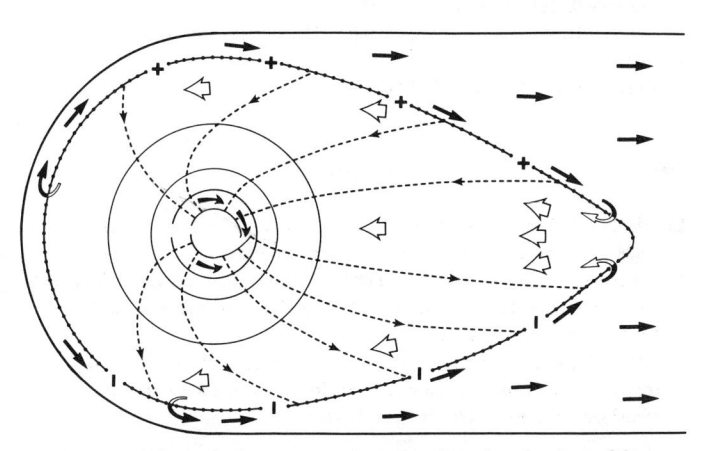

Fig. 7. Flow of plasma in the magnetosphere as projected on the plane of the neutral sheet. Solid arrows indicate low latitude boundary layer and open arrows the central plasma sheet (cps). The shear zone between the boundary layer and cps is the site of space charge leading to the flow of region I Birkeland currents (dashed lines) [after Rostoker and Samson, 1984].

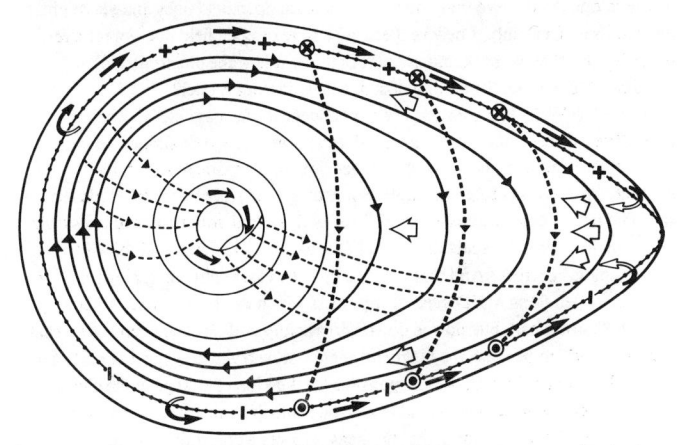

Fig. 8. Projection of ring current (solid lines) and Birkeland currents (dashed lines entering auroral ionosphere) on the plane of the neutral sheet. Dashed lines across the tail signify cross tail current. Open arrows indicate cps plasma sheet flow while solid arrows just inside the magnetopause indicate low latitude boundary layer plasma flow. Curvature in flow at the end of the tail leads to space charge whose discharge leads to the formation of the substorm current wedge [after Rostoker, 1983].

Birn: I would like to comment on your last statement. We have, indeed, checked other possible explanations alternative to the passage of a plasmoid, and the main alternatives that come to mind are either upward and downward motion of the whole thing or a local thickening of the plasma sheet and some wave passing by. And none of the other alternatives is consistent with the observations, for instance, that you see B_z turning northward on both sides of the plasma sheet, and that you, indeed, see strong northward B_z then followed by strong southward B_z in connection with the tailward streaming.

Rostoker: First of all, let's address this boundary that we're talking about - the tail lobe-plasma sheet boundary in which the boundary layer plasma resides. It's not necessarily just a flat interface. As I mentioned, you can have Kelvin-Helmholtz instability at this interface. In fact, I think this is an important feature of what results eventually in pulsation activity as seen down at the earth's surface. And it can extend all the way along the boundary layer-central plasma sheet interface, I'm sure. And if you map back the boundary plasma sheet, you can imagine some nasty irregularities. So that I'm sure when you make the comparison, you just imagine a flat interface. But I think we can be as convoled as the people who make one neutral point, then x it and o it and then multiply the x's and o's in order to explain the observations. I think those kinds of capabilities are also available when one inspects this kind of model with waves propagating along the boundary layer plasma sheet-central plasma sheet interface.

Eastman: Let me note that I have a paper with me that provides some ISEE observations which, I think, strongly support the idea that we do need to to take this kind of thing into consideration. And I have copies available with me, and you can sign up for preprints, if you like. It must be noted that I have yet to find clear flow in the magnetotail in ISEE or IMP data that's not, in some way, obviously associated with the boundary phenomena. And when I can finally find such strong flow, not obviously correlated with boundary phenomena, I intend to publish that.

D. N. Baker: I don't see what gives rise to a quality of suddenness in your model, nor do I see what gives rise to open field line topologies in your model. Can you comment?

Rostoker: Well, There are certainly open field lines. I'm, indeed, a believer in reconnection. I think it certainly is occurring in the front side, and it takes field lines and peels them over the pole and creates a population called plasma mantle.

Baker: within the Plasma sheet, I mean.

Rostoker: Ah well. Within the plasma sheet? Open field lines within the plasma sheet?

Baker: Yes.

Rostoker: I would find it hard to believe that there are such things as open field lines within the plasma sheet, and I know that there are all sorts of cunning ways in which one can believe that you think you're on open field lines and yet might be on closed. Certainly, I believe there are a suite of open field lines swept over the pole, and they go back and are part of this whole wake phenomenon.

Baker: How about the quality of suddenness in your model?

Rostoker: Well, The question is what happens to the magnetotail when the interplanetary field flips northward; it doesn't even have to be sudden - it can start gently gradiating northward and then, for some reason or other, the magnetosphere starts to go unstable. Something obviously happens within the tail. The tail, in essence, has to relieve itself of flux. And you then have to ask the question, how does it decide to do that? One of the ways you could think of is shortening it - correct? And that could happen by essentially imagining this plasma is open to an MHD generator process which can bring it to a screeching halt. Well, not a halt, but slowed down very rapidly and turned around. And that would be sudden because the magnetosphere is now trying to get rid of part of its tail. So I try to envision that as responsible for the kind of change in the character of the vorticity which, then, would lead to an explosive substorm current wedge development. The other thing that happens involves a crosstail current. It flows across the tail and up and around the magnetopause (viz. the traditional magnetotail θ current). I also believe that one way the tail reacts to northward turnings of the IMF is to arrange that some of that current, instead of flowing up along the magnetopause, diverts into what I believe is a large-scale ring current system (Figure 8). This reduces the magnitude of the tailfield pointing earthward or anti-earthward while, at the same time, providing a southward component of B_z in the center of the tail.

Baker: What you describe sounds suspiciously like reconnection to me, but I don't know.

Rostoker: I'm not objecting to reconnection.

Baker: I mean in the plasma sheet.

Axford: Do you just say that it is driven by a combination of viscous interaction and reconnection, no more than that?

Rostoker: I think probably both play a role. People have tried to attack how much viscous can give. And reach the conclusion, I believe, that it's inconsequential, perhaps? I don't know whether it is or not. I'm caught a little bit in here. I was also caught a little bit in the early days because as one imagines this boundary layer flow going on, you imagine the field lines being stretched out, and, you know, how far can you stretch them? Can you really make something that goes out beyond 80 earth radii? However, I'm now finding out the magnetosphere, even on the sides, is like a swiss cheese. The magnetosheath plasma is coming in arbitrarily, depending on the configuration of the IMF with the magnetotail boundary. The magnetotail is flapping this way and that way, apparently exposing itself in different configurations to the incoming field. So I'm not worried too much anymore. Magnetosheath plasma can keep coming in and adding. And it could be considered all as part of the reconnection phenomenon. Just one more comment further to Dan Baker. Again, I don't object to reconnection. I could eventually decide to think of some reconnection phenomena going on at the end of my "separation bubble," although I'm not compelled to at the moment. And I don't object to it closer to the earth either. I'm just worried. Somebody has to tell me why the entire electrojet system around midnight doesn't get knocked all to hell by the passage of the plasmoid. If they could do that. . . .

Hones: I don't see how you can so glibly discount our plasmoid observations; the near-earth observations that implied plasmoids, and now the distant observations by ISEE 3 that show them. For example, in the distant tail, we have the delay time of 30 minutes or so, appropriate to the speed we see a plasmoid going by. We see openness and then closedness of field lines implied by the model, and observed in the energetic electron data. It seems to me that what you're saying is, "Oh well,I can have almost anything if you let me flap the tail around and move it up and down." You can get any kind of signatures. But the point is, the signatures we see are rather explicit and reproducible, and they cannot be realistically interpreted as just random variations of the boundaries.

Rostoker: Ed, I'm not really trying to be that much of a nasty guy. I presented to you something that worries me.

Hones: OK. Well, the answer to that is we've already tested your idea, and it doesn't work.

Rostker: And it is published? Where?

References

Lighthill, M. J., Introductory boundary layer theory, in *Laminar Boundary Layers,* (ed. by L. Rosenhead) p. 46, Clarendon Press, Oxford, 1963.

Rostoker, G., Triggering of expansive phase intensifications of magnetospheric substorms by northward turnings of the interplanetary magnetic field, *J. Geophys. Res. 88,* 6981, 1983.

Rostoker, G. and J. C. Samson, Can substorm expansive phase effects and low frequency Pc magnetic pulsations be attributed to the same source mechanism? *Geophys. Res. Lett.*, in press, 1984.

THE LAST WORDS

V. M. Vasyliunas

Max-Planck-Institut für Aeronomie
D-3411 Katlenburg-Lindau 3, West Germany

Much of what I'm going to say can be subsumed under the title: the meaning of merging or reconnection. When I look back at the controversy which has surrounded the concept for the past thirty years and try to understand historically how it arose, I find it rather remarkable that there has been that much controversy. The word reconnection becomes in some cases almost an emotional symbol, and some people seem to object merely to the name. One can find papers where the concept is violently objected to, but if one examines what the author is doing, one finds he is doing essentially the same physics; he is only absolutely refusing to use that name and insisting that anyone who uses it is wrong. So I would like to, so to speak, demythologize that and discuss what it is we really talk about when we talk about reconnection or merging. Leaving aside the precise legal definitions, I think what we are talking about is very simple. It's a system with a complex magnetic topology, where we have a plasma flow in it of some sort: plasma flow in a complex topology. In the case of the earth, initially there may have been some argument as to whether the topology is really complex or whether all the field lines are just nicely contained in one volume—there was a big battle about that some 15 years ago—but today I think it's generally accepted that the magnetosphere is open; so we do have a complex topology, and of course we know that the solar wind is flowing.

Now, one immediate consequence of looking at the concept in this broad way is related to something which Arons has said. He said that if the reconnection is there, it has no obvious signatures. Of course, if it's not there, it has no obvious signatures either. Conclusion: there are no obvious signatures. By now, it should be clear why there aren't. Reconnection is not a single specific process, like bremsstrahlung or whatever. It is a class of problems; it is something as broad as supersonic flow, for example, or viscous flow. When a system is said to have a complex topology and a flow, that's still very far from a complete specification. So one should not look for a single specific model and should not ask of any of the models, Petschek's or Biskamp's or anybody else's, that they should be universal models; there is no such thing, any more than there is a universal model of supersonic flow past a body. So from this general theoretical point of view, merging or reconnection is simply a class of problems, and you have to look at each specific instance. In the case of the magnetosphere, given the fact that it's open and given the fact that the solar wind flows, there is no question that some sort of reconnection is occurring, but we still have to investigate in detail exactly what the configuration is and what precisely happens where.

A good example for what I consider a sane view of the whole problem is what we heard yesterday afternoon on laboratory plasmas. In fusion devices, one has situations in which a complex topology develops. One then studies, in each individual case, what the topology is, what the field lines do; there's no search for the universal signature of anything or for the universal model. Now in space physics, I think, one reason why there has been this emphasis on expected universality is that the theorists tend to break into groups, as we could see clearly at this meeting. One way is to look at the whole morphological problem; what makes it interesting, of course, is the fact that over much of the region the MHD approximation is valid. But this approximation breaks down in various small regions (above the aurora, for example), and in particular it has to break down in the vicinity of the X line just because of the geometry; theorists very often tend to focus on those small regions because it's interesting, it's unknown, it may lead to remarkable local effects like particle acceleration or whatever. If one looks at the large scale flow again, there's been a big controversy: does it depend on the resistivity or not? One should distinguish clearly three regions, not just two. There is the very small local region around the X line, the diffusion region, where MHD breaks down, and there, of course, everything depends on the resistivity and on the inertia and on whatever else happens. Now if you go to a sufficiently large scale, however, if you really look at the whole magnetosphere, I must say it would be incredible that the very small diffusion region could really control everything on a large scale—there, things will go very much the way dictated by large-scale dynamics. But this need not hold all the way down until one approaches the diffusion region. There may very well be an intermediate region, large-scale in the sense that MHD holds over it (i.e. much bigger than the diffusion region) but still relatively small compared to the entire system; there, it is a fair argument how much the diffusion region really does or does not affect it in detail. OK, so those are the theorists' approaches: you can either concentrate on the small diffusion regions, or deal with this intermediate-scale region (that's most of the classical models), or else discuss the whole large-scale configuration. Now the experimenters, I think, tend to be a little mesmerized by the theorists' emphasis on interesting things that happen near the X line; they then tend to think of reconnection as an almost magical process which somehow is supposed to do all these wonderful things (accelerate particles and so on) and which presumably agrees with some very special models, and whenever these very special models do not seem to predict what is seen, they immediately raise the question "does the whole concept work?" Well, I've been asked several times by experimenters who will come to me and say, I see this and that, what does reconnection predict about it? And my answer always is, "until you tell me more, you have told me nothing." When you say "reconnection", you tell me it's a complex system (which I knew already) and it's got a flow (which I knew already). One has to really think of specific circumstances in each case before one can search for a detailed model. And there's no reason why this concept or any other should really give a good detailed model in advance of observations; the system is just too complex to be all taken in at once. So my main effort was to bring across that we really are talking about a class of problems and not a specific model. We are not simply testing some very simple predictions; we are considering a complex geometry and just have to work it out, both theoretically and observationally.

Then there is this whole argument about circuits and laboratory experiments. The experiments are very interesting, very illuminating, but one has to keep in mind some of the differences between space and the laboratory. One main limitation comes from the fact that in space the MHD approximation holds over most of the system except for small regions, whereas in the laboratory it does not hold outside of your plasma chamber. In a laboratory experiment one can have all kinds of electrodes in the plasma and have also an external circuit with possibly a lot of energy stored in it in magnetic form. And, for example, if I now change the current in the circuit or in the plasma and consider a line integral of E between two points, $\int d\vec{\ell} \cdot \vec{E}$, in the external circuit it is certainly limited by the resistance, inductance, etc. of the circuit and can be quite large if the resistance is low. But in the plasma, however, (and in space even the part of the system external to the magnetosphere is plasma), it is limited by the MHD approximation and becomes simply $\int \vec{B} \times \vec{v} \cdot d\vec{\ell}$ (even if the path crosses the small localized regions where MHD breaks down, the integral is not much affected). So if we want to develop a large value for this quantity, we have to change $\vec{v}$ appropriately, and to do that, we have to exert forces on the plasma; acceleration responds to all applied forces, pressures, $\vec{j} \times \vec{B}$ and so on. In short, in space we always are constrained by the dynamics as to how fast we can change magnetic and electric fields, whereas in the laboratory we're constrained only in the plasma chamber and not outside. The other major difference is that most experiments that were discussed here (at least the simulation experiments like Stenzel's) involve magnetic fields of 10's of gauss to maybe a kilogauss or so; the magnetic force in that case is relatively small and the walls can take up anything in terms of stresses on the chamber. On the other hand, in the magnetosphere the magnetic energy density is generally comparable to the plasma energy density, so it's like doing an experiment with fields of 100's of kilogauss where the copper coils begin to yield: we really have to worry about the self-consistency of the magnetic forces and the mechanical forces in the magnetosphere (we don't in the laboratory experiments discussed here). Now there may be situations in space where that is not the case. In the sun, for example, one can perhaps treat the photosphere as essentially having its own dynamics not much influenced by the magnetic forces, and the same may be true in the earth's upper atmosphere, but not in most of the magnetosphere—there, we really have to think about how to maintain the "walls", if you call them that, the boundaries of the plasma.

So I think that we have now made considerable progress, with an enormous amount of information from many observations on the various phenomena related to plasma flow in the complex topology of the earth's magnetosphere, and we should now appreciate that there is no unique given answer that we are supposed to find, that we really have to study the system in detail. This is just like a study of supersonic flow: for each shape of body there is a different configuration, and if anyone finds that the measurements in a given case are difficult to interpret in terms of one model, it doesn't mean that there's no supersonic flow; it just may mean that I constructed a model for a square object where I should have used a round one. And I think the mutual interplay of laboratory experiments and space observations is essential for progress and very illuminating, provided one keeps in mind the relative limitations of both fields and the limited extent to which one can scale things from one to another. Thank you.

Moore: In terms of having a look at specific cases, what's your opinion on how seriously we may be led astray by just considering two dimensions, rather than all three for the reconnection problem?

Vasyliunas: Well, you have to distinguish two aspects of that. A system may be two-dimensional in the sense that nothing depends on the third dimension and the fields and the flows are in one plane. Those are the classical models. They are highly restricted and are meant to apply at most to the intermediate-scale region I discussed; unless one has a situation which is found to be quasi-two-dimensional, one would not expect them to be particularly applicable. Now two-dimensional may also mean that quantities do not vary in the third dimension, but the fields do have components out of the plane, and that may be a reasonable first approximation to localized regions, say, on the dayside magnetopause. Now the full three-dimensional thing, where everything depends on the three spatial variables, is, of course, essential for large scales and may or may not be essential for intermediate and small scales. I think there is no general answer. The restricted two-dimensional models are probably fairly good in the magnetotail and less good in the front side of the magnetosphere.

Lui: I'd like to comment on the statement that you regard reconnection as a class of problems. I tend to agree with that. One of the reasons for thinking this way is that if one looks at the observations in the tail, one tends to get that feeling. We talk about southward magnetic field in the neutral sheet being associated with substorms, and if one really does a good job and looks at southward B_z in the neutral sheet and looks at the activity at that time, say using some reliable measure of activity like global imaging, one sees that one can have southward B_z in the neutral sheet without any activity, without any auroral activity or energetic particle bursts. So in that sense these magnetic structures are there all the time as a persistent feature in the neutral sheet itself without invoking explosive reconnection type of thing. I would also like to make a suggestion, maybe the magnetic bubbles that we see in the neutral sheet which are not related with substorm can be considered as a class of reconnection, a benign kind of reconnection which doesn't really produce very explosive phenomena. I think in terms of observations, this is important too, in terms of understanding the geometry and the physics of the neutral sheet. And also, I'd like to make a comment since I didn't have time to make it earlier. I'd like to go back to Ian Axford's comment. He made a very strong statement about substorms, and he thinks that substorm is nothing but current interruption, they're the same thing. I tend to disagree with that because if you go back to history and find out how the word substorm originates, it says nothing about magnetic fields. It really comes from the auroral activity that Syun Akasofu has looked at and then, together with the magnetic activity, he defined what a substorm is. I don't think one can jump to the conclusion of the equivalence of these two terms, substorm is equivalent to current interruption. Maybe Ian Axford can explain a little bit.

Axford: I said that current interruption is reconnection.

Lui: Well, I don't think that is necessarily the case either, but it will take more than a minute to argue that point.